Green Energy and Technology

More information about this series at http://www.springer.com/series/8059

Carlos Ricardo Soccol · Satinder Kaur Brar
Craig Faulds · Luiz Pereira Ramos
Editors

Green Fuels Technology

Biofuels

Springer

Editors
Carlos Ricardo Soccol
Department of Bioprocesses Engineering and Biotechnology
Federal University of Paraná
Curitiba
Brazil

Satinder Kaur Brar
Centre for Water, Earth and Environment
Institut national de la recherche scientifique
Québec
Canada

Craig Faulds
Aix Marseille Université
INRA
Marseille
France

Luiz Pereira Ramos
Department of Chemistry
Federal University of Paraná (UFPR)
Curitiba, PR
Brazil

ISSN 1865-3529 ISSN 1865-3537 (electronic)
Green Energy and Technology
ISBN 978-3-319-30203-4 ISBN 978-3-319-30205-8 (eBook)
DOI 10.1007/978-3-319-30205-8

Library of Congress Control Number: 2016946016

Printed on acid-free paper

This Springer imprint is published by Springer Nature
The registered company is Springer International Publishing AG Switzerland

Preface

At the United Nations Climate Conference (COP21) that took place in Paris in December 2015, 195 governments agreed to adopt a global action plan aimed a long-term maintainence of the global average temperature increase to be well below 2 °C above preindustrial levels.

Improving the global policy for the stimulation of the research and the consumption of green fuels can contribute to a quick reduction of CO_2 emission of fossil fuels in next decades.

This book comprises 22 chapters, providing an extensive overview about the state of the art regarding to the current technological developments in green biofuels. Chapter "History and Global Policy of Biofuels" provides a generalized overview of the history of biofuels production and the global policies related to their development and brings to the fore the problem of emissions greenhouse gas (GHG) and the sense that energy is an essential input for global economic growth. In Chapter "Feedstocks for Biofuels" the most important vegetal feedstock and agroindustrial wastes utilized in the production of green biofuels are presented. Chapters "Oil Crops in the Context of Global Biodiesel Production" and "An Overview of Production, Properties and Uses of Biodiesel from Vegetable Oil" present the global market of the main vegetal oils used in the biodiesel production and an overview about their production, properties, uses of biodiesel from vegetable oil, and various technical and economic aspects that require to be made for increasing the production and consumption of this biofuel.

The pretreatment technologies of lignocellulose biomass available to date, with emphasis on those that are already closed to or have eventually reached pre-commercial scale such as steam explosion and/or dilute acid hydrolysis are presented and discussed in Chapter "Pretreatment Processes for Cellulosic Ethanol Production: Processes Integration and Modeling for the Utilization of Lignocellulosics Such as Sugarcane Straw".

The main enzymes involved in cellulose degradation include the classical glycoside hydrolases, namely endoglucanases, cellobiohydrolases, and β-glucosidases, as well as oxidative enzymes, among which cellobiodehydrogenases and the newly discovered lytic polysaccharide monooxygenases are introduced and explored in Chapter "Fungal Enzymatic Degradation of Cellulose". In Chapter "Principles and Challenges Involved in the Enzymatic Hydrolysis of Cellulosic Materials at High Total Solids" the new challenges involved in the production of enzymatic hydrolysates with high sugar concentrations (180–200 g L^{-1}), signaled as the most important factor to economically produce second generation ethanol, are argued.

In Chapters "First Generation Bioethanol" and "Second Generation Bioethanol" the currently technology developed for the production of first- and second generation bioethanol is presented. Chapter "Bioethanol from Soybean Molasses" focuses on the production of bioethanol from soybean molasses at laboratory, pilot, and Industrial scales. Soybean molasses is the main by-product generated during the industrial processing of soybean to produce the soy protein concentrate. It is a rich source of carbohydrates, proteins, and lipids and demonstrated to be a suitable fermentation medium to produce bioethanol, either with *Saccharomyces cerevisiae* or *Zymomonas mobilis*.

Chapter "Bioethanol Wastes: Economic Valorization" describes the most promising technology for the reuse and valorization of the solid, liquid, and gaseous wastes generated during the production of ethanol through the fermentation of sugarcane.

Chapters "General Assessment of the Currently Available Biodiesel Production Technologies"–"Biodiesel and Bioethanol from Microalgae" present a general discussion about available technologies, with a special focus on hydroesterification and biodiesel production from algae. Chapter "Microbial Oil for Biodiesel Production" is a description of biodiesel production (at bench and pilot scale) from oleaginous microorganisms cultivation in alternative substrates for microbial oil production and extraction of lipids. Also, a promising example of microbial oil production from sugarcane juice by yeasts and microalgae and its use as raw material for biodiesel production is presented as a case study.

The research development in biohydrogen is presented in a chronological order in Chapter "Biohydrogen", and anaerobic digestion for biogas production as an evolutionary perspective in the Indian context are discussed in Chapter "Biogas: An Evolutionary Perspective in the Indian Context".

Chapter "Bio-butanol—"A Renewable Green Alternative of Liquid Fuel" from Algae" describes briefly, biobutanol as a fuel and its biochemical production and challenges and a particular emphasis is given on their production from algae as potential substrate.

Pyrolysis process is the thermal degradation of biomass under an inert atmosphere leading to three different products: solid char, liquid biofuel, and fuel gas. This thermochemical process involves complex and multiple reactions. In Chapter "Pyrolysis of Biomass for Biofuel Production", the biomass pyrolysis is studied using the thermogravimetric analysis (TGA) coupled with mass spectrometry (MS), one of the main analytical tools to evaluate the potential of a feedstock.

Life cycle assessment is a powerful tool to analyze the economic efficiency and the environmental impacts of a product, processes, or human activity on the environment. In Chapter "Life-Cycle Assessment of Biofuels", the theory of life cycle assessment, scope and objectives, besides some of the commonly used parameters to evaluate and compare efficiency of different processes are briefly discussed. Then some specificities of biofuel's life cycle are focused, and finally some case studies concerning the most important biofuels such as biodiesel, bioethanol, and biohydrogen, among others, are presented.

The statistics of patent applications is used as indicator for evaluating the technological development in different knowledge areas. Chapter "Patents on Biofuels" covers every aspect of the production process of biofuels, focusing on biotechnological aspects and its most recent trends.

Finally, Chapter "Economic and Environmental Aspects of Biofuels" inspects the economic and environmental impacts of the first generation biofuel, which is the only largely available fuel in the market today. Based on this experience, the process of continuous improvement for the future generations of biofuels is evaluated.

The book would be of special interest for academic, researchers, graduate students, and industry scientists that are working in the area of biofuels.

We would like to thank the authors and the reviewers of the chapters for their cooperation and also for their preparedness and revising the articles in a timed date. We also thank the team from Springer, especially, Dr. Antony Doyle and Vani Gopi, for their collaboration in editing this book.

Curitiba, Brazil — Carlos Ricardo Soccol
Québec, Canada — Satinder Kaur Brar
Marseille, France — Craig Faulds
Curitiba, Brazil — Luiz Pereira Ramos

Contents

About the Editors

Prof. Carlos Ricardo Soccol is leading a research group at the Department of Bioprocesses Engineering and Biotechnology (DEBB) in UFPR. Professor Soccol has experience in bioprocessing engineering, especially in the areas of enzymes, industrial bioprocesses, nutraceuticals and functional foods, applied microbiology, and fermentation technology. The research carried out till date has credited him with 1087 publications including patents (67), books (21), book chapters (110), original papers (317), and research communications in conferences/symposia (572). His research articles have so far been cited 9515 times (ResearchGate database) with Index h=46. It has also led the successful supervision of 57 Ph.D. students, 112 Master's students, and 16 postdoctoral fellows. In 2001, He did Professor HDR in Agri-Foods Biotechnology at Aix Marseille University, Marseille-France. He was visitor professor at École Polytechnique Fédérale de Lausanne, Switzerland (2009), University Lille 1-France (2014) and University Blaise Pascal, Clermont Ferrand-France (2015). He has received several national and international awards, including the Science and Technology award of the Government of Paraná (1996), Best Scientific Achievement of the Year 2001 by Ministry of Sugar of Cuba (MINAZ), The Scopus/Elsevier Award (2009), Dr. Honoris Causa, University Blaise Pascal-France (2010), Outstanding Scientist at 5th International Conference on Industrial Bioprocesses, Taipei, Taiwan, Henri Nestlé Award—Nestlé, Brazil (2014). Since 2014, Prof. Soccol has been the member of the Brazilian Academy of Sciences. He served as the Editor of the *Brazilian Archives of Biology and Technology* during 1997–2015. He is also the scientist IA of CNPq, Brazil. He is a technical and scientific consultant for several companies, agencies and an editorial member of scientific journals.

Prof. Satinder Kaur Brar is leading the research group Bioprocessing and Nano-Enzyme Formulation Facility (BANEFF) at Institut National de la Recherche Scientifique (Eau, Terre et Environnement, INRS-ETE), Québec, Canada. Professor Brar has experience in the development of finished products (formulations) of municipal and industrial wastewater and wastewater sludge based value-added bioproducts, such as enzymes, organic acids, platform chemicals, biocontrol agents, biopesticides, biobutanol, and biohydrogen. She is also working on the fate of endocrine disrupter compounds, pharmaceuticals, nanoparticles, and other toxic organic compounds during value addition of wastewater and wastewater sludge in turn finding suitable biological detoxification technologies. The BANEFF facility has so far led successful supervision of 20 Ph.D. students, six Master's students, and six postdoctoral scholars. She has collaborative programs with several industries in Canada and researchers from Argentina, Spain, Chile, Switzerland, France, Vietnam, China, USA, India, Thailand, Sri Lanka, Mexico, Morocco, Tunisia, and Ivory Coast. Professor Brar has won several national and international recognitions, including, the ASCE State-of-the-Art of Civil Engineering award (2007) for her article, "Bioremediation of Hazardous Wastes—A Review," which was published in the Practice Periodical of Hazardous, Toxic & Radioactive Waste Management—Special issue on Bioremediation; the Rudolf gold medal (2008) for her originality of the article published in Practice Periodical of Hazardous, Toxic & Radioactive Waste Management. She has been elected as member of the College of New Scholars, Artists and Scientists of the Royal Society of Canada in recognition for the emerging generation of Canadian intellectual leadership and outstanding performance in her field, environmental biotechnology. She is also presented with "YWCA women in science" Excellence Award. She is on the editorial board of Brazilian Archives of Biology and Technology Journal, Science of Total Environment and associate editor of Journal of Hazardous, Toxic, and Radioactive Waste (ASCE). She recently initiated a new journal on Nanotechnology for Environmental Engineering with Springer Publishers as Editor-in-Chief. She has won several accolades throughout her professional career including the outstanding young scientist in India in 2002. She has more than 230 research publications which include five books, 40 book chapters, 125 original research papers, 60 research communications in international and national conferences and has registered 2 patents to her credit.

Prof. Craig Faulds is the Director of the Aix Marseille Université-INRA Joint Research Unit "Biodiversity and Biotechnology of Fungi (BBF)" and Professor of Biochemistry at Polytech Marseille in the Luminy Campus of Aix Marseille University. Professor Faulds has experience in lignocellulosic enzymes, especially feruloyl esterases, and the synergistic interaction of polysaccharide and lignin-degrading enzymes, in particular of fungal origin. He has also experience in nutraceuticals and functional foods. He has previously worked at the Institute of Food Research in Norwich, UK (1988–2011), Centro de Investigaciones Biológicas (CSIC), Madrid, Spain (2009–2011) and VTT Technical Research Centre of Finland (2011–2012) before moving to Marseille. He has published to date over 120 peer-reviewed scientific articles, 13 book chapters, and 4 patents. His research articles have so far been cited 5200 times (Thomson Reuters database) with an h-Index=43. Professor Faulds also led the successful supervision and co-supervision of six Ph.D. students and 11 visiting Ph.D. students. He is responsible for the M2 Masters speciality "Biotechnology for Sustainable Development (BIODEV)" within the "Microbiology, Plant Biology and Biotechnology (MVBV)" Masters, Faculty of Science, Aix Marseille University. This is an online "distance learning" course. He is a member of the Editorial Board of Fungal Biology and Biotechnology and has been on the organizing committees of Ferulate 98 (Norwich 1998) and Total Food (Norwich 2004 and 2009).

Dr. Luiz Pereira Ramos is full Professor of Analytical Organic Chemistry at the Federal University of Paraná (UFPR), Department of Chemistry, Curitiba, Paraná, Brazil. He graduated in Chemistry (1982) from the Catholic University of Paraná in 1982, obtained his Master's in Biochemistry from the Department of Biochemistry of the UFPR in 1988 and his Ph.D. in Applied Biology from the University of Ottawa, Canada in 1992 under the supervision of Dr. John (Jack) Saddler. Dr. Ramos is currently the leader of a research group that is actively working on the development of biorefinery strategies for the total conversion of biomass into valuable chemicals, renewable fuels and biobased materials. His research activities involve wood and carbohydrate chemistry, second and third generation biofuels (biodiesel and cellulosic ethanol from conventional and nonconventional feedstocks such as microalgae), biomass fractionation using advanced pretreatment technologies, intensification of bioconversion processes (ultrasound, microwaves), heterogeneous catalysis, and several other topics within the main concept of sustainable biorefineries. As former Head of the UFPR's Graduate Program in Chemistry for several years and permanent staff in two local graduate programs (Chemistry and Bioenergy), he has participated in the successful supervision of 17 Ph.D. students, 41 Master's students, and five postdoctoral fellows. Also, he has been honored as a distinguished professor by the UFPR graduate students in Chemistry for 10 consecutive years. Dr. Ramos participates actively in several networks and collaborative programs involving research institutes and industries from Brazil and abroad (Sweden, Canada, Argentina, Uruguay, Spain, France, Portugal, Denmark, Colombia, United States). As the member of the Brazilian Chemistry Society and of the American Chemistry Society (ACS), he is the recipient of the 2013 6th Edition of the Petrobras Technology Award (supervision of the best dissertation about bioproduct technologies) for his work entitled "Enzymatic hydrolysis of steam-exploded sugarcane bagasse using high total solids and low enzyme loadings", which was published in *Bioresource Technology* in 2015, and received the 2012 General Electric Research Incentive Award for his work with the application of reactive distillation in biodiesel synthesis. He has organized and chaired three international meetings and participated in numerous organizing committees for conferences held in Brazil or abroad. He has authored or co-authored numerous research publications that include two edited books, 17 book chapters, 117 original research papers, 162 extended abstracts plus 154 abstracts in conference proceedings, more than 60 research communications in international and national conferences and 15 patent applications, most of them filed in Brazil. The impact of his published work is reflected in more than 2400 citations and in an H factor of 31 (ISI Webofknowledge, Jan 2016). Currently, he is one of the Associate Editors of the ACS journal *Energy & Fuels* (since 2012), participates in the Editorial Advisory Board of the Bioethanol Journal (edited by De Gruyter), BioResources (Raleigh, USA) and Science and Technology (Uberlândia, Brazil) and has acted as ad hoc consultant for several companies, leading international scientific journals, and both national and international funding agencies.

History and Global Policy of Biofuels

Mariem Ayadi, Saurabh Jyoti Sarma, Vinayak Laxman Pachapur, Satinder Kaur Brar and Ridha Ben Cheikh

Abstract The increasing cost of crude oil, concerns about energy security, greenhouse gas (GHG) emissions, and the realization that energy is an essential input for economic growth have resulted in renewed focus on biofuels. In the last decade, their development has been driven by various governments' policies. Laws and regulations linked to renewable energy were legislated in several countries. Financial supports, tax policies, and the mandatory blending of some biofuels with fossil fuels in transportation sector are the major policies. Thus, the present chapter provides a generalized overview of the history of biofuels production and the global policies related to their development.

Keywords Bioethanol · Biohydrogen · Biogas · Biobutanol · Biooil · History · Policy

1 Introduction

Many countries have started searching for energy stability, especially after the oil crisis in the 1970s which put an end to an abundant and low-cost fuel. In this context, biofuels are considered an attractive alternative from economic, social, and environmental points of view (Cremonez et al. 2015). For the European Union, the primary motivations of biofuels development are to ensure the energy security, to reduce the greenhouse gas (GHG) emissions, and to promote the rural agricultural development (Su et al. 2015). Biofuel is a term used for liquid, gas, and solid fuels which are predominantly produced from biomass; they include biomethanol,

M. Ayadi · R.B. Cheikh
Laboratory of Materials, Optimization and Energy for Sustainability,
National Engineering School of Tunis, BP 37 Le Belvédère, 1002 Tunis, Tunisia

S.J. Sarma · V.L. Pachapur · S.K. Brar (✉)
Centre - Eau Terre Environnement, Institut National de la Recherche Scientifique,
490, Rue de la Couronne, Québec, QC G1K 9A9, Canada
e-mail: satinder.brar@ete.inrs.ca

C.R. Soccol et al. (eds.), *Green Fuels Technology*,
Green Energy and Technology, DOI 10.1007/978-3-319-30205-8_1

bioethanol, biohydrogen, biodiesel, vegetable oils, biooil, bio-char, biogas, bio-synthetic gas named also bio-syngas and Fisher–Tropsch liquids (Demirbas 2008). During the last 10 years, biofuels production has increased in a remarkable way passing from 6.4 in 2003 to 23.4 billion gallons in 2013 (Su et al. 2015). The world leaders in biofuels development and use are Brazil, United States, Germany, France, and Sweden (https://en.wikipedia.org/wiki/Biofuel). The present chapter summarizes the developments of global biofuel initiatives in a chronological order and the policies backing such developments.

2 History of Biofuel Production

2.1 History of Biodiesel Production

Biodiesel is a renewable alternative fuel produced from vegetable oils, animal fats, or waste vegetable oils (Bergmann et al. 2013). The story began in 1890 when the German "Rudolf Diesel" invented the diesel engine. In 1900, one of these engines was powered by peanut oil (http://www.extension.org/pages/27135/history-of-biodiesel#.VhbgAfl_Oko). After Diesel's death in 1913, petroleum fuels became available and cheap reducing the importance of vegetable fuels (http://www.biodiesel.com/biodiesel/history/). After the oil crises of the 1970s, the interest in using vegetable oils increased significantly; however, the problem was that the newer diesel engine could not run on traditional vegetable oils due to their high viscosity in comparison to petroleum diesel fuel (http://www.biodiesel.com/biodiesel/history/). Many methods were proposed including pyrolysis, blending with solvents, and emulsifying the fuel with water or alcohols in order to decrease their viscosity to a level they could be burned properly in the diesel engine (http://www.biodiesel.com/biodiesel/history/); unfortunately, none of these solutions was suitable. Actually, biodiesel patent was granted to the Belgian Charles Chavanne from the University of Brussels who proposed in 1937 the use of ethyl esters derived from the acidic transesterification of palm oil. However, the term biodiesel was mentioned for the first time in 1988 in a Chinese research paper (Cremonez et al. 2015).

2.2 History of Bioethanol Production

Bioethanol is a fuel produced from natural sources or feedstock, e.g., wheat, wood, corn, straw, and sugar beet; it is a way to reduce environmental pollution and crude oil consumption (Demirbas 2008). Figure 1 shows bioethanol production in different countries during 2005 and 2006; where, America is the world's leader of its production (Balat et al. 2009). In the year 2014, global ethanol fuel production was 24,570 million gallons, where USA (14,300) and Brazil (6190) were the two leading

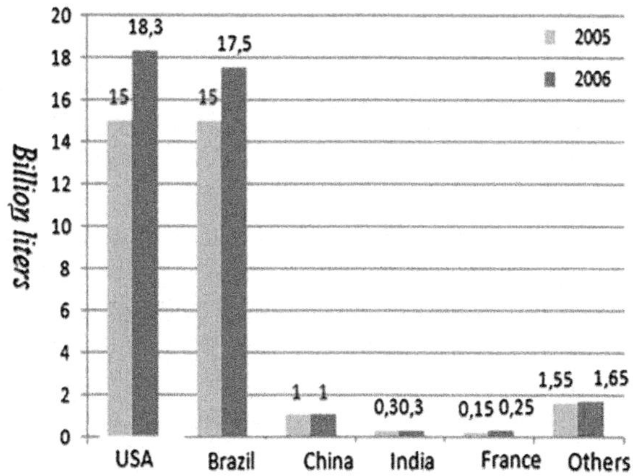

Fig. 1 Bioethanol production worldwide between 2005 and 2006. Data taken from Balat et al. (2009)

producers (http://www.chemistryviews.org/details/ezine/7841381/Alcohol__Not_Just_for_Drinking.html). Bioethanol can be produced by sugars fermentation, in this process, sugar is converted with water into ethanol, carbon dioxide, and water as described in the following equation: (http://fr.scribd.com/doc/21260/The-History-and-Development-of-Bioethanol-as-an-Alternative-Fuel#scribd)

$$C_6H_{12}O_6 + H_2O \rightarrow 2C_2H_5OH + 2CO_2 + H_2O$$

Brazil has successfully produced bioethanol as a fuel for automobile use since the 1970s and it has been produced almost exclusively from sugarcane. Indeed, 1 ton of sugarcane affords 72 l of bioethanol (http://fr.scribd.com/doc/21260/The-History-and-Development-of-Bioethanol-as-an-Alternative-Fuel#scribd). It should be indicated that bioethanol production from sugarcane has been known since 6000 BC. Sugars fermentation into ethanol is one of the oldest organic reactions that mankind learned. Ethanol has been used as an intoxicating agent in the preparation of alcoholic beverages for a long time. Dried ethanol residues were found in China on a 9000-year-old pottery, their isolation as a relatively pure compound was achieved by the Persian Zakariya Razi. In order to increase ethanol yield in beverages, ancient people resorted to distillation, this process was known by the Greeks working in Alexandria, and then Arabs learned it from Alexandrians. In 1796, Johann Tobias Lowitz obtained successfully the pure ethanol by filtering the distilled one, finally, its chemical formula was determined by Nicolas Théodore De Saussure in the early nineteenth century (http://www.aber.ac.uk/en/media/departmental/ibers/pdf/innovations/07/07ch8.pdf).

2.3 *History of Biogas Production*

Biogas is a combustible mixture of gases formed from the anaerobic bacterial decomposition of organic matter; it consists of essentially 50–70 % methane (CH_4),

30–50 % carbon dioxide (CO_2) and may contain small amount of other gases. It has a calorific value around 21–24 MJ/m^3.

As an alternative energy source, biogas is generally used for heating, cooking, lighting, or electricity generation. (Bond et al. 2011). It was suggested that biogas was used for bath water heating in Assyria as early as the tenth century BC and that anaerobic digestion of solid waste was applied in ancient China (He 2010). However, it was reported that harnessing anaerobic digestion of biomass, a process by which microorganisms break down biodegradable material in the absence of oxygen, has started from the nineteenth century when digesters were built in New Zealand and India with a sewage sludge digester built in Exeter, UK to fuel street lamp in 1890 (Bond et al. 2011). In China, commercial use of biogas was attributed to Guorui Luo, as in 1921 he constructed an 8 m^3 biogas tank fed with household waste (He 2010). In Germany, the first sewage treatment plant for biogas feeding was constructed in 1920, thirty years later; the first agricultural biogas plant began operating. During the first half of the 1980s, Chinese government installed more than 7 million digesters in order to overcome the high cost of oil. However, only 4.7 million household biogas digesters were reported. Early in this century, there was a rapid increase in the number of plants especially in 2007 when 26.5 million biogas plants were created in China (Bond et al. 2011).

2.4 *History of Biohydrogen Production*

Biohydrogen (H_2) offers a clean and renewable energy source. Upon utilization, it does not generate carbon-based emissions responsible for environmental pollution and climate change (Levin et al. 2004). Moreover, its energy yield is about 2.75 times greater than hydrocarbon-based fuels (Magnusson et al. 2008). There are different methods to produce biohydrogen including: direct biophotolysis, indirect biophotolysis, photo-fermentation, and dark fermentation (Levin et al. 2004). Dark fermentation is considered the best method for producing H_2 thanks to its ability to produce it at higher rate; in addition, this promising technology can treat a variety of waste streams (Magnusson et al. 2008). In 1939, Hans Gaffron, a German researcher, obtained photochemical and fermentative hydrogen from Algae (Gaffron and Rubin 1942); he reported with his coworker that the green microalgae *Scenedesmus* developed molecular hydrogen under light after being kept in anaerobic and dark conditions (Gaffron and Rubin 1942). Gest et al. studied the photochemical production of molecular hydrogen by growing cultures of photosynthetic bacteria (Gest and Kammen 1949). Aiba et al. (1973) reported that hydrogen gas could be generated from mixing acid fermentation of *Esherichia coli* the butylene glycol fermentation of aerobacter and the butyric acid fermentation of *Clostridium* spp. In 1973, when Benneman et al. studied the hydrogen evolution from water using *Clostridium kluyveri* hydrogenase, he revealed that such evolution could be employed in solar energy conversion (Demirbas 2009). Since 1980s, hydrogen production has been investigated with different anaerobic bacteria, after that, it was integrated with agricultural and industrial activities for the future renewable energy demand (Demirbas 2009).

2.5 *History of Biobutanol Production*

Biobutanol production via biochemical process is known as acetone–butanol–ethanol (ABE) fermentation. It was Pasteur who discovered the biobutanol production via anaerobic bacteria fermentation in 1861. Between 1912 and 1914, Chaim Weizmann isolated *Clostridium acetobutylicum* which is responsible for the production of acetone and butanol; their yield was greater than that given by the previous species (http://www.abercade.ru/en/materials/analytics/339.html). The industrial production of butanol by this species flourished from the first half to the second half of the twentieth century until cheaper butanol was produced from petrochemical derivatives.

After the World War I, butanol demand increased significantly, therefore, many large-scale industrial plants were created in Canada and USA. After 1936, ABE fermentation industries were founded in Egypt, South Africa, China, Japan, and Soviet Union. In 1945, Japan started the butanol production from sugar plants as biofuel for airplanes (Ndaba et al. 2015). By 1960, most of ABE fermentation industries were closed due to the decrease in oil prices which supported petrochemical syntheses (Ndaba et al. 2015). However, Petrochemical route did not resist for a long time when ABE fermentation facilities had spread again in China and Brazil. Nowadays, there are many plants in different countries including USA, UK, Slovakia, and France producing biobutanol for several industrial applications (Ndaba et al. 2015).

2.6 *History of Biooil Production*

Palm oil, an alternative energy source, is the second most widespread oil in the world. It is widely used for biodiesel production and electricity generation in power stations. Moreover, this biooil is largely produced in Malaysia and Indonesia (Oil world 1999). Palm oil is originated from West Africa where it has been used as a principle food crop since 5000 years. Ancient Egyptian tombs indicate that people were buried with palm oil which reflects the prestigious status of this product. With the expansion of the trade across the sea and the industrial revolution in Europe, palm oil commerce expanded dramatically in the international markets leading to industrial lubricants production and candle-making. In addition, with the increase of its demand, Europeans started investing in palm oil production in West Africa and Southeast Asia where the first commercial planting was founded in Malaysia in 1917 (http://theoilpalm.org/history-and-origin/). Then, the cultivation rapidly increased especially in the 1960s under the government agricultural policy which was applied to mitigate dependence on tin and rubber (http://theoilpalm.org/history-of-the-industry/).

Sales et al. produced biodiesel from sunflower oil and ethanol by base catalyzed transesterifiication (Sales 2011). History shows that this plant was cultivated in North America by American–Indian tribes in about 3000 BC. Oil extracted from sunflower was patented in England in 1719 and it was later commercialized in the national market (http://www.botanical-online.com/english/sunflower_history.htm).

Jatropha oil was used for many centuries by crushing seeds to produce basic oil lamps which were then marked in Portugal. However, this oil was abandoned for a long time with the appearance of cheaper paraffin oil. In the nineteenth century, Jatropha reappeared as a key energy plant; crude oil was extracted from the seeds and refined into a biodiesel (http://www.jatropha-bio-fuel.com/jatropha-curcas-l/).

3 Global Policies for Biofuel Production and Use

3.1 Climate Change and GHG Emission Reduction Targets

Fossil fuels are accused of the GHGs released in the atmosphere such as CO, CO_2, CH_4, and NO_2. Vehicles, especially those associated with diesel engine, has caused a rapid growth of GHG emissions resulting in a number of health diseases and a harm to the environment; there are about 22 % of global GHG emissions coming from transport sector. In addition, The International Energy Agency (IEA) estimated that carbon dioxide (CO_2) emissions from this sector will increase by 92 % between 1990 and 2020, moreover, its release in the atmosphere is expected to reach 8.6 billion metric tons from 2020 to 2035 resulting in an increase of the global temperature by 2 °C which may involve the death of hundreds of millions of people (Mofijur et al. 2015).

Biofuel is an alternative which can reduce enormously the dependence on oil in several industrial sectors. Many countries have put their target to exploit biofuels as they have potential to reduce more than 80 % of GHG emissions (Mofijur et al. 2015). This alternative has already covered 2 % of the total transportation fuels and it is expected that it will be more promoted in the near future with technology and researches development (Su et al. 2015).

In America, since the energy consumption is expected to grow 50 % by 2030, there is a big interest in biofuels which become essential to reduce dependence on oil and to ensure a clean energy. In this context, The U.S. Department of Energy (DOE) announced over $1 billion to finance biofuel projects. Integral to this work is the ongoing examination of reducing GHGs (http://www.fanrpan.org/documents/d00533/US_DOE_Biofuels_Myth-v-Fact.pdf). Researches show that biofuels can

Table 1 Variety of GHG emissions by feedstock and type of energy used in processing (Wang et al. 2007)

Fuel	Gasoline	Corn ethanol			Sugarcane ethanol	Cellulosic ethanol
Energy used	Fossil fuels	Current average	Natural gas	Biomass	Biomass	Biomass
GHG emissions reduction (%)	–	19	28	52	78	86

emit the same amount of emissions or more as gasoline; but the difference is that biofuels burn cleaner than gasoline resulting in fewer GHG emissions. Table 1 shows that GHG emissions of fuels depend on the type of feedstock and energy used during the processing. Advanced biofuel is a term identified by The Energy Independence and Security Act of 2007 (EISA 2007) in USA; it presents a biofuel produced from non-corn feedstock generating 50 % lower life cycle GHG emissions of gasoline. Among its acts, EISA 2007 supplied $550 million funding for reducing GHG emissions (Su et al. 2015).

In 2003, the Directive on the Promotion of the Use of Biofuels or Other Renewable fuels for Transport (2003/30/EC) set goals that the use of biofuels in vehicles should reach 2 % in 2005 and 5.75 % in 2010 and only Germany, France, Sweden Austria, and Czechoslovakia achieved successfully these goals (Su et al. 2015).

3.2 Carbon Credit

According to the 'Collins English Dictionary' carbon credit is defined as "a certificate showing that a government or company has paid to have a certain amount of carbon dioxide removed from the environment". One carbon credit is equivalent to one ton of carbon dioxide. This term presents a national and international attempt to mitigate GHG emissions, indeed, countries or groups receive credits once they are able to reduce their GHGs below their emission quota (https://en.wikipedia.org/wiki/Carbon_credit#cite_note-ced-1).

Carbon credits are widely traded in Europe and United states (http://www.biodieselmagazine.com/articles/1874/carbon-credits-offer-opportunity-for-biodiesel-producers/). As emissions reduction in Europe is costly, many companies in Africa are engaged in renewable energy projects such as "biofuels program in Madagascar". This project aims at producing biofuel from Jatropha plant with fewer emissions than fossil fuels; the African company certifies its emission reductions and sells them to a European company to get an additional income stream for the project.

3.3 Policy on Biofuel Feedstock

Feedstock is a term referred to crops or products which can be converted into biofuels or bioenergy or other such products. Their advantages depend on their production, water content, and energy yield (http://www.bioenergywiki.net/Feedstocks). Biomass-based feedstock has three principal elements which are Carbon, Oxygen, and Hydrogen with small percentages of Nitrogen, Sulfur, and ashes. They vary in basic components including cellulose, lignin, hemicellulose, proteins, and triglycerides. In addition, they are characterized by heating value, water content, and specific volume (Churubini et al. 2009). Biomass feedstock is divided into two groups as illustrated in Table 2. Many researchers have classified feedstock into two generations; first feedstock generation includes those which are largely used for biofuel production, the majority of these crops are simultaneously harvested for feed production which presents a real threat for the food security by creating, in another expression, "food versus fuel" conflicts. This category includes sugar, starches, oils, fats, or traditional biomass feed like bamboo stems.

The second generation refers most commonly to cellulose feedstock. Such crops are not widely cultivated or not cultivated as bioenergy source; they are characterized by high potential yields of biofuels. This generation includes grasses and trees. Algae and halophytes, a saltwater plant, are also belonging to the second feedstock generation (http://www.bioenergywiki.net/Feedstocks#Other_second_generation_feedstocks).

In the United States, all the biofuels are currently produced from soybeans and corn. But now, there is an increasing interest in developing new feedstock sources that will not be competitive with food resources. Large quantities of vegetables are left in fields or lost in food processing facilities instead of ending up on consumer's plate. In the United States, it is estimated that the rate of food loss is ranging between 15 and 35 % with a worth of $25–30 billion. This case has brought researchers to think about the exploitation of crop and food wastes for biofuels production (Hacker et al. 2009). Wastewater, rich in organic substance such as starch, is derived from

Table 2 Types of feedstock (http://www.bioenergywiki.net/Feedstocks)

Dedicated feedstock (agriculture, aquaculture, foresty)	Residues (industries and households)
• Sugar crops (e.g., sugar beets, sugarcane) • Strach crops (e.g., corn, wheat, cassava) • Lignocellulosic crops (e.g., wood, switchgrass) • Oil based crops (e.g., palm oil, Jatropha, soy beans, rapessed) • Grasses (e.g., prairie grasses, plants shoots, grass silage) • Marine biomass (e.g., seaweed, micro and macro algae)	• Oil based residues (e.g., animal fat, waste vegetable oil) • Lignocellulosic residues • Organic residues (e.g., manure, organic urban wastes)

food processing with large quantities. Anaerobic digestion was suggested in the presence of a processing plant to produce biogas by capturing methane and hydrogen released from fermentation process. Another study suggested a method to convert starch-rich wastewater into hydrogen gas with a cost of 5 million $ each year (http://www.sciencedaily.com/releases/2003/05/030521092358.htm).

In 2008, 88 % of yard waste and table scraps, 31 % of municipal wastewater, and 26 % of paper mill sludge were landfilled in Quebec. These materials, when landfilled, can decompose and create damage for the environment by the increase of GHG emissions (http://www2.publicationsduquebec.gouv.qc.ca/dynamicSearch/telecharge.php?type=3&file=/Q_2/Q2R35_1_A.HTM). However, recycling these wastes can help in preventing these undesirable effects, creating jobs and energy. Quebec has started banning organic wastes from its landfills, by 2020; it will be prohibited to dispose compostable materials in landfills. Pierre Arcand, the sustainable development minister announced $4 million to support industrial, commercial sectors, and municipalities to manage wastes and to develop biological treatment procedures which help in GHG emissions mitigation. In addition, the government wants to make sure that bioenergy can be produced from organic waste treatment. In July 2012, biocycle.net discussed some challenges such as the possibility of gaining the public attention for organics separation and increasing research linked to organics management (http://www.biocycle.net/2012/10/22/compostingorganicsincanada/).

3.4 Global Policy of Biodiesel Production

Biodiesel, an alternative biofuel, was mostly produced in EU before 2005; later the share started to decline especially with the appearance of United States and Brazil as two important producers in 2013 Fig. 2. Germany is nowadays the largest biodiesel producer with an output of 2.8 million tons in 2011 which presents 35 %

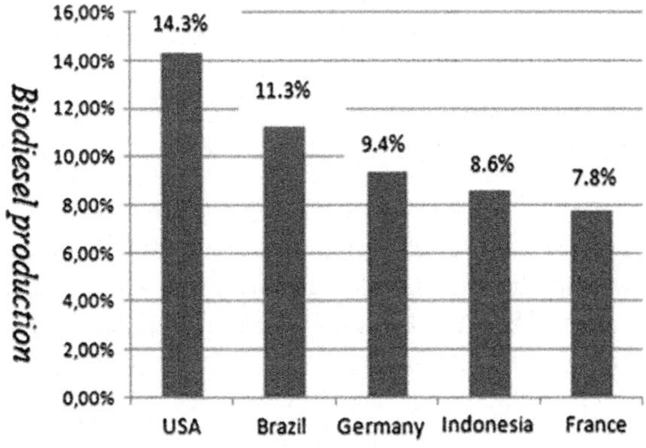

Fig. 2 Leaders of biodiesel production in 2013

of the total European production, and United States is the second largest biodiesel producer with the output of 1100 million gallons as of 2012 (Su et al. 2015).

In last decades, biodiesel production has been driven by government policies. In Argentina, this promising biofuel is granted a financial support once it is sold to the internal market (Sorda et al. 2010). In addition, since 2010, it has been required that gasoline or diesel must contain at least 5 % of biofuel such as biodiesel or bioethanol. The same regulation has been almost applied in Germany since 2009: Biodiesel content in transport diesel has been set at 4.4 % (Sorda et al. 2010). China also provides tax reliefs for biodiesel production in order to promote its development (Su et al. 2015). Many countries in the world have invested in non-food energy plants such as Jatropha as a raw material for biodiesel production. Jatropha oil was developed in 1995 in Brazil and Zimbabwe by the GTE program funded by the German government and Rockefeller foundation (Su et al. 2015). The Indian government has set ambitious biodiesel target: By 2017, 20 % of biodiesel share should be blended with mineral diesel and gasoline. In order to reach this goal, non-edible oil seeds are aimed to be cultivated in marginal lands (Altenburg et al. 2009).

3.5 Global Policy of Bioethanol Production

Brazil is accounted the leadership of bioethanol production and sales in the world. In transport sector, more than 80 % of vehicles are using bioethanol blended fuels (Soccol et al. 2010). The National Ethanol Fuel Program was launched by the Brazilian government in 1970s at the time of oil crisis; series of measures were introduced to improve bioethanol production in order to reduce the oil import and to deal with the decrease of the national sugar price (Su et al. 2015). Three important stages have been experienced by this program. Between 1970 and 1990, to cope with the situation of oil crisis, the government started to encourage the developing of aqueous ethanol. The period ranging from 1990 to 2000 was characterized by the stability in oil prices, Brazil succeeded in developing anhydrous ethanol with initial volume ratio of 20 % ethanol mixed with gasoline, then 22 % in 1993, and 25 % in 2002. From 2000 to today, about 98 % of ethanol is used in transport sector, therefore, ethanol-gasoline cars are emerged in Brazilian market without forgetting that sugarcane and ethanol industry have created 3.6 million jobs and account for 3.5 % of the national GDP (Su et al. 2015).

In Colombia, biofuel policies include mandatory blending a 10 % bioethanol in cities whose population is above 500,000 inhabitants (Sorda et al. 2010). The law was applied in 2005; the aim came true 4 years after when 75 % of total consumed gasoline had 10 % ethanol content (Rutz et al. 2009). By 2020, Colombian government aims to increase the ethanol content up to 25 %. Colombia offers many facilities to support and to encourage bioethanol production: prices of sugarcane from which derived the bioethanol are fixed by the government on the basis of international sugar prices, in addition, bioethanol is exempted from the VAT (Sorda et al. 2010).

3.6 Global Policy of Biogas Production

Biogas is paid more and more attention; it is considered a carbon-neutral energy source. It can be exploited for household use and can be used as vehicle fuel or as a combustible to produce heat and electricity in a combined heat and power (CHP) plant (Engdahl 2010).

Some European countries have taken measures to promote the biogas development. In August 2010, Swedish Energy Agency suggested a strategy for the future biogas production in cooperation with the Swedish Board of Agriculture and the Swedish Environmental Protection Agency. Biogas production was estimated to range between 3 and 4 TWh, out of which, 2.5 TWh could be generated from sludge digestion and waste streams from society, 700 GWh could be achieved from manure digestion (Engdahl 2010). The target of the government is to put an end to GHGs by 2050. As a first step, 40 % of GHG emissions will be reduced by 2020.

Germany is known as the largest producer of biogas in Europe. In 2008, there were about 3891 units of biogas plants which produce about 10 W. The smaller farm-scale plants are used to produce the majority of biogas (Poeschl et al. 2010). German government has set up goals related to energy use. One of them is to exploit 30 % of renewable energy to produce electricity and 14 % to produce heat by 2020. Consequently, biofuels utilization is estimated to reduce 7 % of the total GHGs (Engdahl 2010). The government is encouraging young farmers to invest in biogas sector; many of them see that the future of biogas production is better than the dairy-cattle industry. Projects of community plants occur once a single farm-scale seems not suitable (Holland 2010). In these recent years, there has been a political desire to introduce CHP produced from biogas in the district heating network; many grids remaining from the Soviet period and located in Germany Eastern will be reactivated for heat production (Holland 2010).

Biogas production and consumption is increased significantly in China from 2005 to 2012 (Fig. 3). Chinese government has believed that biogas investment can bring economic benefit and reduce environment pollution, according to statistics, it has invested about 3.8 billion Euros to support biogas development between 2003 and 2012 and now there are more than 91,000 biogas projects in China. By the end of 2020, it is estimated that 80 million household biogas will be set up. The target of government is to build 8000 large-scale biogas projects, consequently, the annual production could reach 45 billion m^3 (https://www.dbfz.de/fileadmin/user_upload/Vortraege/BiogasWorld2014/02_Jiming.pdf).

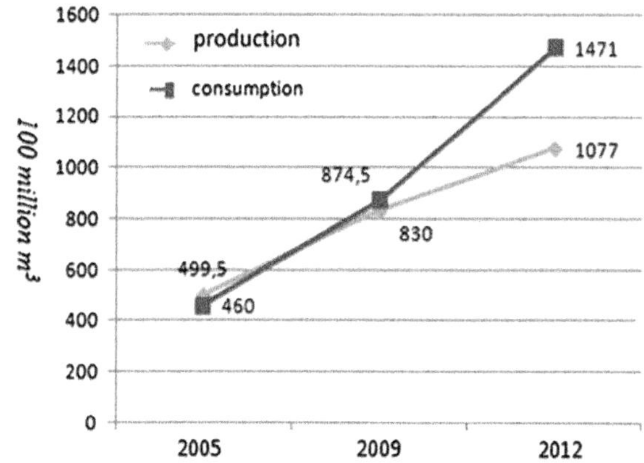

Fig. 3 Biogas production and consumption in China during 2005 and 2012

3.7 Global Policy of Biohydrogen Production

The total annual worldwide hydrogen consumption is ranging between 400 and 500 billion Nm^3 (Demirbas 2009). Currently, most hydrogen is generated from non-renewable resources in particular natural gas, this orientation is not sustainable and generates at least the same amount of carbon dioxide as direct combustion of fossil fuels. Hydrogen can be also produced from biomass by gasification; this method depends strongly on the feedstock cost consequently, and it can be a low-cost method for many countries (Balat et al. 2009).

Hydrogen as a clean energy carrier has a great potential to be an alternative fuel, it is nonpoisonous gas and does not generate pollutants. It can be produced from abundant biomass such as cellulosic materials, starch-based wastes, food or dairy wastes (Chong et al. 2009). Therefore, demand on hydrogen production has been in considerable increase in recent years; European energy policies has showed a great interest for biohydrogen and fuel cell development since they are able to reduce GHG emissions. Consequently, in most European states, hydrogen is exempted from any taxation or taxed at low rates. In March 2007, EU has agreed on the objective to save "20 % of its energy consumption compared to projections for 2020". It is predicted that legislation on energy efficiency (EE) can enhance the development and market introduction of fuel cells which are relatively energy efficient in comparison with gasoline powered by the internal combustion engines (Bleischwitz et al. 2010). Thanks to investment in research, hydrogen-related technologies has increased significantly; many companies in EU, USA, Canada, and Japan are involved in commercialization of biohydrogen technologies (Balat et al. 2009). It is believed that hydrogen production from biomass is economically competitive; however, it presents some challenges as this technology needs a large amount of cooperation and planning at international levels to be more competitive with the other renewable energies.

4 Conclusions

Energy is a fundamental input for economic growth and social progress. To ensure their energy stability, governments have taken measures and designed strategies to support bioenergy development by investing in research or building plants for biofuel production. This chapter presents an overview of the history of biofuels. Likewise, it focuses in their low environmental impact compared to that generated by fossil fuels and summarizes policies of governments towards biofuels which include legislations in feedstock, organic wastes, and issues of clean energy production, trading, distribution, and consumption. However, achieving large-scale changes in energy development requires cooperation at national and international programs.

References

Aiba S, Humphrey AE, Millis NF (1973) Biochemical engineering, vol 2. Academic Press, New York, pp 92–127

Altenburg T, Dietz H, Hahl M, Nikolidakis N, Rosendahl C, Seelige K (2009) Biodiesel in India—value chain organisation and policy options for rural, development. Deutsches Institut für Entwicklungspolitik, Germany

Balat M, Balat M (2009) Political, economic and environmental impacts of biomass-based hydrogen. Int J Hydrogen Energy 34:3589–3603

Bergmann JC, Tupinamba DD, Costa OYA, Almeida JRM, Barreto CC, Quirino BF (2013) Biodiesel production in Brazil and alternative biomass feedstocks. Renew Sustain Energy Rev 21:411–420

Bleischwitz R, Bader N (2010) Policies for the transition towards a hydrogen economy: the EU case. Energy Policy 38:5388–5398

Bond R, Templeton MR (2011) History and future of domestic biogas plants in the developing world. Energy Sustain Dev 15:347–354

Cherubini F, Jungmeier G, Wellisch M, Willke T, Skiadas I, Ree V R, Ed de Jong (2009) Toward a common classification approach for biorefinery systems. Society of Chemical Industry

Chong ML, Sabaratnam V, Shirai Y, Hassan MA (2009) Biohydrogen production from biomass and industrial wastes by dark fermentation. Int J Hydrogen Energy 34:3277–3287

Cremonez PA, Feroldi M, Nadaleti WC, de Rossi E, Feiden A, de Camargo MP, Cremonez FE, Klajn FF (2015) Biodiesel production in Brazil: current scenario and perspectives. Renew Sustain Energy Rev 42:415–428

Demirbas A (2008) Biofuels sources, biofuel policy, biofuel economy and global biofuel projections. Energy Convers Manag 49:2106–2116

Demirbas A (2009) Biohydrogen for future engine fuel demands. Springer-Verlag, London, pp 1–276

Engdahl K (2010) Biogas policies, incentives and barriers—a survey of the strategies of three European Countries, Master thesis

Gaffron H, Rubin J (1942) Fermentative and photochemical production of hydrogen in algae. J General Physiol 26(2):219–240

Gest K, Kamen MD (1949) Studies on the metabolism of photosynthetic bacteria iv. Photochemical production of molecular hydrogen by growing cultures of photosynthetic bacteria. J Bacteriol 58(2):239–245

Hacker J, Mattingly J, Caputo J (2009) Food and crop waste: a valuable biomass feedstock. Environmental and Energy Study Institute, USA

He PJ (2010) Anaerobic digestion: an intriguing long history in China. Waste Manag 30(4): 549–550

Holland H (2010) Ecological farmer owning and operating a 70 kW biogas plant in connection to the farm in Ochsenhausen, Germany

Levin DB, Pitt L, Love M (2004) Biohydrogen production: prospects and limitations to practical application. Int J Hydrogen Energy 29(2004) 173–185

Magnusson L, Islam R, Sparling R, Levin D, Cicek N (2008) Direct hydrogen production from cellulosic waste materials with a single-step dark fermentation process. Int J Hydrogen Energy 33:5398–5403

Mofijur M, Rasul MG, Hyde J, Bhuyia MMK (2015) Role of biofuels on IC engines emission reduction. Energy Procedia 75:886–892

Ndaba B, Chiyanzu S, Marx I (2015) n-Butanol derived from biochemical and chemical routes: a review. Biotechnology reports. doi:10.1016/j.btre.2015.08.001

Oil World (1999) cited in Casson A (2003) Oil-palm, soy beans and critical habitat loss, a review

Poeschl M, Ward S, Owende P (2010) Prospects for expanded utilization of biogas in Germany. Renew Sustain Energy Rev 14(7):1782–1797

Rutz D, Thebaud A, Janssen R, Segura S, Riegelhaupt E, Ballesteros M, Manzanares P, St. James C, Serafini D, Coelho S, Guardabassi P, Aroca G, Soler L, Nadal G, Bravo G (2009) Biofuel policies and legislation in Latin America, BioTop Project. Seventh Framework Programme, European Commission

Sales A (2011) Production of biodiesel from sunflower oil and ethanol by base catalyzed transesterification, Msc thesis

Soccol CR, de Souza Vandenberghe LP, Medeiros ABP, Karp SG, Buckeridge M, Ramos LP et al (2010) Bioethanol from lignocelluloses: status and perspectives in Brazil. Bioresour Technol 101:4820–4825

Sorda G, Banse M, Kemfert C (2010) An overview of biofuel policies across the world. Energy Policy 38:6977–6988

Su Y, Zhang P, Su Y (2015) An overview of biofuels policies and industrialization in the major biofuel producing countries. Renew Sustain Energy Rev 50:991–1003

Wang M, Wu M, Huo H (2007) Life-cycle energy and greenhouse gas emission impacts of different corn ethanol plant types. Environ Res Lett 2(2):024001 (13 pp)

Feedstocks for Biofuels

Adenise Lorenci Woiciechowski, Adriane Bianchi Pedroni Medeiros, Cristine Rodrigues, Luciana Porto de Souza Vandenberghe, Valcineide Oliveira de Andrade Tanobe, Amélio Dall'Agnol, Decio Luiz Gazzoni and Carlos Ricardo Soccol

Abstract Considering the great interest of the countries around the world and the great number of research of many groups from Institutes, Universities, Companies, aiming to allow the use of biomass to produce biofuel, this chapter may give data and basic information about vegetable cultures and organic residues available to generate any kind of biofuel. Concerning to this a general view is given about crops where its main constituent component is sugar, starch, oil, with some principal examples. Besides, data, facilities and examples are given about lignocellulolytic cultures and residues, as far as solid and liquid residues generated in some industrial processes. The data available, includes, amount of each potential energetic source of raw material for biofuel, physic-chemical characteristics, production region, agronomic data, productivity, production, and other information useful for researchers works aiming the production of bioethanol, biodiesel, and biohydrogen.

Keywords Biomass, sugarcane · crop plants · lignocellulosic materials · organic waste

---

A.L. Woiciechowski (✉) · A.B.P. Medeiros · C. Rodrigues · L.P. de Souza Vandenberghe · V.O. de Andrade Tanobe · C.R. Soccol
Federal University of Paraná (UFPR), Curitiba, PR CEP 81531-970, Brazil
e-mail: adenise@ufpr.br

A.B.P. Medeiros
e-mail: adrianebpm@ufpr.br

C. Rodrigues
e-mail: crislabor@gmail.com

L.P. de Souza Vandenberghe
e-mail: lvandenberghe@ufpr.br

V.O. de Andrade Tanobe
e-mail: valcitanobe@gmail.com

C.R. Soccol
e-mail: soccol@ufpr.br

A. Dall'Agnol · D.L. Gazzoni
Embrapa Soybeans, Brazilian Agricultural Research Corporation (Embrapa), Londrina, PR, Brazil

C.R. Soccol et al. (eds.), *Green Fuels Technology*,
Green Energy and Technology, DOI 10.1007/978-3-319-30205-8_2

1 Introduction

In recent decades, many countries have promoted actions for the development of renewable energy; the biofuel have had significant participation in their energy matrixes. The main motivation for the biofuel policies include, for example, reduce dependence on fossil fuels, greenhouse gases reducing and increasing agricultural commodities demand (Ziolkowska 2013). To achieve these goals, besides the economic, environmental, and social conditions, the availability of raw materials production and the potential to produce biofuels must be analyzed.

Considering the alternative technologies involved in renewable energy generation and those commercially viable, only the use of biomass as feedstocks in processes with high efficiency, has the flexibility to supply both the electricity generation sector and the biofuels for transport.

Different biomass has been used in the production of biofuels comprising crop plants, lignocellulosic materials and also organic waste, which include agricultural residues, municipal, and industrial (Table 1). Microalgae and algae is a considered a third-generation feedstocks for biofuels (Sinha and Pandey 2014) .

2 Sugar-Containing Plant Crops

2.1 Sugarcane

The sugarcane is a semi-perennial plant of family grasses originally from Southeast Asia that has long thin stem of the genus *Saccharum* L. The sugarcane cultivated is

Table 1 Different feedstocks used for biofuel production

Feedstocks	Biofuel	Country	Reference
Corn, soybean oil, sorghum	Ethanol, biodiesel	EUA	Koçar and Civas (2013)
Sugarcane, soybean, palm oil	Ethanol, biodiesel	Brazil	Koçar and Civas (2013)
Rapeseed, sunflower, wheatsugar beet, barley, sewage, manure, food wastes, landfill	Ethanol, biodiesel, biogas	EU	Koçar and Civas (2013)
Corn, cassava, sweet potato, rice, jatropha	Ethanol, biodiesel	China	Koçar and Civas (2013)
Corn, wheat	Ethanol	Canada	Koçar and Civas (2013)
Wheat, sugarcane, molasses, palm oil, cotton oil	Ethanol, biodiesel	Australia	Koçar and Civas (2013)
Vinasse wastewater	Biohydrogen	–	Fernandes et al. (2010)
Cheese whey wastewater	Biohydrogen	–	Azbar et al. (2009)

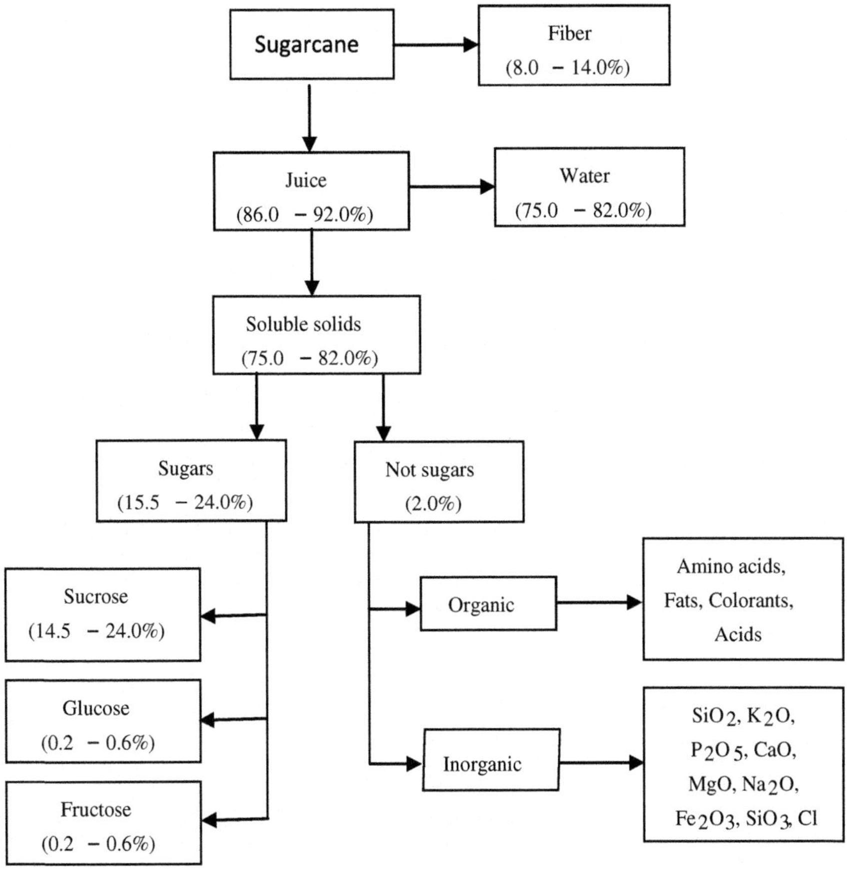

Fig. 1 Sugarcane mass balance (%). *Source* Adapted from Silva (2012)

a hybrid multispecies named *Saccharum* spp. It has a high concentration of sugar in the stalks, which makes up the aerial part of the plant while the sugarcane straw is at their tips and leaves (BNDES and CGEE 2008) is considered the main raw material for the manufacture of sugar and alcohol (ethanol).

The chemical composition of sugarcane varies widely, depending on weather conditions, the physical, chemical, and microbiological soil, variety, maturation stage, and others. In general the stalks are composed 65–75 % water, 8–14 % ashes and 10–17 % sugars (BNDES and CGEE 2008). A more detailed flowchart composition is shown in Fig. 1.

It fits easily in tropical climates, since it requires good amount of rainfall and sunlight. The tropics is privileged from the energy point of view, it presents better conditions for biomass production. In this sense, Brazil is the largest single producer of sugarcane with about 27 % of global production. As for the yield, Peru is first with 32 dry Mg/ha followed by Brazil with 18 dry Mg/ha (Kim and

Dale 2004). The Brazil should achieve average rate of increase in production of 3.25 % until 2018/19, and reap 47.34 million tons of sugarcane. For exports, the estimated volume for 2009 is 32.6 million tons (MAPA 2016).

Different by-products and residues are obtained by ethanol production and are used for energy, animal feed and as fertilizer. Sugarcane bagasse is the major subproduct and is obtained from juice extraction after the crushing of sugar cane. The production of ethanol from sugarcane is more energy efficient than others fonts (corn, sugarbeets, vegetable oils), mainly if sugarcane bagasse is used to produce power for the process and transport (IEA 2007; Lora and Nascimento 2004). Furthermore, using highly efficient boilers, up to 52 % of the bagasse would become available for other uses, such as biofuel production (Botha and Blottnitz 2006).

Sugarcane bagasse is composed essentially of cellulose, hemicellulose, lignin, and extractives. Several research groups studies technologies for sugarcane bagasse use with second-generation biofuels, for example, aiming the selection of sugarcane varieties or to down-regulate lignin biosynthesis in transgenic plants (Petersen et al. 2015). The process normally uses a pretreatment to biomass deconstruction to overcome recalcitrance and conversion to biofuels by enzymatic hydrolysis and subsequent fermentation (Jung et al. 2010).

2.2 *Sugarbeet*

The beets used in the ethanol production are sugar beet (*B. vulgaris*) also known as white beet originally from Europe. It is a plant whose tuber contains a high concentration of sucrose, and although these nonfood beets would not be efficient feedstock for the production of sugar for human consumption, it is one of the main raw materials for the production of biofuel.

Sugar beets are generally grown in the high-altitude region and in temperate climate but due to genetic enhancement, the crop has proven to adapt to various soil and climatic conditions (Içöz et al. 2009). It is recommended that these beets are grown in 3–5 year rotation with other crops to improve soil fertility and manage diseases and nematodes (Ali 2004).

Sugar beets are composed of about 75 % water, 18 % sugar, and 7 % insoluble and soluble materials. Because they have high sugar content beets, they are being considered for biofuels production. Most of the sugar beet is in the form of sucrose but other sugar as maltose, glucose, fructose are present, though these does not interfere with fermentation and distillation for the ethanol production (Haankuku et al. 2015).

Many countries have adopted bioethanol inclusion policies and sugar beet for its high content of sugars and not compete with food, it has been considered with potential for ethanol production. Theoretical ethanol yield gal/ton for sugar beet is 24.8–26.9 competing with other saccharide crops such as sugarcane (15.5–18.6) (Szulczyk et al. 2010). Besides, that could potentially double ethanol production per hectare compared to other feedstocks (corn, cellulose) (Shapouri and Salassi 2006; Panella and Kaffka 2010). Although the sugar beet area has decreased around 20 % between 2007 and 2010 in the EU (Eurostat 2011), USA Energy Independence and Security Act (EISA) of IEA (2007), was considered that sugar beets may be an eligible feedstock for advanced biofuel (NREL 2014).

2.3 Sweet Sorghum

Sweet sorghum is a perennial plant of Andropogoneae tribe and sub-family Panicoidae, Polaes order, *Poaceae* family, genereo *S. S*orghum species (Ratnavathi et al. 2010); is a native of tropical grass countries Africa, the Sudan, Ethiopia. *Sorghum* saccharine size is high, more than three meters, featured mainly due to its sweet and juicy stem as the sugarcane. The panicle (bunch) is open. It produces few grains (seeds).

Sorghum is a versatile crop, since their grain from the stalks and different products can be obtained, such as sugar, ethanol, paper, and other chemical compounds (Ratnavathi et al. 2010). The chemical composition of the juice obtained from stalks may result (Sipos et al. 2009) high ethanol productivity depending on the cultivar. Pereira Filho et al. (2013) noted that most of the characteristic value for the cultivar BR 506, reached 24. 895 L ha^{-1}, followed by BR 505, with 23.286 L ha^{-1}. However, in relation to the other cultivars (BR 505, 507, 501, and 601), the differences in relation to cultivate more productive were, respectively 1.609, 3.846, 4.609, and 8.194 L ha^{-1}.

Sweet sorghum cultivars are characterized by the accumulation of high levels of fermentable carbohydrates (15–23 %) within the stalk (Sarath et al. 2008; Smith et al. 1987). Total fermentable carbohydrates are comprised of three main sugars; sucrose (70 %), glucose (20 %), and fructose (10 %) variation in percentages depends on variety and environmental conditions (Prasad et al. 2007). Sweet sorghum requires less water and contains higher FC levels than corn, making it a favorable biofuel crop for semiarid temperate climate regions (Reddy et al. 2007). Sugar content in the juice increases with maturity, and is low prior to seed development. Sweet sorghum is typically seeded in widely spaced rows (30–40 inches). The ideal seeding rate for most sweet sorghum varieties is 3–4 seeds per linear foot of row with a final stand of 2–3 plants per linear foot of row. If plant populations are too high, the stalks will be spindly and contain less juice (Shoemaker and Bransby 2010).

Because of its agronomic flexibility and productivity, shorter growing cycle and percentage sugars of the same order of sugarcane,(Cunha and Severo Filho 2010) sweet sorghum is viewed as a viable feedstock option for ethanol production in some regions of the world (Davila-Gomez et al. 2011). In addition to industrial and agronomic characteristics, it can be used in the same system for the production of sugarcane, since it has the same physical characteristics (stalks), not requiring handling or to modify the facilities. In some countries (USA and Brazil), it is already being used in conjunction with sugarcane to increase the production of ethanol (Pereira Filho et al. 2013).

3 Starchy Crops

Among biofuels feedstocks, there are the starchy materials such as corn, cassava, wheat, and barley (Balat et al. 2008). However, corn is the most employed feedstock that is significantly used for bioethanol production. Starch materials must pass through an acid and or enzymatic pretreatment so as to produce a high sugar concentration for biofuel production. The following flowchart (Fig. 2) illustrates the main steps of the starchy materials till biofuel production.

3.1 Corn

In the 2013/2014, USA's corn production reached nearly 13.8 billion bushels (351.3 million metric tons) of corn. More than one-third of USA's corn crop is used to feed livestock, 13 % is exported and 40 % is used to produce ethanol. The remainder goes toward food and beverage production (Carter and Miller 2012; EIA 2013).

Corn stover, the residue left in the fields after harvesting corn, has been identified as a near- to mid-term agriculture residue feedstock for the lignocellulose-to-ethanol process. Corn stover has high carbohydrate content, can be collected in a sustainable fashion, and provides economic benefits to the farm community. Corn kernels have starch, which is an α-linked glucose polymer that can be easily broken down to glucose monomers and fermented to ethanol. It has fiber, which encases

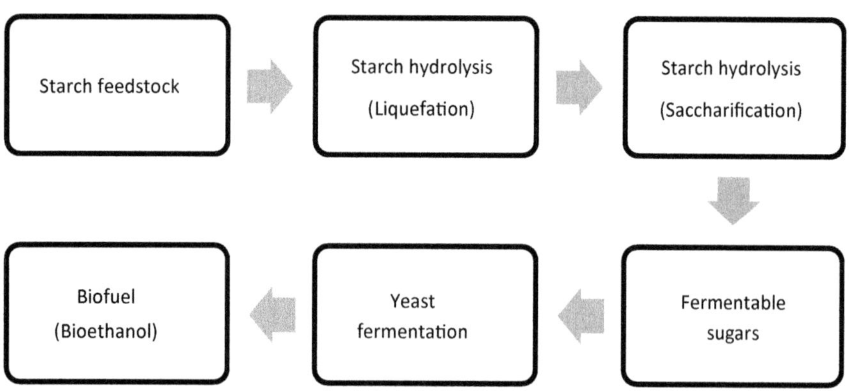

Fig. 2 Schematic diagram of bioethanol production by fermentation process of starch feedstock

Table 2 Corn kernel and corn stover compositions

Corn kernel	% Dry basis	Corn stover	% Dry basis
Starch	72	Cellulose	37.3
Hemicellulose/cellulose	10.5	Galactan/mannan	1.4
Protein	9.5	Xylan	20.6
Oil	4.5	Arabinan	2.1
Sugars	2.0	Lignin	17.5
Ash	1.5	Ash	6.1
		Acetate	2.0
		Extractives	13.0
% Humidity		% Humidity	

Source Watsom and Hamstad 1987

the starch, and about 15 % moisture. The comparative composition of corn kernel and corn stover is presented in Table 2 (Watsom and Hamstad 1987).

Currently, the maximum amount of pure ethanol that can be made from a bushel of corn is 2.74 gallons (98 gallons per ton at 15 % moisture or 115 gallons per dry ton). Yield is primarily dependent on the starch content, which may vary considerably.

Corn stover contains considerable quantities of cellulose, a beta-linked glucose polymer, which is more difficult to break down to glucose monomers than the α-linked polymer in starch. In addition, it contains hemicellulose, which is a more complex polymer of several sugars. The predominant sugars in hemicellulose are xylose and arabinose. These five-carbon sugars can also be fermented to ethanol with the proper microorganism. The maximum theoretical yield from corn stover with the composition is 107 gallons per dry ton (or 91 gallons per ton at 15 % moisture). Around the two sugar polymers is lignin. Lignin has an interesting by-product value and can be sold for different applications.

It is known that 1 acre yields about 130 bushels (3.65 tons at 15 % moisture) of corn, (USDA 2015) and about 1 ton of harvested corn yields 1 dry ton of stover. With an estimated 240 million dry tons of stover produced, the 80 million dry tons available for harvesting is equivalent to 6 billion gallons of ethanol (Glassner et al. 1998).

The U.S. Department of Agriculture (USDA) has a program devoted to the corn ethanol industry. Areas of scientific research address the establishment of new higher-value ethanol coproducts, the development of microbes capable of converting various biomass materials into ethanol, improved processes for the enzymatic saccharification of corn fibers into sugars, and various methods of improving corn ethanol process efficiencies (McLoom et al. 2000).

Fuel ethanol production from corn can be described as a five-stage process: raw material pretreatment, hydrolysis, fermentation, separation and dehydration, and wastewater treatment. The production of bioethanol from starch includes the breakdown of this polysaccharide to obtain an appropriate concentration of fermentable sugars, which are transformed into ethanol by yeasts. After washing, crushing, and milling the corn grains (dry milling process), the starchy material is gelatinized in order to suceptilize the amylose and amylopectin for enzymatic attack in the following liquefaction step. This step is considered as a pretreatment process because of the partial hydrolysis of the starch chains using thermostable α-amylase. The hydrolyzate obtained has reduced viscosity and contains starch oligomers called dextrins. Then, the fermentation process occurs where sugar is immediately assimilated by the yeast *Saccharomyces cerevisiae* in the same reactor and converted into ethanol. The culture broth containing 8–11 % (w/w) ethanol is recovered in a separation step consisting of two distillation columns (Quintero et al. 2008).

3.2 Cassava

Cassava is a shrub with tuberous roots. World production of cassava is around 281 million tones (Mt) a year. Africa contributes to more than half of global supply. Asia encourages the development of cassava crops for industrial and energy purposes. This continent contributes to around a third of world production, with 26 Mt produced by Thailand and 28 Mt by Indonesia. In Latin America, production is around 35 Mt where Brazil dominates with around 70 % of regional production and in third place in world production (Conab 2013).

Cassava is primarily grown for its roots but all of the plant can be used: the wood as a fuel, the leaves and peelings for animal feed and even the stem as dietary salt (UNCTAD 2015). Cassava is used in both human and animal food, in many industrial sectors, particularly in the form of starch, and more recently to produce ethanol. The current market price for fresh cassava roots is based on the food market price. Revenues are based on farm gate prices for fresh cassava roots that fluctuate due to seasonal influences and supply and demand (Van Eijck et al. 2014).

Cassava has starch 59–70 % of starch in its composition (Table 3), which is a polysaccharide comprising solely of glucose monomers that are linked together by glycosidic bonds. It is composed of two types of glucan namely amylose, a linear glucose polymer having only α-1,4 glycosidic linkage and amylopectin, a branched glucose polymer containing mainly α-1,4 glycosidic linkage in a linear part and a few α-1,6 at a branch structure (Sriroth et al. 2012).

Table 3 Cassava composition

Cassava	%
Moisture	59–70
Starch	77–94
Fiber/cell wall materials	1.5–3.7
Protein	1.7–3.8
Lipid	0.2–1.4
Ash	1.8–2.5

Source Breuninger et al. (2009)

Starch granules are less susceptible to enzyme hydrolysis. Upon cooking in excess water, the granular structure of starch is disrupted, making glucose polymers become solubilized and more susceptible to enzyme attacks. At the same time, the starch slurry becomes more viscous. This process is known as gelatinization and the temperature at which starch properties are changed is named as gelatinization temperatures. Different starches have different gelatinization temperatures that lead to different thermal treatment conditions (Swinkels 1998; Thirathumthavorn and Charoenrein 2005).

Cassava is still a small player on the biofuel scenario. In effect, with one ton of cassava, which has a starch content of 30 %, around 280 L can be produced of 96 % pure ethanol (Sriroth et al. 2012). The starch hydrolysis by enzymes is a two-stage process involving liquefaction and saccharification. Liquefaction is a step where starch is degraded by α-amylase, which hydrolyzes only α-1,4 and causes viscosity reduction of starch. Liquefying enzymes usually work at high temperatures (>85 °C) so that the enzyme can help reduce starch paste viscosity during cooking. Dextrins, which are obtained after liquefaction, are further hydrolzyed to glucose by glucoamylase enzyme. These enzymes can hydrolyze both α-1,4 and α-1,6 glycosidic linkage. Glucose is then converted to ethanol by yeast. After fermentation, approximately 10 % (v/v) ethanol are obtained and subjected to distillation and dehydration to remove water and other impurities, yielding anhydrous ethanol (Sriroth et al. 2012).

Nowadays, the production process of bioethanol from starch feedstock is developed to significantly reduce processing time and energy consumption by conducting saccharification and fermentation in a same step. This process is called "Simultaneous Saccharification Fermentation", or SSF process (Sriroth et al. 2012). In this SSF process, the liquefied slurry is cooled down to 32 °C, afterward glucoamylase and yeast are added together. While glucoamylase produces glucose, yeast can use glucose to produce ethanol immediately. No glucose is accumulated throughout the fermentation period (Rojanaridpiched et al. 2003).

4 Oil Seeds

Oilseeds are among the most important crops in international trade. Annually, world consumption of vegetable oils and fats exceeds 300 Mt (USDA 2015d). According to the FAO database (Faostat 2015), world production of vegetable oils increased more than 600 % in 40 years, jumping from 23.6 Mt in 1972/1973 crop season versus 180 Mt in the 2014/2015 crop season being produced mainly in United States (USA), China, Brazil, India, Argentina, and Indonesia. The oil produced is used mainly in nutrition, but also at industry application, fine chemistry, and energy.

There are hundreds of species with potential to provide oil for domestic use or as a raw material for oil chemistry or biodiesel industry. However, few of them have characteristics such as high oil content, well-structured supply chain and production

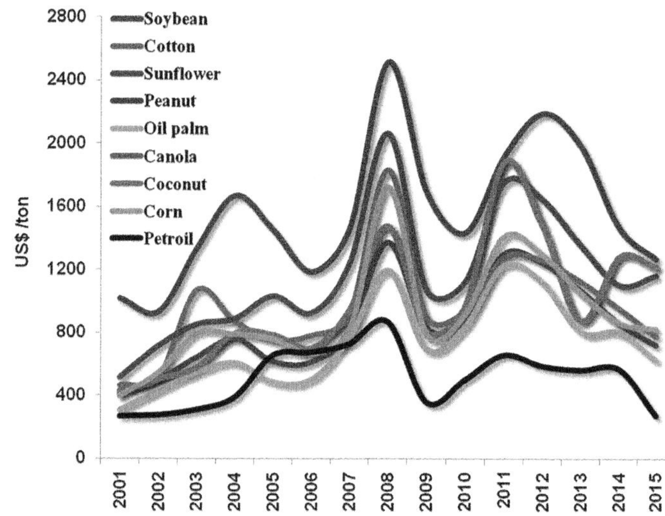

Fig. 3 Vegetable oil and petrol prices on the international market, adjusted to Sept, 2015

technology that justify their large-scale farming. About 80 % of vegetable oil produced worldwide comes from only four oil crops: palm (pulp + almond), soybean, rapeseed (canola), and sunflower. Four other crops account for the next 11 % share: peanut, cotton, coconut, and olive much suitable for nutrition application. Completing the world oil production dozens of other oil producing plants includes corn, castor, flaxseed, sesame, jatropha, jojoba, peanut among others.

As for biodiesel production, with a global production of about 35 billion liters (GL) (REN21 2015), and which demands annually over 30 Mt of vegetable oil and animal fat, four aspects are crucial for a given crop to be considered as a feedstock (Gazzoni et al. 2012): (a) large production; (b) well-organized value chain; (c) insertion as a commodity in the international market; and (d) competitive price, as compared to other oils, but specially against petrol, the fossil energy paradigm.

Figure 3 shows the evolution of market prices of the internationally traded vegetable oils, compared to international petrol prices. Each 1 % biodiesel added to mineral diesel results in the creation of 45,000 jobs, according to estimates of the Ministry of Agrarian Development of Brazil (Abreu et al. 2012).

4.1 Feedstock for Biodiesel Production

Depending on the oil content, yield and harvest of the seeds, the resulting oil volumes obtained from each hectare is variable according to the crop. Figure 4 presents the consolidated world oil production for the last 54 years, representing an eightfold increase.

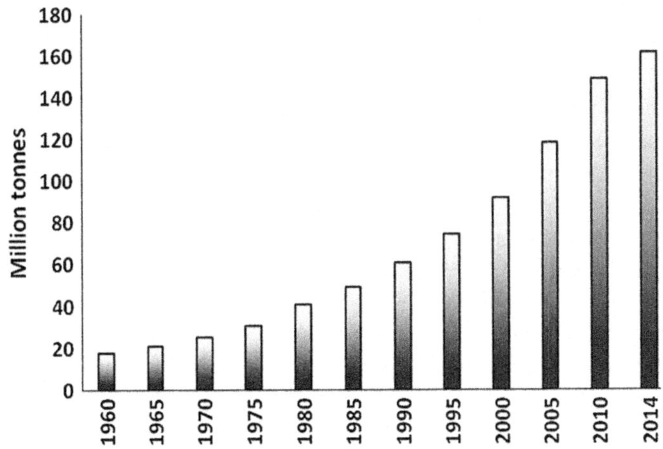

Fig. 4 World oil production from major oil crops

Oils have different characteristics, according to the crop. Oils are used for different purposes at industrial level (nutrition, oilchemistry, bioenergy, etc.). Its average composition included mainly saturated and unsaturated fatty acids. The four most important oil crops cultivated and traded in the international market, representing over 90 % of the global biodiesel feedstock are analyzed below.

4.1.1 Soybean (*Glycine Max* (L.) Merrill)

Soybean is one of the most important crops in the world, especially due to the high quality of its protein meal. Oil content (18–20 %) in the seed is lower than protein (36–40 %), but due to the large amount of soybean meal demanded to feed meat animals, the resulting oil volume is significant (EMBRAPA 1994).

In 2014, about 45 Mt of soybean oil was produced worldwide, being second only behind palm oil. In fact, soybean oil leads the vegetable oil production until last decade, and might be the leader again, in the near future. In 2013, Argentina cultivating 19.42 Mha with productivity of 2539 kg/ha produced 49.31 Mt, Brazil in 27.91 Mha, with a productivity of 2929 kg/ha produced 81.72 Mt and USA in 30.7 Mha, yielding 2915 kg/ha produced 89.48 Mt of soybeans. (FAOSTAT database).

Nowadays, in addition to the food market for humans and pets nutrition, new markets like bioenergy and oil chemistry, extend the horizons of soybean demand, increasing annual growing rates since 1990, when global production (108 Mt) was just one-third of the current (315 Mt) production. The mean annual growth during the last 20 years surpassed 8 Mt (Faostat 2015).

4.1.2 Oil Palm (*Elaeis Guineensis*)

Originally from the Gulf of Guinea, west central Africa, it is also known as African palm (dendê, in Brazil). Even known and used for millennia, its commercial cultivation started on the first decade of the twentieth century, in Malaysia. Palm is recognized as the oil crop that produces the largest amount of oil per hectare, and is responsible for 57 Mt of the global production of 170 Mt of vegetable oils (www.statista.com). Along with soybean oil, it accounts for over 60 % of the world vegetable oil production, but, considering the 2015 average oil yield of each crop, one hectare of palm oil yields the same amount of oil as 10 hectares of soybeans. (Corley and Tinker 2003)

The high oil yield allow palm oil to occupy only 8 % of the world area cultivated with oil crops, but providing almost a third of vegetable oil produced globally (Faostat 2015). Due to its tropical origin, palm oil is cultivated at humid tropics, as well as Southeastern Asian, Northwestern South America, and part of Central America. Presently, Asian countries account for nearly 90 % of its cultivated area. Indonesia, Malaysia, and Thailand are the major producers (FAOSTAT 2015). The largest importers are also located in Asia (China and India). In 2013, according to FAOSTAT database, Indonesia, Malaysia, Thailand, Nigeria, and Colombia produced, respectively, 26.9, 19.22, 1.97, 0.96, and 0.95 Mt of palm oil fruit.

The palm oil is used for bioenergy production, but the largest use is in the nutrition segment (industrial frying, chocolates, pasta, margarine, vegetable creams, cookies, ice cream), and in cosmetics industry (beauty products, shampoos, detergents, and soaps). The energetic balance (input/output) of biodiesel from palm oil is very favorable, sometimes reaching up to 1:8, according to Gazzoni et al. (2008).

The area presently used for palm oil cultivation in Southeastern Asia countries was formerly occupied by native forest, causing intense deforestation, collapsing the rainforest in countries like Indonesia within a decade.

In contrast, Brazil has the world largest reserve of suitable land for palm cultivation, estimated to be around 50 Mha (Müller 1980), but cultivates only 0.16 Mha (Faostat 2015), due to the restrictions imposed by the Brazilian environmental legislation for Amazonian lands, which restricts to 20 % the amount of area of a given farm, preserving 80 % of the biome (Müller and Furlan Junior 2001). So, Brazil is the nineth palm oil producer (0.37 Mt), resulting in continuous import of palm oil.

Palm oil is a perennial and huge oil production plant. However, its residues after oil extraction have only marginal or no commercial value. The main uses for palm oil residues are organic fertilizer or electricity generation by burning the waste. Palm oil fruits produce the palm oil itself, extracted from the pulp; and palm kernel oil, extracted from the fruit kernel. The oil fraction constitutes about 22 % of the weight of the palm bunch, and only 3 % is palm kernel oil. Lauric acid is almost absent in palm oil, the predominant component of palm kernel oil (Ramos et al. 2009). In 2013, Indonesia, Malaysia and Thailand produced 26.9, 19.22, and 1.97 Mt of oil palm, respectively, and 3.06, 2.27 and 0.18 Mt of Kernel oil. Then Nigeria and Colombia produced 0.96 and 0.95 Mt of palm oil and 0.51 and 0.08 Mt of kernel oil, respectively (FAOSTAT database).

4.1.3 Canola (Rapeseed) (*Brassica Napus* L.)

Rapeseed belongs to the Brasscicaceae family (formerly Cruciferae), the same family as mustard, broccoli, or cauliflower. The canola name results from a contraction of "CANadian Oil Low Acid", a variety of rapeseed modified in the 1970s by traditional breeding, by Canadian scientists from the University of Manitoba selected varieties, which oil has low erucic acid (toxic for humans and animals). The rapeseed differs from canola due to high levels of erucic acid and glucosinolate present in the grains.

Brassica oilseed varieties are among the oldest plants cultivated by humanity, with documentation of its use in India 4000 years ago, and in China and Japan 2000 years ago. Because of its lubricant properties, there was a high demand for rapeseed oil during World War I to supply the increasing number of steam engines in naval and merchant ships. After the war, the lubricant demand declined sharply, and other uses for the oil were developed (USDA 2015a).

Presently, canola is the leading group of varieties grown worldwide as rapeseed. The oil is the main product of canola, although its meal is also highly valued for the formulation of animal feed, because of the high protein content. According to De Mori et al. (2014), the oil content of canola seeds is high (38–45 %) and the volume of oil produced worldwide is surpassed only by palm and soybean oil.

Low amounts of unsaturated fatty acids are found in canola oil, being palmitic (16:0) the one with higher content (4 %). The major fatty acid found in canola is the mono-unsaturated oleic (18:1) (63 %), followed by polyunsaturated linoleic (18:2) (20 %) and linolenic (18:3) (9 %) (Ramos et al. 2009). High prices of canola oil make biodiesel from canola costly for the market and for supporting public policies. As for energy balance of biodiesel from canola oil, considering meal utilization, it was concluded that for each input energy unit along the life cycle 2.9 energy units are obtained; when considering only oil production (not computing energy on the meal) this relationship decreases to 1:1.4 (Gazzoni et al. 2009).

In 2014/15, world production of canola was 72 Mt of grains allowing the extraction of 26 Mt of oil, representing 16 % of global vegetable oil production (Faostat 2015). The leading production region is the European Union (24.0 Mt), followed by China (14.7 Mt), Canada (14.45 Mt), India (7.5 Mt) and Japan (2.0 Mt). Canola grain contains around 40 % of oil; Canola is more adapted to mild temperature regions, distant from the Equator (USDA 2015a). In 2013, according to FAOSTAT database, Europe, Canada, and China produced 9.91, 2.83 and 5.6 Mt of canola oil.

4.1.4 Sunflower (*Helianthus Annuus* L.)

The center of origin of sunflower is the region comprising Southwest USA and Northern Mexico, from where it disseminated to the rest of the continent. Its most likely domestication occurred in that region, where there is evidence of its cultivation by North American Indians over 3000 years ago (Lentz et al. 2001).

Russia was largely responsible for the spread of sunflower as a worldwide economically important crop. The importance of sunflower as edible oil source, only emerged by the 1920s. However, it was after World War II that sunflower aroused to the front line of oil crop production (USDA 2015b).

The global sunflower cultivated area in 2014 was, approximately, 18 Mha, with an overall production of 40 Mt of grain, 16 Mt of oil, and 17 Mt of meal (Faostat 2015), ranking fourth among the most important oils and meal production, globally. Sunflower vegetable oil is about 7.5 % of world production, behind palm (34 %), soybeans (30 %), and canola (16 %). The oil content of the seeds is about 45 %, consumed almost completely as edible oil for its excellent quality, while protein content situates on the range of 28–32 % (Leite et al. 2005; USDA 2015b).

According to Ungaro (2000), sunflower requires insensitive photoperiod and can be cultivated from the vicinity of the equator to latitudes above 40°. The optimum temperatures for proper plant growth are between 27 and 28 °C, but develops quite satisfactorily from 8 to 34 °C, being a good second summer crop (off-season) and an agronomic important option for rotation with soybeans, corn, and wheat.

The oil is rich in unsaturated fatty acids, like the monounsaturated oleic (18:1), with 16 % and the polyunsaturated linoleic, with 72 %; major saturated fatty acids are palmitic (16:0), with 6 % and stearic (18:0), with 4 % (Ramos et al. 2009).

Sunflower is used as ornament plant and its meals are used for feeding domestic bees, and sillage (animal fodder). The nutritional quality of sunflower oil is similar to the canola oil, being highly suitable for biodiesel production (Leite et al. 2005)

Regarding to energy efficiency of biodiesel production from sunflower oil, Gazzoni et al. (2005), using Life Cycle Analysis techniques, determined that with the whole grain destination (meal for nutrition, oil for biodiesel), 2.69 units energy were obtained from each energy unit input to the system. This relation was reduced when meal was not considered then each unit of input energy represented 1.61 units obtained from biodiesel use.

4.1.5 Minor and Potential Oil Crops

A series of species are used locally, even regionally, for oil production in small scale. Some are directed for self-consumption, either for human or animal nutrition, for soaps or energy production. Represents less than 5 % of the world oil production are restricted to commercial or purposes niches many of them based on native production and extractive systems but with median to high oil content, and a theoretical potential for oil production. Its commercial development depends on (a) possibility of production of over 500 kg/ha of oil, in order to compete with major oil crops; (b) domestication of the species; (c) establishment of production systems; (d) organization of the productive chain connecting growers, suppliers, processors, industry, and consumers. Among others, besides cotton and peanut, potential oil crops include, castor, oil radish, flaxseed, sesame, safflower, crambe, tucuman, oiticica, tung, pequi, jatropha, jojoba.

5 Lignocellulosic Wastes

Second-generation biofuels produced from (larger) feedstocks from lignocellulosic materials include cereal straw, forest residues, bagasse, and purpose-grown energy crops such as vegetative grasses and short rotation forests (Demirbas 2009). Among these sources for biofuel production, the percentage of sugar is variable, as well as the conversion processes used in the production of biofuel. A few companies in European Community (Gnansounou 2010) and USA have operated pilot plants to make cellulosic ethanol but no commercial amounts of the fuel are being made (Banerjee et al. 2010).

The main routes for obtaining bioethanol from lignocellulosic sources comprise several steps: pretreatment for delignification and release the cellulose and hemicellulose fractions; hydrolysis of cellulose and hemicellulose to fermentable sugars obtained (glucose, xylose, galactose, mannose, arabinose) (Sarkar et al. 2012). Furthermore, due to the high cost of enzyme, the current fuel grade ethanol produced from lignocellulosic material is still not able to compete with gasoline. In a contemporary process of lignocellulosic ethanol which is being worked out for more than 2–3 decades is not yet materialized into a viable technology. The permissible cost of enzymes is 15–30 cents/gallon of ethanol which is still not a reality (Menon and Rao 2012). Lignocellulosic materials could produce up to 442 billion liters per year of bioethanol (Balat 2011).

Among the main waste generated in the world and Brazil is sugarcane bagasse, rice hulls, oat hulls, straw and cob and corn husks, which have in their chemical pulp composition, hemicellulose, lignin, and other compounds (Sarkar et al. 2012).

5.1 *Sugarcane Bagasse*

The solid waste generated after processing the sugarcane is called bagasse. The chemical composition is 40 % cellulose, 25 % hemicellulose, 20 % lignin, and 10 % of other chemical compounds; it can be estimated that a ton of the pulp to produce approximately 300 L of ethanol (Halling and Simms-Boore 2008).

Sugarcane planted area in Brazil grew by 7.56 % per year during the last decade. The state of São Paulo was responsible for 55.3 % of all Brazilian sugarcane planted area in 2010, appreciating even more arable land values (Meyer et al. 2013).

The Brazilian Energy Plan scenarios estimate a mass sugarcane bagasse offering to be used only for second-generation ethanol around 7.0 $\times$ 106 tons year-1 for 2015, and 25.9 $\times$ 106 tons year-1 for 2030 (Hofsetz and Silva 2012).

Average productivity of sugarcane in Brazil is 85 tons per hectare; each ton of processed cane generated about 140 kg of straw and 140 kg of bagasse (dry basi), i.e., 12 tons of straw and 12 tons of bagasse. Assuming that the conversion of glucose to ethanol is complete, then full use of sugarcane (thatched, straw, and bagasse) can significantly increase ethanol production per hectare, from the current 7000 L to about 14,000 L. Sugarcane straw is 15 % of the weight of the stalks of sugar can ripe, or 12 % when seca. 13.29. In energy terms is the straw that is one-third of the potential energy of sugarcane that is currently underutilized (Santos et al. 2012)

Currently, 6000–7000 L of ethanol is produced from one hectare of sugarcane—not including the bagasse. When bagasse can be utilized for ethanol production, the output is likely to double to 12,000–15,000 L per hectare (Halling and Simms-Boore 2008).

5.2 Rice Husk

Rice (Oryza sativa) is a herbaceous plant included in the class Liliopsida (Monocotyledon), order Poales, Poaceae family, genus Oryza. It is one of the cereals produced and consumed in the world, characterized as staple food for over half the world's population. The annual rice production is approximately 606 Million tons. In this scenario, Brazil participates with 13.140.900t (2.17 % of world production)) (FAO 2015). Global production for 2015/16 is up from last month due to larger crops in China, the Philippines, and Mali, but remains at its lowest level in 4 years (FAO 2015).

Rice husk (RH), which is part of the rice paddy (rice grain), is a by-product of the rice milling process that involves the separation of the husk and bran (the outer layer of the rice grain) from the edible portion. Global production of RH is very significant and falls in the range of tens of millions of tons per annum. This presents an attractive opportunity to utilize such waste material for further processing particularly for the conversion into bioethanol. Typically about 50 % of the husk produced in a rice mill is burnt onsite to produce steam to drive the mechanical milling machinery (Abbas and Ansumali 2010)

Rice husk is composed mainly of cellulosic sugars. Being a lignocellulosic material, RH also contains lignin, which is present in up to 20 % of the husks. After gasification, RH ash is produced containing a useful secondary product—silica (SiO_2). Silica has been shown to be present in RH ash in high quantities varying from 15.30 to 24.60 % (Abbas and Ansumali 2010)

Rice straw is one of the abundant lignocellulosic waste materials in the world. It is annually produced about 731 million tons which is distributed in Africa (20.9 million tons), Asia (667.6 million tons), Europe (3.9 million tons), America (37.2 million tons), and Oceania (1.7 million tons). This amount of rice straw can potentially produce 205 billion liters bioethanol per year, which is the largest amount from a single biomass feedstock (FAO 2015).

5.3 Corn Stover

Corn (*Zea mays* L.) is a plant belonging to the family Gramineae/Poaceae. It is a monocotiledone slender stem, which can reach two meters in height (Thompson and Tyner 2014).

Corn stover consists in the different parts of the plant, which are the cobs, husks, stalks, leaves, and tassel (Thompson and Tyner 2014; Qureshi et al. 2010). Corn stover was reported as an average level of pulp (33–43 %), hemicellulose (20–34.5 %), lignin (8–14.1 %), protein (5 %), ash (4 %) (Aguiar and Ferraz 2011).

The potential amount of bioethanol derived from corn stover could replace 42:1 GL of gasoline used in a midsize passenger vehicle fueled by E85 (a mixture of 85 % ethanol/15 % of gasoline by volume), or about 3.8 % of world annual gasoline consumption (Kim and Dale 2004).

The United States is predominantly a producer of bioethanol derived from corn. Feedstock availability is not expected to be a constraint for bioethanol production over the next decade. Corn is expected to remain the predominant feedstock in the United States, although its share likely will decline modestly by 2015 (Balat 2011). In US, corn ethanol is currently the predominant biofuel, and is already using over 30 % of the corn produced (ERS 2010) though over 90 % of waste (corn stover) are left in the field (Kim and Dale 2004).

The current US stover yield (average from 2006 to 2010) was 7.3 Mg ha^{-1}. The annual total production was 237 Tg at present (2006–2010) and is projected to be 261 Tg in 2022 and 303 Tg in 2050 with an assumption of no changes in the total harvested area. Of the stover production, the cobs account for about 18 % (Tan et al. 2012).

5.4 Wheat Straw

Wheat (*T. aestivum*) is the world's most widely grown crop, cultivated in over 115 nations under a wide range of environmental conditions (Talebnia et al. 2010). Over the past 100 years, the yields of wheat have been increased and annual global production of dry wheat in 2008 was estimated to be over 650 Tg.

The overall chemical composition of wheat straws could slightly differ depending on wheat species, soil, and climate conditions. Cellulose, hemicellulose, and lignin content of wheat straw are in the range of 33–40, 20–25, and 15–20 (%w/w), respectively (Prasad et al. 2007).

The straw produced might be left on the field, plowed back into the soil, burned or even removed from the land depending on the decision made by landowner. Disposal of wheat straw by burning has been practiced for a long time. In recent years however, this practice has been challenged due to increased concern over the health effects of smoke from burning fields. Thus, finding an alternative way for disposal of surplus wheat straw is of high interest and an immediate necessity (Kerstetter and Lyons 2001).

6 Other Solid and Liquid Wastes

The organic waste from urban activity, rural, and mostly agricultural industry has been submitted to anaerobic digestion process. While reducing the pollution potential of waste, these processes provides end products as biogas or hydrogen. The production of biohydrogen and biogas from these sources is considered a promising solution for the energy demand (Lin et al. 2012).

Many studies have been conducted and projects have been developed at different scales to the development of the digestion process (Cortez et al. 2011). The kind of feedstocks is one of the factors related to the biodigester performance. The volatile solid content, the lignin content, and the C/N ratio of the waste influence the level of biological activity and consequently the production rates.

6.1 Wastewater

The wastewater from sewage or resulting from industrial processes is traditionally discarded by the industry as waste but can be used as feedstock to produce biohydrogen and biogas.

Many kinds of wastewaters are being studied in order to establish the process and performance parameters of fermentative biohydrogen production. Due to its low productivity and yield, the biohydrogen production on commercial scale is still developing and needs studies to become viable (Lin et al. 2012).

6.1.1 Vinasse

Vinasse is the main liquid stream from the first-generation ethanol production process. It is collected from the bottom of ethanol distillation columns. Due to its high level of organic compounds and nutrients, vinasse is a potential pollutant. In Brazil, sugarcane processing plants generally generate from 10 to 15 L of vinasse per liter of produced ethanol. More than 320 billion m^3 of vinasse were produced in 2014/2015 (UNICA 2015).

This residue has been tested as feedstock for biohydrogen and biogas. Fernandes et al. (2010) found vinasse as the highest potential feedstock for hydrogen production among other wastewater tested. The hydrogen yield was 25 mmol H_2/g COD.

The vinasse characteristics are dependent on the raw material. In the case of sugarcane vinasse, its composition also varies according to the fermentation feedstock. Bioethanol are mainly produced from sugarcane juice and/or molasses or corn (Moraes et al. 2015). The sugarcane vinasse characteristics are presented in Table 4.

Table 4 Physicochemical characteristics of sugarcane vinasse

Characteristic	
pH	3.8–5.0
Total solids (g/L)	21–85
Soluble solids (g/L)	4–31
Non-soluble solids (g/L)	3–13
COD (mg/L)	15,000–27,000
Water (%)	89–96
Organic matter in total solids (%)	70
Nitrogen (g/L)	1.0–3.5
Phosphorus (g/L)	0.4–4.0
Potassium (g/L)	9.0–13.0
Magnesium (g/L)	0.8–1.5

Adapted from Sydney (2013)

6.1.2 Glycerol

Glycerol is a feedstock for the industrial production of many products with commercial interests. However, when it comes from the production of biodiesel, the generated glycerin has a very low commercial value, primarily due to the impurities it contains.

Glycerol has become one of the most inexpensive and abundant carbon sources for microorganisms, since this is the main residue of the biodiesel production worldwide. Many workers have used crude glycerol from biodiesel process as a feedstock for biohydrogen production (Table 5). The conversion of glycerol to high energy fuels, such as the biohydrogen is an interesting and innovative alternative. It has been reported higher yields than those obtained with the conversion of sugars (Gonzalez et al. 2008).

Table 5 Biohydrogen production using crude glycerol as feedstock in different bioreactor systems

Bioreactor type	Hydrogen yield	Reference
120 mL serum bottles containing 40 mL media	2.73 ± 0.14 mol-H_2/mol glycerol	Ngo et al. (2011)
500 mL serum bottles with 250 mL of media	0.31 mol-H_2 mol/glycerol	Priscilla et al. (2009)
Packed-bed reactor of 60 mL working volume	63 mmol-H_2/L. h	Ito et al. (2005)
2 L glass flasks, with 1 L of liquid volume	200 ml-H_2/g COD	Bruna et al. (2010)
Bio-electrochemical two-compartment reactor	0.77 mol-H_2/mol glycerol	Sakai and Yagishita (2007)
Single-chamber membrane reactor	0.41 ± 0.1 m^3-H_2/m^3. d	Selembo et al. (2009)
125 mL serum bottles	4 mol-H_2/mol glycerol	Guillaume and Patrick (2009)

Adapted from: Sarma et al. (2012)

Table 6 Characteristics of palm oil mill effluent

Parameter	Values
pH	4–5
BOD (mg/L)	25,000
COD (mg/L)	55,000–60,000
Total Solids (mg/L)	40,500
Oils and grease (mg/L)	4000
Alkalinity ($CaCO_3$) (mg/L)	50–150

Source Ahmad et al. (2003)

6.1.3 Pome

POME (palm oil mill effluent) is the aqueous effluent from the production of biodiesel from palm oil and can be used for the production of biogas in anaerobic digester (Poh et al. 2010). The extraction process of oil from palm required 5–7.5 tons of water for each ton of oil. About 50 % of this water result as palm oil effluent (Ahmad et al. 2003). Due to its high content of phosphorous, carbon and nitrogen, this wastewater has highly negative environmental impact, and must be properly treated before disposal in water bodies (Poh et al. 2010).The anaerobic digestion of POME to produce biogas or biomethane has been studied (Table 6).

6.2 Urban Solid Wastes

The conversion of municipal solid waste to biofuel has become increasingly popular in recent years as a sustainable technology. In many industrialized countries around the world many facilities have operated in industrial-scale. In Edmonton, a Canadian city, 100,000 tons of municipal waste per year are converted into biofuels and chemicals. Also in San Francisco and Portland, in North America, 80–85 % of the residential organic waste is collected and composted (CleanTechnica 2014).

Different components of municipal solid waste determine the biogas and methane production potential. Getahun et al. (2014) found the highest biogas and methane yield with a mixed waste composed with fruit waste (15 %), food waste (12 %), yard waste (23 %), and paper waste (4 %). They attributed this due to its optimum C/N ratio (25:1) and good nutrient composition for the growth of methanogenic bacteria.

7 Conclusion and Perspectives

The diversity of renewable raw materials and residues used as feedstocks for biofuel production, combined with new technologies that have been developed, enable the future of this renewable energy source.

Biofuels include a very wide range of products, including bioethanol, biodiesel, biogas, biomethanol, biohydrogen, among others. The most common are bioethanol and biodiesel. Biodiesel is produced mainly from oil plants. For bioethanol production, the most interesting feedstocks are plants of rapid growth and annual collection, rich in simple sugars or easily hydrolysable. Sugar cane, beets, sweet *sorghum*, and cereals (corn, wheat, maize, cassava, etc.) are the most used.

Actually, microbial lipids, particularly single cell oils produced by oleaginous microorganisms have been used as potential raw material for biodiesel production due to their similar fatty acids compositions to vegetable oil. There is much interest in fuels produced from algae and a number of facilities are in the demonstration stage or commercial scale (Janssen et al. 2013). The process facility is generally colocated with a ethanol facility and utilizes carbon dioxide from the ethanol facility in its algae production process.

References

Abbas A, Ansumali S (2010) Global potential of rice husk as a renewable feedstock for ethanol biofuel production. Bioenerg Res 3:328–334

Abreu YV, Oliveira HR, Leal JEC (2012) Biodiesel no Brasil em Três Hiatos: Selo Combustível Social, Empresas e Leilões. Eumed, Univ. Malaga (Esp) 213

Aguiar A, Ferraz A (2011) Revisão: Mecanismos envolvidos na biodegradação de materiais lignocelulósicos e aplicações tecnológicas correlatas. Quim Nova 34:1729–1738

Ahmad AL, Ismail S, Bhatia S (2003) Water recycling from palm oil mill effluent (POME) using membrane technology. Desalination 157:87–95

Ali MB (2004). Characteristics and production costs of us sugar beet farms, United States Department of Agriculture Statistical Bulletin, SB974–8

Azbar N, Cetinkaya Dokgoz FT, Keskin T, Korkmaz KS, Syed HM (2009) Continuous fermentative hydrogen production from cheese whey wastewater under thermophilic anaerobic conditions. Int J Hydrogen Energy 34:7441–7447

Balat M, Balat H, Cahide OZ (2008) Progress in bioethanol processing. Prog Energ Combust 34:551–573

Balat M (2011) Production of bioethanol from lignocellulosic materials via the biochemical pathway: a review. Energy Convers Manag 52:858–875

Banerjee S, Mudlia RS, Sem R, Giri B (2010) Commercializing lignocellulosicbioethanol: technology bottlenecks and possible remedies Biofuels. Bioprod Bioref 4:77–93

BNDES and CGEE (Org) (2008) Bioetanol de cana-de-açúcar: energia para o desenvolvimento sustentável. Rio de Janeiro: Bndes, 316 DC (http://www.cei.org/pdf/5774.pdf)

Botha T, Von Blottnitz H (2006) A comparison of the environmental benefits of bagasse-derived lectricity and fuel ethanol on a life-cycle basis. Energy Policy 34:2654–2661

Breuninger WF, Piyachomkwan K, Sriroth K (2009). Tapioca/cassava Starch: Production and use. In: BeMiller J, Whistler R (eds) Starch chemistry and technology, 3rd edn. Academic press, New York. p 544

Bruna SF, Guilherme P, Fernanda RA, Nora K, Saavedra A, Marcelo Z (2010) Potential to produce biohydrogen from various wastewaters. Energy Sustain Develop 14(2):143

Carter C, Miller HI (2012) Corn for food, not fuel. http://www.nytimes.com/2012/07/31/opinion/corn-for-food-not-fuel.html?_r=0. Accessed 8 Nov 2015

CleanTechnica (2014) First industrial-scale municipal solid waste to biofuel facility opens. http://cleantechnica.com/2014/06/09/first-industrial-scale-municipal-solid-waste-biofuel. Accessed 15 Jan 2016

Conab (2013) http://www.conab.gov.br/OlalaCMS/uploads/arquivos/13_09_12_17_38_24_5_mandioca_.pdf

Corley RHV, Tinker PB (2003) The oil palm. 5th edn. Blackwell, Oxford, p 627

Cortez LAB, Silva A, Jordan RA, Castro, LR (2011) Biodigestão de Efluentes In: Cortez LAB. Lora EES., Gómez EO (Org) (eds) Biomassa para Energia, Unicamp, Campinas. p 493–529

Cunha SP; Severo Filho WA (2010) Avanços tecnológicos na obtenção de etanol a partirde sorgo sacarino (Sorghum bicolor (L.) Moench). TECNO-LÓGICA, Santa Cruz do Sul. 14, n. 2, 69–75, jul./dez

Davila-Gomez AC, Chuck-Hernandez AE, Perez-Carrilloa WL, Rooney BSO (2011) Serna-Saldivar Evaluation of bioethanol production from five different varieties of sweet and forage sorghums (*Sorghum bicolor* (L) Moench). Ind Crops Prod 33:611–616

De Mori C, Tomm GO, Ferreira PEP (2014) Aspectos econômicos e conjunturais da cultura da canola no mundo e no Brasil. Passo Fundo: Embrapa Trigo, 2014. p 36. html. (Embrapa Trigo. Documentos online, 149)

Demirbas MF (2009) Biorefineries for biofuel upgrading: A critical review. Appl Energy 86: S151–S161

EIA (2013) http://www.eia.gov/todayinenergy/detail.cfm?id=11551. Acessed 8 Nov

Embrapa (1994) Tropical soybean improvement and production. Food and Agriculture Organization of the United Nations (FAO), Rome 253 p

ERS (United States Department of Agriculture Economic Research Service), Feed Grains Database, 2010. Available at www.ers.usda.govS

Eurostat (2011) Statistical data. Available on http://epp.eurostat.ec.europa.eu/portal/page/portal/eurostat/home/S

FAO—Food and Agriculture Organizations of United Nations, Statistic Division (2015) Available at http://faostat3.fao.org/download/Q/%2A/E

FAOSTAT (2015) Faostat database. Available at www.faostat.fao.org

Fernandes BS, Peixoto G, Albrecht FR, del Aguila NKS, Zaiat M (2010) Potential to produce biohydrogen from various wastewaters. Energy Sustain Dev 14:143–148

Gazzoni DL, Avila MT, Felici PHN, Ralisch R (2008) Balanço energético do biodiesel de dendê. In: CONGRESSO BRASILEIRO DE AGROENERGIA, 1, SIMPÓSIO INTERNACIONAL DE BIOCOMBSTÍVEL, 1., Uberlândia. Anais. Uberlândia: UFU

Gazzoni DL, Azurdia I, Blanco G, Estrada C, Macedo IC (2012) Sustainable energy in Latin America and the Caribbean: potential for the future. Available at http://www.icsu.org/latin-america-caribbean/publications/reports-and-reviews/sustainable-energy/energy_english.pdf. Retrieved on 15 Sept 2015

Gazzoni DL, Borges JLB, Avila MT, Felici PHN (2009) Balanço energético da cultura da canola para a produção de biodiesel. Espaço Energia. 11:24–28

Gazzoni DL, Felici PHN, Coronato RMS, Ralisch R (2005) Balanço energético das culturas de girassol e soja pra produção de biodiesel. Biomassa e Energia 2(4):259–265

Getahun T, Gebrehiwot M, Ambelu A, Van Gerven T, Van der Bruggen B (2014) The potential of biogas production from municipal solid waste in a tropical climate Environ Monit Assess. 186:4637–4646

Glassner D, Hettenhaus J, Schechinger T (1998) Corn stover collection project. BioEnergy 98. Expanding bioenergy partnerships: proceedings, vol 2, Madison, WI, pp 1100–1110

Gnansounou E (2010) Production and use of lignocellulosic bioethanol in Europe: Current situation and perspectives. Bioresour Technol 11:4842–4850

Gonzalez R, Murarka A, Dharmadi Y, Yazdani SS (2008) A new model for the anaerobic fermentation of glycerol in enteric bacteria: Trunk and auxiliary pathways in Escherichia coli. Metab Eng 10:234–245

Guillaume SP, Patrick CH (2009) High yield conversion of a crude glycerol fraction from biodiesel production to hydrogen by photofermentation. Bioresour Technol 100:3513–3517

Haankuku C, Epplin FM, Kakani VG (2015) Industrial sugar beets to biofuel: Field to fuel production system and cost estimates. Biomass Bioenergy 80:267–277

Halling P, Simms-Borre P (2008) Overview of lignocellulosic feedstock conversion into ethanol—focus on sugarcane bagasse. Int Sugar J 110(1311)

Hofsetz K, Silva MA (2012) Brazilian sugarcane bagasse: Energy and non-energy consumption. Biomass Bioenergy 46:564–573

IEA (2007) Energy Technology Essentials: biofuel Production, International Energy Agency http://www.iea.org/techno/essentials2.pdf

Içöz E, Tugrul MK, Saral A, Içöz E (2009) Research on ethanol production and use from sugar beet in Turkey. Biomass Bioenergy 33:1–7

Ito T, Nakashimada Y, Senba K, Matsui T, Nishio N (2005) Hydrogen and ethanol production from glycerol-containing wastes discharged after biodiesel manufacturing process. J Biosci Bioeng 100:260–265

Janssen R, Turhollow AF, Rutz D, Mergner R (2013) Production facilities for second-generation biofuels in the USA and the EU—current status and future perspectives. Biofuels, Bioprod Biorefin 7:647–665

Jung S, Foston M, Sullards MC, Ragauskas AJ (2010) Surface characterization of dilute acid pretreated *Populus deltoides* by ToF-SIMS. Energy Fuels 24:1347–1357

Kerstetter JD, Lyons JK (2001) Wheat straw for ethanol production in washington: a resource, technical, and economic assessment. Washington State University Cooperative Extension. http://www.energy.wsu.edu/documents/renewables/WheatstrawForEthanol.pdf

Kim S, Dale BE (2004) Global potential bioethanol production from wasted crops and crop residues. Biomass Bioenergy 26:361–375

Koçar G, Civas N (2013) An overview of biofuels from energy crops: Current status and future prospects. Renew Sustain Energy Rev 28:900–916

Leite RMVB de C, Brighenti AM, Castro C de (eds) (2005) Girassol no Brasil. Londrina: Embrapa Soja. p 613

Lentz DL, Pohl ED, Pope KO, Wyatt AR (2001) Prehistoric sunflower (*Helianthus annuus* L.) domestication in Mexico. Econ Botany 55(3):370–376

Lin CY, Lay CH, Sen B, Chu CY, Kumar G, Chen CC, Chang JS (2012) Fermentative hydrogen production from wastewaters: a review and prognosis. Int J Hydrogen Energy 37:15632–15642

Lora EES, Nascimento MAR (2004) Geração Termelétrica: Planejamento, Projeto e Operação. Rio de Janeiro, Editora Interciência. vol 2, p 1296

MAPA (2016) http://www.agricultura.gov.br/vegetal/culturas/cana-de-acucar. Acessed in Jan 2016

McLoom A, Taylor F, Yee W, Ibsen K, Wooley R (2000) Determining the cost of producing ethanol from corn starch and lignocellulosic feedstocks. National Renewable Energy Laboratory. http://www.nrel.gov/docs/fy01osti/28893.pdf. Retrieved on July 2016

Menon V, Rao M (2012) Trends in bioconversion of lignocellulose: Biofuels, platform chemicals & biorefinery concept. Prog Energy Combust Sci 38:522–550

Meyer PM, Rodrigues PHM, Millne DD (2013) Impact of biofuel production in Brazil on the economy, agriculture, and the environment. Annimal Frontiers 3(2):28–37

Moraes BS, Marcelo Zaiat M, Bonomi A (2015) Anaerobic digestion of vinasse from sugarcane ethanol production in Brazil: Challenges and perspectives. Renew Sustain Energy Rev 44:888–903

Müller AA (1980) A cultura do dendê. Belém, PA: EMBRAPA-CPATU. 24 p. il. (EMBRAPA-CPATU. Miscelânea, 5)

Muller AA, Furlan Junior J (eds) (2001) Agronegócio do dendê: uma alternativa social, econômica e ambiental para o desenvolvimento sustentável da Amazônia. Belém: Embrapa Amazônia Oriental, 2001. 288p. il

National Renewable Energy Laboratory (NREL) (2014) State and Local Governments: renewable Fuel Standards, [Updated 08.09.14; cited 10.02.15] Available from http://www.nrel.gov/tech_deployment/state_local_governments/basics_fuel_standards.html

Ngo TA, Kim MS, Sim SJ (2011) High-yield biohydrogen production from biodiesel manufacturing waste by Thermotoga neapolitana. Int J Hydrogen Energy 36:5636–5642

Panella L, Kaffka SR (2010) Sugar beet (*Beta vulgaris* L) as a biofuel feedstock in the United States. In: Eggleston G (ed), Sustainability of the sugar and sugar ethanol industries. ACS symposium series, vol 105, American Chemical Society, Washington, DC, pp 163–175

Pereira Filho IA, Parrella RAC, Moreira JAA, May A, Souza VF, Cruz J (2013) Avaliação de cultivares de sorgo sacarino (*Sorguhm bicolocr* (l.) Moench) em diferentes densidades de semeadura visando a características importantes na produção de etanol Revista Brasileira de Milho e Sorgo. 12, n.2, 118–127

Petersen AM, Melamu Rethabi, Knoetze JH, Görgens JF (2015) Comparison of second-generation processes for the conversion of sugarcane bagasse to liquid biofuels in terms of energy efficiency, pinch point analysis and Life Cycle Analysis. Energy Convers Manag 91:292–301

Poh PE, Yong W-J, Chong MF (2010) Palm oil mill effluent (POME)—characteristic in high crop season and the applicability of high-rate anaerobic bioreactors for the treatment of POME. Ind Eng Chem Res Washington 49:11732–11740

Prasad SA, Singh N Jain, Hoshi HC (2007) Ethanol production from sweet sorghum syrup for utilization as automotive fuel in India. Energy Fuel 21:2415-2420

Priscilla AS, Joe MP, Wallis AL, Bruce EL (2009) Enhanced hydrogen and 1,3-propanediol production from glycerol by fermentation using mixed cultures. Biotechnol Bioeng 104: 1098–1106

Quintero JA, Montoya MI, Sánchez OJ, Giraldo OH, Cardona CA (2008) Fuel ethanol production from sugarcane and corn: Comparative analysis for a Colombian case. Energy 33:385–399

Qureshi N, Saha BC, Hecto RRE, Dienb B, Hughes S, Liu S, Iten L, Bowman MJ, Sarath G, Cott AM (2010) A Production of butanol (a biofuel) from agricultural residues: Part II—Use of corn stover and switchgrass hydrolysates. Biomass and Bio energy. 34:566–571

Ramos MJ, Fernández CM, Casas A, Rodríguez L, Pérez A (2009) Influence of fatty acid composition of raw materials on biodiesel properties. Bioresource Tech 100(1):262–268

Ratnavathi CV, Suresh K, Vijay-Kumar BS, Pallav IM, Komala VV, Seetharam N (2010) Study on genotypic variation for etanol protution from sweet sorghum juice. Biomass Bioenergy 34 (7):947–952

Reddy BVS, Kumar AA, Ramesh S (2007) Sweet sorghum: a water saving bio-energy crop. Patancheru-502 324. Int Crops Res. Institute for the Semi-Arid Tropics

REN 21 (2015) Global Status Report. Available at www.ren21.net. Retrieved on 7 Oct 2015

Rojanaridpiched C, Kosintarasaenee S, Sriroth K, Piyachomkwan K, Tia S, Kaewsompong S, Nitivararat M (2003) Development of ethanol production technology from cassava chip at a pilot plant scale. National Research Council of Thailand

Santos FA, Queiróz JH, Colodett EJL, Fernandes SA, Guimarães VM, Rezende ST (2012) Potencial da palha de cana-de-açúcar para produção de etanol. Quim Nova 35(5):1004–1010

Sarath G, Mitchell RB, Sattler SE, Funnell D, Pederson JF, Graybosch RA, Vogel KP (2008) Opportunities and roadblocks in utilizing forages and small grains for liquid fuels. J Ind Microbiol Biotechnol 35:343–354

Sarma SJ, Brar SK, Sydney EB, Le Bihan Y, Buelna G, Soccol CR (2012) Microbial hydrogen production by bioconversion of crude glycerol: A review. Int J Hydrogen Energy 37: 6473–6490

Sarkar N, Ghosh SK, Bannerje SE, Aikat K (2012) Bioethanol production from agricultural wastes: An overview. Renewable Energy 37:19–27

Selembo PA, Perez JM, Lloyd WA, Logan BE (2009) High hydrogen production from glycerol or glucose by electrohydrogenesis using microbial electrolysis cells. Int J Hydrogen Energy 34:5373–5381

Shapouri H, Salassi ME (2006) The Economic Feasibility of Ethanol Production from Sugar in the United States. USDA-ers, Washington DC, p 69 July

Shoemaker CE, Bransby DI (2010) The role of sorghum as a bioenergy feedstock chapter 9 the role of sorghum as a bioenergy feedstock. p 149–159

Silva OC (2012) Uma avaliação da produção de bioetanol no contexto da agricultura familiar a partir da experiência do COOPERBIO. Graduate Energy Program (2012). USP—São Paulo University, Brazil

Sinha P, Pandey A (2014) Biohydrogen production from various feedstocks by *Bacillus firmus* NMBL-03. Int J Hydrogen Energy 39:758–7525

Sipos B, Réczey J, Somorai Z, Kádár Z, Dienes D, Réczey K (2009) Sweet sorghum as feedstock for ethanol production: enzymatic hydrolysis of steam-pretreated bagasse. Appl Biochem Biotechnol 153:151–162

Sakai S, Yagishita T (2007) Microbial production of hydrogen and ethanol from glycerol-containing wastes discharged from a biodiesel fuel production plant in a bio-electrochemical reactor with thionine. Biotechnol Bioeng 98(2):340–348

Smith GA, Bagby MO, Lewellan RT, Doney DL, Moore PH, Hills FJ, Campbell LG, Hogaboam GJ, Coe GE, Freeman K (1987) Evaluation of sweet sorghum for fermentable sugar production potential. Crop Sci 27:788–793

Sriroth K, Wanlapatit S, Piyachomkwan K (2012) Cassava Bioethanol, Bioethanol. In: Marco Aurelio Pinheiro Lima (ed), ISBN: 978-953-51-0008-9, InTech, Available from http://www.intechopen.com/books/bioethanol/-cassava-bioethanol

Sydney EB (2013) Valorization of vinasse as broth for biological hydrogen and volatile fatty acids production by means of anaerobic bacteria. Other. Universit'e Blaise Pascal - Clermont-Ferrand II

Szulczyk KR, McCarl BA, Cornforth G (2010) Market penetration of ethanol. Renew Sustain Energy Rev 14:394–403

Swinkels JJM (1998) Industrial starch chemistry. AVEBE Brochure, The Netherlands

Talebnia F, Karakashev D, Angelidaki I (2010) Production of bioethanol from wheat straw: An overview on pretreatment, hydrolysis and fermentation. Bioresource Technol 101:4744–4753

Tan Z, Liu S, Bliss N, Tieszen LL (2012) Current and potential sustainable corn stover feedstock for biofuel production in the United States. Biomass Bioenergy 47:372–386

Thompson JL, Tyner WE (2014) Corn stover for bioenergy production: cost estimates and farmer supply response. Biomass Bioenergy 62:166–173

Thirathumthavorn D, Charoenrein S (2005) Thermal and Pasting Properties of Acid-Treated Rice Starch. Starch/Starke 57:217–222

UNCTAD (2015) Commodity profile—cassava. http://www.unctad.info/en/Infocomm/AACP-Products/COMMODIRY-PROFILE—Cassava. Acessed in 8 Nov 2015

Ungaro MRG (2000) Cultura do girassol. Campinas: IAC, 2000. p 36 (IAC. Boletim Técnico, 188)

UNICA (2015) Brazilian Sugarcane Industry Association. http://www.unica.com.br/. Accessed 4 Jan 2016

USDA (2015) National Agriculture Statistics Service. http://www.afdc.energy.gov/data/10339. Accessed 8 Nov

USDA (2015a) Canola. Available at http://www.ers.usda.gov/topics/crops/soybeans-oil-crops/canola.aspx. Retrieved on 26 Sept 2015

USDA (2015b) Sunflowerseed. Available at http://www.ers.usda.gov/topics/crops/soybeans-oil-crops/sunflowerseed.aspx. Retrieved on 26 Sept 2015

USDA (2015d) Oilseeds. Available at http://www.fas.usda.gov/data/oilseeds-world-markets-and-trade. Retrieved on 11 Oct 2015

Van Eijck J, Batidzirai B, Faaij A (2014) Current and future economic performance of first and second generationbiofuels in developing countries. Appl Energy 135:115–141

Watsom S, Ramstad P (1987) Corn: Chemistry and technology, american association of cereal chemists, Inc., St. Paul, MN.

Ziolkowska JR (2013) Evaluating sustainability of biofuels feedstocks: a multi-objective framework for supporting decision making. Biomass Bioenergy 59:425–440

Oil Crops in the Context of Global Biodiesel Production

Decio Luiz Gazzoni and Amélio Dall'Agnol

Abstract There are hundreds of plants with potential to produce oil for biodiesel industry but only a few are largely cultivated, being the same that have been used for other purposes, especially for nutrition. Along the last 10 years, palm, soybean, canola, and sunflower accounted for 80 % of the world vegetable oil production, or 90 % if peanut, cotton, coconut, and olive are also included. Similarly, almost all countries are potential producers of oil but USA, China, Brazil, India, Argentina, and Indonesia account for about 70 % of the 180 Mton produced worldwide, and for almost all of the vegetable oil traded in the international market. In order to consider an oil crop as a trusty supplier of raw material for the biodiesel industry, it must fulfill the following criteria: (a) must be produced on a large scale; (b) should belong to a well-organized supply chain; (c) be considered a commodity in the international market and its oil be competitive in price with not only other vegetable oils but also with petrol; (d) its by-products obtained besides the oil should also have a steady demand on the domestic and international market. The world biodiesel production strongly accelerated within 2000–2009, reducing the rates of expansion after 2009 because of the world financial crisis (2008–2010). Currently (2016), biodiesel global production is estimated to be slightly over 35 Mton, awaiting new stimuli to reaccelerate considering the well-known environmental and social benefits of biodiesel, which can overcome the disadvantage of the higher cost as compared to mineral diesel.

Keywords Vegetable oils · Oilseeds · Palm oil · Soybean · Sustainability · Public policies

D.L. Gazzoni (✉) · A. Dall'Agnol
Embrapa Soybeans, Brazilian Agricultural Research Corporation (Embrapa), Londrina, PR, Brazil
e-mail: decio.gazzoni@embrapa.br

C.R. Soccol et al. (eds.), *Green Fuels Technology*,
Green Energy and Technology, DOI 10.1007/978-3-319-30205-8_3

1 Introduction

Oilseeds are among the most important crops in international trade. Annually, world consumption of vegetable oils and animal fats exceeds 300 Mton (USDA 2015d). According to the FAO database (FAOSTAT 2015), world production of vegetable oils has grown dramatically in recent decades, increasing more than 600 % in roughly 40 years, jumping from 23.6 Mton in 1972/1973 crop season to 180 Mton in the 2014/2015 crop season. More than 70 % of the world production of vegetable oil is concentrated in just six countries: United States (USA), China, Brazil, India, Argentina, and Indonesia. Nutrition is by far the major market, but general industry application, fine chemistry, and energy are also demanding increasing amounts of vegetable oils in recent years.

There are hundreds of species with potential to provide oil for domestic use or as a raw material for oil chemistry or biodiesel industry. However, few of them have characteristics that justify their large-scale farming, chiefly: high-oil content, well-structured supply chain, and production technology at the cutting edge. Many oilseeds are only economically viable because their coproducts after oil extraction (ex., soybean meal and cotton fiber) are highly demanded in the market, sometimes even higher than the oil itself.

Currently, about 80 % of vegetable oil produced worldwide comes from only four oil crops: palm (pulp + almond), soybean, rapeseed (canola), and sunflower. Four other crops account for the next 11 % share: peanut, cotton, coconut, and olive. Then, dozens of other oil producing plants share less than 10 % of the world oil production. Among which it is worth mentioning some with different potential degrees for future expansion, depending on the availability of adequate production technology and a well-structured production chain: castor, flaxseed, sesame, safflower, forage turnip, crambe, tucuman, rubs, buriti, macaúba, indaiá, açaí, gerivá, patauá, cotieira, oiticica, nhandiroba, tung, pequi, jatropha, jojoba, tingui, among others.

As for biodiesel production, which demands annually over 20 Mton of vegetable oil and animal fat, four aspects are crucial for a given crop to be considered as a feedstock, according to Gazzoni et al. (2012): (a) large production; (b) well-organized value chain; (c) insertion as a commodity in the international market; and (d) competitive price, as compared to other oils, but specially against petrol, the fossil energy paradigm.

Figure 1 shows the evolution of market prices of the internationally traded vegetable oils, compared to international petrol prices. Usually, vegetable oils do not compete with petrol prices, but other advantages should be taken into account, like being more environmentally friendly, the creation of additional jobs evenly distributed, and more business opportunities, among others.

In some oilseeds, the protein fraction is more important to the market than the oil, as for soybeans. Others, such as cotton, are grown primarily for the production of fiber, being oil clearer a byproduct; peanuts, sesame, coconut, and açaí are

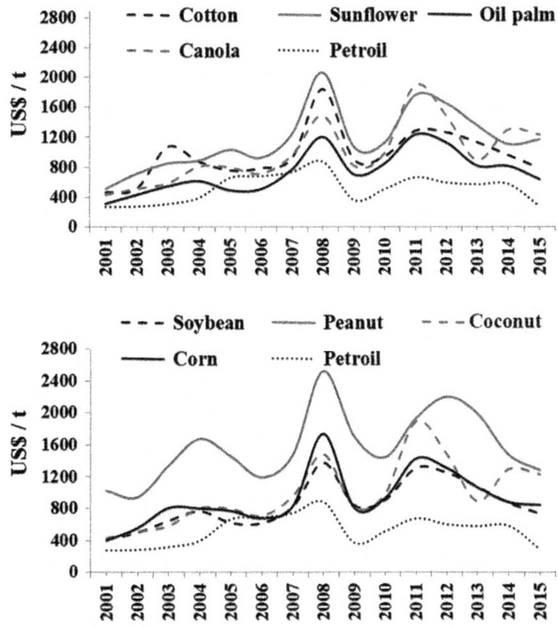

Fig. 1 Vegetable oil and crude oil prices in the international market, adjusted to Oct, 2015

cultivated to meet demands for direct human consumption, and only marginally for oil extraction.

The global biodiesel production is about 35 billion liters (GL), according to 2015 statistics (REN 21 2015). The major feedstocks for its production are palm, soybeans, and rapeseed (canola), with some participation of animal fat and sunflower. As a rule, each country uses its more abundant raw material to produce biodiesel, so that USA, Brazil, and Argentina largely relies on soybean oil; Indonesia and Malaysia use palm oil; and the European Union counts on rapeseed (canola). Animal fat also constitutes an important source of raw material for biodiesel production, being the main raw material in China (pork fat). In Brazil, about 20 % of the biodiesel is obtained from animal fat (mainly tallow), usually blended with biodiesel obtained from vegetable oil, to meet legal and technical specifications.

Biodiesel global production increased dramatically in the first decade of the present century, when oil prices hovered around US$100.00/barrel, with a peak of US$142.00/barrel in 2008,[1] surfing the wave of the pursuing of a more sustainable energy matrix. As an average for all locations and raw materials, biodiesel yields 93 % more energy than invested on its production (Hill 2006), not accounting for solar radiation energy captured by plants through photosynthesis.

During the first decade of the twenty-first century, the Land Use Change/Indirect Land Use Change theory(LUC/ILUC) was very popular, blaming biofuels as

[1]www.commoditycharts.com/commodities/Energies.

responsible for carbon debt and for competition with food production (Fargione et al. 2008; Searchinger et al. 2008), leading to higher food prices. A review by Gazzoni (2014a) concluded that the model described by the LUC/ILUC theory did not fit to actual agricultural production data, according to the most recent studies and FAO statistics.

A case study of the Brazilian biofuels production and use demonstrated that from 2007 to 2011 the use of biodiesel in Brazil accounted for avoided emissions up to 16 Mton of CO_2 (Gazzoni 2014b). Due to the uncertainty in the scientific literature regarding the ecological benefits of biofuels, Davis et al. (2009) proposed that providing new information on biogeochemistry and plant physiology, ecologists, and plant scientists could increase the accuracy of Life Cycle Analysis for biofuel production systems.

As for the moment, the vegetable oil for biodiesel production cost exceeds that of mineral diesel (see Fig. 1). Following the financial crisis of the end of last decade, and the recent reduction on petrol prices, biodiesel production has stabilized, waiting for new stimuli to resume former production increase rates, which includes public policies supporting its production and use. Nevertheless, besides environmental benefits, it should be considered that social gains partially help offsetting its higher costs, as biodiesel production generates much more jobs than the petrol chain. Each 1 % biodiesel added to mineral diesel results in the creation of approximately 45,000 jobs, according to estimates of the Ministry of Agrarian Development of Brazil (Abreu et al. 2012).

2 Feedstock for Biodiesel Production

Even though several plants can produce oil, stored on grains or fruits, only a few of them are actually commercially important, traded in the international market and constituting important feedstock for industrial purposes. Table 1 presents the global area cultivated with oil crops while Table 2 details the area cultivated with the most important oil crops worldwide, including ones that are not used for biodiesel production due to unsuitable oil characteristics.

The competitiveness of any crop, including the oil crops, largely relies on its yield. Table 3 shows the evolution of the average global yield of several oil crops, including the ones not suited for the biodiesel industry, whereas the Table 4 presents the grain production of those oil crops. It is important to consider the different

Table 1 Worldwide cultivated area and grain production of the major oil crops

	1960	1965	1970	1975	1980	1985	1990	1995	2000	2005	2010	2014
Area (Mha)	98	108	116	127	145	155	164	185	198	224	239	265
Production (Mton)	106	126	145	174	201	253	282	328	373	468	536	634

Source FAOSTAT database

Table 2 Global area cultivated with selected oil crops worldwide, in million hectares (Mha)

Crop	1960	1965	1970	1975	1980	1985	1990	1995	2000	2005	2010	2014
Castor	1.2	1.5	1.5	1.4	1.5	1.7	1.7	1.2	1.8	1.6	1.5	1.5
Coconuts	5.3	6.0	6.7	7.4	8.8	9.4	10.0	10.7	10.8	11.2	11.8	12.1
Cottonseed	31.9	33.7	34.1	32.5	34.3	33.4	33.1	35.5	31.8	35.0	31.8	36.9
Peanuts	16.6	19.8	19.5	20.0	18.4	18.5	19.8	22.0	23.2	24.0	25.5	25.9
Hempseed	0.3	0.4	0.3	0.3	0.2	0.2	0.1	0.0	0.0	0.0	0.0	0.0
Rapeseed (Canola)	6.3	7.1	8.2	9.9	11.0	14.8	17.6	23.8	25.8	27.7	32.2	38.4
Safflower	0.8	0.8	1.0	1.3	1.3	1.4	1.2	1.2	0.8	0.8	0.8	0.8
Sesame seed	5.0	5.6	5.9	5.8	6.3	6.9	6.1	6.7	7.2	7.5	8.3	9.5
Soybeans	23.8	25.8	29.5	38.8	50.6	53.1	57.2	62.5	74.4	92.6	102.8	113.3
Sunflower	6.7	7.5	8.7	9.2	12.4	14.8	17.0	20.9	21.2	23.2	23.1	26.0
Tallowtree	0.0	0.0	0.0	0.0	0.0	0.3	0.3	0.3	0.4	0.4	0.4	0.4
Tung	0.1	0.1	0.1	0.1	0.1	0.3	0.2	0.2	0.2	0.2	0.2	0.2

Source FAOSTAT database

Table 3 Average global yield for selected oil crops, in kilograms per hectare (kg/ha)

Crop	1960	1965	1970	1975	1980	1985	1990	1995	2000	2005	2010	2014
Castor	470	547	563	571	504	707	818	876	777	943	1136	1235
Coconuts	4532	4247	3922	4149	3681	3825	4353	4649	4759	5126	5094	5103
Cottonseed	862	1063	1042	1113	1206	1523	1639	1599	1668	2005	2153	1979
Peanuts	849	801	922	956	920	1134	1169	1298	1494	1602	1677	1777
Hempseed	241	265	270	308	281	347	559	1187	1612	3973	3128	3114
Jojoba	0	0	0	0	0	337	340	1000	800	300	380	417
Rapeseed (Canola)	573	741	816	887	979	1304	1387	1435	1529	1806	1864	2000
Safflower	432	627	688	782	703	643	690	712	757	700	810	827
Sesame seed	286	279	340	297	277	336	388	381	385	466	529	506
Soybeans	1129	1228	1480	1657	1600	1906	1896	2031	2169	2318	2578	2684
Sunflower	1023	1059	1149	1068	1099	1270	1333	1255	1250	1323	1363	1800
Tallowtree						2590	2299	2309	2343	2342	2398	2457
Tung	7967	3898	10,536	10,797	8033	2059	2227	2943	3236	2696	2873	2745

Source FAOSTAT database

Table 4 Grain production for selected oil crops worldwide, in million ton

Crop	1960	1965	1970	1975	1980	1985	1990	1995	2000	2005	2010	2014
Castor	0.6	0.8	0.8	0.8	0.8	1.2	1.4	1.1	1.4	1.5	1.7	1.9
Coconuts	23.8	25.5	26.3	30.8	32.2	35.9	43.5	49.7	51.2	57.4	60.3	62.0
Cottonseed	27.5	35.8	35.6	36.2	41.4	50.9	54.2	56.7	53.1	70.1	68.6	73.0
Peanuts	14.1	15.8	18.0	19.1	16.9	20.9	23.1	28.6	34.7	38.5	42.7	45.2
Hempseed	0.1	0.1	0.1	0.1	0.1	0.1	0.0	0.0	0.0	0.1	0.1	0.1
Jojoba	0.0	0.0	0.0	0.0	0.0	0.0	0.0	0.0	0.0	0.0	0.0	0.0
Rapeseed (Canola)	3.6	5.2	6.7	8.8	10.8	19.2	24.4	34.2	39.5	50.0	60.1	72.5
Safflower	0.3	0.5	0.7	1.0	0.9	0.9	0.8	0.8	0.6	0.6	0.6	0.6
Sunflower	6.8	8.0	10.0	9.9	13.7	18.9	22.7	26.3	26.6	30.8	31.5	4.6.6
Sesame	1.4	1.6	2.0	1.7	1.7	2.3	2.4	2.5	2.8	3.5	4.4	4.8
Soybeans	26.9	31.7	43.7	64.2	81.0	101.2	108.5	127.0	161.3	214.6	265.0	276.4
Tallowtree	0.6	0.6	0.7	0.7	0.7	0.8	0.8	0.8	0.8	0.9	0.9	1.0
Tung	0.5	0.3	0.6	0.7	0.5	0.5	0.5	0.5	0.5	0.4	0.5	0.5

Source FAOSTAT database

uses of the vegetable oil, which include the most important one—gastronomy and nutrition industry—but also other industrial (not food), lubricants, fine chemistry, cosmetics and hygiene, besides the energy industry.

Depending on the oil content of the seeds, the resulting oil volumes obtained from each hectare is variable according to the crop, but also depends on the total grain harvest of a given crop.

The world vegetable oil production since 1960 with a projection to 2020 is given in Fig. 2, while the percent share of each one of the major feedstocks of the world vegetable oil production, in 5-year time scale between 1960 and 2015, is shown in Fig. 3.

Table 5 illustrates the evolution of world oil production of different oil crops and Table 6 states the characteristics of the major vegetable oils.

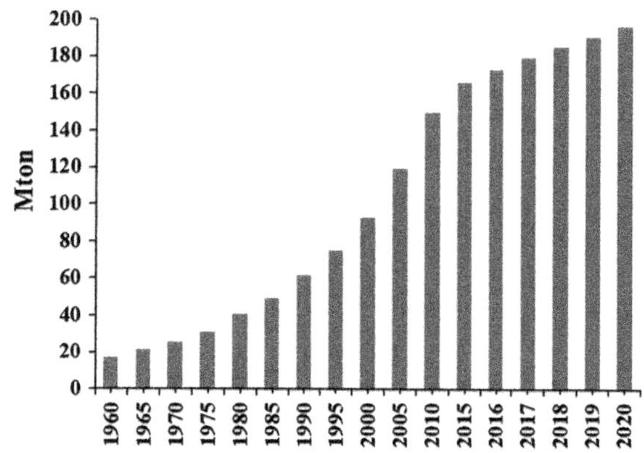

Fig. 2 World vegetable oil production

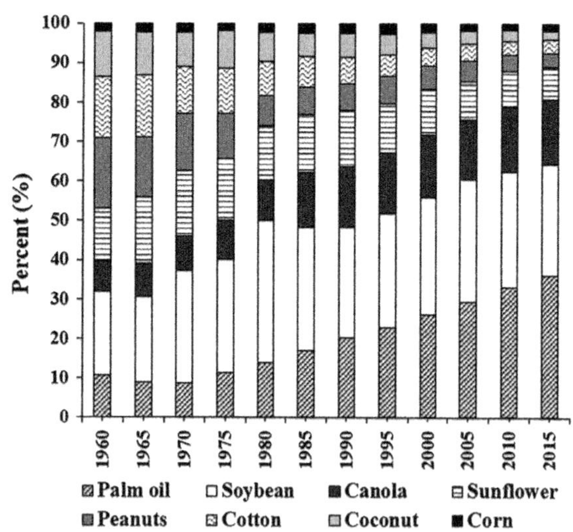

Fig. 3 Share of the major vegetable oils

Table 5 Vegetable oil production from selected oil crops, in million ton

Source	1960	1965	1970	1975	1980	1985	1990	1995	2000	2005	2010	2014
Corn	0.3	0.4	0.5	0.5	0.8	1.1	1.4	1.8	1.9	2.1	2.3	2.9
Soybean	3.0	3.8	6.3	7.8	13.2	14.1	15.9	20.1	25.6	34.2	40.7	42.7
Peanut	2.5	2.7	3.2	3.1	2.7	3.1	3.8	4.7	5.1	5.6	5.6	5.2
Coconut	1.6	1.9	1.9	2.6	2.7	2.6	3.4	3.7	3.4	3.4	3.9	3.2
Palm	1.5	1.6	1.9	3.1	5.1	7.6	11.4	15.9	22.2	32.3	45.8	54.4
Palm kernel	0.5	0.5	0.5	0.6	0.7	1.2	1.7	2.1	2.8	4.4	5.6	6.7
Olive	1.4	1.2	1.4	1.8	2.0	1.7	1.5	1.7	2.5	2.6	3.3	2.8
Sunflower	1.9	3.0	3.7	4.2	5.1	6.6	8.1	8.8	9.8	10.7	12.6	12.6
Safflower	0.1	0.1	0.2	0.2	0.3	0.2	0.2	0.2	0.2	0.1	0.1	0.1
Rapeseed (Canola)	1.1	1.5	1.9	2.7	3.7	6.2	8.6	10.7	13.5	16.8	22.8	24.7
Sesame	0.4	0.5	0.6	0.5	0.5	0.7	0.7	0.7	0.7	0.9	1.1	1.1
Cotton	2.2	2.8	2.6	3.1	3.2	3.5	3.8	3.8	3.8	4.9	4.8	5.1
Linseed	0.9	1.0	1.0	0.7	0.7	0.7	0.7	0.7	0.7	0.6	0.5	0.6

Source FAOSTAT database

Table 6 Characteristics of the major vegetable oils

Feedstock	Saturated	Monounsaturated		Polyunsaturated			Smoke point (°C)
		Total	Oleic	Total	Linolenic (ω-3)	Linoleic (ω-6)	
Canola (rapeseed)	7	64		29	10	10	204
Coconut	91	6	6	3		2	177
Corn	15	30	28	55	1	58	232
Cottonseed	28	19	19	53	1	54	216
Flaxseed/linseed	8	13	11	79	64	15	107
Olive	14	72		14	2	15	193
Palm	49	42	40	9		10	235
Peanut	18	50	48	32		32	225
Safflower (>70 % linoleic	8	17		75			210
Safflower (high oleic)	8	77		15			210
Soybean	16	26	24	58	7	50	238
Sunflower (<60 % linoleic)	10	45	45	40		40	227
Sunflower (>70 % oleic)	9	84		4			227
Cottonseed hydrogenated	94	2	1				
Palm hydrogenated	48	41	8				
Soybean hydrogenated	21	74	1	1			

Sources USDA Rel 27 (http://www.ars.usda.gov/Services/docs.htm?docid=24912); Ivanov et al. (2010), Katragadda et al. (2010)

By far, gastronomic and nutritional uses (salad dress, industrial food, cooking, etc.) is the largest market for vegetable oil, followed by other industrial uses, including fine chemistry. In spite of being used as fuel since the beginning of the twentieth century, only during last decade the energy market expanded its share, specially based on public policies incentivizing its production and use.

It is noteworthy to observe that soybeans led the global vegetable oil production up to the beginning of the twenty-first century, when it was surpassed by palm oil production. The high oil yield by unit of area obtained from palm oil plantations is the major drive for the expansion of this crop, especially in Southeastern Asia.

Oils have different characteristics, according to the crop. Depending on the range of its use, oils can be considered as multipurpose or have a narrow market, with quite specific purpose (nutrition, oil chemistry, bioenergy, etc.).

The major oil crops, in the sense of being the most cultivated and traded in the international market, and representing over 90 % of the global biodiesel feedstock, are analyzed below.

2.1 *Soybean* (**Glycine max** *(L.) Merrill)*

2.1.1 Soybean History

Soybean has been cultivated in China since 5000 years BC, considered the most important legume in ancient Chinese culture (Merril 1931). By the first century BC to the Age of Discovery (fifteen and sixteenth centuries), soy was introduced in several Asian countries (Japan, Indonesia, Philippines, Vietnam, Thailand, Malaysia, Burma, Nepal, and northern India). These regions are considered today as the secondary center of soybean dispersion (Hymowitz 1983). Although considered a sacred grain and extensively used in the diet of the East for thousands of years, its introduction in the West happened only during the eighteenth century (Bretschneider 1882).

The first report on soybean crops in the United States dates from 1765, having first been tested for use as silage and green manure (Piper and Morse 1910). Until 1941, the area cultivated as a forage crop or green manure was greater than the one dedicated to grain production. At the peak of its use as forage, more than 2 million hectares were grown in the USA for that purpose. However, since the 1950s, the main use toggled from fodder to grains, and by the early 1960s the "soy fodder" had disappeared from the USA fields (USDA 2015e).

During the first three decades of the twentieth century, soybean production on a large scale, was still confined to the East (China, Indonesia, Japan, and Korea), in latitudes near or above 35°N. China continued to be the major global producer until the mid-50s, when it was surpassed by the USA. Since the 1960s, the area and soy production increased dramatically, not only in the USA but also in Brazil and Argentina (Myasaka and Medina 1981; Dall'Agnol et al. 2007).

China, once the largest world producer (up to mid-1950s), nowadays is the largest soybean importer, absorbing in excess of 60 % of the soybeans internationally traded in 2015. Brazil is the largest nonprocessed grain exporter and Argentina leads the sales of meal and oil, as well as biodiesel (Gazzoni 2013).

2.1.2 Soybean Expansion in Latin America

The exceptional high price of soybeans in the world market in the mid-70s—when soybeans reached the highest value of all time (US$1249/ton of grains in 1973, adjusted price to 2015[2]) was the main driver of the rapid expansion of its cultivation in the Mercosur (regional common market) region. Since 2010, countries of this economic block are leading the world soy production, with a market share of 53 %, and a total production of 168 Mton. These countries are Brazil (95.5 Mton), Argentina (60.5 Mton), Paraguay (8.5 Mton), and Uruguay (3.5 Mton) (FAOSTAT 2015).

Soybean production started showing socioeconomic importance in Brazil from the mid-50s, in Argentina and Paraguay from the 1970s and Uruguay only recently gained importance as a soybean producer. Even though, it is already the main item on the export basket of this country, as also happens in Brazil, Argentina, and Paraguay.

While the crop was confined to temperate and subtropical regions, Brazilian pioneer growers depended on soybean cultivation technology imported from the USA, especially, the varieties. However, when the cultivation of soybean shifted to Brazilian tropical regions, the imported varieties did not grow properly. It was necessary a comprehensive program for developing local technology, mainly varieties adapted to low latitude conditions, as well as other techniques like soil management and fertilization or pest management (Myasaka and Medina 1981; Câmara 2000; Dall'Agnol et al. 2007). Nowadays, soybean is grown with similar efficiency from the south to the extreme north of the South America, with consistently higher yields in the tropics than those obtained in the subtropics, even higher than the ones obtained in the traditional cropping area of the USA (Congresso Brasileiro de Soja 2006). Such facts brought economic development and social well-being to a previously poor and under habited regions, largely because land was undervalued, as there was no adequate technology for extensive cropping, according to Gazzoni (2013).

Soybean is a milestone in the agro-industrial development of Mercosur countries. Its influence is so deep that two phases of regional agriculture are clearly differentiated, before and after 1970. Until that date, the prevailing cropping system in the region was the subsistence agriculture, for own or local consumption. When growers started cropping soybeans, they were obliged to face modern agriculture

[2]http://archives.chicagotribune.com/1973/02/15/page/58/article/soybean-prices-surge-to-peaks.

strategies and to be connected to the international market, which led to a chain of unprecedented changes in regional agriculture.

Extensive soybean fields were largely responsible for accelerating mechanization of the farms; the transport system had to quickly modernize; the growing demand forced agriculture to open a new agricultural frontier; growers had to professionalize their farm management; private organizations had to enhance their international trade skills; an overall and accelerated process of development covered all soybean regions. Due to soybean influence, other crops, like corn, wheat, and cotton, also experienced a quick revolution on their cultivation and management, as well as created solid new big business on animal production, like the modern poultry and pork value chains (Dall'Agnol et al. 2007; Gazzoni 2013).

2.1.3 Soybean Grain and Oil Production

Soybean is one of the most important crops in the world, ranked among the four top producing and traded grains, which also include corn, wheat, and rice. While the other grains are important due to its carbohydrate content (especially starch), the demand of soybean is driven by the high quality of its protein meal, consisting a key raw material for meat production. In this sense, the soybean oil can be considered a byproduct of the soybean processing.

Oil content (18–20 %) in the seed is lower than protein (36–40 %), but given the large amount of soybean meal demanded to feed meat producing animals, the resulting oil volume is significant (EMBRAPA 1994). Soy scientists, especially breeders, refer the difficulties to increase the oil content on soybeans, as a result of inappropriate cross links with protein content as well as with the crop yield. Considering present average soybean yield and oil content, ca. 600 kg of oil can be obtained out of each hectare of cultivated with soybean.

Oil represents 18–20 % of the soybean seed weight. Triacylglycerols represent over 94 % of the lipid fraction, followed by phospholipids (3.7 %), unsaponifable matter (1.5 %), sterols (0.24 %), tocopherols (0.12) and free fatty acids (0.5 %), referred by Hammond (2005) as a typical soybean oil composition.

In order to extract the oil, the soybean is crashed, adjusted for moisture content, heated to between 60 and 88 °C, rolled into flakes, and solvent-extracted with hexanes. The oil is then refined, blended for different applications and, sometimes, hydrogenated. Soybean oils, both liquid and partially hydrogenated, are traded as vegetable oil or are constituents for a wide variety of processed foods.

The residue remaining from oil extraction (soybean meal) is used in the nutrition industry for animal feed. Soybean is one of the most important protein sources (36–40 %). According to Hammond et al. (2005), lysine (2.6 %), threonine (1.5 %), cysteine (0.7 %), and methionine (0.6 %) are the most common amino acids found on soybean meal.

As an average, soybean oil has 17 % of saturated fat, 24 % of monounsaturated fat, and 59 % of polyunsaturated fat (Poth 2001). According to Ivanov et al. (2010), the major unsaturated fatty acids in soybean oil triglycerides are the polyunsaturated alpha-linolenic acid (C-18:3), with 7–10 %, and linoleic acid (C-18:2), with 51 %. The monounsaturated oleic acid (C-18:1) represents 23 %. Soybean oil also contains saturated fatty acids like 10 % palmitic (C-16:0) and 4 % stearic (C-18:0). Hammond et al. (2005) mention lineolate (54.5 %), oleate (22.9 %), linolenate (23 %), palmitate (10.6 %), and stearate (4.1 %) as being the most common methyl esters found on typical soybean oils, while myristate, palmitoleate, arachidate, gondoate, behenate, and lignocerate are also present, but at lower concentrations. According to the same authors, the average values for saponification and iodine are 190.4 and 132.7, respectively. Other components of soybean seeds are carbohydrates and ashes, with 29.4 and 4.6 %, respectively, expressed on dry weight basis.

Gazzoni et al. (2005), using Life Cycle Analysis methodology, determined that the relationship between input and output energy, considering the whole soybean grain (oil for biodiesel plus meal for other uses) was 1:3.38. Considering only the oil fraction for biodiesel production the relation was 1:1.12.

Global soybean area and production for the last 53 years are shown on Fig. 4. In 2014, about 45 Mton of soybean oil was produced worldwide, being second only to palm oil. In fact, soybean oil led the vegetable oil production until last decade, and might be the leader again, in the near future. In 1960, the world production was roughly 25.5 Mton, 92 % concentrated in the USA (59 %) and China (33 %).

Currently, the USA remains at the forefront of soy production, with 108 Mton (34.3 %), but Latin America, led by Brazil (98.6 Mton) and Argentina (60.5 Mton) overcome with great advantage (49.3 vs. 34.3 %) the USA production, turning the region the main center of global soybean production. These three countries account for about 84 % of the global harvest of 2014/15 season, meaning 264 Mton out of the world total of 315 Mton. Detailed statistics regarding soybean area, yield, and production for the major producing countries, are shown in Table 7.

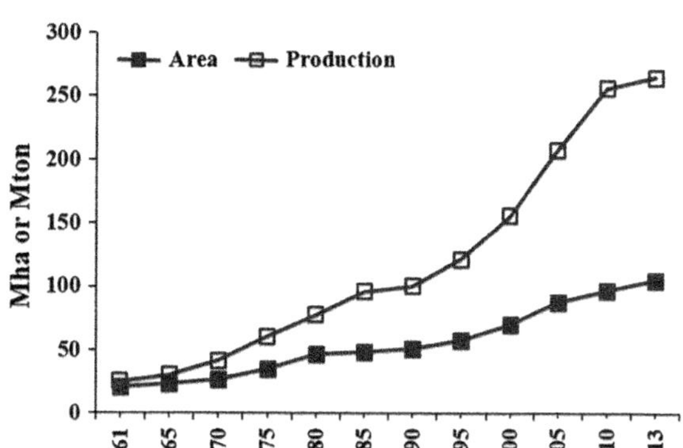

Fig. 4 Global soybean area and production

Table 7 Soybean area, yield and production for the most important countries

Year	Argentina			Bolivia			Brazil			Canada			India		
	Area (Mha)	Yield (kg/ha)	Production (Mton)	Area (Mha)	Yield (kg/ha)	Production (Mton)	Area (Mha)	Yield (kg/ha)	Production (Mton)	Area (Mha)	Yield (kg/ha)	Production (Mton)	Area (Mha)	Yield (kg/ha)	Production (Mton)
1961							0.24	1127	0.27	0.09	2103	0.18	0.01	455	0.01
1965	0.02	1035	0.02				0.43	1212	0.52	0.11	2038	0.22	0.02	435	0.01
1970	0.03	1032	0.03				1.32	1144	1.51	0.14	2085	0.28	0.03	438	0.01
1975	0.36	1363	0.49	0.01	1262	0.01	5.82	1699	9.89	0.16	2324	0.37	0.09	979	0.09
1980	2.03	1724	3.50	0.04	1267	0.05	8.77	1727	15.16	0.28	2489	0.69	0.61	727	0.44
1985	3.27	1988	6.50	0.06	1376	0.08	10.15	1800	18.28	0.41	2499	1.01	1.34	764	1.02
1990	4.96	2157	10.70	0.14	1623	0.23	11.49	1732	19.90	0.48	2610	1.26	2.56	1015	2.6
1995	5.93	2045	12.13	0.43	2031	0.87	11.68	2200	25.68	0.82	2783	2.29	5.04	1012	5.1
2000	8.64	2331	20.14	0.62	1941	1.20	13.64	2400	32.73	1.06	2548	2.70	6.42	822	5.28
2005	14.03	2729	38.29	0.94	1799	1.69	22.95	2230	51.18	1.17	2708	3.16	7.71	1073	8.27
2010	18.13	2905	52.68	1.09	1558	1.69	23.33	2948	68.76	1.48	2942	4.35	9.55	1333	12.74
2013	19.42	2539	49.31	1.24	1896	2.35	27.91	2929	81.72	1.82	2857	5.20	12.2	979	11.95

Year	China			Paraguay			USA			Uruguay		
	Area (Mha)	Yield (Kg/ha)	Production (Mton)	Area (Mha)	Yield (Kg/ha)	Production (Mton)	Area (Mha)	Yield (Kg/ha)	Production (Mton)	Area (Mha)	Yield (Kg/ha)	Production (Mton)
1961	10.01	626	6.26	0.01	1.615	0.01	10.93	1.690	18.47			
1965	8.64	719	6.21	0.01	1.925	0.02	13.94	1.651	23.01			
1970	8.02	1.094	8.78	0.03	1.459	0.04	17.1	1.794	30.68			
1975	7.03	1.038	7.30	0.15	1.465	0.22	21.7	1.942	42.14	0.01	1.719	0.02
1980	7.23	1.101	7.97	0.48	1.130	0.54	27.44	1.783	48.92	0.04	1.216	0.05
1985	7.73	1.361	10.51	0.72	1.631	1.17	24.92	2.292	57.13	0.01	1.449	0.02

(continued)

Table 7 (continued)

Year	China			Paraguay			USA			Uruguay		
	Area (Mha)	Yield (Kg/ha)	Production (Mton)	Area (Mha)	Yield (Kg/ha)	Production (Mton)	Area (Mha)	Yield (Kg/ha)	Production (Mton)	Area (Mha)	Yield (Kg/ha)	Production (Mton)
1990	7.56	1.455	11.01	0.9	1.994	1.79	22.87	2.292	52.42	0.03	1.298	0.04
1995	8.13	1.662	13.51	0.74	3.008	2.21	24.91	2.376	59.17	0.01	1.824	0.02
2000	9.31	1.656	15.41	1.18	2.533	2.98	29.3	2.561	75.05	0.01	764	0.01
2005	9.59	1.704	16.35	1.97	2.024	3.99	28.83	2.896	83.5	0.28	1.838	0.51
2010	8.52	1.771	15.08	2.67	2.793	7.46	31.01	2.922	90.61	0.86	2.317	2
2013	6.79	1.760	11.95	3.08	2.950	9.09	30.7	2.915	89.48	1.21	2.667	3.2

Source FAOSTAT database

2.1.4 Soybean in the Near Future

Demand for soybean remains strong chiefly because of the continuous growing need for protein meal, both for human consumption and animal feed. Nowadays, in addition to the food market for both commercial animal production, humans, and pet's nutrition, new markets like bioenergy and oil chemistry, extend the horizons of soybean demand, leveraging breakneck annual growing rates since 1990, when global production (108 Mton) was just one third of the current (315 Mton) production. The mean annual growth during the last 20 years surpassed 8 Mton (FAOSTAT 2015).

A close look into the future make it clear that a growing world population, as well as the increasing in longevity and income per capita of this population, with consequent changes in food habits and consumption patterns, will keep the demand growing at a steady pace, similar to the process of the last 50 years (Dall'Agnol et al. 2007). This forecasted continuous growing for soybean meal demand result in increasing soybean oil production, helping to assure the supply of vegetable oil for the biodiesel industry.

Prospective scenarios for soybean in the medium term (Fig. 5) show that Brazil and Argentina will capture most of the incremental market for the next two decades, given the depletion of the North American, Chinese, and Indian agricultural frontier. Among current competitors, Brazil is the one with the best comparative advantages, like abundant land, favorable climate to produce throughout the year, technology in state of the art, modern businesspersons, and entrepreneurs, but needs satisfactorily solve the issues encompassed in the so-called "Brazil cost" to ensure market leadership. Among the Brazil cost restrictions, ones linked to storage, transportation, and ports are the most challenging ones.

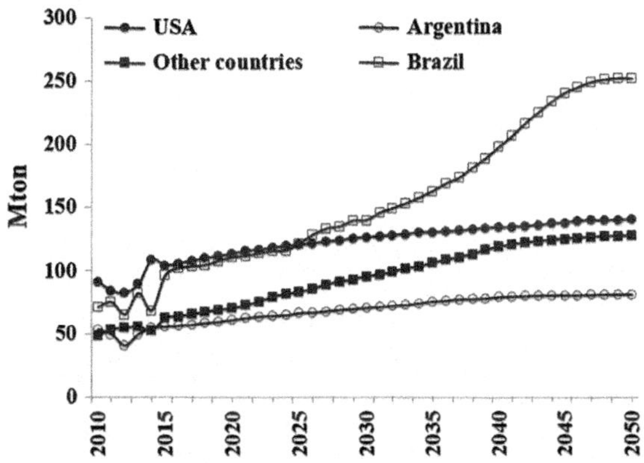

Fig. 5 Forecast of the major soybean producing countries

2.2 *Oil Palm (*Elaeis guineensis*)*

2.2.1 Origin and Highlights

Originally from the Gulf of Guinea, west central Africa, it is also known as African palm and dendê (only in Brazil). Although known and exploited for millennia in Africa, its commercial cultivation is relatively new, starting on the first decade of the twentieth century, in Malaysia. Palm is recognized as the oil crop that produces the largest amount of oil per hectare, which supports its leadership on the world vegetable oil production (Corley and Tinker 2003).

Palm oil is responsible for 57 Mt of the global production of 180 Mt of vegetable oils. Along with soybean oil, it accounts for over 50 % of the world vegetable oil production but, considering the 2015 average oil yield of each crop, one hectare of palm oil yields approximately the same amount of oil as 10 ha of soybean.

The high oil yield allows palm oil to occupy only 8 % of the world area cultivated with oil crops, while providing almost a third of vegetable oil produced globally (FAOSTAT 2015).

According to Cornet (2001), due to its tropical origin, palm oil is quite suitable for cultivation in the humid tropics of the original region, as well as southeastern Asia, northwestern South America and part of Central America. Presently, Asian countries account for nearly 90 % of its cultivated area, being Indonesia, Malaysia, and Thailand the major producers, according to the FAO database (FAOSTAT 2015). As well as the major producers, the largest importers are also located in Asia (China and India).

Indonesia, given its monumental production, is using part of its oil to produce biodiesel, with mandates for adding 5 % palm oil biodiesel in petrodiesel in 2006, 10 % in 2010, and 25 % in 2025 (REN 21 2015). The energy balance (input/output) of biodiesel from palm oil is very favorable, sometimes reaching up to 1:8, according to Gazzoni et al. (2008). The palm oil is used for bioenergy production, but the largest use is found in the nutrition segment (industrial frying, chocolates, pasta, margarine, vegetable creams, cookies, ice cream), and in cosmetics industry (beauty products, shampoos, detergents, and soaps).

2.2.2 Palm Oil Production

Figure 6 illustrates the evolution of the palm oil production worldwide, and the major palm oil producing countries are shown in Table 8.

Besides the leading Southern Asia countries (Indonesia, 26.9 Mton; Malaysia, 19.2 Mton; and Thailand, 1.97 Mton), Nigeria (0.96 Mton), and Colombia (0.95 Mton), plus over 40 other countries produce palm oil (Fig. 6). Up to the 1970s, Malaysia was the major producer of palm oil, with more than half of world production. In the last 40 years, Indonesia's production has skyrocketed from 0.7 Mton, in 1980, to 31 Mton, in 2014.

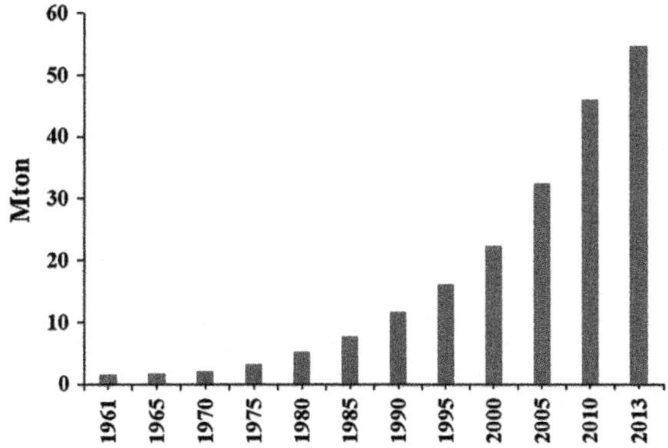

Fig. 6 Global palm oil fruit production

A significant proportion of the area presently used for palm oil cultivation in Southeastern Asia countries was formerly occupied by native forest, causing intense deforestation. Environmentalists advert that if deforestation proceeds at this pace, there will be no rainforests in countries like Indonesia within a decade, which would jeopardize the survival of Sumatra tiger, Asian rhino and orangutan. According to the environmentalist NGO Greenpeace, every year Indonesia loses 620,000 ha of rainforest, making it one of the largest emitters of greenhouse gases on the planet. This fact, associated with the loss of biodiversity could undermine the future of millions of Indonesians who depend on the forests for their food, shelter and livelihoods (Greenpeace 2015).

In contrast, Brazil has the world largest reserve of suitable land for palm cultivation, estimated to be around 50 Mha (Müller 1980), but cultivates only 0.16 Mha (FAOSTAT 2015). Largely, this is due to the restrictions imposed by the Brazilian environmental legislation for Amazonian lands, which restricts to 20 % the amount of area of a given farm that can undergo any kind of economic exploitation, imposing that more than 80 % of the biome should be preserved (Müller and Furlan Junior 2001). As a consequence, in spite of having the largest potential area for oil palm cultivation, Brazil is 9th among palm oil producers (0.37 Mt) and such a small production does not meet the country's needs, resulting in continuous import of palm oil, reason why the amount dedicated to biodiesel production is very small.

As for the future palm oil production, it should be taken into account that vast areas of the Brazilian tropical rainforests were cleared in the 1960s and 1970s, for the establishment of national integration highways. These areas are currently occupied by degraded pastures with low nutrition levels, and might be reinserted on profitable and sustainable business through the palm cultivation (Müller and Furlan Junior 2001). This land use change could provide new opportunities for employment and income to thousands of poor small farmers established on the banks of these highways, given adequate market and government incentives are put in place.

Table 8 Palm oil fruit production of the major countries, in Mton

Country	1961	1965	1970	1975	1980	1985	1990	1995	2000	2005	2010	2013
Indonesia	0.15	0.16	0.22	0.40	0.72	1.24	2.41	4.48	7.00	11.86	21.96	26.90
Malaysia	0.09	0.15	0.43	1.26	2.57	4.13	6.09	7.81	10.84	14.96	16.99	19.22
Thailand	0.00	0.00	0.00	0.00	0.02	0.09	0.23	0.37	0.58	0.78	1.29	1.97
Nigeria	0.67	0.69	0.49	0.50	0.65	0.62	0.73	0.86	0.90	1.17	0.97	0.96
Colombia	0.00	0.00	0.03	0.04	0.07	0.13	0.25	0.39	0.52	0.67	0.75	0.95

Source FAOSTAT database

Table 9 Oil of palm and oil of kernel produced by the major countries, in Mton

Year	Colombia		Indonesia		Malaysia		Nigeria		Thailand	
	Palm	Kernel	Palm	Kernel	Palm	Kernel	Palm	Kernel	Palm	Kernel
1961			0.15		0.09		0.67			
1965			0.16		0.15	0.01	0.69	0.02		
1970	0.03		0.22		0.43	0.03	0.49	0.05		
1975	0.04		0.40	0.02	1.26	0.11	0.50	0.06		
1980	0.07	0.01	0.72	0.04	2.57	0.22	0.65	0.08	0.02	0.00
1985	0.13	0.01	1.24	0.12	4.13	0.51	0.62	0.15	0.09	0.01
1990	0.25	0.02	2.41	0.31	6.09	0.83	0.73	0.16	0.23	0.02
1995	0.39	0.03	4.48	0.43	7.81	1.04	0.86	0.24	0.37	0.04
2000	0.52	0.05	7.00	0.72	10.84	1.39	0.90	0.20	0.58	0.05
2005	0.67	0.07	11.86	1.46	14.96	1.84	1.17	0.57	0.78	0.07
2010	0.75	0.07	21.96	2.36	16.99	2.01	0.97	0.51	1.29	0.13
2013	0.95	0.08	26.90	3.06	19.22	2.27	0.96	0.51	1.97	0.18

Source FAOSTAT database

Palm oil is a huge oil producing plant and Table 9 shows the evolution of the palm oil and kernel oil from major producing countries. However, unlike most of the major oil crops, its residues after oil extraction have only marginal or no commercial value. The main uses for palm oil residues are organic fertilizer or electricity generation by burning the waste. Two types of oils are obtained from the palm oil fruits: the palm oil itself, extracted from the pulp; and palm kernel oil, extracted from the fruit kernel. The oil fraction constitutes about 22 % of the weight of the palm bunch, and only 3 % is palm kernel oil. Lauric acid is almost absent in palm oil, being the predominant component of palm kernel oil (Ramos et al. 2009).

Palm oil is composed almost by 50:50 saturated:unsaturated fatty acids. Major saturated are palmitic (44 %), followed by stearic (4 %), while oleic (monounsaturated, 37 %), and linoleic (polyunsaturated, 9 %) are the major unsaturated fatty acids found on palm oil (Ramos et al. 2009).

2.2.3 Special Requirements

Müller and Furlan Junior (2001) point out that the establishment of a palm plantation is expensive, being an investment of long maturity and late paying back, taking 4–6 years for the first harvest. During this period, the crop does not generate income, unless it is consortiated with other food or fiber crops, in between palm lines, like cassava, pineapple, papaya, banana, or even pastures. This is important not only to ensure food for self-consumption, but also to allow an extra income from the sale of the remaining production.

For this reason, small growers need official or private support to withstand the heavy crop establishment costs and to survive during the initial period of the project (Müller and Furlan Junior 2001). Moreover, it is paramount the existence of a processing industry in the surroundings of the crop plantation, because the fruit demands rapid processing after harvest, and the low value of the fruit does not allow transport over long distances (Müller 1980).

*2.3 Canola (Rapeseed) (***Brassica napus** *L.)*

2.3.1 History and Highlights

Rapeseed belongs to the Brassicaceae family (formerly Cruciferae), the same family as mustard, broccoli, or cauliflower. *Brassica napus* is the result of an interspecific cross between *Brassica campestris* and *Brassica oleracea*. The canola name results from a contraction of "**CAN**adian **O**il **L**ow **A**cid," a variety of rapeseed modified in the early 1970s by traditional breeding, through which Canadian scientists from the

University of Manitoba selected varieties, which oil has low erucic acid (toxic for humans and animals). Its bran has very low glucosinolates content (antinutritional components), making both excellent alternatives for humans (oil) and animals (cake) consumption (Downey and Harvey 1963). The rapeseed, in turn, differs from canola because high levels of erucic acid and glucosinolates are present in the grains (Cultura da colza 1980).

Brassica oilseed varieties are among the oldest plants cultivated by humanity, with documentation of its use in India 4000 years ago, and in China and Japan 2000 years ago, but *B. napus* use is more recent, and first records are restricted to the Mediterranean region (Prakash and Hinata 1980). Its use in northern Europe for oil lamps dates to the thirteenth century (Snowdon 2007), but a larger use was limited until the development of steam power, when machinists found rapeseed oil clung to water- and steam-washed metal surfaces better than other lubricants. Because of its lubricant properties, there was a high demand for rapeseed oil during World War I to supply the increasing number of steam engines in naval and merchant ships. The war demand used all the European and Asian rapeseed oil available, creating a critical shortage, giving the opportunity for Canada to expand its rapeseed production. After the war, the lubricant demand declined sharply, and other uses for the oil were developed (USDA 2015a).

Presently, canola is the leading group of varieties grown worldwide as rapeseed. The oil is the main product of canola, although its meal is also highly valued for the formulation of animal feed, because of the high-protein content. According to De Mori et al. (2014), the oil content of canola seeds is high (38–45 %) and the volume of oil produced worldwide is surpassed only by palm and soybean oil, and the meal is second only to soybeans.

Low amounts of unsaturated fatty acids are found in canola oil, being palmitic (16:0) the one with higher content, normally 4 %. The major fatty acid found in canola is the mono unsaturated oleic (18:1) with 63 %, followed by polyunsaturated linoleic (18:2) with 20 %, and linolenic (18:3) with 9 % (Ramos et al. 2009). Due to its favorable fatty acid profile, doctors and nutritionists indicate canola and sunflower oils as the best composition of fatty acids for people interested in healthy diets.

In canola oil are found high amount of omega-3, vitamin E, monounsaturated fats, and the lowest saturated fat content of all vegetable oils. Perhaps this is the reason why the demand exceeds supply and the market value exceeds the price of soybean oil. High prices of canola oil make biodiesel from this source rather costly for the market and for supporting public policies. As for the total energy balance of biodiesel from canola oil, considering the utilization of its meal, it was concluded that for each energy unit input along the life cycle (from feedstock production to biodiesel consumption), 2.9 energy units are obtained; when considering only oil production (not computing energy on the meal), this relationship decreases to 1:1.4 (Gazzoni et al. 2009).

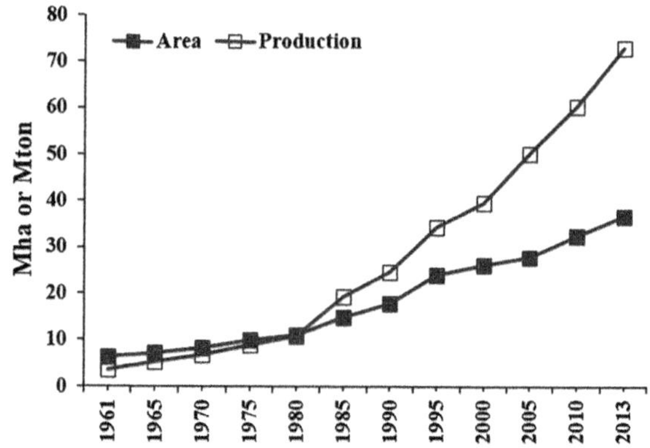

Fig. 7 World area and production of canola

2.3.2 Canola Production

World area production of canola is shown on Fig. 7, and Table 10 details the canola production parameters for the leading producing regions or countries

In 2014/15, world production of canola was 72 Mton of grains, allowing the extraction of 26 Mton of oil, representing 16 % of global vegetable oil production (FAOSTAT 2015). The leading production region is the European Union (24.0 Mton), followed by China (14.7 Mton), Canada (14.45 Mton), India (7.5 Mton), and Japan (2.0 Mton).

Canola is more adapted to mild temperature regions, distant from the Equator. In these locations, the cropping window is very narrow, not favoring crops such as soybean or corn. In regions of more severe climates, canola is seeded previously to the formation of snow and remains dormant until the spring when germinates after soils thawing, completing the cycle with approximately 85 days (USDA 2015a). This system is more profitable (20–30 % higher yields) than the canola seeded in the spring, after the melting of the snow.

Canola grain are rich in oil content (around 40 %; USDA 2015a), leading to large amounts of oil produced worldwide, as demonstrated on Fig. 8, being the aggregated data detailed by each producing region or countries on Table 11.

In regions where canola is cultivated during the spring either corn, soybeans, or cotton may result more profitable. Canola would be an excellent crop rotation alternative for soybeans and corn, but as it is very susceptible to the attack of a disease known as *sclerotinia*, the rotation is not recommended because the pest inoculum is build up during canola cycle and negatively affect soybean yield (De Mori et al. 2014). Furthermore, producing canola requires appropriated machines, especially harvesters and seed machines, due to the very small size of its grains, in order to provide adequate sowing and avoid harvesting losses.

Table 10 Canola area, yield and production for the most important countries

Year	Europe			Canada			China			India			Japan		
	Area (Mha)	Yield (kg/ha)	Production (Mton)	Area (Mha)	Yield (kg/ha)	Production (Mton)	Area (Mha)	Yield (kg/ha)	Production (Mton)	Area (Mha)	Yield (kg/ha)	Production (Mton)	Area (Mha)	Yield (kg/ha)	Production (Mton)
1961	0.61	1546	0.95	0.29	885	0.25	1.47	263	0.39	2.88	467	1.35	0.19	159	0.27
1965	0.84	1898	1.59	0.58	883	0.51	1.84	604	1.11	2.91	507	1.47	0.09	1403	0.13
1970	1.05	1869	1.97	1.64	999	1.64	1.45	666	0.97	3.17	493	1.56	0.02	1470	0.03
1975	1.29	2057	2.65	1.83	1006	1.84	2.31	665	1.54	3.68	612	2.25	0.01	1568	0.01
1980	1.66	2265	3.77	2.08	1194	2.48	2.84	839	2.39	3.47	412	1.43	0.01	1649	0.01
1985	2.57	2462	6.32	2.78	1257	3.50	4.49	1248	5.61	3.99	771	3.07	0.01	1711	0.01
1990	3.62	2553	9.23	2.53	1291	3.27	5.50	1264	6.96	4.97	831	4.13	0.01	1739	0.01
1995	4.27	2534	10.81	5.27	1221	6.44	6.91	1416	9.78	6.06	950	5.76	0.01	1795	0.01
2000	4.62	2542	11.74	4.86	1483	7.21	7.49	1519	11.38	6.03	960	5.79	0.01	2123	0.01
2005	5.46	3023	16.51	5.18	1832	9.48	7.28	1793	13.05	7.32	1038	7.59	0.01	2038	0.01
2010	8.94	2598	23.24	6.85	1865	12.77	7.37	1775	13.08	5.58	1184	6.61	0.01	1276	0.01
2013	9.31	2749	25.58	8.01	2240	17.94	7.52	1923	14.46	6.34	1233	7.82	0.01	1208	0.01

Source FAOSTAT database

Fig. 8 World production of canola oil from 1961 to 2013

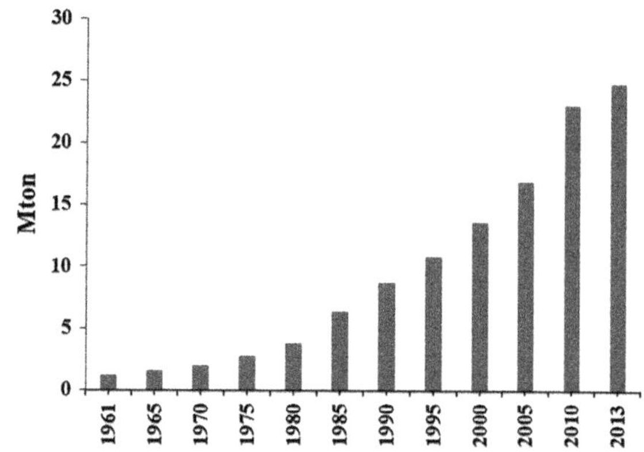

Table 11 Oil of canola produced by the major countries or regions, in Mton

Year	Europe	Canada	India	Japan	China
1961	0.34	0.01	0.40	0.12	0.11
1965	0.53	0.02	0.44	0.10	0.30
1970	0.68	0.07	0.48	0.14	0.29
1975	0.91	0.14	0.69	0.29	0.42
1980	1.38	0.42	0.42	0.41	0.80
1985	2.37	0.50	0.94	0.59	1.58
1990	3.37	0.58	1.37	0.75	2.14
1995	3.61	1.15	1.76	0.79	2.73
2000	4.51	1.30	1.79	0.91	3.64
2005	5.79	1.29	2.35	0.93	4.65
2010	9.37	2.50	2.05	0.99	5.39
2013	9.91	2.83	2.31	1.04	5.60

Source FAOSTAT database

2.4 *Sunflower (***Helianthus annuus** *L.)*

2.4.1 History and Highlights

The center of origin of sunflower is the region comprising southwest USA and northern Mexico, from where it disseminated to the rest of the continent. Its most likely domestication occurred in that region, based on evidences of its cultivation by North American Indians over 3000 years ago (Lentz et al. 2001).

Upon the discovery of America, the Spaniards introduced sunflower in Spain as an ornamental plant, from where it spread to the rest of Europe. In the eighteenth century, the sunflower reached Eastern Europe, presently the main world producing region, led by Ukraine (10.0 Mton) and Russia (9.0 Mton), followed by the European Union (8.8 Mton), Argentina (2.5 Mton), and Turkey (1.2 Mton) (FAOSTAT 2015).

Russia was largely responsible for the spread of sunflower as a worldwide economically important crop. By 1880, after being improved by Russian agronomists, the sunflower was reintroduced into the USA, where it was initially used as fodder. The importance of sunflower as an edible oil source only emerged by the 1920s. However, it was after World War II that sunflower aroused to the front line of the international oil crop production worldwide (USDA 2015b).

2.4.2 Production and Use

The global sunflower cultivated area in 2014 was, approximately, 18 Mha, with an overall production of 40 Mton of grain, being 16 Mton of oil, and 17 Mton of meal (FAOSTAT 2015), ranking fourth among the most important oils and meal production, globally. Sunflower accounts for about 7.5 % of world production of vegetable oil, behind palm (34 %), soybeans (30 %), and canola (16 %).[3] The oil content of the grains is about 45 %, consumed almost completely as edible oil for its excellent quality, while grain protein content range from 28 to 32 % (de Leite et al. 2005; USDA 2015b).

Figure 9 displays the evolution of sunflower area and production worldwide, while Table 12 details the history of sunflower area, yield, and production, for the major producing countries. In this Table, data for Russia and Ukraine are absent until 1995, as they were aggregated under the common name of Soviet Union on the FAO database.

According to Ungaro (2000), sunflower requires soil with good content of potassium and phosphorus, being more tolerant to drought than other major grains, because of its deeper root system. It is insensitive to photoperiod and can be cultivated from the vicinity of the equator to latitudes above 40°. Temperatures around 27 °C are considered optimum for proper plant growth, but it develops quite satisfactorily from 8 to 34 °C, reason why can be grown as a second summer crop (off-season). The crop is also an agronomic important option for rotation with soybeans, corn, and wheat.

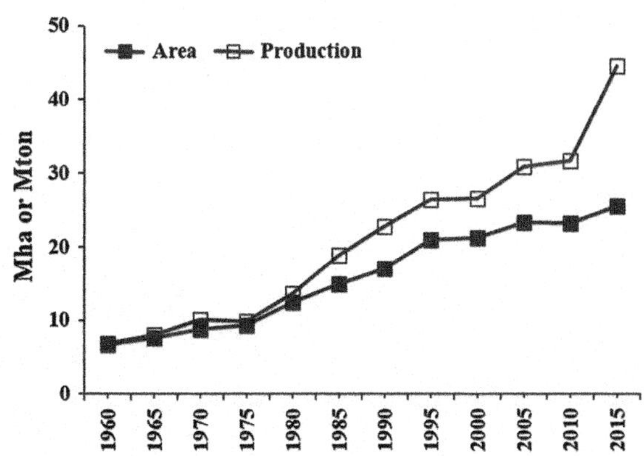

Fig. 9 Global sunflower area and production

[3]www.statista.com.

Table 12 Sunflower area, yield and production for the most important countries

Year	Argentina			Russia			Turkey			Ukraine			USSR		
	Area (Mha)	Yield (kg/ha)	Production (Mton)	Area (Mha)	Yield (kg/ha)	Production (Mton)	Area (Mha)	Yield (kg/ha)	Production (Mton)	Area (Mha)	Yield (kg/ha)	Production (Mton)	Area (Mha)	Yield (kg/ha)	Production (Mton)
1961	0.90	651	0.59				0.12	822	0.10				4.22	1127	4.75
1965	1.02	746	0.76				0.16	1002	0.16				4.87	1119	5.45
1970	1.35	846	1.14				0.36	1042	0.38				4.78	1286	6.14
1975	1.01	728	0.73				0.42	1168	0.49				4.05	1234	4.99
1980	1.86	890	1.65				0.58	1304	0.75				4.34	1065	4.62
1985	2.36	1441	3.40				0.64	1245	0.80				4.05	1298	5.26
1990	2.69	1451	3.90				0.71	1204	0.86				4.65	1376	6.40
1995	2.95	1963	5.80	4.13	1018	4.20	0.59	1539	0.90	2.01	1425	2.86			
2000	3.48	1746	6.07	4.37	896	3.91	0.54	1476	0.80	2.84	1217	3.46			
2005	1.92	1905	3.66	5.41	1190	6.44	0.57	1723	0.98	3.69	1276	4.71			
2010	1.50	1493	2.23	5.58	959	5.34	0.64	2058	1.32	4.53	1496	6.77			
2013	1.62	1916	3.10	6.80	1550	10.53	0.61	2498	1.52	5.09	2170	11.05			

Source FAOSTAT database

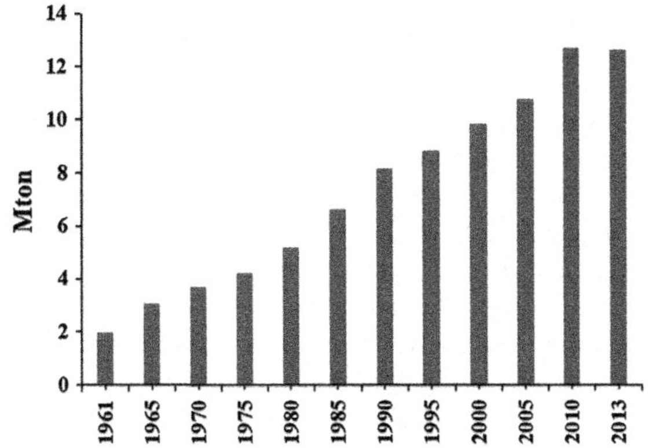

Fig. 10 World sunflower oil production

From sunflower grains, it is extracted high-quality oil and the remaining meal is excellent for animal nutrition. The oil is rich in unsaturated fatty acids, like the monounsaturated oleic (18:1), with 16 % and the polyunsaturated linoleic, with 72 %; major saturated fatty acids are palmitic (16:0), with 6 % and stearic (18:0), with 4 % (Ramos et al. 2009).

The sunflower meal contains about 50 % protein and is rich in sulfur amino acids, allowing a perfect integration with soybean meal, which is rich in lysine and low in sulfur amino acids. A mixture of both would provide ideal balanced food for animal nutrition (de Leite et al. 2005). According to these authors, besides oil and meal production, sunflower is an important feeding source for domestic and native bees (honey production), as well as an ornamental plant and for silage (animal fodder). Sunflower seeds are great for feeding birds and for edible oil production, and used as a lubricant. The nutritional quality of sunflower oil is similar to the canola oil, being highly suitable for biodiesel production. Figure 10 displays the historical series of the world sunflower oil production.

Regarding energy efficiency of biodiesel production from sunflower oil, Gazzoni et al. (2005), using Life Cycle Analysis techniques, determined that when the whole grain destination was considered (meal for nutrition, oil for biodiesel), 2.69 energy units were obtained from each energy unit input to the system. This relation was reduced when meal was not considered, and then each unit of input energy generated 1.61 units released by biodiesel combustion.

2.5 *Cotton (*Gossypium hirsutum *L.)*

2.5.1 History and Uses

Iqbal et al. (2001) refers cotton as one of the oldest plants domesticated by man, known for more than 8000 years, with records of its use to about 4000 years ago. The most likely center of origin of cotton is India, although some kind of cotton is found on all continents (Iqbal 1997), including 40 native species found in subtropical and tropical regions, some of which are used for commercial production of textile fibers. The most common species used for fiber production are *Gossypium hirsutum* (USA and Australia), *Gossypium arboreum* and *Gossypium herbaceum* (Asia), and *Gossypium barbadense* (Egypt) (USDA 2015c).

Cotton is a tropical crop, but has broad adaptation and can be grown on latitudes ranging from 0° to 40° (Cia et al. 1999), growing well in various soil types, but the plant root needs a well-oxygenated environment, for what the cropped area cannot be compacted. It is considered quite tolerant to median water deficit, requiring good soil moisture during the growing season and relatively dry weather during ripening and harvesting (Beltrão 1999).

The cottonseed contains medium levels of oil (15–18 %), used both for industrial purposes (hygiene/cosmetic industry and bioenergy) and for domestic consumption (fried foods and margarine) (USDA 2015c). Cotton oil contains both saturated fatty acids like palmitic (20 %) and stearic (2 %), and unsaturated ones as oleic (35 %) and linoleic (42 %) (Ramos et al. 2009). The meal resulting after cottonseed processing contains from 20 to 25 % protein and is directed to animal feed, preferably mixed with other proteinaceous cakes, because of the presence of gossypol, a toxic substance found in cottonseed meal (Cia et al. 1999).

2.5.2 Cotton Production

Cotton is a crop favored by a well-structured production chain, with extensive technological expertise and comprehensive coverage of research institutions and networks, readily available to solve any technological problem (Cia et al. 1999). Cotton varieties differ in the size of fiber (short, medium, and long), plant height (tall and short), and the cycle length (early: 120/150 days or late: 150/180 days). The perennial cotton—which is a tree—depends on manual harvest, being restricted to small farms. Approximately, 90 % of world production corresponds to annual cotton with early cycle (Beltrão 1999).

The cotton production cost is high because the plant is a suitable host for several pests, demanding a large number of pesticide applications, turning its production cost one of the most expensive among the major crops.

Cotton is widely grown, being present in over 80 countries, occupying an area in excess of 30 Mha (2014/15 season), producing 45 Mt of cottonseed (Fig. 11) and 26.3 Mt of cotton lint (total of 71.3 Mt).

The major driver for cotton production in the world is the fiber, used mainly for textile applications, besides other minor industrial uses. It should be taken into consideration that cotton market is quite unstable and competitive, partially because of the limited demand of its fiber, which competes with other natural, but chiefly with synthetic fibers.

The large number of producing countries contributes for an unstable market, making it easy to replenish low global stocks as a reaction to the stimulus of good market prices. This is one of the reasons why, despite the large number of

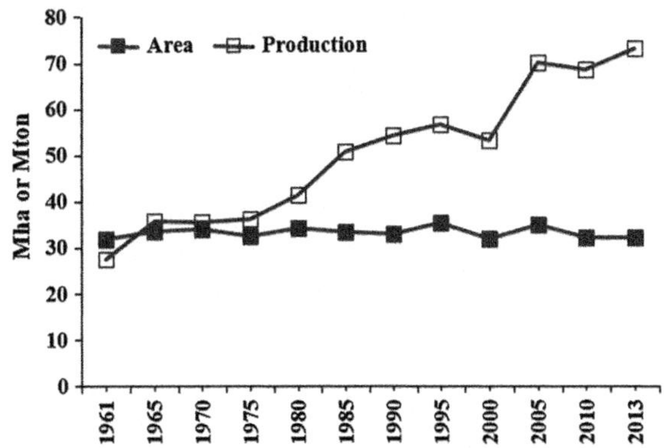

Fig. 11 Global cotton area and production

producing countries, not all of them are present on the market every other year, due to production problems, mainly climatic and phytosanitary constraints, increasing production costs. So, the competition status of the countries changes, depending on the production amount and its costs and on the market price.

Table 13 details the cotton area, yield, and production for the most important producing countries, while Fig. 12 presents the recent history of cotton oil production. According to the FAO database (FAOSTAT 2015), the major producers of cotton lint are India (6.51 Mton), China (6.48 Mton), USA (3.55 Mton), Pakistan (2.31 Mton), and Brazil (1.51 Mton). The plume international trade, amounting 7.67 Mton, are led by the USA (2.33 Mton) and followed by India (0.98 Mton), Brazil (0.87 Mton), Australia (0.63 Mton), and Uzbekistan (0.61 Mton). China, despite being the second largest producer, is also the number one cotton fiber importer, followed by the East Asian countries, Europe, Bangladesh, and Pakistan.

2.6 *Peanut (*Arachis hypogea *L.)*

2.6.1 History and Highlights

Wild peanuts are common plants along South America (mainly Brazil, Paraguay, Bolivia, and northern Argentina), between latitudes 10 and 30°S, with its most probable center of origin located in the Chaco region (Kochert et al. 1996). It belongs to the botanic family Fabaceae, the same as beans, peas, and soybean. Peanut plants are classified into four groups, according to differential characteristics: Runner, Spanish, Valencia, and Virginia, being the first three upright and early types; the latter is creeping and have a longer cycle (Beasley and Baldwin 2015).

According to archaeological documentation referred by Jones (2007), there is evidence of its consumption since 3800 BC. The cultivation and dispersion of peanut began with the Indians, spreading it to various regions of Latin America. In

Table 13 Cotton area, yield and production for the most important countries

Year	Brazil			India			Pakistan			USA			China		
	Area (Mha)	Yield (kg/ha)	Production (Mton)	Area (Mha)	Yield (kg/ha)	Production (Mton)	Area (Mha)	Yield (kg/ha)	Production (Mton)	Area (Mha)	Yield (kg/ha)	Production (Mton)	Area (Mha)	Yield (kg/ha)	Production (Mton)
1961	2.02	631	1.28	7.72	358	2.76	1.40	697	0.97	6.33	1349	8.53	3.87	621	2.40
1965	2.33	583	1.36	7.94	392	3.12	1.56	801	1.25	5.51	1593	8.78	5.00	1259	6.30
1970	4.30	455	1.95	7.61	392	2.98	1.75	932	1.63	4.51	1309	5.91	4.99	1368	6.83
1975	3.88	451	1.75	7.46	486	3.63	1.85	833	1.54	3.56	1328	4.73	4.95	1443	7.14
1980	3.70	453	1.68	7.82	516	4.04	2.11	1017	2.14	5.35	1211	6.48	4.92	1652	8.12
1985	3.59	796	2.86	7.53	615	4.64	2.36	1544	3.65	4.14	1863	7.71	5.14	2420	12.44
1990	1.90	1009	1.92	7.44	703	5.23	2.66	1845	4.91	4.75	1851	8.79	5.59	2420	13.52
1995	1.19	1218	1.45	9.04	756	6.83	3.00	1804	5.41	6.48	1561	10.11	5.42	2638	14.30
2000	0.80	2508	2.01	8.58	598	5.13	2.93	1871	5.48	5.28	1814	9.58	4.04	3279	13.25
2005	1.26	2904	3.67	8.68	1133	9.83	3.10	2141	6.64	5.59	2305	12.88	5.06	3387	17.14
2010	0.83	3555	2.95	11.14	1594	17.76	2.69	2088	5.61	4.33	2188	9.47	4.85	3694	17.91
2013	0.94	3621	3.42	11.70	1617	18.91	2.81	2225	6.24	3.05	2498	7.63	4.35	4356	18.93

Source FAOSTAT database

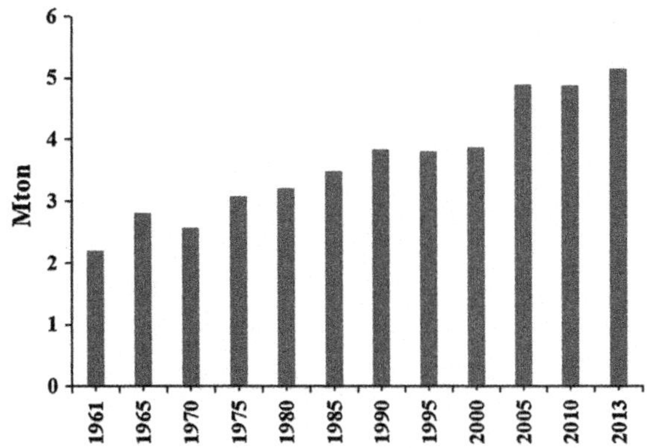

Fig. 12 World cotton oil production

the eighteenth century, it was introduced in Europe. In the nineteenth century, it was introduced to Africa, from Brazil, and to Asia, from Peru.

Peanut is one of the oil crops with highest oil fraction, ranging from 45 to 50 % (Mercer et al. 1990). It is highly prized in the market and appreciated for human consumption, and can be used in the pharmaceutical, cosmetics, and in the production of biodiesel. The meal quality is comparable to soybean meal making it highly valued for animal feed (dos Santos et al. 2013).

2.6.2 Production

The peanut plant grows well at temperatures between 20 and 30 °C, throughout the cycle. The plant prefers loamy sandy soil, well fertilized. Like all legumes, peanut does not tolerate high soil acidity, requiring liming when appropriate. It requires good availability of calcium for the formation of pods, as well as phosphorus, for grain formation. Nitrogen can be made available by the inoculation of grains with N-fixing bacteria of the *Bradhyrizobium* genus prior to sowing. There are limitations of appropriate machinery for the harvest process, normally carried out manually (dos Santos et al. 2013).

Peanut is highly susceptible to the attack of microorganisms-producing mycotoxins, particularly aflatoxin, depreciating its commercial value (Pitt and Hocking 2006). The inadequate management of humidity and temperature during peanut harvest, transport, and storage favors this attack. These microorganisms can survive in plant debris and infect subsequent crops. For this reason, it is recommended to avoid continuous peanut cultivation in the same area. Rotation with other crops is highly desirable. In addition to providing a good meal to feed pigs and poultry from the beans, its shoots can provide hay or quality silage for feeding cattle.

Peanut world area has been stable around 30 Mha for the last 55 years, while the production jumped from less than 30 Mt to over 70 Mt (Fig. 13). The largest producers are China, India, Nigeria, USA, and Brazil (Table 14). Major uses of

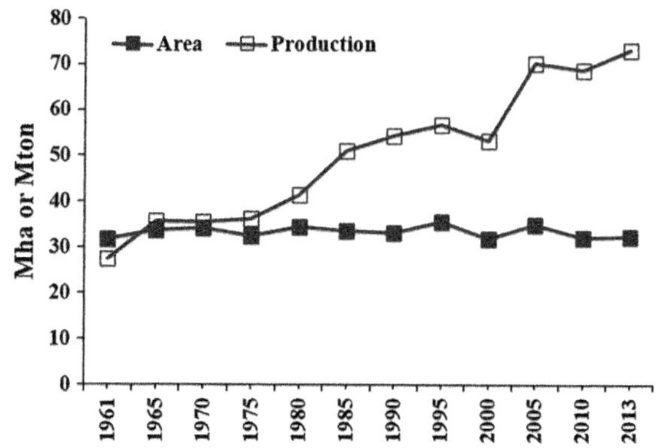

Fig. 13 Global peanut area and production

peanut are for oil production, human food, and animal feed (dos Santos et al. 2013). World peanut oil production is displayed on Fig. 14.

2.7 *Minor Oil Crops*

A series of species are used locally, even regionally, for oil production in small scale. Some are directed for self-consumption, either for human or animal nutrition, for elaborating soaps or producing energy. Those minor crops represent less than 5 % of the world oil production, and are restricted to commercial or purposes niches. Two species of minor oil crops are described below.

2.7.1 Castor (*Ricinus communis* L.)

The center of origin of the castor bean is undefined, as both India and Ethiopia are mentioned as its center of origin (Anjami 2012). It belongs to the family Euphorbiaceae, the same of cassava, rubber, and jatropha. The plant shows broad adaptation, being cultivated or naturally occurring on latitudes from 0° to 40°, with prevailing temperatures between 20 and 30 °C, requiring annual rainfall between 500 and 1500 mm (Abreu et al. 2012). It is recognized as a suitable crop for semi-arid regions, because of its relative tolerance to drought (Carvalho 2005).

According to Azevedo and Beltrão (2007), under dry conditions, castor yields are very low, but there are situations where castor is the only cash crop for peasants living on semi-arid regions, even though the plant is more productive on well-drained, deep, non-compacted and fertile soils, with pH on the range 6.0–7.0 (Rodrigues Filho 2000).

The area and the world production of castor beans is approximately 1.5 Mha/Mt, being India responsible for over 50 % of this production, followed by China and Brazil, and the castor oil production is around 0.5 Mt (FAOSTAT 2015). Its market

Table 14 Peanut area, yield and production for the most important countries

Year	Brazil			China			India			Nigeria			USA		
	Area (Mha)	Yield (kg/ha)	Production (Mton)	Area (Mha)	Yield (kg/ha)	Production (Mton)	Area (Mha)	Yield (kg/ha)	Production (Mton)	Area (Mha)	Yield (kg/ha)	Production (Mton)	Area (Mha)	Yield (kg/ha)	Production (Mton)
1961	0.44	1339	0.58	1.30	889	1.15	6.89	725	4.99	1.49	1052	1.57	0.57	1329	0.75
1965	0.54	1374	0.74	1.95	1054	2.05	7.70	554	4.26	2.23	887	1.98	0.58	1862	1.08
1970	0.67	1386	0.93	1.80	1264	2.27	7.33	834	6.11	1.87	846	1.58	0.59	2276	1.35
1975	0.35	1281	0.44	1.94	1218	2.36	7.22	935	6.75	1.42	323	0.46	0.61	2874	1.74
1980	0.31	1543	0.48	2.39	1542	3.69	6.80	736	5.01	0.56	837	0.47	0.57	1844	1.04
1985	0.19	1756	0.34	3.37	2003	6.75	7.12	719	5.12	0.59	1046	0.62	0.59	3148	1.87
1990	0.08	1655	0.14	2.94	2187	6.43	8.31	904	7.51	0.71	1649	1.17	0.74	2223	1.63
1995	0.09	1796	0.17	3.85	2684	10.33	7.52	1007	7.58	1.77	894	1.58	0.61	2558	1.57
2000	0.10	1793	0.18	4.88	2972	14.52	6.56	988	6.48	1.93	1500	2.90	0.54	2740	1.48
2005	0.14	2317	0.32	4.68	3073	14.40	6.74	1187	7.99	2.19	1590	3.48	0.66	3351	2.21
2010	0.09	2772	0.26	4.55	3454	15.71	5.86	1410	8.27	2.79	1362	3.80	0.51	3712	1.89
2013	0.12	3222	0.39	4.65	3659	17.02	5.25	1804	9.47	2.36	1271	3.00	0.42	4496	1.89

Source FAOSTAT database

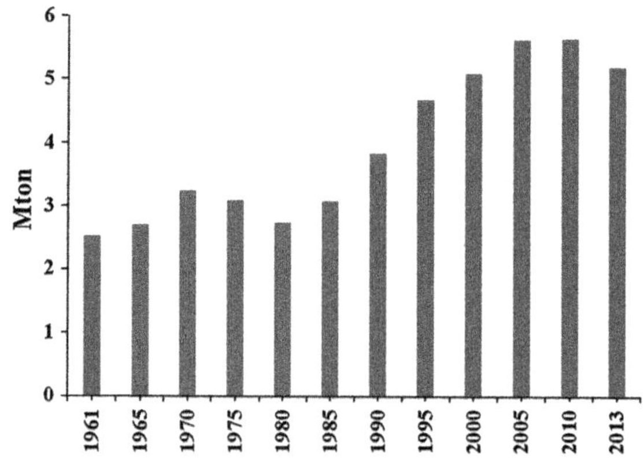

Fig. 14 World peanut oil production

is narrow and limited, reason why a production much higher than this amount can lead to exaggerated stocks and low market prices.

Severino et al. (2006) mention that castor yields are low when compared to major oil crops, but the seeds are rich in oil (45–52 %). Although there are cultivars of annual cycle, the most commonly cultivars grown worldwide are late ripening (180–240 days), requiring manual harvest, which is one of the limitations for the crop expansion. Ogunniyi (2006) mention that ricinoleic, a monounsaturated (18:1) omega-9 fatty acid, represents up to 90 % of the seed oil obtained from mature castor beans. It differs from oleic acid due to the presence of a hydroxyl radical linked to the 12th carbon of the chain. Ricinoleic is not an edible fatty acid, but has multiple uses in the industry (manufacturing of paints, varnishes, soaps, detergents, insecticides, fungicides, bactericides, candles, synthetics, plastics, pharmaceuticals, specialty greases, etc.), according to Ogunniyi (2006).

The high proportion of ricinoleic acid on the castor oil is largely responsible for its low viscosity and for the formation of polymers on the combustion chambers of engines, limiting its use for biodiesel production, unless blended with biodiesel obtained from feedstocks with oils of higher viscosity. It is a much sued oil to lubricate high-speed engines (aircraft, rockets, ships), not changing its characteristics whether used in high or low temperatures. Castor meal has no commercial value because it is toxic to animal feed, being generally used as organic fertilizer, due to its effectiveness in controlling soil nematodes.

2.7.2 Oil Radish (*Raphanus sativus* L. var. *oleiferus* Metzg)

Originally from Asia, it is one of the oldest species exploited for oil production (Wang et al. 2015). Although the oil radish is a plant whose seeds are rich in oil, its major use is as a winter cover plant, for crop rotation and for feed. The plant belongs to the family Brassicaceae (formerly Cruciferae), the same as crambe, canola and mustard.

According to Hernani and Henn (1995), this species shows large adaptability to different climates and soils, being a very vigorous plant, with pivoting and aggressive root system, able to break through extremely dense soil layers and/or compacted, at depths greater than 2.50 m. It grows fast, exerting high suppressive effect on weeds. Sixty days after sowing, the oil radish covers about 70 % of the land surface; its biomass has easy and rapid decomposition due to the low carbon/nitrogen ratio (C/N), providing, instantly nutrients to subsequent crops (CATI 2001). Produces between 20 and 35 t/ha of biomass, 3–8 t/ha of dry matter, and 500–1500 kg/ha of seeds, resulting in 150–500 kg/ha of oil.

Oliveira et al. (2011) stated that the oil radish is a rustic plant that grows well in poor soils, either in cold or hot places, indifferent to low (0 m) or high (1000 m) altitudes. The plant requires the presence of moisture in the soil during implantation and early development, but during the rest of the cycle shows median tolerance to drought and frost, being adequately cropped during fall and winter. It is quite resistant to pests and diseases and does not demand soil preparation. Despite its tolerance to soils with aluminum saturation and high acidity, the plant increases the green mass and grain production when cultivated on fertile soils.

The oil content of the seeds is relatively high (32–42 %) but, due to its low grain yield, the oil production is small and is not edible (Wang et al. 2015). The oil market is restricted to industrial uses and its production chain is deficient.

2.8 *Potential Oil Crops*

There are innumerous plant species with median to high oil content, with a theoretical potential for oil production. Some are source of oil on extractive systems, based on native formation. Its commercial development depends on (a) possibility of production of over 500 kg/ha of oil, in order to compete with major oil crops; (b) domestication of the species; (c) establishment of sound production systems; (d) organization of the productive chain connecting growers, suppliers, processors, industry and consumers (Gazzoni et al. 2012). Among others, potential oil crops include flaxseed, sesame, safflower, crambe, tucuman, rubs, buriti, macaúba, indaiá, açaí, gerivá, patauá, cotieira, oiticica, nhandiroba, tung, pequi, jatropha, jojoba, and tingui. Two examples of potential oil crops are described below.

2.8.1 Crambe (*Crambe abyssinica*)

This species is native from the Mediterranean region and has been cultivated on several regions as central and west Asia, Europe, USA, and South America (Weiss 2000). The plant belongs to the family Brassicaceae (formerly Cruciferae), the same as turnip, mustard, and canola. Until recently, it was only used as fodder. However, given its rusticity, precocity (90–120 days) and high potential to produce oil

(26–38 % content in seeds) (Meier and Lessman 1971), it has been investigated as a potential oil crop, aiming biodiesel production, in spite of its low productivity.

Weiss (2000) describes crambe as an annual, herbaceous plant, about one meter high. The oil is inedible because of the presence of erucic acid (60 %), being useful as a raw material for the manufacture of plastic films, nylon, adhesives, anticorrosive, and lubricating products, which are traditionally dependent on rapeseed oil.

According to Dalchiavon et al. (2012), crambe shows lower production costs when compared to soybean, sunflower or canola, and potential for winter cultivation as it can withstand temperatures as low as 4 °C below zero, being relatively tolerant to drought. Crambe grows better on well-drained soil with a pH between 6 and 7 (White and Higgins 1966).

The oil extracted from crambe seeds is used as an industrial lubricant, a corrosion inhibitor, and as an ingredient in the manufacture of synthetic rubber. The oil contains 50–60 % erucic acid, a long chain fatty acid, which is used in the manufacture of plastic films, plasticizers, nylon, adhesives, and electrical insulation (Oplinger et al. 2015). The authors refer that crambe is being promoted in the USA as a new domestic source of erucic acid, primarily obtained from imported rapeseed oil. Supplies of industrial rapeseed are less-plentiful since the development of varieties (canola) that have no erucic acid content, in contrast with crambe oil that contains 8–9 % more erucic acid than industrial rapeseed oil.

Crambe meal contains 25–35 % protein when the pod is included and 46–58 % protein when the pod is removed, with a well-balanced amino acids content (Hesketh et al. 1963). Defatted crambe meal is a protein supplement for livestock feeds (Oplinger et al. 2015) and its use has been approved by the FDA for beef cattle rations for up to 5 % of the daily intake. Nevertheless, the meal has not been approved for nonruminant feeds due to the presence of glucosinolates, broken down during digestion to harmful products that depress the appetite and can cause liver and kidney damage.

Untreated oil-free crambe meal may contain up to 10 % thioglucosides (McGhee et al. 1965), which is toxic to nonruminant animals, such as hogs and chickens (Van Etten et al. 1965, 1969). However, subjecting whole seed to moist heat before processing can deactivate the enzyme, and the glucosinolates remain intact through the oil extraction process, according to Oplinger et al. (2015).

2.8.2 Jatropha (*Jatropha curcas*)

This plant belongs to the family Euphorbiaceae, the same as the castor bean and cassava, and its center of origin is located in Mexico (Dias et al. 2012). It is a tree of rapid growth, whose average height is two to three meters, but can reach up to five meters, under special conditions of climate and soil. It takes 3–4 years for commercial harvesting and production may extend from 40 to 100 years.

The plant has been traditionally used as a living fence, from where fruits are harvested for oil extraction (Carvalho et al. 2009). Jatropha seeds contains 25–40 % oil (average of 37.5 %) and its use has been restricted to self-consumption on production sites (farms), for energy purposes or for soap production. As it happens with castor, the meal resulting from oil extraction is highly toxic to animals, and cannot be used as feed, unless it is detoxified (Dias et al. 2007).

During the 1990s and up to the first decade of the twenty-first century, there was a global wave of incentives for using jatropha seeds as feedstock for biodiesel production, with several private and public initiatives aiming to establish large commercial jatropha plantations in Asia (mainly China, India, Indonesia, Malaysia, and others), Africa and Latin America. All these initiatives failed, due to the absence of feasible production systems (commercial varieties, recommendations for plant nutrition and for controlling several pests hosted by the plant), low productivity, large period until first commercial harvest, and the high demand for labor force, especially for harvesting.

The fruit ripening of jatropha does not occur at the same time, but extends for 3–4 months, exacerbating the requirement of manpower. In addition, the high toxicity of the jatropha meal prevents its use as animal feed and even its use as organic fertilizer may pose environmental hazards. In this case, only the jatropha oil would have commercial value, making it impossible to compete with major oil crops, like soybean, canola or sunflower, whose meal is highly demanded in the market, or cotton oil, supported by the commercialization of the lint.

References

Abreu YV, Oliveira HR, Leal JE. C (2012) Biodiesel no Brasil em Três Hiatos: Selo Combustível Social, Empresas e Leilões. Eumed, Univ. Malaga (Esp), 213 p

Anjami K (2012) Castor genetic resources: a primary gene pool for exploitation. Ind Crop Prod 35 (1):1–14

Beasley J, Baldwin J (2015) Peanuts cultivars and description. The University of Georgia College of Agricultural & Environmental Sciences. Available at http://www.caes.uga.edu/commodities/fieldcrops/peanuts/production/cultivardescription.html. Retrieved on 27 Sept 2015

Bretschneider W (1882) Botanicum Sinicum. Notes on Chinese botany from native and western sources. J Roy Astron Soc N China Br NS 16–17:18–23

Camara GMS (ed) (2000) Soja: tecnologia da produção II. Piracicaba, ESALQ/LPV 450p

Carvalho BCL (2005) Manual do cultivo da mamona. EBDA, Salvador, 65 p (il.)

CATI (2001) Nabo forrageiro: adubação verde para inverno. CATI, Campinas. 1 folder (CATI. Responde, 25)

Cia E, Freire EC, dos Santos WJ (eds) (1999) Cultura do algodoeiro. Piracicaba: POTAFOS, 286 p

Congresso Brasileiro de Soja 4 (2006) Londrina. Resumos. Londrina: Embrapa Soja, 202 p. Organizado por Odilon Ferreira Saraiva, Simone Ery Grosskopf

Corley RHV, Tinker PB (2003) The oil palm, 5th edn. Blackwell Science Ltd., Oxford, 627 p

Cornet D (2001) Oil Palm—*Elaeis guineensis* Jacq. In: Raemaekers RH (ed) Crop production in Tropical Africa. Directorate General for International Cooperaction, Brussels, pp 769–797

Cultura da colza (1980) Fert Informa 3(28):4–5

Dalchiavon MP, Santos RF, de Souza SNM, Bassegio D, Rossetto C, Bauerman HB (2012) Growth analysis of *Crambe abyssinica* according to the variation in planting density. Acta Iguazu 1(3):33–43

Dall'Agnoll A, Roessing AC, Lazzarotto JJ, Hirakuri MH, Oliveira AB (2007) O complexo agroindustrial da soja brasileira. Circular Técnica, 43, Embrapa Soja, Londrina, PR, Brasil, 11 p

Davis SC, Anderson-Teixeira KJ, Delucia EH (2009) Life-cycle analysis and the ecology of biofuels. Trends Plant Sci 14(3):140–146

de Azevedo DMP, de Beltrão NEM (eds) (2007) O Agronegócio da mamona no Brasil. 2 ed. rev. amp. Campina Grande: Embrapa Algodão; Brasília, DF: Embrapa Informação Tecnológica. 506 p

de Beltrão NEM (Org.) (1999) O agronegócio do algodão no Brasil. Brasília, DF: Embrapa Comunicação para a Transferência de Tecnologia; Campina Grande: Embrapa Algodão, 2:1–2, 491p

de Carvalho BCL, Oliveira EAS, Leite VM, Dourado VV (2009) Informações técnicas para o cultivo do pinhão-manso no estado da Bahia. EBDA, Salvador, p 79

de Leite RMVBC, Brighenti AM, de Castro C (eds) (2005) Girassol no Brasil. Embrapa Soja, Londrina, 613 p

De Mori C, Tomm GO, Ferreira PEP (2014) Aspectos econômicos e conjunturais da cultura da canola no mundo e no Brasil. Embrapa Trigo, Passo Fundo, 36 p (html) (Embrapa Trigo. Documentos online, 149)

Dias LA, Leme LP, Laviola BG, Pallini A, Pereira OL, Dias DCFS, Carvalho M, Manfio CE, AS, de Sousa LCA, de Oliveira TS, Pretti LA (2007) Cultivo de pinhão manso (*Jatropha curcas* L.) para produção de óleo combustível. L. A. S. Dias, Viçosa, MG, 40 p

Dias LA, Missio RF, Dias DC (2012) Antiquity, botany, origin and domestication of *Jatropha curcas* (Euphorbiaceae), a plant species with potential for biodiesel production. Genet Mol Res 11(3):2719–2728

dos Santos RC, Freire RMM, de Lima LM (eds) (2013) O agronegócio do amendoim no Brasil. 2. ed. rev. e ampl. Brasília, DF: Embrapa, 585 p. il. color. Autores: Alessandra Pereira Fávero, Ana Cristina Miranda Brasileiro, Ana Heloneida de Araújo Morais, Ana Maria Rauen de Olvieira Miguel, Bill Jorge Costa, David John Bertioli, Denizart Bolonhezi, Francisco José Alves Fernandes Távora

Downey RK, Harvey BL (1963) Methods of breeding for oil quality in rape. Can J Plant Sci 43(3):271–275

Embrapa (1994) Tropical soybean improvement and production. Food and Agriculture Organization of the United Nations (FAO), Rome 253 p

FAOSTAT (2015) FAOSTAT database. Available at www.faostat.fao.org

Fargione J, Hill J, Tilman D, Polasky S, Hawthorne P (2008) Land clearing and the biofuel carbon debt. Science 319(5867):1235–1238

Gazzoni DL (2013) A sustentabilidade da soja no contexto do agronegócio brasileiro e mundial. Embrapa Soja Série Documentos 344, 50 p

Gazzoni DL (2014a) O impacto do uso da terra na sustentabilidade dos biocombustíveis. Embrapa Soja Série Documentos 347, 82 p

Gazzoni DL (2014b) Balanço de emissões de CO_2 por biocombustíveis no Brasil: histórico e perspectivas. Embrapa Soja Série Documentos 334, 107 p

Gazzoni DL, Felici PHN, Coronato RMS, Ralisch R (2005) Balanço energético das culturas de girassol e soja pra produção de biodiesel. Biomassa e Energia 2(4):259–265

Gazzoni DL, Avila MT, Felici PHN, Ralisch R (2008) Balanço energético do biodiesel de dendê. In: CONGRESSO BRASILEIRO DE AGROENERGIA, 1., SIMPÓSIO INTERNACIONAL DE BIOCOMBSTÍVEL, 1., Uberlândia. Anais. UFU, Uberlândia

Gazzoni DL, Borges JLB, Avila MT, Felici PHN (2009) Balanço energético da cultura da canola para a produção de biodiesel. Espaço Energia 11:24–28

Gazzoni DL, Azurdia I, Blanco G, Estrada C, Macedo IC (2012) Sustainable energy in Latin America and the Caribbean: potential for the future. Available at http://www.icsu.org/latin-america-caribbean/publications/reports-and-reviews/sustainable-energy/energy_english.pdf. Retrieved on 15 Sept 2015

Greenpeace (2015) Indonesia forest threats. Available at http://www.greenpeace.org/international/en/campaigns/forests/asia-pacific/threats/. Retrieved on 26 Sept 2015

Hammond EG, Johnson LA, Su C, Wang T, White PJ (2005) Soybean oil. In: Bailey's industrial oil and fat products. Available at https://pharosproject.net/uploads/files/cml/1360080729.pdf. Retrieved on 12 Sept 2015

Hernani LC, Henn R (1995) Nabo forrageiro. Dourados: EMBRAPA-CPAO (1 folder)

Hesketh HR, Creger CR, Crouch JR (1963) *Crambe abyssinica* meal as a protein source for broilers. Poultry Sci 42:1276

Hill J, Nelson E, Tilman D, Polasky S, Tiffany D (2006) Environmental, economic, and energetic costs and benefits of biodiesel and ethanol biofuels. PNAS 103(30):11206–11210

Hymowitz T (1983) History of soy—soybeans: the success story. Available at http://www.nsrl.uiuc.edu/aboutsoy/history3.html. Retrieved on 11 July 2015

Iqbal MJ, Aziz N, Saeed NA, Zafar Y, Malik KA (1997) Genetic diversity evaluation of some elite cotton varieties by RAPD analysis. Theor Appl Genet 94(1):139–144

Iqbal MJ, Reddy OUK, El-Zik KM, Pepper AE (2001) A genetic bottleneck in the evolution under domestication of upland cotton *Gossypium hirsutum* L. examined using DNA fingerprinting. Theor Appl Genet 103(4):547–554

Ivanov DS, Jovanka D, Sredanović A (2010) Fatty acid composition of various soybean products. J Inst Food Tech Novi Sad 37(2):65–70

Jones J (2007) Earliest-known evidence of peanut, cotton and squash farming found. Anthropoly.net. Available at http://anthropology.net/2007/06/28/earliest-known-evidence-of-peanut-cotton-and-squash-farming-found/. Retrieved on 27 Sept 2015

Katragadda HR, Fullana AS, Sidhu S, Carbonell-Barrachina ÁA (2010) Emissions of volatile aldehydes from heated cooking oils. Food Chem 120:59–65

Kochert G, Stalker HT, Gimenes M, Galgaro L, Lopes CR, Moore K (1996) RFLP and cytogenetic evidence on the origin and evolution of allotetraploid domesticated peanut, *Arachis hypogaea* (Leguminosae). Am J Bot 83(10):1282–1291

Lentz DL, Pohl ED, Pope KO, Wyatt AR (2001) Prehistoric sunflower (*Helianthus annuus* L.) domestication in Mexico. Econ Bot 55(3):370–376

McGhee JE, Kirk CD, Mustakas GC (1965) Methods for determining thioglucosides in *Crambe abyssinica*. J Am Oil Chem Soc 42:889–891

Meier VD, Lessman KJ (1971) Estimation of optimum field plot shape and size for testing yield in *Crambe abyssinica* Hochst. Crop Sci 11(5):648–650

Mercer LC, Wynne JC, Young CT (1990) Inheritance of fatty acid content in peanut oil. Peanut Sci 17(1):17–21

Merrill ED (1931) The phytogeography of cultivated plants in relation to the assumed pre-Columbian-Eurasian-American contacts. Am Anthropologist 33:375–382

Miyasaka S, Medina JC (eds) (1981) A soja no Brasil. ITAL, Campinas, 1062 p

Müller AA (1980) A cultura do dendê. Belém. EMBRAPA-CPATU, PA, 24 p (il.) (EMBRAPA-CPATU. Miscelânea, 5)

Muller AA, Furlan Junior J (eds) (2001) Agronegócio do dendê: uma alternativa social, econômica e ambiental para o desenvolvimento sustentável da Amazônia. Embrapa Amazônia Oriental, Belém, 288 p (il.)

Ogunniyi DS (2006) Castor oil: a vital industrial raw material. Biores Tech 97(9):1086–1091

Oliveira AdS, de Carvalho MLM, Nery MC, Oliveira JA, Guimaraes RM (2011) Seed quality and optimal spatial arrangement of fodder radish. Sci Agricola 68(4):417–423

Oplinger ES, Oelke EA, Kaminski AR, Putnam DH, Teynor TM, Dolll JD, Kelling KA, Durgan BR, Noetzel DM (2015) Crambe. In: Altenative crops manual. Available at https://www.hort.purdue.edu/newcrop/afcm/crambe.html. Retrieved on 17 Oct 2015

Piper CV, Morse WJ (1910) The soybean: history, varieties and field studies. USDA Bur Pl Ind Bull 197:1–84

Pitt JI, Hocking AD (2006) Mycotoxins in Australia: biocontrol of aflatoxin in peanuts. Mycopathologia 162:233–243

Poth U (2001) Drying oils and related products. Ullmann's encyclopedia of industrial chemistry. Wiley-VCH Verlag GmbH & Co. KGaA

Prakash S, Hinata K (1980) Taxonomy, cytogenetics and origin of crop Brassicas, a review. Opera Bot 55:1–57

Ramos MJ, Fernández CM, Casas A, Rodríguez L, Pérez A (2009) Influence of fatty acid composition of raw materials on biodiesel properties. Bioresour Tech 100(1):262–268

REN 21 (2015) Global status report. Available at www.ren21.net. Retrieved on 7 Oct 2015

Rodrigues Filho A (2000) Cultura da mamona. Belo Horizonte, EMATER, MG, 18 p (il.) (Agricultura: culturas)

Searchinger T, Heimlich R, Houghton RA et al (2008) Use of US croplands for biofuels increases greenhouse gases through emissions from land-use change. Science 319:1238–1240

Severino LS, Nóbrega MB de, Gonçalves NP, Eguia MTJ (2006) Viagem à Índia para prospecção de tecnologias sobre mamona e pinhão manso. Campina Grande, Embrapa Algodão, 56 p (Embrapa Algodão. Documentos, 153)

Snowdon R, Lühs N, Friedt W (2007) Oilseed rape. In: Kole C (ed) Genome mapping and molecular breeding in plans, vol 2. Springer, Berlin, pp 55–114

Ungaro MRG (2000) Cultura do girassol. IAC, Campinas, 36 p (IAC. Boletim Técnico, 188)

USDA (2015a) Canola. Available at http://www.ers.usda.gov/topics/crops/soybeans-oil-crops/canola.aspx. Retrieved on 26 Sept 2015

USDA (2015b) Sunflowerseed. Available at http://www.ers.usda.gov/topics/crops/soybeans-oil-crops/sunflowerseed.aspx. Retrieved on 26 Sept 2015

USDA (2015c) Cotton. Available at http://www.ers.usda.gov/topics/crops/cotton-wool/background.aspx. Retrieved on 26 Sept 2015

USDA (2015d) Oilseeds. Available at http://www.fas.usda.gov/data/oilseeds-world-markets-and-trade. Retrieved on 11 Oct 2015

USDA (2015e) Soybeans & Oil crops. Available at http://www.ers.usda.gov/topics/crops/soybeans-oil-crops/background.aspx. Retrieved on 13 Oct 2015

Van Etten CH, Daxenbichler ME, Peters JE, Wolff IA, Booth AN (1965) Seed meal from *Crambe abyssinica*. J Agric Food Chem 13:24–27

Van Etten CH, Daxenbichler ME, Wolff IA (1969) Natural glucosinolates (thioglucosides) in foods and feeds. J Agric Food Chem 17:483–491

Wang Q, Zhang L, Zheng P (2015) Genetic diversity and evolutionary relationship analyses within and among Raphanus species using EST-SSR markers. Mol Breeding 62:35–62

Weiss, EA (2000) Oil seed crops. World Agriculture Series, Cab Direct, 373 p

White GA, Higgins JJ (1966) Culture of crambe: a new industrial oilseed crop. Agricultural Research Service, USDA. ARS Production Research Report. 95

An Overview of Production, Properties, and Uses of Biodiesel from Vegetable Oil

Arindam Sinha Roy, Akoijam Chingkheihunba and Kannan Pakshirajan

Abstract The search for alternative fuels is continuously increasing owing to the ever growing energy demand worldwide. Vegetable oils and their derivatives (particularly methyl esters), commonly referred to as biodiesel, are prominent candidates as alternative to diesel fuel. In addition to its advantages as a renewable and domestic fuel resource, biodiesel use reduces emission of environmental pollutants. Moreover, engine performance and fuel economy due to biodiesel are nearly identical compared to those with conventional fuels. It can even be used directly in most diesel engines without requiring extensive engine modifications. Recent research studies have shown that it has advanced from being a purely experimental fuel to initial stages of commercialization. However, large scale application of this biofuel is still limited due to its economics, combustion value, emissions and low-temperature properties. This chapter presents an overview of history, different sources, properties and uses of biodiesel. Besides, this work deals with vegetable oil as an efficient feed stock for biodiesel production and different factors influencing the production of biodiesel from vegetable oil. Also, the economics of biodiesel as a transportation fuel is discussed and compared with that of the conventional petro-diesel.

Keywords Biodiesel · Vegetable oil · Biofuel · Production · Properties · Economics

1 Introduction

The demand for robust supply of energy from various sources is a result of high economic growth for several decades in most of the developing countries across the globe. The main drivers of increasing primary oil demand are ever-increasing

A.S. Roy · A. Chingkheihunba · K. Pakshirajan (✉)
Department of Bioseceinces and Bioengineering,
Indian Institute of Technology Guwahati, Guwahati 781039, Assam, India
e-mail: pakshi@iitg.ernet.in

C.R. Soccol et al. (eds.), *Green Fuels Technology*,
Green Energy and Technology, DOI 10.1007/978-3-319-30205-8_4

growth of industrialization and population, the need for mobility and peoples' aspirations for improved living conditions (Goldemberg and Johansson 2004). For example, the transport sector relies almost entirely on oil supplies for fuel. In the last 30 years, global energy use has increased almost 70 %, but this growth is uneven, because developing countries have almost tripled their energy consumption, while in industrialized countries it has increased by 21 % (Brown et al. 2000). The oil requirement is projected to rise by about 1.0 % per year, reaching approximately 105 million barrels per day (mb/d) level by 2030 (Fulton et al. 2004).

Several factors, such as in increase energy prices, increased market volatility (particularly during 2008–2009 and 2012–2013), growing energy demand, political crisis in the oil rich middle east countries, heavy dependence of many countries on imported oil, growing concerns about the environmental impact of fossil fuels have created an energy crisis (Goldemberg 2007). This has led to the search for alternative sources that allow a diverse offering and increased life span of existing resources (Fulton et al. 2004; Wang et al. 2006). In this background there is a growing strong interest and support for the biofuels in many parts of the world. Hundreds of scientific articles and various other reports from around the world dealing with vegetable oil-based alternative diesel fuels ("biodiesel") have appeared in print. They have advanced from being purely experimental fuels to initial stages of commercialization (Shay 1993). Nevertheless, various technical and economic aspects require further improvement of these fuels. The development of alternate energy such as bioethanol and biodiesel has allowed the energy crops gain importance every day, with greater strength in agricultural and energy policies in both industrialized and developing countries, and even contributed to the research and development activity of cars and engines that run on such biofuels (Alternative Fuels Committee of the Engine Manufacturers Association 1995).

The contribution of biofuels as an alternative energy source is currently very small, but this may change as the global production of biofuels has been growing rapidly in recent years, more than tripling from about 18 billion liters [10 million tons of oil equivalent (MToe)] in 2000 to about 60 billion l (42 MToe) in 2008. Supply is dominated by bioethanol, which accounted for approximately 84 % of total biofuel production in 2008 (Mandil and Shihab-Eldin 2010). But the major controversy for bioethanol is the diversion of food crop and crop land for production biofuel which causes shortage of food as well as increase in the food prices. Also the moral and ethical aspect of the use of food crops for ethanol production is into much controversy (Goldemberg 2007). In contrast, the biodiesel can be produced from alcoholysis of any oil or fat through the transerterification process, employing a catalyst which can be homogeneous, heterogeneous, or enzymatic. The biomass source for biodiesel can be edible oils like rapeseed, coconut, soybean, palm, etc., nonedible oils like Jatropha, Pongamia, etc., low-cost waste products such as waste cooking oil, soapstock, grease, etc. It is also important to note that from the transesterification process of biodiesel synthesis, value added by-product glycerol is derived (Fulton et al. 2004). Despite this, it is important to recognize that biofuels such as ethanol and/or biodiesel will not be able to end the petroleum oil dependency of developed/industrialized countries, because there will not be enough land and water to meet the energy requirements of the automotive industry.

2 Biodiesel

2.1 History

The use of vegetable oils in diesel engines is nearly as old as the diesel engine itself. The inventor of the diesel engine, Rudolf Diesel, reportedly used groundnut (peanut) oil as a fuel for demonstration purposes in 1900. Though his initial engine experiments were not successful, but by the time he showed his engine at the World Exhibition in Paris in 1897, his engine was running on 100 % peanut oil. Much earlier to this discovery in 1853, scientists E. Duffy and J. Patrick, conducted the transesterification of a vegetable oil (Nitske and Wilson 1965). Diesel's ideas on agriculture and his invention provided the foundation for a society fuelled with clean, renewable, locally grown fuel. The inventors expected their invention to run on fuel derived from plants, but cheap petroleum proved to be more popular at that time.

Looking into the origin of biodiesel fuel, the earlier engines worked so smoothly on earthnut or peanut oil that only a few people were aware of it. These engines were then worked on vegetable oil without making any alterations. The French government at the time was toying with the idea of testing the applicability to power production of the Arachide or earthnut. Diesel himself was supportive of the idea as he had conducted related tests. During the 1920s, diesel engine manufacturers decided to alter their engines utilizing the lower viscosity of the fossil fuel, best known as petrodiesel, rather than such biomass vegetable oil fuel. Some further modifications were carried out on the use of vegetable oils in diesel engines during the third and fourth decades of twentieth century.

During these times vegetable oils were used occasionally as diesel substitutes in some cases, but usually only in emergency situations for instance, shortage of petrodiesel. In 1937, G. Chavanne in Belgium was granted a patent for a "Procedure for the transformation of vegetable oils for their uses as fuels" Belgian Patent 422,877. This patent described the alcoholysis of vegetable oils using ethanol. This probably is the first account of the production of what is known as "biodiesel" today in the biodiesel fuel history (Meher et al. 2006). Since then there has not been much work on the improvement of biodiesel-based fuel engines, all researches were confined only in for the academic purpose. The fuel and energy crises of the late 1970s and early 1980s as well as accompanying concerns about the depletion of the world's nonrenewable resources provided the incentives to seek alternatives to conventional, petroleum-based fuels (Friedrich 2004). In this context, vegetable oils as fuel for diesel engines were remembered. Again recently, there has been a renewed focus on vegetable oils and animal fats to make biodiesel. Researchers are focusing on improving transesterification process, use of different regenerable catalyst, low-cost alternative sources of oil/fat, etc.

Numerous different vegetable oils have been tested as biodiesel. Often the vegetable oils investigated for their suitability as biodiesel are those which occur abundantly in the country of testing. Therefore, soybean oil is of primary interest as biodiesel source in the United States while many European countries are concerned with rapeseed oil, and countries with tropical climate prefer to utilize coconut oil or palm oil. Other vegetable oils, including sunflower, safflower, etc., have also been investigated. Furthermore, other sources of biodiesel studied include animal fats and used or waste cooking oils. Sources of biodiesel with some emphasis on developing countries have been discussed (Shay 1993).

2.2 Different Sources of Biodiesel

The different sources of biodiesel can be grouped into mainly three categories; they are (i) vegetable oil feedstock (edible and nonedible), (ii) animal fats, and (iii) waste sources. Vegetable oils as raw materials for production of biodiesel have the following advantages, such as: liquid in nature, portability, availability, renewability, higher heat value (about 88 % of no. 2 diesel fuel), lower sulfur content, biodegradability. But there are certain disadvantages as well such as: higher viscosity, lower volatility, the reactivity of unsaturated hydrocarbon chains, etc. (Demirbas 2006). To overcome these disadvantages, vegetable oils are converted to biodiesel fuel by transesterification with methanol.

Vegetable oils like soybean, rapeseed, canola, rice oil, safflower, groundnut, coconut, oat, sorghum oil are edible oils that have been successfully tested to produce biodiesel. Due to higher prices of these edible vegetable oils, compared to diesel fuel, and their uses in dietary, they have restricted use in biofuel applications. Because of this, the focus has been shifted to waste vegetable oils and nonedible crude vegetable oils as biodiesel sources. In this regard, nonedible oils such as these form *Jatropha curcas*, neem, palm, mahua, Castor, *Pongamia glabra*, tobacco seed, tall, etc., have shown great potential in biodiesel production. Table 3 summarizes the type of feedstock for biodiesel production and their current status in detail.

Geography and climate have played an active role in the selection of feedstock for biodiesel production. For example, palm oil is used dominantly in tropical countries like Malaysia, rapeseed/canola oil is primarily used in Europe, and soybean oil is used in the United States (Bajpai and Tyagi 2006). In India, jatropha, neem, pongamia have great potential as they are natural habitat of India, grows on infertile lands, it can withstand severe conditions, and can be cultivated as a part of the approach for reclaiming the spoiled lands.

Animal fats such as lard, tallow, fishoil, poultry fat, etc., are lesser explored as biodiesel sources. Animal fats are also triglycerides like vegetable oil, but there is a subtle difference between oils and animal fats, fats are generally unsaturated (carbon–carbon double bonds), whereas fats are saturated (all single bonds) (Nelson and Schrock 2006). Generally, animal fats are solid at room temperature and due to their chemical structure and physical property animal fats are less suitable for biodiesel application. But due to their high availability and low cost they should be explored further for biodiesel production.

Waste cooking oil is another source for biodiesel production; the easy availability and cheap price of waste cooking oils makes biodiesel from this source highly competitive in price with petroleum fuel. They are basically vegetable oil or derived from animal fat and hence conventional transesterification processes can be used for biodiesel production from used cooking oil. Waste cooking oil and trap grease contain 5–30 % and 40–100 % of free fatty acids, respectively, making it a good alternative source (Fan and Burto 2009). Large quantity of waste cooking oil is generated worldwide; for example, restaurants in the US alone produce about 300 million US gallons of waste cooking oil annually, in China the amount of waste cooking oil is around 4.5 million tons each year. The major problem with waste cooking oil is its water content, most biodiesel production processes can tolerate up to 1 % water in the feedstock, even this small quantity of water will increase soap formation and measurably affect the transesterification process (Zheng et al. 2006; Ma and Hanna 1999).

2.3 *Properties and Uses*

The physical and chemical properties of biodiesel varies greatly depending upon the feedstock used, catalyst, and type of alcohol (methanol, ethanol, propanol, etc.) used for production (Fulton et al. 2004). The important properties based on which biodiesel is characterized are: kinematic viscosity, specific gravity, cold flow properties, flash point, cetane number, iodine value, lubricity, and oxidative stability. Other than these properties, water and sediment content, total ash, total glycerin, ester content, phosphorous content, sulfur content are also considered for a given biodiesel (Bajpai and Tyagi 2006; Dmytryshyn et al. 2004). Among these properties kinematic viscosity, specific gravity, and flash point are related to fuel efficiency and determine the power output of biodiesel. Depending upon these properties the blending of biodiesel is decided. The water content and amount of contaminant also determine the heating value of biodiesel as a fuel. The net heating value for biodiesel is 118,296 Btu/gal which is 8.5 % lower than no. 2 diesel (petrodiesel), and its density is about 0.880 g/cm^3, 3.5 % more than no. 2 diesel (Anonymous 2004).

Other properties though directly impact the fuel efficiency and combustion characteristic of biodiesel are important to determine the nature of biodiesel as a clean fuel by measuring its emission characteristic. Ash content, sulfur content, phosphorous content, etc., are such aspects of biodiesel important to determine the environmental impact of biodiesel. Typically, ash content for biofuels is lower than for most coals and petrodiesel, and sulfur content is much lower than for many fossil fuels (Anonymous 2002). Unlike coal ash, which may contain toxic metals and other trace contaminants, biomass ash may be used as a soil amendment to help replenish nutrients removed by harvest, and hence biodiesel is considered as a cleaner fuel compared to fossil fuels or coal-based fuels (Alternative Fuels Committee of the Engine Manufacturers Association 1995). Table 1 describes the specification for biodiesel as a transportation fuel.

Table 1 Specification of biodiesel for transportation purpose in the US and EU (Demirbas 2009; Jääskeläinen 2009; ACEA 2009; ASTM 2002)

Property	US specification (ASTM D6751-12)			EU specification (EN 14214:2012)		
	Limits	Units	Test	Limits	Units	Test
Kinematic viscosity	1.9–6.0	mm^2/s	D445	3.5–5.0	mm^2/s	EN ISO 3104
Density	–	–	–	860–900	kg/m^3	EN ISO 3675 EN ISO 12185
Flash point	93, min	°C	D93	101, min	°C	EN ISO 2719
Cetane number	47, min		D613	51.0, min		EN ISO 5165
Water and sediment	0.050, max	% vol	D2709	500, max	mg/kg	EN ISO 12937
Total contamination,	–	–	–	24, max	mg/kg	EN 12662
Ester content	–	–	–	96.5 %, min		EN 14103
Distillation temperature (% vol recovered)	90 %: 360 °C, max	%	D1160			
Sulfur, (by mass)	Two grades: S15 15 S500 0.05 %, max	ppm	D5453	10.0	mg/kg	EN ISO 20846 EN ISO 20884 EN ISO 13032
Sulfated ash	0.020, max	% mass	D874	0.020, max	% mass	ISO 3987
Carbon residue on 10 % distillation residue	0.050, max	%wt	D4530	–	–	–
Acid number	0.50, max	mg KOH/g	D664	0.50, max	mg KOH/g	EN 14104
Iodine value	–	–	–	120, max	GIod/100 g	EN 14111 EN 16300
Oxidation stability	3, min	h min	EN 14112	8, min	h min	EN 14112

(continued)

Table 1 (continued)

Property	US specification (ASTM D6751-12)			EU specification (EN 14214:2012)		
	Limits	Units	Test	Limits	Units	Test
Phosphorous	0.001, max	%wt	D4951	4.0, max	mg/kg	EN 14107 pr EN 16294
Free glycerin	0.020, max	%wt	D6584	0.02, max	%wt	EN 14105 EN 14106
Total glycerin	0.240, max	%wt	D6584	0.25, max	%wt	EN 14105
Group I metals: (Na + K)	5, max	mg/kg	EN 14538	5.0, max	mg/kg	EN 14108 EN 14109 EN 14538
Group II metals: (Ca + Mg)	5, max	mg/kg	EN 14538	5.0, max	mg/kg	EN 14538
Monoglycerides, diglycerides and triglycerides	MG 0.40, max	%wt	D6584	MG 0.70, max DG 0.20, max TG 0.20, max	%wt	EN 14105
Alcohol content	0.2 methanol, max	%wt	EN14110	0.20 methanol, max	%wt	EN 14110

Biodiesel can be used in similar way like petrodiesel as transport fuel, for heating purposes or to run generators. Biodiesel is generally used in pure form (named B100) or may be blended with petroleum diesel; 2 % biodiesel (B2), 5 % biodiesel (B5), and 20 % biodiesel (B20) (Wang et al. 2006). The use of biodiesel in pure or blended form to run vehicles is well reported. Though some car manufacturers have instructed their customers not to use biodiesel, there are a large number of manufactures either upgrading or modifying their engine configuration to suit biodiesel (Anonymous 2014). Biodiesel has also been used even to fly aircraft with a growing number of airlines conducting trials or flying commercial flights using biodiesel (Anonymous 2009; Air Transport Action Group 2009).

3 Biodiesel from Vegetable Oils

A lot of efforts are currently being made worldwide to find alternative fuels to meet the present and future demands of energy, without causing further global warming effects. Such an alternative fuel should be comparable to the conventional fuels with respect to various desired fuel characteristics and properties. Currently, a variety of substances and their natural sources are being investigated as potential alternatives for fossil fuels, especially petroleum-derived fuels.

Vegetable oil in the pure form also called "straight vegetable oils" (SVOs), or in blends with conventional fuels have attained significant importance as an alternative to the conventional fuels. Use of pure vegetable oils in low-speed diesel engines, such as those of large ships neither produce net carbon dioxide nor generate sulfur oxides (Espadafor et al. 2009). At present appropriate engine modifications are required to minimize its high NOx emissions. Pure sunflower oils preheated at 75 °C give the same heat release curve and gas cylinder pressures as those of diesel in a direct injection engine with a reduced emission of carbon and smoke opacity by 2.05 and 4 %, respectively (Canakci et al. 2009). Re-refining crude sunflower oil improves fuel performance and appears more promising than raw sunflower oil. However, there are problems associated with the long-term use such as carbonization in the engine (Canakci et al. 2009).

An alternative to conventional fuels or pure vegetable oils is the use of appropriate blends of vegetable oils and diesel. A 30 % blend of putranjiva oil (oil obtained from the seeds of *Putranjiva roxburghii*) with diesel when used in Ricardo Variable Compression Diesel Engine gives a performance equivalent to that of pure diesel with significant reduction in the emission of CO, NOx, and smoke particulate (Haldar et al. 2009). Deccan hemp oil blends with diesel at 25 and 50 % can be used as a substitute for diesel without any engine modification, although it has a disadvantage of larger emissions than diesel alone (Hebbal et al. 2006). Pyrolysis oil obtained from castor oil seeds can be blended with diesel in diesel engines (Figueiredo et al. 2009). A blend of diethyl ether (DEE) with orange oil has a higher brake efficiency, peak cylinder pressure, and heat release rate in comparison to diesel or blends of orange oil and diesel. In addition, the DEE and orange oil blend has lower emissions of hydrocarbons (HCs), CO, and smoke, but leads to higher emissions of NOx (Purushothaman and Nagarajan 2009).

Biodiesel is one of the most promising future fuels, comprises of mono alkyl esters of long chain fatty acid derived from vegetable oil, waste frying oils, animal fats, etc. One of the major advantages of biodiesel is that it can be used in an internal combustion engine without any engine modification(s). Biodiesel use can result in a substantial reduction in unburned hydrocarbon, CO, and particulate matter.

3.1 Characteristics of Vegetable Oil Affecting Their Suitability for Use as Biodiesel

A number of physical and chemical characteristics of plant oil, such as heating value, pour point, cloud number, flash point, iodine number, viscosity, density, and cetane number influence their suitability as a fuel. Table 2 describes the productivity and the fuel characteristics of some common biodiesel feedstock. Heating

Table 2 Productivity and fuel characteristics of biodiesel from common feedstock (Atabani et al. 2012, 2013; Issariyakul and Dalai 2014; Moka et al. 2014; Pinzi et al. 2009; Demirbas 2003)

Plants	Oil content (%)	Yield (L/ha/year)	Saponification value	Iodine value	Cetane number	Viscosity (mm^2/s) (°C)	Low heating value (kJ/kg)	Carbon residue (%)
Rapeseed	38–46	1190	179	94–120	37.6	37 (38 °C)	39,709	0.30
Soybean	15–20	446	190	117–143	37.9	32.6 (38 °C)	39,623	0.27
Sunflower	44–51	952	189-190	110–143	37.1	37.1 (38 °C)	39,575	0.23
Palm	30–60	5950	200-246	35–61	42	39.6 (38 °C)	36,553	–
Cottonseed	18–25	325	195	90–140	41.8	33.5 (38 °C)	39,468	0.24
Peanut	36–56	1059	199	80–106	41.8	39.6 (38 °C)	39,782	0.24
Coconut	63–65	2689	256	6–12	70	–	–	–
Neemseed	20–30	–	–	–	47	30 (30 °C)	39,399	–
Jatropha	35–60	1892	–	–	42.5	49.9 (38 °C)	39,774	–
Linseed	40–44	478	189	168–204	34.6	27.2 (38 °C)	39,307	0.22
Karanja	27–39	225–2250	–	–	58	120 (20 °C)	37,100	–
Olive	45–70	1212	192	75–94	49.3	29.4 (–)	39,700	–
Corn	48	172	195	103–140	37.6	34.9 (38 °C)	39,500	0.24
Sesame	50	696	188	104–120	40.2	35.5 (38 °C)	39,349	0.25
Castor	53	1413	202	126–152	42.3	25 (38 °C)	39,500	–
Mahua	–	–	–	–	45	16.9 (30 °C)	30,248	–
Safflower	–	779	–	–	41.3	31.3 (38 °C)	39,517	0.25
Crambe	–	–	–	–	44.6	53.6 (38 °C)	40,482	0.23

value or the heat of combustion is the amount of heat energy released by the combustion of a unit value of fuels, and is a measure of the energy content of a fuel. It is determined in a standard bomb calorimeter, and its value for most plants range from 39,310 kJ/kg for linseed oil to 40,480 kJ/kg for crambe oil, which is comparable to that of diesel fuel (45,340 kJ/kg) (Filemon 2010; Atabani et al. 2012). Flash point is the minimum temperature at which the fuel will ignite on application of an ignition source, and can predict their possible fire hazards during transportation, handling, and storage. The flash point of plant oil ranges from 240 °C for linseed oil to 277 °C for corn, which are much higher than that of diesel oil, which is only 52 °C (Filemon 2010; Atabani et al. 2012). Iodine value indicates the degree of saturation of oil, and is measured in grams of iodine absorbed by 100 ml of a given oil sample. Lower the iodine number, the fuels are more combustible and more efficient, but has the disadvantage of having higher melting point and are usually solid at ambient temperature. Therefore, biodiesel produced from low iodine value oil might only be suitable for use in warm tropical countries. In contrast, biodiesel made from oils with high iodine value, such as linseed oil, soyabean oil, and sunflower oil, should be stored appropriately and used quickly to avoid its oxidation and polymerization. Further due to its lower melting point, they are suitable for use in cold weather.

Viscosity or the measure of the thickness of the oil is another important parameter since it affects injector lubrication and fuel atomization (Filemon 2010; Atabani et al. 2012). Highly viscous oil tends to form larger droplets on injection which can cause poor combustion, increased exhaust smoke and emissions, and may not provide sufficient lubrication for the precision fit of the fuel injection fit and can cause leakage and increased tear (Filemon 2010; Atabani et al. 2012). Since the viscosity of a fluid decreases with increasing temperature and plant oils have kinematic viscosities ranging from 31 mm^2/s for safflower oil to 54 mm^2/s for crambe oil, preheating and proper processing are required to reduce the kinematic viscosity to a level close to that of diesel fuel (3–5 mm^2/s) (Filemon 2010; Atabani et al. 2012).

Cetane number is the relative measure between the beginning of injection and autoignition of the fuel and is generally specific to a particular engine being used. Therefore, fuels with higher cetane number will have a lower ignition delay period than the fuel with a lower cetane number and diesel engines operate better with fuels with a cetane number of about 50 or higher (Filemon 2010; Atabani et al. 2012). Vegetable oils, such as linseed oil and rapeseed oil have relatively low cetane number compared to that of palm stearine, palm kernel oil, and palm olein, and, therefore lead to difficulty in engine starting and produce noise and thick exhaust smoke.

Other important characteristics of plant oil, which affects the suitability of its use as a fuel, are ash content, which is the measure of the amount of metal contained in the fuel; and biofuels usually have lower ash content than most coals (Filemon 2010; Atabani et al. 2012). The ash content is important for the heating value, as heating value decreases with increasing ash content; sulfur percentage is the percentage by weight of sulfur in fuel and in biofuels it is much lower than in many fossil fuels; potassium percentage which is the percentage by weight of potassium in fuel; cloud point and melt point are the measure of the temperature at which the fuel solidify and melts, respectively, are important parameters while operating any engine to prevent waxing of fuel in fuel lines and fuel tanks.

The major component of vegetable oils is triacylglycerol (TAG) or triglyceride (TG), which is a molecule, composed of three esters of fatty acid chain (acyl group) attached to the glycerol backbone (glycerol group). The major difference between various vegetable oils is the type of fatty acids attached in the triglyceride molecule. Fatty acid composition is of utmost importance as it determines fuel properties of biodiesel derived from corresponding vegetable oils. Fatty acid composition also determines the degree of saturation/unsaturation and molecular weight of vegetable oils. The most common fatty acids in vegetable oil are lauric acid ($C_{12}H_{24}O_2$), myristic acid ($C_{14}H_{28}O_2$), palmitic acid ($C_{16}H_{32}O_2$), stearic acid ($C_{18}H_{36}O_2$), oleic acid ($C_{18}H_{34}O_2$), linileic acid ($C_{18}H_{32}O_2$), and linolenic acid ($C_{18}H_{30}O_2$) (Filemon 2010; Atabani et al. 2012).

3.2 *Feedstock for Vegetable Oil*

Lipid feedstock, such as vegetable oil, animal fats and from microalgae, including cyanobacteria, is currently used for the production of biodiesel. The important parameters to consider a biodiesel feedstock is the amount of oil that can be produced or extracted per unit area of land planted to the oil containing crop. Among various plant oils yield per unit area of cultivated land, oil palm produces the highest amount of oil per hector of cultivated land, yielding 5000 kg oil per hectare, followed by coconut with 2260 kg oil per hectare, jatropha typically produces 1590 kg oil per hectare (Filemon 2010). In addition to the oil productivity per hectare, overall productivity must be assessed taking into consideration other needed agricultural inputs such as water, fertilizer, labor, and energy requirement particularly for harvesting and processing. Therefore, in practice, soybean and rapeseed dominate the world oilseed production with oil content of 21 and 35 %, respectively (Issariyakul and Dalai 2014).

The availability and production of vegetable oil as lipid feedstock also depends on the climatic conditions with the use of rapeseed oil feedstock in European countries and Canada, soybean oil in the United States, and palm oil in tropical countries such as Indonesia and Malaysia (Issariyakul and Dalai 2014; Sharma and Singh 2009). The ability of brassica crop (rapeseed, canola, mustard) to tolerate low temperature allows them to be cultivated in cold climate regions. These seeds have high oil (40 %) content and the dominant fatty acids include oleic acid (C18:1), linoleic acid (C18:2), and erucic acid (C22:1) (Röbbelen 1990; USDA-FAS 2016; Williams 2005; Wang 2002).

In terms of total production and international trade, soybean is the world's largest oilseed and its oil content range from 15 to 22 % depending on environmental conditions during seeds maturity and the dominant fatty acids are oleic acid

(C18:1) and linoleic acid (C18:2) (Röbbelen 1990; USDA-FAS 2016). Although palm is the most efficient oil-producing plant per area per year but due to lesser availability they have received considerable attention, except in tropical countries. Palm oil is either derived from mesocarp or palm kernel inside the seed (palm kernel oil) and is more saturated than soybean oil and rapeseed oil (Pantzaris and Basiron 2002; Lin 2002). The dominant fatty acids in palm oil and palm kernel oil are palmitic (C16:0), stearic (C18:0), oleic (C18:1), and linoleic (C18:2) lauric (C12:0), myristic (C14:0), oleic (C18:1) acids (Röbbelen 1990; USDA-FAS 2016; Williams 2005). Fractionation of palm oil at ambient temperature (25–30 °C) yields palm olein or oleic-rich oil (liquid fraction) and palm stearin or stearic-rich oil (solid fraction).

Sunflower is one of the most ancient oilseed species as its cultivation can be traced back to 3000 B.C. Sunflower was once the world top-rank oil-producing plant prior to the advent of soybean boom after World War II. The oil content in sunflower seeds range from 40 to 50 % with a yield of 280–700 kg/hector/year. Oleic (C18:1) and linoleic (C18:2) are the major fatty acid constituents of sunflower oil (Röbbelen 1990; USDA-FAS 2016).

Rice oil are extracted from rice bran using extruder, expander, and expeller to form a bran flakes or pallet followed by solvent extraction and contains triacylglyceride (TAG) with palmitic (C16:0), oleic (C18:1), and linoleic (C18:2) as the major fatty acids (Orthoefer 2005). Diacylglyceride (DAG), monoacylglyceride (MAG), and sterols are present as minor constituent. Due to their ability to be grown on noncultivable and degraded wasteland they are considered as one of the most promising feedstocks for biodiesel production.

Although jatropha plant has low nutritional requirements, cultivation of Jatropha under acidic soil requires additional nutrients such as calcium and magnesium due to its preference for alkaline soil. Oil derived from jatropha is nonedible due to curcin, a toxic compound, found in the seeds. Oil content ranges from 35 to 40 % in seed and 50–60 % in kernel with oleic (C18:1) and linoleic (C18:2) as its major fatty acids (Misra and Murthy 2011).

Karanja is an oil seed-bearing tree native to humid and subtropical environments. It is highly tolerant to salinity and can be cultivated on degraded wasteland on a variety of soil types ranging from clay to sandy or stony. The oil droplets extracted from karanja appear yellowish orange to brown and are not edible due to the presence of toxic flavonoids. Oil content varies from 9 to 46 % with oleic (C18:1) and linoleic (C18:2) as its major fatty acids (Kumar and Sharma 2011; Meher et al. 2006).

There are large numbers of plants that produces oils, that can be processed to produce biodiesel, which can be used as a diesel substitute or blend. Most of this oil, such as soybean oil, coconut oil, and palm oil are also used as human or animal food or in the production of various types of cosmetics and pharmaceuticals, but increasing amount are now being processed for the production of biodiesel. Detailed list of different feedstocks for biodiesel production and their current status is presented in Table 3.

Table 3 Different feedstock categories for biodiesel production and their current status (Atabani et al. 2012; Issariyakul and Dalai 2014; Moka et al. 2014)

Feedstock categories	Feedstock	Current status as biodiesel feedstock
Edible vegetable oil	Soybeans, rapeseed, safflower, rice bran, barley, sesame, groundnut, sorghum, wheat, corn, coconut, canola, peanut, palm and palm kernel, sunflower	Edible vegetable oil accounts for 95 % of the world's biodiesel feedstock Relatively high oil content and gives the highest oil yield per area per year as compared to other oils
		Creates food versus fuel crises and the destruction of ecosystem of arable land
		Rise in the price of vegetable oil affects the economic viability of biodiesel industries
		Currently, edible oil is not feasible for the long-term supply as biodiesel feedstock
Nonedible vegetable oil	*Jatropha curcas*, Mahua, Pongamia, Camelina, Cotton seed, Karanja, Cumaru, *Cynara cardunculus*, *Abutilon muticum*, Neem, Jojoba, Passion seed, Moringa, Tobacco seed, Rubber seed tree, Tall, Coffee ground, Nagchampa, *Croton megalocarpus*, *Pachira glabra*, *Aleurites moluccana*, *Terminalia belerica*	Nonedible oil plants are gaining worldwide attention due to their wide occurrence and adaptability
		Cultivable in wasteland not suitable for food crops thus eliminates the competition for food and are eco-friendly
		Jatropha is considered as one of the most promising feedstocks for biodiesel production
Waste or recycled oil	–	Use of waste or recycle oil can reduce the feedstock cost reaching up to 45 % of direct production in addition to various environmental benefits
		Although waste cooking oil has a great potential for the production of biodiesel, their collection infrastructure and logistics is a problem
Animal fats	Pork lard beef tallow poultry fat fish oil chicken fat	The use of animal fats eliminates the need to dispose them but they are not plentiful to satisfy the global energy demand

(continued)

Table 3 (continued)

Feedstock categories	Feedstock	Current status as biodiesel feedstock
Microorganisms	Photosynthetic microalgae and cyanobacteria	Represents a very promising feedstock because they are more efficient than conventional crop plants and the oil yield can be 25 times higher than the yield of oil palm and 250 times the amount of soybean
		Can provide a source of renewable biodiesel to meet the global demand for transport fuels
		Ability to sequester carbon from flue gas gives the opportunity to use greenhouse gases as feedstock for the growth of algae
		High production cost and the requirement of desirable strains with high oil productivity and effective large-scale bioreactors

3.3 *Factor Affecting the Production of Biodiesel*

Biodiesel production involves a number of different types of physical and chemical steps depending on the kind of feedstock. The first step is the extraction of the crude oil from fruits, seeds, or parts of oil containing plants, and usually involves mechanical press extraction and solvent extraction. Crude plant oil extract is much more viscous than conventional diesel fuel (11–17 times thicker) and has very different chemical properties and combustion characteristics to those of conventional diesel fuel and is not suitable for direct use.

Biodiesel is produced either by the transesterification of TAG or the esterification of free fatty acid, which also reduces the viscosity of vegetable oil (Figs. 1 and 2). These reactions consume one mole of alcohol for every mole of ester produced, i.e.,

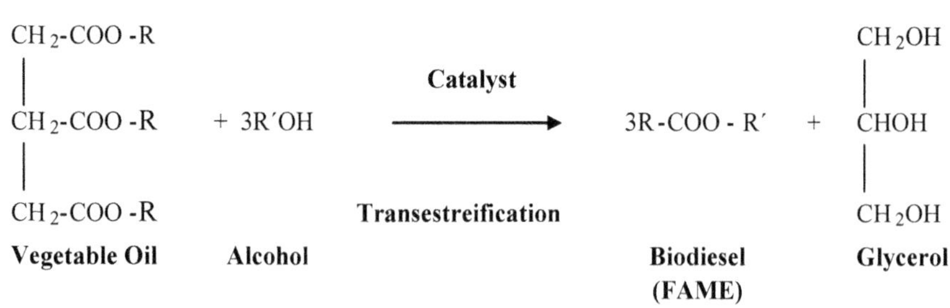

Fig. 1 Transesterification reaction for vegetable oil conversion to biodiesel

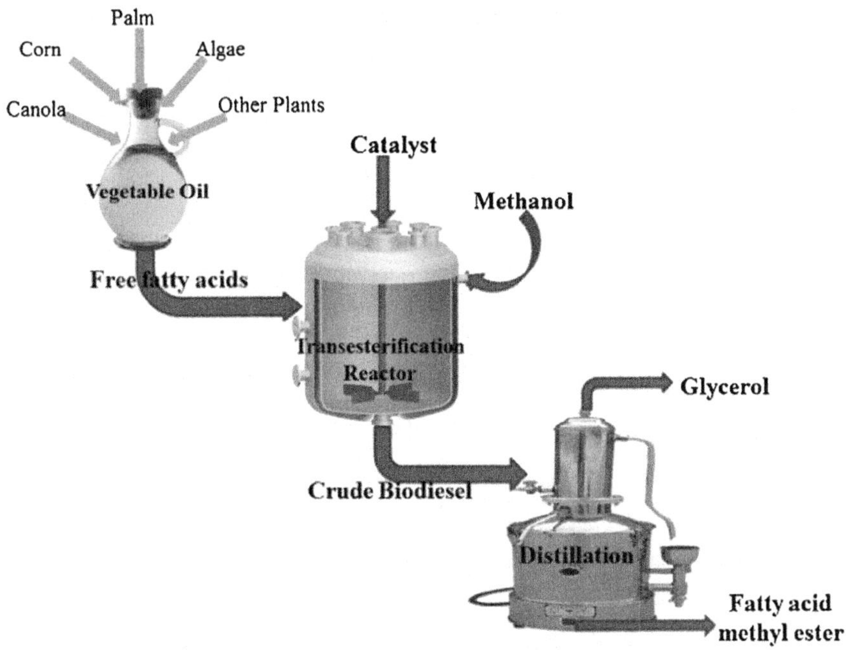

Fig. 2 Schematic of biodiesel production process from vegetable oils

three moles of alcohol to produce three moles of ester and one mole of glycerol for every mole of TAG consumed. The efficiency of transesterification reaction is influenced by various parameters, which includes the type of alcohol, alcohol to oil molar ratio, water content, reaction temperature, reaction duration, and the type of catalyst (Isariyakul and Dalai 2014). The type of catalyst used in transesterification is the most important parameters, which determine the outcome of the biodiesel production.

Selection of catalyst depends on the type and quality of feedstock, and base catalyst is the most commercially used due to high yield, short reaction time, and low reaction temperature requirement, whereas acid catalyst is used when the feedstock contains a high concentration of free fatty acids and water. Solid catalyst and enzymatic catalysis are also subjected to investigation, and further research and development is needed to realize their potential applications.

Transesterification under supercritical conditions involves the use of extreme temperature and pressure, and, therefore, are susceptible to polymerization, thereby purification of the desired product becomes difficult (D'Ippolito et al. 2007). Commercially, homogeneous base catalysis is most commonly used and offers a high reaction yield (97 % or more) in a short time (10 min to 1 h) with mild reaction temperatures (25–70 °C) (Isariyakul and Dalai 2014). The most common homogeneous base catalysts are hydroxides and alkoxides of alkali metals such as NaOH, KOH, $NaOCH_3$, $KOCH_3$. Homogeneous base catalysis is limited to anhydrous feedstock with acid value lower than 1 and acid catalyst is more suitable if the feedstock contains a high amount of free fatty acid and water.

Homogeneous acid catalysis reduces saponification by the esterification of the free fatty acid but has the disadvantage of requiring a high reaction temperature and a long reaction time. H_2SO_4, H_3PO_4, HCl, BF_3, and CF_3CO_2H are the commonly used homogeneous acid catalysts. In contrast to homogeneous catalysis, heterogeneous catalyst can be regenerated and reused, rendering continuous production of biodiesel. Heterogeneous acid catalyst is most promising for biodiesel production due to simple biodiesel purification step and their ability to handle low quality feedstock with high FFA content and water.

Due to the reversible nature of the transesterification reaction an excess of alcohol is usually used in order to shift the reaction to the product side. The optimum ratio of alcohol to oil to achieve a maximum conversion for an alkali catalyzed depends on the quality of oil and the type of vegetable oil. A maximum of 92–98 % conversion was reported using alcohol to oil ranging from 6:1 to 15:1 with various biodiesel feedstocks in an alkali-catalyzed reaction (Karmee and Chadha 2005; Leung and Guo 2006; Enciner et al. 2002; Rashid and Anwar 2008). Acid-catalyzed reaction, however, requires higher alcohol to oil molar ratio (30:1–245:1) when compared to alkali-catalyzed reaction (Freedman et al. 1986; Canakci and Van Gerpen 1999; Bhatti et al. 2008; Zheng et al. 2006).

Increasing the alcohol used in transesterification above the optimum ratio increases the polarity of the mixture and the solubility of glycerol promoting the reverse reaction, thereby reducing the ester yield. The most commonly used alcohol for transesterification reaction is methanol due to its economical benefit although TAG is sparsely soluble in methanol. This immiscibility behavior referred to as mass transfer resistance or mass transfer limitation can be overcome by rigorous mechanical stirring, an aid of cosolvent, the use of super critical conditions, and the use of other techniques such as microwave and ultrasonic (Isariyakul and Dalai 2014). Other alcohols such as ethanol, propanol, and butanol are also explored to improve the mass transfer of TAG. The use of ethanol alleviates the initial mass transfer hereby increasing the initial rate of reaction but with a reduce rate of reaction due to lower reactivity of ethoxide compared to methoxide (Sridharan and Mathai 1974).

The nature of the starting materials also affects the ester yield and glyceride conversion in alkali-catalyzed process. Soap formation results from the reaction between the catalyst and FFA, and also from the hydrolysis of glycerides to form soaps and glycerol when FFA and water are present in the feedstock. The hydrolysis of ester to form FFA is also enhanced in the presence of water, thereby reducing the ester yield. Comparison of the different oil extraction methods and description of various transesterification reactions are presented in Tables 4 and 5, respectively.

Table 4 Comparison of various oil extraction methods (Achten et al. 2008; Mahanta and Shrivastava 2012; Atabaniab et al. 2012)

Methods	Characteristics	Advantages
Mechanical extraction	Achieved with the use of manual ram press or an engine driven screw press. Requires filtration and degumming of the extracted oil. Oil yields range from 60 to 80 %	The cost conventional method with minimum capital investment
Solvent extraction	Particle size, the type of liquid chosen, temperature, and agitation of the solvent affects the process of leaching	Economical for large-scale production of biodiesel. Highest oil yield
Enzymatic extraction	Oil extraction achieved by used of suitable enzyme in a controlled reaction temperature, pH, and reaction time	Environmental friendly and limits the production of volatile organic compounds

3.4 Economics: An Estimate of Biodiesel Production Cost

There are several reports that estimate the cost and feasibility of industrial scale production of biodiesel taking into account various feedstock and operation scales. The construction and the expansion of existing knowledge has estimated a total production of 1.7×10^9 L (US) and 114 million L (Europe) of biodiesel by the leading biodiesel producers of the world (European Biodiesel Board 2004). Globally, biodiesel production increased from 8.4 million ton in 2007 to 20 million ton in 2010. It is expected to reach 150 million ton by 2020 (Moka et al. 2014).

Among the different types of reaction configurations, the choice of chemical technology to employ for the production of biodiesel depends on the choice of feedstock and its quality which in turn influences the overall cost and feasibility. The cost of production including the cost of feedstock and its conversion to biodiesel is estimated from US$0.30/L ($1.14/gal) to US$0.69/L ($2.62/gal) for fuel produced from soybeans and rapeseed, respectively (Haas et al. 2006). A production cost of US$0.42/L ($1.58/gal) was estimated from refined, bleached, and deodorized soy oil in a small pilot scale plant (190/L, batch process), excluding the cost of feedstock and the profit from the sale of coproduct glycerol and the capital cost of production (Canakci and Van Gerpen 2001).

In the overall economics of commercial biodiesel production, the cost of feedstock comprises a very substantial portion and in case of biodiesel from soy oil it constitutes about 88 % of the overall production cost (American Biofuels Association & Information Resources, Inc 1994; Bender 1999; Graboski and McCormick 1998). This highlights the need for the development of technologies to allow the use of lower value feedstock. Contemporary process simulation software designing could assist in choosing appropriate feedstock, chemical process, plant capacity, and design thereby determining the overall economic feasibility of a proposed operation.

The current price of bulk petroleum diesel is considerably lower than the cost of biodiesel production, and this substantial price difference contributed largely by cost of feedstock highlight the potential value of low cost alternatives in improving

Table 5 Characteristics of various transesterification reactions (Fulton et al. 2004; Goldemberg 2007; Wang et al. 2006)

Transesterification reactions	Catalyst	Reaction temperature (K)	Reaction time	Methyl ester yield (%)	Post-transesterification processing	Advantages	Disadvantages
Alkali catalysis	NaOH, $NaOCH_3$, $KOCH_3$, KOH, NaMeO, K_2CO_3	303–338	30–60 min	~96	Energy intensive removal of glycerol and the catalyst. Wastewater treatment	Faster and economical, high conversion efficiency, cost effective method	Soap formation and reduction in ester yield in the presence of Free fatty acid
Acid catalysis	Sulfuric, hydrochloric, ferric sulfate, phosphoric and organic sulfonic acid	338	3–48 h	~90	Energy intensive removal of glycerol and the catalyst. Wastewater treatment	More tolerant than alkaline catalysts for vegetable oils having high free fatty acids and water. High conversion efficiency and cost effective	Lower yield compared to alkali catalysis
Enzyme catalysis	Lipase catalysts such as diazomethane CH2N2	–	–	–	No complex operations for the recovery of glycerol and the elimination of catalyst and soap	Tolerance for the free fatty acid level of the feedstock	Lipases are very expensive
Supercritical method	–	523–573	7–15 min	98	Recovery of methanol is easier and environmental friendly	Very short reaction time. High conversion efficiency	High cost of operation

the economic viability of biodiesel. Although biodiesel has a number of advantages over the conventional fuels, they still have a long way to go before they can be considered as economically viable alternatives to the conventional fuels. In addition, the use of oil seeds as raw materials for the production of biodiesel requires a large acreage of land to meet demand, whereas in case of animal fats a large capital is required to feed animals, which will be used for raw oil production.

The use of oleaginous microorganisms to produce raw materials is also considered, and is attracting a number of researchers for the development of a feasible biofuels technology. Notwithstanding the above limitations, the use of biodiesel is gradually catching up with conventional diesel and petrol although it has yet to reach its full commercial potential in the developing countries. Comparative techno-economic analysis shows that the cost of biodiesel from waste cooking oils is lower than that of fresh oils. Life cycle cost analysis for 50,000 ton of palm biodiesel production plant is likely to bring the palm oil derived biodiesel cost at par with the fossil fuel cost, but to achieve that an even marginal subsidy should be provided by government.

4 Biodiesel Versus Diesel as a Transportation Fuel: Advantages and Disadvantages

The transportation sector has been the focus of economic analyses and cost comparison studies for conventional versus alternative fuels. Biodiesel has the advantages of being portable, readily available, renewable, higher combustion efficiency, lower sulfur and aromatic content, higher cetane number, and biodegradable (Moka et al. 2014; Haas et al. 2006; Canakci and Van Gerpen 2001; American Biofuels Association & Information Resources, Inc 1994; Bender 1999; Graboski and McCormick 1998; Park et al. 2010). Compared to diesel, biodiesel has a higher viscosity, lower energy content, high cloud point and pour point, higher NO_X emission, lower engine speed and power, injector coking, engine compatibility, high price, and higher engine wear (Balat and Balat 2010). Biodiesel offers safety during transportation, handling, and storage due to their higher flash point, which in turns requires higher temperature to ignite the fuel on application of an ignition source. The presence of electronegative element oxygen makes biodiesel slightly more polar and viscous than diesel fuel. This in turn lowers the heating value of biodiesel, and it generally has a lower heating value (LHV), (12 % less than No. 2 diesel fuel) (Ohadi and Garis 2005).

Preheating palm oil methyl ester significantly improves the brake power output and exhaust emission and studies on soy oil methyl ester in a direct injection diesel engine shows relevant combustion parameters such as ignition delay, peak pressure, and rate of pressure rise was close to those observed for diesel combustion at the same engine load, speed, timing, and nozzle diameter. Soy oil methyl esters also exhibited comparable ignition delay to that of diesel fuel with drastic reduction in the emission of NOx and slightly lower CO emission (Balat and Balat 2010; Ohadi and Garis 2005). The emission of sulfate is negligible due to the near absence of sulfur in biodiesel thus helps to reduce the problem of acid rain due to transportation fuels. Biodiesel reduces unregulated emissions of aromatic hydrocarbons, smoke, and particulate matter due to oxygenated nature of biodiesel where more oxygen is available for burning and reducing hydrocarbon emission in the exhaust.

Biodiesel blends up to 20 % with petroleum-based diesel can be used in nearly all diesel engine, and the most common blends are B2 (2 % biodiesel and 98 % petroleum diesel), B5 (5 % biodiesel and 95 % petroleum diesel), and B20 (20 % biodiesel and 80 % petroleum diesel). Some of the drawbacks for using biodiesel blends include the problem with fuel freezing in cold weather; reduced energy density and degradation of fuel under prolong storage. Biodiesel also has higher gel point and cloud point than petrodiesel, thus require heating of storage tank, especially in cooler climate. Biodiesel also produces higher NOx emission due to the higher cetane rating and oxygen content of the fuel. Another challenge in using biodiesel is the hydrophobic nature of the fuel, and the presence of water can cause a number of problems such as:

- reduces the heat of combustion leading to more smoke and harder starting,
- corrosion of fuel pumps, injector pumps, and fuel lines,
- formation of ice crystal in cold climates and gelling of the fuels thereby decreasing the flow properties,
- plugging of fuel system by microbial growth due to the presence of water.

5 Conclusions

With the depletion of fossil fuels and the rising levels of GHGs contributed substantially from the transportation sector, biofuel provides a renewable and environmental-friendly alternative to the conventional fossil fuels. Among the several alternatives that can be used as transportation fuels, methyl esters of vegetable oil have several advantages and are recommended for use as a substitute for petroleum-based diesel. Although the use of biodiesel has gone up drastically during the last decades, in order to successfully compete with conventional fuels, improvement in terms of their proprieties, production efficiency as well as end user suitability is needed. Since the use of particular vegetable oil feedstocks affects the overall cost and the characteristics of the resulting biodiesel, selection of appropriate vegetable oil and the production technology is vital for the advancement of biodiesel industries. At present, biodiesel derived from vegetable oil is not economically feasible because it is more expensive than petroleum fuels but the demand for biodiesel as conventional diesel additive will likely increase in the coming years with an increase in petroleum prices and uncertainties concerning their availability. With the magnitude of resources and efforts being put in the development of an economically feasible biodiesel technology, biodiesel derived from vegetable oil will ensure a clean and green environment in the future.

References

ACEA (2009) Biodiesel guidelines. European Automobile Manufacturers Association, Brussels, Belgium, Mar 2009. http://www.acea.be/images/uploads/files/20090423_B100_Guideline.pdf

Achten WMJ, Verchit L, Mathijs Franken YJ, Singh E, Aerts VP, Muys RB (2008) Jatropha bio-diesel production and use. Biomass Bioenergy 32(12):1063–1084

Air Transport Action Group (2009) Beginner's guide to aviation biofuels. http://www.cphcleantech.com/media/2450696/beginnersguide.pdf

Alternative Fuels Committee of the Engine Manufacturers Association (1995) Biodiesel fuels and their use in diesel engine applications

American Biofuels Association & Information Resources, Inc. (1994) Biodiesel—a technology performance and regulatory review. National SoyDiesel Development Board, Jefferson City, MO, p 118

Anonymous (2002) A comprehensive analysis of biodiesel impacts on exhaust emissions. US Environmental Protection Agency EPA420-P-02-001

Anonymous (2004) Biodiesel handling and use guidelines. US Department of Energy, US Department of Energy DOE/GO-102004–1999

Anonymous (2009) Tomorrow's solution for today's environment. http://www.gomedic.com/applications.html

Anonymous (2014) Your ultimate biodiesel guide. Usage of Biodiesel Fuel. http://www.berkeleybiodiesel.org/usage-of-biodiesel-fuel-a.html

ASTM (2002) Standard specification for biodiesel fuel (B100) blend stock for distillate fuels. American Society for Testing and Materials, D6751-02

Atabani AE, Silitonga AS, Badruddina IA, Mahlia TMI, Masjuki HH, Mekhilef S (2012) A comprehensive review on biodiesel as an alternative energy resource and its characteristics. Renew Sust Energy Rev 1(1):2070–2093

Atabani AE, Silitonga AS, Ong HC, Mahlia TMI, Masjuki HH, Badruddina IA, Fayaza H (2013) Non-edible vegetable oils: a critical evaluation of oil extraction, fatty acid compositions, biodiesel production, characteristics, engine performance and emissions production. Renew Sustain Energy Rev 18:211–245

Bajpai D, Tyagi VK (2006) Biodiesel: source, production, composition, properties and its benefits. J Oleo Sci 55(10):487–502

Balat M, Balat H (2010) Progress in biodiesel processing. Appl Energy 87(6):1815–1835

Bender M (1999) Economic feasibility review for community-scale farmer cooperatives for biodiesel. Bioresour Technol 70:81–87

Bhatti HN, Hanif MA, Qasim M, Rehman A (2008) Biodiesel production from waste tallow. Fuel 87:2961–2966

Brown RC, Liu Q, Norton G (2000) Catalytic effect observed during the co-gasification of coal and switchgrass. Biomass Bioenergy 18:499–506

Canakci M, Van Gerpen J (1999) Biodiesel production via acid catalysis. Trans ASAE 42 (5):1203–1210

Canakci M, Van Gerpen J (2001) A pilot plant to produce biodiesel from high free fatty acid feedstocks. Paper No. 016049, 2001 ASAE Annual International Meeting, Sacramento, CA

Canakci M, Ozsezen AN, Turkcan A (2009) Combustion analysis of preheated crude sunfower oil in an IDI diesel engine. Biomass Bioenergy 33:760–767

D'Ippolito SA, Yori JC, Iturria ME, Pieck CL, Vera CR (2007) Analysis of a two-step, noncatalytic, supercritical biodiesel production process with heat recovery. Energy Fuels 21:339–346

Demirbas A (2003) Biodiesel fuels from vegetable oils via catalytic and non-catalytic supercritical alcohol transesterifications and other methods: a survey. Energy Convers Manage 44:2093–2109

Demirbas A (2006) Biodiesel production via non-catalytic SCF method and biodiesel fuel characteristics. Energy Convers Manage 47:2271–2282

Demirbas A (2009) Progress and recent trends in biodiesel fuels. Energy Convers Manag 50 (1):14–34
Dmytryshyn SL, Dalai AK, Chaudhari ST, Mishra HK, Reaney MJ (2004) Synthesis and characterization of vegetable oil derived esters: evaluation for their diesel additive properties. Bioresour Technol 92:55–64
Enciner JM, Gonzalez JF, Rodriguez JJ, Tajedor A (2002) Biodiesels fuel from vegetable oils: transesterification of *Cynara cardunculus* L. oils with ethanol. Energy Fuels 16:443–450
Espadafor FJ, García MT, Villanueva JB, Gutiérrez JM (2009) The viability of pure vegetable oil as an alternative fuel for large ships. Transp Res Part D 14:461–469
European Biodiesel Board (2004) Biodiesel production statistics. Available from http://www.ebb.eu.org/stats.php
Fan X, Burto R (2009) Recent development of biodiesel feedstocks and the applications of glycerol: a review. Open Fuels Energy Sci J 2:100–109
Figueiredo MKK, Romeiro GA, Avila LA, Damasceno RN, Franco AP (2009) The isolation of pyrolysis oil from castor seeds via a low temperature conversion (LTC) process and its use in a pyrolysis oil–diesel blend. Fuel 88:2193–2198
Filemon J (2010) Biofuels from plant oils. Available from http://www.aseanfoundation.org/documents/books/biofuel.pdf
Freedman B, Butterfeld RO, Pryde EH (1986) Transesterification kinetics of soybean oil. J Am Oil Chem Soc 63(10):1375–1380
Friedrich MS (2004) A worlwide review of the commercial production of biodiesel—a technological, economic and ecological investigation based on case studies. Department of Technology and Sustainable Product Management (ITNP), Wien, http://itnp.wu-wien.ac.at/archiv_doc/BAND41_FRIEDRICH.PDF
Fulton L, Howes T, Hardy J (2004) Biofuels for transport: an international perspective. Iternational Energy Agency, Paris
Goldemberg J (2007) ethanol for a sustainable energy future. Science 315:808–810
Goldemberg J, Johansson TB (2004) World energy assessment overview: update, United Nations Devlopment Programme
Graboski MS, McCormick RL (1998) Combustion of fat and vegetable oil derived fuels in diesel engines. Prog Energy Combust Sci 24:125–164
Haas MJ, McAloon AJ, Yee WC, Foglia TA (2006) A process model to estimate biodiesel production costs. Bioresour Technol 97(4):671–678
Haldar SK, Ghosh BB, Naga A (2009) Utilization of unattended *Putranjiva roxburghii* non-edible oil as fuel in diesel engine. Renew Energy 34:343–347
Hebbal OD, Reddy KV, Rajagopal K (2006) Performance characteristics of a diesel engine with Deccan hemp oil. Fuel 8:2187–2194
Isariyakul T, Dalai AK (2014) Biodiesel from vegetable oils. Renew Sustain Energy Rev 31:46–471
Issariyakul T, Dalai AK (2014) Biodiesel from vegetable oils. Renew Sust Energy Rev 31: 446–471
Jääskeläinen H (2009) Biodiesel standards & properties. www.DieselNet.com. Copyright © Ecopoint Inc. http://www.dieselnet.com/tech/fuel_biodiesel_std.php
Karmee SK, Chadha A (2005) Preparation of biodiesel from crude oil of *Pongamia pinnata*. Bioresour Technol 96:1425–1429
Kumar A, Sharma S (2011) Potential non-edible oil resources as biodiesel feedstock: an Indian perspective. Renew Sustain Energy Rev 15(4):1791–1800
Leung DYC, Guo Y (2006) Transesterification of neat and used frying oil: optimization for biodiesel production. Fuel Process Technol 87:883–890
Lin SW (2002) Palm oil. In: Gunstone FD (ed) Vegetable oils in food technology composition, properties and uses. Blackwell Publishing, CRC Press LLC, Boca Raton, FL, USA
Ma F, Hanna MA (1999) Biodiesel production: a review. Bioresour Technol 70:1–15
Mahanta P, Shrivastava A (2012) Technology development of bio-diesel as an energy alternative. Available from http://www.newagepublishers.com/samplechapter/001305.pdf

Mandil C, Shihab-Eldin A (2010) assessment of biofuels potential and limitations. International Energy Forum
Meher LC, Sagar DV, Naik SN (2006) Technical aspects of biodiesel production by transesterification—a review. Renew Sustain Energy Rev 10:248–268
Misra RD, Murthy MS (2011) Jatropa-the future fuel of India. Renew Sustain Energy Rev 15 (2):1350–1359
Moka S, Pande M, Rani M, Gakhar R, Sharma M, Rani J, Bhaskarwar AN (2014) Alternative fuels: an overview of current trends and scope for future. Renew Sustain Energy Rev 32: 697–712
Nelson RG, Schrock MD (2006) Energetic and economic feasibility associated with the production, processing and conversion of beef tallow to a substitute diesel fuel. Biomass Bioenergy 30:584–591
Nitske WR, Wilson CM (1965) Rudolf diesel: pioneer of the age of power. University of Oklahoma Press, Oklahoma, p 139
Ohadi M, Garis D (2005) Oil price increases and the energy industry: what role for the majors? Soc Pet Eng. doi:10.2118/93755-MS
Orthoefer FT (2005) Rice bran oil. In: Shahidi F (ed) Bailey's industrial oil & fat products, vol 2, 6th ed. Wiley, New Jersey
Pantzaris TP, Basiron Y (2002) The lauric (coconut and palmkernel) oils. In: Gunstone FD (ed) Vegetable oils in food technology composition, properties and uses. Blackwell Publishing, CRC Press LLC, Boca Raton, FL, USA
Park SH, Kim HJ, Sik Lee CS (2010) Fuel spray and exhaust emission characteristics of an undiluted soybean oil methyl ester in a diesel engine. Energy Fuels 24(11):6172–6178
Pinzi S, Garcia IL, Lopez-Gimenez FZ, Luque de Castro MD, Dorado G, Dorado MP (2009) The ideal vegetable oil-based biodiesel composition: a review of social, economical and technical implications. Energy Fuels 23:2325–2341
Purushothaman K, Nagarajan G (2009) Experimental investigation on a C.I. engine using orange oil and orange oil with DEE. Fuel 88:1732–1740
Rashid U, Anwar F (2008) Production of biodiesel through optimized alkaline-catalyzed transesterification of rapeseed oil. Fuel 87:265–273
Röbbelen G (1990) Mutation breeding for quality improvement—a case study for oilseed crops. Mutat Breed Rev 6:1–44
Sharma YC, Singh B (2009) Development of biodiesel: current scenario. Renew Sust Energy Rev 13:1646–1651
Shay EG (1993) Diesel fuel from vegetable oils: status and opportunities. Biomass Bioenergy 4:227–242
Sridharan R, Mathai IM (1974) Transesterification reaction. J Sci Ind Res 33:178–186
USDA-FAS (2016) Oilseeds: world markets and trade. The United State Department of Agriculture—Foreign Agricultural Service Website http://www.fas.usda
Wang T (2002) Soybean oil. In: Gunstone FD (ed) Vegetable oils in food technology composition, properties and uses. Blackwell Publishing, CRC Press LLC, Boca Raton
Wang YD, Al-Shemmeri T, Eames P, McMullan J, Hewitt N, Huang Y (2006) An experimental investigation of the performance and gaseous exhaust emissions of a diesel engine using blends of a vegetable oil. Appl Thermal Eng 26:1684–1691
Williams MA (2005) Recovery of oils and fats from oilseeds and fatty materials. In: Shahidi F (ed) Bailey's industrial oil & fat products, vol 5, 6th ed. Wiley, New Jersey
Zheng S, Kates M, Dubé MA, McLean DD (2006) Acid-catalyzed production of biodiesel from waste frying oil. Biomass Bioenergy 30:267–272

Pretreatment Processes for Cellulosic Ethanol Production: Processes Integration and Modeling for the Utilization of Lignocellulosics Such as Sugarcane Straw

Danilo Ribeiro de Lima, Marcos Henrique Luciano Silveira, Luis Del Rio and Luiz Pereira Ramos

Abstract Pretreatment is the key step for a viable and efficient cellulosic ethanol production process and, for this reason, it must be very selective in avoiding polysaccharide degradation and inhibitors formation. This work provides a brief overview of the leading pretreatment technologies available to date, with emphasis on those that are already closed to or eventually reached commercial scale such as steam explosion and/or dilute acid hydrolysis. Details are also given with regard to the fundamental effects of pretreatment on the chemical composition and organizational structure of the plant cell wall. Furthermore, the impact of steam explosion and enzymatic hydrolysis on the overall capital cost of cellulosic ethanol production has been determined in light of the following process integration approaches using sugarcane straw as the reference material: simultaneous saccharification and co-fermentation (SSCF), separated hydrolysis and co-fermentation (SHCF) and separated hydrolysis and fermentation (SHF). As a result, cellulosic ethanol produced from SSCF, SHCF, and SHF processes resulted in capital cost estimates of $1.66, $1.75, and $2.23 per liter of ethanol produced. The difference among these values is related to the easiness with which different unit operations are harmonized in a sustainable and fully operational biorefinery unit.

D.R. de Lima · M.H.L. Silveira
Center of Sugarcane Technology—CTC, Piracicaba, SP, Brazil

D.R. de Lima
Process and Product Development Technology Center,
Fibria Celulose SA, Jacareí, SP, Brazil

L. Del Rio
FPInnovations, 570 Boul St. Jean, Pointe-Claire, QC H9R 3J9, Canada

L.P. Ramos (✉)
Department of Chemistry, Research Center of Applied Chemistry/INCT Energy and Environment (INCT E&A), Federal University of Paraná, Curitiba, PR, Brazil
e-mail: luiz.ramos@ufpr.br

C.R. Soccol et al. (eds.), *Green Fuels Technology*,
Green Energy and Technology, DOI 10.1007/978-3-319-30205-8_5

Keywords Cellulosic ethanol · Pretreatment · Process integration · Modeling · Sugarcane straw

1 Biomass Structure

The constant increase in the world energy demand and the on-going environmental concerns about the use of fossil fuels has led many countries and research institutes to investigate and implement the production and use of liquid biofuels. In this scenario, lignocellulosic biomass has emerged as an option for several industrial applications such as in the case of cellulosic ethanol. However, it is well known that the plant cell wall has a very well-organized structural arrangement that complicates its use for the production of fuels and chemicals through chemical or biological processes.

Biomass recalcitrance relies primarily on the close interaction that exists among the three main components of the cell wall: cellulose, hemicellulose, and lignin. Lignin is a noncellulosic polyphenolic material whose biosynthesis results from the oxidative combinatorial coupling of several resonant forms of radicals that are generated from *p*-hydroxy-cinnamic alcohols with different degrees of methoxylation and these originate the following lignin substructures or building blocks: the hydroxyphenyl or H units (from *p*-coumaryl alcohol), the guaiacyl of G units (from coniferyl alcohol), and the syringyl or S units (from sinapyl alcohol) (Fig. 1a). The ratio among these building blocks varies in the plant kingdom, as well as among species of the same genus and among individuals of the same species. As result, its structure contains a great variety of chemical bonds, characterizing a rather diverse and nonuniform hydrophobic structure whose properties depend on several factors including the experimental conditions used for extraction and purification (Fig. 1b). This complex polyphenolic biopolymer can be fractionated or chemically modified in many ways to produce fuels, chemicals, adsorbents, and polymeric materials. For this reason, lignin represents a promising alternative feedstock for the development of sustainable biorefineries (Laurichesse and Avérous 2014). Besides, lignin is the most abundant natural macromolecule after cellulose, representing 20–30 % of lignocellulosic biomass produced on Earth.

Differently from lignin, hemicelluloses correspond to a great family of heteropolysaccharides that are often highly acetylated and covalently linked to lignin. Most hemicelluloses are either branched or fully decorated along the main chain by different side groups including organic acids, carbohydrate residues, and cinnamic acid derivatives. Hence, compared to cellulose (see below), hemicelluloses have a lower molecular mass and a lower molecular organization, resulting in polymeric matrix with a greater amorphous character. In general, these biopolymers are more susceptible to thermal, chemical, and biological degradation than cellulose.

Hemicellulose chains are strongly associated to cellulose by hydrogen bonding and contain cinnamic acids linked to the main backbone that may or may not be directly connected to lignin. When hemicelluloses are submitted to acid hydrolysis, different monosaccharides are released depending on their chemical composition. These include D-mannose, D-galactose, D-xylose, D-glucose, D-glucuronic acid,

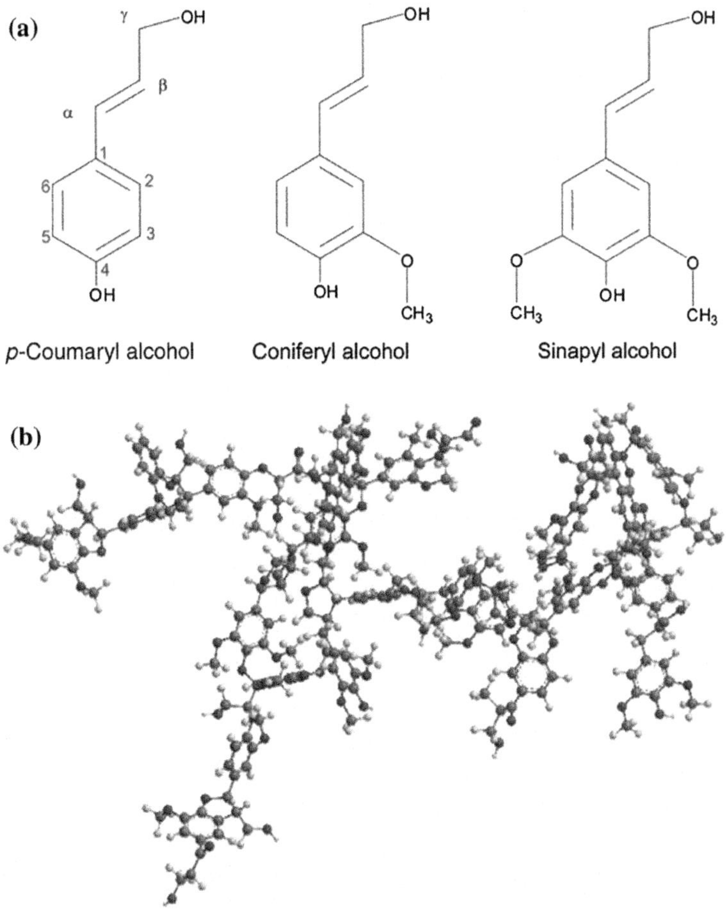

Fig. 1 **a** Structure of the alcohol precursors of lignin; **b** 3D model of the lignin structure as proposed by Fengel and Wegener (1989)

L-arabinose, 4-*O*-methyl-D-glucuronic acid (Ramos 2003). These biopolymers are more flexible and prone to fill empty spaces in the cell wall aggregate, therefore contributing to the strength and cohesiveness of the plant cell was architecture. Figure 2 shows a structural model of grassy hemicellulose with the main chain composed of β-(1→4)-xylan.

The most important polysaccharide of the plant cell wall is cellulose, which represents 40–50 wt.% (dry basis) of all biomass found on Earth (Pérez and Mazeau 2005). Differently than hemicelluloses, cellulose is a linear β-(1→4)-glucan that is able to establish a hydrogen bonding network with adjacent chains, forming planar and/or lamellar structures of great cohesiveness (Fig. 3). Such interactions result in highly ordered crystalline regions that are interrupted by regions of lower molecular organization (amorphous regions) (Matthews et al. 2006). Because of their high level of molecular organization, the resulting

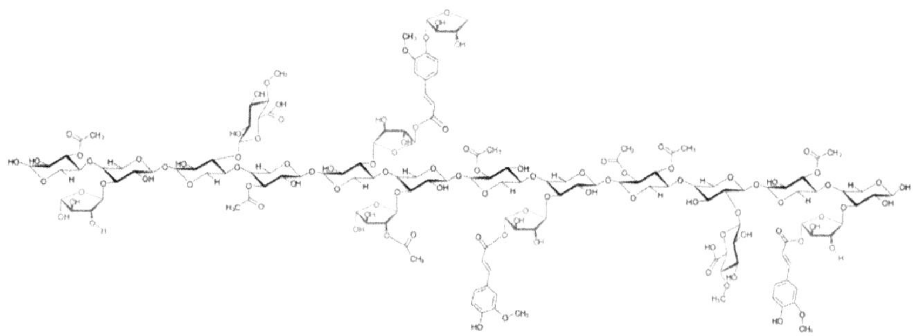

Fig. 2 Structural representation of an acetylated arabino-feruloyl-xylan fragment, with anhydroxylose in *black*, anhydroarabinose in *red*, 4-*O*-methylglucuronic acid in *green* and acetyl groups in *blue* (color figure online)

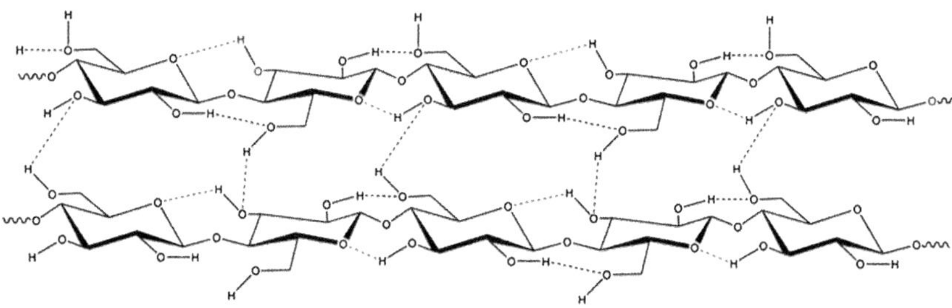

Fig. 3 Structural model of cellulose, with inter-chain hydrogen bonds in *red* and intra-chain hydrogen bonds in *black* and *blue* (color figure online)

composite is a very stable aggregate whose interaction with external agents is limited to its available surface area (Ramos 2003).

In association with the major components of the biomass, minor components are also present in lignocellulose matrix. These components are collectively referred to as extractives and are often responsible for certain plant characteristics, such as color, smell, natural resistance to rot, flavor, and abrasive properties, which are attributed to the presence of ash that may or may not be originated from soil contamination during harvesting (D'Almeida 1988). The organic fraction of biomass extractives may contain fats, waxes, fatty acids, alcohols, steroids, low molar mass hydrocarbons, terpenes, lignans, stilbenes, and flavonoids, among others.

2 Cellulosic Ethanol Production

Because of the high level of chemical association that is exists in the lignocellulosic matrix, a pretreatment step is required to increase its accessibility to hydrolases (Silveira et al. 2015a). By doing so, these materials become suitable for the production of second generation biofuels, particularly ethanol and other oxygenated compounds. In this case, at least four main steps must be carried out in optimal arrangement: (1) biomass pretreatment; (2) enzymatic hydrolysis; (3) monosaccharide fermentation; and (4) ethanol distillation. Therefore, the integration of such processes is of utmost importance to mitigate processing costs.

The most usual routes to achieve high accessibilities in plant biomass are through hemicellulose and/or lignin removal. Pretreatment technologies based on acid hydrolysis promote hemicellulose removal and, with this, a C5 liquid fraction is generated, leaving a solid substrate with relatively high cellulose and lignin contents. As a result, the cellulosic material acquires high susceptibility to enzymatic hydrolysis even in the presence of high lignin contents of 30–40 % (Tomás-Pejó et al. 2008).

When both enzymatic hydrolysis and fermentation steps are carried out at their optimal conditions, the process is referred to as separated hydrolysis and fermentation (SHF). The advantage of this strategy is that both processes are carried out at their optimal conditions but both capital and operational costs are higher due to the need of carrying out both hydrolysis and fermentation separately (Zhu et al. 2015). However, the integration of these two steps can lead to a considerable reduction in process costs (Palmqvist and Hahn-Hägerdal 2000). In this case, four integration approaches can be adopted: (1) simultaneous saccharification and fermentation (SSF); (2) co-fermentation of pentoses and hexoses (CF); (3) simultaneous saccharification and co-fermentation (SSCF); and (4) the consolidated bioprocess (CBP) (Taherzadeh and Karimi 2007).

In the SSF approach, the monosaccharides released by enzymatic hydrolysis are rapidly consumed by fermentation and their inhibitory effects on enzyme performance are considerably reduced if not eliminated. However, hydrolysis must be performed under nonoptimal conditions unless a thermotolerant or thermophilic organism is used for fermentation (Karimi et al. 2006). The CF approach takes place when the C6 fraction coming from enzymatic hydrolysis and the C5 fraction coming from an acid-catalyzed pretreatment are fermented together using either a genetically modified organism or a mix culture containing both C5 and C6 fermenting organisms. However, the co-fermentation of C5 and C6 is still a challenge, even though great progresses have been achieved in the last decade. The SSCF process is a combination of SSF and CF, meaning that the fermentation step must be carried out by a C5/C6 fermenting organism. Finally, the CBP was initially defined as the direct one-pot biological conversion of native biomass into fuels and chemicals. In this case, the employed organism secretes a powerful enzyme cocktail that is able to deconstruct the plant cell wall architecture, producing reducing sugars that are immediately metabolized to produce ethanol by co-fermentation (Horisawa et al. 2015).

Pretreatment is considered the key step for the biochemical production of cellulosic ethanol. Besides, pretreatment responds for a considerable share of the overall process production cost (Singh et al. 2015).

3 Biomass Pretreatment Technologies

Efficient pretreatment technologies for cellulosic ethanol production must be able to operate at high total solids, release highly digestible solids, minimize sugar losses, generate reduced amounts of hydrolysis/fermentation inhibitors, and operate in low to moderate cost reactors employing appropriate construction materials that are able to withstand highly corrosive chemical environments (Yang and Wyman 2012). A variety of pretreatment processes have been developed to date and these can be classified as physical, chemical, biological, or two or more combinations of these (Silveira et al. 2015a). Table 1 shows an overview of the winner pretreatment technologies, with their main characteristics and effects on the biomass chemistry and process yields.

Recently, ionic liquids (ILs) have been applied as solvents for cellulose dissolution and also as an option for biomass pretreatment. In this case, the biomass is dissolved and lignin and hemicelluloses are partially or totally removed after the addition of an anti-solvent that precipitates cellulose (Swatloski et al. 2002). This pretreatment technology results in a very susceptible substrates for hydrolysis, normally requiring very low enzyme loadings for achieving high glucose yields (da Costa Lopes et al. 2013). Also, low chemical modifications in lignin are observed and no inhibitors are formed. However, care must be taken with the inhibitory effect that ILs may have on hydrolysis and fermentation. On the other hand, green solvents such as ILs are still too expensive for industrial applications, therefore requiring efficient strategies for their recovery and reuse, particularly when high ILs to biomass ratios of 10:1–20:1 are used for pretreatment (Dibble et al. 2011).

Likewise ILs, supercritical fluids, such as carbon dioxide ($scCO_2$) increase substrate accessibility without modifying the lignin component or generating hydrolysis and/or fermentation inhibitors. Another positive aspect of $scCO_2$ is that xylooligosaccharides are produced as water soluble extracts and this can improve the economics of the overall pretreatment process since these are valuable chemicals for several industrial applications (Morais et al. 2014). However, this technology operates at high pressures and temperatures and both capital and operational costs are very high.

A new pretreatment technology has been developed by combining ILs with $scCO_2$ (Silveira et al. 2015b). By doing so, it was possible to reduce the usual 20:1 IL to biomass ratio to 1:1 or less by introducing ethanol as a pretreatment co-solvent. High pretreatment efficiencies were obtained with regard to delignification (around 40 %) and susceptibility to hydrolysis at very low enzyme loadings. Besides, high IL recoveries were obtained in the ethanol extract at the optimal pretreatment conditions. Although promising for green chemistry applications, this technology was only tested in bench scale and its viability in large scale has yet to be demonstrated.

Another alternative for a selective biomass pretreatment is the use of microorganisms for lignin degradation (Lee et al. 2007). Although positive in many ways, particularly due to its low energy requirements, less corrosion issues, and high

Table 1 Major aspects of pretreatment technologies for cellulosic ethanol production

Effect on the process	Pretreatment technology[a]										
	Milling	US	Acid	Alkaline	Organosolv	SCF	STEX	AFEX	LHW	IL	Biological
Increases of accessible surface area	H	M	M	M	H	L	H	H	M	H	H
Cellulose decrystallization	H	L	L	H	H	L	M	H	L	H	L
Hemicellulose removal	L	M/H[b]	H	M	M	M	H	M	M/H[b]	H	H
Lignin removal	L	L/M[b]	L	H	H	H	L	H	L/M[c]	H	H
Generation of inhibitors	L	L/H[c]	H	H	H	L	H	H	L	H	L
Temperature	H	M	M	M	H	H	H	H	H	M	L
Carbohydrate losses	M/H[b]	M/H[b]	H	H	M	L	M	L	L	L	L
Time consuming	L	M	M	M	M	M	L	L	M	M	H
Enzymatic digestibility	H	H	H	H	H	M	H	H	H	H	H
Changes in lignin structure	L	L	H	H	L	M	H	H	M/H[b]	H	H
Glucan conversion (%)	89	92	32	87	97	80	92	85	80	90	37

[a]*US* Ultrasound; *SCF* supercritical fluids; *STSE* Steam Explosion; *AFEX* Ammonia Fiber Expansion; *LHW* liquid hot water; *IL* ionic liquid
[b]Depends on the presence of a chemical catalyst (e.g. mineral acids)
[c]When carried out in alkaline medium

selectivity for delignification, long pretreatment times is are needed, extra care must be taken with biological contamination and some polysaccharide may be lost together with lignin (Silveira et al. 2015a).

Lignin removal is the probably the ideal way to produce substrates with high glucan contents and high susceptibility to enzymatic hydrolysis. Alkaline washing has been the most widely pretreatment method used for this purpose (Silveira et al. 2015a). Developments with sodium hydroxide and lime have led to pretreatment processes with a relatively low capital cost, low inhibitors formation and high glucose yields after enzymatic hydrolysis (Wyman et al. 2005; Cheng et al. 2010; Aita et al. 2011). However, such alkaline pretreatments complicate the recovery of lignin in useful form and generate substrates that need to be washed extensively until the pH is adjusted for enzymatic hydrolysis and/or fermentation. Also, there is a need for recycling chemicals in a similar way to what is currently done in *kraft* mills for pulp and paper. By contrast, other alkaline pretreatment processes are ready for industrial application and these may have a strong impact on the viability of biorefineries. One of such methods is named AFEX for ammonia fiber expansion. This pretreatment increases biomass surface area, produces a very low amount of inhibitors (if any), modifies the crystalline structure of cellulose, and generates a lignin stream with great potential for its conversion to fuels, chemicals, and materials. Another advantage is the easiness with which ammonia can be recovered and recycled. However, AFEX requires the use of large amounts of ammonia at high pressures as well as the use of expensive unit operations such as an ammonia compressor that increases the total capital cost for implementation. For this reason, AFEX is still not a fully deployable commercial technology. Nevertheless, a commercial Dupont Cellulosic Ethanol unit in USA utilizes a mild alkaline pretreatment process based on dilute ammonia (NREL 2015).

Acids have been often applied for biomass pretreatment and these generate highly accessible substrates for hydrolysis on the basis of hemicellulose removal (Palmqvist and Hahn-Hägerdal 2000). However, depending on the pretreatment severity, high polysaccharide losses are achieved with the subsequent formation of furan compounds that are inhibitory to hydrolysis and fermentation (Ramos 2003). Pretreatment technologies based on acid hydrolysis are leading the first pilot scale trials and demonstration plants for cellulosic ethanol production but other technologies such as AFEX represent promising solutions to debottleneck this important step of the production process. The use of concentrated mineral acids does not require enzymes for hydrolysis but severe detoxification treatments are required for optimal fermentation (Von Sivers and Zacchi 1995). Besides, corrosion and the inevitable formation of hazardous residues are immediate consequences of this pretreatment technology. As a result, high capital and operational costs are usually attached to the development of such acid pretreatment. For this reason, dilute acid pretreatments have been intensively studied to lower these capital costs (Vancov and McIntosh 2012). With this process, enzymatic hydrolysis is facilitated and less chemicals are required in the process, producing less inhibitory compounds and increasing the sustainability of the overall pretreatment process.

In order to mitigate capital costs and reduce the use of harsh chemicals, auto-catalytic processes (hot water extraction, auto-hydrolysis, and hydrothermal pre-treatments) have emerged as the most promising alternative technologies for biomass pretreatment (Ehara and Saka 2005). However, to compensate the absence of chemical catalysts, higher operating temperatures, and residence times are required for optimal performance. Besides, by performing a simple biomass cooking, higher enzyme loading are usually required for enzymatic hydrolysis and higher amounts of water are consumed compared to some other pretreatment technologies (Knez et al. 2014). Hot water pretreatment is economically attractive because no exogenous catalyst is required and the reactor construction has a relatively low capital cost, such as in the case of horizontal tubular pretreatment vessels. Although it results in high recovery of pentose sugars and in the generation of low inhibitory chemicals, the hot water pretreatment is a water and energy demanding process that is difficult to reach out to commercial scale.

The combination of physical and chemical pretreatment processes is recognized as the most effective strategy to meet the cost and performance requirements of a suitable pretreatment technology (Balat et al. 2008; Li et al. 2011). For this reason, steam explosion is the most widely investigated pretreatment method for biomass deconstruction. In this case, biomass is cooked under high-pressure steam prioritizing hemicellulose removal by acid hydrolysis and this is followed by an adiabatic expansion when the reactor content is released to and subsequently collected from a stainless-steel cyclone. By doing so, sheering effects are promoted that decrease the substrate particle size and improve the rheology of the substrate slurry (Ramos 2003). Other positive aspects of this technology include its ability to remove hemicelluloses and redistribute lignin without impeding the successful enzymatic hydrolysis of the steam-treated substrate at high total solids and relatively low enzyme loadings. Also, the hemicellulose component is recovered in the water soluble C5 stream but mostly in the oligomeric form. However, the addition of an acid catalyst may reduce the accumulation of oligosaccharides while decreasing the temperature and time requirements for optimal pretreatment (Aguiar et al. 2013). On the other hand, being an acid-catalyzed pretreatment process, steam explosion inevitably releases inhibitory compounds from the chemical modification of lignin (mostly phenolic acids) and from the hydrolysis and dehydration of hemicellulose sugars (mostly acetic acid and furan compounds, respectively) and these must be controlled and/or overcome to ensure the achievements of high hydrolysis and/or fermentation yields at the end of the process.

The use of a two-step pretreatment process can maximize sugar recovery by pretreating biomass at low temperature to solubilize hemicellulose sugars in the first step and then subject it to higher temperatures in the second step to improve the cellulose accessibility to enzymes. Even though it may offer advantages such as hydrolysis at low enzyme loadings and high ethanol yields through CF or SSCF technologies, a techno-economic evaluation is necessary to determine the real benefits of including an additional steam explosion step (Galbe and Zacchi 2007). POET-DSM in Emmetsburg (Iowa, USA), Beta Renewables in Crescentino (Italy), Granbio in Alagoas (Brazil) and Abengoa in Hugoton (Kansas, USA) utilize variations of the steam explosion technology (Abengoa 2011; Evans 2014; LuxResearch 2014).

The alkaline version of steam explosion consists in the use of liquid anhydrous ammonia instead of steam. Differently from steam explosion, this technology provides cellulose swelling and reduction of crystallization as well as a sheering effect. Besides, no hemicellulose removal occurs and substrates with high total solids are generated (Harun et al. 2013).

By contrast to chemical pretreatments, where the main goal is to increase the accessibility of cellulose to cellulases via the partial removal of lignin and hemicelluloses, mechanical pretreatments preserve the majority of the biomass components. Mechanical milling has a very high energy demand and this represents the main barrier for its scale-up (Da Silva et al. 2010). As the main effects, milling reduces the substrate crystallinity while increases its accessibility to the enzymes without formation of inhibitory compounds to hydrolysis and fermentation. Nonetheless, because of the high degree of lignin retention, the pretreated solids typically require very high enzyme loadings to achieve effective saccharification of both cellulose and hemicelluloses (Mooney et al. 1998; Batalha et al. 2015). Consequently, a "chemi-mechanical" approach involving the treatment of corn stover with sodium hydroxide prior to mechanical treatment in the form of disk refining has been the subject of a recent study (Chen et al. 2014).

In the same vein, the Canadian research institute FPInnovations has recently developed a proprietary process known as the TMP-Bio process, which converts lignocellulosic biomass into various value-added bioproducts (Yuan et al. 2015). Briefly, the process includes a unique mild biomass treatment technology composed of a mild chemical treatment and mechanical refining followed by enzymatic hydrolysis to produce a sugar solution composed of glucose and xylose and a solid fraction composed primarily of near-native lignin (hydrolysis lignin). Advantages of this process include ease of implementation into the existing thermomechanical and chemi-thermomechanical (TMP and CTMP respectively) pulp mill infrastructure and the ability to ferment the solution without detoxification.

Over the past 2 years, a TMP-Bio pilot plant able to operate at the 200 kg per week scale was successfully started in Pointe-Claire (Quebec, Canada). This pilot plant allows researchers at FPInnovations to conduct enzymatic hydrolysis at greater than 20 wt.% biomass total solids, and to obtain carbohydrate conversion yields exceeding 90 %. Moreover, during a 3-month pilot plant campaign, sugar solutions at concentrations of 120–140 g/L (as well as 1.5 tons of hydrolysis lignin) were produced. Current work is focused on developing novel applications for the hydrolysis lignin fraction.

Most if not all of the pretreatment technologies presented so far require biomass conveying from unpressurized to pressurized environments and this may be rather costly and energy intensive. Hence, the development of simple but effective large-scale equipment to transfer biomass in such conditions is extremely necessary for facilitating the scale-up of pretreatment technologies to a commercial scale.

4 Biomass Feeding for Pretreatment Reactors

Feeding is critical for biomass utilization processes such as cellulosic ethanol production in modern biorefineries. Biomass properties, such as mean size, distribution, shape, density, moisture content, compressibility, and content of abrasive materials can affect adversely the feeding operation hindering the uniform and continuous flow of the feed material (Basu 2010). Plant operation at high pressure leads frequently to biomass feeding failure and further challenges in establishing reliable feeding (Elliot 1989; Cummer and Brown 2002). A variety of conveyors, feeders, and storage vessels have been utilized in this industry to achieve smoothly feeding and accurate feed rate control; biomass feeders such screw feeders, rotary valves, slurry pumping, batch reactors and piston feeders are commonly employed for pressurized reactor systems (Rautalin and Wilen 1992; Berglin et al. 2012).

In this context, screw feeders have been widely used to transfer biomass to pressurized systems. As screw feeders are volumetric devices, pliable and compressible feedstocks such as sugarcane bagasse result in different mass delivered per unit time. Other factors affecting the screw feeder volumetric capacity are screw speed, screw flight diameter, shaft diameter, and the fullness of the screw (Rautalin and Wilen 1992). Although the solid friction on screw flights and casing surface determine the equipment efficiency, frictional effect of abrasive materials present in the biomass leads to equipment erosion. In addition, biomass plug formation (necessary to seal the feeder system) and excessive accumulation inside a section beyond the screw flight result in clogging. Companies such Raizen and Granbio reported difficulties related to feeding due to the utilization of biomass with a high content of abrasive materials (Mizutani 2015; Scharr 2015). For instance, Granbio faced erosion problems in pretreatment equipment systems related to inorganic impurities presented in sugarcane straw. According to the company, debugging and debottlenecking processes are still required in order to identify pretreatment problems and troubleshooting. Raizen reported 30 wt.% nameplate capacity in the first 6 months after its cellulosic ethanol unit start-up as well as pretreatment equipment erosion and corrosion. To be cost effective and technically viable, feeding systems should handle appropriate biomass types (size, shape, moisture content, and presence of impurities, among others), ensure seal, resist back pressure, and avoid bridging and blockage in the feeder. For example, short pitch screws result in better mixing and higher screw fullness (Tsai and Lin 1994). Besides that large internal clearance results in efficiency loss while small clearance leads to jamming and severe mechanical wear between screw flight and casing surface, especially, if abrasive material is present in the biomass. Many adjustments are necessary to adapt screw feeders to modern biorefineries and although screw feeders have worked in pilot scale processing units, they are difficult to handle in industrial operations, being sometimes inefficient and economic unfeasible, particularly for herbaceous feedstocks in pressurized systems (Dai et al. 2012).

Other systems used for transferring pressurized biomass are rotary valves, slurry pumping and batch reactors. Rotary valve transfers material from unpressurized to pressurized pockets located between the rotor and the feeder case, sealing the system and resisting back pressure. The feeder pocket is discharge by blowing with steam (Rautalin and Wilen 1992). Recently, modified rotary valves have been used to convey biomass at high pressures in biorefineries because of processing advantages such as easy flow rate control and less contact between biomass and valve casing, which is achieved by controlling the position of the conical rotor resulting in less erosion and maintenance (IEA 2014). However, disadvantages are associated with incomplete pocket discharge caused by sticky materials, steam leakages by wear and clearance between the rotor and the rotary valve frame (Rautalin and Wilen 1992).

Slurry pumping is a viable method to transfer biomass to a pressurized pre-treatment vessel using commercial off-the-shelf equipment. Biomass feedstocks such as pine wood chips and corn stover particles are mixed with water to form paste-type non-Newtonian slurry at 15 wt.% total solids and pumped up to 150–200 bar (Berglin et al. 2012). However, robust and energy demanding pumps are required as well as large amounts of water to reduce the biomass total solids coming out of pretreatment to around 15 wt.%; nevertheless, erosion in pump and other moving parts may be frequent. Batch reactors operating in parallel can be used to transfer biomass in pressurized system. Such reaction systems require no special equipment and keep the same level of production of a continuous process; however, it requires higher capital costs as mentioned before. Once the biomass is conveyed to the feed system it may not be fed directly into the pretreatment equipment because of the presence of foreign abrasive materials. Biomass may be screened or washed prior feeding and major drawbacks are, however, the high capital cost, energy demand, and utilities consumption that are involved in these unit operations.

5 Enzymatic Hydrolysis of Pretreated Lignocellulose

Once a substrate with high accessibility is generated, the enzymatic hydrolysis takes place using a cellulolytic complex acting in synergy as exposed in Fig. 4. In this case, the main enzymes involved in saccharification are endo-β-(1→4)-glucanases

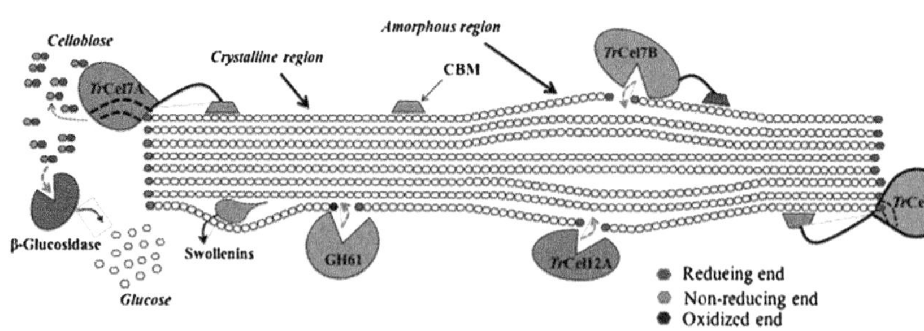

Fig. 4 Enzymes involved in the hydrolysis of cellulose (adapted from Andreaus et al. 2014)

(EnG), the exo-β-(1→4)-glucanases (ExG), or celobioidrolases, and β-(1→4)-glucosidases (βG) (Vinzant et al. 2001). However, other auxiliary enzymes are involved in this process, such as xylanases, pectinases. and feruloil esterases (Decker et al. 2009). Specifically, the cellulose biodegradation starts by the random break down of amorphous cellulose by the action of EnG that leads to the formation of a new reducing and nonreducing chain ends. These ends are the starting loci for ExGs that are able to progressively solubilize cellulose releasing cellobiose as the main reaction product. As a final step, βGs break-down cellobiose to fermentable glucose (Silveira et al. 2014). Furthermore, it is important to mention the contribution of family 61 of glycosyl hydrolases (GH61) that are able to act on crystalline cellulose by an oxidative mechanism promoting amorphogenesis and boosting the enzyme synergy involved in cellulose conversion (Eibinger et al. 2014). In addition, swollenins and expansins are also known as auxiliary proteins for their amorphogenesis effect on the cellulose structure (Arantes and Saddler 2010).

6 Economic Analysis of Cellulosic Ethanol Production

There are several economic analyses in the literature for lignocellulosic ethanol process from different feedstocks and conversion technologies (Wooley et al. 1999; Aden et al. 2002; Eggeman and Elander 2005; Kazi et al. 2010b; Dias et al. 2011; Albarelli et al. 2014). Various studies indicate that feedstock cost and ethanol price are the main economic drivers for cellulosic ethanol projects and both variables are correlated. Other variables that affect the economic feasibility of biomass conversion are pretreatment technology, solid contents of enzymatic hydrolysis and process streams, onsite production or the use of commercial off-the-shelf enzymes, segregated or simultaneous saccharification and co-fermentation, co-generation, wastewater treatment, vinasse concentration, and maturity of different technology scenarios.

Reports from National Renewable Energy Laboratory (NREL) (Wooley et al. 1999; Aden et al. 2002; Kazi et al. 2010a; Humbird et al. 2011) provided process models that included all details of major cellulosic ethanol unit operations. Indeed NREL developed process models that provide a base case for many further techno-economic studies and cost estimates of ethanol production. While several researchers have developed process models for cellulosic ethanol using different methodologies to calculate capital and operational cost, NREL studies are presented here because only results from similar process modeling frameworks can be compared on a consistent basis (Table 2).

Depending on its complexity (technology or multistep process), pretreatment can easily represent one of the most costly unit operation in the cellulosic ethanol biorefinery. According to Table 2, the cost of all four pretreatment scenarios are in the range of 4–29 % of the total installed equipment cost with hot water pretreatment being the lowest and two-stage dilute acid pretreatment being the highest. For

Table 2 Cost ratio (%) of typical biorefinery unit operations for cellulosic ethanol production, considering all equipment already installed

Pretreatment technology	Pretreatment total solids (%)	Hydrolysis and fermentation (%)	Distillation (%)	Waste water treatment (%)	Cogeneration and others (%)	References
Acid-catalyzed steam explosion	19	10	10	8	53	Wooley et al. (1999)
Acid-catalyzed steam explosion	13	13	10	21	43	Humbird et al. (2011)
Hot water extraction	4	19	20	1	56	Kazi et al. (2010b)
Two-stage dilute acid	29	6	17	3	45	Kazi et al. (2010b)
Ammonia fiber extraction (AFEX)	20	15	18	1	46	Kazi et al. (2010b)

instance, acid pretreatment utilization results in effective conversion of hemicellulose into pentose sugars but also formation of inhibitory compounds, equipment corrosion, and high operational and maintenance costs (Wyman 1996). In addition, the dilute acid pretreatment process requires long retention time for detoxification by overliming, which requires large and expensive reaction vessels. Nevertheless, the two-stage dilute acid pretreatment reduces the need of enzymes for saccharification, one of the major operational costs in other pretreatments. Companies like Arkenol (now BlueFire) and Massada Corp utilize concentrated acid pretreatments and no commercial use of dilute acid pretreatment has been noticed in modern biorefineries (Taherzadeh and Karimi 2007; Silveira et al. 2015a).

Many researchers have used simulation process models for cellulosic ethanol production and integration. Wooley et al. (1999) developed a process model using Aspen Plus for cellulosic ethanol production from wood chips using dilute acid pretreatment that provided a base case for many other techno-economic studies. Aden et al. (2002) provided an updated version of the process model developed by Wooley et al. (1999) utilizing co-current dilute acid pretreatment and enzymatic hydrolysis of corn stover; this process model included details of operations, such as feed handling, product recovery, wastewater treatment, and cogeneration from this non-forest agricultural residue.

Sugarcane biorefinery has been extensively studied (Dias et al. 2011; Furlan et al. 2012; Macrelli et al. 2012; Albarelli et al. 2014). Dias et al. (2011) concluded that cellulosic ethanol production competes with electricity generation if sugarcane straw and efficient biomass conversion technologies are available. Furlan et al. (2012) stated that sugarcane biorefinery in Brazil is currently not economically feasible for ethanol production due to the local electricity market prices, even though a sugar mill and cellulosic ethanol biorefinery selling ethanol are closer to feasibility than the conventional sugar mill selling electricity to the grid.

7 Simulation of Processes Integration

This study compares the techno-economic performance of a cellulosic ethanol biorefinery operating as a stand-alone unit and located in a sugarcane straw biomass hub. Process models recently developed (Murphy et al. 2013) using the software Superpro Designer were adapted to reflect the sugarcane straw feedstock composition and different saccharification and fermentation processes were used to estimate the cellulosic ethanol production cost. A brief description of the simulation used and main modeling assumption are described below. Flow diagrams representing the proposed cellulosic ethanol production processes are shown in Figs. 5, 6 and 7.

Sugarcane straw is an abundant agricultural residue in Brazil. A typical variety of sugarcane harvested in Brazil generates around 140 kg dry basis of straw and its composition is show in Fig. 8.

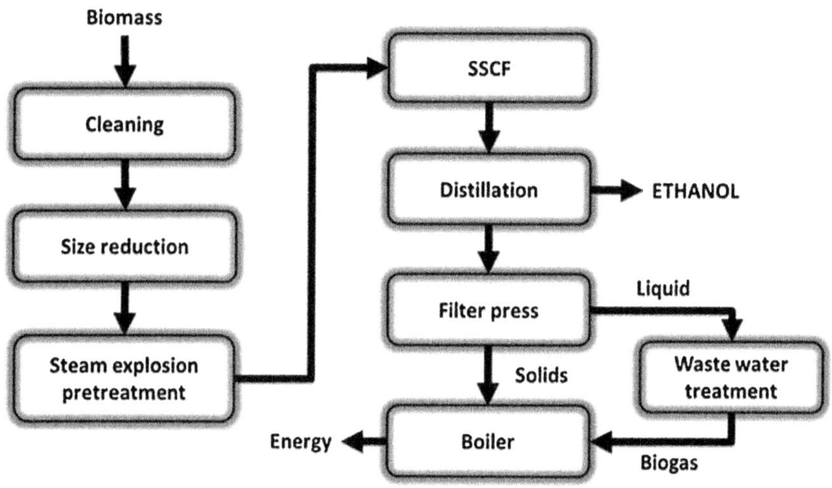

Fig. 5 Flow diagram representing the SSCF (simultaneous saccharification and co-fermentation) cellulosic ethanol production process

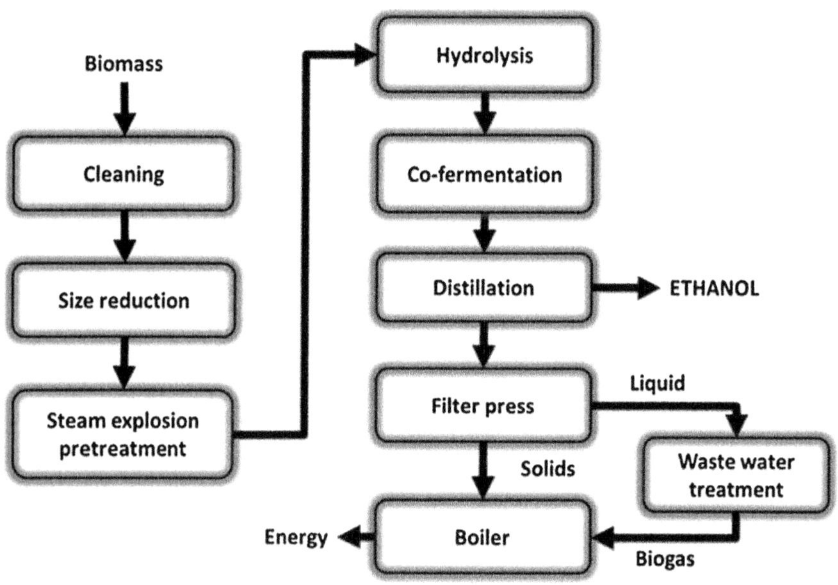

Fig. 6 Flow diagram representing the SHCF (separated hydrolysis and co-fermentation) cellulosic ethanol production process

Sugarcane straw used in the simulation process is initially cleaned in order to remove undesirable biomass fractions, such as sugarcane chunks, trash, and inorganic impurities, such as soil, stones, and other debris. The removal of biomass

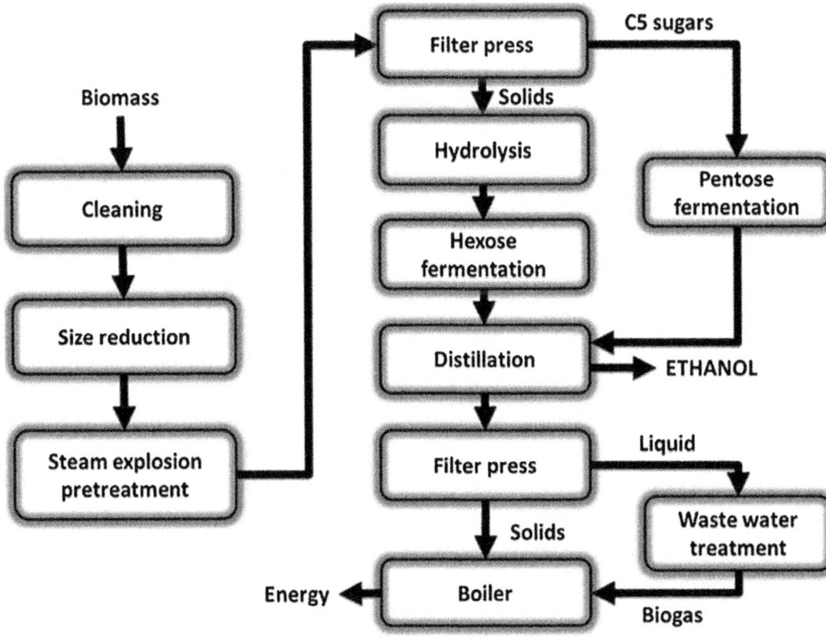

Fig. 7 Flow diagram representing the SHF (separated hydrolysis and fermentation) cellulosic ethanol production process

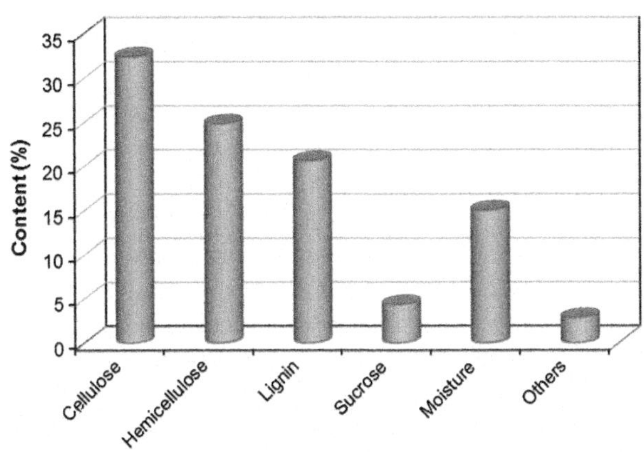

Fig. 8 Sugarcane straw composition (wt%) according to Milanez et al. (2015)

chunks is necessary to select and transfer to the pretreatment unit only the fraction that is effectively useful while inorganic impurities must be removed because they affect the equipment by erosion and abrasion. In the next step, biomass has its size reduced by a shredding machine (knife mill) in order to form a uniform biomass sample.

The pretreatment is performed immediately after the feedstock handling step. As steam explosion is the pretreatment technology most widely used by commercial players today, it has been adopted in this work as well. Kumar and Murthy (2011) developed simulation modeling of dilute acid, dilute alkali and hot water pretreatment methods. High-pressure steam is injected into the pretreatment reactor loaded with the sugarcane straw and pretreated under the desired experimental conditions of pressure and residence time; then, the system is suddenly depressurized to a flash tank. After steam explosion, the unwashed pretreated material may be directly sent to (1) simultaneous saccharification and co-fermentation (SSCF), (2) separated hydrolysis and co-fermentation (SHCF), or (3) to a filter press to remove the pentose liquor (C5 stream) to be fermented in a separated vessel, with the remaining solid stream being hydrolyzed and then fermented to fuel ethanol (hexose sugars only). Enzymatic hydrolysis is adopted in this study because it is favored over acid hydrolysis due to mild operating conditions, high sugar yields, low capital investment and low maintenance cost. Pretreated sugarcane straw was hydrolyzed using commercial enzymes at an enzyme loading of 15 FPU/g cellulose while the engineered yeast *Zymomonas mobilis* was used to ferment hexoses and pentoses into ethanol.

The fermented slurry was stored in beer well and sent to a continuous distillation process consisting of a combination of columns and molecular sieves. The first distillation column separates ethanol as overhead vapors and the bottom effluent containing lignin, proteins and other non-fermentable materials, which is filtered to result in two separate streams: the solids containing lignin are combusted in fluidized bed combustor for steam generation and the liquid stream is treated in the wastewater treatment unit. The ethanol enriched vapor stream is transferred to further downstream processing (rectification and stripper columns, molecular sieves) to produce anhydrous ethanol, which is then denatured by addition of gasoline. In Table 3 the process conditions and efficiencies used in this study are presented in detail (Kumar and Murthy 2011).

Costs of all equipment, utilities, and other consumables were taken from Kumar and Murthy (2011). SuperPro Designer estimates the additional cost of installation, piping, electrical, insulation, design work, and buildings for the industrial facility (direct costs). Other costs, such as engineering costs and construction costs (accounted as indirect costs), contractors' fees, contingency costs, and start-up costs as well as project life and depreciation followed the same methodology proposed by Kumar and Murthy (2011). Equipments added for SHCF (hydrolysis vessels) and SHF (filter belt press, hydrolysis, and fermentation vessels) were designed and had their capital cost estimated by SuperPro Designer. Currency used for this simulation was the US dollar.

Cellulosic ethanol production capacities using 200,000 metric tons/year of sugarcane straw were calculated as 53.2, 53.2 and 43.5 million litres for plants using SSCF, SHCF and SHF, respectively. However, the SHF approach was restricted to hexose fermentation with non-genetically modified microorganisms. Although this is a simple process that allows sugar conversion into ethanol in short

Table 3 Process conditions and efficiencies used in this study for a biomass feed rate of 771.8 dry ton/day (adapted from Kumar and Murthy 2011 and Dias et al. 2011)

Parameters	Value	Unit
Pretreatment		
Temperature	180	C
Pressure	11	bar
Residence time	15	min
Solids loading	30	%
Cellulose to glucose	5	%
Xylan to xylose	70	%
Lignin to soluble lignin	5	%
Xylose to furfural	15	%
Glucose to HMF	15	%
SSCF		
Temperature	35	C
Enzyme loading	15	FPU/g cellulose
Time	5	days
Cellulose to glucose	70	%
Xylan to xylose	80	%
Glucose to ethanol	95	%
Xylose to ethanol	70	%
SHCF and SHF		
Hydrolysis		
Temperature	50	C
Enzyme loading	15	FPU/g cellulose
Time	2	days
Cellulose to glucose	70	%
Xylan to xylose	80	%
Fermentation		
Time SHCF	3	days
Time hexose SHF	8	hr
Time pentose SHF	3	days
Glucose to ethanol*	90	%
Xylose to ethanol*	80	%
Ethanol recovery	98.8	%

*For SHCF fermentation yields were the same as SSCF

periods of time using yeasts such as *Saccharomyces cerevisiae*, pentose sugars (C5) are not converted to any value added-products. Hence, the C5 stream should be sold as by-product but this scenario was not considered in the SHF approach adopted in this work. SSCF and SHCF consisted in hexose and pentose co-fermentation resulting in higher ethanol outputs.

Ethanol produced from SSCF, SHCF, and SHF processes resulted in capital cost estimations of $1.66, $1.75, and $2.23 per L of cellulosic ethanol produced. Kumar and Murthy (2011) obtained $1.71 per L using SSCF and tall fescue as feedstock.

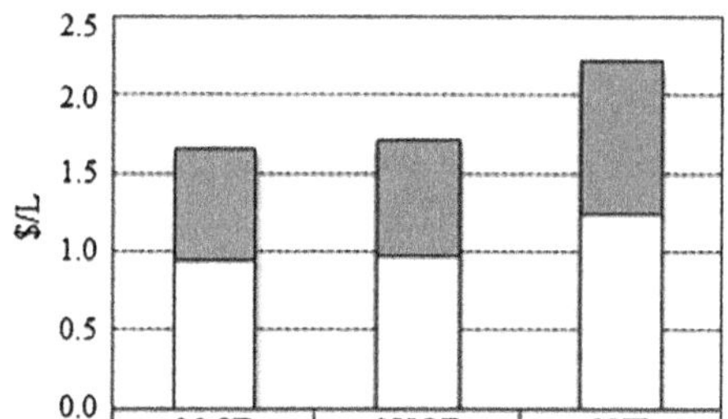

	SSCF	SHCF	SHF
OPEX (S/L)	0.86	0.90	1.18
Other Costs (S/L)	0.72	0.74	0.98
Equipment Cost (S/L)	0.94	0.97	1.24

Fig. 9 Capital cost (equipment and other costs) and OPEX for different cellulosic ethanol production processes

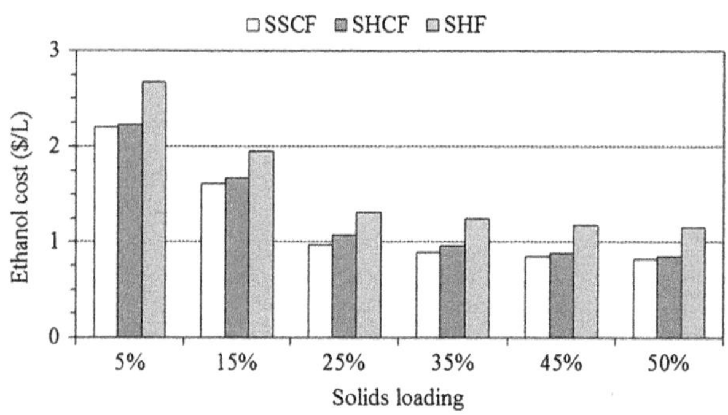

Fig. 10 Effect of solids loading on the cost of different cellulosic ethanol production processes

SHCF required additional hydrolysis reactors compared with SSCF and this resulted in higher capital expenditure. SHF requires much higher capital cost due to the need of a filter press to separate the pentose liquor from the pretreated solid stream; thus, an additional hydrolysis reactor is necessary to convert cellulose into glucose as well extra fermentation vessel to ferment pentoses.

The pentose liquor stream may require detoxification but this was not considered in this study because non-catalyzed steam explosion was used for hemicellulose solubilization. In the SHF process, both yeast and enzymes can work at their optimal temperature and biological contamination is mitigated. In addition, the conventional yeast *Saccharomyces cerevisiae* can ferment hexoses and minimize risks. However, SHF is an intensive capital cost process and the investment required to carry it out may not compensate the risks. The break-down of installed equipment costs and the operational expenditure (OPEX) are shown in Fig. 9.

A sensitivity analysis was performed for the biomass feeding (Fig. 10). As described previously, this is an important aspect of modern biorefineries as many companies are reporting technical complications when attempting to feed biomass into the pretreatment vessel. While Kumar and Murthy (2011) modeled pretreatment at 30 wt.% total solids, the sensitivity analysis on ethanol cost was performed in this study by changing the pretreatment feeding from 5 to 50 wt.% total solids. As biomass is washed to remove trash and others debris, equipment erosion is not analyzed in this sensitivity study.

By increasing the total solids from 5 to 25 wt.%, the ethanol production cost decreased by 56, 52, and 51 % for processes using SSCF, SHCF, and SHF, respectively (Fig. 9). However, above 25 wt.% total solids, the ethanol production cost reduced at a slower pace. Steam explosion pretreatment reactors operate at solid loading around 35 wt.% and any variation in biomass moisture content can be adjusted directly into the pretreatment system (water addition/drainage). However, solid loadings below 25 wt.% are not usual for this kind of pretreatment and higher capital expenditures are necessary because of the need larger equipment designs. Finally, pumping at 15 wt.% total solids as described before may not be economically viable to be implemented in cellulosic ethanol biorefineries.

Acknowledgments The authors are grateful to the National Research Council of Brazil (CNPq, grants: 551404/2010-8; 311554/2011-3), Araucaria Foundation (grants 3572013-35356 and 2742013-35145) (Brazil) and the INCT Energy and Environment (INCT E&A) for the financial support to carry out this study. M.H.L.S. is also grateful to CAPES (Brazil) for providing a scholarship to carry out his graduate studies.

References

Abengoa (2011) Abengoa bioenergy hybrid of kansas. http://www.kcc.state.ks.us/energy/kwrec_09/presentations/B1_Robb.pdf. Accessed 5 Sept 2015

Aden A, Ruth M, Ibsen K, et al (2002) Lignocellulosic biomass to ethanol process design and economics utilizing co-current dilute acid prehydrolysis and enzymatic hydrolysis for corn stover. Other Inf PBD 1 Jun 2002 Medium: ED; Size: 154 pages. doi:NREL/TP-510–32438

Aguiar RS, Silveira MHL, Pitarelo AP et al (2013) Kinetics of enzyme-catalyzed hydrolysis of steam-exploded sugarcane bagasse. Bioresour Technol 147:416–423. doi:10.1016/j.biortech.2013.08.067

Aita G, Salvi D, Walker MS (2011) Enzyme hydrolysis and ethanol fermentation of dilute ammonia pretreated energy cane. Bioresour Technol 102:4444–4448. doi:10.1016/j.biortech.2010.12.095

Albarelli JQ, Ensinas AV, Silva M (2014) Product diversification to enhance economic viability of second generation ethanol production in Brazil: the case of the sugar and ethanol joint production. Chem Eng Res Des 92:1470–1481. doi:10.1016/j.cherd.2013.11.016

Andreaus J, Olekszyszen DN, Henrique M, Silveira L (2014) Processing of cellulosic textile materials with cellulases. Cellul Other Naturally Occurring Polym 661:11–19

Arantes V, Saddler JN (2010) Access to cellulose limits the efficiency of enzymatic hydrolysis: the role of amorphogenesis. Biotechnol Biofuels 3:4. doi:10.1186/1754-6834-3-4
Balat M, Balat H, Öz C (2008) Progress in bioethanol processing. Prog Energy Combust Sci 34:551–573. doi:10.1016/j.pecs.2007.11.001
Basu P (2010) Biomass handling. In: Basu P (ed) Biomass gasification and pyrolysis: practical design and theory. Elsevier, Amsterdam, pp 269–299
Batalha RAL, Han Q, Jammel H et al (2015) Production of fermentable sugars from sugarcane bagasse by enzymatic hydrolysis after autohydrolysis and mechanical refining. Bioresour Technol 180:97–105. doi:10.1016/j.biortech.2014.12.060
Berglin EJ, Enderlin CW, Schimidt AJ (2012) Review and assessment of commercial vendors/options for feeding and pumping biomass slurries for hydrothermal liquefaction. Pacific Northwest National Laboratory, Springfield
Chen X, Shekiro J, Pschorn T et al (2014) A highly efficient dilute alkali deacetylation and mechanical (disc) refining process for the conversion of renewable biomass to lower cost sugars. Biotechnol Biofuels 7:98. doi:10.1186/1754-6834-7-98
Cheng YS, Zheng Y, Yu CW et al (2010) Evaluation of high solids alkaline pretreatment of rice straw. Appl Biochem Biotechnol 162:1768–1784. doi:10.1007/s12010-010-8958-4
Cummer KR, Brown RC (2002) Ancillary equipment for biomass gasification. Biomass Bioenergy 23:113–128. doi:10.1016/S0961-9534(02)00038-7
D'Almeida MLO (1988) Composição química de materiais lignocelulósicos. Celulose e Papel. In: Composição química de materiais lignocelulósicos. Celulose e Papel, 2nd edn. Instituto de Pesquisas do Estado de São Paulo, São Paulo, Brazil
Da Costa Lopes AM, João KG, Rubik DF et al (2013) Pre-treatment of lignocellulosic biomass using ionic liquids: wheat straw fractionation. Bioresour Technol 142:198–208. doi:10.1016/j.biortech.2013.05.032
Da Silva ASA, Inoue H, Endo T et al (2010) Milling pretreatment of sugarcane bagasse and straw for enzymatic hydrolysis and ethanol fermentation. Bioresour Technol 101:7402–7409. doi:10.1016/j.biortech.2010.05.008
Dai J, Cui H, Grace JR (2012) Biomass feeding for thermochemical reactors. Prog Energy Combust Sci 38:716. doi:10.1016/j.pecs.2012.04.002
Decker SR, Siika-aho M, Viikari L (2009) Enzymatic depolymerization of plant cell wall hemicelluloses. In: Himmel ME (ed) Biomass recalcitrance: deconstructing the plant cell wall for bioenergy. Blackwell Publishing Ltd, Oxford, pp 352–373
Dias MOS, Cunha MP, Jesus CDF et al (2011) Second generation ethanol in Brazil: can it compete with electricity production? Bioresour Technol 102:8964–8971. doi:10.1016/j.biortech.2011.06.098
Dibble DC, Li C, Sun L et al (2011) A facile method for the recovery of ionic liquid and lignin from biomass pretreatment. Green Chem 13:3255–3264. doi:10.1039/C1GC15111H
Eggeman T, Elander RT (2005) Process and economic analysis of pretreatment technologies. Bioresour Technol 96:2019–2025. doi:10.1016/j.biortech.2005.01.017
Ehara K, Saka S (2005) Decomposition behavior of cellulose in supercritical water, subcritical water, and their combined treatments. J Wood Sci 51:148–153. doi:10.1007/s10086-004-0626-2
Eibinger M, Ganner T, Bubner P et al (2014) Cellulose surface degradation by a lytic polysaccharide monooxygenase and its effect on cellulase hydrolytic. J Biol Chem 289:35929–35938. doi:10.1074/jbc.M114.602227
Elliot TC (1989) Standard handbook of power plant engineering, 1st edn. McGraw-Hill, New York
Evans JH (2014) Update on project LIBERTY, the POET-DSM advanced biofuels cellulosic ethanol biorefinery in Emmetsburg Iowa. In: Proceedings of 36th symposium on biotechnology for fuels and chemicals
Fengel D, Wegener G (1989) Wood: chemistry, ultrastructure, reactions. Walter de Gruyter, Berlin
Furlan FF, Costa CBB, Fonseca GDC et al (2012) Assessing the production of first and second generation bioethanol from sugarcane through the integration of global optimization and

process detailed modeling. Comput Chem Eng 43:1–9. doi:10.1016/j.compchemeng.2012.04.002

Galbe M, Zacchi G (2007) Pretreatment of lignocellulosic materials for efficient bioethanol production. Adv Biochem Engin/Biotechnol 108:41–65. doi:10.1007/10_2007_070

Harun S, Balan V, Takriff MS et al (2013) Performance of AFEXTM pretreated rice straw as source of fermentable sugars: the influence of particle size. Biotechnol Biofuels 6:40. doi:10.1186/1754-6834-6-40

Horisawa S, Ando H, Ariga O, Sakuma Y (2015) Direct ethanol production from cellulosic materials by consolidated biological processing using the wood rot fungus *Schizophyllum commune*. Bioresour Technol 197:37–41. doi:10.1016/j.biortech.2015.08.031

Humbird D, Davis R, Tao L et al (2011) Process design and economics for biochemical conversion of lignocellulosic biomass to ethanol process design and economics for biochemical conversion of lignocellulosic biomass to ethanol. National Renewable Energy Laboratory-NREL, Golden, Colorado

IEA (2014) Tecnologia Etanol Celulósico E2G-CTC [Online]. http://www.iea.org/media/technologyplatform/workshops/brazilnov2014/TheTechnologicalBoundaryRDforEthanol2G.pdf. Accessed 10 May 2015

Karimi K, Emtiazi G, Taherzadeh MJ (2006) Ethanol production from dilute-acid pretreated rice straw by simultaneous saccharification and fermentation with *Mucor indicus*, *Rhizopus oryzae*, and *Saccharomyces cerevisiae*. Enzym Microb Technol 40:138–144. doi:10.1016/j.enzmictec.2005.10.046

Kazi FK, Fortman J, Anex R (2010a) Techno-economic analysis of biochemical scenarios for production of cellulosic ethanol. National Renewable Energy Laboratory-NREL, Golden, Colorado

Kazi FK, Fortman JA, Anex RP et al (2010b) Techno-economic comparison of process technologies for biochemical ethanol production from corn stover. Fuel 89:S20–S28. doi:10.1016/j.fuel.2010.01.001

Knez Z, Markočič E, Hrnčič MK et al (2014) High pressure water reforming of biomass for energy and chemicals: a short review. J Supercrit Fluids 96:46–52. doi:10.1016/j.supflu.2014.06.008

Kumar D, Murthy GS (2011) Impact of pretreatment and downstream processing technologies on economics and energy in cellulosic ethanol production. Biotechnol Biofuels 4:27. doi:10.1186/1754-6834-4-27

Laurichesse S, Avérous L (2014) Chemical modification of lignins: towards biobased polymers. Prog Polym Sci 39:1266–1290. doi:10.1016/j.progpolymsci.2013.11.004

Lee JW, Gwak KS, Park JY et al (2007) Biological pretreatment of softwood *Pinus densiflora* by three white rot fungi. J Microbiol 45:485–491

Li C, Cheng G, Balan V et al (2011) Influence of physico-chemical changes on enzymatic digestibility of ionic liquid and AFEX pretreated corn stover. Bioresour Technol 102:6928–6936. doi:10.1016/j.biortech.2011.04.005

LuxResearch (2014) GranBio begins operations at its 22 MGY commercial cellulosic ethanol facility in Brazil. http://www.luxresearchinc.com/content/granbio-begins-operations-its-22-mgy-commercial-cellulosic-ethanol-facility-brazil. Accessed 9 Oct 2015

Macrelli S, Mogensen J, Zacchi G (2012) Techno-economic evaluation of 2nd generation bioethanol production from sugar cane bagasse and leaves integrated with the sugar-based ethanol process. Biotechnol Biofuels 5:22. doi:10.1186/1754-6834-5-22

Matthews JF, Skopec CE, Mason PE et al (2006) Computer simulation studies of microcrystalline cellulose Iβ. Carbohydr Res 341:138–152. doi:10.1016/j.carres.2005.09.028

Milanez AY, Nyko D, Valente MS et al (2015) BNDES Setorial 41:237–294. Available at https://web.bndes.gov.br/bib/jspui/handle/1408/4283

Mizutani P (2015) O Cenário macroeconômico e a busca pela competitividade para as energias renováveis. In: Ethanol Summit. São Paulo, Brazil, July 6–7

Mooney CA, Mansfield SD, Touhy MG, Saddler JN (1998) The effect of initial pore volume and lignin content on the enzymatic hydrolysis of softwoods. Bioresour Technol 64:113–119. doi:10.1016/S0960-8524(97)00181-8

Morais ARC, Mata AC, Bogel-Łukasik R (2014) The CO_2-assisted autohydrolysis of wheat straw. Green Chem 16:238–246. doi:10.1039/C3GC41870G
Murphy L, Bohlin C, Baumann MJ et al (2013) Product inhibition of five *Hypocrea jecorina* cellulases. Enzym Microb Technol 52:163–169. doi:10.1016/j.enzmictec.2013.01.002
NREL (2015) Science central to success of new biofuels projects: DuPont-NREL partnership delivered key innovations for large scale cellulosic ethanol facility in Iowa. http://www.nrel.gov/news/features/2015/16468. Accessed 1 Jan 2015
Palmqvist E, Hahn-Hägerdal B (2000) Fermentation of lignocellulosic hydrolysates. II: inhibitors and mechanisms of inhibition. Bioresour Technol 74:25–33. doi:10.1016/S0960-8524(99)00161-3
Pérez S, Mazeau K (2005) Conformation, structures, and morphologies of cellulose. In: Polysaccharides: structural diversity and functional versatility. Marcel Dekker, New York, pp 41–68
Ramos LP (2003) The chemistry involved in the steam treatment of lignocellulosic materials. Quim Nova 26:863–871
Rautalin A, Wilen C (1992) Feeding biomass into pressure. Technical Research Center of Finland (VTT), Espoo
Scharr A (2015) Os problemas da primeira usina de etanol celulósico do Brasil. In: NovaCana. http://www.novacana.com/n/etanol/2-geracao-celulose/problemas-primeira-usina-etanol-celulosico-brasil-070415/. Accessed 10 Apr 2015
Silveira MHL, Siika-aho M, Kruus K, et al (2014) The essential role of plant cell wall degrading enzymes in the success of biorefineries: current status and future challenges. In: da Silva SS, Kumar AC (eds) Biofuels in Brazil. Springer International Publishing, Berlin, pp 151–172
Silveira MHL, Morais ARC, da Costa Lopes AM et al (2015a) Current pretreatment technologies for the development of cellulosic ethanol and biorefineries. ChemSusChem 8:3366–3390. doi:10.1002/cssc.201500282
Silveira MHL, Vanelli BA, Corazza ML, Ramos LP (2015b) Supercritical carbon dioxide combined with 1-butyl-3-methylimidazolium acetate and ethanol for the pretreatment and enzymatic hydrolysis of sugarcane bagasse. Bioresour Technol 192:389–392. doi:10.1016/j.biortech.2015.05.044
Singh J, Suhag M, Dhaka A (2015) Augmented digestion of lignocellulose by steam explosion, acid and alkaline pretreatment methods: a review. Carbohydr Polym 117:624–631. doi:10.1016/j.carbpol.2014.10.012
Swatloski RP, Spear SK, Holbrey JD, Rogers RD (2002) Dissolution of cellose with ionic liquids. J Am Chem Soc 124:4974–4975. doi:10.1021/ja025790m
Taherzadeh MJ, Karimi K (2007) Enzyme-based hydrolysis processes for ethanol from lignocellulosic materials: a review. BioResources 2:707–738. doi:10.15376/biores.2.4.707-738
Tomás-Pejó E, Oliva JM, Ballesteros M, Olsson L (2008) Comparison of SHF and SSF processes from steam-exploded wheat straw for ethanol production by xylose-fermenting and robust glucose-fermenting Saccharomyces cerevisiae strains. Biotechnol Bioeng 100:1122–1131. doi:10.1002/bit.21849
Tsai W, Lin C (1994) On the mixing of granular materials in a screw feeder. Powder Technol 80:119–126. doi:10.1016/0032-5910(94)02839-7
Vancov T, McIntosh S (2012) Mild acid pretreatment and enzyme saccharification of *Sorghum* bicolor straw. Appl Energy 92:421–428. doi:10.1016/j.apenergy.2011.11.053
Vinzant TB, Adney WS, Decker SR et al (2001) Fingerprinting *Trichoderma reesei* hydrolases in a commercial cellulase preparation. Appl Biochem Biotechnol 91–93:99–107. doi:10.1385/ABAB:91-93:1-9:99
Von Sivers M, Zacchi G (1995) A techno-economical comparison of three processes for the production of ethanol from pine. Bioresour Technol 51:43–52. doi:10.1016/0960-8524(94)00094-H

Wooley R, Ruth M, Sheehan J et al (1999) Lignocellulosic biomass to ethanol process design and economics utilizing co-current dilute acid prehydrolysis and enzymatic hydrolysis current and futuristic scenarios. National Renewable Energy Laboratory—NREL, Golden, Colorado

Wyman CE (1996) Handbook on bioethanol: production and utilization. Taylor & Francis, Washington

Wyman CE, Dale BE, Elander RT et al (2005) Coordinated development of leading biomass pretreatment technologies. Bioresour Technol 96:1959–1966. doi:10.1016/j.biortech.2005.01.010

Yang B, Wyman CE (2012) Pretreatment: the key to unlocking low-cost cellulosic ethanol. Biofuels Bioprod Biorefin 6:246–256. doi:10.1002/bbb

Yuan Z, Browne T, Zhang X (2015) Biomass fractionation process for bioproducts. Patents

Zhu J-Q, Qin L, Li W-C et al (2015) Simultaneous saccharification and co-fermentation of dry diluted acid pretreated corn stover at high dry matter loading: overcoming the inhibitors by non-tolerant yeast. Bioresour Technol 198:39–46. doi:10.1016/j.biortech.2015.08.140

Fungal Enzymatic Degradation of Cellulose

Marie Couturier, Chloé Bennati-Granier, Mateus Barbian Urio, Luiz Pereira Ramos and Jean-Guy Berrin

Abstract In nature, filamentous fungi are potent degraders of cellulose as they are able to produce a high number and broad variety of cellulases with complementary catalytic activities. These enzymes include notably classical glycoside hydrolase activities, i.e., endoglucanases, cellobiohydrolases, and β-glucosidases. Oxidative enzymes are also involved in cellulose deconstruction, such as the newly discovered lytic polysaccharide monooxygenases (LPMOs), and auxiliary nonenzymatic proteins are involved in substrate targeting and loosening. In this chapter, the actions of the enzymatic partners are described, as well as their kinetics and the interactions between cellulases and with non cellulase enzymes (i.e., synergism). Because recalcitrant cellulose is still a challenge to date, strategies to discover new efficient biocatalysts from fungal biodiversity are also presented here.

Keywords Cellulose · Cellulase · Fungi · Sugar oxidation · Saccharification · Synergies · Auxiliary activities

M. Couturier · C. Bennati-Granier · J.-G. Berrin (✉)
INRA, UMR1163 Biodiversité et Biotechnologie Fongiques,
13288 Marseille, France
e-mail: jean-guy.berrin@univ-amu.fr

M. Couturier · C. Bennati-Granier · J.-G. Berrin
Aix Marseille Université, UMR1163 Biodiversité et Biotechnologie Fongiques,
13288 Marseille, France

M. Couturier · C. Bennati-Granier · J.-G. Berrin
Polytech'Marseille, UMR1163 Biodiversité et Biotechnologie Fongiques,
13288 Marseille, France

M.B. Urio · L.P. Ramos
Department of Chemistry, Research Center in Applied Chemistry (CEPESQ),
Federal University of Paraná (UFPR), P. O. Box 19032
Curitiba, PR 81531-990, Brazil

L.P. Ramos
INCT Energy & Environment (INCT E&A),
Federal University of Paraná, Curitiba, Brazil

C.R. Soccol et al. (eds.), *Green Fuels Technology*,
Green Energy and Technology, DOI 10.1007/978-3-319-30205-8_6

1 Introduction

Lignocellulosic biomass is the largest renewable source of carbohydrates on Earth and cellulose is its main component. Cellulose is a homopolymer of β-1,4 linked glucose, organized in linear microfibrils that form very recalcitrant crystalline-like structures. In the plant cell wall, cellulose is tightly intermeshed with the other components, hemicellulose, lignin and pectin, making the whole structure extremely recalcitrant to microbial attack.

In the past decades, the deconstruction of the plant cell wall has become a major challenge for many industrial applications, including production of biofuels, biomaterials, and high value products. In particular, the access to cellulose and its hydrolysis into monomers and oligomers is still a bottleneck that has been mobilizing research efforts.

In nature, microorganisms are potent degraders of lignocellulose which they use as energy source. In particular, filamentous fungi play a key role in recycling nutrients in forest ecosystems. They are extremely well adapted for the degradation of biomass and as such are able to produce a high number and broad variety of enzymes with complementary catalytic activities to degrade cellulose-rich materials (Couturier et al. 2012; Sigoillot et al. 2012). Such enzymes include the classical glycoside hydrolases, namely, endoglucanases, cellobiohydrolases, and β-glucosidases, as well as oxidative enzymes, among which cellobiodehydrogenases and the newly discovered lytic polysaccharide monooxygenases (LPMOs). Filamentous fungi have adopted different strategies to perform efficient degradation of cellulosic biomass.

2 Enzymes Involved in Cellulose Degradation

2.1 The CAZy Classification

Enzymes involved in carbohydrate deconstruction are grouped in the carbohydrate-active enzyme (CAZy) classification based on comparison of their amino acid sequence, three-dimensional structure and catalytic mechanism [www.cazy.org; www.cazypedia.org; (Lombard et al. 2014)]. The CAZy database gathers the enzymes involved in the modification of carbohydrates into several groups, Glycoside Hydrolases (GH) that cleave glycosidic bonds, Glycosyl Transferases (GT) which form new glycosidic bonds, Polysaccharide Lyases (PL) which cleave uronic acid-containing polysaccharide chains, Carbohydrate Esterases (CE) that allow deacylation of polysaccharide chains (Henrissat et al. 1991). Auxiliary Activity (AA) enzymes have been added more recently (Levasseur et al. 2013). Most of AA enzymes are oxidoreductases acting on lignin and carbohydrates and among them four families have been recently described as LPMOs. Finally, Carbohydrate-Binding Modules (CBM) are noncatalytic modules appended to enzymes which are involved in substrate targeting. In October 2015, the CAZy database included 135 GH families, 16 CE families, 13 AA families, and 71 CBM families.

2.2 *The Classical Cellulose-Acting Enzymes*

Historically, a system of three complementary enzymatic activities has been described as being in charge of cellulose degradation: endoglucanases, cellobiohydrolases, and β-glucosidases (For a review see Payne et al. 2015; Fig. 1). They are able to hydrolyze the β-1,4 covalent bonds that connect glucose units in the cellulose chains and act synergistically with different specificities. Accordingly, their structural organization and catalytic mechanisms allow for the accommodation of corresponding substrates.

Endoglucanases (EG, endo-1,4 β D-glucanases, EC 3.2.1.4) randomly cleave β-1,4 bonds in amorphous areas of cellulose chains and generate new reducing and nonreducing ends. They are classified in several CAZy families, namely, GH5, GH6, GH7, GH9, GH12, GH44, GH45, and GH74. Endoglucanases display a variety of structures, such as β jelly roll as *Aspergillus niger* family GH12 *An*EglA (1KS4, Khademi et al. 2002) or $(\beta/\alpha)_8$ barrel as *Trichoderma reesei Tr*Cel5A (3QR3, Lee et al. 2011) as well as two possible catalytic mechanisms with retention of configuration or with inversion of configuration (Davies and Henrissat 1995). However, to accommodate cellulose chains, endo-acting cellulase structures have in common a large cleft containing the catalytic amino acids (Davies and Henrissat 1995).

Cellobiohydrolases (CBHs, cellulose 1,4-β-cellobiosidases, EC 3.2.1.91), are processive enzymes which release cellobiose from either reducing (GH7 CBHs) or nonreducing ends (GH6 CBHs) of cellulose fragments released by endoglucanases. GH6 CBHs display an inverting catalytic mechanism, whereas GH7 CBHs use a

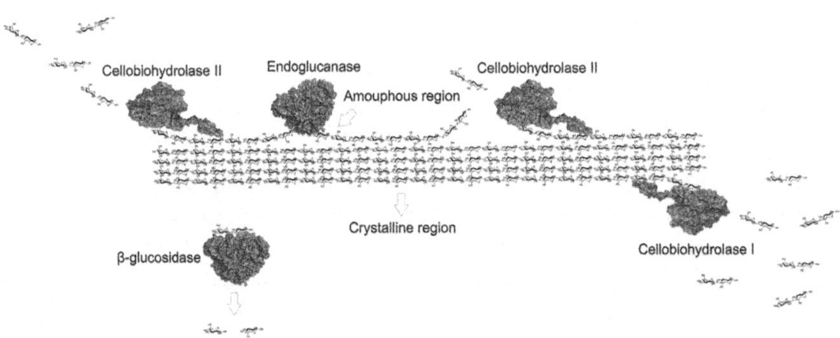

Fig. 1 Illustration depicting the hydrolysis of cellulosic materials using endoglucanases, exoglucanases, and β-glucosidases

retaining mechanism. Three-dimensional structure examples include the Basidiomycetes *Coprinopsis cinerea* Cel6A and Cel6C CBHs (3VOG and 3A64, respectively, Tamura et al. 2012) and *Phanerochaete chrysosporium* *Pc*Cel7D (1GPI, Munoz et al. 2001). CBHs harbor cleft- or tunnel-bearing structures which allow the enzyme to slide on cellulose chain for the next cleavage while the product is being released.

β-glucosidases are the third partner of the cellulase system and catalyze the cleavage of cellobiose or cello-oligomers into glucose. They are characterized by a pocket-containing topology that allows optimal detection of the nonreducing extremity and leads to the cleavage of a single sugar unit. Because of this topology, β-glucosidases are nonprocessive enzymes, since the substrate has to be released after each cleavage event to allow the new glucose unit to exit the pocket. In the CAZy database, β-glucosidases are grouped in families GH1 and GH3. A few fungal β-glucosidase structures have been solved, among which the ones of *T. reesei*, the family GH1 *Tr*Bgl2 (3AHY) and family GH3 *Hj*Cel3A (3ZYZ), the latter being the most abundant β-glucosidase in *T. reesei* enzyme cocktails (Jeng et al. 2011; Karkehabadi et al. 2014). Both exoglucanases and β-glucosidases are strongly inhibited by their reaction products cellobiose and glucose, respectively (Teugjas and Väljamäe 2013).

2.3 Oxidative Enzymes Involved in Cellulose Deconstruction

Complementary to their typical hydrolytic cellulases, fungi have developed oxidative degradation enzymes. These enzymes have been recently identified and described as LPMO enzymes (Quinlan et al. 2011; Vaaje-Kolstad et al. 2010; Harris et al. 2010). LPMOs are classified into four auxiliary activity (AA) families, AA9 (formerly GH61), AA10 (formerly CBM33), AA11, and AA13 of the Carbohydrate-Active enZyme database (CAZy; http://www.cazy.org; Levasseur et al. 2013). The AA10 family contains mainly enzymes of bacterial and viral origin that cleave cellulose and chitin mostly at the C1 position (Forsberg et al. 2011; Hemsworth et al. 2013). The LPMOs classified in the AA11 and AA13 families, respectively, cleave chitin and starch and share important structural features with the two previously characterized families (Vu et al. 2014a; Leggio et al. 2015; Hemsworth et al. 2014). This section will focus mainly on the AA9 family containing only fungal LPMOs active on lignocellulose although much of what is known about the fungal cellulolytic LPMOs is likely applicable across the LPMO superfamily.

In fungi, these enzymes have first been classified into the GH61 family after one member of the family, *T. reesei* EGL4, was reported displaying a weak endoglucanase activity (Saloheimo et al. 1997; Karlsson et al. 2001). However, there was described as “weak endoglucanases” as the activity was several orders of magnitude lower than what had been observed in other endoglucanases. In 2008, the first reported structure of the *T. reseei* Cel61B (Karkehabadi et al. 2008) suggested another activity for those enzymes. Its structure closely resembles to the CBP21 protein (AA10 formerly CBM33), a chitin-binding protein from the bacterium *Serratia marcescens*. This enzyme was obtained few years earlier and had been proposed to enhance chitin degradation through a non-catalytic mechanism (Vaaje-Kolstad et al. 2005).

In the last few years, GH61 have drawn increasing attention because of their «stimulating» effect on cellulase cocktails for biomass conversion (Harris et al. 2010). Other structure and biochemical study have revealed their oxidative mechanism, described their active site and highlighted some important structural features (Quinlan et al. 2011; Harris et al. 2010; Vu et al. 2014b; Kittl et al. 2012; Beeson et al. 2012; Li et al. 2012; Phillips et al. 2011). In August 2015, family AA9 includes 301 members among which 7 members have had their three-dimensional structure solved (Karkehabadi et al. 2008; Wu et al. 2013; Quinlan et al. 2011; Harris et al. 2010; Borisova et al. 2015; Li et al. 2012). These analyses revealed a structural β-sandwich fold of typically 8–10 β-strands with a flat surface where binding with the substrate occurs mostly via stacking interactions with planar aromatic residues. A type II copper ion exposed at the surface is coordinated a "histidine brace" formed by two highly conserved histidine residues, one of which corresponds to the N-terminal histidine, and one tyrosine (Langston et al. 2011; Li et al. 2012). Fungal AA9 LPMOs are secreted enzymes and can contain post-translational modifications. One of the most unusual is the methylation of the N-terminal histidine at the imidazole Nε. This modification is found only in fungal LPMOs and its role is unclear and still under debate.

The oxidation of glucose units has been described mostly at the C1 or C4 position (Beeson et al. 2012; Phillips et al. 2011; Bennati-Granier et al. 2015; Li et al. 2012; Vu et al. 2014b), but a few studies suggested oxidation of the C6 position as well (Bey et al. 2013; Quinlan et al. 2011). AA9 LPMOs are classified into three groups, depending on their regioselective mode of action: type 1 LPMOs will oxidize at C1 and release soluble oligosaccharides with an aldonic acid at their reducing end; type 2 LPMOs will oxidize at C4 and release ketoaldose at the nonreducing end; and type 3 will oxidize at both C1 and C4 and release a mixture of alodnic acid and ketoaldose. AA9 LPMOs require a reducing cofactor for activity, such as ascorbic acid (Forsberg et al. 2011; Quinlan et al. 2011), fragment of lignins (Dimarogona et al. 2012), or enzymes like the cellobiose dehydrogenase (CDH) (Langston et al. 2011; Bey et al. 2013; Phillips et al. 2011). CDHs and AA9 LPMOs are often cosecreted in fungal cultures (Poidevin et al. 2014; Navarro et al. 2014). A clear indication of the synergy was obtained when it was shown that the combination of *Thermoascus auranticus* AA9 and *Humicola isolens* CDH greatly enhanced cellulose degradation (Langston et al. 2011).

Although no structural complex with their substrates is available, binding of the substrate may occur via aromatic-carbohydrate interactions. Indeed, some aromatic residues on the substrate-binding surface are conserved and the spacing matches the spacing between glucose subunits in cellulose (Harris et al. 2010; Li et al. 2012; Wu et al. 2013). Some structural differences have been observed among the different

AA9 LPMOs characterized. More AA9 LPMO members need to be characterized in order to identify the molecular determinants involved in their substrate specificity. For instance, two AA9 LPMOs have been recently shown to act on soluble cello-oligosaccharides, i.e., *Nc*LPMO9c and *Pa*LPMO9H (Isaksen et al. 2014; Bennati-Granier et al. 2015).

2.4 Ancillary Proteins

2.4.1 Carbohydrate-Binding Modules

Cellulolytic enzymes can be associated with non-catalytic modules among which the CBM are an important group (for an extensive review, see Várnai et al. 2014). CBMs play a role for substrate targeting and binding and often increase the overall catalytic activity of the enzyme especially on crystalline substrates. Based on their topology, CBMs have been grouped in three structural and functional groups by Boraston et al. (2004): type-A or surface-binding CBMs, type-B or glycan-chain binding CBMs, and type-C, or small sugar binding CBMs. Cellulose-acting enzymes are typically associated with type-A CBMs, which present a flat surface exposing aromatic residues allowing the interaction with cellulose chains. Another classification of CBMs is found in the CAZy database, in which CBMs are classified based on structure and binding specificity (www.cazy.org; Lombard et al. 2014). Among the 71 CBM families, three families gather CBMs identified in fungal cellulases that show binding to cellulose: CBM1, CBM6, and CBM63. CBM1 family comprises most of the modules associated with fungal cellulose-acting enzymes. These CBMs are approximately 40 residues long, and can be located either at the N- or C-terminus of the catalytic module, alone or in a multimodular organization (Guillén et al. 2010).

In fungal genomes, the number of identified CBM1 modules varies, from none in some brown-rot fungi such as *Postia placenta* and *Fomitopsis pinicola* to more than 30 in some white-rot species such as *Phanerochaete chrysosporium* and *Bjerkandera adusta*. In white-rot fungi, the distribution of CBM1s among the different families of cellulases is heterogeneous, with some families such as GH7 CBH and AA9 LPMOs being often found as single modules, whereas GH5 endoglucanase and GH6 CBH are associated with CBM1s. In brown-rot fungi, the cellulose degradation system does not rely on cellulases and accordingly the number of associated CBM1s is also smaller, with most GH5 endoglucanases and AA9 LPMOs being found as single domains.

2.4.2 Expansins

Expansins are another type of non-catalytic proteins that can play a role in cellulose degradation (For a recent review, see Liu et al. 2015). The presumed mechanism is a disruption of hydrogen bonding between cellulose microfibrils or between cellulose and other cell wall polysaccharides leading to an enhanced accessibility of cellulases to cellulose chains (Saloheimo et al. 2002; McQueen-Mason and Cosgrove 1994). The expansin-like protein from *T. reesei*, *Tr*Swo1, revealed a capability for disruption of cellulose fibers in cotton or filter paper without yielding any detectable reducing sugars (Saloheimo et al. 2002). The resolution of its structure revealed that *Tr*Swo1 has a CBM1 N-terminal, a linker region, and an expansin-like C-terminal domain (20 % identity), which in expansins are similar in structure and sequence to the catalytic site of family GH45 (Saloheimo et al. 2002). More recently, Andberg et al. (2015) proved the hydrolytic activity of *Tr*Swo1. The mode of action of this enzyme is similar to both endo- and exoglucanases. Hence, *Tr*Swo1 could reduce the viscosity of reaction environments containing barley β-glucan, hydroxyethyl cellulose, and carboxymethyl cellulose (or typical endoglucanases substrates) at a consistency of 1 %. On the other hand, when the composition of barley β-glucan hydrolysates were investigated, cellobiose was the main reaction product with no evidence for intermediates, while for a typical endoglucanase (Cel5A), cellopentaose and cellohexaose were predominantly released. *Tr*Swo1 presented a limited activity on barley β-glucan, since only 1.2 % of dry mass was solubilized in either 15 s or 24 h of hydrolysis. It was suggested that the *Tr*Swo1 mode of action involved an initial attack in the middle of cellulose chain and a subsequent processive action along the chain releasing cellobiose. Hydrolysis of barley β-glucan was probably stopped when *Tr*Swo1 came across a β-(1→3) glycosidic bond and possibly stayed permanently bound at this substrate site. Expansin-related proteins have been identified in both Basidiomycetes and Ascomycetes and a few have been characterized from *Aspergillus fumigatus* (Chen et al. 2010), *B. adusta* (Quiroz-Castañeda et al. 2011), *Schizophyllum commune* (Tovar-Herrera et al. 2015). Studies have investigated the activity of expansins and expansin-like proteins in cellulase cocktails and they revealed that expansins enhance cellulose degradation. For instance, Gourlay et al. (2013) observed that the SWOI addition on a steam pretreated corn stover promoted cellulose and hemicellulose solubilisation primarily to their corresponding oligomers. The authors attributed this nonhydrolytic effect to the release of preexisting oligomers that were bound to the substrate surface, but small concentrations of glucose and xylose were also released in the substrate hydrolysate.

3 Strategies to Improve Cellulose Degradation

3.1 Combination and Synergism of Cellulolytic Enzymes

Although the combination of enzymes from the different families (glycoside hydrolases and oxidases) is theoretically enough to carry out complete conversion of cellulose into monomers, complex kinetics, cellulose crystallinity as well as product inhibition leads to a limited degradation efficiency in practice. The kinetics of cellulose degradation by the different enzymatic partners are complex and many models have been proposed to understand the activity of cellulases on cellulose over the course of degradation. Bansal et al. (2009), have summarized in a review the sequence of steps involved in cellulose degradation from the adsorption of endoglucanase and cellobiohydrolase onto their substrate to detachment from the chain and hydrolysis of cellobiose into glucose by beta-glucosidases. In this sequence of events, many factors impact kinetics, causing an overall decreasing rate of reaction over time. Nonproductive binding of cellulases on other components of lignocellulosic substrates has been extensively studied, by FPLC (Gao et al. 2014), colorimetric quantification (Guo et al. 2014), or quartz crystal microbalance (Rahikainen et al. 2013) such as nonproductive cellulase binding, enzyme deactivation, and mostly substrate depletion, and product inhibition. Substrate cristallinity is also cited as a factor for decreasing reaction rate, since amorphous regions are hydrolysed first and the more crystalline regions remain as recalcitrant, resulting in an increase of crystalline fraction of cellulose over time (Chen et al. 2007). This model is called the two-phase substrate model and reflects the physical complexity of the cellulose, which affect both accessibility and reactivity. For an extensive review of models and parameters involved in cellulose degradation kinetics, see Bansal et al. (2009).

Cellulolytic enzymes with different specificities exhibit synergistic action on fibers, simultaneous action of multiple enzyme components resulting in a significantly higher hydrolysis yield than the sum of the hydrolysis yields of the individual enzyme components. This phenomenon called synergism has been described more than twenty years ago. Endo–exo and exo-exo synergisms have been distinguished. An example of endo-exo synergy in fungal cellulases occurs between *T. reesei* endoglucanase *Tr*Cel5A (EGII) and its cellobiohydrolase partner *Tr*Cel7A (CBHI) (Medve et al. 1998). Different types of CBHs working together can also lead to a synergetic degradation of cellulose, such as TrCel7A (CBHI) working on the reducing end of cellulose chains and TrCel6A (CBHII) which acts on nonreducing ends. Real-time visualization of crystalline cellulose degradation by *T. reesei* CBHs was performed using high-speed atomic force microscopy (Igarashi et al. 2011). TrCel7A molecules were observed to slide unidirectionally along the crystalline cellulose surface but at one point exhibited collective halting analogous to a traffic jam. Changing the crystalline polymorphic form of cellulose by means of an ammonia treatment increased the apparent number of accessible lanes on the crystalline surface and consequently the number of moving cellulase molecules. Treatment of this bulky crystalline cellulose simultaneously or separately with TrCel6A resulted in a remarkable increase in the proportion of mobile enzyme molecules on the surface.

While endoglucanases increase the available sites for exoglucanases, β-glucosidases decrease the exoglucanase inhibition by converting cellobiose into glucose (Wood 1985). Multiple factors affect the synergy between cellulases. These include the specific activity of the enzymes, the ratio between them, the enzyme loading and the chemical composition and structure of cellulosic substrates.

Cellulolytic glycoside hydrolases also exhibit strong synergism with their oxidative counterparts, LPMOs. This synergy was first described in 2010, when the actual effect of LPMO was still under question (Harris et al. 2010), but their addition to a cellulase cocktail allowed to significantly decrease enzyme loadings to hydrolyse cellulosic biomass. AA9 LPMOs seem to exhibit activity on both amorphous and crystalline cellulose, whereas endoglucanases presents no activity against the latter. This might be the reason why AA9 LPMOs display a high synergy with cellulases and may provide half of the enzymatic loading required for hydrolysis (Harris et al. 2010, 2014).

3.2 *Fungal Accessory Enzymes*

Another type of synergy involving cellulases and noncellulolytic enzymes can increase fungal degradation of cellulose: because of the complex structure of lignocellulose, efficient cellulose degradation also depends on accessory activities to allow access for cellulases to cellulosic fibers.

Cellulose microfibrils are associated with some hemicelluloses which hampers access for cellulolytic enzymes. In particular, whereas arabinoxylan and (1,3)(1,4)-β-glucan do not interact strongly with cellulose (Mikkelsen et al. 2015), other components of the cell wall such as softwood mannan has been suggested being tightly associated with cellulose fibrils (Åkerholm and Salmén 2001). Accordingly, addition of mannanases to *T. reesei* cellulolytic cocktails led to a synergistic effect and an increase of glucose release from nonpretreated softwood in saccharification assays (Couturier et al. 2011). Such synergies have also been described between pectinases and cellulases (Zhang et al. 2013) for the hydrolysis of steam-exploded hemp and confirms the spatial contacts between pectin and cellulose that have been described in primary cell wall in several studies (Cosgrove 2014; Wang et al. 2015). Xylanases (Hu et al. 2011) and a xyloglucanase (Benko et al. 2008) have demonstrated a synergistic effect when employed in combination with cellulases on specific substrates, leading to improved conversion of cellulose compared to cellulases alone. Gao et al. (2011) demonstrated that, by adding endoxylanases to a cellulolytic enzyme system, the glucose release from AFEX (ammonia fiber expansion) pretreated corn stover increased from 56 to 83 % after 24 h hydrolysis. Selig et al. (2008) achieved an 84 % improvement in the enzymatic hydrolysis of hot water pretreated corn stover by adding an endoxylanase, a ferulic acid esterase and an acetyl xylan esterase to a the cellobiohydrolase Cel7A. These authors also observed that the resulting synergistic effect is more evident when low Cel7A loadings are used.

Tabka et al. (2006) studied the effects of adding xylanases, feruloyl esterases, and laccases on the hydrolysis of dilute sulphuric acid impregnated steam-exploded wheat straw. The addition of hemicellulases caused an enhancement in the substrate glucose yield. On the other hand, the addition of laccase promoted the cleavage of covalent bonds in lignin, showing a negative effect that was associated to the inhibition of cellulases by the accumulation of phenolic compounds in the substrate hydrolysates.

4 Perspectives in Cellulose Hydrolysis Through the Exploration of Fungal Biodiversity

The most studied cellulolytic system to date is probably that of the ascomycete fungus *T. reesei*, largely used in industry and engineered for decades to be used in biomass hydrolysis applications. *T. reesei* genome was sequenced in 2008 (Martinez et al. 2008) and revealed a relatively reduced set of GHs in general and of cellulases in particular, with 5 endoglucanase genes, 2 cellobiohydrolase genes, and 15 β-glucosidase genes. Only 3 LPMO genes of family AA9 are also encoded by *T. reesei* genome. Despite the presence of all types of cellulase activities, *T. reesei* enzyme cocktails are not able to achieve a complete degradation of cellulose. In the past few years, the search for novel CAZymes has been expanded to the exploration of fungal strains from tropical forests (Berrin et al. 2012), marine environment (Arfi et al. 2013) and pathogens (Couturier et al. 2012). For example, the maize pathogen *Ustilago maydis* was identified as a good source of enzymes for improvement of *T. reesei* cellulolytic capabilities. *U. maydis* genome revealed one of the smallest sets of genes that encode for CAZymes with only 95 glycoside hydrolases (Kämper et al. 2006), but investigation of its secretome highlighted a significant fraction of putative oxidoreductases that are potentially involved in the depolymerisation of lignocellulose. The authors suggested that *U. maydis* oxidoreductases could participate in the depolymerisation of lignocellulose via the formation of highly reactive oxidants.

Classically, cellulolytic systems of white-rot and brown-rot fungi have been opposed. Cellulose degradation of white-rot mostly rely on GH6, GH7, LPMOs, and numerous CBM1s allowing anchoring of catalytic modules on crystalline cellulose. On the other hand, brown-rot fungi system barely contains cellulases and mostly use nonenzymatic processes based on hydroxyl radical produced by Fenton reaction (Martinez et al. 2009). However, recent work suggests that the separation of brown-rot and white-rot fungi in two distinct groups might have been an oversimplification and that some fungi display intermediate modes of action (Floudas et al. 2012, 2015; Riley et al. 2014).

A wealth of fungal genomics and postgenomics (transcriptomics and secretomics) information has been generated in the last few years. More than 265 fungal genomes (more than 90 corresponding to basidiomycetes) are publically available. These studies constitute a solid basis to identify the main players involved in the degradation of cellulosic biomass through comparative–omic studies. Complex portfolios of fungal enzymes are secreted in response to environment and growth substrates. The study of fungal secretomes from the scope of their different lignocellulosic biomass degradation strategies and lifestyles would facilitate their use in the treatment of lignocellulose as carbon feedstock for biofuel production and further biorefinery processes (Alfaro et al. 2014).

References

Alfaro M, Oguiza JA, Ramírez L et al (2014) Comparative analysis of secretomes in basidiomycete fungi. J Proteomics 102:28–43

Åkerholm M, Salmén L (2001) Interactions between wood polymers studied by dynamic FT-IR spectroscopy. Polymer 42(3):963–969

Andberg M, Penttilä M, Saloheimo M (2015) Swollenin from *Trichoderma reesei* exhibits hydrolytic activity against cellulosic substrates with features of both endoglucanases and cellobiohydrolases. Biores Technol 181:105–113

Arfi Y, Chevret D, Henrissat B et al (2013) Characterization of salt-adapted secreted lignocellulolytic enzymes from the mangrove fungus *Pestalotiopsis* sp. Nat Commun 4:1810

Bansal P, Hall M, Realff MJ et al (2009) Modeling cellulase kinetics on lignocellulosic substrates. Biotechnol Adv 27(6):833–848

Beeson WT, Phillips CM, Cate JH et al (2012) Oxidative cleavage of cellulose by fungal copper-dependent polysaccharide monooxygenases. J Am Chem Soc 134(2):890–892

Benko Z, Siika-aho M, Viikari L et al (2008) Evaluation of the role of xyloglucanase in the enzymatic hydrolysis of lignocellulosic substrates. Enzyme Microb Technol 43(2):109–114

Bennati-Granier C, Garajova S, Champion C et al (2015) Substrate specificity and regioselectivity of fungal AA9 lytic polysaccharide monooxygenases secreted by *Podospora anserina*. Biotechnol Biofuels 8(1):90

Berrin JG, Navarro D, Couturier M et al (2012) Exploring the natural fungal biodiversity of tropical and temperate forests toward improvement of biomass conversion. Appl Environ Microbiol 78(18):6483–6490

Bey M, Zhou S, Poidevin L et al (2013) Cello-oligosaccharide oxidation reveals differences between two lytic polysaccharide monooxygenases (Family GH61) from *Podospora anserina*. Appl Environ Microbiol 79(2):488–496

Boraston AB, Bolam DN, Gilbert HJ et al (2004) Carbohydrate-binding modules: fine-tuning polysaccharide recognition. Biochem J 382:769–781

Borisova AS, Isaksen T, Dimarogona M et al (2015) Structural and functional characterization of a lytic polysaccharide monooxygenase with broad substrate specificity. J Biol Chem 290(38):22955–22969

Chen Y, Stipanovic AJ, Winter WT et al (2007) Effect of digestion by pure cellulases on crystallinity and average chain length for bacterial and microcrystalline celluloses. Cellulose 14:283–293

Chen X, Ishida N, Todaka N et al (2010) Promotion of efficient Saccharification of crystalline cellulose by *Aspergillus fumigatus* Swo1. Appl Environ Microbiol 76(8):2556–2561

Cosgrove DJ (2014) Re-constructing our models of cellulose and primary cell wall assembly. Curr Opin Plant Biol 22:122–131

Couturier M, Haon M, Coutinho PM et al (2011) *Podospora anserina* hemicellulases potentiate the trichoderma reesei secretome for saccharification of lignocellulosic biomass. Appl Environ Microbiol 77(1):237–246

Couturier M, Navarro D, Olivé C et al (2012) Post-genomic analyses of fungal lignocellulosic biomass degradation reveal the unexpected potential of the plant pathogen *Ustilago maydis*. BMC Genom 13(1):57

Davies G, Henrissat B (1995) Structures and mechanisms of glycosyl hydrolases. Structure 3(9):853–859

Dimarogona M, Topakas E, Olsson L et al (2012) Lignin boosts the cellulase performance of a GH-61 enzyme from *Sporotrichum thermophile*. Biores Technol 110:480–487
Floudas D, Binder M, Riley R et al (2012) The paleozoic origin of enzymatic lignin decomposition reconstructed from 31 fungal genomes. Science 336:1715–1719
Floudas D, Held BW, Riley R et al (2015) Evolution of novel wood decay mechanisms in Agaricales revealed by the genome sequences of *Fistulina hepatica* and *Cylindrobasidium torrendii*. Fungal Genet Biol 76:78–92
Forsberg Z, Vaaje-Kolstad G, Westereng B et al (2011) Cleavage of cellulose by a CBM33 protein. Protein Sci 20(9):1479–1483
Gao D, Uppugundla N, Chundawat SP et al (2011) Hemicellulases and auxiliary enzymes for improved conversion of lignocellulosic biomass to monosaccharides. Biotechnol Biofuels 4(1):5
Gao D, Haarmeyer C, Balan V et al (2014) Lignin triggers irreversible cellulase loss during pretreated lignocellulosic biomass saccharification. Biotechnol Biofuels 7(1):175
Gourlay K, Hu J, Arantes V et al (2013) Swollenin aids in the amorphogenesis step during the enzymatic hydrolysis of pretreated biomass. Biores Technol 142:498–503
Guillén D, Sánchez S, Rodríguez-Sanoja R (2010) Carbohydrate-binding domains: multiplicity of biological roles. Appl Microbiol Biotechnol 85:1241–1249
Guo F, Shi W, Sun W et al (2014) Differences in the adsorption of enzymes onto lignins from diverse types of lignocellulosic biomass and the underlying mechanism. Biotechnol Biofuels 7(1):38
Harris PV, Welner D, McFarland KC et al (2010) Stimulation of lignocellulosic biomass hydrolysis by proteins of glycoside hydrolase family 61: structure and function of a large, enigmatic family. Biochem 49(15):3305–3316
Hemsworth GR, Henrissat B, Davies GJ et al (2014) Discovery and characterization of a new family of lytic polysaccharide monooxygenases. Nat Chem Biol 10(2):122–126
Harris PV, Xu F, Kreel NE et al (2014) New enzyme insights drive advances in commercial ethanol production. Curr Opin Chem Biol 19:162–170
Hemsworth GR, Taylor EJ, Kim RQ et al (2013) The copper active site of CBM33 polysaccharide oxygenases. J Am Chem Soc 135(16):6069–6077
Henrissat B, Vegetales M, Grenoble F (1991) A classification of glycosyl hydrolases based sequence similarities amino acid. Biochem J 280:309–316
Hu J, Arantes V, Saddler JN (2011) The enhancement of enzymatic hydrolysis of lignocellulosic substrates by the addition of accessory enzymes such as xylanase: is it an additive or synergistic effect? Biotechnol Biofuels 4(1):36
Igarashi K, Uchihashi T, Koivula A et al (2011) Traffic jams reduce hydrolytic efficiency of cellulase on cellulose surface. Science 333(6047):1279–1282
Isaksen T, Westereng B, Aachmann FL et al (2014) A C4-oxidizing lytic polysaccharide monooxygenase cleaving both cellulose and cello-oligosaccharides. J Biol Chem 289(5): 2632–2642
Jeng WY, Wang NC, Lin MH et al (2011) Structural and functional analysis of three β-glucosidases from bacterium *Clostridium cellulovorans*, fungus *Trichoderma reesei* and termite *Neotermes koshunensis*. J Struct Biol 173(1):46–56
Kämper J, Kahmann R, Bölker M et al (2006) Insights from the genome of the biotrophic fungal plant pathogen *Ustilago maydis*. Nature 444(7115):97–101
Karkehabadi S, Hansson H, Kim S et al (2008) The first structure of a glycoside hydrolase family 61 member, Cel61B from *Hypocrea jecorina*, at 1.6Å Resolution. J Mol Biol 383(1):144–154
Karkehabadi S, Helmich KE, Kaper T et al (2014) Biochemical characterization and crystal structures of a fungal family 3 β-glucosidase, Cel3A from *Hypocrea jecorina*. J Biol Chem 289 (45):31624–31637
Karlsson J, Saloheimo M, Siika-Aho M et al (2001) Homologous expression and characterization of Cel61A of *Trichoderma reesei* (EGIV). Eur J Biochem 268:6498–6507

Khademi S, Zhang D, Swanson SM et al (2002) Determination of the structure of an endoglucanase from *Aspergillus niger* and its mode of inhibition by palladium chloride. Acta Crystallogr D Biol Crystallogr 58(4):660–666

Kittl R, Kracher D, Burgstaller D et al (2012) Production of four *Neurospora crassa* lytic polysaccharide monooxygenases in *Pichia pastoris* monitored by a fluorimetric assay. Biotechnol Biofuels 5(1):79

Langston JA, Shaghasi T, Abbate E et al (2011) Oxidoreductive cellulose depolymerization by the enzymes cellobiose dehydrogenase and glycoside hydrolase 61. Appl Environ Microbiol 77(19):7007–7015

Lee TM, Farrow MF, Arnold FH et al (2011) A structural study of *Hypocrea jecorina* Cel5A. Prot Science 20:1935–1940

Levasseur A, Drula E, Lombard V et al (2013) Expansion of the enzymatic repertoire of the CAZy database to integrate auxiliary redox enzymes. Biotechnol Biofuels 6(1):41

Li X, Beeson WT 4th, Phillips CM et al (2012) Structural basis for substrate targeting and catalysis by fungal polysaccharide monooxygenases. Structure 20(6):1051–1061

Liu X, Ma Y, Zhang M (2015) Research advances in expansins and expansion-like proteins involved in lignocellulose degradation. Biotechnol Lett 37(8):1541–1551

Lo Leggio L, Simmons TJ, Poulsen JC et al (2015) Structure and boosting activity of a starch-degrading lytic polysaccharide monooxygenase. Nat Commun 6:5961

Lombard V, Golaconda Ramulu H, Drula E et al (2014) The carbohydrate-active enzymes database (CAZy) in 2013. Nucleic Acids Res 42(D1):D490–D495

Martinez D, Berka RM, Henrissat B et al (2008) Genome sequencing and analysis of the biomass-degrading fungus *Trichoderma reesei* (syn. Hypocrea jecorina). Nat Biotechnol 26 (5):553–560

Martinez D, Challacombe J, Morgenstern I et al (2009) Genome, transcriptome, and secretome analysis of wood decay fungus *Postia placenta* supports unique mechanisms of lignocellulose conversion. Proc Natl Acad Sci (USA) 106(6):1954–1959

McQueen-Mason S, Cosgrove DJ (1994) Disruption of hydrogen bonding between plant cell wall polymers by proteins that induce wall extension. Proc Natl Acad Sci (USA) 91(14):6574–6578

Medve J, Karlsson J, Lee D et al (1998) Hydrolysis of microcrystalline cellulose by cellobiohydrolase I and endoglucanase II from *Trichoderma reesei*: adsorption, sugar production pattern, and synergism of the enzymes. Biotechnol Bioeng 59(5):621–634

Mikkelsen D, Flanagan BM, Wilson SM et al (2015) Interactions of arabinoxylan and (1,3)(1,4)-β-glucan with cellulose networks. Biomacromol 16(4):1232–1239

Munoz IG, Ubhayasekera W, Henriksson H et al (2001) Family 7 cellobiohydrolases from *Phanerochaete chrysosporium*: crystal structure of the catalytic module of Cel7D (CBH58) at 1.32 Å resolution and homology models of the isozymes. J Mol Biol 314:1097

Navarro D, Rosso MN, Haon M et al (2014) Fast solubilization of recalcitrant cellulosic biomass by the basidiomycete fungus *Laetisaria arvalis* involves successive secretion of oxidative and hydrolytic enzymes. Biotechnol Biofuels 7(1):143

Payne CM, Knott BC, Mayes HB et al (2015) Fungal cellulases. Chem Rev 115(3):1308–1448

Phillips CM, Beeson WT, Cate JH et al (2011) Cellobiose dehydrogenase and a copper-dependent polysaccharide monooxygenase potentiate cellulose degradation by *Neurospora crassa*. ACS Chem Biol 6(12):1399–1406

Poidevin L, Berrin JG, Bennati-Granier C et al (2014) Comparative analyses of *Podospora anserina* secretomes reveal a large array of lignocellulose-active enzymes. Appl Microbiol Biotechnol 98(17):7457–7469

Quinlan RJ, Sweeney MD, Lo Leggio L et al (2011) Insights into the oxidative degradation of cellulose by a copper metalloenzyme that exploits biomass components. Proc Natl Acad Sci USA 108(37):15079–15084

Quiroz-Castañeda RE, Martínez-Anaya C, Cuervo-Soto LI et al (2011) Loosenin, a novel protein with cellulose-disrupting activity from *Bjerkandera adusta*. Microb Cell Fact 10:8

Rahikainen JL, Martin-Sampedro R, Heikkinen H et al (2013) Inhibitory effect of lignin during cellulose bioconversion: the effect of lignin chemistry on non-productive enzyme adsorption. Biores Technol 133:270–278

Riley R, Salamov AA, Brown DW et al (2014) Extensive sampling of basidiomycete genomes demonstrates inadequacy of the white-rot/brown-rot paradigm for wood decay fungi. Proc Natl Acad Sci USA 111(27):9923–9928

Saloheimo M, Nakari-Setälä T, Tenkanen M et al (1997) cDNA cloning of a *Trichoderma reesei* cellulase and demonstration of endoglucanase activity by expression in yeast. Eur J Biochem 249(2):584–591

Saloheimo M, Paloheimo M, Hakola S et al (2002) Swollenin, a *Trichoderma reesei* protein with sequence similarity to the plant expansins, exhibits disruption activity on cellulosic materials. Eur J Biochem 269(17):4202–4211

Selig MJ, Knoshaug EP, Adney WS et al (2008) Synergistic enhancement of cellobiohydrolase performance on pretreated corn stover by addition of xylanase and esterase activities. Biores Technol 99(11):4997–5005

Sigoillot JC, Berrin JG, Bey M et al (2012) Fungal strategies for lignin degradation. In: Jouanin L, Lapierre C (eds) Advances in botanical research. Elsevier, Amsterdam, pp 263–308

Tabka MG, Herpoël-Gimbert I, Monod F et al (2006) Enzymatic saccharification of wheat straw for bioethanol production by a combined cellulase xylanase and feruloyl esterase treatment. Enzyme Microb Technol 39(4):897–902

Tamura M, Miyazaki T, Tanaka Y et al (2012) Comparison of the structural changes in two cellobiohydrolases, *Cc*Cel6A and *Cc*Cel6C, from *Coprinopsis cinerea*—a tweezer-like motion in the structure of *Cc*Cel6C. FEBS J 279:1871–1882

Teugjas H, Väljamäe P (2013) Product inhibition of cellulases studied with 14C-labeled cellulose substrates. Biotechnol Biofuels 6(1):104

Tovar-Herrera OE, Batista-García RA, Sánchez-Carbente Mdel R et al (2015) A novel expansin protein from the white-rot fungus *Schizophyllum commune*. PLoS ONE 10(3):e0122296

Vaaje-Kolstad G, Horn SJ, van Aalten DM et al (2005) The non-catalytic chitin-binding protein CBP21 from *Serratia marcescens* is essential for chitin degradation. J Biol Chem 280 (31):28492–28497

Vaaje-Kolstad G, Westereng B, Horn SJ et al (2010) An oxidative enzyme boosting the enzymatic conversion of recalcitrant polysaccharides. Science 330(6001):219–222

Várnai A, Mäkelä MR, Djajadi DT et al (2014) Carbohydrate-binding modules of fungal cellulases: occurrence in nature, function, and relevance in industrial biomass conversion. Adv Appl Microbiol 88:103–165

Vu V, Beeson WT, Span EA et al (2014a) A family of starch-active polysaccharide monooxygenases. Proc Natl Acad Sci USA 111(38):13822–13827

Vu V, Beeson WT, Phillips CM et al (2014b) Determinants of regioselective hydroxylation in the fungal polysaccharide monooxygenases. J Am Chem Soc 136(2):562–565

Wang T, Park YB, Cosgrove DJ et al (2015) Cellulose-pectin spatial contacts are inherent to never-dried *Arabidopsis thaliana* primary cell walls: evidence from solid-state NMR. Plant Physiol 168(3):871–884

Wood TM (1985) Properties of cellulolytic enzyme systems. Biochem Soc Trans 13(2):407–410

Wu M, Beckham GT, Larsson AM et al (2013) Crystal structure and computational characterization of the lytic polysaccharide monooxygenase GH61D from the basidiomycota fungus *Phanerochaete chrysosporium*. J Biol Chem 288(18):12828–12839

Zhang J, Pakarinen A, Viikari L (2013) Synergy between cellulases and pectinases in the hydrolysis of hemp. Biores Technol 129:302–307

Principles and Challenges Involved in the Enzymatic Hydrolysis of Cellulosic Materials at High Total Solids

Douglas H. Fockink, Mateus B. Urio, Luana M. Chiarello, Jorge H. Sánchez and Luiz Pereira Ramos

Abstract The enzymatic hydrolysis of plant polysaccharides is a key unit operation for the production of cellulosic ethanol. However, hydrolysates with sugar concentrations as high as 180–200 g L^{-1} (18–20 °Brix) must be produced for a successful ethanol fermentation and this can only be achieved at high total solids. With this, a significant decrease in both capital and production costs is achieved. In addition, less water is needed, effluent generation is minimized and the cost of wastewater treatment is reduced. At high total solids, the rheology of biomass slurries exhibits large apparent viscosities and yield stresses that increase with the volume fraction of the insoluble solids, creating both mass and heat transfer limitations at various levels within the body of the fiber suspension. In this context, this chapter reviews the most recent developments in enzymatic hydrolysis for the production of high sugar concentrations using high total solids and low enzyme loadings. This chapter also reviews several strategies to overcome these rheological problems, such as fed-batch feeding and the addition of chemical additives that are able to decrease the effects of extremely high initial viscosities, thus facilitating the substrate liquefaction while decreasing the unproductive binding of enzymes. Lastly, a brief discussion is given about the impact of different impellers setups on the hydrolysis rate, since adequate mixing capacity and low energy consumption are key factors in designing bioreactors for lignocellulose processing.

Keywords Cellulosic materials · Enzymatic hydrolysis · Cellulases · High total solids · Substrate rheology

D.H. Fockink · M.B. Urio · L.M. Chiarello
Department of Chemistry, Research Center in Applied Chemistry (CEPESQ), Federal University of Paraná, Curitiba, PR, Brazil

J.H. Sánchez
Pulp and Paper Research Group, Department of Chemical Engineering, Universidad Pontificia Bolivariana—UPB, P.O. Box 56006, Medellín, Colombia

L.P. Ramos (✉)
Department of Chemistry, Research Center of Applied Chemistry/INCT Energy and Environment (INCT E&A), Federal University of Paraná, Curitiba, PR, Brazil
e-mail: luiz.ramos@ufpr.br

C.R. Soccol et al. (eds.), *Green Fuels Technology*,
Green Energy and Technology, DOI 10.1007/978-3-319-30205-8_7

1 Introduction

Lignocellulose has been considered one of the most important resources for the sustainable production of liquid biofuels such as ethanol (Demirbaş 2001; Himmel et al. 2007). For instance, in the case of energy crops such as sugarcane, the average ethanol production from one autonomous distillery may be boosted by 39 % if the C6 sugars from cane bagasse are used for fermentation, with additional yields being expected from the use of C5 sugars as well (Ramos et al. 2015). However, to achieve this goal, lignocellulosic materials must be submitted to a pretreatment method to open up the structure of the cell wall, therefore, exposing plant polysaccharides to the subsequent steps of hydrolysis and fermentation (Silveira et al. 2015).

The hydrolysis of plant polysaccharides can be carried out by chemical or biochemical routes involving acids or enzymes, respectively. The enzymatic route is preferable because, compared to acid hydrolysis, it can be carried out at milder conditions, usually at 45–50 °C and pH values between 4.8 and 5.2 (Sun and Cheng 2002). Besides, it does not lead to yield losses due to carbohydrate dehydration and present no corrosion problems to the unit operations involved in the process (Balat 2011). On the other hand, the cost of enzymes still remains as the main obstacle for the full economic viability of the overall conversion process (Kim and Kim 2014). Nevertheless, several demonstration plants have been built worldwide and the first commercial units have started their operations within the last few years. While this opens up a new era in this field, studies are still ongoing to develop better pretreatment methods, advanced enzyme systems, and energy efficient process configurations to reduce costs and increase the profitability of this biofuel production process.

As shown in Chapter "Fungal Enzymatic Degradation of Cellulose", three main classes of enzymes are required for hydrolysis: endo-β-(1→4)-glucanases, exo-β-(1→4)-glucanases (or cellobiohydrolases) and β-(1→4)-glucosidases (Ladisch et al. 1983; Sun and Cheng 2002). At first, endoglucanases breakdown linkages in regions of low crystallinity, producing new reducing and nonreducing chain ends. Then, cellobiohydrolases I and II remove mostly cellobiose from reducing and nonreducing chain ends, respectively. Finally, β-glucosidases convert cellobiose and other low molar mass oligosaccharides to glucose while decreasing the end-product inhibition of cellobiohydrolases (Walker and Wilson 1991; Bhat and Bhat 1997; Arantes and Saddler 2010). As a result, a high degree of synergy is required among these enzymes if an efficient enzymatic hydrolysis of β-(1→4)-glucans (mostly cellulose) is to be achieved (Ng et al. 2011).

While β-glucosidases are more active against soluble oligosaccharides, particularly cellobiose (Sørensen et al. 2013), endo and exoglucanases act on cellulosic substrates (e.g., crystalline and amorphous cellulose) and, for this reason, they are called "true" cellulases (Sukharnikov et al. 2012). Their mode of action depends on the progressive adsorption and diffusion over the insoluble substrate surface. Therefore, some cellulases exhibit a carbohydrate-binding module (CBM) in their structures (Mello and Polikarpov 2014). In addition, numerous studies have shown that non-hydrolytic ancillary proteins such as expansin-like proteins and lytic polysaccharide monooxygenases (LPMOs) have an essential role in boosting the enzymatic hydrolysis of cellulose (Saloheimo et al. 2002; Arantes and Saddler 2010; Harris et al. 2010).

Recently, studies on the enzymatic hydrolysis of lignocellulosics have been oriented to the use of high total solids, defined as the initial concentration at which little or no free water is present in the reaction environment (Kristensen et al. 2009b). This way, high sugar and ethanol concentrations can be obtained by enzymatic hydrolysis and fermentation, further improving the distillation efficiency and the productivity of this liquid biofuel (Jørgensen et al. 2007b). However, such option has several consequences in both hydrolysis rates and yields as discussed below.

2 Factors Affecting Enzymatic Hydrolysis

The enzymatic hydrolysis of lignocellulosic materials is limited by several factors and these can be divided in two groups: enzyme-related and substrate-related factors.

2.1 *Enzyme-Related Factors*

Several enzyme-related factors are known to influence the rate and extend of enzymatic hydrolysis. Enzyme inhibition may arise from the accumulation of glucose and cellobiose (end-product inhibition) or from the release pretreatment by-products, such as furan compounds, phenolic acids, aliphatic organic acids, and xylo-oligomers. Also, the thermal and/or shear denaturation of enzymes and their nonproductive adsorption on lignin and lignin-carbohydrate complexes are also highly influential and may partially compromise the overall hydrolysis yield (Ramos et al. 1992; Tengborg et al. 2001b; Xiao et al. 2004; Rosgaard et al. 2007; Wang et al. 2009). Needless to say, all of these effects are even more critical when operating at high total solids.

It is widely known that glucan conversion decreases when enzymatic hydrolysis is carried out at high total solids and this phenomenon has been referred to as "the solids effect" by Kristensen et al. (2009a). Cara et al. (2007) observed this effect during hydrolysis of hydrothermally treated olive tree prunings using 15 FPU (filter paper units) g^{-1} of dry substrate. The conversion at 72 h decreased linearly from 76.2 to 49.9 % when the substrate total solids were increased from 2 to 30 wt.%, respectively. By contrast, the final glucose concentration increased from around 5–60 g L^{-1}, respectively.

Kristensen et al. (2009a) investigated the factors that are responsible for the solids effect, such as changes in substrate chemical composition, end-product inhibition, low availability of water, and cellulase adsorption. This study was conducted with filter paper (a pure cellulose substrate) to avoid the influence of lignin adsorption or lignin-derived inhibitors. Compared to lignin-containing cellulosic materials, similar trends in reaction conversion were observed when hydrolysis was carried out at high total solids.

High levels of glucose and cellobiose accumulation are probably the most important reason for a gradual loss in enzyme performance (Andrić et al. 2010; Puri et al. 2013). For instance, Kristensen et al. (2009a) observed different conversions of 64.5 and 38.6 % after 48 h of enzymatic hydrolysis at 5 and 20 wt.% total solids, respectively. However, when 50 g L^{-1} of glucose were added in the reaction beginning, both reaction systems resulted in similar glucose yields after 48 h of hydrolysis (29.7 and 26.3 % for 5 and 20 wt.% total solids, respectively). These authors suggested that enzymes are inhibited to a similar extent once a certain glucose concentration is reached in the reaction medium.

The water content is also influential at high total solids due to its role as reagent and solvent (Zaccai 2004). Kristensen et al. (2009a) investigated the effect of water-to-enzyme ratio by replacing 25 % of the buffer by oleyl alcohol to keep almost the same sample viscosity and, by doing so, the solids-to-water ratio was increased from 20 to 25 wt.%. After 40 h of hydrolysis, a 5.6 % decrease in total glucose release was observed when the corresponding increase in solids-to-water ratio usually promotes a yield loss of about 12 % or more. Hence, there was no direct correlation between the reduction in water content and the corresponding decrease in reaction conversion.

It is known that the enzyme performance is strongly associated with their adsorption rates onto the surface of the cellulosic material (Kyriacou et al. 1988). In addition, the presence of accumulated end-products such as glucose and cellobiose influences the enzyme adsorption profile (Kumar and Wyman 2008). In this way, the last factor investigated by Kristensen et al. (2009a) was the correlation between different initial total solids and cellulase adsorption, which was determined by changes in the total substrate nitrogen content. A negative linear correlation between the initial total solids and total cellulase adsorption was observed after 24 h of hydrolysis.

The degree of synergy is another important parameter that affects the enzymatic hydrolysis of cellulosic materials. Studies with purified cellulases have shown that synergy is dependent on the relative proportion and activity of key enzyme components as well as on the enzyme-to-substrate ratio and some substrate physical/chemical properties. At high total solids, the supplementation of cellulase preparations with other enzymes such xylanases and β-glucosidases can promote higher levels of synergy among enzymes. Another way to improve the yield of enzymatic hydrolysis at high total solids is the use of higher enzyme dosages but this inevitably reflects in higher production costs (Modenbach and Nokes 2013). Even so, one must keep in mind that these variables do not have a linear correlation; hence, doubling the dosage will not produce the same effect on hydrolysis yields.

On the basis of these factors, the complexity of performing enzymatic hydrolysis at high total solids was clearly evidenced, as well as the need for more detailed optimization studies of such biomass conversion process. Several strategies have been used to improve both substrate and enzyme performances at high total solids. For instance, a number of methods have been developed to reduce enzyme inhibition. Hydrolysis can be carried out while soluble sugars are removed by ultrafiltration or simultaneous saccharification and fermentation (SSF). In the SSF process, the end-product inhibition is decreased because the amount of sugars that is released by enzymatic hydrolysis is immediately fermented to ethanol. However, if a thermotolerant yeast is not available, hydrolysis cannot be carried out in its optimal temperature of 45–50 °C (Antil et al. 2015) and ethanol production may be limited by its inhibitory effect on yeast growth (Bisson 1999; Muller et al. 2007).

2.2 Structural Features

The substrate-related factors refer to properties such as porosity, surface area, particle size, degree of polymerization, hemicellulose content, lignin content, lignin/hemicellulose chemical composition, and crystallinity, as well as to how these propertied change during a time-course hydrolysis. Moreover, by performing enzymatic hydrolysis at high total solids, both mass and heat transfers are compromised as a result of the high viscosity of the fiber suspension, particularly at the early stages of hydrolysis (Yun et al. 2001).

Several authors have suggested that amorphous cellulose, due to its looser molecular organization and larger porosity, is more susceptible to enzymatic hydrolysis than its crystalline form. Enzymatic hydrolysis of cellulose is typically 3–30 times faster for amorphous cellulose compared that of crystalline cellulose (Zhang and Lynd 2004). However, several enzyme complexes are reported to catalyze the hydrolysis of both amorphous and crystalline cellulose to soluble sugars such as glucose and cellobiose (Chandra et al. 2007).

The degree of crystallinity has been considered one of the most important factors to explain the apparent recalcitrance of lignocellulosic materials to bioconversion (Puri 1984; Rivers and Emert 1988; Chang and Holtzapple 2000). When pretreatment is able to reduce cellulose crystallinity and increase the available surface area, it will most certainly have a positive effect on the rate and extent of enzymatic hydrolysis (Zhang and Lynd 2004). For instance, Yoshida et al. (2008) observed that higher crystallinities led to lower hydrolysis yields of cellulosic substrates derived from *Miscanthus sinensis*, indicating that amorphous cellulose hydrolyses faster than crystalline cellulose.

Lignin can also inhibit hydrolysis by blocking the access of cellulases to the cellulose component and by irreversibly binding enzymes by hydrophobic interactions. However, the distribution and composition of lignin is as important as the concentration of lignin in terms of enzyme accessibility and digestibility (Mooney et al. 1998). The lignin distribution over cellulose fibers prevents their swelling and compromises the substrate recognition by the hydrolytic enzymes. Different types of lignin, particularly after pretreatment, present different reactivities and their hydrophobicity is critical for the overall substrate accessibility as well. For instance, guaiacyl lignins from conifers (softwoods) usually have a higher degree of condensation that restricts substrate accessibility and swelling, compared to herbaceous and hardwood lignins (Ramos et al. 1992; Mooney et al. 1998). Therefore, lignin removal often cause a dramatically increase in hydrolysis rate (McMillan 1994; Sun and Cheng 2002).

Delignification and hemicellulose deacetylation remove barriers to enzymatic hydrolysis. Therefore, an effective pretreatment process must result in the complete deacetylation of plant polysaccharides and the reduction of at least 10 % of the biomass lignin content (Chang and Holtzapple 2000). Kim and Holtzapple (2006) studied the effect of these structural features on the enzyme digestibility of corn stover. Oxidative lime pretreatment lowered the acetyl and lignin content to produce substrates with high digestibility, regardless of their crystallinity index.

Cellulosic substrates with low lignin content usually present high accessibilities, therefore requiring less enzyme dosages for optimal hydrolysis. However, a complete delignification of biomass is difficult due to the distribution of highly hydrophobic lignin in the cell wall. Also, depending on the reaction conditions, lignin fragments tend to react with one another and with carbohydrate derivatives to produce compounds of high molecular mass that are detrimental to enzymatic hydrolysis (Balat 2011). To overcome this limitation, proteins (albumin), nonionic surfactants (Tween 80), and polymers (polyethylene glycol) have been added to the reaction mixture in order to minimize the unproductive adsorption of enzymes, particularly those that are caused by hydrophobic interactions with lignin and lignin-carbohydrate complexes (Kaar and Holtzapple 1998; Kim et al. 2003; Qing et al. 2010).

The relationship between substrate concentration and enzyme loading can also affect the rate and extent of hydrolysis of plant polysaccharides (Ramos et al. 1992). High substrate concentrations directly interfere with the efficiency of the process by obstructing the mass transfer phenomena and maximizing the loss of catalytic activity by shear effects. Also, the concentration of hydrolysis products (cellobiose and glucose) is increased in the reaction medium, and this excels the limits of end-product inhibition as mentioned before (Huang and Penner 1991; Penner and Liaw 1994; Sun and Cheng 2002). Ramos et al. (1992) demonstrated that such effects are real at increased substrate concentrations even when the ratio between substrate and enzymes is kept constant. Also, the relationship between enzyme loading and hydrolysis efficiency is not linear but, in general, the greatest the enzyme loading, the highest the reaction conversion up to a limit when substrate saturation is reached. Hence, the use of high enzyme loadings not only results in competition for the most accessible substrate sites but also increases the nonspecific and/or unproductive binding of cellulase enzymes (Xiao et al. 2004).

3 Process-Related Factors Affecting the Enzymatic Hydrolysis at High Total Solids

When operating at high total solids, a significant decrease in both capital and production costs can be achieved by reducing the size of critical equipment (reactors, storage tanks, and distillation column) and minimizing the energy requirement for heating and cooling during distillation (Mohagheghi et al. 1992; Jørgensen et al. 2007a; Roche et al. 2009b; Yang et al. 2011). In addition, less water is needed, effluent generation is minimized and the cost of wastewater treatment is reduced (Modenbach and Nokes 2013).

In general, the distillation is viable if the fermentation broth contains more than 4 wt.% of ethanol. Hence, the concentration of fermentable sugars must be at least 80 g L^{-1}. Considering that most pretreated lignocellulosic materials contain 60 wt.% of glucans, this would require the use of at least 15 wt.% total solids, a 90 % glucan conversion during enzymatic hydrolysis and a final ethanol yield of 95 % after fermentation (Zhao et al. 2013). Economic evaluations suggested that an increase from 5 to 8 wt.% in total solids may reduce the total ethanol production cost by about 20 % (Galbe et al. 2007; Zhao et al. 2011). However, cellulosic slurries are highly hygroscopic and difficult to handle at solids concentrations exceeding 10–15 wt.% (Lynd 1996). Zhang et al. (2010) found that the energy required for mixing slurries of pretreated corn stover increased one order of magnitude when the total solids was increased from 15 to 30 wt.% (from 79.5 to 1009.2 MJ t^{-1} slurry, respectively). Table 1 illustrates the wide variety of operating conditions that have been studied for enzymatic hydrolysis at high total solids.

When pretreated substrates are present at total solids below 4 wt.%, the fibrous materials are suspended in the abundance of free water and the resulting suspension is easy to be mixed and transferred. There is a minimum amount of fiber flocs or fiber network formation at low total solids, and pulps dispersed as single fibers or small fiber aggregates facilitate the even and thorough distribution of enzymes on the fiber surface (Osawa and Schuerch 1963; Nutt et al. 1993; Switzer and Klingenberg 2004). However, once the substrate consistency increases up to 8 wt.%, a greater degree of fiber interactions occurs and this leads to a substantial increase in the strength of the fiber network. At consistencies even higher (20–40 wt.%), the liquid volume is lower than or equal to the interparticle void volume. Dense suspensions are formed and multiple body interactions prevail in these systems (Coussot 2005).

A further complicating aspect of biomass slurries is that biomass can absorb water. As the total solids approach 20 wt.%, the liquid fraction becomes fully absorbed into the biomass and may cause the bulk to become unsaturated (i.e., absence of a free bulk water continuous phase) (Hodge et al. 2009). At this point, portions of the void volume contain air instead of liquid and the biomass now behaves as a wet granular material. Once there is no free water in the system, the apparent viscosity of the mixture increases and both mixing and handling of the fibrous material becomes much more difficult (Modenbach and Nokes 2013). In this case, the enzymes can only reach the inter-floc spaces but not the intra-floc voids

Table 1 Batch strategies for enzymatic hydrolysis at high total solids (adapted from Modenbach and Nokes 2013)

Material	Pretreatment	Total solids (wt.%)	Enzyme loading	Hydrolysis conditions	Glucan conversion (%)	Reference
Olive tree pruning biomass	Liquid hot water	20	15 FPU g^{-1} of dry substrate	72 h at 50 °C and 150 rpm	64	Cara et al. (2007)
		30			50	
	Steam explosion	20			55	
		30			40	
Corn stover	Steam explosion	30	30 FPU g^{-1} glucan	72 h at 50 °C and 150 rpm	60	Yang et al. (2011)
Corn stover	Steam explosion	15	20 FPU g^{-1} of dry substrate	96 h at 50 °C and 220 rpm	75	Lu et al. (2010)
		20			74	
		25			74	
		30			73	
Corn stover	Dilute acid	20	12 FPU g^{-1} glucan	168 h at 48 °C and 4 rpm	77	Knutsen and Liberatore (2010)
Wheat straw	Steam	20	7 FPU g^{-1} of dry substrate	96 h at 50 °C and 6.6 rpm	60	Jørgensen et al. (2007b)
		30			42	
		40			35	
Corn stover	Dilute acid	28	22 FPU g^{-1} glucan	168 h at 45 °C and 130 rpm	73	Hodge et al. (2008)
Sweet sorghum bagasse	Liquid hot water	20	30 FPU g^{-1} glucan	72 h at 50 °C and 100 rpm + 0.175 mL Tween 80 g^{-1} of dry substrate	60	Wang et al. (2012)
Cassava bagasse	Dilute acid	15	20 FPU g^{-1} of dry substrate	72 h at 50 °C	65	Ma et al. (2011)
		20			56	
		25			50	
Sugarcane bagasse	Steam explosion	20	4.5 FPU g^{-1} of dry substrate	72 h at 50 °C and 200 rpm	69	Ramos et al. (2015)

where most of the substrate is located and this affects the enzymatic hydrolysis efficiency considerably (Viamajala et al. 2009).

Mixing problems are very difficult to solve during enzymatic hydrolysis at high total solids. Therefore, reactors with different impellers setups, such as peg mixer, helical, Rushton, plate-and-frame, double-curved-blade, pitched-blade, and anchor have been tested to address some of these problems. For instance, some studies have been conducted at 15 wt.% total solids or higher and the resulting hydrolysis yields were significantly improved by exploiting different reaction designs (Fan et al. 2003; Humbird et al. 2011). Roller bottle reactors have been used to circumvent the challenging rheology problems of thick fiber slurries and these are able to give 2.4 times higher biomass conversion than shake flasks at 30 wt.% total solids under otherwise identical hydrolysis conditions (Mohagheghi et al. 1992; De Bari et al. 2002; Varga et al. 2004; Roche et al. 2009a).

3.1 Mass Transfer Considerations

Water content is essential to the hydrolysis of lignocellulosic materials as it is the medium through which enzymes and reaction products diffuse, as well as being a reactant in the hydrolysis of glycosidic bonds. The enzyme transfer to the fiber surface involves three steps: (i) convection in the liquid phase in which most fibers are dispersed, (ii) dissolution in the water layer surrounding the individual fibers, and (iii) diffusion to the reaction site. According to the Osawa and Schuerch (1963) model (Fig. 1), enzymes are transported to the fiber reaction site by convection across the mobile water layer (d1) and by diffusion across the immobile water layer (d2) immediately surrounding the fibers (Osawa and Schuerch 1963; Bouchard et al. 1995). At low consistency and under the external force of shaking or agitation, the mass transfer resistance associated with forced convection across the d1-layer is negligible, and the diffusion across the d2-layer is then the controlling step.

As the substrate total solids is increased, the mobile layer is gradually reduced and it can be assumed that, at medium consistencies (about 10 wt.%), no mobile water is present anymore. Therefore, the thickness of the immobile water layer is

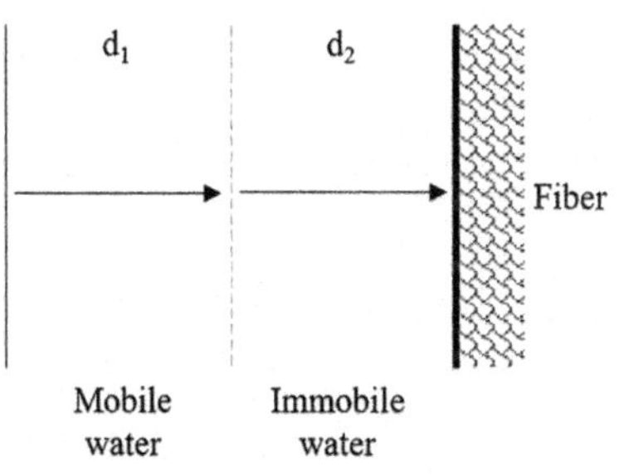

Fig. 1 Mass transfer process model (adapted from Osawa and Schuerch 1963)

controlling the mass transfer rate in accordance to the *film theory*, where the mass transfer rate is inversely proportional to the layer depth, $k_c = D/\delta$. It has been suggested that, in a high-intensity mixing system, fluidization of a fiber suspension effectively sets the d2 layer in motion and changes the environment so that the mass can be transported by convection instead of through the more sluggish diffusion process. In the high consistency range (above 20 %), most of the water is stored within the fiber and only a thin mobile water layer covers the fiber, thus decreasing the enzyme diffusion path length toward the fiber. However, due to the disappearance of the mobile layer, enzymes cannot freely disperse to all fiber sites and the enzyme may end up being concentrated in a smaller area of the fiber aggregates (Reeve and Earl 1986; Laxen et al. 1990; Kappel et al. 1994).

Dispersion of enzymes at high total solids is accomplished by high shear mixing, where the particle-liquid mass transfer depends on the thickness of the fiber suspension as well as on the turbulence structure of the stirring tank reactor (STR). Thus, an increase in biomass total solids also increases the suspension viscosity (see next section), leading to mass transfer and mixing limitations. Unfortunately, mass transfer in pulp suspensions have been examined only for gas-liquid mixing (Rewatkar and Bennington 2000). The convective coefficient k_L for liquid-particle mass transfer in stirred tanks can be calculated by an empirical equation (Eq. 1) given by Pangarkar et al. (2002):

$$k_{\mathrm{L}} = 5.31 \times 10^{-5}\left(\frac{N}{N_{\mathrm{S}}}\right)\left(\frac{\mu_{\mathrm{L}}}{\rho_{\mathrm{L}} D_{\mathrm{m}}}\right)^{-0.53} \quad (1)$$

where N is the impeller speed, N_{S} is the minimum impeller speed for complete suspension of the particles, μ_{L} is the viscosity of the liquid, ρ_{L} is the density of the liquid, and D_{m} is the molecular diffusivity of solute in the liquid. A correlation proposed by Zwietering (1958) for N_{S} is given by Eq. 2:

$$N_{\mathrm{S}} = S v^{0.1} d_{\mathrm{p}}^{0.2}\left(\frac{g\Delta\rho}{\rho_{\mathrm{L}}}\right)^{0.45}\frac{x^{0.13}}{D^{0.85}} \quad (2)$$

where S is a function of type of impeller, v is the kinematic viscosity (m^2/s), d_{p} is de particle diameter (m), g is the gravitational acceleration (m/s^2), x is the particle's weight percentage, and D is the impeller diameter (m).

At high total solids, the suspension can be considered as a porous medium in which the species transport is controlled by diffusion. Accordingly, the effective diffusion of enzymes and products can be determined by Eq. 3:

$$D_{\mathrm{eff}} = \frac{\varepsilon}{\tau} D_{\mathrm{m}} \quad (3)$$

where D_{eff} is effective diffusivity, ε is suspension porosity, and τ is tortuosity given by Eq. 4 (Roberts et al. 2011):

$$\tau = \left(\frac{1 - 0.037}{\varepsilon - 0.037}\right)^{0.661} \quad (4)$$

Thus, as the solid content increases, a decrease in porosity is observed, leading to a decrease in the effective diffusion (Roberts et al. 2011).

3.2 *Rheology of Biomass Slurries*

An efficient conversion during hydrolysis at high total solids requires adequate and uniform distribution of heat and enzymes within the biomass slurry as well as a good mass transport of products into the bulk phase to prevent localized accumulations that could lead to enzyme inhibition (Viamajala et al. 2009). At high total solids, the rheology of biomass slurries exhibit strong non-Newtonian flow properties, with large apparent viscosities and yield stresses that increase with the volume fraction of the insoluble solids (Knutsen and Liberatore 2009; Roche et al. 2009a; Stickel et al. 2009; Ehrhardt et al. 2010; Wiman et al. 2011; Palmqvist et al. 2015). This creates both mass and heat transfer limitations at various levels within the body of the fiber suspension. Also, slurries need to be transported to different unit operations along the process and, at high total solids, these mixing and transport issues become challenging because slurries are too thick and paste-like to be pumped through (Hodge et al. 2008, 2009; Roche et al. 2009b; Viamajala et al. 2009; Samaniuk et al. 2011).

The high viscosity is not only due to the presence of relatively high contents of insoluble materials, but also a result of the high water binding capacity of the cellulosic substrate. Most lignocellulosic materials must undergo a pretreatment process to increase fiber porosity by removing part of the hemicellulose and lignin components, thereby improving the accessibility of the plant polysaccharides to the hydrolytic enzymes (Thompson et al. 1992; Rosgaard et al. 2007).

Empirically, the rheological properties of slurries undergo dynamic and dramatic changes as result of cellulolytic activity because various chemical bonds within the biomass solid phase structure are hydrolyzed while components are solubilized into the liquid phase (Dasari and Berson 2007; Rosgaard et al. 2007; Viamajala et al. 2009; Samaniuk et al. 2011; Palmqvist et al. 2015). However, the unproductive binding of cellulases to the lignin component, as demonstrated for *Trichoderma reesei* cellulases, may increase at high substrate loadings, especially as the hydrolysis proceeds and the amount of cellulose decreases in the reaction mixture (Palonen et al. 2004). Therefore, a fundamental understanding of the rheological properties of highly concentrated biomass slurries needs to be developed and used to design appropriate reaction and pumping systems (Viamajala et al. 2009; Wiman et al. 2011).

Rheology studies were carried out by Viamajala et al. (2009) using steam-exploded corn stover that was pre-impregnated with dilute sulfuric acid.

Rheology measurements were made using plate/plate geometry, and each slurry was subjected to a logarithmic increase in shear rates from 0.1 to 10 s^{-1} to capture the non-thixotropic flow behavior. The flow curves were obtained under continuous shear using total solids ranging from 10 to 40 wt.% for small (20 mesh) and large (80 mesh) particle sizes. The measured apparent viscosities increased with increasing total solids such that the rheograms obtained at these conditions shifted upward in the positive *y*-direction. However, after increasing to a certain level, the curves appear to "stack" on top of each other. For example, in the case of the 80 mesh slurries, significant shifts in the flow curves are seen as the total solids increase from 12.5 to 22.5 wt.% but the curves almost overlap for concentrations between 25 and 32.5 wt.%. Hence, an increase in apparent viscosity is observed with increasing total solids but only up to a certain point after which there is no further increase. These slurries exhibited a pseudoplastic or shear-thinning behavior in the range of shear rates tested in this study.

Others studies with pretreated corn stover slurries also reported a similar pseudoplastic behavior (Pimenova and Hanley 2003, 2004). While the exact mechanism leading to pseudoplasticity in biomass slurries is unknown, Sato (1995) suggested that the substrate particles interact to form a three-dimensional network and the progressive breakdown of this network under shear results in lowering the apparent viscosity and shear-thinning behavior. It is expected that many of the principles of shear-thinning that has been hypothesized for other types of suspensions would also apply to these biomass slurries.

Similar behavior was observed for steam-treated sugarcane bagasse slurries at 10–30 wt.% total solids. Rheology measurements also were made using the plate/plate geometry and through an increase in shear rates. The results are shown in Fig. 2 where the biggest increases in apparent viscosities were observed until 20 wt.% for both substrates produced (authors' unpublished data).

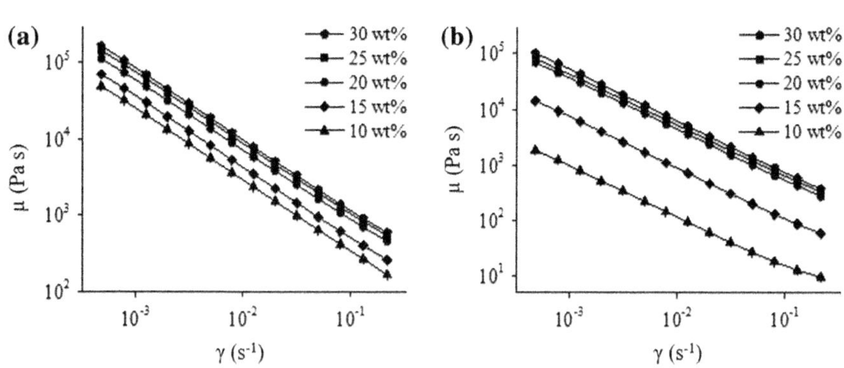

Fig. 2 Apparent viscosity as a function of shear rate for sugarcane bagasse slurries pretreated by **a** autohydrolysis and **b** sulfuric acid-catalyzed steam explosion

3.3 Process Configurations

A favorable strategy to overcome the rheology problems is to use fed-batch, by which the pretreated substrate is fed continuously or by scheduled additions to minimize the nonuniformity of the system. The scheduled addition of the solid substrate has been referred to as multistep fed-batch to differentiate it from the continuous fed-batch (Rudolf et al. 2005; Rosgaard et al. 2007; Liu et al. 2010; Yang et al. 2010; Zhang et al. 2010; Zhao et al. 2013).

Various fed-batch strategies have been applied to the enzymatic hydrolysis and/or SSF of various pretreated substrates to overcome the effects of extremely high initial viscosities, to avoid the unproductive binding of enzymes and to achieve high sugar and/or ethanol concentrations at the end of the process (Borden et al. 2000; Varga et al. 2004; Du et al. 2014). A fed-batch regime also facilitates the substrate liquefaction and this maintains a level of free water to ensure a fast diffusion of enzymes and products (Modenbach and Nokes 2013). Table 2 illustrates the variety of applications using fed-batch configurations.

Zhang et al. (2012) applied the fed-batch approach for the conversion of sugarcane bagasse and wheat straw after alkaline delignification. The pretreated biomass was fed into the reactor at 9, 8, 7, and 6 wt.% total solids over a 48 h reaction course to achieve final total solids of 30 wt.%. However, the total amount of enzymes was loaded together with the first step of substrate addition. For wheat straw, the highest glucan conversion of 60 % was obtained after the first feeding and this was attributed to the low total solids and high enzyme loading at this reaction stage. However, the glucan conversion decreased afterwards together with the enzyme:substrate ratio, reaching a total glucan conversion of only 39 % after 72 h of hydrolysis. A slightly different conversion profile was observed for sugarcane bagasse. For this, the glucan conversion continued to increase over the course of hydrolysis with the exception of the last feeding stage (6 wt.% total solids at 48 h), when a sharp decrease was observed. Nevertheless, the conversion increased again 24 h after the feeding, leading to a final glucan conversion of 55 %. Differences in the effect of pretreatment on the lignocellulose chemistry may have led to differences in glucose yields between the two substrates. Alkaline delignification increased the surface roughness in both cases, allowing for a better access of the enzymes to the glucan (mostly cellulose) component. Sugarcane bagasse produced substrates with a rougher and more fragmented surface than wheat straw.

Ma et al. (2011) used a fed-batch strategy to achieve a 25 wt.% total solids for the hydrolysis of dilute acid pretreated cassava bagasse. However, enzymes were added either all at once at the beginning of the reaction or progressively with each substrate addition. At total solids of 25 %, the batch reaction reached a glucan conversion of about 50 %, whereas the fed-batches with a single enzyme addition and multiple enzyme additions achieved conversions of 75 and 84 %, respectively. These results are similar to those reported in other fed-batch studies (Hodge et al. 2009; Yang et al. 2011), indicating that, under the right conditions, fed-batch

Table 2 Fed-batch strategies for enzymatic hydrolysis at high total solids (adapted from Modenbach and Nokes 2013)

Material	Pretreatment	Solids loading	Enzyme loading	Hydrolysis conditions	Glucan conversion (%)	Reference
Wheat straw	Sodium hydroxide	9 + 8 + 7 + 6 wt.% at 0, 8, 24, 48 h	9.6 FPU g^{-1} of dry substrate[a]	144 h at 50 °C and 120 rpm	35	Zhang et al. (2012)
Sugarcane bagasse					51	
Sweet sorghum bagasse	Liquid hot water	10 + 5 + 5 wt.% at 0, 24, 48 h	30 FPU g^{-1} glucan[b]	120 h at 50 °C and 100 rpm + 0.175 mL Tween 80 g^{-1} of dry substrate	60	Wang et al. (2012)
		15 + 7.5 + 7.5 wt.% at 0, 24, 48 h			54	
Sugarcane bagasse	Sodium hydroxide	12 + 7 + 7 + 7 wt.% at 0, 6, 12, 24 h	10 FPU g^{-1} of dry substrate[a]	120 h at 50 °C and 150 rpm	60	Gao et al. (2011)
		12 + 9 + 7 + 5 wt.% at 0, 6, 12, 24 h			55	
Barley straw	Steam	5 + 5 + 5 wt.% at 0, 6, 24 h	7.5 FPU g^{-1} of glucan[a]	72 h at 50 °C and 57 rpm	65	Rosgaard et al. (2007)
		10 + 5 wt.% at 0, 24 h			68	
Barley straw	Steam	5 + 5 + 5 wt.% at 0, 6, 24 h	7.5 FPU g^{-1} of glucan[b]	72 h at 50 °C and 57 rpm	64	Rosgaard et al. (2007)
		10 + 5 wt.% at 0, 24 h			69	
Cassava bagasse	Dilute acid	10 + 7.5 + 7.5 wt.% at 0, 6, 12 h	20 FPU g^{-1} of dry substrate[a]	72 h at 50 °C	74	Ma et al. (2011)
			20 FPU g^{-1} of dry substrate[b]		84	

[a]Enzyme was added at the reaction beginning
[b]Enzyme was applied with each substrate application

systems may be a plausible solution for achieving higher conversion rates when hydrolysis is performed at high total solids.

Rosgaard et al. (2007) reported different strategies for both batch and fed-batch hydrolysis of steam-treated barley straw, including variations in the sequential addition of substrate as well as substrate plus fresh enzyme. Three reactions with fixed substrate loadings of 5, 10, and 15 wt.% were performed and two reactions starting at 5 and 10 wt.% were supplied with additional substrate after 6 and 24 h ("5 + 5 + 5 wt.%") and 24 h ("10 + 5 wt.%"), respectively, to increase the substrate total solids to a final 15 wt.%. The addition of fresh enzyme with each substrate addition was used to maintain a constant enzyme:substrate ratio throughout the whole reaction time, as opposed to fed-batch feeding schemes where all of the enzyme is added together with the first substrate addition. In these cases, the effective enzyme:substrate ratio decreases with the subsequent addition of fresh substrate. Not surprisingly, the fed-batch schemes that received the full enzyme loading at the beginning produced higher glucose yields during the first few hours as compared to the fed-batch reactions that received fresh enzyme at each substrate addition. As result of these high initial enzyme:substrate ratios, the viscosity decreased in a rate much faster than that of reactions carried out with stepwise enzyme loading. When the final substrate addition was added at 24 h, the subsequent measurement at 48 h showed that the viscosity had increased only slightly in both reactions: 80 and 240 mPa s for the 5 + 5 + 5 wt.% and 10 + 5 wt.%, respectively, compared to 90 mPa s for the 15 wt.% total solids. The viscosity of these samples continued to decrease to levels similar to those obtained with full substrate loading at the beginning of the reaction, which ranged from 20 mPa s (5 wt.%) to 85 mPa s (10 + 5 wt.%). However, the extent of the hydrolysis reaction was not affected by the method of enzyme loading as the final glucose concentrations (62–67 g L^{-1}) were not different for fed-batch reactions that were carried out with and without stepwise enzyme loading.

Lower viscosities are often touted as an advantage of fed-batch systems over batch systems because mixing becomes easier as viscosity decreases. The viscosities of the fed-batch systems reported by Rosgaard et al. (2007) were lower than those of the batch systems but no benefits were observed with regard to glucose production because the batch system at 15 wt.% total solids resulted in higher glucose production (78 g L^{-1}) after 72 h of hydrolysis. The hydrolysis performance of the fed-batch systems was impacted by the stepwise addition of the substrate. There was a decrease in hydrolysis rates and these were never fully recovered, resulting in lower final yields than the corresponding batch systems.

Surfactants and polymers are believed to form a hydrated layer on the lignin surface, presenting a steric hindrance to the unproductive binding of cellulases. As a result, more enzymes are available for cellulose hydrolysis (Eriksson et al. 2002). Polymers and surfactants might also disrupt the lignocellulose structure by removing lignin or amorphous cellulose, reinforcing biomass swelling, and increasing cellulose accessibility (Helle et al. 1993; Kaar and Holtzapple 1998; Li et al. 2012). Besides substrate-related mechanisms, an effect on enzyme stability has been suggested as well. Surfactants and polymers can protect enzymes from thermal denaturation, impede macromolecular aggregation, and help enzyme desorption from strong binding sites (Helle et al. 1993; Kaar and Holtzapple 1998).

Chemical additives can also modify the rheology of biomass slurries. The bulk rheological properties of these suspensions are mainly due to frictional forces between fiber particles and "hooking" between kinked and curled fibers. By generating steric repulsive forces, these additives reduce the surface friction between fibers, mitigate the formation of flocs, and reduce the viscosity of fiber suspension (Beghello and Lindström 1998; Kerekes 2006).

Knutsen and Liberatore (2010) investigated the effect of 18 different chemical additives on slurry rheology and hydrolysis rates for pretreated corn stover. In general, surfactants added to lignocellulosic slurries at 2 wt.%, including cetylpyridinium chloride (CPCl), cetyl trimethylammonium bromide (CTAB), sodium dodecylbenzene sulfonate (NaDBS), and sodium dodecyl sulfonate (SDS), positively affected the rheological properties of the slurry by reducing the viscosity by nearly four-fold compared to the viscosity of the unmodified slurry. Tween 20 reduced the yield stress by 30–40 %. However, slight decreases in the extent of enzymatic hydrolysis were observed but CPCl and CTAB did not affect the resulting hydrolysis rates.

The effects of lignosulfonates (SXSL) and long-chain fatty alcohols (LFAs) on the rheology and enzymatic hydrolysis of high total solids corncob slurries were investigated by Lou et al. (2014). The application of 2.5 wt.% SXSL increased the 72 h substrate enzymatic digestibility from 31.7 to 54.0 % but it also increased the slurry yield stress, making the slurry difficult to stir and pump. n-Octanol ($C_8H_{18}O$) and n-decanol ($C_{10}H_{22}O$) improved the rheological properties of the thick slurry and were able to counteract the negative effect of SXSL. In addition, $C_8H_{18}O$ and $C_{10}H_{22}O$ clearly enhanced the enzymatic hydrolysis of thick corncob slurries regardless of the SXSL addition.

A mechanism was proposed to explain the observed negative effect of SXSL and the positive effect of LFAs on the rheological properties of biomass slurries. Initially, water adsorption by hydrogen bonding creates a hydration layer on the surface of the cellulosic material except on more hydrophobic regions where lignin is located. Lignosulfonates adsorb on the lignified surface by $\pi - \pi$ interactions involving aromatic rings (Deng et al. 2012) and hydrophobic binding (Lou et al. 2013). The sulfonic acid groups of lignosulfonates remain oriented toward the aqueous phase, making the surface hydrophilic. Thus, water adsorption is triggered to form a hydration layer where enzymes do not adsorb unproductively. As the free water in the slurry is reduced, its rheological properties are aggravated with a concomitant increment in yield stress and complex viscosity. On the other hand, lignosulfonate can also be adsorbed on the cellulose surface to disrupt the hydrogen bonding network of thick fiber slurries. The resulting negatively charged substrate particles disperse more easily, therefore improving the rheology of the fiber slurry (Lou et al. 2014).

LFAs are composed of a long alkyl chain and a terminal hydroxyl group that binds strongly to cellulose, hemicellulose, lignin, and lignosulfonate by hydrogen bonding. The alkyl chain in LFAs make the lignocellulose surface more hydrophobic and destroy the hydration layers around cellulose and lignosulfonate, freeing water molecules and improving the rheological properties of thick fiber slurries (Lou et al. 2014).

3.4 *Effect of Different Impellers*

When high total solids are used, the enzymatic hydrolysis of cellulose becomes dramatically slower (Lau and Dale 2009). Although the lower process performance at high total solids is due to several factors, material mixing and mass transfer play a significant role (Stickel et al. 2009). Approximately 12–15 wt.% total solids represent the upper limit at which solids can be mixed effectively in stirred-tank reactors (Hodge et al. 2009). In addition, the use of low enzyme loadings due to cost considerations excels even further the need for a good material mixing. Sufficient mixing capacity, low energy consumption, and advanced cellulase systems are therefore, the key factors for designing bioreactors for lignocellulose processing (Du et al. 2014).

The power consumption for mixing is determined by the rheology of the material (Wiman et al. 2011). The mixing energy requirements during SSF can be as high as 60 % of the energy consumption for producing cellulosic ethanol at high total solids, a figure that would preclude the need for economic feasibility studies of large-scale processes. In fact, this is an area for further improvements through the application of advanced engineering principles (Zhang et al. 2010; Palmqvist et al. 2015).

Mixing clearly affects the rate of enzymatic hydrolysis (Tengborg et al. 2001a). Roche et al. (2009b) verified that the hydrolysis yield increased because the enzymes were more uniformly distributed when the samples were thoroughly mixed. Studies carried out in shake flasks with phosphoric acid impregnated steam-exploded cane bagasse showed that, using 20 wt.% total solids and the same enzyme loading, there was an increase in the final sugar concentration when the agitation was increased from 150 to 200 rpm (Ramos et al. 2015).

The method used for mixing the fiber slurry has a substantial impact on the hydrolysis rate of lignocellulosic materials. For instance, Zhang et al. (2009) observed a significant reduction in the liquefaction time when hydrolysis at high total solids (20 wt.%) were performed in a lab scale peg mixer rather than in a shake flasks. The mixer used in this study was a 9 L reactor fitted with a rotating shaft with pegs extending out radially (Table 3, entry 1). Peg mixers are commonly used in the pulp and paper industry, which may utilize solids loadings up to 35 wt.% (Zhang et al. 2009). Liquefaction occurred after 1 h of hydrolysis in the peg mixer, whereas shake flasks took much longer to reach the same effect. The decrease in liquefaction time was attributed to the effective mixing provided by the peg mixer and the breaking down of the large fiber network that tends to occur at fiber loadings above 8 wt.%. The hydrolysis performed in the peg mixer resulted in 144

Table 3 Different types of impellers studied for use with high-solids enzymatic hydrolysis

Entry	Impeller		Reference
	Type	Format	
1	Peg mixer		Zhang et al. (2009)
2	Helical	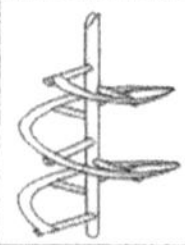	Zhang et al. (2010)
3	Rushton	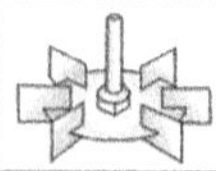	Zhang et al. (2010)
4	Plate-and-frame	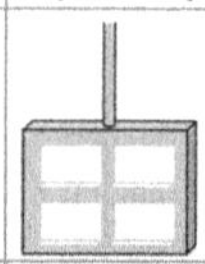	Wang et al. (2012)
5	Double-curved-blade	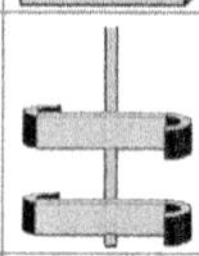	Wang et al. (2012)
6	Pitched-blade		Palmqvist et al. (2015)
7	Anchor		Palmqvist et al. (2015)

and 158 g L^{-1} of glucose from unbleached hardwood and organosolv pretreated poplar, respectively.

The effects of both helical (Table 3, entry 2) and Rushton paddle (Table 3, entry 3) impellers at solids loadings up to 30 wt.% were investigated by Zhang et al. (2010). The helical impeller performed better than the Rushton impeller with regard to every aspect tested. The substrate feeding rate into the reactor was adjusted so that a liquefied slurry could be maintained throughout the feeding period. The helical impeller provided a better mixing because the feeding period was completed more than 2 h sooner than that of the Rushton impeller. The helical impeller also consumed less power and resulted in higher ethanol concentrations (51.0 g L^{-1} vs. 43.9 g L^{-1}) and productivities. At 30 wt.% total solids (prior to inoculation with the fermentative organism), the Rushton impeller required nearly 40 W kg^{-1} corn stover (CS) before decreasing it to ~29 W kg^{-1} CS after 72 h of SSF. The helical impeller required ~8 W kg^{-1} CS and ~1 W kg^{-1} CS prior to inoculation and after 72 h, respectively. Lastly, the mixing efficiency of the helical impeller was better than that of the Rushton impeller.

Other impeller geometries were also tested by Wang et al. (2012). A plate-and-frame impeller (Table 3, entry 4) and a double-curved-blade impeller (Table 3, entry 5) were tested at various speeds and 100 rpm resulted in the best conversion efficiencies for both geometries. However, the plate-and-frame impeller outperformed the double-curved-blade impeller by nearly 18 %, indicating that the geometry of the impeller can have an important effect on hydrolysis efficiency. These authors suggested that the plate-and-frame impeller provides a more consistent mixing regime throughout the deepness of the reaction vessel, whereas the axial flow induced by the double-curved-blade impeller was a function of the actual distance from the blades.

Palmqvist et al. (2015) verified different impeller setups for enzymatic hydrolysis of Norway spruce with the Cellic CTec2 complex from Novozymes. A dual pitched-blade impeller (Table 3, entry 6) was used in a configuration similar to that used in demonstration plants, as well as a wide anchor impeller (Table 3, entry 7) that creates a different flow pattern inside the reactor chamber. The results for three different agitation rates showed that, regardless of impeller configuration, the highest level of agitation (200 rpm) achieved the highest glucan conversion by enzymatic hydrolysis in lab scale.

From the aforementioned reactors used for enzymatic hydrolysis at high total solids, there are several suggestions to improve the mixing of highly viscous fiber slurries. Free-fall mixing relies on gravity to effectively mix the slurry, which consumes less energy than a stirred-tank reactor providing a similar degree of mixing. An effective mixing regime can greatly depend on the impeller geometry, as the shape of an impeller can cause large differences in speed and shear effects at various impeller slurry interfaces throughout the reactor. High shear rates can also disrupt the adsorption of cellulase onto the cellulosic material or to even cause the denaturation of the cellulolytic enzymes (Kaya et al. 1996; Cao and Tan 2004).

The use of thick slurries in a biomass to ethanol conversion process is likely to take place under low shear rates. For example, during dilute acid pretreatment in pilot scale reactors at the National Renewable Energy Laboratory (NREL), biomass is fed to a steam pressurized reactor using a screw feeder operating at maximum speed of 55 rpm (Schell et al. 2003). The downstream enzymatic hydrolysis is also envisioned to be carried out at a low shear rate because the reaction is slow and does not require a continuous vigorous mixing to achieve the mass and heat transfer needed for hydrolysis to succeed (Goto et al. 1986; Sato 1995; Turian et al. 1997; Goudoulas et al. 2003; Pimenova and Hanley 2003; He et al. 2004; Houchin and Hanley 2004; Pimenova and Hanley 2004; Stickel and Powell 2005).

Many studies involving enzymatic hydrolysis at high total solids have embraced the idea of using horizontal rather than vertical reactors. Gravitational or free-fall mixing provides many advantages over typical vertical stirred-tank reactors and are used in other industrial processes that require mixing of highly viscous slurries, such as peanut butter, ketchup, and concrete (Roche et al. 2009a). The horizontal orientation minimizes particle settling and local accumulation of reaction products within the reactor, as well as ensuring a better enzyme distribution in the bulk. These reactors are also easily scalable from bench to both pilot and industrial scales. Power requirements are lower for horizontal reactors equipped with paddles over vertical stirred tanks that provide the same level of effective mixing (Dasari and Berson 2007).

A horizontally oriented rotating drum was utilized for the enzymatic hydrolysis of steam pretreated wheat straw at 40 wt.% total solids and an enzyme loading of 7 FPU g^{-1} of dry substrate. This study found that cellulose and hemicellulose conversion decreased from 90–33 % to 70–35 %, respectively, with the increase in solids loading from 2–40 wt.% but the reactor provided adequate mixing as evidenced by the high glucose concentration (86 g kg^{-1}) (Jørgensen et al. 2007b).

Hydrolysis studies carried out by Dasari and Berson (2007) utilized an 8 L scraped surface bioreactor for the hydrolysis of dilute acid pretreated corn stover. The reactor was constructed from a cylinder made of Pyrex glass with aluminum lids fitted over the ends to be employed for enzymatic hydrolysis at high total solids and to facilitate scale-up studies from laboratory-scale shake flasks. An adjustable speed, rotating shaft with attached rubber-tipped stainless steel blades was inserted into the reactor. Three sampling ports were located along the length of the reactor. Compared to shake flasks, the horizontal reactor was able to increase the glucose yield by approximately 10 % at a 25 wt.% total solids loading.

Du et al. (2014) applied a horizontal rotating reactor (HRR) and a vertical stirred-tank reactor (VSTR) for the saccharification of pretreated corn stover at high total solids. The high initial viscosity of the substrate slurry hindered the effective blending and slowed the yields of enzymatic hydrolysis. To overcome this problem, a fed-batch enzymatic hydrolysis was performed in both HRR and VSTR by adding the substrate and/or enzymes gradually so that the viscosity of the fiber suspension was maintained constant. The glucose concentration was always higher in the HRR than in the VSTR for the same substrate and enzyme feeding strategy.

Enzymatic hydrolysis in demonstration scale was developed by Jørgensen et al. (2007b) using a reactor with a 280 L total volume. Several features were added to the pilot scale drum reactor, as well as to the small scale glass reactor, to address issues associated with high total solids. The horizontal orientation of both reactors takes advantage of free-fall mixing, eliminating the need for mechanical mixing. Evaluation of a range of mixing speeds (3.3–11.5 rpm) resulted in no significant differences in cellulose conversion over the tested range but the energy input for mixing was significantly reduced as compared to vertically oriented stirred tank reactors. In addition to free-fall mixing, a rotating shaft affixed with paddles supplied additional mixing capabilities, as the shaft in the pilot scale reactor can be programmed to change rotational direction twice per minute. The paddles also provide a scrapping action that removes lignocellulosic materials from the reactor walls, improving the heat transfer between the biomass and the reactor walls.

4 Conclusion

A fundamental understanding about the enzymatic hydrolysis of lignocellulosics at high total solids is crucial to improve the efficiency of cellulosic ethanol production. The enzyme complex has an important role in this process, particularly due to the impact of several inhibitory effects on both enzymes and fermenting microorganisms. Increasing sugar and ethanol yields are obtained at high total solids and this could contribute to a more economically feasible process if compared to operations at low total solids. However, hydrolysis at high substrate concentrations are limited by the lack of available water and these high viscosities create both mass and heat transfer limitation that translate into difficulties with mixing and handling. To overcome this rheology problem, fed-batch hydrolysis systems and/or the use of chemical additives can minimize the nonuniformity of these systems. Finally, the method used for mixing the fiber slurry has a substantial impact on the hydrolysis rate and different impellers can be applied for this purpose.

Acknowledgments This work was partially supported by CNPq (grants: 551404/2010-8; 311554/2011-3) and Araucaria Foundation (grants 3572013-35356 and 2742013-35145) (Brazil). D.H.F., M.B.U. and L.M.C. are grateful to CAPES (Brazil) for providing scholarships to carry out their graduate studies. Finally, the authors wish to acknowledge the support of the CYTED (Red Provalor, grant 312RT0456) for funding of a student exchange between UFPR and UPB.

References

Andrić P, Meyer AS, Jensen PA, Dam-Johansen K (2010) Reactor design for minimizing product inhibition during enzymatic lignocellulose hydrolysis: I. Significance and mechanism of cellobiose and glucose inhibition on cellulolytic enzymes. Biotechnol Adv 28(3):308–324. doi:10.1016/j.biotechadv.2010.01.003

Antil PS, Gupta R, Kuhad RC (2015) Simultaneous saccharification and fermentation of pretreated sugarcane bagasse to ethanol using a new thermotolerant yeast. Ann Microbiol 65(1):423–429. doi:10.1007/s13213-014-0875-2

Arantes V, Saddler JN (2010) Access to cellulose limits the efficiency of enzymatic hydrolysis: the role of amorphogenesis. Biotechnol Biofuels 3:4. doi:10.1186/1754-6834-3-4

Balat M (2011) Production of bioethanol from lignocellulosic materials via the biochemical pathway: a review. Energy Convers Manage 52(2):858–875. doi:10.1016/j.enconman.2010.08.013

Beghello L, Lindström T (1998) The influence of carboxymethylation on the fiber flocculation process. Nord Pulp Pap Res J 13(4):269–273. doi:10.3183/NPPRJ-1998-13-04-p269-273

Bhat MK, Bhat S (1997) Cellulose degrading enzymes and their potential industrial applications. Biotechnol Adv 15(3–4):583–620. doi:10.1016/S0734-9750(97)00006-2

Bisson LF (1999) Stuck and sluggish fermentations. Am J Enol Vitic 50(1):107–119

Borden JR, Lee YY, Yoon H-H (2000) Simultaneous saccharification and fermentation of cellulosic biomass to acetic acid. Appl Biochem Biotechnol 84–86(1–9):963–970. doi:10.1385/ABAB:84-86:1-9:963

Bouchard J, Nugent HM, Berry RM (1995) The role of water and hydrogen ion concentration in ozone bleaching of kraft pulp at medium consistency. Tappi J 78(1):74–82

Cao Y, Tan H (2004) The effect of shear field on the hydrolysis of cellulose. J Macromol Sci Part B Phys 43(6):1115–1121. doi:10.1081/MB-200040726
Cara C, Moya M, Ballesteros I, Negro MJ, González A, Ruiz E (2007) Influence of solid loading on enzymatic hydrolysis of steam exploded or liquid hot water pretreated olive tree biomass. Process Biochem 42(6):1003–1009. doi:10.1016/j.procbio.2007.03.012
Chandra RP, Bura R, Mabee WE, Berlin A, Pan X, Saddler JN (2007) Substrate pretreatment: the key to effective enzymatic hydrolysis of lignocellulosics? In: Olsson L (ed) Biofuels, vol 108. Advances in biochemical engineering/biotechnology. Springer, Berlin, pp 67–93. doi:10.1007/10_2007_064
Chang VS, Holtzapple MT (2000) Fundamental factors affecting biomass enzymatic reactivity. Appl Biochem Biotechnol 84–86(1–9):5–37. doi:10.1385/ABAB:84-86:1-9:5
Coussot P (2005). Rheometry of pastes, suspensions, and granular materials: applications in industry and environment. Wiley, Hoboken. doi:10.1007/s00397-006-0118-y
Dasari RK, Berson RE (2007) The effect of particle size on hydrolysis reaction rates and rheological properties in cellulosic slurries. In: Mielenz JR, Klasson KT, Adney WS, McMillan JD (eds) Applied biochemistry and biotecnology, vol 137–140. ABAB Symposium. Humana Press, Nashville, pp 289–299. doi:10.1007/978-1-60327-181-3_26
De Bari I, Viola E, Barisano D, Cardinale M, Nanna F, Zimbardi F, Cardinale G, Braccio G (2002) Ethanol production at flask and pilot scale from concentrated slurries of steam-exploded aspen. Ind Eng Chem Res 41(7):1745–1753. doi:10.1021/ie010571f
Demirbaş A (2001) Biomass resource facilities and biomass conversion processing for fuels and chemicals. Energy Convers Manage 42(11):1357–1378. doi:10.1016/S0196-8904(00)00137-0
Deng Y, Feng X, Yang D, Yi C, Qiu X (2012) π-π stacking of the aromatic groups in lignosulfonates. BioResources 7(1):1145–1156. doi:10.15376/biores.7.1.1145-1156
Du J, Zhang F, Li Y, Zhang H, Liang J, Zheng H, Huang H (2014) Enzymatic liquefaction and saccharification of pretreated corn stover at high-solids concentrations in a horizontal rotating bioreactor. Bioprocess Biosyst Eng 37(2):173–181. doi:10.1007/s00449-013-0983-6
Ehrhardt MR, Monz TO, Root TW, Connelly RK, Scott CT, Klingenberg DJ (2010) Rheology of dilute acid hydrolyzed corn stover at high solids concentration. Appl Biochem Biotechnol 160 (4):1102–1115. doi:10.1007/s12010-009-8606-z
Eriksson T, Börjesson J, Tjerneld F (2002) Mechanism of surfactant effect in enzymatic hydrolysis of lignocellulose. Enzyme Microb Technol 31(3):353–364. doi:10.1016/S0141-0229(02)00134-5
Fan Z, South C, Lyford K, Munsie J, van Walsum P, Lynd LR (2003) Conversion of paper sludge to ethanol in a semicontinuous solids-fed reactor. Bioprocess Biosyst Eng 26(2):93–101. doi:10.1007/s00449-003-0337-x
Galbe M, Sassner P, Wingren A, Zacchi G (2007) Process engineering economics of bioethanol production. In: Olsson L (ed) Biofuels, vol 108. Advances in biochemical engineering/biotechnology. Springer, Berlin, pp 303–327. doi:10.1007/10_2007_063
Gao D, Uppugundla N, Chundawat S, Yu X, Hermanson S, Gowda K, Brumm P, Mead D, Balan V, Dale B (2011) Hemicellulases and auxiliary enzymes for improved conversion of lignocellulosic biomass to monosaccharides. Biotechnol Biofuels 4:1. doi:10.1186/1754-6834-4-5
Goto S, Nagazono H, Kato H (1986) The flow behavior of fiber suspensions in Newtonian fluids and polymer solutions. Rheol Acta 25(3):246–256. doi:10.1007/BF01357952
Goudoulas TB, Kastrinakis EG, Nychas SG (2003) Rheological aspects of dense lignite-water suspensions; time dependence, preshear and solids loading effects. Rheol Acta 42(1–2):73–85. doi:10.1007/s00397-002-0257-8
Harris PV, Welner D, McFarland KC, Re E, Navarro Poulsen J-C, Brown K, Salbo R, Ding H, Vlasenko E, Merino S, Xu F, Cherry J, Larsen S, Lo Leggio L (2010) Stimulation of lignocellulosic biomass hydrolysis by proteins of glycoside hydrolase family 61: structure and function of a large, enigmatic family. Biochemistry 49(15):3305–3316. doi:10.1021/bi100009p
He M, Wang Y, Forssberg E (2004) Slurry rheology in wet ultrafine grinding of industrial minerals: a review. Powder Technol 147(1–3):94–112. doi:10.1016/j.powtec.2004.09.032

Helle SS, Duff SJB, Cooper DG (1993) Effect of surfactants on cellulose hydrolysis. Biotechnol Bioeng 42(5):611–617. doi:10.1002/bit.260420509

Himmel ME, Ding S-Y, Johnson DK, Adney WS, Nimlos MR, Brady JW, Foust TD (2007) Biomass recalcitrance: engineering plants and enzymes for biofuels production. Science 315 (5813):804–807. doi:10.1126/science.1137016

Hodge DB, Karim MN, Schell DJ, McMillan JD (2008) Soluble and insoluble solids contributions to high-solids enzymatic hydrolysis of lignocellulose. Bioresour Technol 99(18):8940–8948. doi:10.1016/j.biortech.2008.05.015

Hodge DB, Karim MN, Schell DJ, McMillan JD (2009) Model-based fed-batch for high-solids enzymatic cellulose hydrolysis. Appl Biochem Biotechnol 152(1):88–107. doi:10.1007/s12010-008-8217-0

Houchin TL, Hanley TR (2004) Measurement of rheology of distiller's grain slurries using a helical impeller viscometer. In: Finkelstein M, McMillan J, Davison B, Evans B (eds) Proceedings of the twenty-fifth symposium on biotechnology for fuels and chemicals. Humana Press, Breckenridge, pp 723–732. doi:10.1007/978-1-59259-837-3_57

Huang X, Penner MH (1991) Apparent substrate inhibition of the *Trichoderma reesei* cellulase system. J Agric Food Chem 39(11):2096–2100. doi:10.1021/jf00011a042

Humbird D, Davis R, Tao L, Kinchin C, Hsu D, Aden A, Schoen P, Lukas J, Olthof B, Worley M, Sexton D, Dudgeon D (2011) Process design and economics for biochemical conversion of lignocellulosic biomass to ethanol. Technical Report NREL/TP-5100-47764. National Renewable Energy Laboratory, Golden

Jørgensen H, Kristensen JB, Felby C (2007a) Enzymatic conversion of lignocellulose into fermentable sugars: challenges and opportunities. Biofuels Bioprod Biorefin 1(2):119–134. doi:10.1002/bbb.4

Jørgensen H, Vibe-Pedersen J, Larsen J, Felby C (2007b) Liquefaction of lignocellulose at high-solids concentrations. Biotechnol Bioeng 96(5):862–870. doi:10.1002/bit.21115

Kaar WE, Holtzapple MT (1998) Benefits from Tween during enzymic hydrolysis of corn stover. Biotechnol Bioeng 59(4):419–427. doi:10.1002/(SICI)1097-0290(19980820)59:4<419:AID-BIT4>3.0.CO;2-J

Kappel J, Brauer P, Kittel FP (1994) High-consistency ozone bleaching technology. TAPPI J 77 (6):109–116

Kaya F, Heitmann JA, Joyce TW (1996) Deactivation of cellulase and hemicellulase in high shear fields. Cellul Chem Technol 30(1–2):49–56

Kerekes RJ (2006) Rheology of fibre suspensions in papermaking: an overview of recent research. Nord Pulp Pap Res J 21(5):598–612. doi:10.3183/NPPRJ-2006-21-05-p598-612

Kim S, Holtzapple MT (2006) Effect of structural features on enzyme digestibility of corn stover. Bioresour Technol 97(4):583–591. doi:10.1016/j.biortech.2005.03.040

Kim TH, Kim TH (2014) Overview of technical barriers and implementation of cellulosic ethanol in the U.S. Energy 66:13–19. doi:10.1016/j.energy.2013.08.008

Kim TH, Kim JS, Sunwoo C, Lee YY (2003) Pretreatment of corn stover by aqueous ammonia. Bioresour Technol 90(1):39–47. doi:10.1016/S0960-8524(03)00097-X

Knutsen JS, Liberatore MW (2009) Rheology of high-solids biomass slurries for biorefinery applications. J Rheol 53(4):877–892. doi:10.1122/1.3143878

Knutsen JS, Liberatore MW (2010) Rheology modification and enzyme kinetics of high solids cellulosic slurries. Energy Fuels 24(5):3267–3274. doi:10.1021/ef100140g

Kristensen J, Felby C, Jorgensen H (2009a) Yield-determining factors in high-solids enzymatic hydrolysis of lignocellulose. Biotechnol Biofuels 2(1):11. doi:10.1186/1754-6834-2-11

Kristensen JB, Felby C, Jørgensen H (2009b) Determining yields in high solids enzymatic hydrolysis of biomass. Appl Biochem Biotechnol 156(1–3):127–132. doi:10.1007/s12010-008-8375-0

Kumar R, Wyman CE (2008) An improved method to directly estimate cellulase adsorption on biomass solids. Enzyme Microb Technol 42(5):426–433. doi:10.1016/j.enzmictec.2007.12.005

Kyriacou A, Neufeld RJ, MacKenzie CR (1988) Effect of physical parameters on the adsorption characteristics of fractionated *Trichoderma reesei* cellulase components. Enzyme Microb Technol 10(11):675–681. doi:10.1016/0141-0229(88)90059-2

Ladisch MR, Lin KW, Voloch M, Tsao GT (1983) Process considerations in the enzymatic hydrolysis of biomass. Enzyme Microb Technol 5(2):82–102. doi:10.1016/0141-0229(83)90042-X

Lau MW, Dale BE (2009) Cellulosic ethanol production from AFEX-treated corn stover using *Saccharomyces cerevisiae* 424A(LNH-ST). Proc Natl Acad Sci 106(5):1368–1373. doi:10.1073/pnas.0812364106

Laxen T, Ryynanen H, Henricson K (1990) Medium-consistency ozone bleaching. Pap Puu 75 (5):504–507

Li J, Li S, Fan C, Yan Z (2012) The mechanism of poly(ethylene glycol) 4000 effect on enzymatic hydrolysis of lignocellulose. Colloids Surf B 89:203–210. doi:10.1016/j.colsurfb.2011.09.019

Liu K, Lin X, Yue J, Li X, Fang X, Zhu M, Lin J, Qu Y, Xiao L (2010) High concentration ethanol production from corncob residues by fed-batch strategy. Bioresour Technol 101(13):4952–4958. doi:10.1016/j.biortech.2009.11.013

Lou H, Wang M, Lai H, Lin X, Zhou M, Yang D, Qiu X (2013) Reducing non-productive adsorption of cellulase and enhancing enzymatic hydrolysis of lignocelluloses by noncovalent modification of lignin with lignosulfonate. Bioresour Technol 146:478–484. doi:10.1016/j.biortech.2013.07.115

Lou H, Wu S, Li X, Lan T, Yang D, Pang Y, Qiu X, Li X, Huang J (2014) Improving rheology and enzymatic hydrolysis of high-solid corncob slurries by adding lignosulfonate and long-chain fatty alcohols. J Agric Food Chem 62(33):8430–8436. doi:10.1021/jf502534s

Lu Y, Wang Y, Xu G, Chu J, Zhuang Y, Zhang S (2010) Influence of high solid concentration on enzymatic hydrolysis and fermentation of steam-exploded corn stover biomass. Appl Biochem Biotechnol 160(2):360–369. doi:10.1007/s12010-008-8306-0

Lynd LR (1996) Overview and evaluation of fuel ethanol from cellulosic biomass: technology, economics, the environment, and policy. Annu Rev Energy 21(1):403–465. doi:10.1146/annurev.energy.21.1.403

Ma X, Yue G, Yu J, Zhang X, Tan T (2011) Enzymatic hydrolysis of cassava bagasse with high solid loading. J Biobased Mater Bioenergy 5(2):275–281. doi:10.1166/jbmb.2011.1138

McMillan JD (1994) Pretreatment of lignocellulosic biomass. In: Himmel ME, Baker JO, Overend RP (eds) Enzymatic conversion of biomass for fuels production, vol 566. American Chemical Society, Washington, pp 292–324. doi:10.1021/bk-1994-0566.ch015

Mello BL, Polikarpov I (2014) Family 1 carbohydrate binding-modules enhance saccharification rates. AMB Express 4:36. doi:10.1186/s13568-014-0036-9

Modenbach AA, Nokes SE (2013) Enzymatic hydrolysis of biomass at high-solids loadings—a review. Biomass Bioenergy 56:526–544. doi:10.1016/j.biombioe.2013.05.031

Mohagheghi A, Tucker M, Grohmann K, Wyman C (1992) High solids simultaneous saccharification and fermentation of pretreated wheat straw to ethanol. Appl Biochem Biotechnol 33(2):67–81. doi:10.1007/BF02950778

Mooney CA, Mansfield SD, Touhy MG, Saddler JN (1998) The effect of initial pore volume and lignin content on the enzymatic hydrolysis of softwoods. Bioresour Technol 64(2):113–119. doi:10.1016/S0960-8524(97)00181-8

Muller JL, Protti KL, Machado MdS, Lacerda LLVd, Bresolin TMB, Podlech PS (2007) Comparação do crescimento de *Saccharomyces boulardii* em fermentador por batelada tipo air lift e shaker. Cienc Tecnol Aliment (Campinas, Braz) 27 (4):688–693. doi:10.1590/S0101-20612007000400003

Ng IS, Tsai S-W, Ju Y-M, Yu S-M, T-hD Ho (2011) Dynamic synergistic effect on *Trichoderma reesei* cellulases by novel β-glucosidases from *Taiwanese fungi*. Bioresour Technol 102 (10):6073–6081. doi:10.1016/j.biortech.2010.12.110

Nutt WE, Griggs BE, Eachus SW, Pikulin MA (1993) Developing an ozone bleaching process. Tappi J 76(3):115–123

Osawa Z, Schuerch C (1963) The action of gaseous reagents on cellulosic materials. I: zonization and reduction of unbleached kraft pulp. Tappi J 46(2):79–89

Palmqvist B, Kadić A, Hägglund K, Petersson A, Lidén G (2015) Scale-up of high-solid enzymatic hydrolysis of steam-pretreated softwood: the effects of reactor flow conditions. Biomass Conv Bioref. doi:10.1007/s13399-015-0177-3

Palonen H, Tjerneld F, Zacchi G, Tenkanen M (2004) Adsorption of *Trichoderma reesei* CBH I and EG II and their catalytic domains on steam pretreated softwood and isolated lignin. J Biotechnol 107(1):65–72. doi:10.1016/j.jbiotec.2003.09.011

Pangarkar VG, Yawalkar AA, Sharma MM, Beenackers AACM (2002) Particle–liquid mass transfer coefficient in two-/three-phase stirred tank reactors. Ind Eng Chem Res 41(17):4141–4167. doi:10.1021/ie010933j

Penner MH, Liaw E-T (1994) Kinetic consequences of high ratios of substrate to enzyme saccharification systems based on *Trichoderma* cellulase. In: Himmel ME, Baker JO, Overend RP (eds) Enzymatic conversion of biomass for fuels production, vol 566. American Chemical Society, Washington, pp 363–371. doi:10.1021/bk-1994-0566.ch018

Pimenova NV, Hanley TR (2003) Measurement of rheological properties of corn stover suspensions. Appl Biochem Biotechnol 106(1–3):383–392. doi:10.1385/ABAB:106:1-3:383

Pimenova NV, Hanley TR (2004) Effect of corn stover concentration on rheological characteristics. Appl Biochem Biotechnol 114(1–3):347–360. doi:10.1385/ABAB:114:1-3:347

Puri VP (1984) Effect of crystallinity and degree of polymerization of cellulose on enzymatic saccharification. Biotechnol Bioeng 26(10):1219–1222. doi:10.1002/bit.260261010

Puri D, Heaven S, Banks C (2013) Improving the performance of enzymes in hydrolysis of high solids paper pulp derived from MSW. Biotechnol Biofuels 6(1):107. doi:10.1186/1754-6834-6-107

Qing Q, Yang B, Wyman CE (2010) Impact of surfactants on pretreatment of corn stover. Bioresour Technol 101(15):5941–5951. doi:10.1016/j.biortech.2010.03.003

Ramos LP, Breuil C, Saddler JN (1992) Comparison of steam pretreatment of eucalyptus, aspen, and spruce wood chips and their enzymatic hydrolysis. Appl Biochem Biotechnol 34–35 (1):37–48. doi:10.1007/BF02920532

Ramos LP, da Silva L, Ballem AC, Pitarelo AP, Chiarello LM, Silveira MHL (2015) Enzymatic hydrolysis of steam-exploded sugarcane bagasse using high total solids and low enzyme loadings. Bioresour Technol 175:195–202. doi:10.1016/j.biortech.2014.10.087

Reeve DW, Earl PF (1986) Mixing gases, water and pulp in bleaching. Tappi J 69(7):84–88

Rewatkar VB, Bennington CPJ (2000) Gas-liquid mass transfer in low- and medium-consistency pulp suspensions. Can J Chem Eng 78(3):504–512. doi:10.1002/cjce.5450780309

Rivers DB, Emert GH (1988) Factors affecting the enzymatic hydrolysis of municipal-solid-waste components. Biotechnol Bioeng 31(3):278–281. doi:10.1002/bit.260310314

Roberts KM, Lavenson DM, Tozzi EJ, McCarthy MJ, Jeoh T (2011) The effects of water interactions in cellulose suspensions on mass transfer and saccharification efficiency at high solids loadings. Cellulose 18(3):759–773. doi:10.1007/s10570-011-9509-z

Roche C, Dibble C, Stickel J (2009a) Laboratory-scale method for enzymatic saccharification of lignocellulosic biomass at high-solids loadings. Biotechnol Biofuels 2(1):28. doi:10.1186/1754-6834-2-28

Roche CM, Dibble CJ, Knutsen JS, Stickel JJ, Liberatore MW (2009b) Particle concentration and yield stress of biomass slurries during enzymatic hydrolysis at high-solids loadings. Biotechnol Bioeng 104(2):290–300. doi:10.1002/bit.22381

Rosgaard L, Andric P, Dam-Johansen K, Pedersen S, Meyer A (2007) Effects of substrate loading on enzymatic hydrolysis and viscosity of pretreated barley straw. Appl Biochem Biotechnol 143(1):27–40. doi:10.1007/s12010-007-0028-1

Rudolf A, Alkasrawi M, Zacchi G, Lidén G (2005) A comparison between batch and fed-batch simultaneous saccharification and fermentation of steam pretreated spruce. Enzyme Microb Technol 37(2):195–204. doi:10.1016/j.enzmictec.2005.02.013

Saloheimo M, Paloheimo M, Hakola S, Pere J, Swanson B, Nyyssönen E, Bhatia A, Ward M, Penttilä M (2002) Swollenin, a *Trichoderma reesei* protein with sequence similarity to the plant expansins, exhibits disruption activity on cellulosic materials. Eur J Biochem 269 (17):4202–4211. doi:10.1046/j.1432-1033.2002.03095.x

Samaniuk JR, Wang J, Root TW, Scott CT, Klingenberg DJ (2011) Rheology of concentrated biomass. Korea-Aust Rheol J 23(4):237–245. doi:10.1007/s13367-011-0029-z

Sato T (1995) Rheology of suspensions. J Coat Technol 67:69–79

Schell DJ, Farmer J, Newman M, McMillan JD (2003) Dilute-sulfuric acid pretreatment of corn stover in pilot-scale reactor. Appl Biochem Biotechnol 105(1–3):69–85. doi:10.1385/ABAB:105:1-3:69

Silveira MHL, Morais ARC, da Costa Lopes AM, Olekszyszen DN, Bogel-Łukasik R, Andreaus J, Pereira Ramos L (2015) Current pretreatment technologies for the development of cellulosic ethanol and biorefineries. ChemSusChem 8 (20):3366–3390. doi:10.1002/cssc.201500282

Sørensen A, Lübeck M, Lübeck PS, Ahring BK (2013) Fungal beta-glucosidases: a bottleneck in industrial use of lignocellulosic materials. Biomolecules 3(3):612–631. doi:10.3390/biom3030612

Stickel JJ, Powell RL (2005) Fluid mechanics and rheology of dense suspensions. Annu Rev Fluid Mech 37(1):129–149. doi:10.1146/annurev.fluid.36.050802.122132

Stickel JJ, Knutsen JS, Liberatore MW, Luu W, Bousfield DW, Klingenberg DJ, Scott CT, Root TW, Ehrhardt MR, Monz TO (2009) Rheology measurements of a biomass slurry: an inter-laboratory study. Rheol Acta 48(9):1005–1015. doi:10.1007/s00397-009-0382-8

Sukharnikov LO, Alahuhta M, Brunecky R, Upadhyay A, Himmel ME, Lunin VV, Zhulin IB (2012) Sequence, structure, and evolution of cellulases in glycoside hydrolase family 48. J Biol Chem 287(49):41068–41077. doi:10.1074/jbc.M112.405720

Sun Y, Cheng J (2002) Hydrolysis of lignocellulosic materials for ethanol production: a review. Bioresour Technol 83(1):1–11. doi:10.1016/S0960-8524(01)00212-7

Switzer LH, Klingenberg DJ (2004) Flocculation in simulations of sheared fiber suspensions. Int J Multiphase Flow 30(1):67–87. doi:10.1016/j.ijmultiphaseflow.2003.10.005

Tengborg C, Galbe M, Zacchi G (2001a) Influence of enzyme loading and physical parameters on the enzymatic hydrolysis of steam-pretreated softwood. Biotechnol Prog 17(1):110–117. doi:10.1021/bp000145+

Tengborg C, Galbe M, Zacchi G (2001b) Reduced inhibition of enzymatic hydrolysis of steam-pretreated softwood. Enzyme Microb Technol 28(9–10):835–844. doi:10.1016/S0141-0229(01)00342-8

Thompson DN, Chen H-C, Grethlein HE (1992) Comparison of pretreatment methods on the basis of available surface area. Bioresour Technol 39(2):155–163. doi:10.1016/0960-8524(92)90135-K

Turian RM, Ma TW, Hsu FLG, Sung DJ (1997) Characterization, settling, and rheology of concentrated fine particulate mineral slurries. Powder Technol 93(3):219–233. doi:10.1016/S0032-5910(97)03274-9

Varga E, Klinke HB, Réczey K, Thomsen AB (2004) High solid simultaneous saccharification and fermentation of wet oxidized corn stover to ethanol. Biotechnol Bioeng 88(5):567–574. doi:10.1002/bit.20222

Viamajala S, McMillan JD, Schell DJ, Elander RT (2009) Rheology of corn stover slurries at high solids concentrations—effects of saccharification and particle size. Bioresour Technol 100 (2):925–934. doi:10.1016/j.biortech.2008.06.070

Walker LP, Wilson DB (1991) Enzymatic hydrolysis of cellulose: an overview. Bioresour Technol 36(1):3–14. doi:10.1016/0960-8524(91)90095-2

Wang K, Jiang J-X, Xu F, Sun R-C (2009) Influence of steaming explosion time on the physic-chemical properties of cellulose from *Lespedeza stalks* (*Lespedeza crytobotrya*). Bioresour Technol 100(21):5288–5294. doi:10.1016/j.biortech.2009.05.019

Wang W, Zhuang X, Yuan Z, Yu Q, Qi W, Wang Q, Tan X (2012) High consistency enzymatic saccharification of sweet sorghum bagasse pretreated with liquid hot water. Bioresour Technol 108:252–257. doi:10.1016/j.biortech.2011.12.092

Wiman M, Palmqvist B, Tornberg E, Lidén G (2011) Rheological characterization of dilute acid pretreated softwood. Biotechnol Bioeng 108(5):1031–1041. doi:10.1002/bit.23020

Xiao Z, Zhang X, Gregg DJ, Saddler JN (2004) Effects of sugar inhibition on cellulases and β-glucosidase during enzymatic hydrolysis of softwood substrates. Appl Biochem Biotechnol 115(1–3):1115–1126. doi:10.1385/ABAB:115:1-3:1115

Yang M, Li W, Liu B, Li Q, Xing J (2010) High-concentration sugars production from corn stover based on combined pretreatments and fed-batch process. Bioresour Technol 101(13):4884–4888. doi:10.1016/j.biortech.2009.12.013

Yang J, Zhang X, Yong Q, Yu S (2011) Three-stage enzymatic hydrolysis of steam-exploded corn stover at high substrate concentration. Bioresour Technol 102(7):4905–4908. doi:10.1016/j.biortech.2010.12.047

Yoshida M, Liu Y, Uchida S, Kawarada K, Ukagami Y, Ichinose H, Kaneko S, Fukuda K (2008) Effects of cellulose crystallinity, hemicellulose, and lignin on the enzymatic hydrolysis of *Miscanthus sinensis* to monosaccharides. Biosci Biotechnol Biochem 72(3):805–810. doi:10.1271/bbb.70689

Yun S-I, Jeong C-S, Chung D-K, Choi H-S (2001) Purification and Some Properties of a β-Glucosidase from *Trichoderma harzianum* Type C-4. Biosci Biotechnol Biochem 65 (9):2028–2032. doi:10.1271/bbb.65.2028

Zaccai G (2004) The effect of water on protein dynamics. Philos Trans R Soc B 359(1448):1269–1275. doi:10.1098/rstb.2004.1503

Zhang Y-HP, Lynd LR (2004) Toward an aggregated understanding of enzymatic hydrolysis of cellulose: noncomplexed cellulase systems. Biotechnol Bioeng 88(7):797–824. doi:10.1002/bit.20282

Zhang X, Qin W, Paice MG, Saddler JN (2009) High consistency enzymatic hydrolysis of hardwood substrates. Bioresour Technol 100(23):5890–5897. doi:10.1016/j.biortech.2009.06.082

Zhang J, Chu D, Huang J, Yu Z, Dai G, Bao J (2010) Simultaneous saccharification and ethanol fermentation at high corn stover solids loading in a helical stirring bioreactor. Biotechnol Bioeng 105(4):718–728. doi:10.1002/bit.22593

Zhang Y, Liu Y-Y, Xu J-L, Yuan Z-H, Zhuang X-S (2012) High solid and low enzyme loading based saccharification of agricultural biomass. BioResources 7(1):345–353. doi:10.15376/biores.7.1.0345-0353

Zhao X, Song Y, Liu D (2011) Enzymatic hydrolysis and simultaneous saccharification and fermentation of alkali/peracetic acid-pretreated sugarcane bagasse for ethanol and 2,3-butanediol production. Enzyme Microb Technol 49(4):413–419. doi:10.1016/j.enzmictec.2011.07.003

Zhao X, Dong L, Chen L, Liu D (2013) Batch and multi-step fed-batch enzymatic saccharification of Formiline-pretreated sugarcane bagasse at high solid loadings for high sugar and ethanol titers. Bioresour Technol 135:350–356. doi:10.1016/j.biortech.2012.09.074

Zwietering TN (1958) Suspending of solid particles in liquid by agitators. Chem Eng Sci 8 (3):244–253. doi:10.1016/0009-2509(58)85031-9

First Generation Bioethanol

Emmanuel Bertrand, Luciana P.S. Vandenberghe, Carlos Ricardo Soccol, Jean-Claude Sigoillot and Craig Faulds

Abstract At the beginning of 2016, first generation bioethanol still contributes to the majority of the 25 billion of gallons' bioethanol produced worldwide, with the United States and Brazil producing approximately 85 % of the global production predominantly based on corn and sugarcane, respectively. However, concerns over the long-term sustainability of first generation bioethanol, such as the impacts on land use, water resource, the potential contamination of soils with the distillation residues, and the competition for food and feed production is frequently highlighted. Current fuel ethanol research and development strives to minimize these negative externalities. The fundamental role that process design plays during the development of cost-effective technologies is evaluated through the modification of the major pathways in first generation ethanol synthesis. In this context, the central role that better performing enzymes and microorganisms play in the intensity and integration of the process, such as the typical example of simultaneous saccharification and fermentation from starchy material in first generation facilities is acknowledged. Compensating ethanol production costs by the integrated valorization of energy and by-products for feed and green chemistry in a typical

E. Bertrand · J.-C. Sigoillot · C. Faulds
UMR 1163 Biodiversité et Biotechnologie Fongiques,
INRA, 13288 Marseille, France

E. Bertrand · J.-C. Sigoillot · C. Faulds
UMR 1163 Biodiversité et Biotechnologie Fongiques, Aix Marseille Université,
13288 Marseille, France

E. Bertrand · J.-C. Sigoillot · C. Faulds (✉)
Polytech' Marseille, UMR 1163 Biodiversité et Biotechnologie Fongiques,
13288 Marseille, France
e-mail: craig.faulds@univ-amu.fr

L.P.S. Vandenberghe · C.R. Soccol
Bioprocess Engineering and Biotechnology Department, Federal University of Parana,
81531-990 Curitiba, PR, Brazil

C.R. Soccol et al. (eds.), *Green Fuels Technology*,
Green Energy and Technology, DOI 10.1007/978-3-319-30205-8_8

biorefinery concept are striking outputs of the first generation ethanol real scale experiment. Finally, rather than a mistake, first generation bioethanol should be considered as the first step that made it possible to gain the necessary experience for the successful implementation of the future greener generations biofuels from the field to the tank, starting with second generation lignocellulosic that is now coming on the market. In this context, integrated biorefineries are a promising way to diversify the usable feedstocks, leading to reduced facilities size and optimized supply-chains, to valorize more efficiently bagasse's from sugarcane and corn stover or even to exploit the potential of microalgae to capture the carbon dioxide that is produced during the fermentation steps. Major stakeholders in bioenergy production are taking advantage of the large-scale successful development of first generation bioethanol, using the most promising processing schemes for next generation facilities, although the industry is still facing uncertainties with respect to its economic viability and longevity.

Keywords Sugarcane · Sugar beet · Corn · Cassava · Enzymatic treatments · Process engineering · High gravity fermentation · Integrated biorefineries

1 Introduction

Global population growth, projected to exceed 9 billion by 2050, will raise the average calorie intake thus pushing productivity from already scarce arable land to its limit. At the same time, the energy demand in developing nations is expected to increase by 84 % over the same period, with nearly one-third of this additional fuel probably needing to come from alternative renewable sources such as biofuels (Graham-Rowe 2011; Dutta et al. 2014). First generation ethanol (1G ethanol) processes utilize either soluble sugars or starch. In 2014 there were more than 200 starch-based bioethanol plants operating in the USA, with an average capacity of 260,000 m^3 ethanol produced per year from corn (maize) and sorghum (www.ethanolproducers.com). Figure 1 shows the global ethanol production by country or region, over the period 2007–2014. The United States is the world's largest producer of bioethanol, producing over 14 billion gallons in 2014 alone with more than 40 % of the US corn crop is being used to produce ethanol. Together, the U.S. and Brazil produce 83 % of the world's ethanol, which globally amounts to around 21 million m^3 ethanol produced from sugarcane and 60 million m^3 from corn and other grains (REN21 2012; Dutta et al. 2014; AFDC 2015). The fuels generated from these raw materials are readily used in today's petrol engines. However, there are country-specific mandates for blending biofuels, as there are concerns about possible food versus fuel conflicts of interest in land use.

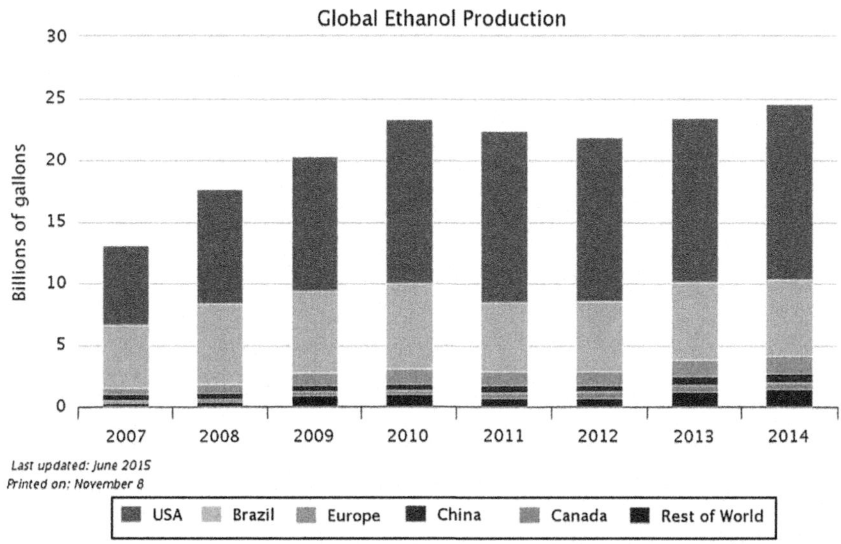

Fig. 1 Global bioethanol production from 2007 to 2014. *Source* Renewable Fuels Association, Ethanol Industry Outlook 2008–2015 reports (AFDC 2015)

2 Feedstocks for First Generation Bioethanol

Bioethanol feedstocks can be classified into three types: (i) sucrose-containing feedstocks (e.g., sugar beet, sweet sorghum, and sugarcane), (ii) starchy materials (e.g., wheat, corn, and barley), and (iii) lignocellulosic biomass (e.g., wood, straw, and grasses) (Balat et al. 2008). The availability of feedstocks for bioethanol can vary considerably from season to season and depending up on geographic locations. The changes in the price of feedstocks can highly affect the production costs of bioethanol (Yoosin and Sorapipatana 2007). Another point to consider is that, the major feedstocks for first generation biofuels are the sources of food, which may cause certain competition. Only 2 % of world's arable land is used to grow biomass feedstock for first generation biofuel production (OECD/IEA 2008), which may contribute to the increase of commodity prices for food and animal feeds. However, direct or indirect impact of biofuels on food price hike remains inconclusive.

1G bioethanol production in different producing countries and the main feedstock used is described in Table 1. With an increasing instability in petroleum prices, many countries have decided to direct their energy policy toward the use of biofuels. This imposes the production of crops such as maize, sugar beet, and others that can supply the demand for bioethanol, without conflict with food production. Cereal grains are the most abundant crops used at present for the generation of 1G ethanol. The infrastructure for growing, harvesting, and processing maize is well established, and the conversion of corn starch and corn syrups into ethanol is a relatively simple process. Corn grains contain approximately 65–76 % w/w starch,

Table 1 World's first generation ethanol production from different feedstocks

Country/continent	Major feedstock sugar and starchy crops	Ethanol production per year (billion liters)	Costs (US$/L)
Asia			–
China	Molasses, Sweet Sorghum	–	0.32, 0.29
Thailand	Cassava	1.0	0.18
Europe			
Belgium	Wheat	0.4	–
EU	Cereal and Sugar beet	4.5	–
France	Sugar beet	1.0	0.60–0.68
Spain	Barley, Wheat	0.4	–
Sweden	Wheat	–	0.40–0.45
Poland	Rye	0.2	0.55–0.65
North America			
US	Corn/Maize	50.3	0.25–0.40
Canada	Wheat/Cereal	1.8	–
South America			
Brazil	Sugarcane	25.5	0.16–0.22
Argentina	Sugarcane	0.5	–
Oceania			
Australia	Sugarcane	0.3	–

Source Modified from Gupta and Verma (2015), Haankuku et al. (2015)

while wheat (66–82 %), barley (55–74 %), sorghum (68–80 %), oat (45–69 %), and rice (74–85 %) are also rich sources of 1G raw materials.

Sugarcane is the second most used raw material for the production of bioethanol. The majority of the world's sugarcane is grown in Brazil. Unlike cereals, which produce starch as the source of fermentable sugars, sugarcane produces directly sugar, and so does not require an initial heating step prior to fermentation. As with maize, the infrastructure for the production, harvesting, and processing of sugarcane is well established.

Palm oil (SE Asia), sugarcane (Brazil), and sweet sorghum (China) appear to be the most sustainable crop for the generation of bioenergy as these crops make the most efficient use of land, water, nitrogen, and energy resources (de Vries et al. 2010). In comparison, maize (USA) and wheat (NE Europe) were poor performers for ethanol production, while rapeseed (NW Europe), cassava (Thailand), sugar beet (NW Europe), and soybean (USA, S. America) take a more intermediate position in feedstock sustainability factors. In temperate climates, first generation ethanol from maize and wheat appear not to be sustainable, as they are not fully meeting their primary goals of reducing fossil fuel consumption and GHG emissions.

Biofuels facilities can be broadly separated into two based on the feedstock, and thus the technologies required to produce the bioethanol. For instance, some of the major producers include ADM Hamburg AG (Germany), a subsidiary of Archer Daniels Midland, producing bioethanol for transportation from food processing. LS9, Inc (USA), which uses synthetic biology to produce ethanol from sugarcane syrup, but is also investigating processes for sorghum and lignocellulosic 2G feedstock. Proalcool (Brazil) is the National State-sponsored company for supporting the production of bioethanol from sugarcane. Jilin Fuel Ethanol and the Henan Tianguan Alcohol Chemical Group Co (China) have plants with the capacity to produce 1.3 million tons of ethanol per year.

2.1 Sugarcane 1G Bioethanol

Sugarcane represents two-third of world sugar production and one-third is from sugar beet (Linoj et al. 2006). They are the most promising sources for bioethanol production (UNCTAD 2015). Both are produced in geographically distinct regions. Sugarcane is grown in tropical and subtropical countries, while sugar beet is only grown in temperate climate countries. (Balat et al. 2008). Unlike cereals, which produce starch as the source of fermentable sugars, sugarcane produces directly sugar, and so does not require an initial heating step prior to fermentation. As with maize, the infrastructure for the production, harvesting, and processing of sugarcane is well established. Brazil is the largest producer of sugarcane worldwide with 632 billion tons (Unica 2015). The center-south region of Brazil accounts for almost 80 % of its feedstock production (Zarrilli 2006). Sugarcane was chosen as the substrate for ethanol production due to its great adaptation to the Brazilian soil, the weather conditions and high sucrose juice composition. Due to this great interest, agricultural and technological studies were intensified. This fact, led Brazil to a very favorable position in terms of energy security. The sugarcane yields for 2015 and 2020 are estimated to be 79 and 84 t cane/ha, respectively, with an average increase rate of 1.3 % per year (Wang et al. 2014).

The National Alcohol Program—ProAlcool, created by the government of Brazil in 1975 resulted in less dependency on fossil fuels (Rosillo-Calle and Cortez 1998; Soccol et al. 2005). In this way, the Brazilian government started its policy to substitute gasoline with sugarcane alcohol. The program saw the participation of several politicians and military, sugarcane producers, researchers, the alcohol industry, and the media. The use of a mixture of ethanol and gasoline (gasohol) to fuel common cars was then intensified. The addition of 25 % ethanol to gasoline reduced the import of 550 million barrels' oil and CO_2 emission by 110 million tons (Soccol et al. 2010). In March 2003, the introduction of flex fuel vehicles (FFVs) revitalized the Brazilian car industry. FFVs can use various mixtures of alcohol and gas, thus allowing the consumers to react to the different prices signals of the two markets (Hira and Oliveira 2009). Presently, Brazil has more than 90 % of flex fuel vehicles in its fleet (ANFAVEA 2015).

Nowadays, around 39.4 % of the Brazilian energy matrix is renewable and 157 % is derived from sugarcane (Unica 2015). Brazil has a land area of 851 million hectares, of which 54 % are preserved, such as the Amazon rainforest (350 million hectares). From the land available for agriculture (340 million hectares), only 2.59 % is used to produce sugarcane, representing 10 million hectares, showing a great expansion potential for this crop (Udop 2015; Unica 2015). Brazil ethanol production is entirely based on the fermentation of simple sugars extracted from harvested sugarcane stem either in autonomous distilleries or in annexed plants co-located with sugar mills that coproduce ethanol and crystalline sugar (Seabra et al. 2011; Wang et al. 2014). As one of the worlds' largest ethanol producers, Brazil has used sugarcane as feedstock to produce over 28 billion liters of ethanol in 2014/2015 (Unica 2015), which is destined for fuel. Currently, there are 403 bioethanol production units installed in the country (Udop 2015). Amongst these, 392 units are located in the south, southeast, and center-west and only five units in the Amazon region. However, an expansion of the ethanol production to 104 billion liters in 2025 will necessitate the reduction of production costs to sustain the transportation from more distant areas within Brazil to internal and external markets. In addition, advanced technology can provide better environmental performance and greater productivity per unit of land. However, this is also almost always bringing additional costs. A hectare of sugarcane can produce about 6000 liters of ethanol (Cerqueira Leite et al. 2009). Around 70 % of the ethanol production costs correspond to the raw materials (IBGE 2008).

2.2 Sugar Beet 1G Bioethanol

Based on the USA Energy Independence and Security Act (EISA) of 2007, sugar beets (*Beta vulgaris* L.) may be an eligible feedstock for advanced biofuel provided that production and conversion to biofuel meets the 50 % greenhouse gas reduction threshold required for advanced biofuel designation (Congress U.S. 2007; NREL 2014; Haankuku et al. 2015). The new energy strategy for Europe starting from 2011 to 2020 has been discussed in the European Union institutions (European Commission 2010; European Parliament 2010). This strategy has to be in line with the Lisbon Treaty to guide long-term emission- reduction goals, the so-called 20–20–20. To achieve energy and climate goals, the potential of bioenergy is a key issue. The main inputs in the production of bioethanol in the EU are sugar beet, wheat, corn, or barley (Salazar-Ordóñez et al. 2013). Wheat and sugar beet are frequently used in Northwestern Europe, while corn is employed in Central Europe and Spain, where barley is also often used. Thirty percent of bioethanol is produced from sugar beets (Agrosynergie 2011); around 24 % of the total production of this crop has been destined to bioethanol in the EU for the past three years (Eurostat 2011).

Sugar beets are tuber crops composed of about 75 % water, 18 % sugar (mainly sucrose), and 7 % insoluble and soluble materials (which are required to be at low levels). Unlike conventional sugar beets that are bred to produce sugar for table use, biofuel feedstock industrial beets are specialized nongrade varieties bred for total sugar production (Haankuku et al. 2015). Some alternatives are being examined in order to reduce bioethanol production costs from sugar beet. New sugar beets varieties and multi-effect evaporation processes were proposed as the major factors in the future cost reduction. Although the costs of direct fermentation of sugar beet juice (adjust the sugar content by adding molasses) is lower than the process using sugar beet juice concentration, the multi-effect evaporation enables a high-sugar fermentation and saves distillation and equipment costs (Ruan et al. 2001). At the same time, it also reduces the microbial infection of the squeeze juice. Part of impregnated water and diluted water are the wastes from the distillation tower. Water can be recycled in the production process and therefore reduce emissions (Zhou et al. 2011). In addition, with this method, separating sugar beet pulps before fermentation improves the equipment utilization of fermentation and distillation, saves energy consumption and makes the comprehensive utilization of sugar beet pulps much easier. Enrichment process preserves the sugar, which will be able to extend the production period in ethanol plants.

2.3 Corn 1G Bioethanol

The U.S. Department of Agriculture (USDA) has a program devoted to the corn ethanol industry. Areas of scientific research address the establishment of new higher value ethanol coproducts, the development of microbes capable of converting various biomass materials into ethanol, improved processes for the enzymatic saccharification of corn fibers into sugars, and various methods of improving corn ethanol process efficiencies (McAloon et al. 2000). In the 2013/2014 USA's corn production reached nearly 13.8 billion bushels (351.3 million tons) of corn and roughly 11 % of the production was exported to more than 100 different countries. More than one-third of USA's corn crop is used to feed livestock, 13 % is exported and 40 % is used to produce ethanol. The remainder goes toward food and beverage production. Federal renewable-fuel standards require the blending of 13.2 billion gallons of corn ethanol with gasoline in 2012. This required 4.7 billion bushels of corn, which corresponds to forty percent of the annual crop (Carter and Miller 2012; EIA 2013).

Fuel ethanol production from corn can be described as a five-stage process: raw material pretreatment, hydrolysis, fermentation, separation and dehydration, and wastewater treatment. The production of bioethanol from starch includes the breakdown of this polysaccharide to obtain an appropriate concentration of fermentable sugars, which are transformed into ethanol by yeasts. After washing, crushing, and milling the corn grains (dry milling process), the starchy material is gelatinized in order to make the amylose and amylopectin susceptible for enzymatic attack in the following liquefaction step. This step is considered as a pretreatment process because of the partial hydrolysis of the starch chains using thermostable α-amylase. The hydrolysate obtained has reduced viscosity and contains starch oligomers called dextrins. Then, the fermentation process occurs where sugar is immediately assimilated by the yeast *Saccharomyces cerevisiae* in the same reactor and converted into ethanol. The culture broth containing 8–11 % (w/w) ethanol is recovered in a separation step consisting of two distillation columns (Quintero et al. 2008).

2.4 Cassava 1G Bioethanol

Cassava is a shrub with tuberous roots. It is the third source of food calories in tropical countries after rice and corn. Cassava is used in both human and animal food, in many industrial sectors, particularly in the form of starch, and more recently to produce ethanol. Cassava is primarily grown for its roots but all of the plant can be used: the wood as a fuel, the leaves and peelings for animal feed, and even the stem as dietary salt (UNCTAD 2015). World production of cassava is around 281 million tons (Mt) a year. Africa contributes to more than half of the global supply. Asia encourages the development of cassava crops for industrial and energy purposes. This continent contributes to around one-third of the world production, with 26 Mt produced by Thailand and 28 Mt by Indonesia. In Latin America production is around 35 Mt where Brazil dominates with around 70 % of regional production and in third place in world production (Conab 2013). Cassava is still a small player on the biofuel scenario. For example, cassava roots, which have a starch content of 30 % (w/w), can generate 180 liters pure ethanol (96 %) per ton of raw material (500–4000 liters per hectare per year) (Larkin et al. 2004).

3 Enzymes for the First Generation Bioethanol

The starch hydrolysis by enzymes is a two-stage process involving liquefaction and saccharification. Liquefaction is a step where starch is degraded by α-amylase, which hydrolyzes only α-1,4 bonds between glucose units and causes a reduction in starch viscosity. Liquefying enzymes usually work at high temperatures (>85°C) so that the enzyme can help reduce starch paste viscosity during cooking. Dextrins, which are obtained after liquefaction, are further hydrolyzed by limit dextrins/pullulanases which can hydrolyze both α-1,4 and α-1,6 glycosidic linkage and then to glucose by glucoamylase. Glucose is then subsequentially converted to ethanol by yeast fermentation. After fermentation, approximately 10 % (v/v) ethanol is obtained and subjected to distillation and dehydration to remove water and other impurities, yielding anhydrous ethanol (Sriroth et al. 2012).

Biotech companies such as Dyadic, Amryis, and Gevo have focused on developing enzymatic solutions for the high-value steps. While enzyme production is considered to be an expensive step, accounting for nearly 50 % of the costs of 2G cellulosic ethanol production, the cost of enzyme production is being driven down by the manufacturers, such as Novozymes, Dupont, and Dyadic, from $2 per gallon in 2010 to approx. $0.30 in 2015, together with much more efficient, feedstock specific pretreatment process development. This has also led to onsite enzyme production facilities provided by Dupont, Dyadic, and DSM. Dupont, who opened the world's largest bioethanol plant in Iowa (USA) in 2015 have developed enzyme technologies for the production of more than 68 billion liters of 1G ethanol per year from corn cobs, stems, and leaves. Enzyme companies are constantly trying to improve their products for 1G bioethanol production, where even the smallest improvement in hydrolysis can lead to an extra 1–2 % increase in ethanol production. Recent developments in the two main starch-degrading enzymes, α-amylase and glucoamylase, encouraged by the availability of various fungal and bacterial genome sequences, are moving in the direction of more robust products operating at pH 5 or below, and with limited supplementary calcium requirement (Harris et al. 2014). A number of enzymes, as described below, are required for the production of 1G ethanol.

3.1 α-Amylases

α-Amylases (EC 3.2.1.1) are 1,4-α-D-glucan glucanohydrolases that catalyze the cleavage of internal α-1,4-glycosidic bonds in starch in a random manner, releasing dextrins and gluco-oligosaccharides with the reducing groups liberated in the α-configuration. The α-1,4 bonds close to the α-1,6 branch points in amylopectin are resistant to hydrolysis by α-amylases. Prolonged hydrolysis of the amylopectin with α-amylases yield limit dextrins. Most a-amylases belong to the CAZy family GH13 (Lombard et al. 2014), together with pullulanases, cyclomaltodextrinases, and trehalose-6-phosphate hydrolase. This classification is based on the direct relationship between sequence and folding similarities. Some α-amylases belong to GH Family 57. The sequences of α-amylases from different origins have very few discernable similarities but the catalytic mechanism requires three catalytic residues (two Asp and one Glu), as well as residues involved in substrate binding. These residues are all found in four highly conserved regions. Despite the low sequence similarity, α-amylases from different sources display remarkably similar tertiary folding, with a $(\alpha/\beta)_8$ barrel central core. The enzyme contains a calcium-binding site, similar to that of other amylolytic enzymes, and the removal of the calcium leads to irreversible inhibition. Cereal α-amylases contain starch granule binding sites, which interact with a host of different substrates, from granular starch to cyclodextrin, through two consecutive Trp residues (Lundgard and Svensson 1987). This feature is not present in animal of microbial α-amylases. The SPEZYME® line of α-amylases from Dupont claim to offer robust liquefaction of starch and viscosity reduction over a range of temperatures and pH (http://biosciences.dupont.com/industries/biofuels/bioethanol-from-starch/). Proteinaceous α-Amylase inhibitors have been identified in cereal grains, which are implemented in plant defense and endogenous enzyme regulation (Sancho et al. 2003; Nielsen et al. 2004).

The presence of such inhibitors in the raw materials utilized for 1G production may have a serious impact on the efficiency of the process.

3.2 *β-Amylases*

β-Amylases (EC 3.2.1.2) are exo-acting hydrolases removing successive β-anomeric maltose units from α-1,4-glucans. such as starch and glycogen. They belong to the GH 14 family of CAZy. Sweet potato contains a high level of this enzyme as soluble protein in its tubers, with only trace amounts of α-amylase. As with α-amylase, β-amylases adopt a large $(\alpha/\beta)_8$ barrel central core, with a catalytic pocket containing the two catalytic Glu residues, compared to α-amylases which have their catalytic mechanism in a long cleft open at both ends. Substrate binding causes a structural shift where a flexible loop moves 11 Å upon maltose binding, effectively closing over the substrate like a hinged lid (Rockey et al. 2000; Kang et al. 2005). A single enzyme can release several maltose molecules in a phenomenon where the β-amylase "slides" on the substrate (Ishikawa et al. 2007). Such enzymes can be used to make maltose syrups for further processing into fermentable sugars for 1G ethanol production. More resistant starch can be formed through the action of β-amylases (Luckett and Wang 2012).

3.3 *Glucoamylase*

Glucoamylase (GA; EC 3.2.1.3), also known as amyloglucanases, is an inverting exo-acting multi-domain enzyme which attacks starch from the nonreducing end to produce glucose. The catalytic domain of GA belongs to CAZy family GH15 and has a $(\alpha/\alpha)_6$ barrel structure. The active site is described as a well of ~ 10 Å deep by ~ 15 Å wide so the substrate must penetrate deep into the well before cleavage can occur. This means that the cleaved glucose and the remaining chain must leave the well before the next reaction can proceed. To overcome this, filamentous fungi, such as *Aspergillus* sp., produce very large amounts of GA. Hydrolysis occurs via multichain attacks, but at high glucose concentrations, GA reforms all the glycosidic bonds that it hydrolyses, condensing glucose to form isomaltose, isomaltotriose, and other derivatives (Nikolov et al. 1989). Purified GA has been used to make glucose syrups from maltodextrins produced by the action of α-amylases. A standard industrial saccharification process using GA starts with dextrins of DP 10–15. The most studied GAs are those from *Aspergillus awamori* and *A. niger*, and are composed of three separate structures: a catalytic and a starch-binding domain separated by a rigid, highly glycosylated linker (Kramer et al. 1993). The binding domain can be proteolytically cleaved to leave only the catalytic domain, which is then incapable of acting on granular starch. The intact multi-domain molecule can degrade the whole starch granule. The starch-binding domain belongs to CBM20 family of the CAZy database (Lombard et al. 2014) and is ~108 residues long, extending from the C-terminus of the linker. This binding domain has two binding sites for starch and related substrates, where a Trp in each side has been implicated in the interaction between the parallel strands of the amylosic double helix (Morris et al. 2005). Saccharification of liquefied starch-containing substrates requires addition of glucoamylase (e.g., Spirizyme® from Novozyme) to ensure maximum conversion of dextrins to glucose. New generation glucoamylase preparations have been developed to work directly on the corn-fiber matrix to degrade trapped starches down to glucose. More than 100 glucoamylases with a huge diversity (40–50 % identity) were recently cloned and characterized by Novozymes with respect to ethanol stability, activity in high-density solids and preference for branched dextro-oligosaccharides (Harris et al. 2014).

3.4 Pullulanases or Limit Dextranases

Pullulanases (EC 3.2.1.41) or limit dextranases (EC 3.2.1.142) catalyze the hydrolysis of the α-(1,6)-D-glucosidic linkage in amylopectin. They are usually more active on dextro-oligosaccharides than on polymeric starch. This enzyme, together with the amylases and GA help bring about the complete degradation of starch to glucose and maltose. They belong to the GH13 family and are distributed widely amongst microorganisms (mainly thermophilic bacteria and archaea) and plants. A conformational difference around the active site cleft together with domain organization determines the different substrate specificities between pullulanases and the other α-1,6-glucan debranching enzyme, isoamylase (Mikami et al. 2006).

3.5 Lytic Polysaccharide Monooxygenase (LPMO)

Very recently, a lytic polysaccharide monooxygenase (LPMO) specifically acting on starch has been reported (Harris and Wogulis. 2010; Horn et al. 2012; Lo Leggio et al. 2015). This enzyme belongs to the AA13 family of the CAZy database, are metalloproteins with a histidine-ligated mononuclear copper and in one case have been associated with a starch-binding CBM20 module (Horn et al. 2012; Lo Leggio et al. 2015). A highly starch-specific AA13 showed moderate activity on retrograded starch, degrading it through oxidation at the C1 position to aldonic acids dependent on the presence of copper and the reducing cofactor cysteine, and acted in synergy with β-amylase to release maltose (Lo Leggio et al. 2015). The starch-acting LPMO has a conserved central β-sandwich core with the active site common to other fungal LPMOs, such as the AA9s (see Couturier et al. Chapter "Fungal Enzymatic Degradation of Cellulose" of this book), with the active site presented to the solution in a shallow groove along the protein surface which leads to the bound copper. Lo Leggio and coworkers postulate that this difference in structure to AA9s is likely to accommodate the more contoured surface of retrograded starch.

3.6 Phytases

Phytases can be included in the list of enzymes involved in 1G ethanol production through its ability to stabilizing the α-amylase by degrading phytic acid that could detach vital calcium ions from the α-amylase. The phosphate moieties of phytic acid are able to bind di- and trivalent metal ions such as calcium, magnesium, zinc, and iron. Phytases (EC 3.1.3.8 and 3.1.3.26) de-esterify the phosphate groups from phytate, and are most commonly used in animal feed to date. Two main groups of phytases exist, hence the two EC numbers. Plant-derived enzymes belong to the 3-Phytase (EC 3.1.3.8) group, while most microbial ones belong to the 6-phytase group (EC 3.1.3.26). The number indicates the position of the ester bond on the substrate. Fungal phytases are active in the acid-neutral region, while bacterial ones are more active in the neutral-alkaline range. They hydrolyze phytate and other phosphoesters in a two-step mechanism (Ping-Pong) involving a covalent phosphorylated histidine adduct enzyme intermediate (Ostanin et al. 1992).

3.7 Proteases

Proteases (also called peptidases) have been typically used in the manufacturing of bioethanol as a way of degrading protein present in the raw material providing free amino nitrogen for yeast growth, thus replacing the need to add an exogenous source such as urea during fermentation (Lei et al. 2013; Vidal et al. 2009) but can also be used to breakdown starch-gluten complexes through weakening the endosperm-associated protein matrix encapsulating the starch granule (Wang et al. 2009), thereby providing further accessibility of the starch to the α-amylases and glucoamylases (Alvarez et al. 2010). This increased the fermentation rate and ethanol yield in a dry-grind ethanol production process (Johnston and McAloon 2014). This in turn increases the specific gravity of the mash and improves germ recovery. Proteases are of two kinds: exoproteases and endoproteases. The exoproteases remove one amino acid from the protein chain at a time and can be further subclassified into aminopeptidases and carboxypeptidases. Aminopeptidases cleave the amino acids from their amino (N-) terminus while the carboxypeptidases cleave from the carboxy (C-) terminus. The carboxypeptidases are further subdivided based on their active site mechanism, e.g., metallo-carbopeptidases, serine carboxypeptidases, and cysteine carboxypeptidases. Exoproteases are generally not commercially available but are present in enzymatic cocktails or culture supernatants. Endoproteases act randomly along the polypeptide chain and are subdivided into four classes differing in their catalytic mechanism: serine proteases (e.g., chymotrypsin, trypsin, subtilisins), cysteine proteases (e.g., papain, ficin, bromelain), aspartic proteases (e.g., pepsin) and metalloproteases (e.g., thermolysin, neutral proteases). Acidic and metalloproteases activate a water molecule which then performs a nucleophilic attack on the peptide bond resulting in hydrolysis, while serine, threonine, and cysteine proteases uses a nucleophilic residue and a catalytic triad to form intermediate complexes between the enzyme and the substrate, releasing one part of the product, then activated water performs the second catalytic step to release the second half of the product, regenerating the free enzyme. The addition of and endoprotease from *A. niger* (Genencor International/DuPont GC 100) and an exoprotease from *A. oryzae* (Novozyme 50045) was shown to result in a higher ethanol concentration (mean 0.3–1.8 % v/v) and lower DDGS yield compared to a no protease control (Wang et al. 2009). The addition of an acid protease with or without urea during the fermentation step was calculated to decrease overall process costs by $0.01/L (Johnston and McAloon 2014).

Most of these enzymes are present in commercially available blends or as Trade names from the major enzyme producers. Novozymes attributes 18 % of its US turnover to selling enzymes for 1G starch-to-ethanol. A recent product, Avantec® is designed to make more corn starch available for hydrolysis and thus fermentation through reducing viscosity and so allowing refining plants to run at higher solid loading or higher run rates. The Liquozyme® ranges are enzymatic preparations with low pH tolerance and high thermostability, which reduces the starch down to an optimal dextrin profile. Viscozyme® was designed specifically for cereal crop utilization to degrade the β-glucan and arabinoxylan and other cereal-specific components, which can lead to high viscosity in the process. High viscosity limits the dry feedstock you can add to the process, increasing water consumption, and affecting downstream processing, such as the efficiency of separation, evaporation, and heat exchange, and thus lowering ethanol yield. This allows a smoother flow of the liquefaction process.

Fouling of the bioreactors costs a 1G biorefinery plant time and money. Fouling reduces the efficiency of the heat exchangers and normally will involve the use of additional sulphuric acid and/or hydroblasting to remove the material accumulated in the reactors. Enzymatic treatments have been designed to be added during the fermentation stage to avoid metal chelation and fouling due to the unfermented material present after liquefaction and saccharification.

4 Overview of Processes for the First Generation Bioethanol

Currently, industrial production of first generation bioethanol is made from agricultural products rich in starch or sucrose which are readily fermentable. These sugars serve as an energy reserve for plants and as such are stored in specific tissues for each plant. Two general processes could be highlighted: one using sugar crops and another based on starchy plants.

4.1 Sugar Beet and Sugarcane Processing

Sugarcane, with its widespread use in Brazil, is the sugar producing plant most used for the production of bioethanol. In 2010, sugarcane ethanol represented a third of the ethanol production worldwide (Linoj et al. 2006). But it grows only in tropical areas. Sugar beet is much better suited to Europe and temperate climate. There is also a third plant used to a lesser extent, sweet sorghum. It comes from Africa, but is adaptable to temperate climate and has the advantage of having reduced water needs to grow. The sugar-containing plants have the drawback of being rich in water and therefore cannot be stored over time for later use. Indeed, their sugar content decreases quickly after uprooting or cutting.

Sugarcane contains 12–17 % total sugars on a wet-weight basis with 68–72 % moisture (90 % sucrose and 10 % glucose or fructose). The average extraction efficiency to produce cane juice by crushing is approximately 95 % and the remaining solid residue is cane fiber (bagasse). In factories that only produce ethanol, the cane juice is heated up to 110 °C to reduce microbial contamination, decanted, sometimes concentrated by evaporation and then fermented. In combined sugar–ethanol plants (annexed distilleries), sucrose crystals that are formed after cane juice concentration are removed by centrifugation, leaving a syrup (molasses) that contains up to 65 % w/w sugars. Both sugarcane juice and molasses (after adjusting the sugar concentration) normally contain sufficient minerals and organic nutrients to be immediately suitable for ethanol production by fermentation with Saccharomyces cerevisiae (Wheals et al. 1999). Sugar beets can be used directly in dedicated plants for ethanol production or processed for sugar production. In such a case, clear juices, syrups, and molasses obtained from the clarification steps, evaporation, and crystallization can be used for ethanol production. Table 2 shows the average compositions of these juices.

The maximum theoretical yield of glucose to ethanol conversion is 0.511 g of ethanol per gram of glucose corresponding to the so-called Gay–Lussac equation giving 2 mol of ethanol by mole of glucose. For sucrose, this yield is 0.538 g/g, one mole of sucrose giving 4 mol of ethanol. The actual yield in industrial units is around of 90 % of these maxima. Clear juices have the required sugar concentration to obtain fermented juice with ethanol concentration of 10–15 % (v/v). A high alcohol content is necessary to limit energy consumption and to optimize ethanol recovery during the distillation step (Bai et al. 2008). The upper limit depends on the sugar concentration before fermentation and the greater or lesser adaptation of the strains to high levels of ethanol. In France, the average production of sugar beet is around 70 T/ha, with a sugar content of 17 % (on wet basis) representing an average production of 12 T of sugar/ha (Data from SNSF). Figure 2 illustrates the process of ethanol production from sugar beet described hereafter.

Table 2 Composition of different juices produced during sugar beet processing (adapted from Reiss 2012)

	Diffusion juice	Clear juice	Syrup	Molasses
Dry matter (% m/m)	14.70	14.50	58.80	80.80
Sucrose (% m/m)	12.85	13.13	53.00	49.20
Total nitrogen (% m/m)	0.13	0.13	0.14	1.82
Reducing compounds (% m/m)	0.07	0.01	0.47	0.86
Ashes (% m/m)	0.28	0.34	1.85	9.86
pH	6.3	9.25	7.27	6.98

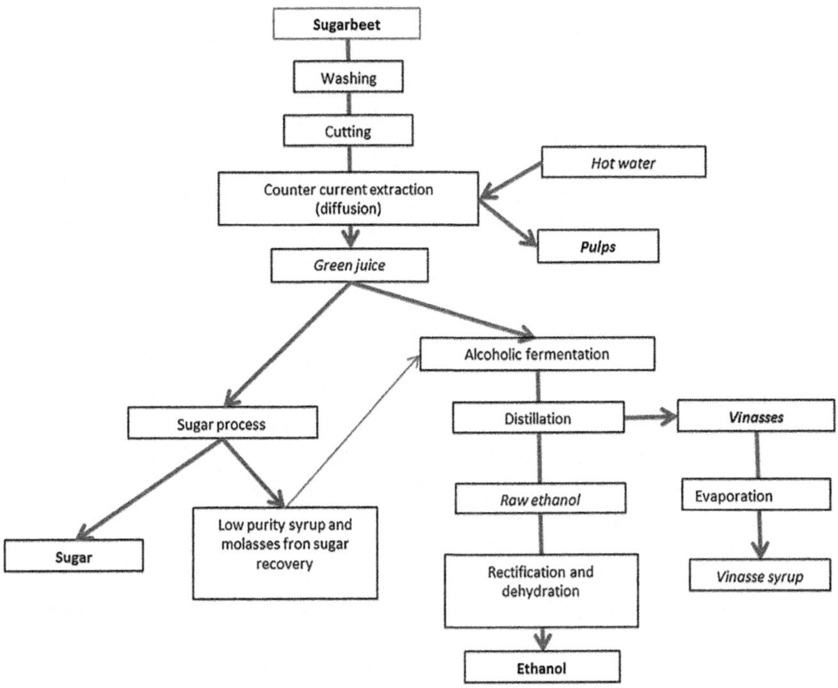

Fig. 2 Diagram of ethanol production from sugar beets. Alcohol should be produced directly from green juice produced by counter-current extraction or from syrups and molasses coming from crystal sugar production (adapted from Reiss 2012)

Ethanol production from sugar beet begins, as for sugar production, with the washing of roots and cutting them into chips (thin strips of 5–6 cm long) in a root cutter. The shape of chips is optimized for good extraction of sugar in water. The chips are then transported to the diffusion step, a counter-current extraction by hot water. The enriched water, the "diffusion juice" is recovered at diffuser head and chips "exhausted" out tail diffuser in the form of pulps. After extraction, pulps contain about 92 % water. Much of this water is separated from the pulp by pressing or dewatering and recycled. Dehydrated pulps are used in animal feed. The extraction juice called also green juice can be used directly for the fermentation by *Saccharomyces cerevisiae* during the sugar campaign (September to January in Europe). Fermented juice containing 10–14 % v/v of ethanol is called beer or wine. The next step is the distillation of beer to recover ethanol.

4.1.1 Purification

The purification of diffusion juice aims to remove some of the insoluble impurities they contain. A lime treatment (liming) results in the precipitation of impurities. Under the action of the lime, mineral acids and number of organic materials, such as pectins and proteins are converted into insoluble salts. This is followed by a double carbonation (adding CO_2) which serves to precipitate the lime remaining in the juice. These insoluble salts and calcium carbonate form a precipitate entraining the removal of some soluble impurities. Liming leads the virtual elimination of iron. Iron can catalyze oxidation reactions leading to the color formation and conducing to the formation of grayish-white sugars (Borges et al. 2012), which is undesirable for sugar production. The liquid fraction, called clear juice is then recovered. Carbonated lime and precipitated impurities are filtered off.

4.1.2 Evaporation

The purified juice still contains 85 % water. Evaporation allows the thin juice concentrate to obtain a syrup with a concentration close to saturation. Evaporation takes place in a multiple "effects" evaporator (generally four successive evaporators). Every evaporator uses the vapor produced by the precedent and the pressure in each evaporator is reduced to compensate for the reduction in temperature of vapors and to decrease correlatively the boiling point. Moreover, the low temperature avoids the cooking of sugar at a stage dedicated to evaporate the juice.

4.1.3 Crystallization

Crystallization involves the separation of sucrose (as crystals) from impurities that remain in the concentrated juice. The crystallization is carried out in 2 or 3 stages called "jets." Each jet is made of a proper crystallization stage, mixing, and centrifugation. The concentrated juice is heated and stirred in large boilers operating under partial vacuum. Its concentration continues and very fine sugar crystals are introduced therein to initiate crystallization (coarsening of crystals). The mixture syrup-crystals obtained then passes into a stirring tank to cool while continuing the crystallization. Finally, it is centrifuged in turbines or centrifuges to recover crystals. Drains still contain sugar as well as impurities that were not removed during the purification. Drains from the third jet are molasses, still rich in sugar but hardly extractable due to the high glucose and fructose content. Molasses are used in fermentation industries and in animal feed. Sugar plants with an annex distillery usually operate only 2 jets and route the drains of the second jet directly at the distillery. Pulps and vinasses (bottom residues of distillation) produced during the process can be valued in animal feeding or by anaerobic digestion. They are mainly recycled at different stages of the process (Reiss 2012).

4.2 Sweet Sorghum Processing

Sweet sorghum is a C4 crop in the grass family belonging to the genus *Sorghum bicolor* L. Moench which also includes grain and fiber sorghum and is characterized by a high photosynthetic efficiency. The average productivity is around 49.7 t/ha of fresh stem producing roughly 63 % of juice (Gnansounou et al. 2005). The total dissolvable content of juice is around 17.9 % (°Brix) with 69.5 % of sucrose (Woods 2001). In addition to sugar, the juice contains other compounds and impurities, which have to be eliminated before crystallization to obtain white sugar. Furthermore, sweet sorghum sugars consist of 85 % sucrose, 9 % glucose, and 6 % fructose—on average—and only sucrose may readily be converted to white sugar (Woods 2000). Juice purification is operated by liming (addition of lime milk and precipitation by carbon dioxide). The lime milk precipitates and captures the impurities in the raw juice. The settled solids (mainly calcium carbonate and nonsugars) from the clarifier are filtered and sent to the spent lime storage area, while the filtrate is again saturated in a second carbonation station. The purified juice obtained after the subsequent filtration is called thin juice and is thickened in a multi-effect evaporator into thick juice. High-pressure steam produced in the boiler provides the energy for evaporation, and the condensed steam is returned to the boiler or used as technical water. The thin juice that has been diluted with water during extraction and purification enters the evaporating station with an average sugar content of 15 % while the thick juice leaving the evaporator contains approximately 70 % sugar.

4.3 Corn and Starchy Grains

Today, most fuel ethanol is produced from corn by either the dry-grind (67 %) or the wet mill (33 %) process. The two processes differ with respect to complexity, associated capital costs, the numbers and types of coproducts produced, and the flexibility to produce different kinds of primary products. The principal differences between the ethanol dry-grind process and the wet mill process are the feedstock preparation steps and the numbers and types of coproducts recovered. Once the starch has been recovered the process of converting it to fuel ethanol and recovering the ethanol is similar in both wet mill and dry-grind facilities (Bothast and Schlicher

2004). Production of ethanol from starch needs its depolymerisation to obtain a glucose syrup suitable for fermentation. The hydrolysis of starch may be considered as a first and key step in corn and other starchy plants processing for bioethanol production. The main role of this step is to effectively provide the conversion of two major starch polymers: amylose, a mostly linear α-D-(1–4)-glucan and branched amylopectin, α-D-(1–4)-glucan, which has α-D-(1–6) linkages at the branch points, into fermentable sugars that could subsequently be converted to ethanol by yeasts or bacteria. Recent advances in the developing of thermostable α-amylases, the starch liquefying enzymes which catalyze the hydrolysis of internal α-D-(1–4)-glucosidic linkages in starch in a random manner and effective glucoamylases, the starch saccharifying enzymes which catalyze the hydrolysis of α-D-(1–4) and α-D-(1–6)-glucosidic bonds of starch from the nonreducing ends giving glucose as the final product, have led to commercial establishment of the so called 'two enzyme cold process' (Baras et al. 2002). The traditional thinning agent used in starch technology was acid (hydrochloric or oxalic acids, pH 2 and 140–150 °C for 5 min). The introduction of thermostable α-amylases has meant milder processing conditions (Aiyer 2005). The formation of by-products is reduced and the main advantages of this process are lower energy consumption and a lower content of non-glycosidic impurities and thus much better suitability for ethanol production.

4.3.1 Wet Milling Process

In the wet milling process, illustrated in Fig. 3, grains are dipped into an aqueous solution containing sulphuric acid, which facilitates the separation of the different components including starch, fiber, gluten, and germ. A grinding is then realized to separate these components. The germ is removed from the kernel and corn oil is extracted from the germ. The remaining germ meal is added to fibers and the hull to form corn gluten feed. Gluten is also separated to become corn gluten meal, a high-protein animal feed. A starch solution is separated from the solids and the starch so obtained is then liquefied and saccharified by enzymatic way to yield corn syrup that can be processed for ethanol production (Bothast and Schlicher 2004). However, secondary reactions occur due to the acid used during the stage of soaking (Sanchez and Cardona 2008). Furthermore, during this process several compounds of the plant are extracted. The composition of juices is thus very variable.

4.3.2 Dry Milling Process

In the case of "dry milling," illustrated Fig. 4, grains are cleaned and crushed in ball milling apparatus or hammer mill. The flour obtained is hydrolysed in two stages by enzymatic way. The first stage, realized by means of one α-amylase is called liquefaction. The second stage is called saccharification and uses a glucoamylase (GA). This ends in the formation of a syrup of glucose. This syrup will then be used

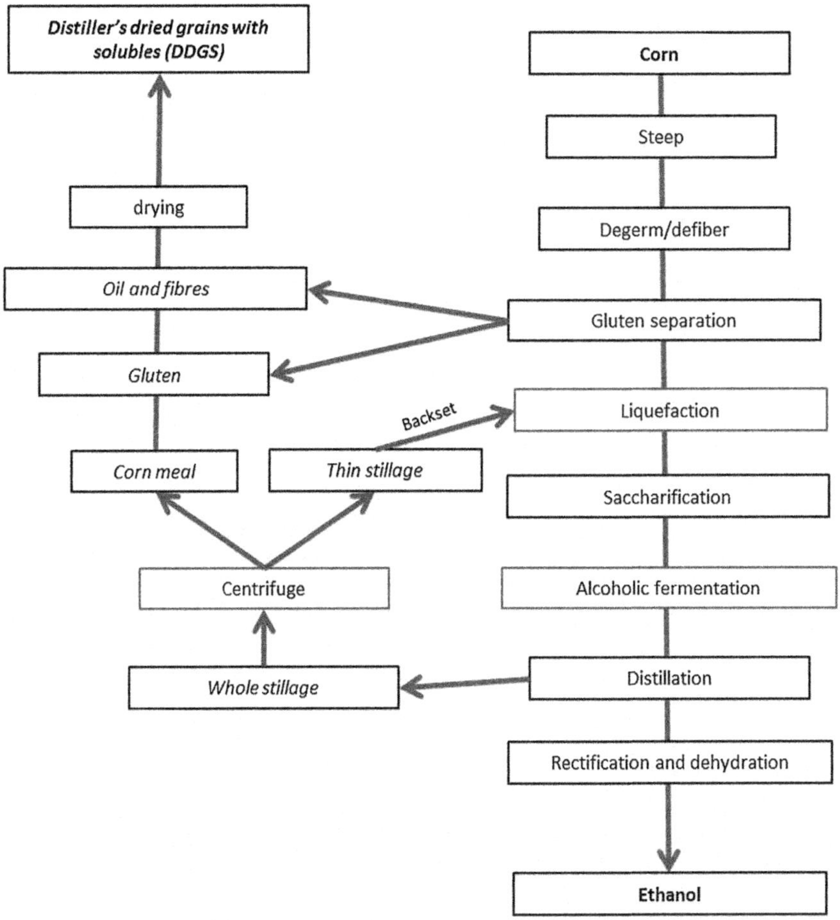

Fig. 3 Schematic representation of the wet mill process for starchy materials (adapted from Reiss 2012 and Bothast and Schlicher 2004). After steeping, crushing of hydrated grains gives fiber and oil from germs, proteins (gluten) and starch. Associated to corn meal recovered after distillation, these products are known as Distiller's dry grain with solubles (DDGS)

to produce ethanol by fermentation. Spent grains and cheap wines will be valued in animal feed. The stages of saccharification and of fermentation can be associated to decrease the duration of the process. This allows the glucose release at the desired rate in the medium by the regulation of the amylolytic activity.

4.3.3 Liquefaction, Saccharification, and Fermentation

The ground corn is first sent to a slurry tank along with process water, thermostable alpha-amylase, ammonia, and lime. After the slurry is prepared, the mixture

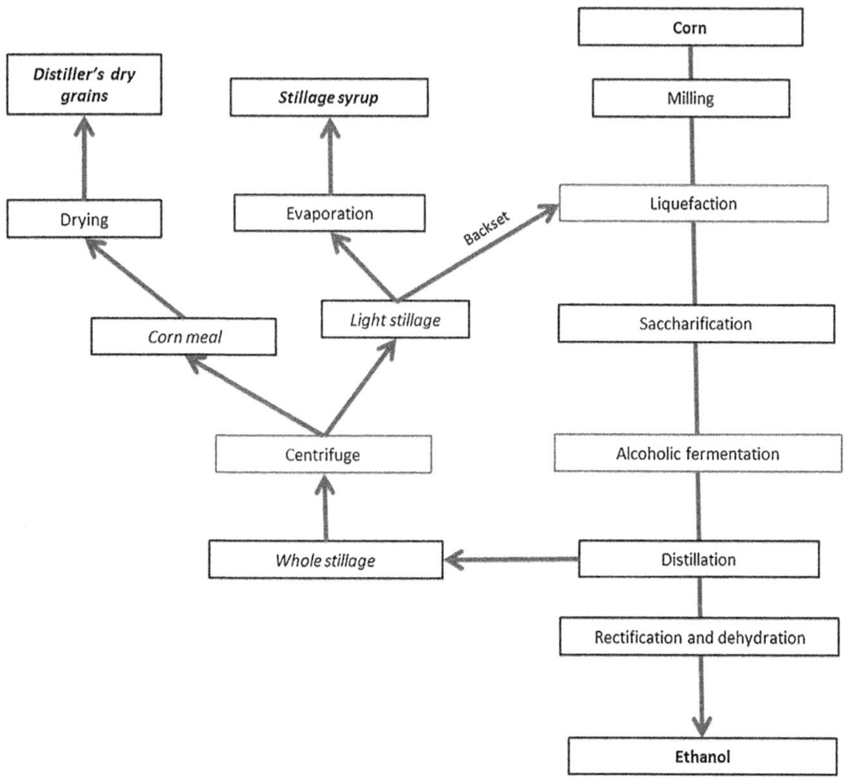

Fig. 4 Schematic representation of the dry grind process (adapted from Reiss 2012 and Bothast and Schlicher 2004). All grains components follow the entire process yielding only one coproduct: Distiller's dried grains (DDG)

undergoes liquefaction, where starch is gelatinized using a "jet-cooker" (steam injection heater) and hydrolysed with thermostable alpha-amylase into oligosaccharides also known as dextrins. During the gelatinization step, there is a sharp rise in the slurry viscosity that is rapidly decreased as the alpha-amylase hydrolyses the starch. Liquefaction is done at pH 6.5 and is initially held for 60 min at 88 °C with agitation. The output from the initial liquefaction step is combined with "backset," a recycled stream taken from the liquid portion of the "stillage" separated by centrifugation after the distillation step. The backset provides critical nutrients for the yeast later in fermentation.

These combined streams are "cooked" (i.e., held at 110 °C for 15 min), cooled, and then transferred to the saccharification tank. The resulting solution contains mainly dextrins, shorts oligosaccharides. Addition of glucoamylase converts dextrins into glucose. During this incubation at a temperature of 60 °C, almost all of the dextrins are converted to glucose. Glucoamylase continues to be active and can further hydrolyse during fermentation if there are any remaining dextrins.

Following the saccharification reaction, the slurry is transferred to the fermentation vessel and cooled at the yeast's optimal temperature (around 30 °C) prior to yeast addition. Often, ammonium sulfate or urea is added as a nitrogen source for the growth of yeast. Proteases can also be added. They break down the corn proteins to free amino acids, which serve as an additional source of nitrogen for the yeast. The fermentation requires 48–72 h to reach a final ethanol concentration of 10–12 %. The pH of the beer declines during the fermentation below pH 4, because of the carbon dioxide formed during the ethanol fermentation. This decrease in pH is important both for increasing the activity of glucoamylase and inhibiting the growth of contaminating bacteria. Either batch, fed-batch or continuous fermentation systems may be used, although batch processing is more common (Bothast and Schlicher 2004).

4.4 Cassava

After harvesting, the roots are chopped into chips for drying. Chips are usually sun dried. Dry chips are packed in bags and can be stored for months. Their starch content is more than 65 % (Sriroth et al. 2012). However, during storage, the starch yields decreases somewhat, depending on storage temperature: typically, 5 % reduction of starch yield is observed after 8 months of storage (Abera and Rakshit 2004). Another advantage of chips is the easy transportation. A big advantage of cassava over many other traditional crops is that it can be grown and harvested throughout the year. This results in a constant supply of cassava to the ethanol production facility in contrast to more seasonally crops. As for other starchy materials, the process described in Fig. 5 is carried out with two distinguishable technologies: wet milling process and dry-grinding process. Currently, most new facilities use the dry grinding process. The wet milling process starts with soaking the cassava chips in an acid to soften the material which results in the separation of starch from other components. The fibers are recovered in several separation steps. Next, the starch and protein are separated. In this process the streams are fractionated and several coproducts can be recovered. Most streams are recovered before the fermentation step. The dry grinding process starts with grinding the chips. This is done by hammer mills or roller mills. Next the ground material is mixed with water, cooked and mixed with enzymes. Cassava starch has a lower gelatinization temperature and offers a higher solubility for amylases in comparison to corn starch (Sanchez and Cardona 2008). This process produces only one coproduct that is separated at the end of the whole process, after fermentation, distillation, and drying: distiller dried grains with solubles. This is mostly used as animal feed. The use as animal feed is, however, limited due to the high fiber content.

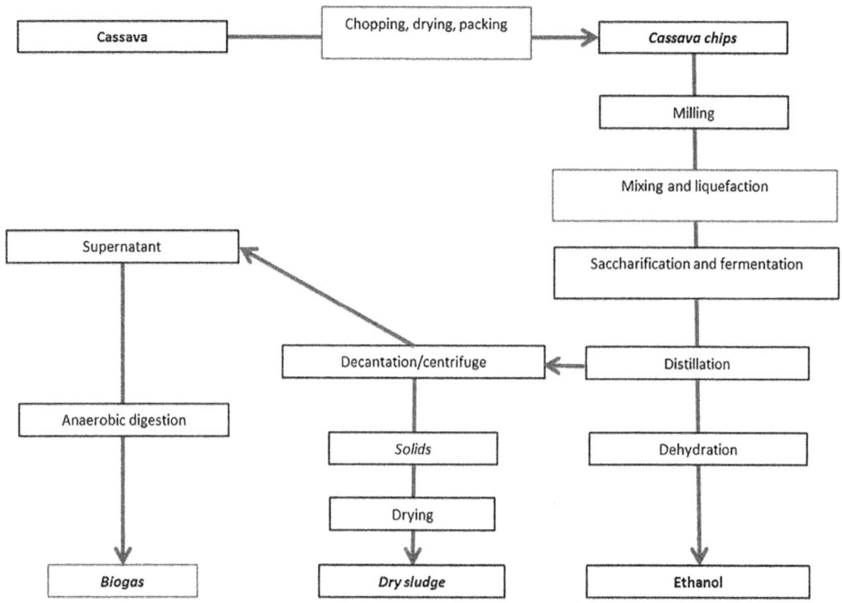

Fig. 5 Cassava processing for ethanol production. In a first step, starch-containing roots are chopped and dried. The resulting chips are further processed in ethanol facilities either by wet or dry milling

4.4.1 Fermentation

Ethanol fermentation can be realized classically in batch bioreactors of high volumes, typically around 500 m^3. These types of cultures have no constant yields throughout the process because of the change of medium composition during the fermentation. The cultures in batch impose time-outs during the filling and during the draining of tanks, their cleaning, and sterilization. The semi-continuous techniques called fed-batch are often used. These processes are said semi-continuous and allow limiting the sugar concentration in the medium as it is added in a progressive manner, limiting inhibition by the substrate or by the fermentation products which can be removed in the same way. Some precursors could also be added when it is necessary allowing a very precise regulation of the strain metabolism (Echegaray et al. 2000). The continuous systems are opened systems in which the cellular population is constantly maintained in a stable environment and a state of balanced growth, by removing continuously a part of the culture and by replacing it by fresh medium. Classically, they can be operated in chemostat or turbidostat mode.

To increase the productivity and decrease the cost of the production of ethanol, many researches are performed on high-density fermentation (Das Neves et al. 2006). These fermentations are said at high density because the fermentation medium contains more than 250 g/L of sugar, which in theory allows to obtain more than 15 % (v/v) of ethanol instead of 10–12 % generally obtained in most of the distilleries. The high-density fermentations possess numerous advantages. Indeed, these fermentations allow increasing production capacities without any modification of the structure of production (Bvochora et al. 2000; Puligundla et al. 2011). However, high-density cultures are very sensitive to temperature, concentration and cellular viability, medium composition, oxygen concentration. Furthermore, strains selected for strong concentrations in ethanol must be used (D'Amore 1992).

Nowadays, the production process of bioethanol from starch feedstock is developed to significantly reduce processing time and energy consumption by conducting saccharification and fermentation in a same step. This process is called "Simultaneous Saccharification Fermentation," or SSF process (Sriroth et al. 2012). In this SSF process, the liquefied slurry is cooled down to 32 °C, afterward glucoamylase and yeast are added together. While glucoamylase produces glucose, yeast can use glucose to produce ethanol immediately. No glucose is accumulated throughout the fermentation period (Rojanaridpiched et al. 2003).

4.4.2 Distillation and Ethanol Recovery

The fermentations presented above allow the obtaining of wines containing between 10 and 12 % (v/v) of ethanol in the case of sugar plants and until 18 % for the starchy plants, it is thus necessary to separate the ethanol of the water contained in beers. The mixture obtained from fermentation is not a water-ethanol binary system, even if it represents the main part, but a complex mixture containing volatile secondary products of the fermentation as aldehydes, esters, methanol, or higher alcohols possessing more than two carbons. The presence of these secondary products is regulated, for fuel alcohol, by the US or European standards.

The first step in ethanol recovery is the beer column, which recovers nearly all of the ethanol produced during fermentation in the distillate. An almost equal amount of water is also distilled that must be separated from the ethanol in the next stage of rectification/stripping. To obtain anhydrous ethanol, two stages are necessary after the distillation. The first part of this process is intended to extract the head products (aldehydes, ethyl acetate). The second part, is intended to concentrate the alcohol and to eliminate the tails (superior alcohols). Finally, the last part eliminates the methanol contained in the alcohol. The second stage for the obtaining of the pure ethanol consists in eliminating the residual water. Indeed, by distillation, one can obtain only a composition near the azeotrope composition, which is around 96 % (v/v) of ethanol for 4 % (v/v) of water. The alcohol so produced can serve then directly as biofuel. But if this one must be mixed with gasoline it is necessary to add a stage of dehydration to obtain anhydrous alcohol. The solution used in industry is the use of molecular sieve allowing the separation of ethanol from water according to the existing size difference between these two molecules.

5 Microorganisms for the First Generation Bioethanol

Technological development can help to diminish the environmental impact and the prices of the ethanol fuels. Numerous research has been conducted in order to obtain better fermentation conditions including organisms, low cost substrates, and process optimization to achieve optimal environmental conditions (Siqueira et al. 2008). Many microbial species are able to metabolize sugars and convert it into ethanol. However, only a few have proven sufficient efficiencies to be deployed at industrial scale. The yeast *Saccharomyces cerevisiae* and the bacteria *Zymomonas mobilis,* are the two main microorganisms traditionally found in first generation facilities. Although *Zymomonas mobilis* gave better yields, lower biomass production, and does not require any addition of oxygen contrary to *Saccharomyces cerevisiae*, it is also more sensitive to the environmental contaminations, it only uses a limited range of substrates and the produced biomass is not reusable for feeding purposes. Consequently, *Saccharomyces cerevisiae* is most of the time the microorganism of choice for the large-scale bioethanol production.

5.1 Metabolic Pathways Towards Ethanol Production

Three metabolic behaviors can be considered in yeasts. They are depending on the way the carbon source is used to produce the energy necessary for the cellular machinery. This type of metabolic behavior is strongly dependent on the environmental conditions such as what are the sugars available as a substrate, what is the local oxygen level available for the yeast and on the yeast itself. Purely oxidative yeasts will never produce ethanol, yeasts sensitive to the oxygen concentration will produce ethanol only in the case of an oxygen limitation and yeasts sensitive to the glucose content will also produce ethanol in the case of an excess of glucose even if oxygen is present. This last effect is known as the Crabtree effect. *Saccharomyces cerevisiae* is the typical example of a Crabtree positive yeast having an oxidative metabolism in the presence of oxygen at very low glucose concentrations (0.1–0.5 $g{\cdot}L^{-1}$ depending on the strain) and a mixed metabolism when the glucose concentration increases above that threshold. Soluble glucose penetrates into the yeast cell and is converted by a series of enzymatic reactions into pyruvate according to two major pathways. Pyruvate is further converted to carbon dioxide, energy, and eventually into ethanol. Some of the released energy is used by the yeast cells to support their growth and maintenance reactions during the fermentation. The rest of the energy is converted into heat and must be taken out of the fermenter or it will cause a temperature increase. Ethanol and carbon dioxide are taken out of the yeast cells.

5.2 *Conversion of Glucose to Pyruvate*

The conversion of glucose (or fructose) into pyruvate inside the cytosol of the yeast is known as the Embden-Meyerhof-Parnas pathway (EMP). This path is common for both aerobic and anaerobic conditions and includes 11 individual enzymatic steps. For instance, Cheng (2009) proposed a detailed description of the EMP pathway. Besides the EMP pathway, a second carbohydrate breakdown pathway is widely used among bacteria. It was discovered by Entner and Doudourof in *Pseudomonas saccharophilia.* Glucose-6-Phosphate is first dehydrogenated into 6-phosphogluconate by the glucose-6-phosphate dehydrogenase and this is further converted by the 6-gluconate dehydratase and the 2-keto-3-deoxy-6-phosphate-gluconate aldolase into one molecule of pyruvate and one molecule of 3-phosphoglyceraldehyde. The 3-phosphoglyceraldehyde can be further oxidized to pyruvate by the enzymes of the EMP pathway. *Zymomonas mobilis* degrades sugars into pyruvate with this pathway.

5.3 *Fermentative Pathway*

In the absence of oxygen, since there is no other electron acceptor *Saccharomyces cerevisiae* has a fermentative metabolism. Under these conditions, the energy is only produced via the EMP pathway. Pyruvate is then catalyzed by a pyruvate decarboxylase into acetaldehyde (with the release of one carbon dioxide) and finally acetaldehyde is handled by the alcohol dehydrogenase and reduced into ethanol. This further allows the recovery of the reduced cofactors NADH, H^+ into its oxidized form NAD^+. In addition to ethanol, other by-products are formed. The most important is glycerol. The purpose of the glycerol production pathway is to balance the redox balance in response to the biomass production associated with the fermentation reaction. The theoretical ethanol yield is 0.51 g ethanol per gram of glucose consumed while the biomass yield is approximately 0.10–0.12 g biomass per gram of glucose consumed. However, maintenance reactions, synthesis of the cellular infrastructure and the formation of secondary compounds (glycerol, acetic acid, reserve substances) limit this efficiency to approximately 90 % of its theoretical value.

5.4 *Oxidative Pathway*

The oxidative metabolism of glucose result in the complete degradation of the molecule into water and carbon dioxide with the simultaneous production of energy through the successive involvement of the EMP pathway, the conversion of pyruvate into acetyl-CoA through the pyruvate dehydrogenase reaction, the production of reduced cofactors ($NADH,H^+$ and $FADH_2$) into the tricarboxylic acid cycle (TCA cycle), the conversion of these reduced cofactors into a proton gradient and finally into adenosine triphosphate in the oxidative phosphorylation pathway. In this metabolic configuration, the biomass yields for *Saccharomyces cerevisiae* are approximately of 0.45–0.50 g biomass per gram of glucose without any production of ethanol.

5.5 Crabtree Effect

In order to explain the origin of the Crabtree effect, two main hypotheses have been considered. However, the exact origin of this saturation mechanism is not clearly established yet. The respiratory capacity might be limited due to the repression of some specific enzymes responsible for the transportation of the reducing power from the cytosol into the mitochondria by glucose. Thus, the decrease in the cell capacity to reoxidize NADH into NAD^+ will favor the ethanol production. As the conversion of pyruvate into ethanol produces little energy in comparison to the oxidative phosphorylation, the glycolytic flow will increase to meet the cellular needs. The metabolic saturation might be located at the pyruvate node. A too high glycolytic flux might progressively induce an over-accumulation of pyruvate and cause its redirection toward the production of ethanol and the others by-products. The Crabtree effect is not energetically interesting for the yeast as it has for first consequence a much lower biomass to glucose yield. However, the secondary consequence is a drastic increase of the growth rate from 0.08–0.15 h^{-1} to 0.4–0.45 h^{-1}. This energetic waste tendency turns out to be advantageous in the case of a competition for the substrate between microorganisms.

5.6 High Gravity Fermentation

The minimal ethanol concentration step that is necessary to achieve an economically viable process has been estimated at approximately 60 $g{\cdot}L^{-1}$ at the end of the fermentation step. This threshold is due to the high energetic costs for the evaporation and rectification unit operations (Fig. 2) of the ethanol process. In the current industrial processes, yields were estimated to be around 90–92 % of the maximal Gay–Lussac theoretical yield (being 0.511 g ethanol produced per g of glucose consumed). The productivity is approximatively of 2 kg ethanol produced per cubic meter of fermentation medium per hour. At the laboratory scale, the optimization of the fermentation conditions such as medium composition, vitamins and nitrogen feeding strategies, aeration, thermal settings, and strain selection makes it possible to enhance these performances significantly. For instance, it has been possible to obtained final concentrations with ethanol content higher than 180 $g.L^{-1}$ (Thomas and Ingledew 1992) or productivity records equivalent to the production of 3.5 kg

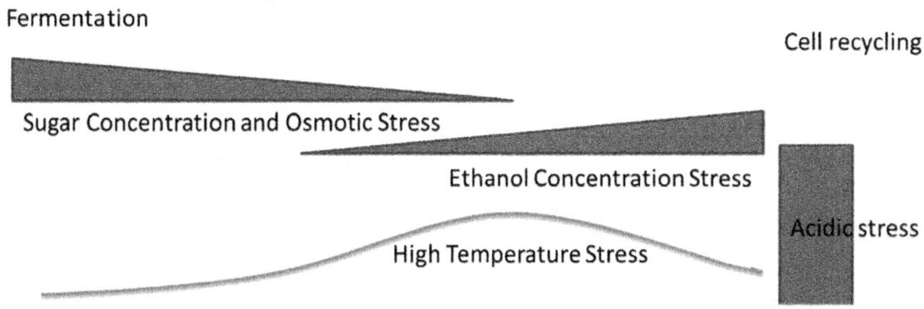

Fig. 6 Various environmental stresses that are traditionally occurring during the alcoholic fermentation (adapted from Della-Bianca et al. 2013; Koppram and Olsson 2014)

ethanol per cubic meter of fermentation medium per hour in a 45 h long fed-batch process (Alfenore et al. 2004). These high substrate concentration processes (>300 $g{\cdot}L^{-1}$) are not only essential for the first generation bioethanol but also for the successful implementation of a cost competitive second generation. Several technological advantages have to be taken into consideration such as a drastic decrease in the water and energy needs (Mussatto and Roberto 2004). However, the implementation of these conditions causes high loading of raw materials (up to 20 % mass concentration) and implies mixing conditions and increased amount of fermentation inhibitors. In fact, the environmental conditions found in the industrial production processes of first and second generation are far from the optimal physiological condition for yeast. It is therefore necessary that high gravity and very high gravity fermentations uses robust industrial strains to deal with these unfavorable environments (Koppram and Olsson 2014).

Figure 6 represents the various stresses that a yeast cell can encounter during single batch fermentation.

According to Della-Bianca et al. (2013), high-sugar concentration at the beginning of the fermentation and high ethanol concentration during the last stage, pH variations, high temperature, and the presence of toxic compounds are the most significant stresses. Furthermore, in the case, of a SSF configuration, the yeast strains have to be active near to the optimal activity of the amylase (pH 7) or cellulases (pH 6, 40–50 °C). It is also important to take into consideration the risk of contamination by bacteria during the fermentation process. In fact, in order to achieve low costs ethanol yields the fermentation is carried out nonaseptically in most of the facilities. It can deviate carbon away from ethanol production due to the bacterial cells metabolites and have detrimental effects on yeast performances (Basso et al. 2014). In this context, yeasts cells are usually recycled between two fermentations runs with the addition of sulfuric acid in order to reduce the bacterial contamination (Della-Bianca et al. 2013). Because they are commonly facing simultaneously or sequentially, a wide variety of stresses conditions, it is generally admitted that industrial strains are better and faster to adapt at these stresses (Pizarro et al. 2008). It is also usually believed that under selective industrial conditions, fermentation requires the use of industrial wild strains (Albers and Larsson 2009).

For instance, ethanol plants in Brazil are traditionally using selected baker's yeast as a starter. However, these yeasts were unable to survive the many recycling processes and were progressively replaced by indigenous yeasts strains during the season (Basso et al. 2014). This has adverse effects on the antifoam consumption, on the ethanol yield and consequently on the global performance of the whole process.

5.7 Strain Selection and Improvement

Given the large volumes of ethanol being produced at the facility scale, an improvement of the fermentation yields of only 1 % will have huge consequences on the overall profitability and environmental impact of the installation. Consequently, research efforts are now focusing on the improvement of the fermentation steps, including more resistant strains to the various stresses encountered during the alcoholic fermentation. Further efforts are also made toward the diversification of the usable carbon sources and on the improvement of the yields. One last focus is the consolidation of processes making it possible to carry out two or many unit operations in a single vessel in order to reduce investments and operating costs. Several strategies such are genome shuffling (Wang et al. 2014; Snoek et al. 2015), transcription machinery engineering, random mutagenesis have been developed with the aim to modify specific metabolic pathways and substrate transport systems in laboratory strains (Pereira 2014; Steensels et al. 2014). Results obtained by Pagliardini (2010) clearly demonstrate that the metabolic engineering of microorganism for the production of interesting molecules including ethanol is very difficult to implement as soon as it changes the central carbon metabolism. In fact, the major interconnections between the different metabolic pathways through energetics and redox couplings make it almost impossible to change one or the other pathway without affecting the system as a whole. In addition, certain metabolic pathways are leading to the production of molecules possessing fundamentals physiological roles. Changing these pathways can consequently affect the behavior of the mutant strains. The author was able to produce mutants that exhibited a reduction of up to 80 % in the glycerol production in comparison with the wild strain. The ethanol to glucose consumptions ratios increased from two to five percent depending on the mutant strain. For some of the mutants a decrease in the production of other organic acids and biomass was also observed. However, growth rates as well as ethanol tolerance and the ability to handle osmotic stresses were drastically reduced in the mutant strains in particular in anaerobic conditions making these new strains hardly usable for industrial purposes.

A second strategy aims at gaining and integrating physiological knowledge for strain improvements. The microflora of traditional and industrial fermentation processes is the potential source for discovering natural microbial strains with the desired physiological properties in order to resist better to the environmental stresses they are facing. Wild yeasts could be found for example in industrial alcoholic fermentation processes for beverage or bioethanol production plants. Once the right host strain is selected, metabolic engineering approaches might be used in a second step (Pereira 2014).

5.8 Resistance to Ethanol Stresses

Cot (2006) studied the physiological adaptation of yeasts strains to very high ethanoic content. According to the author, the key toward high ethanol content lies in the yeast's ability to maintain a high metabolic activity as long as possible. The selected yeast should demonstrate the best ability to withstand stressful conditions induced by ethanol accumulation. Ethanol is known to induce disturbances in the membranes (Walker-Caprioglio et al. 1990). However, in the case of very high gravity fermentations, this can be the cause not only of modifications of the activities of the membrane transporters but also of an irreversible partial or total loss of the membrane integrity. Lipid analysis clearly shows a correlation between the phospholipids content and the cellular viability (Cot 2006). The adaptation of yeast must depend on its ability to maintain its phospholipids content (Alexandre et al. 1994). Monitoring the gene expression reveals the establishment of a general stress response that has already started during the growth phase and reached its paroxysms at approximately 100 grams ethanol per litter when ethanol started to be produced without any production of biomass. Transcription and translation-related genes are downregulated and some genes necessary to enter the stationary phase are on the contrary being induced (SSA3, HSP12…). Carbohydrates and lipids reserves are accumulating, cells are less budding and close to a quiescent physiological state (Cot 2006). According to Herman (2002), quiescent cells are more resistant to ethanoic stresses. However, the mechanisms leading to this quiescent state remain hypothetical. As quiescent states are often found after an essential nutrient deficiency (Gray et al. 2004) and as ethanol is reported to inhibit many active transportation systems (Walker-Caprioglio et al. 1990), a possible mechanism would be that high ethanoic concentrations are responsible for active transportation systems damages and therefore causing indirectly nutritional starvation. In the same temporal windows, the overexpression of genes coding for active transportation systems was also reported. Under high ethanoic pressure, the cell population divides into two or more subpopulations that have different properties: quiescent and nonquiescent cells. Quiescent cells are usual daughter cells without any budding scar; they are accumulating glycogen and have different transcriptomic profiles that those of the nonquiescent cells (Cot 2006).

5.9 *Resistance to Temperature Stresses*

The temperature is known to affect the membrane fluidity (Kim et al. 2006). It may even cause an increase in permeability and ion leakage (Pipper 1995). It has been reported that the membrane composition might change in order to preserve its fluidity (Suutari et al. 1990). The conformation of proteins is also affected and may cause denaturation and aggregations. A large number of heat-shock genes might be induced in order to express chaperones proteins activities and prevent the denaturation of proteins (Morano et al. 1998). Postmus (2011) demonstrated that glycolytic flux is increased in C- and N-limited chemostats at higher temperatures and that the energetic cost for maintaining a proper protein folding is higher at 38 C than at 30 C. He also observed that transient increases of glycolytic flux are occurring immediately after an increase of the cultivation temperature in order to restore the balance between growth and maintenance and to produce ATP at a level that matches with the new energetic needs. The author reported that it might even cause a metabolic switch from respiratory to fermentative metabolism at high temperature. According to the author, the morphology of the mitochondria was severely perturbed. The author hypothesizes that in the 30–37 °C range, the energetic needs for the maintenance increases and at higher temperature, the energetic yield of the respiratory metabolism was not sufficient to support the investments in the mitochondria and causes the shift toward fermentative metabolism.

6 Environmental Assessment of First Generation Bioethanol

Environmental studies on large-scale production of first generation biofuels have shown that the net energy output is generally favorable to bioethanol. Similarly, greenhouse gas emissions are reduced in comparison to fossil fuels. This general assessment must, however, be strongly nuanced when the good agricultural practices are not respected. The gains expected by the technological process improvements are significant. However, they do not constitute the most promising source for further environmental savings (Benoist 2009). In the case of ethanol from wheat, the potential benefit achievable with technical savings is estimated at +0.24 MJ produced per MJ of fossil fuel consumed (−38 %) and at −11.6 g CO_2 equivalent for greenhouse gases emissions (−21 %). A more efficient valuation of the agricultural coproducts is the key to improve the life cycle analyses balance sheets whatever the bioethanol production pathway being considered. When these coproducts are recovered energetically as heat and electricity cogeneration, this allows an important reduction of greenhouse gas emissions potential. However, this gain is hampered by the significant decrease in the apparent productivity of the cultures. It is therefore appropriate to consider the apparent productivity of crops that can include the effects on land use of the by-products generated by these crops, especially their use in the animal feed sector. Consequently, the environmental evaluations must be considered not only in terms of MJ produced at the ethanol plant but rather in terms of kilometers traveled per MJ of fuel consumed and ideally per hectare of crops mobilized.

First generation bioethanol has been pointed out since 2008 because of its possible competition with food use and because of its impact on the local biodiversity (Koh and Ghazoul 2008). It this context, the use of alternative crops that are able to use on marginal lands (such as salted soils) that are not usable for the cultivation of conventional edible crops) is a topic of interest toward better environmental balances of first generation biofuels. More recently, concerns are also raising about the efficiency of the biofuels policies with respect to the climate change mitigation. The major assumption is that biofuels replace the consumption of fossil fuels. However, recent evidences suggested that because an increased use of biofuels will lower oil prices and consequently result in an increase of the crude oil consumption. This so-called rebound effect of biofuels can significantly lower the effectiveness of the biofuel policies in reducing the greenhouse gases emissions (Smeets et al. 2014). Financial mechanisms need to be set on the international trade system in order to offset this rebound effect. This is a major point in the international negotiations on climate change (De Perthuis and Trotignon 2015). Would it be possible to set a unique price for the carbon emissions or are we heading towards a global economy with variable ecological balances depending on the differentiated national development needs (Gollier and Tirole 2015)? Nevertheless, a sustainable and economically viable development of the green energy including bioethanol will probably never be possible without any satisfactory international agreement.

7 Toward Integrated First-, Second-, and Third Generation Biorefineries

Due to the high investment, costs, and uncertainties about the second generation bioethanol process (see Sigoillot and Faulds, Chapter "Second Generation Bioethanol" of this book), one solution is to consider an integration of first and second generation plants (Lennartsson et al. 2014) and even integrated third generation (Maranduba et al. 2015). This strategy is schematically represented in Fig. 7. In order to be functional, it requires the adaptation of the fermentation step to new substrates. Again, yeasts are at the center of this feedstock diversification. In the case of the second generation processes, the crucial point is the ability to convert xylose into ethanol without being negatively affected by inhibitory compounds such as furfural or acetic acids that are produced during the various pretreatments of the biomass. In the case of the third generation many new substrates such as Mannose, Rhamnose, Uronic acid, Glucuronic acid, or even N-acetylglucosamine may be used as a carbon sources. *Pichia stipites, Pichia angoraphorae, Candida shehatae, or Zimomonas palmae* have already proven their potential for fermenting part of these

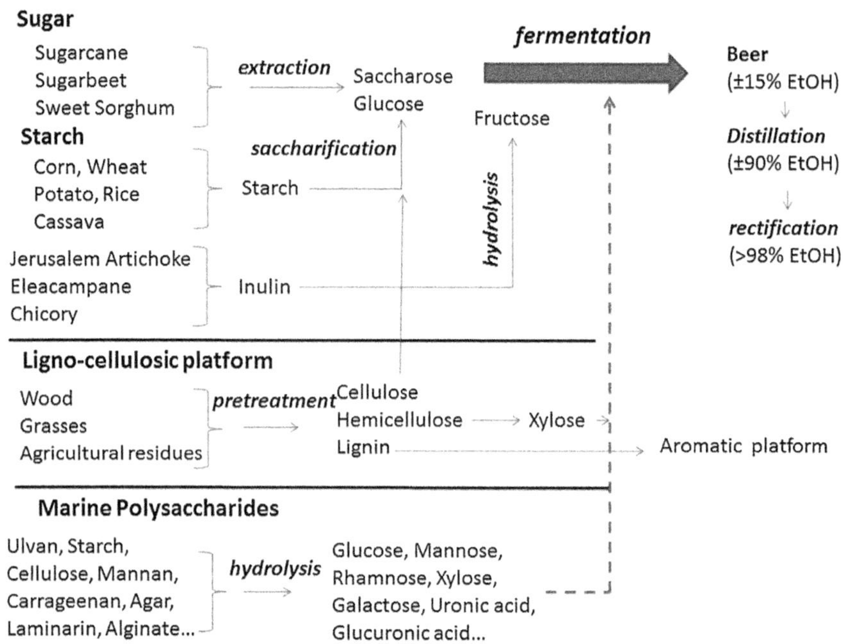

Fig. 7 Substrate diversification strategies for the integration of first, second, and third generation feedstocks into a single integrated biorefinery concept (adapted from Souza et al. 2015)

new substrates and to be more tolerant to inhibitory compounds (Gonçalves et al. 2015; Trivedi et al. 2015).

First generation grain mills currently dominate the world market through the production of Distiller's Dried Grains and Solubles (DDGS), predominantly a high-protein product, which is sold as low cost animal feed. The 1G refineries are usually located close to the production fields and so would evoke low transportation costs for the incorporation of straw, corncobs or bran as 2G raw materials. The combination of the high yielding 1G ethanol process would require less conversion of the 2G feedstock in order to make the biorefinery profitable. Sugarcane provides an attractive raw material feedstock for such a process, with a combined plant being projected to potentially outperform a 1G plant if all the plant is used, e.g., leaves and pentoses (Macrelli et al. 2014). Recent life cycle analysis assessment on the integrated first, second, and even third generation bioethanol facilities prove its potential in terms of reduction of the environmental weaknesses of the first generation facilities (Souza et al. 2015; Maity et al. 2014; Maity 2015a, b). Fungi have probably received the least attention in the biofuel arena. However, there are many studies, which show that fungi may make it easier to process biofuels from plant feedstock. Food grade fungi, such as *Rhizopus* sp., *Aspergillus oryzae, Fusarium vebebatum, Neurospora intermedia,* or *Monoacus purpureus*, already used in the manufacturing and processing of human food, can be used to utilize the remaining pentose sugars and so improve the overall economy of the first generation process (Lennartsson et al. 2014). In addition, those fungi are less prone to inhibition by the inhibitors generated by the process, which can interfere with the yeast metabolic processes. The fungus is easily removed from the liquid broth and can then be dried as a feed source, thus contributing to the improved economics of an integrated processing plant.

8 Conclusion and Perspectives

First generation bioethanol is produced at large industrial scale and has proven its economic viability. The United States and Brazil are the two leaders for production with the predominant use of corn and sugarcane respectively. Despites of the many concerns on the long-term sustainability of the first generation bioethanol, such as the impacts on land use change, the water use, the potential contamination of soils with the distillation residues, and the competition for food and feed production, many potentials routes are arising in order to make this production greener. In this context, integrated biorefineries are a promising way (i) to diversify the feedstocks usable, leading to reduced facilities size and optimized supply chains, (ii) to valorize more efficiently bagasse's from sugarcane and corn stover or even (iii) to exploit the potential of microalgae to capture the carbon dioxide that is produced during the fermentation steps and use it for example after transformation as the biodiesel that is needed for the agricultural machineries. This is the opportunity to take advantage of this large-scale successful deployment to get experience for the development of the most promising processing schemes for the next generation facilities that are still facing uncertainties with respect to their economic viability.

References

Abera S, Rakshit SK (2004) Effect of dry cassava chip storage on yield and functional properties of extracted starch. Starch 56(6):232–240

AFDC (2015) Renewable Fuels Association, Ethanol Industry Outlook 2008-2015 reports. http://www.afdc.energy.gov/data/10331. Accessed 8 Nov 2015

Agrosynergie (2011) Evaluation des measures de la politique agricole commune relatives au secteur du sucre. http://ec.europa.eu/agriculture/eval/...2011/exec_sum_fr.pdf. Accessed 19 Jan 2016

Aiyer PV (2005) Amylases and their applications. Afr J Biotechnol 4(13):1525–1529

Albers E, Larsson C (2009) A comparison of stress tolerance in YPD and industrial lignocellulose-based medium among industrial and laboratory yeast strains. J Ind Microbiol Biotechnol 36(8):1085–1091

Alexandre H, Rousseaux I, Charpentier C (1994) Relationship between ethanol tolerance, lipid composition and plasma membrane fluidity in *Saccharomyces cerevisiae* and *Kloeckera apiculata*. FEMS Microbiol Lett 124:17–22

Alfenore S, Cameleyre X, Benbadis L et al (2004) Aeration strategy: a need for very high ethanol performance in Saccharomyces cerevisiae fed-batch process. Appl Microbiol Biotechnol 63 (5):537–542

Alvarez MM, Pérez-Carrillo E, Serna-Saldivar SO (2010) Effect of decortication and protease treatment on the kinetics of liquefaction, saccharification, and ethanol production from sorghum. J Chem Technol Biotechnol 85:1122–1130

ANFAVEA (2015) http://www.virapagina.com.br/anfavea2015/. Accessed 08 Nov 2015

Bai FW, Anderson WA, Moo-Young M (2008) Ethanol fermentation technologies from sugar and starch feedstocks. Biotechnol Adv 26:89–105

Balat M, Balat H, Cahide OZ (2008) Progress in bioethanol processing. Prog Energ Combust 34:551–573

Baras J, Gacesa S, Pejin D (2002) Ethanol is a strategic raw material. Chem Ind 56:89–10

Basso TO, Gomes FS, Lopes ML et al (2014) Homo- and heterofermentative lactobacilli differentially affect sugarcane-based fuel ethanol fermentation. Ant van Leeuwenhoek Int J Gen Mol Microbiol 105(1):169–177

Benoist A (2009) Eléments d'adaptation de la méthodologie d'analyse de cycle de vie aux carburants végétaux: cas de la première génération. PhD. Ecole Nationale des Mines de Paris, France

Borges EP, Lopes ML, Amorim H (2012) Impact of sugar cane juice chemical composition on clarification and VHP sugar quality. Int Sugar J 114(1364):552–558

Bothast RJ, Schlicher MA (2004) Biotechnological processes for conversion of corn into ethanol. Appl Microbiol Biotechnol 67:19–25

Bvochora JM, Read JS, Zvauya R (2000) Application of very high gravity technology to the cofermentation of sweet stem sorghum juice and sorghum grain. Ind Crop Prod 11:11–17

Carter CA, Miller HI (2012) Corn for food, not fuel. http://www.nytimes.com/2012/07/31/opinion/corn-for-food-not-fuel.html?_r=0. Accessed 08 Nov 2015

Cerqueira Leite RC, Leal MRLVL, Cortez LAB et al (2009) Can Brazil replace 5% of the 2025 gasoline world demand with ethanol? Energy 34:655–661

Cheng JJ (2009) Biological process for ethanol production. Biomass Renew Energy Processes 209–269

Conab (2013) http://www.conab.gov.br/OlalaCMS/uploads/arquivos/13_09_12_17_38_24_5_mandioca_pdf

Congress US, Renewable Fuels, Consumer Protection, and Energy Efficiency Act of 2007 (2007) Section 102, Subtitle A, H.R. 6 (EAS)

Cot M (2006) Etudes physiologiques de l'adaptation et de la résistance de la levure Saccharomyces cerevisiae au cours de la production intensive d'éthanol. PhD. Institut National des Sciences Appliquées de Toulouse

D'Amore T (1992) Improving yeast fermentation performance. J Inst Brew 98:375–382

Das Neves MA, Kimura T, Shimizu N, Shiiba K (2006) Production of alcohol by simultaneous saccharification and fermentation of low-grade wheat flour. Braz Arch Biol Technol 49 (3):481–490

Della-Bianca BE, Basso TO, Stambuk BU et al (2013) What do we know about the yeast strains from the Brazillian fuel ethanol industry? Appl Microbiol Biotechnol 97(3):979–991

De Perthuis C, Trotignon R (2015) Le Climat à quel prix?. La Négociation climatique. Editions Odile Jacob, Paris

de Vries SC, van de Ven GWJ, van Ittersum MK et al (2010) Resource use efficiency and environmental performance of nine major biofuel crops, processed by first generation conversion techniques. Biomass Energy 34:588–601

Dutta K, Daverey A, Jih-Gaw Lin J-G (2014) Evolution retrospective for alternative fuels: first to fourth generation. Renew Energy 69:114–122

Echegaray O, Carvalho J, Fernandes A et al (2000) Fed-batch culture of *Saccharomyces cerevisiae* in sugarcane blackstrap molasses: invertase activity of intact cells in ethanol fermentation. Biomass Bioenergy 19:39–50

EIA (2013) http://www.eia.gov/todayinenergy/detail.cfm?id=11551. Accessed 08 Nov

European Commission (2010) Energy 2020: a strategy for competitive, sustainable and secure energy. COM (2010) 639 final, Brussels

European Parliament (2010) A new energy strategy for Europe 2011-2020. P7_TA (2010)0441, Brussels

Eurostat (2011) Statistical data. http://epp.eurostat.ec.europa.eu/portal/page/portal/eurostat/home/S

Gnansounou E, Dauriat A, Wyman CE (2005) Refining sweet sorghum to ethanol and sugar: economic trade-offs in the context of North China. Biores Technol 96(9):985–1002

Gonçalves FA, dos Santos ES, de Macedo GR (2015) Alcoholic fermentation of Saccharomyces cerevisiae, Pichia stipitis and Zymomonas mobilis in the presence of inhibitory compounds and seawater: Fermentation in the presence of inhibitors. J Basic Microbiol 55(6):695–708

Gollier C, Tirole J (2015) Negotiating effective institutions against climate change. Econ Energy Environ Pol 4(2):5–28

Graham-Rowe D (2011) Beyond food versus fuel. Nature 474:S6e8

Gray JV, Petsko GA, Johnston GC et al (2004) “Sleeping beauty”: quiescence in *Saccharomyces cerevisiae*. Microbiol Mol Biol Rev 68:187–206

Gupta A, Verma JP (2015) Sustainable bio-ethanol production from agro-residues: A review. Renew Sust Energy Rev 41:550–567

Haankuku C, Epplin FM, Kakani VG (2015) Industrial sugar beets to biofuel: field to fuel production system and cost estimates. Biomass Bioen 80:267–277

Harris PV, Wogulis M (2010) Polypeptides having amylolytic enhancing activity and polynucleotides encoding the same. Patent No. WO/2010/059413

Harris PV, Xu F, Kreel NE et al (2014) New enzyme insights drive advances in commercial ethanol production. Curr Opin Chem Biol 19:162–170

Herman PK (2002) Stationary phase in yeast. Curr Opin Microbiol 5:602–607

Hira A, Oliveira LG (2009) No substitute for oil? How Brazil developed its ethanol industry. Energ Policy 37:2450–2456

Horn SV, Vaaje-Kolstad G, Westereng B et al (2012) Novel enzymes for the degradation of cellulose. Biotechnol Biofuels 5:45

IBGE (2008) Brazilian Institute of Geography and Statistics [Instituto Brasileiro de Geografia e Estatística—IBGE]. Evolução da Produtividade da Cana-de-Açúcar no Brasil. www.ibge.gov.brS

Ishikawa K, Nakatani H, Katsuya Y et al (2007) Kinetic and structural analysis of enzyme sliding on a substrate: multiple attack in β-amylase. Biochem 46:792–798

Johnston DB, McAloon AJ (2014) Protease increases fermentation rate and ethanol yield in dry-grind ethanol production. Biores Technol 154:18–25

Kang Y-N, Tanabe A, Adachi M et al (2005) Structural analysis of threonine 342 mutants of soybean b-amylase: role of a conformational change of the inner loop in the catalytic mechanism. Biochem 44:5106–5116

Kim I-S, Moon H-Y, Yun H-S et al (2006) Heat shock causes oxidative stress and induces a variety of cell rescue proteins in Saccharomyces cerevisiae KNU5377. J Microbiol (Seoul, Korea) 44(5):492–501

Koppram R, Olsson L (2014) Combined substrate, enzyme and yeast feed in simultaneous saccharification and fermentation allow bioethanol production from pretreated spruce biomass at high solids loading. Biotechnol Biofuel 7. doi:10.1186/1754-6834-7-54

Kramer GFH, Gunning AP, Morris VJ et al (1993) Scanning tunneling microscopy of *Aspergillus niger* glucoamylase. J Chem Soc, Faraday Trans 89:2595–2602

Koh LP, Ghazoul J (2008) Biofuels, biodiversity, and people: Understanding the conflicts and finding opportunities. Biol Conserv 141(10):2450–2460

Larkin S, Ramage J, Scurlock J (2004). Bioenergy. In: Boyle G (ed) Renewable energy. Oxford University Press, p 135

Lei H, Zheng L, Wang C, Zhao H et al (2013) Effects of worts treated with proteases on the assimilation of free amino acids and fermentation performance of lager yeast. Int J Food Microbiol 161:76–83

Lennartsson PR, Erlandsson P, Taherzadeh MJ (2014) Integration of the first and second generation bioethanol process and the importance of by-products. Biores Technol 165:3–8
Linoj KNV, Dhavala P, Goswami A et al (2006) Liquid biofuels in South Asia: resources and technologies. Asian Biotechnol Develop Rev 8:31–49
Lo Leggio L, Simmons TJ, Poulsen J-CN et al (2015) Structure and boosting activity of a starch-degrading lytic polysaccharide monooxygenase. Nature Comms 6:5961
Lombard V, Ramulu HG, Drula E et al (2014) The carbohydrate-active enzymes databases (CAZy) in 2013. Nucleic Acids Res 42:490–495
Luckett CR, Wang Y-J (2012) Effects of β-amylolysis on the resistant starch formation of debranched corn starches. J Agric Food Chem 60:4751–4757
Lundgard R, Svensson B (1987) The four major forms of barley a-amylase. Purification, characterization and structural relationship. Carlsberg Res Commun 52:313–326
Macrelli S, Galbe M, Wallberg O (2014) Effects of production and market factors on ethanol profitability for an integrated first and second generation ethanol plant using the whole sugarcane as feedstock. Biotech Biofuels 7:26
Maity JP, Bundschuh J, Chen C-Y et al (2014) Microalgae for third generation biofuel production, mitigation of greenhouse gas emissions and wastewater treatment: present and future perspectives—a mini review. Energy 78:104–113
Maity SK (2015a) Opportunities, recent trends and challenges of integrated biorefinery: part I. Renew Sustain Energy Rev 43:1427–1445
Maity SK (2015b) Opportunities, recent trends and challenges of integrated biorefinery: Part II. Renew Sustain Energy Rev 43:1446–1466
Maranduba HL, Robra S, Nascimento IA et al (2015) Reducing the life cycle GHG emissions of microalgal biodiesel through integration with ethanol production system. Biores Technol 194:21–27
McAloon A, Taylor F, Yee W et al (2000) Determining the cost of producing ethanol from corn starch and lignocellulosic feedstocks. http://www.nrel.gov/docs/fy01osti/28893.pdf. Accessed 08 Nov 2015
Mikami B, Iwamoto H, Malle D et al (2006) Crystal structure of pullulanase: evidence for parallel binding of oligosaccharides in the active site. J Mol Biol 359:690–707
Morano KA, Liu PC, Thiele DJ (1998) Protein chaperones and the heat shock response in *Saccharomyces cerevisiae*. Curr Opin Microbiol 1(2):197–203
Morris VJ, Gunning AP, Faulds CB et al (2005) AFM images of complexes between amylose and *Aspergillus niger* glucoamylase mutants, native and mutant starch binding domains: a model for the action of glucoamylase. Starch 57:1–7
Mussatto SI, Roberto IC (2004) Alternatives for detoxification of diluted-acid lignocellulosic hydrolyzates for use in fermentative processes: a review. Biores Technol 93:1–10
Nielsen PK, Bønsager BC, Fukuda K et al (2004) Barley a-amylase/subtilisin inhibitor: structure, biophysics and protein engineering. Biochim Biophys Acta 1696:157–164
Nikolov ZL, Meagher MM, Reilly PJ (1989) Kinetics, equilibria, and modeling of the formation of oligosaccharides from D-glucose by *Aspergillus niger* glucoamylase. Biotechnol Bioeng 34:694–704
NREL (2014) Renewable Energy Data Book, US Dept of Energy. www.nrel.gov/docs/fy16osti/64720.pdf. Accessed 19 Jan 2016
OECD/IEA (2008) From 1st to 2nd generations of biofuel technologies, overview of current industry and RD&D activities
Ostanin K, Harms EH, Stevis PE (1992) Overexpression, site directed mutagenesis and mechanism of *Escherichia coli* acid phosphatase. J Biol Chem 267:22830–22836
Pagliardini J (2010) Optimisation du rendement de production de bioéthanol chez *Saccharomyces cerevisiae* par minimisation de la synthèse du glycérol: approche intégrée de génie métabolique et microbiologique. PhD. Toulouse University, France
Pereira LFB (2014) Bioethanol—robust production strains for process identification. PhD. Univesidade do Minho, Portugal

Pipper PW (1995) The heat-shock and ethanol stress responses of yeast exhibit extensive similarity and functional overlap. FEMS Yeast Res 134(2–3):121–127

Pizarro FJ, Jewett MC, Nielsen J et al (2008) Growth temperature exerts differential physiological and transcriptional responses in laboratory and wine strains of *Saccharomyces cerevisiae*. Appl Environ Microbiol 74(20):6358–6368

Postmus J (2011) The physiological response of *Saccharomyces cerevisiae* to temperature stress. PhD. Amsterdam University, The Netherlands

Puligundla P, Smogrovicova D, Obulam VSR, Ko S (2011) Very high gravity (VHG) ethanolic brewing and fermentation: a research update. J Ind Microbiol Biotechnol 38:1133–1144

Quintero JA, Montoya MI, Sánchez OJ et al (2008) Fuel ethanol production from sugarcane and corn: comparative analysis for a Colombian case. Energy 33:385–399

Reiss J (2012) Intensification de la brique «fermentation alcoolique» de substrats betteraviers pour la production d'éthanol. PhD. Toulouse University, France

REN21 (2012) Renewables 2012? Global Status Report, REN21 Secretariat, Paris

Rockey WM, Laederach A, Reilly PJ (2000) Automated docking of α-(1→4)- and α-(1→6)-linked glucosyl trisaccharides and maltopentaose into the soybean β-amylase active site. Proteins Struct Funct Genet 40:299–309

Rojanaridpiched C, Kosintarasaenee S, Sriroth K et al (2003) Development of Ethanol Production Technology from Cassava Chip at a Pilot Plant Scale. National Research Council of Thailand

Rosillo-Calle F, Cortez LAB (1998) Towards ProAlcool II—a review of the Brazilian bioethanol programme. Biomass Bioenergy 14:115–124

Ruan Q, Chen WB, Huang SH (2001) The mathematics model and matrix method of complex cocurrent multi-effect evaporation. Eng Sci 3:36–41

Salazar-Ordóñez M, Pérez-Hernández PP, Martín-Lozano JM (2013) Sugar beet for bioethanol production: an approach based on environmental agricultural outputs. Energy Policy 55: 662–668

Sanchez OJ, Cardona CA (2008) Trends in biotechnological production of fuel ethanol from different feedstocks. Biores Technol 99:5270–5295

Sancho AI, Faulds CB, Svensson B et al (2003) Cross-inhibitory activity of cereal protein inhibitors against a-amylases and xylanases. Biochim Biophys Acta 1650:136–144

Seabra JEA, Macedo IC, Chum HL et al (2011) Life cycle assessment of Brazilian sugarcane products: GHG emissions and energy use. Biofuels Bioprod Biorefin 5:519–532

Siqueira PF, Karp SG, Carvalho JC et al (2008) Production of bio-ethanol from soybean molasses by Saccharomyces cerevisiae at laboratory, pilot and industrial scales. Biores Technol 99:8156–8163

Smeets E, Tabeau A, van Berkum S et al (2014) The impact of the rebound effect of the use of first generation biofuels in the EU on greenhouse gas emissions: a critical review. Renew Sustain Energy Rev 38:393–403

Snoek T, Picca Nicolino M, Van den Bremt S et al (2015) Large-scale robot-assisted genome shuffling yields industrial Saccharomyces cerevisiae yeasts with increased ethanol tolerance. Biotech Biofuels 8(1):32

Soccol CR, Vandenberghe LPS, Costa B et al (2005) Brazilian biofuel program: an overview. J Sci Ind Res 64:897–904

Soccol CR, Vandenberghe LPD, Medeiros ABP et al (2010) Bioethanol from lignocelluloses: status and perspectives in Brazil. Biores Technol 101(13):4820–4825

Souza SP, Gopal AR, Seabra JEA (2015) Life cycle assessment of biofuels from an integrated Brazilian algae-sugarcane biorefinery. Energy 81:373–381

Sriroth K, Wanlapatit S, Piyachomkwan K (2012) Cassava Bioethanol, Bioethanol, Prof. Marco Aurelio Pinheiro Lima (Ed.), ISBN: 978-953-51-0008-9, InTech. http://www.intechopen.com/books/bioethanol/-cassava-bioethanol

Steensels J, Snoek T, Meersman E et al (2014) Improving industrial yeast strains: exploiting natural and artificial diversity. FEMS Microbiol Rev 38(5):947–995

Suutari M, Liukkonen K, Laakso S (1990) Temperature adaptation in yeasts: the role of fatty acids. J Gen Microbiol 136(8):1469–1474

Thomas KC, Ingledew WM (1992) Production of 21% (v/v) ethanol by fermentation of very high gravity (VHG) wheat mashes. J Ind Microbiol 10(1):61–68
Trivedi N, Gupta V, Reddy C et al (2015) Marine macroalgal biomass as a renewable source of bioethanol. In: Kim S-K, Lee C-G (eds) Marine Bioenergy. CRC Press, pp 197–216
Udop (2015) União dos Produtores de Bioenergia http://www.udop.com.br/index.php?item=safras. Accessed 08 Nov 2015
Unica (2015) União da Indústria da Cana de Açúcar. http://www.unicadata.com.br/historico-de-producao-e-moagem.php. Accessed 08 Nov 2015
UNCTAD (2015) Commodity profile—Cassava. http://www.unctad.info/en/Infocomm/AACP-Products/COMMODIRY-PROFILE—Cassava/. Accessed 08 Nov 2015
Vidal BC Jr, Rausch KD, Tumbleson M et al (2009) Protease treatment to improve ethanol fermentation in modified dry gring corn processes. Cereal Chem 86:323–328
Walker-Caprioglio HM, Casey WM, Parks LW (1990) *Saccharomyces cerevisiae* membrane sterol modifications in response to growth in the presence of ethanol. Appl Environ Microbiol 56:2853–2857
Wang P, Johnston DB, Rausch KD et al (2009) Effects of protease and urea on a granular starch hydrolyzing process for corn ethanol production. Cereal Chem 86:319–322
Wang L, Raul Quiceno R, Price C et al (2014) Economic and GHG emissions analyses for sugarcane ethanol in Brazil: looking forward. Renew Sustain Energy Rev 40:571–582
Wheals AE, Basso LC, Alves DMG et al (1999) Fuel ethanol after 25 years. Trends Biotechnol 17 (12):482–487
Woods J (2000) Integrating sweet sorghum and sugarcane for bioenergy: modelling the potential for electricity and ethanol production in SE Zimbabwe. PhD thesis. King's College, London
Woods J (2001) The potential for energy production using sweet sorghum in southern Africa. Energy Sustain Dev 1:31–38
Yoosin S, Sorapipatana C (2007) A study of ethanol production cost for gasoline substitution in Thailand and its competitiveness: thammasat. Int J Sci Technol 12:69–80
Zarrilli S (2006) The emerging biofuels market: regulatory, trade and development implications. UNCTAD Intergovernmental Expert Meeting on BioFuels, Geneva, November 3 2006
Zhou GT, Jiang ZM, Dong XL et al (2011) Recycling and refining of alcohol derived from waste beer separated from spent yeast. J Am Soc Brew Chem 69(3):158–162

Second Generation Bioethanol

Jean-Claude Sigoillot and Craig Faulds

Abstract Second generation bioethanol, i.e., ethanol from lignocellulosic biomass is envisaged as a renewable source of transport fuel in the next years to complement or replace the first generation and the fossil fuel. At present, the most economic solution seems to be the enzymatic way that produces fermentable sugars from cellulose and hemicellulose. To make this process commercially viable, several improvements are needed to enhance the process. Currently, stocks of biomass are constituted by available agro-industrial residues such as corn stover and straws, wood and wood processing residues and dedicated crops such as miscanthus and other grasses. The biological process for converting the lignocellulose to fuel ethanol requires: a pretreatment to liberate cellulose and hemicellulose from their complex with lignin, a depolymerization of the carbohydrate polymers to produce free sugars, and a fermentation of mixed hexose and pentose sugars to produce ethanol. Pretreatment must be cost effective and must be adapted to the type of biomass. They must not produce high amount of inhibitors such as 5-hydroxymethyl furfural, even if different technique are currently available to detoxify fermentation broths. Saccharification by enzymes and fermentation should be conducted separately (SHF) or simultaneously (SSF). Integration of the process the development of cofermentation of pentoses produced through the pretreatment of the biomass (SSCF). Progress in genetic engineering allows us to consider the development of Consolidated BioProcesses (CBP) in which a single microbial strain is able to ferment polysaccharides to produce ethanol in one step. The economic viability of ethanol recovery from the fermentation mash depends of the overall process, as

J.-C. Sigoillot · C. Faulds (✉)
UMR1163 Biodiversité et Biotechnologie Fongiques, INRA,
13288 Marseille, France
e-mail: craig.faulds@univ-amu.fr

J.-C. Sigoillot · C. Faulds
UMR1163 Biodiversité et Biotechnologie Fongiques, Aix Marseille Université,
13288 Marseille, France

J.-C. Sigoillot · C. Faulds
Polytech' Marseille, UMR1163 Biodiversité et Biotechnologie Fongiques,
13288 Marseille, France

C.R. Soccol et al. (eds.), *Green Fuels Technology*,
Green Energy and Technology, DOI 10.1007/978-3-319-30205-8_9

distillation heat balance is dependant of ethanol concentration. The future of ethanol from biomass is widely dependent on the oil price and on the political will of the different countries.

Keywords Bioethanol · Biomass · Pretreatment · SHF · SSF · CBP

1 Introduction

Renewable carbon termed as "biomass" seems the most promising way to produce energy in the near future. It would complement solar, wind, and other intermittent energy sources in the renewable energy mix of the future. Although wood and crop residues have been used ancient times as a source of energy and combustible materials, and are used in modern heat and energy producing plants as fuelwood, wood chips and pellets, etc., one of the most promising applications is to process biomass to obtain secondary biofuels such as ethanol that can be used in vehicles and various industrial processes. The secondary biofuels are further divided into first, second and third generation biofuels on the basis of raw material and technology used for their production (Nigham and Singh 2011).

Nearly all fuel ethanol is currently produced in first generation plants by fermentation of glucose from starch or sucrose from sugar cane and sugar beet (Rosillo-Calle and Cortez 1998; See Bertrand et al. Chapter "First Generation Bioethanol" of this book). The main ethanol producers are the USA and Brasil (Table 1). Starch-containing biomass includes corn, wheat, oats, rice, potato, and cassava. On a dry basis, corn, wheat, sorghums and other grains contain around 60–75 % (wt/wt) of starch, hydrolyzable to hexose with a significant weight increase

Table 1 On going projects for bioethanol production from lignocellulosic biomasses

Country	Ethanol production (million liters)
USA	60536.2
Brasil	37323.3
China	9350
EU	7625.1
India	2397.8
Canada	1512.5
Thailand	1110.9
Argentina	625.9
Australia	434.8

All are using enzymatic hydrolysis and fermentation except Ineos Bio process based on gazeification and bacterial ethanol production from syngas
Ethanol production for biofuel in 2015
Source OECD-FAO Agricultural outlook 2013–2022

(stoechiometrically the starch to hexose ratio is 9:10) and these offer a good resource in many fermentation processes (Jackman 1987). Therefore, agronomic residues such as corn stover (corn cobs and stalks), sugarcane waste, wheat or rice straw, forestry, and paper mill discards, the paper portion of municipal waste and dedicated energy crops such as miscanthus and switch grass are sources of polysaccharides, mainly cellulose which could be converted in sugars (Lin and Tanaka 2006).

Plant cell walls contain three major polymers: cellulose (an insoluble linear unbranched homopolysaccharide consisting of glucose subunits linked via β-1,4-glycosidic linkages), hemicellulose (non-cellulosic polysaccharides including mainly xylans, mannans, and glucans) and lignin (an intricate polyphenolic structure). The total complex of these polymers is often referred to as lignocellulose (Comb and Hatzis 1996). Cellulosic materials represent the most abundant global source of biomass and have been largely underexploited. The global production of plant biomass, of which over 90 % is lignocellulose, amounts to about 200×10^9 tons per year, where about $8–20 \times 10^9$ tons of the primary biomass remains potentially accessible (Lin and Tanaka 2006). Conversion of cellulose and other polysaccharides into fermentable sugars could be achieved by both chemical and biological processes. Acid hydrolysis of plant lignocellulosic biomass has been known since 1819 (Harris and Beglinger 1946) and was developed during World War II. Again, it was the war-time efforts by researchers at the U.S. Army Quartermaster Research Center in Natick, Massachusetts, that led to the discovery of cellulose hydrolyzing enzymes (Reese 1976). Cellulases were identified as the cause of degradation of cellulose in the cotton fabrics comprising the army uniforms in tropical arenas. The culture that was isolated as one of the potent cellulases producers was *Trichoderma viride* which was later re-named *Trichoderma reesei* in honor of Dr. Reese (Katzen and Tsao 2000).

The development of strain hyperproducing cellulases in several countries has allowed the design of industrial-scale biological processes. The biological process for converting the lignocellulose to fuel ethanol requires: a pretreatment to liberate cellulose and hemicellulose from their complex with lignin, a depolymerization of the carbohydrate polymers to produce free sugars, and a fermentation of mixed hexose and pentose sugars to produce ethanol. The general process is schematized in Fig. 1. It includes the two main ways (Simultaneous Saccharification and Fermentation, SSF and Separate Hydrolysis and Fermentation, SHF), the use of C5 hydrolysates to produce cellulose and the ethanol recovery and dehydration. Extensive research has been carried out in this field for decades (Yu and Zhang 2004), and the first demonstration plant using lignocelluloses feedstocks has been in operation in Canada since April 2004 (Tampier et al. 2004). Since this date, several projects were developed by different companies and are currently operational (Table 2). However, despite the work done, the industrial scale up of this process appears to be still hindered by technological issues or by the lack of a biomass refinery approach in which ethanol is one of several products. In fact, because raw material cost comprises more than 20 % of the production cost (Brown et al. 2001; Kaylen et al. 2000;

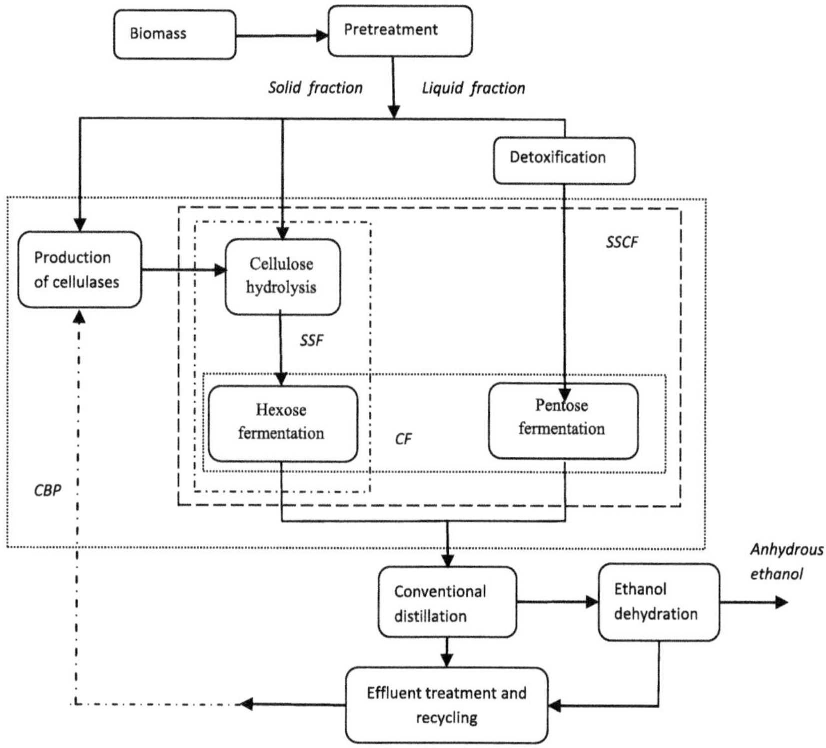

Fig. 1 Schematic diagram of lignocellulosic ethanol production displaying the main process steps: pretreatment, detoxification, production of cellulases from liquid or solid fractions of pretreated biomass and/or waste recycling, cellulose hydrolysis, fermentation of hexoses and pentoses, ethanol recovery and dehydration. From inner to external rectangle: *SSF* includes hydrolysis and fermentation; *CF* co-fermentation of hexoses and pentoses, *SSCF* simultaneous saccharification and co-fermentation, *CBP* consolidated bioprocess where a single organism is able to achieve all the process. Adapted from Cardona et al. (2007)

Zhuang et al. 2001), the optimization of the cellulose conversion should be accomplished by correct management and utilization of all process streams.

In general, prospective lignocellulosic materials for fuel ethanol production can be divided into six main groups: crop residues (cane bagasse, corn stover, wheat straw, rice straw, rice hulls, barley straw, sweet sorghum bagasse, olive stones and pulp), hardwood (aspen, poplar), softwood (pine, spruce), cellulose wastes (newsprint, waste office paper, recycled paper sludge), herbaceous biomass (alfalfa hay, switchgrass, reed canary grass, coastal Bermudagrass, thimothy grass), and municipal solid wastes (MSW). The composition of most of these materials can be found elsewhere (e.g., Sun and Cheng 2002). The main processing challenge in the ethanol production from lignocellulosic biomass is the feedstock pretreatment which allows cellulose accessibility to enzymes or chemicals.

Table 2 Ethanol production (million litres) for biofuel in several countries in 2015

Company	Location	Production (kt/y)	Biomass
Beta-renewable Mossi-Ghisolfi Novozyme	Crescentino (Italy)	35	Wheat and rice straws, *Arundo donax*
Abengoa bioenergy	Hugoton, KS	70	Corn stover, wheat straw, milo stubble and switchgrass
POET-DSM	Emmetsburg, IO	50–70	Corn crop residues
IOGEN RAIZEN (Shell)	Costa Pinto (Brasil)	30	Bagasse, wheat straw
INEOS	Vero Beach, FLA	24	Gaseification of all organic residues
IMBICON	Kalundborg (Denmark)	30	Straw
DUPONT	Nevada	80	Corn stover

Source OECD-FAO Agricultural outlook 2013–2022

2 Biomass: Availability and Potential Applications

Due to the structure of lignocellulosic complexes, an adapted pretreatment is required for their degradation leading to the removal of lignin, the partial or total hydrolysis of the hemicelluloses and the decrease in the fraction of crystalline cellulose related to the amorphous one, which is more easily hydrolyzed in the subsequent steps (Cardona and Sanchez 2007). Lignocellulosic waste materials obtained from energy crops, wood and agricultural residues represent the most abundant global source of renewable biomass (Lin and Tanaka 2006). Among the agricultural residues, wheat straw is the largest biomass feedstock in Europe and the second largest in the world after rice straw (Kim and Dale 2004).

2.1 Wheat Straw

In France, for example, wheat represents the main grown crop, producing roughly 22 Mt of straw annually. Straw is mainly used for cattle breeding. Disposal of wheat straw by burning has been practiced for a long time. In recent years, however, this practice has been challenged due to increased concern over the health effects of smoke from burning fields (Kerstetter and Lyons 2001). Burning of wheat straw results in large amounts of air pollutants including particulate matter (PM10), CO and NO_2 (Li et al. 2008). Thus, finding an alternative way for disposal of surplus wheat straw is of high interest and an immediate necessity. New agricultural practices taking into account the carbon balance of soils favor the landfill and decrease the availability, leaving only 8 Mt for energy production. However, as straw density is very low, transport on long seems not economically viable. More generally, biomass availability at a short distance around the plant is a key parameter for biorefinery implantation. Consequently, a careful study of the availability of resources is essential and is conditioning plant implantation. Kumar et al. (2005) have proposed an interesting way for integrating the feedstock transport to the ethanol production facility and the saccharification process named simultaneous transport and saccharification. These authors consider that the enzymatic hydrolysis of corn stover can be carried out in pipelines during its transport; the hydrolysed corn stover could directly enter the ethanol fermentation plant. Another parameter is that annual harvest campaign involves the constitution of large stocks to allow operation of the plant throughout the year submitted to climatic hazards. Diversification of biomass should overcome the problem allowing optimal operation of the plant throughout the year and avoiding climatic hazards.

2.2 Rice Straw

Rice straw is one of the most abundant lignocellulosic waste materials in the world. In terms of total production, rice is the third most important grain crop in the world behind wheat and corn. Every kilogram of grain harvested is accompanied by production of 1–1.5 kg of the straw (Maiorella 1985). Rice straw has several characteristics that make it a potential feedstock for fuel ethanol production. It has high cellulose and hemicelluloses content that can be easily hydrolyzed into fermentable sugars. In terms of chemical composition, the straw predominantly contains cellulose (32–47 %), hemicellulose (19–27 %) and lignin (5–24 %).

2.3 Corn Stover

Corn stover is considered as one of the potential biomass and most studied material too. According to Kadam and McMillan (2003), about 80–100 dry tons of corn stover/year can be used for ethanol production in the US. It has been estimated that approximately 256 million dry tons of corn stover will be available in the year 2030 due to collection technologies improvement and a steady yield increase (Karunanithy and Muthukumarappan 2009). Several pretreatment were compared to enhance ethanol production from corn stover (Uppugundia et al. 2014).

2.4 Grasses

Perennial grasses show some ecological advantages in comparison with annual crops. Perennial rhizomatous grasses, as miscanthus (*Miscanthus* × *giganteus*) and giant reed (*Arundo donax* L.) are generating much interest in Europe, as new sources of biomass for energy production. In fact miscanthus and giant reed need to have a limited soil management (planting and related tillage), reducing risk of soil erosion and determining a likely increase in soil carbon content and in biodiversity. Moreover, due to the recycling of nutrients by their rhizome systems, perennial grasses have a low demand for nutrient inputs and since they have few natural pests, they may also be produced without pesticide use. It is frequently asserted that plants convert only 0.1 % of solar energy into biomass, therefore requiring unacceptable amounts of land for production of fuel feedstocks. The C4 perennial grass *M. giganteus* has proved a promising biomass crop in Europe, while switchgrass (*Panicum virgatum*) has been tested at several locations in North America (Heaton et al. 2008). Angelini et al. (2009) found that the energetic average yields of *A. donax* and *M. giganteus* grown in central Italy during a 12 year cycle of culture was 637 GJ/Ha/y (14 TEP) and 467 GJ/Ha/y (10 TEP), respectively. In the climatic condition of Central Italy, giant reed was characterized by higher dry yield than miscanthus (37.7 vs. 28.7 t/Ha averaged from 2 to 12 years of growth).

2.5 Wood

Wood from several tree species must also be considered for bioethanol production. Wood is used in several ways. Power plants using wood chips instead of fossil fuels are of growing interest, which implies a competition for wood supply between processes for "green" energy. Traditional use for pulp and paper production is still large consumers of wood, competing also for available resources. The main advantage of wood is its high density compared to grasses and straw, allowing cheaper transportation costs and storage facilities. The more promising specie is poplar, as well in USA than in Europe or Asian countries (Littlewood et al. 2014). The genus *Populus* comprises 25–35 species of deciduous plants native to the Northern Hemisphere. Common names used for the different species include poplar, aspen, and cottonwood. Yields of first-generation hybrid poplar planted on croplands in the Lake States of the USA have been estimated to be in the range of 7.9–11.8 dry tons ha^{-1} $year^{-1}$. The reported yield is slightly lower on corn lands in Minnesota, with values ranging from 7.7 to 9.9 dry tons ha^{-1} $year^{-1}$ for hybrid poplar species in North America (Sannigrahi and Ragauskas 2009).

Short rotation production seems a promising cultural way as poplar produces fast growing shoots from the stump after the first cut. These new stalks should be harvested after two years (Short-term rotation, STR), or annually in very short-term rotation. Nominal yield is estimated to be 14 tons/ha/year. This is comparable to that of switchgrass (14 tons/ha/year) and much higher than corn stover (8.4 tons/ha/year) and wheat straw (6 tons/ha/year).

Softwood is also a promising source of biomass for biorefinery use. Pine and spruce, contain around 43–45 % cellulose, 20–23 % hemicellulose and 28 % lignin. The hemicellulose is mainly made up of mannose, which is a hexose that can be fermented by normal baker's yeast, and the content of pentoses is only around 6–7 % of the total wood. Theoretically, around 410 l of ethanol can be produced per metric ton dry raw material using only the hexose fraction and 455 l if all carbohydrates are considered (Galbe and Zacchi 2002). However, Softwoods are generally recognized as being much more refractory than hardwoods or agricultural residues. This is due to the fact that softwoods have a more rigid structure and contain more lignin. Also, the content of acetylated groups is lower than in hardwoods and autohydrolysis cannot occur to the same extent. Acid-catalyzed steam pretreatment is therefore the best available pretreatment method.

3 Pretreatments

The aim of the pretreatment is the removal of lignin and hemicellulose, the reduction of crystalline cellulose and the increase in the porosity of the materials. Additionally, the pretreatment should improve the formation of sugars or the ability to form them during the succeeding enzymatic hydrolysis, and avoid the formation of inhibitors for subsequent hydrolysis and fermentation processes. The main constraints are minimizing sugar degradation and the formation of inhibitors (furanic and phenolic compounds), and limiting the consumption of chemicals, energy and water, and the production of wastes (Galbe and Zacchi 2007). For the pretreatment of lignocellulosics, several physical, physical–chemical, chemical and biological processes have been proposed and developed (Sun and Cheng 2002). The most studied pretreatments adapted to different biomasses are summarized in Tables 3 and 4 (adapted from Sanchez and Cardona 2008). Accessibility of cellulose to cellulases is a parameter that allows comparing the pretreatment efficacy. It is often estimated by the lignin or hemicelluloses removal and glucan digestibility (Mosier et al. 2005). However, cellulose accessibility seems to be more important in the final yield than extended delignification (Rollin et al. 2010). A chimeric non-hydrolytic protein containing a cellulose-binding module (CBM; see Couturier et al. Chapter "Fungal Enzymatic Degradation of Cellulose" of this book, for further details of CBMs) and a green fluorescent protein moiety was developed allowing a direct determination of the accessibility of cellulose (Hong et al. 2007; Zhu et al. 2009).

3.1 Physical Methods

In all processes, a first step of size reduction is necessary, to allow a better diffusion of chemicals and/or enzymes, which is the main limiting step of biomass conversion. Size diminution is the best way to enhance reagent penetration and reduce operation time (or increase conversion yield). However, the advantages of size reduction are balanced by energy consumption. As indicated above, materials can be comminuted by a combination of chipping, grinding and milling to reduce

Table 3 Compilation of physical and physical–chemical pretreatment of biomasses

Methods	Procedure/agents	Examples of pretreated materials	References
Physical methods: Mechanical comminution	Chipping, grinding, milling	Wood and forestry wastes (hardwood), Straw, corn stover, cane bagasse, Timothy, alfalfa	Sun and Cheng (2002)
Pyrolysis	$T > 300$ °C, then cooling and condensing	Wood, waste cotton, corn stover	Khiyami et al. (2005), Sun and Cheng (2002), Yu and Zhang (2004)
Physical–chemical methods: Steam explosion	Saturated steam at 160–290 °C, $P = 0.69$–4.85 MPa for several sec or min, then decompression until atm. pressure	Poplar, aspen, eucalyptus Softwood (Douglas fir) Bagasse, corn stalk, wheat straw, rice straw, barley straw, sweet sorghum bagasse, *Brassica carinata* residue, olive stones Timothy grass, alfalfa, reed canary grass	Ballesteros et al. (2001, 2002, 2004), Belkacemi et al. (1997, 2002), De Bari et al. (2002), Hamelinck et al. (2005), Lynd et al. (2002), Nakamura et al. (2001), Negro et al. (2003), Soderstrom et al. (2003), Sun and Cheng (2002)
Liquid hot water (LHW)	Pressurized hot water, $P > 5$ MPa, $T = 170$–230 °C, 1–46 min; solids load <20 %	Bagasse, corn stover, Alfalfa fiber	Ballesteros et al. (2002), Laser et al. (2002), Lynd et al. (2002), Negro et al. (2003), Sreenath et al. (2001)
Ammonia fiber explosion (AFEX)	1–2 kg ammonia/kg dry biomass, 90 °C, 30 min, $P = 1.12$–1.36 MPa	Aspen wood Bagasse, wheat straw, barley straw, rice hulls, corn stover Switchgrass, coastal Bermuda grass, alfalfa	Dale et al. (1996), Lynd et al. (2002), Sun and Cheng (2002)
CO_2 explosion	4 kg CO_2/kg fiber, $P = 5.62$ MPa	News print Alfalfa Recycled paper	Sun and Cheng (2002)

Adapted from Sanchez and Cardona (2008). Treatment parameters are summarized and types of biomass and corresponding references are indicated

Table 4 Compilation of chemical treatments

Methods	Procedure/agents	Examples of pretreated materials	References
Ozonolysis	Ozone, room temperature and pressure	Poplar sawdust Bagasse, wheat straw, cotton straw, green hay, peanut	Sun and Cheng (2002)
Dilute-acid hydrolysis	0.75–5 % H_2SO_4, HCl, or HNO_3, P = 1 MPa; continuous process for low solids loads (5–10 wt % dry substrate/mixture): T = 160–200 °C; batch process for high solids loads (10–40 wt% dry substrate/mixture): T = 120–160 °C	Bagasse, corn stover, wheat straw, rye straw, rice hulls Switchgrass, Bermudagrass	Hamelinck et al. (2005), Lynd et al. (2002), Martinez et al. (2000), Rodrıguez-Chong et al. (2004), Saha et al. (2005a, b), Schell et al. (2003), Sun and Cheng (2002), Wooley et al. (1999)
Concentrated-acid hydrolysis	10–30 % H_2SO_4, 170–190 °C, 1:1.6 solid–liquid ratio Acid recovery is required 21–60 % peracetic acid, silo-typesystem	Poplar, sawdust Bagasse	Cuzens and Miller (1997), Teixeira et al. (1999)
Alkaline hydrolysis	Dilute NaOH, 24 h, 60 ° C; $Ca(OH)_2$, 4 h, 120 °C; it can be complemented by adding H_2O_2 (0.5–2.15 vol.%) at lower temperature (35 °C)	Hardwood, Bagasse, corn stover, straws with low lignin content (10–18 %), cane leaves	Hamelinck et al. (2005), Kaar and Holtzapple (2000), Lynd et al. (2002), Sun and Cheng (2002), Teixeira et al. (1999)
Oxidative delignification	Peroxidase and 2 % H_2O_2, 20 °C, 8 h	Bagasse	Sun and Cheng (2002)
Wet oxidation	1.2 MPa oxygen pressure, 195 °C, 15 min; addition of water and small amounts of Na_2CO_3 or H_2SO_4	Corn stover, wheatstraw	Bjerre et al. (1996), Varga et al. (2004)
Organosolvprocess	Organic solvents (methanol, ethanol, acetone, ethylene glycol, triethylene glycol) or their mixture with 1 % of H_2SO_4 or HCl; 185–198 °C, 30–60 min, pH = 2.0–3.4	Poplar wood, Mixed softwood (spruce, pine, Douglas fir)	Lynd et al. (2002), Pan et al. (2005), Rezzoug and Capart (1996), Sun and Cheng (2002)
Biological methods: Fungal pretreatment	Brown-, white- and soft-rot fungi	Corn stover, wheat straw	Sun and Cheng (2002), Tengerdy and Szakacs (2003)

Adapted from Sanchez and Cardona (2008). Treatment parameters are summarized and types of biomass and corresponding references are indicated cellulose crystallinity. The size of the materials is usually 10–30 mm after chipping and 0.2–2 mm after milling or grinding. Vibratory ball milling has been found to be more effective in breaking down the cellulose crystallinity of spruce and aspen chips and improving the digestibility of the biomass than ordinary ball milling (Millett et al. 1979). The power requirement of mechanical comminution of agricultural materials depends on the final particle size and the biomass characteristics (Cadoche and Lopez 1989).

Extrusion process is a novel and promising physical pretreatment method for biomass conversion to ethanol production. In extrusion, the materials are subjected to heating, mixing and shearing, resulting in physical and chemical modifications during the passage through the extruder. Screw speed and barrel temperature are believed to disrupt the lignocellulose structure causing defibrillation, fibrillation, and shortening of the fibers, and, in the end, increasing accessibility of carbohydrates to enzymatic attack (Karunanithy and Muthukumarappan 2009). Compression and expansion cycle in the extruder should allow biomass to be soaked with chemicals or enzymes (Liu et al. 2013). The different bioreactor parameters must be taken into account to achieve the highest efficiency in the process. Application of enzymes during extrusion process is being considered as a promising technology for ethanol production.

3.2 *Physical–Chemical Methods*

Steam Explosion

Physical–chemical pretreatment methods are considerably more effective than physical. Steam explosion is the most commonly used method for pretreatment of lignocellulosic materials. In this method, chipped biomass is treated with high-pressure saturated steam and then the pressure is swiftly reduced, which makes the materials undergo an explosive decompression. Steam explosion is typically initiated at a temperature of 160–260 °C (corresponding pressure 0.69–4.83 MPa) for several seconds to a few minutes. The reaction is then stopped by sudden decompression to atmospheric pressure. When steam is allowed to expand within the lignocellulosic matrix it separates the individual fibers. The process causes hemicellulose degradation (“flash” hydrolysis) and lignin transformation due to high temperature, thus increasing the potential of cellulose hydrolysis. In some extent, hemicelluloses hydrolysis release the lignin located between hemicelluloses layers. The most important factors affecting the effectiveness of steam explosion are particle size, temperature, residence time and the combined effect of both temperature (*T*) and time (*t*), which is described by the severity factor $R_0 : R_0 = t \cdot e^{(T-100)/14.75}$ (Alvira et al. 2010), where *t* represent the treatment time and *T* the temperature. Alfani et al. (2000) used log (R_0) between 3.5 and 4.5 for wheat straw pretreatment, obtaining a correlated decrease in hemicellulose content. The steam explosion efficiency can be enhanced using acidic conditions, such as presoaking the crude feedstock for 8 h at 60 °C in a 0.2 N sulphuric acid solution corresponding to a 3.85 % weight of acid to dry material ratio (Ballerini et al. 1994).

Ammonia Fiber Explosion

Ammonia fiber explosion (AFEX) is an alkaline thermal pretreatment which exposes the lignocellulosic materials at relatively high temperature and pressure treatment followed by rapid pressure release (Holtzapple et al. 1990). The moderate temperatures (60–100 °C) are significantly less than that of the steam explosion process, meaning less energy input and overall costs associated with the process. Pressures exceeding 12 atm are required for operation at ambient temperature (Lynd et al. 1999). Ground, pre-wetted lignocellulosic material at a moisture content of 15–30 % is placed in a pressure vessel with liquid ammonia (NH_3) at a loading of about 1–2 kg NH_3/kg dry biomass. The residence time can be altered from low (5–10 min) to moderate (30 min) lengths depending on the degree of saturation needed for the type of biomass. When released to atmospheric temperature, the rapid expansion of the ammonia gas causes swelling of the biomass feedstock, creating a disruption in the lignin-carbohydrate linkage, hemicellulose hydrolysis and ammonolysis of glucuronic cross-linked bonds, and partial decrystallization of the cellulose structure, all leading to a higher accessible surface area for enzymatic attack (Brodeur et al. 2011). This method does not produce inhibitors of the downstream processes and small particle size is not required for efficacy. AFEX can achieve greater than 90 % conversion of cellulose and hemicellulose to fermentable sugars for a variety of lignocellulosic materials including alfalfa, barley straw, corn residue, wheat straw, rice straw, corn fiber, sugarcane bagasse, switchgrass, coastal bermudagrass, and rye grass straw (Sendich et al. 2008). However, against aspen chips, which contain higher lignin content than sugar cane bagasse, for example, the AFEX process is less effective (Sun and Cheng 2002). This pretreatment does not remove lignin or any other substances from the biomass; however, the lignin-carbohydrate complexes are cleaved, and the lignin is deposited on the surfaces of the material possibly causing blockage of cellulases to cellulose (Brodeur et al. 2011). The AFEX process requires efficient ammonia recovery to be economical due to the high cost of ammonia. Ammonia recovery and recycle is feasible despite of its high volatility (Teymouri et al. 2005) but the associated complexity and costs of ammonia recovery may be significant regarding commercial potential of the AFEX pretreatment (Eggeman and Elander 2005; Mosier et al. 2005).

Liquid Hot Water treatment (LHW)

The liquid hot water method uses compressed hot liquid water to hydrolyze the hemicellulose (Neves et al. 2007). Much like the steam explosion process, liquid hot water (LHW) pretreatment uses water at elevated temperatures and high pressures to maintain its liquid form in order to promote disintegration and separation of the lignocellulosic matrix. It is a hydrothermal pretreatment method which releases high fraction of hemicellulosic sugars in the form of oligomers (Sarkar et al. 2012).

The treatment generally occurs at temperatures of 170–230 °C and pressures above 5 MPa over lengths of time ranging from a few minutes up to an hour. This process uses many of the same features that steam explosion employs, primarily autohydrolysis, without the rapid decompression, utilizing flow through reactors of varying configurations or batch techniques with the latter being the primary emphasis at the laboratory scale (Negro et al. 2003; Brodeur et al. 2011). However, this process also contributes to the production of small amounts of undesired degrading compounds such as furfural, carboxylic acid, that are very toxic to ethanol fermentation as they inhibit microbial growth (Talebnia et al. 2010). The solubilized product, consisting primarily of oligosaccharides derived from hemicellulose (88–98 % removal from solid fraction) and lignin (35–60 % of total starting material) and a minor amount of cellulose (4–15 %), without acid or chemical requirement, it is an environmentally attractive and economically interesting method. Yu et al. (2010) studied two-step liquid hot water treatment of *Eucalyptus grandis* and obtained a xylose recovery of 86.4 %. Maximal glucose yield of 70–76 % corresponding to 80 % of xylan removal from soybean straw was obtained through combined liquid hot water and alkaline treatments (Wana et al. 2011).

3.3 Chemical Methods

Chemical pretreatments employ different chemical agents as ozone, acids, alkalis, peroxide and organic solvents. Inorganic acids as H_2SO_4 and HCl have been preferably used for biomass pretreatment. The main objective of acid pretreatment is to solubilize the hemicelluloses fraction of the biomass and to make the cellulose more accessible to enzymes (Ishizawa et al. 2007; Alvira et al. 2010). High hydrolysis yield have been reported when pretreating the lignocelluloses with diluted sulphuric acid (Lee et al. 1999; Mosier et al. 2005). However, sulphuric acid treatment has some process limitation such as corrosion, neutralization before fermentation producing gypsum which is not easily recycled and inhibitor production in the liquid phase together with other soluble constituents, mainly sugars (Yang and Wyman 2008). Several processes were recently developed such as organosolv process or use of ionic liquid (Li et al. 2013) but further research are needed to improve these pretreatment before they can be applied at industrial scale.

3.4 Biological Methods

Wood decay fungi have been extensively studied for lignocelluloses degradation. Brown rot, white rot and soft rot fungi are considered for enzyme production. White-rot, particularly, and in a lesser extent brown rot fungi are producing lignin degrading enzymes such as peroxidases and laccases. Several attempts have been made to design biological pretreatments allowing the degradation of lignin and hemicelluloses. Such a biological pretreatment has low energy requirements and mild environmental conditions. However, most of these processes are too slow limiting its application at industrial level. Moreover, fungal strains are generally using cellulose for growth, producing a decrease of ethanol yield. White-rot fungi were screened to select fast growing strain that degrade mainly lignin and grow preferentially on hemicelluloses and conserving the cellulose (Salvachua et al. 2011; Zhou et al. 2015). This could be achieved by using laccase producing strains working with "natural" mediators such as lignin derived phenols occurring naturally during the lignin degradation process (Fillat et al. 2010). Storage period can give the opportunity of such treatment which needs a long time to be efficient.

4 Detoxification of Lignocellulose Hydrolysates

It is well-known that thermo-chemical pretreatment of lignocelluloses, e.g., dilute acid hydrolysis and steam explosion, can release not only the fermentable pentose and hexose sugars, but also various compounds which are inhibitory to microorganisms and lead to apparent reduction in fermentation yield and productivity. In general, there are three major groups of inhibitors: furan derivatives, weak acids, and phenolic compounds (Almeida et al. 2007). The action of these three groups of inhibitor is variable, depending on the fermentative microorganisms and operating conditions. Moreover, this effect is largely strain dependant, and several strains have been selected for their resistance to inhibitors. Microorganisms considered for ethanol production from lignocellulose include both bacteria and yeasts. Many studies were done on the bacteria *Zymomonas mobilis* which uses the Entner-Doudoroff pathway and offers potentially higher yields than the other organisms. Genetically modified *Escherichia coli* are also considered due to their well-known genetic engineering. However, the yeast *Saccharomyces cerevisiae* has proved to be more robust than bacteria, both with respect to tolerance to the end product ethanol and to other compounds present in hydrolysates. By far, the most widely used organism in the existing fermentation industry is the yeast *S. cerevisiae*. More over the Baker's yeast *S. cerevisiae* is a GRAS (Generally Regarded as Safe) microorganism commonly used in industrial wine making, brewing and baking processes for the production of ethanol and CO_2 from fermentable sugars (Van Zyl et al. 1988). Several strains of *S. cerevisiae* show tolerance to furans due to their ability to convert HMF and furfural to less harmful compounds (Palmqvist and Hahn-Hagerdal 2000; Diaz de Villegas 1992). HMF is reduced to 2,5-bis-hydroxymethylfuran (HMF alcohol) under aerobic and anaerobic conditions (Almeida et al. 2007).

The detoxification can be chemical, physical, or biological. The most commonly used methods for detoxification of hydrolyzates before fermentation are: evaporation, solvent extraction, overliming with calcium hydroxide, activated charcoal, ion exchange resins, and enzymatic detoxification. Comparisons of different methods for detoxification, or conditioning, indicate that they differ significantly with respect to effects on hydrolysate chemistry and fermentability (Larsson et al. 1999; Cantarella et al. 2004). In comparisons of detoxification methods, treatment with calcium hydroxide (overliming) has emerged as one of the most efficient methods. In many cases, overliming also seems to be the most economical choice (Ranatunga et al. 2000). Although biotechnical methods are very promising in a longer perspective, they are seldom compared to conventional methods, such as alkaline detoxification (Jönsson et al. 2013).

5 Hydrolysis and Fermentation

5.1 *Acid Hydrolysis*

Historically, industrial cellulose digestion has been made with acid hydrolysis. Concentrated acid process was developed in Germany prior to World War II. Technology was also developed in Germany utilizing a dilute sulfuric acid percolation process to hydrolyze and extract pentoses and hexoses from wood waste. Several installations were built prior to and during the war. Examples are the modified Bergius process (40 % HCl) operated during World War II in Germany, and the more recently modified Scholler processes (0.4 % H_2SO_4) in the former Soviet Union, Japan and Brazil (Keller 1996). Acid hydrolysis can be performed with several types of acids, including sulphurous, sulphuric, hydrochloric, hydrofluoric, phosphoric, nitric and formic acid. These acids may be either concentrated or diluted. Processes involving concentrated acids are operated at low temperature and give high yields (e.g., 90 % of theoretical glucose yield), but the large amount of acids used causes problems associated with equipment corrosion and energy-demanding acid recovery (Jones and Semrau 1984). For these reasons, dilute acid hydrolysis was more attractive with a relatively low acid consumption. However, high temperatures were required to achieve acceptable rates of conversion of cellulose to glucose, and high temperatures also increase the rates of hemicellulose sugar decomposition and equipment corrosion. The maximum yield of glucose was obtained at high temperature and short residence time, but even under these conditions the glucose yield was only between 50 and 60 % of the theoretical value (Wyman 1996), far from the results obtained with enzyme. As a consequence, ongoing projects develop biological processes.

5.2 *Enzymatic Hydrolysis*

Enzymes specialized in breaking up the β-1-4-glycosidic bonds of glucan are collectively called cellulases (see Couturier et al. Chapter “Fungal Enzymatic

Degradation of Cellulose" of this book). Since the initial work of Reese on *Trichoderma*, many enzymatic cocktails have been developed and commercialized with improved performances. Optimum temperature of cellulases is generally above 50 °C and their thermal stability allow residence time of 48 h or more for the hydrolysing step. To obtain optimal performances, separate hydrolysis and fermentation processes (SHF) were first designed, as yeast strains were generally mesophilic with optimal growth temperature in the range 30–37 °C. However, cellulases responsible for enzymatic hydrolysis of pretreated cellulosic biomass are strongly inhibited by hydrolysis products: glucose and short cellulose chains. One way to overcome cellulase inhibition is to ferment the glucose to ethanol as soon as it appears in solution. Simultaneous saccharification and fermentation (SSF) combines enzymatic hydrolysis with ethanol fermentation to keep the concentration of glucose low and avoid inhibitory effects. However, difference in temperature optima for hydrolysis (45–50 °C) and fermentation (28–35 °C) implies reduced performances of enzymatic cocktails resulting in low rates of cellulose hydrolysis, which is the stage limiting the rate of alcohol production counterbalanced by an increase of enzyme loading and subsequently operating costs (Ballesteros et al. 2004; Jeffries and Jin 2000; Jeffries and Shi 1999).

5.3 SHF Versus SSF

The advantage of SHF is the ability to carry out each step under optimal conditions, i.e., enzymatic hydrolysis at 45–50 °C and fermentation at about 30 °C. It is also possible to run fermentation in continuous mode with cell recycling. However, end-product inhibition of the activity of certain key enzymes and contamination problems are associated with this process due to accumulation of glucose at a temperature allowing contaminant growth (Talebnia et al. 2010).

The idea of performing the enzymatic hydrolysis and fermentation simultaneously was put forward by Gauss et al. in a patent from 1976. The authors stated that the glucose yield in a traditional separate enzymatic hydrolysis (using enzymes produced by the fungus *T. reesei*) was low, probably due to end-product inhibition of the hydrolysis by glucose and cellobiose. The authors could, however, show that they obtained a higher overall ethanol yield when using SSF, which they attributed to the removal of glucose and cellobiose by the fermentation, and the consequent release of end-product inhibition. The term SSF (the abbreviation SSF is often used also for *solid state fermentation*) was not used by the authors at the time, but became the common notation for this process within just a few years from the original invention.

As already mentioned, an important advantage with SSF compared to SHF is the reduction of end-product inhibition by sugars formed in the hydrolysis which are metabolized by fermenting organisms (Margeot et al. 2009). The fermentation product ethanol also inhibits hydrolysis, but to a lesser extent than cellobiose or glucose and, moreover inhibit contaminant growth when a minimum concentration is reached (Hoztapple 1990). However, also the SSF process may suffer from incomplete hydrolysis of the solid lignocellulosic fraction. Except for inhibition by end-products or other components (Wu and Lee 1997), this can be due to enzyme deactivation, unproductive enzyme adsorption (Ooshima et al. 1990), decreasing availability of chain ends (Zhang and Lynd 2004) and increasing crystallinity with conversion of pretreated cellulose (Zhang et al. 1999). Bearing in mind that the sugars are more inhibitory for conversion process than ethanol is, SSF can reach higher rates, yields and ethanol concentrations in comparison with SHF (Wyman et al. 1992; Koppram and Olsson 2014). SSF offers an easier operation and a lower equipment requirement than the sequential process since no hydrolysis reactors are needed; moreover, the presence of ethanol in the broth makes that the reactioning mixture be less vulnerable to the action of undesired microorganisms (Wyman 1996). Nevertheless, SSF has the inconvenient that the optimal conditions for hydrolysis and fermentation are different, which implies a difficult control and optimization of process parameters (Claassen et al. 1999). In addition, larger amounts of exogenous enzymes are required.

Other promising integration alternative is the inclusion of the pentose fermentation in the SSF, process called simultaneous saccharification and co-fermentation (SSCF).

5.4 Pentoses Fermentation

For utilization of lignocelluloses for bioethanol production, both pentoses and hexoses released by pretreatment and enzymatic hydrolysis should be fermented (Talebnia et al. 2010). Simultaneous saccharification and co-fermentation (SSCF) was the first attempt to use C5 fraction produced by the pretreatment. Another way is to use it to produce cellulases with *T. reesei* (Bischof et al. 2013). Ballerini et al. (1994) were producing hydrolytic enzymes with *T. reesei* CL847 fed with a mix of lactose from whey and C5 fraction from steam explosion pretreatment. Naturally xylose-fermenting yeasts, such as *Pichia stipitis* and *Candida shehatae* (Prior et al., 1989) could potentially be advantageous to use in SSF of materials with high xylan contents (Olofsson et al. 2008). However, their tolerance to inhibitory compounds in undetoxified lignocellulose hydrolyzates is rather low (Van Zyl et al. 1988; Roberto et al. 1991). The main "competitors" to the yeast have been the bacteria *Z. mobilis* and genetically engineered *E. coli*, but they had the same limitations. Since *S. cerevisiae* cannot utilize pentoses such as xylose, genetic engineering of the organism is necessary (Hahn-Hägerdal et al. 2007; Tomas-Pejo 2008). Recently, Sanda et al., (2011) have produced a recombinant *S. cerevisiae* strain that ferments xylose in the presence of fermentation inhibitors (30 mM acetate and 20 mM formate).

Sasaki et al. (2015) designed membrane reactors allowing increased ethanol production during simultaneous saccharification and co-fermentation of rice straw by xylose-fermenting *S. cerevisiae*. Hydrothermal pretreatment of rice straw gives a cellulose-rich solid fraction and a liquid fraction rich in hemicellulose-derived sugars (Alvira et al. 2010). Most hemicellulose-derived sugars exist in oligomeric form, which is adequate for utilizing the liquid fraction after membrane separation process by nanofiltration (NF) and ultrafiltration (UF) under decreased pressure (Sasaki et al. 2014). Membrane separation has the advantage of low energy consumption (He et al. 2014) and could selectively concentrate sugars by removing fermentation inhibitors such as carboxylic acids, furfurals and lignin in the liquid fraction (Sasaki et al. 2014). Demeke et al. (2013) have modified the industrial strain *S. cerevisiae* Ethanol Red by inserting expression cassette containing 13 genes including *Clostridium phytofermentans* XylA, encoding D-xylose isomerase (XI), and enzymes of the pentose phosphate pathway. After DNA shuffling and evolutionary engineering, they obtained a strain (GS1.11-26) which produced 32 % more ethanol than the parent strain Ethanol Red, due to efficient D-xylose utilization.

The logic culmination of reaction integration for the transformation of biomass into ethanol is the consolidated bioprocessing (CBP), known also as direct microbial conversion (DMC). The key difference between CBP and the other strategies of biomass processing is that only one microbial community is employed both for the production of cellulases and fermentation, i.e., cellulase production, cellulose hydrolysis, and fermentation are carried out in a single step (Xu et al. 2009) (see Fig. 1). Wyman (1992) indicates that in most of the studies on CBP, the bacterium *Clostridium thermocellum* is used for enzyme production, cellulose hydrolysis, and glucose fermentation, whereas the co-fermentation using *Clostridium thermosaccharolyticum* allows the simultaneous conversion of pentoses obtained from hemicellulose hydrolysis into ethanol. In particular, the CBP using *C. thermocellum* showed a substrate conversion 31 % higher than a system using *T. reesei* and *S. cerevisiae.*

Progress in genetic engineering and massive sequencing of genomes has introduced a change in strategy and organisms involved (Grigoriev et al. 2011). Two strategies are equally chosen. The native cellulolytic strategy involves engineering naturally occurring cellulolytic microorganisms to improve product-related properties, such as yield and titer; the recombinant cellulolytic strategy involves engineering of non-cellulolytic organisms that exhibit high product yields and titers so that they express a heterologous cellulase system that enables cellulose utilization. Each strategy involves considerable uncertainty, and different strategies could prove advantageous for different products. (Lynd et al. 2005). Using the first strategy, Shaw et al. (2008) developed an engineered strain of *Thermoanaerobacterium saccharolyticum* producing ethanol at high yield without appreciable amount of organic acids owing to the deletion of genes involved in their formation. The second strategy was used by Sakamoto et al. (2011) to construct a recombinant *S. cerevisiae* that not only hydrolyzed hemicelluloses by codisplaying endoxylanase from *T. reesei*, xylosidase from *Aspergillus oryzae*, and β-glucosidase from *Aspergillus aculeatus* but that also assimilated xylose through the expression of xylose reductase and xylitol dehydrogenase from *P. stipitis* and xylulokinase from *S. cerevisiae*. The recombinant strain successfully produced ethanol from rice straw hydrolysate and yielded 0.41 g ethanol/g of total sugars in rice straw hydrolysate which corresponded to 82 % of the theoretical yield.

6 Ethanol Recovery and Dehydration

The product stream from fermentation is called beer or wine depending on the country's most favorite beverage. As fermenting microorganisms can tolerate ethanol concentration as high as 16 % v/v (i.e., so called Ethanol Red *S. cerevisiae* commercial strain), the actual limitation of alcoholic degree is generally due to biomass loading. To obtain a beer containing 8 % ethanol by weight, which is the minimum level for an economically viable distillation, biomass suspension at the beginning of hydrolysis/fermentation step should be around 16 % by weight. The rheology of suspension limits biomass loading and depends mainly of the water retention capacity of the biomass. If all the water is absorbed by the substrate, enzymes could hardly operate. A liquefaction step is generally added at the process, either SHF or SSF, in which a first quantity of hydrolytic enzymes is used and pretreated biomass introduced in a sequential mode (Fed-batch) at enzyme optimal temperature or adapted to high temperatures (Pakarinen et al. 2014; Liu et al. 2014) in order to avoid rheology limitations (Zhang et al. 2010).

6.1 Distillation and Dehydration

For transportation fuel use, ethanol is generally blended with gasoline at various concentrations. For mixtures with gasoline water-free (anhydrous) ethanol is required. There is a common problem in the mixture or azeotrope, at 95.6 % by weight (97.2 % by volume) with water at a temperature of 78.15 °C, which makes it impossible to separate ethanol–water in a single distillation column. The first step is to recover the ethanol in a distillation or beer column, where most of the water remains with the solids part. The product (37 % ethanol) is then concentrated in a rectifying column to a concentration just below the azeotrope (95 %). Such a technology was used for centuries to obtain ethanol from wines or other alcoholic beverages. The resulting ethanol is further dehydrated in order to achieve anhydrous ethanol by employing azeotropic distillation, extractive distillation, liquid–liquid extraction, adsorption, or some complex hybrid separation methods (Huang et al. 2008).

6.2 Azeotropic Distillation

The binary ethanol-water may form with a third volatile compound a ternary azeotrope with a boiling point lower than the ethanol-water azeotrope. Phase separation at the condenser level allows to discarding an aqueous phase and recycling a binary mixture containing ethanol and the third compound which play the role of entrainer. Anhydrous ethanol is recovered at the bottom of the distillation column. The three commonly used entrainers for breaking binary ethanol–water azeotropes by heterogeneous azeotropic distillation are benzene (Wasylkiewicz et al. 2003), toluene (Feng et al. 2000) and cyclohexane (Gomis et al. 2005). Benzene was the traditional carrier in heterogeneous azeotropic distillation for ethanol dehydration. However, it has been substituted by other solvents because of its carcinogenic effect. Currently, cyclohexane is one of the most used entrainers for this separation. Composition of the ternary azeotrope is 76 % cyclohexane, 17 % ethanol and 7 % water, with a boiling point of 62.1 °C. Azeotropic distillation allows the recovery of a cyclohexane-ethanol mixture containing less than 50 ppm of water (Gomis et al. 2005).

6.3 Extractive Distillation

Extractive distillation uses a selective high boiling solvent which alter the activity coefficients and hence increase the separation factor (Nakamura et al. 2001). The third component added as separating agent can be liquid solvent, ionic liquid, dissolved salt, a mixture of volatile liquid solvent and dissolved salt, or hyper-branched polymer. One of the most commonly used extractive solvents in extractive distillation for ethanol dehydration is ethylene glycol. (Meirelles et al. 1992), but use of gazoline seems very useful for transport fuel production Extensive study of alcohol dehydration by extractive distillation and other processes was reviewed by Huang et al. (2008).

6.4 Adsorption

Zeolites molecular sieve (type 3A and 4A) are widely employed in separating ethanol–water mixture. 3A zeolite molecular sieves which has a nominal pore size of 3 Angstroms can be used for dehydration of polar liquids such as ethanol (Huang et al. 2008). Water molecules, with an approximate molecular diameter of 0.28 nm, can easily penetrate the pores of the molecular sieve adsorbent, while ethanol, with an approximate molecular diameter of 0.44 nm are excluded. Zeolite molecular sieves are highly selective, but water is very strongly adsorbed and high temperatures and/or low pressures are required to regenerate them. Other adsorbents, mainly biobased adsorbents are considered such as natural corncobs, natural and activated palm stone and oak (Al-Ashesh et al. 2004). They are generally less selective than zeolite, but they need less energy to be regenerd. In some case, the feedstock can be used, and further processed without need of regeneration step.

7 Perspectives

Regardless of the feedstock, the final ethanol selling prize must be competitive with that for gasoline. Thus, profit margins in ethanol production processes are low, and returns on capital are uncertain due to the fluctuations in the oil price. Numerous studies on bioethenol production were initiated after the first oil crisis and were abandoned when oil price fell. However, environment concerns, global change due to CO_2 level increase makes essential the energy transition and the production of renewable fuel. On a technical point of view, progresses on genetic engineering and mass genome sequencing must allow the obtention of industrial microoganisms adapted to each biomass source and able to ferment both hexoses and pentoses, increasing ethanol yield and over all profitability of the process. However, within a biorefinery concept, valorization of the residual fraction, mainly composed of lignin, generally burned to ensure the energetic balance could enhance the sustainability of the process.

References

Al-Asheh S, Ganat F, Al-Lagtah N (2004) Separation of ethanol–water mixtures using molecular sieves and biobased adsorbents. Chem Eng Res Des 82:855–864

Alfani F, Gallifuoco A, Saporosi A et al (2000) Comparison of SHF and SSF processes for the bioconversion of steam-exploded wheat straw. J Ind Microbiol Biotechnol 25:184–192

Almeida JR, Modig T, Petersson A et al (2007) Increased tolerance and conversion of inhibitors in lignocellulosic hydrolysates by *Saccharomyces cerevisiae*. J Chem Technol Biotechnol 82 (4):340–349

Alvira P, Tomás-Pejó E, Ballesteros M et al (2010) Pretreatment technologies for an efficient bioethanol production process based on enzymatic hydrolysis: a review. Bioresour Technol 101(13):4851–4861

Angelini LG, Ceccarini L, Nassi N et al (2009) Comparison of *Arundo donax* L. and *Miscanthus* x *giganteus* in a long-term field experiment in Central Italy: analysis of productive characteristics and energy balance. Biomass Bioenergy 33:635–643

Ballerini D, Desmarquest JP, Pourquié P et al (1994) Ethanol production from lignocellulosics: large scale production and economy. Bioresour Technol 50:17–23

Ballesteros I, Oliva JM, Saez F et al (2001) Ethanol production from lignocellulosic byproducts of olive oil extraction. Appl Biochem Biotechnol 91–93:237–252

Ballesteros I, Oliva JM, Negro MJ et al (2002) Enzymic hydrolysis of steam exploded herbaceous agricultural waste (*Brassica carinata*) at different particule sizes. Process Biochem 38:187–192

Ballesteros M, Oliva JM, Negro MJ et al (2004) Ethanol from lignocellulosic materials by a simultaneous saccharification and fermentation process (SSF) with *Kluyveromyces marxianus* CECT 10875. Process Biochem 39:1843–1848

Belkacemi K, Turcotte G, Savoie P et al (1997) Ethanol production from enzymatic hydrolyzates of cellulosic fines and hemicellulose-rich liquors derived from aqueous/steam fractionation of forages. Ind Eng Chem Res 36:4572–4580

Belkacemi K, Turcotte G, Savoie P (2002) Aqueous/steam-fractionated agricultural residues as substrates for ethanol production. Ind Eng Chem Res 41:173–179

Bischof R, Fourtis L, Limbeck A et al (2013) Comparative analysis of the *Trichoderma reesei* transcriptome during growth on the cellulase inducing substrates wheat straw and lactose. Biotech Biofuels 6:127
Bjerre AB, Olesen AB, Fernqvis T et al (1996) Pretreatment of wheat straw using combined wet oxidation and alkaline hydrolysis resulting in convertible cellulose and hemicellulose. Biotechnol Bioeng 49:568–577
Brodeur G, Yau E, Badal K et al (2011) Chemical and physicochemical pretreatment of lignocellulosic biomass: a review. Enzym Res. doi:10.4061/2011/787532
Brown RC, Radlein D, Piskorz J (2001) Pretreatment processes to increase pyrolytic yield of levoglucosan from herbaceous feedstocks. In: Bosell JJ (ed) American chemical society symposium series no. 784. American Chemical Society, Washington DC, pp 123–134
Cadoche L, López GD (1989) Assessment of size reduction as a preliminary step in the production of ethanol from lignocellulosic wastes. Biol Wastes 30(2):153–157
Cantarella M, Cantarella L, Gallifuoco A et al (2004) Comparison of different detoxification methods for steam-exploded poplar wood as a substrate for the bioproduction of ethanol in SHF and SSF. Proc Biochem 39:1533–1542
Cardona CA, Sanchez OJ (2007) Fuel ethanol production: process design trends and integration opportunities. Bioresour Technol 98:2415–2457
Claassen PAM, Van Lier JB, Contreras AL et al (1999) Utilisation of biomass for the supply of energy carriers. Appl Microbiol Biotechnol 52(6):741–755
Comb N, Hatzis A (1996) Development of an epi-fluorescence assay for monitoring yeast viability and pretreatment hydrolysate toxicity in the presence of lignocellulosic solids. Appl Biochem Biotechnol 57(58):649–657
Cuzens JC, Miller JR (1997) Acid hydrolysis of bagasse for ethanol production. Renewable Energy 10(2/3):285–290
Dale BE, Leong CK, Pham TK et al (1996) Hydrolysis of lignocellulosics at low enzyme levels: application of the AFEX process. Bioresour Technol 56:111–116
De Bari I, Viola E, Barisano D et al (2002) Ethanol production at flask and pilot scale from concentrated slurries of steam-exploded aspen. Ind Eng Chem Res 41:1745–1753
Demeke MM, Dietz H, Li Y et al (2013) Development of a D-xylose fermenting and inhibitor tolerant industrial *Saccharomyces cerevisiae* strain with high performance in lignocellulose hydrolysates using metabolic and evolutionary engineering. Biotechnol Biofuels 6:89
Diaz de Villegas ME, Villa P, Guerra M et al (1992) Conversion of furfural into furfuryl alcohol by *Saccharomyces cerevisiae*. Acta Biotechnol 12:351–354
Eggeman T, Elander RT (2005) Process and economic analysis of pretreatment technologies. Bioresour Technol 96:2019–2025
Feng G, Fan LT, Friedler F (2000) Synthesizing alternative sequences via a P-graph-based approach in azeotropic distillation systems. Waste Manage 20:639–643
Fillat A, Colom JF, Vidal T (2010) A new approach to the biobleaching of flax pulp with laccase using natural mediators. Bioresour Technol 101(11):4104–4110
Galbe M, Zacchi G (2002) A review of the production of ethanol from softwood. Appl Microbiol Biotechnol 59:18–628
Galbe M, Zacchi G (2007) Pretreatment of lignocellulosic materials for efficient bioethanol production. Adv Biochem Eng Biotechnol 108:41–65
Gauss WF, Suzuki S, Takagi M (1976) Manufacture of alcohol from cellulosic materials using plural ferments. US Patent 3,990,944
Gomis V, Font A, Pedraza R et al (2005) Isobaric vapor–liquid and vapor–liquid–liquid equilibrium data for the system water + ethanol + cyclohexane. Fluid Phase Equilib 235:7–10
Grigoriev IV, Cullen D, Goodwin SB et al (2011) Fueling the future with fungal genomics. Mycology 2(3):192–209
Hahn-Hägerdal B, Karhumaa K, Fonseca C et al (2007) Towards industrial pentose-fermenting yeast strains. Appl Microbiol Biotechnol 74(5):937–953

Hamelinck CN, van Hooijdonk G et al (2005) Ethanol from lignocellulosic biomass: techno-economic performance in short-, middle- and long-term. Biomass Bioenergy 28: 384–410

Harris EE, Beglinger E (1946) Madison wood sugar process. Ind Eng Chem 38:896–904

He J, Wu A, Chen D et al (2014) Cost-effective lignocellulolytic enzyme production by *Trichoderma reesei* on a cane molasses medium. Biotechnol Biofuels 7:43

Heaton EA, Dohleman FG, Long SP (2008) Meeting US biofuel goals with less land: the potential of Miscanthus. Glob Change Biol 14(9):2000–2014

Holtzapple M, Cognata M, Shu Y et al (1990) Inhibition of *Trichoderma reesei* cellulase by sugars and solvents. Biotechnol Bioeng 36(3):275–287

Hong J, Ye X, Zhang Y-H (2007) Quantitative determination of cellulose accessibility to cellulase based on adsorption of a nonhydrolytic fusion protein containing CBM and GFP with its applications. Langmuir 23:12535–12540

Huang HJ, Ramaswamy S, Tschirner UW et al (2008) A review of separation technologies in current and future biorefineries. Sep Purif Technol 62(1):1–21

Ishizawa CI, Davis MF, Schell DF et al (2007) Porosity and its effect on the digestibility of dilute sulfuric acid pretreated corn stover. J Agric Food Chem 55:2575–2581

Jackman EA (1987) Industrial alcohol. In: Bu'lock JD, Christiansen B (eds) Basic biotechnology. Academic, London, pp 309–336

Jeffries TW, Jin YS (2000) Ethanol and thermotolerance in the bioconversion of xylose by yeasts. Adv Appl Microbiol 47:221–268

Jeffries TW, Shi NQ (1999) Genetic engineering for improved xylose fermentation of yeasts. Adv Biochem Eng Biotechnol 65:117–161

Jones JL, Semrau KT (1984) Wood hydrolysis for ethanol production—previous experience and the economics of selected processes. Biomass 5(2):109–135

Jönsson LJ, Alriksson B, Nilvebrant NO (2013) Bioconversion of lignocellulose: inhibitors and detoxification. Biotechnol Biofuels 6(1):16

Kaar WE, Holtzapple MT (2000) Using lime pretreatment to facilitate the enzymic hydrolysis of corn stover. Biomass Bioenergy 18:189–199

Kadam KL, McMillan JD (2003) Availability of corn stover as a sustainable feedstock for bioethanol production. Bioresour Technol 88:17–25

Karunanithy C, Muthukumarappan K (2009) Influence of extruder temperature and screw speed on pretreatment of corn stover while varying enzymes and their ratios. Appl Biochem Biotechnol 162(1):264–279

Katzen R, Tsao GT (2000) A view of the history of biochemical engineering. In: Scheper T (ed) History of modern biotechnology II. Springer, Berlin

Kaylen ML, Van Dyne D, Choi YS et al (2000) Economic feasibility of producing ethanol from lignocellulosic feedstocks. Bioresour Technol 72:19–32

Keller FA (1996) Integrated bioprocess development for bioethanol production. In: Wyman CE (ed) Handbook on bioethanol: production and utilization. Taylor & Francis, Bristol, pp 351–379

Kerstetter JD, Lyons JK (2001) Logging and agricultural residue supply curves for the Pacific Northwest. Washington State University Energy Program. US Department of Energy. Contract #DE-FC01-99EE50616

Khiyami MA, Pometto AL III, Brown RC (2005) Detoxification of corn stover and corn starch pyrolysis liquors by *Pseudomonas putida* and *Streptomyces setonii* suspended cells and plastic compost support biofilms. J Agric Food Chem 53:2978–2987

Kim S, Dale BE (2004) Global potential bioethanol production from wasted crops and crop residues. Biomass Bioenergy 26:361–375

Koppram R, Olsson L (2014) Combined substrate, enzyme and yeast feed in simultaneous saccharification and fermentation allow bioethanol production from pretreated spruce biomass at high solids loadings. Biotechnol Biofuels 7:54

Kumar A, Cameron JB, Flynn PC (2005) Pipeline transport and simultaneous saccharification of corn stover. Bioresour Technol 96(7):819–829

Larsson S, Reimann A, Nilvebrant N-O et al (1999) Comparison of different methods for the detoxification of lignocellulose hydrolysates of spruce. Appl Biochem Biotechnol 77:91–103

Laser M, Schulman D, Allen SG (2002) A comparison of liquid hot water and steam pretreatments of sugar cane bagasse for bioconversion to ethanol. Bioresour Technol 81:33–44

Lee CK, Low KS, Gan PY (1999) Removal of some organic dyes by acid-treated spent bleaching earth. Environ Technol 20(1):99–104

Li L, Wang Y, Zhang Q et al (2008) Wheat straw burning and its associated impacts on Beijing air quality. Sci China Ser D Earth Sci 51:403–414

Li C, Tanjore D, He W (2013) Scale-up and evaluation of high solid ionic liquid pretreatment and enzymatic hydrolysis of switchgrass. Biotechnol Biofuels 6:154

Lin Y, Tanaka S (2006) Ethanol fermentation from biomass resources: current state and prospects. Appl Microbiol Biotechnol 69:627–642

Littlewood J, Guo M, Boerjan W et al (2014) Bioethanol from poplar: a commercially viable alternative to fossil fuel in the European Union. Biotechnol Biofuels 7:113

Liu C, van der Heide E, Wang H et al (2013) Alkaline twin-screw extrusion pretreatment for fermentable sugar production. Biotechnol Biofuels 6:97

Liu ZH, Qin L, Zhu JQ et al (2014) Simultaneous saccharification and fermentation of steam-exploded corn stover at high glucan loading and high temperature. Biotechnol Biofuels 7:167

Lynd R, Wyman CE, Gerngross TU (1999) Biocommodity engineering. Biotechnol Prog 15: 777–793

Lynd LR, Weimer PJ, van Zyl WH et al (2002) Microbial cellulose utilization: fundamentals and biotechnology. Microbiol Mol Biol Rev 66(3):506–577

Lynd LR, van Zyl WH, McBride JE et al (2005) Consolidated bioprocessing of cellulosic biomass: an update. Curr Opin Biotechnol 16:577–583

Maiorella BL (1985) Ethanol. Comprehensive biotechnology, vol 3. Pergamon, Oxford, pp 861–909

Margeot A, Hahn-Hagerdal B, Edlund M et al (2009) New improvements for lignocellulosic ethanol. Curr Opin Biotechnol 20:372–380

Martinez A, Rodriguez ME, York SW et al (2000) Effects of $Ca(OH)_2$ treatments ("Overliming") on the composition and toxicity of bagasse hemicellulose hydrolysates. Biotechnol Bioeng 69 (5):526–536

Meirelles A, Weiss S, Herfurth H (1992) Ethanol dehydration by extractive distillation. J Chem Technol Biotechnol 53:181–188

Millett MA, Effland MJ, Caulfield DF (1979) Influence of fine grinding on the hydrolysis of cellulosic materials-acid vs. enzymatic. Adv Chem Ser 181:71–89

Mosier N, Wyman C, Dale B et al (2005) Features of promising technologies for pretreatment of lignocellulosic biomass. Bioresour Technol 96(6):673–686

Nakamura Y, Sawada T, Inoue E (2001) Enhanced ethanol production from enzymatically treated steam-exploded rice straw using extractive fermentation. J Chem Technol Biotechnol 76: 879–884

Negro MJ, Manzanares P, Ballesteros I et al (2003) Hydrothermal pretreatment conditions to enhance ethanol production from poplar biomass. Appl Biochem Biotechnol 105–108:87–100

Neves MA, Kimura T, Shimizu N et al (2007) State of the art and future trends of bioethanol production. Dyn Biochem Process Biotechnol Mol Biol 1–13 (Global Science Books)

Nigham PS, Singh A (2011) Production of liquid biofuels from renewable resources. Prog Energy Combust Sci 37:52–68

Olofsson K, Bertilsson M, Lidén G (2008) A short review on SSF—an interesting process option for ethanol production from lignocellulosic feedstocks. Biotechnol Biofuels 1:7

Ooshima H, Burns DS, Converse AO (1990) Adsorption of cellulase from *Trichoderma reesei* on cellulose and lignacious residue in wood pretreated by dilute sulfuric acid with explosive decompression. Biotechnol Bioeng 36(5):446–452

Pakarinen A, Haven MO, Djajadi DT (2014) Cellulases without carbohydrate-binding modules in high consistency ethanol production process. Biotechnol Biofuels 7:27

Palmqvist E, Hahn-Hagerdal B (2000) Fermentation of lignocellulosic hydrolysates. II: inhibitors and mechanisms of inhibition. Bioresour Technol 74:25–33

Pan X, Arato C, Gilkes N et al (2005) Biorefining of softwoods using ethanol organosolv pulping: preliminary evaluation of process streams for manufacture of fuel-grade ethanol and co-products. Biotechnol Bioeng 90(4):473–481

Prior BA, Kilian SG, du Preez JC (1989) Fermentation of D-xylose by the yeasts *Candida Shehatae* and *Pichia Stipitis*. Process Biochem 24:21–32

Ranatunga TD, Jervis J, Helm RF et al (2000) The effect of overliming on the toxicity of dilute acid pretreated lignocellulosics: the role of inorganics, uronic acids and ether-soluble organics. Enzym Microb Technol 27(3):240–247

Reese ET (1976) History of cellulase program at Natick. Biotechnol Bioeng Symp 6:9–30

Rezzoug S-A, Capart R (1996) Solvolysis and hydrotreatment of wood to provide fuel. Biomass Bioenergy 11(4):343–352

Roberto IC, Lacis LS, Barbosa MFS et al (1991) Utilization of sugar cane bagasse hemicellulosic hydrolysate by *Pichia stipitis* for the production of ethanol. Process Biochem 26(1):15–21

Rodrıguez-Chong A, Ramırez JA, Garrote G et al (2004) Hydrolysis of sugar cane bagasse using nitric acid: a kinetic assessment. J Food Eng 61(2):143–152

Rollin JA, Zhu Z, Sathitsuksanoh N et al (2010) Increasing cellulose accessibility is more important than removing lignin: a comparison of cellulose solvent-based lignocellulose fractionation and soaking in aqueous ammonia. Biotechnol Bioeng 108:22–30

Rosillo-Calle F, Cortez L (1998) Towards proalcohol II: a review of the Brazilian bioethanol programme. Biomass Bioenergy 14:115–124

Saha BC, Iten LB, Cotta MA et al (2005a) Dilute acid pretreatment, enzymatic saccharification, and fermentation of rice hulls to ethanol. Biotechnol Prog 21:816–822

Saha BC, Iten LB, Cott MA et al (2005b) Dilute acid pretreatment, enzymatic saccharification and fermentation of wheat straw to ethanol. Process Biochem 40:3693–3700

Sakamoto T, Hasunuma T, Hori Y et al (2011) Direct ethanol production from hemicellulosic materials of rice straw by use of an engineered yeast strain codisplaying three types of hemicellulolytic enzymes on the surface of xylose-utilizing Saccharomyces cerevisiae cells. J Biotechnol 158:203–210

Salvachúa D, Prieto A, Lopez-Abelairas M et al (2011) Fungal pretreatment: an alternative in second-generation ethanol from wheat straw. Bioresour Technol 102:7500–7506

Sanchez OJ, Cardona CA (2008) Trends in biotechnological production of fuel ethanol from different feedstocks. Bioresour Technol 99:5270–5295

Sanda T, Hasunuma T, Matsuda F (2011) Repeated-batch fermentation of lignocellulosic hydrolysate to ethanol using a hybrid *Saccharomyces cerevisiae* strain metabolically engineered for tolerance to acetic and formic acids. Bioresour Technol 102:7917–7924

Sannigrahi P, Ragauskas AJ (2009) Poplar as a feedstock for biofuels: a review of compositional characteristics. Biofuels Bioprod Biorefin 4:209–226

Sarkar N, Ghosh SK, Bannerjee S et al (2012) Bioethanol production from agricultural wastes: an overview. Renewable Energy 37(1):19–27

Sasaki K, Tsuge Y, Sasaki D et al (2014) Optimized membrane process to increase hemicellulosic ethanol production from pretreated rice straw by recombinant xylose-fermenting *Saccharomyces cerevisiae*. Bioresour Technol 169:380–386

Sasaki K, Tsuge Y, Sasaki D et al (2015) Mechanical milling and membrane separation for increased ethanol production during simultaneous saccharification and co-fermentation of rice straw by xylose-fermenting *Saccharomyces cerevisiae*. Bioresour Technol 185:263–268

Schell DJ, Farmer J, Newman M et al (2003) Dilute sulfuric acid pre-treatment of corn stover in pilot-scale reactor. Investigation of yields, kinetics, and enzymatic digestibilities of solids. Appl Biochem Biotechnol 105(1–3):69–85

Sendich EN, Laser M, Kim S et al (2008) Recent process improvements for the ammonia fiber expansion (AFEX) process and resulting reductions in minimum ethanol selling price. Bioresour Technol 99:8429–8435

Shaw AJ, Jenney FE, Adams MW et al (2008) End-product pathways in the xylose fermenting bacterium, *Thermoanaerobacterium saccharolyticum*. Enzym Microb Technol 42(6):453–458

Soderstrom J, Pilcher L, Galbe M et al (2003) Combined use of H_2SO_4 and SO_2 impregnation for steam pretreatment of spruce in ethanol production. Appl Biochem Biotechnol 105:127–140

Sreenath HK, Koegel RG, Moldes AB et al (2001) Ethanol production from alfalfa fiber fractions by saccharification and fermentation. Process Biochem 36:1199–1204

Sun Y, Cheng J (2002) Hydrolysis of lignocellulosic materials for ethanol production: a review. Bioresour Technol 83(1):1–11

Talebnia F, Karakashev D, Angelidaki I (2010) Production of bioethanol from wheat straw: an overview on pretreatment, hydrolysis and fermentation. Bioresour Technol 101(13):4744–4753

Tampier M, Smith DW, Bibeau EL et al (2004) Identifying environmentally preferable uses for biomass resources: phase 2 report: life-cycle emission reduction benefits of selected feedstock-to product threads. Envirochem Inc., Vancouver

Teixeira LC, Linden JC, Schroeder HA (1999) Alkaline and peracetic acid pretreatments of biomass for ethanol production. Appl Biochem Biotechnol 77–79:19–34

Tengerdy RP, Szakacs G (2003) Bioconversion of lignocellulose in solid substrate fermentation. Biochem Eng J 13:169–179

Teymouri F, Laureano-Perez L, Alizadeh H et al (2005) Optimization of the ammonia fiber explosion (AFEX) treatment parameters for enzymatic hydrolysis of corn stover. Bioresour Technol 96(18):2014–2018

Tomas-Pejo E, Oliva JM, Ballesteros M et al (2008) Comparison of SHF and SSF processes from steam-exploded wheat straw for ethanol production by xylose-fermenting and robust glucose-fermenting *Saccharomyces cerevisiae* strains. Biotechnol Bioeng 100(6):1122–1131

Uppugundia N, da Costa Sousa L, Chundawat S et al (2014) A comparative study of ethanol production using dilute acid, ionic liquid and AFEX™ pretreated corn stover. Biotechnol Biofuels 7:72

Van Zyl C, Prior BA, du Preez JC (1988) Production of ethanol from sugar cane bagasse hemicellulose hydrolyzate by *Pichia stipitis*. Appl Biochem Biotechnol 17:357–369

Varga E, Klinkle HB, Reczey K et al (2004) High solid simultaneous saccharification and fermentation of wet oxidized cornstover to ethanol. Biotechnol Bioeng 88(5):567–574

Wana C, Zhou Y, Li Y (2011) Liquid hot water and alkaline pretreatment of soybean straw for improving cellulose digestibility. Bioresour Technol 102:6254–6259

Wasylkiewicz SK, Kobylka LC, Castillo FJL (2003) Synthesis and design of heterogeneous separation systems with recycle streams. Chem Eng J 92:201–208

Wooley R, Ruth M, Glassner D, Sheejan J (1999) Process design and costing of bioethanol technology: a tool for determining the status and direction of research and development. Biotechnol Prog 15:794–803

Wu Z, Lee YY (1997) Inhibition of the enzymatic hydrolysis of cellulose by ethanol. Biotechnol Lett 19(10):977–979

Wyman CE (1996) Ethanol production from lignocellulosic biomass: overview. In: Wyman CE (ed) Handbook on bioethanol: production and utilization. Taylor & Francis, Bristol, pp 1–18

Wyman CE, Spindler DD, Grohmann K (1992) Simultaneous saccharification and fermentation of several lignocellulosic feedstocks to fuel ethanol. Biomass Bioenergy 3(5):301–307

Xu Q, Singh A, Himmel ME (2009) Perspectives and new directions for the production of bioethanol using consolidated bioprocessing of lignocellulose. Curr Opin Biotechnol 20: 364–371

Yang B, Wyman CE (2008) Pretreatment: the key to unlocking low-cost cellulosic ethanol. Biofuels Bioprod Biorefin 2:26–40

Yu ZS, Zhang HX (2004) Ethanol fermentation of acid-hydrolyzed cellulosic pyrolysate with Saccharomyces cerevisiae. Bioresour Technol 93:199–204

Yu G, Yano S, Inoue H et al (2010) Pretreatment of rice straw by a hot-compressed water process for enzymatic hydrolysis. Appl Biochem Biotechnol 160(2):539–551

Zhang Y-HP, Lynd LR (2004) Toward an aggregated understanding of enzymatic hydrolysis of cellulose: noncomplexed cellulose systems. Biotechnol Bioeng 8(7):797–824

Zhang S, Wolfgang DE, Wilson DB (1999) Substrate heterogeneity causes the nonlinear kinetics of insoluble cellulose hydrolysis. Biotechnol Bioeng 66(1):35–41

Zhang MJ, Wang F, Su R et al (2010) Ethanol production from high dry matter corncob using fed-batch simultaneous saccharification and fermentation after combined pretreatment. Bioresour Technol 101:4959–4964

Zhou S, Raouche S, Grisel S et al (2015) Solid-state fermentation in multi-well plates to assess pretreatment efficiency of rot fungi on lignocellulose biomass. Microb Biotechnol 8(6): 940–949

Zhu Z, Sathitsuksanoh N, Vinzant T et al (2009) Comparative study of corn stover pretreated by dilute acid and cellulose solvent-based lignocellulose fractionation: enzymatic hydrolysis, supramolecular structure, and substrate accessibility. Biotechnol Bioeng 103(4):715–724

Zhuang XY, Zhang HX, Yang JZ et al (2001) Preparation of levoglucosan by pyrolysis of cellulose and its citric acid fermentation. Bioresour Technol 79:63–66

Bioethanol from Soybean Molasses

Susan Grace Karp, Adenise Lorenci Woiciechowski, Luiz Alberto Junior Letti and Carlos Ricardo Soccol

Abstract Bioethanol is an important biofuel produced by the fermentation of sugars that, together with biodiesel, biobutanol, biomethane, and biohydrogen, composes the range of alternatives to the fossil fuels. Soybean molasses is the main by-product generated at the industrial processing of soybean to produce the soy protein concentrate. It is a rich source of carbohydrates, proteins, lipids, and minerals that could be used as a fermentation medium for microbial growth. The aim of this chapter is to describe studies on the development of processes to produce bioethanol from soybean molasses, focusing on the use of different microorganisms, fermentation scales, and pretreatment strategies. *Saccharomyces cerevisiae* and *Zymomonas mobilis* presented interesting ethanol yields and productivities at laboratory scale; however, considering the adequateness to industrial facilities, only the yeast-based process was scaled-up. At pilot scale, an average ethanol yield of 44.13 % over the total initial sugars was achieved, which represented an ethanol yield of 129.2 kg, or 163.6 L of absolute ethanol per ton of dry molasses. The average productivity for the pilot scale fed-batch process was 7.882 g/Lh. After scale-up to an industrial plant, one ton of soybean molasses (dry basis) yielded 162.7 L of absolute ethanol and 3.729 tons of vinasse, a by-product containing 19.5 % solids that had to be concentrated to be employed as an energy source. The pretreatment of soybean molasses by acid and enzymatic hydrolyses provided increases in the ethanol yield over total initial sugars to 62 and 68 %, respectively, and reduced the concentration of residual sugars.

Keywords Bioethanol · Soybean molasses · Fermentation · *Saccharomyces cerevisiae* · *Zymomonas mobilis* · Hydrolysis · Scale-up

---

S.G. Karp
Positivo University, Curitiba, Brazil

A.L. Woiciechowski · L.A.J. Letti · C.R. Soccol (✉)
Federal University of Paraná, Curitiba, Brazil
e-mail: soccol@ufpr.br

C.R. Soccol et al. (eds.), *Green Fuels Technology*,
Green Energy and Technology, DOI 10.1007/978-3-319-30205-8_10

Abbreviations

BOD	Biochemical oxygen demand
Disac	Disaccharides
FAT	Fundação André Tosello (André Tosello Foundation)
Fru	Fructose
Gal	Galactose
Glu	Glucose
NRRL	Northern Regional Research Laboratory
Raf	Raffinose
SPC	Soy protein concentrate
Sta	Stachyose
T. S.	Total sugars
UFPR	Universidade Federal do Paraná (Federal University of Paraná)

Nonstandard Symbols

S	Substrate
X	Biomass
P	Product
r_s	Substrate consumption rate
r_x	Biomass production rate
r_p	Product formation rate
$Y_{X/S}$	Biomass yield from substrate
$Y_{P/S}$	Product yield from substrate

1 Introduction

The development of integrated technologies for industrial processing, using renewable resources as feedstock with minimized land use and environmental impact are essential objectives to be pursuit by the industry in the current global scenario. This sustainable production chain of bio-based marketable chemicals forms the concept of a biorefinery. According to Cherubini (2010), renewable carbon-based raw materials for biorefinery are provided from four different sectors: agriculture, forestry, industries, and households (municipal solid waste and wastewaters) and aquaculture (algae and seaweeds).

Soybean (*Glycine max*) is one of the most important agricultural products in the world. It is a rich source of vegetable protein and is mainly cultivated for oil and protein extraction. The average composition of soybean includes protein (43 %), lipids (21 %), fibers (4 %), minerals (6 %), and carbohydrates (26 %), with sugars, such as mono and disaccharides (sucrose, glucose, and fructose) representing around 50 % of the total carbohydrates, and the oligosaccharides raffinose and stachyose representing the other half (Siqueira 2007).

The United States is the leader soybean producer and exporter. Soybean production for 2015/2016 is projected at 3916 billion bushels, or 106.6 million metric tons. Brazilian production of soybean for the harvest of 2015 is estimated in 96.7 million tons, which represents around 30 % of the estimated global production. Soybean cultivation in Brazil corresponds to a harvested area of almost 32 million hectares, the average expected yield being calculated as 3.0 t/ha (USDA 2015; IBGE 2015).

The primary processing of soybean to obtain oil and protein is important to the producer country, for promoting economic development and generating employment. In order to extract the oil, the grains are crushed, and the oil is extracted with an appropriate solvent. The remaining solid fraction is the de-oiled meal, constituted by proteins, sugars, and minerals. Soybean meal can be processed as shown in Fig. 1, producing two other fractions: the protein concentrate and the molasses composed basically of sugars.

Bioethanol is an important biofuel produced by the fermentation of sugars that, together with biodiesel, biobutanol, biomethane, and biohydrogen, represents an alternative to the fossil fuels. The interesting properties of ethanol as an energy source have led to intense research, especially, focusing on obtaining efficient microorganisms, low-cost fermentation substrates, optimal fermentation conditions, and on developing engines that are adapted to ethanol-gasoline mixtures as well as to pure ethanol (Soccol et al. 2005). The most important resources for bioethanol production include sugarcane, corn, and beet. However, other carbohydrate rich materials, such as sorghum, cassava, and soybean residues could be used for this purpose.

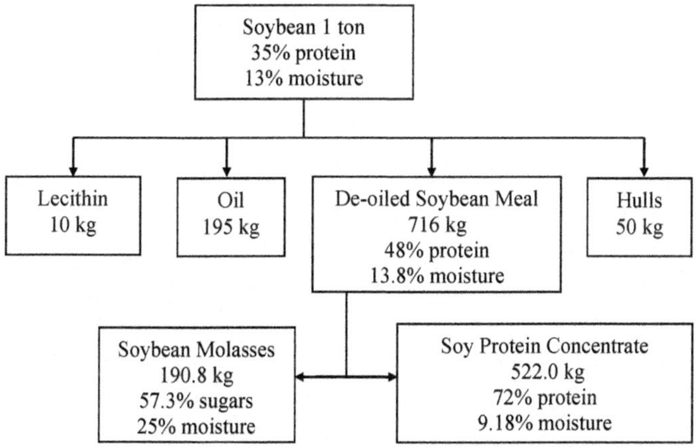

Fig. 1 Products of the industrial processing of soybean. *Source* Adapted from Siqueira et al. (2007)

2 Production of Soy Protein Concentrate and Soybean Molasses

The soy protein concentrate (SPC) is prepared from dehulled and defatted soybeans by removing most of the water-soluble and nonprotein constituents. It contains at least 65–67 % protein on a moisture-free basis. SPC is basically composed of proteins and insoluble carbohydrates (Endres 2001). It can be produced by precipitation with an alcoholic solution, precipitation with an acid solution and water washing with heat denaturing.

The process of alcoholic precipitation is based on the ability of solutions of aliphatic alcohols (methanol, ethanol, and isopropanol) to extract the fraction of soluble sugars, without dissolving proteins from dehulled and defatted soybeans in a batch or a continuous process. After the extraction of the sugars, the alcohol is recovered and reused and the final product is dried (Hettiarachchy and Kalapathy 1997; Lusas and Riaz 1995; Peisker 2001).

In the process of precipitation with acid solution, dehulled and defatted soybeans are suspended in an acid solution at pH 4.5 (isoelectric pH of the soy protein) to remove soluble carbohydrates. This is the pH where most of the proteins precipitate, and then they are separated by centrifugation. More commonly, the concentrate is neutralized to make it more soluble and functional, depending on the final destination, and the final product is dried (Hettiarachchy and Kalapathy 1997; Lusas and Riaz 1995; Peisker 2001).

In the process of water washing with heat denaturation, the proteins of defatted soybeans are denatured by exposure to water vapor, making them water insoluble. Soluble carbohydrates are removed by washing with hot water and the final product is dried (Hettiarachchy and Kalapathy 1997; Lusas and Riaz 1995).

Soybean molasses is the main by-product generated by the production of SPC. It is a brown-colored viscous material whose composition is presented at Table 1. Considering that the molasses contains carbohydrates as major components, it could

Table 1 Average physicochemical composition of the soybean molasses

Component	% in dry basis
Stachyose	18.6
Raffinose	9.68
Sucrose	28.4
Glucose	0.243
Fructose	0.127
Galactose	0.254
Total carbohydrates	57.3
Proteins	9.44
Lipids	21.2
Fibers	5.7
Ash	6.36

Source Siqueira et al. (2008)

be used as a carbon source for fermentative processes, opening a wide range of possibilities for its use in a soybean biorefinery. The soybean molasses in its raw form is generally sold for low prices, for use as ingredient in animal feed.

Many research groups have been working on processes for the biological conversion of soybean molasses and other soybean products and residues into biochemicals of high value, including bioethanol (Siqueira et al. 2008; Letti et al. 2012; Romão et al. 2012; Silva et al. 2012; Long and Gibbons 2013), lactic acid (Karp et al. 2011), enzymes (Sanada et al. 2009; Weingartner 2010), xanthan gum (Soccol et al. 2009), culture medium for self-contained biological indicators (Dlugokenski et al. 2011), edible mushrooms (Chimilovski et al. 2011), bioinsecticides (Melo et al. 2009), and oligopeptides (Rojas et al. 2014).

3 Bioethanol Production from Non-hydrolyzed Soybean Molasses at Laboratory, Pilot, and Industrial Scales

A process for bioethanol production from soybean molasses by *Saccharomyces cerevisiae* was developed at the Bioprocess Engineering and Biotechnology Department, Federal University of Paraná (UFPR), in partnership with a Brazilian soybean-processing company, at laboratory, pilot, and industrial scales (Siqueira et al. 2008). A process for bioethanol production by *Zymomonas mobilis* using the same substrate (soybean molasses) was developed by Letti et al. (2012) at laboratory scale.

3.1 Microorganism Selection

Yeasts, particularly *S. cerevisiae*, are the most commonly and traditionally used microorganisms for ethanol fermentation. However, ethanol producing bacteria such as *Z. mobilis* can also present interesting results in terms of yield and productivity. Table 2 presents the results of a screening test to select the most appropriate strains for bioethanol production from soybean molasses.

Ten *S. cerevisiae* strains, from the culture collection of the Bioprocess Engineering and Biotechnology Department (UFPR), were assayed on soybean molasses containing 30 % (w/v) of soluble solids (222 g/L of total sugars). Also, two strains of the bacteria *Z. mobilis*, named NRRL 806 (available at the National Center for Agricultural Utilization Research, formerly Northern Regional Research Laboratory, USA) and FAT (available at the André Tosello Foundation, São Paulo, Brazil), were evaluated on soybean molasses containing 15 and 10 % (w/v) of soluble solids (112 and 75 g/L of total sugars).

The *S. cerevisiae* strain LPB-SC presented the highest ethanol yield (41.5 % of the theoretical yield from the total initial sugars) among yeasts and the highest

Table 2 Screening of *Saccharomyces cerevisiae* and *Zymomonas mobilis* strains for bioethanol production from soybean molasses

Strain	Total initial sugars (g/L)	Total residual sugars (g/L)	Ethanol (g/L)	Yield[a] over initial sugars (%)	Productivity (g/Lh)
Saccharomyces cerevisiae					
LPB 1	222	135.2	38.9	34.3	1.62
LPB 2		125.0	35.9	31.6	1.50
LPB 3		131.5	37.0	32.6	1.54
LPB 4		135.6	42.9	37.8	1.79
LPB 5		134.9	37.5	33.0	1.56
LPB 6		118.3	33.2	29.3	1.38
LPB-SC		126.9	47.1	41.5	1.96
LPB-MA		145.0	30.6	27.0	1.27
LPB-JP		121.7	30.5	26.9	1.27
Zymomonas mobilis					
NRRL806	112	–	25.8	45.1	1.61
FAT		–	17.8	31.1	1.11
NRRL806	75	–	15.4	40.2	0.963
FAT		–	12.1	31.6	0.756

Concentrations and yields after 24 h (*S. cerevisiae*) and 16 h (*Z. mobilis*) of fermentation
Source Adapted from Siqueira et al. (2008), Letti (2007)
[a]Percentage of the maximum theoretical yield: 51.1 % (w/w) of sugar content; yield's average standard deviation was 1.41

ethanol productivity (1.96 g/Lh) among all microorganisms. This strain was chosen for the subsequent yeast fermentation tests with soybean molasses. The *Z. mobilis* strain NRRL 806 produced 25.8 g/L of ethanol from the soybean molasses containing 15 % (w/v) of soluble solids, which represented the highest yield among all strains (45.1 %), against 17.8 g/L produced by the strain FAT. The strain NRRL 806 also produced more ethanol from the soybean molasses containing 10 % (w/v) of soluble solids (15.4 g/L against 12.1 g/L produced by the strain FAT), so it was selected for the subsequent bacterial fermentations (Letti 2007).

3.2 *Bioethanol Production by* S. cerevisiae*: Kinetics Under Optimized Conditions*

Fermentation tests were conducted at laboratory scale to define the optimal conditions for ethanol production by the strain *S. cerevisiae* LPB-SC. The soybean molasses provided the necessary nutrients for growth, and there was no need to supplement the fermentation medium with inorganic salts ($MgSO_4$, NH_4NO_3) or yeast extract. The molasses was fermented with an initial concentration of 30 % (w/v) of soluble solids (or 30 °Brix), to provide a minimum ethanol concentration

Table 3 Kinetic and yield parameters of bioethanol production from soybean molasses

Time (h)	S (g/L)	X (g/L)	P (g/L)	r_S (dS/dt)	r_X (dX/dt)	r_P (dP/dt)	$Y_{X/S}$, % (r_x/r_s)	$Y_{P/S}$, % (r_p/r_s)
0	182.0	6.00	0.400	–	–	–	–	–
1	176.1	6.05	3.20	5.9	0.05	2.8	0.847	47.5
2	163.6	6.18	8.80	12.5	0.13	5.6	1.04	44.8
3	140.5	6.31	19.6	23.1	0.13	10.8	0.563	46.7
4	123.9	6.49	27.0	16.6	0.18	7.4	1.08	44.6
5	102.1	6.59	36.4	21.8	0.10	9.4	0.459	43.1
6	87.6	6.72	43.0	14.5	0.13	6.6	0.897	45.5
Average				15.7	0.12	8.08	0.815	45.4

Note Average standard deviations were 4.77 % for sugar, 2.10 % for biomass and 5.24 % for ethanol concentrations

Source Adapted from Siqueira et al. (2008)

S substrate; *X* biomass; *P* product; r_s substrate consumption rate; r_x biomass production rate; r_p product formation rate; $Y_{X/S}$ biomass yield from substrate; $Y_{P/S}$ product yield from substrate

that would be economical for the distillation process. For an initial biomass concentration of 3 × 10^8 viable cells/mL, the fermentation finished in 6 h.

Table 3 presents the kinetics, under optimized conditions, of the batch fermentation of soybean molasses to produce bioethanol, conducted in a bench-scale bioreactor (8 L of total capacity). The ethanol yield was 45.4 % from consumed substrate, which represented 88.8 % of the theoretical maximum (51.1 %, since 48.9 % are converted into CO_2). Considering only the initial substrate concentration (182 g/L), ethanol yield was 45.8 % of the theoretical maximum. This was equivalent to 134.1 kg or 169.8 L of absolute ethanol per ton of dry molasses, since one ton of molasses contained 573 kg of sugar on moisture-free basis (Siqueira et al. 2008).

The high residual substrate concentration (87.6 g/L) indicated that almost 50 % of the soybean molasses' sugars were not fermented by the strain LPB-SC. It is known that *S. cerevisiae* produces intra and extracellular invertase, the latter being responsible for cleaving sucrose into glucose and fructose, monomers that are assimilated and converted into ethanol (Zech and Görisch 1995). However, another enzyme is necessary for an organism to metabolize the complex sugars of soybean molasses (stachyose and raffinose), which are formed by fructose, glucose, and galactose monomers linked by β-1,2 and α-1,6 bonds. This enzyme, called α-galactosidase, was probably not produced by *S. cerevisiae* LPB-SC.

3.3 *Bioethanol Production by* Z. mobilis

The production of bioethanol from soybean molasses by *Z. mobilis* NRRL 806 was optimized at laboratory scale as a function of the soybean molasses concentration (the sole carbon and nitrogen source), pH and period of previous aerobic phase. The optimal parameters were soybean molasses concentration of around 200 g/L of soluble solids, pH between 6.0 and 7.0, and the period of previous aerobic phase did not provide significant effect. Kinetic tests were performed in Erlenmeyer flasks and in a 6 L bench-scale bioreactor and the yields of sugar conversion to ethanol were 78.3 and 96.0 % of the theoretical maximum (yields from consumed sugars), with ethanol productions of 24.2 and 29.3 g/L, respectively. Maximum ethanol concentration in the bioreactor was achieved after 8 h. The microorganism was able to consume almost all the fructose, glucose and sucrose contents, but was not able to uptake galactose. Complex sugars with α-1,6 bonds were not metabolized (Letti et al. 2012).

Industrial processes for ethanol production usually require higher concentrations of ethanol at the end of fermentation, in order to make the distillation process economic. However considering that *Z. mobilis* presented good yields of sugar conversion, other studies should be developed before scale-up, especially focusing on continuous processes.

3.4 Bioethanol Production at Pilot Scale

The process of bioethanol production from soybean molasses by *S. cerevisiae* was scaled-up. The pilot plant for ethanol production had the capacity to produce 1 m^3 of fermented broth per day, with two tanks of 1 m^3 capacity (one for the preparation of the must and the other for fermentation) and a circulation system through plate heat exchangers to provide constant temperature. The inoculum was prepared with fresh pressed yeast, which was added at the beginning of each cycle.

The soybean molasses contains significant concentrations of proteins and lipids that increase the viscosity and surface tension, and although viscosity should decrease while ethanol concentration increases, the intense formation of CO_2 bubbles strongly favors foam formation (Togrul and Arslan 2004).

The process was initially operated in batch mode but a great amount of foam was generated. Besides the addition of antifoam and dispersant agents, the alternative chosen to control foam and CO_2 formation was to change the operating mode to fed-batch, which is considered one of the most useful systems for economical ethanol production (Roukas 1996). This is the most common operation mode with cell recycle used in Brazil, also called modified Melle-Boinot process (Soccol et al. 2005). Another advantage of the fed-batch system is that the intermittent feeding of the substrate prevents inhibition and catabolite repression, improving the productivity of fermentation by maintaining a low substrate concentration (Prasad et al. 2007).

Table 4 presents the results of 11 fermentation cycles conducted at the pilot scale plant, two batches and nine fed-batches.

Table 4 Results of pilot scale fermentations for bioethanol production with different concentrations of soybean molasses (°Brix or % w/v of soluble solids)

Cycle/Op. mode[a]	° Brix	Initial sugar (g/L)	Final sugar (g/L)	Ethanol (g/L)	Time (h)	Productivity (g/Lh)	Yield[b] (%)
1/B[c]	20.0	116.6	54.00	26.10	6	4.350	43.80
2/B	30.0	251.8	118.2	58.60	6	9.767	45.54
3/FB[d]	21.0	166.9	59.1	39.28	7	5.611	46.06
4/FB	22.0	168.9	79.2	41.92	5	8.384	48.57
5/FB	23.0	185.0	89.9	43.72	5	8.744	46.25
6/FB	21.0	182.1	83.3	40.01	4	10.000	43.00
7/FB	21.0	189.5	76.9	40.50	6	6.750	41.82
8/FB	21.5	190.7	85.0	40.81	6	6.802	41.88
9/FB	22.0	180.8	83.2	41.14	5	8.228	44.53
10/FB	23.0	173.2	80.0	38.00	4	9.500	42.93
11/FB	30.0	230.7	74.37	48.41	7	6.916	41.06
Average productivity/yield						7.882[e]	44.13

Notes Initial biomass concentration was 3×10^8 cells/mL
Source Adapted from Siqueira et al. (2008)
[a]Operational mode
[b]Yield over total initial sugars
[c]B—Batch
[d]FB—Fed-batch
[e]For the fed-batches only

An average yield of 44.13 % over the total initial sugars was equivalent to an ethanol yield of 129.2 kg, or 163.6 L of absolute ethanol per ton of dry molasses. The average productivity, calculated only for the fed-batches, was 7.882 g/Lh.

3.5 *Bioethanol Production at Industrial Scale*

The process of bioethanol production from soybean molasses by *S. cerevisiae* was finally transferred to an industrial plant with a production capacity of 10 m^3 hydrated ethanol per day. At this stage, the biomass was recovered through centrifugation after fermentation, and then treated with sulfuric acid to pH 2.2, for 2 h, in order to avoid flocculation and inactivate the weak cells, thus preparing the inoculum for the next fermentation cycle.

The major problem identified after starting-up the plant was the contamination with bacteria, specifically Gram-positive bacilli. The problem was solved with the addition of antibiotic, together with the previous treatment of the medium (removal of insoluble solids by centrifugation). This mechanical pretreatment of the medium is important to avoid flocculation of insoluble solids in yeast biomass, thus increasing the efficiency of inoculum's acid treatment, and to preserve the equipment from encrustation.

Table 5 Results obtained at laboratory, pilot and industrial scales, concerning to the alcoholic fermentation process with soybean molasses

	Capacity	Ethanol yield (L/ton[a])
Laboratory scale	10 L/day	169.8
Pilot scale	1 m^3/day	163.6
Industrial scale	10 m^3 ethanol/day	162.7

Source Soccol et al. (2013)
[a]Liters of absolute ethanol per ton of dry molasses

After around 50 fermentation cycles, the operational parameters were adjusted and the maximum capacity of the industrial plant (10,000 L of ethanol per day) was achieved. The mass balance of industrial scale production indicated that one ton of soybean molasses (dry basis) yielded 162.7 L of absolute ethanol and 3.729 tons of a liquid waste called vinasse, produced by the distillation of the fermented broth. The yields of bioethanol production obtained at different production scales are summarized at Table 5.

The high amount of vinasse generated at the bioethanol production plant (almost 23 kg per L of absolute ethanol), containing 19.5 % solids, represents a considerable disposal problem with potential environmental impact.

The treatment of the sugarcane vinasse has been one of the most challenging issues in the industrial production of bioethanol in Brazil, because of its high biochemical oxygen demand (BOD) values, ranging from 30 to 60 gO_2/L (Navarro et al. 2000). The BOD of the soybean vinasse was determined as 77.2 gO_2/L.

This high organic charge is an obstacle for the treatment of the vinasse as a common effluent. The initial alternative was to recover it by concentration through evaporation and burning in the industrial boiler, thus producing steam for industrial activities. Also, other bioprocesses have been developed considering that this wastewater could be used as a raw material of high nutritional value for some microorganisms (Sanada et al. 2009; Karp et al. 2011; Dlugokenski et al. 2011). Table 6 presents the complete physicochemical composition of the soybean vinasse.

Table 6 Average physicochemical composition of the soybean vinasse from bioethanol production

Component	% in dry basis
Stachyose	11.09
Raffinose	22.07
Sucrose	0
Glucose	0
Fructose	0
Galactose	1.84
Total carbohydrates	35.0
Proteins	13.3
Lipids	27.8
Fibers	14.6
Ash	9.24

Source Adapted from Karp et al. (2011)

4 Bioethanol Production from Hydrolyzed Soybean Components at Laboratory Scale

The hydrolysis of the complex oligosaccharides present in soybean molasses could be an alternative to increase fermentation yield and, consequently, decrease the organic charge of the vinasse. Siqueira (2007) reported an increase of 20 % in ethanol production when the raw material was pre-hydrolyzed with the enzyme α-galactosidase. However, the high cost of the enzyme was pointed as an obstacle for its utilization at industrial scale for the production of bioethanol, which is a relatively low priced product.

Rojas et al. (2014) evaluated the soybean hulls as raw material to produce oligopeptides after the proteolysis of hull's proteins, and to produce ethanol after acid hydrolysis of the lignocellulosic fraction and fermentation of the resulting simple sugars. The proteolysis resulted in a pool of oligopeptides with small size, more than 80 % lower than 6.5 kDa, while the fermentation of the simple sugars with *S. cerevisiae* yielded around 90 % of the theoretical maximum (productivity of 0.2 g/Lh of ethanol).

Romão et al. (2012) evaluated the effect of acid hydrolysis on the ethanol yield from soybean molasses by *S. cerevisiae*. The experimental conditions, evaluated using a factorial design, were the type of acid (sulfuric, hydrochloric, and nitric acids), the absolute pressure (from 101.325 to 202.65 kPa) and the pH (from 3 to 5). The best result of ethanol production, 54 % of yield over initial sugar content, was achieved when the pH was set at 4 and the absolute pressure at 151.988 kPa, in the presence of nitric acid. The fermentation conditions were further optimized and a 62 % yield over initial sugars was reached after 14 h of fermentation.

Silva et al. (2012) evaluated the effect of enzymatic hydrolysis of soybean molasses to produce ethanol. In the non-hydrolyzed medium, the yield over total initial sugars was 47.4 %, the productivity was 3.61 g/Lh and the residual sugar concentration was 69.9 g/L, starting from 330 g/L of soybean molasses. When the enzymatic hydrolysis with α-galactosidase was employed, the yield over initial sugars increased to 68.1 % and the residual sugar concentration decreased to 24.9 g/L, from an initial concentration of 157.8 g/L.

Long and Gibbons (2013) developed a different process of hydrolysis using commercial cellulase, β-glucosidase, and pectinase to treat soybean molasses and soy solubles (by-product of soy protein isolate production), at various solid loading rates (33, 50, 60, 75, and 80 %, w/v). The hydrolysates were then fermented for 96 h using *S. cerevisiae* NRRL Y-2034 and *Scheffersomyces stipitis* NRRL Y-7124. Maximum ethanol concentrations for *S. cerevisiae* and *S. stipitis* were 39 and 28 g/L, respectively, obtained in soy solubles with 50 % solids and without the addition of enzymes. From soybean molasses, maximum concentrations for *S. cerevisiae* and *S. stipitis* were 12 g/L with 60 % solids in the presence of enzymes

Table 7 Concentration of sugars, in g/L, after acid hydrolysis of soybean molasses with hydrochloric, sulfuric, and phosphoric acids in different concentrations (N), at 121 °C for 15 min

Control		Sta	Raf	Disac	Glu	Fru + Gal	T. S.
		38.8	13.6	53.6	3.0	4.0	113.0
HCl	0.25 N	0.3	23.8	17.0	27.9	41.3	110.3
	0.5 N	0.2	0.1	4.3	33.4	44.6	82.5
	1.0 N	0	0	3.3	26.8	19.9	50.1
	2.0 N	0	0	2.6	20.8	15.0	38.4
H_2SO_4	0.1 N	8.5	21.0	16.4	19.9	30.0	95.9
	0.2 N	0.2	2.0	8.5	32.8	46.1	89.5
	0.5 N	0	0	3.2	33.6	43.7	80.4
	1.0 N	0	0	1.1	24.1	24.6	49.8
H_3PO_4	0.1 N	36.3	14.1	54.3	4.1	5.3	11.0
	1.0 N	0	24.8	11.1	24.2	35.3	95.4

Source Letti (2007)
Sta stachyose; *Raf* raffinose; *Disac* disaccharides; *Glu* glucose; *Fru + Gal* fructose plus galactose; *T. S.* total sugars

and almost 6 g/L with 33 % solids and without enzymes, respectively. Additional pretreatment to hydrolyse the α-1,6 galactosidic bonds could have improved the process yields.

Letti (2007) tested different conditions for acid hydrolysis of soybean molasses, aiming at the release of fermentable sugars to produce ethanol. The best results were obtained with hydrochloric acid at 0.5 N, as shown in Table 7. The author also evaluated the enzymatic hydrolysis with a commercial α-1,6 galactosidase, and the production of ethanol was enhanced by around 33 % when compared to a control assay. The major drawback of substituting acid hydrolysis by enzymatic hydrolysis was considered the higher cost of the enzyme, as mentioned previously.

5 Conclusion

The soybean molasses demonstrated to be a suitable fermentation medium to produce bioethanol, either with *S. cerevisiae* or *Z. mobilis*. The process involving yeast fermentation is better established in industry and was successfully scaled-up to pilot and industrial plants. Challenges in bioethanol production from soybean molasses include the enhancement of fermentation yield by promoting the hydrolysis of non-fermentable sugars in an economically feasible way, and the investigation of alternatives for the destination of the soybean vinasse, a liquid residue generated in significant amount.

References

Cherubini F (2010) The biorefinery concept: using biomass instead of oil for producing energy and chemicals. Energy Convers Manage 12:1412–1421

Chimilovski JS, Habu S, Teixeira RFB, Thomaz-Soccol V, Noseda MD, Pandey A, Soccol CR (2011) Antitumoral activities of exopolysaccharide of *Grifola frondosa* produced by submerged fermentation using sugarcane and soybean molasses as carbon source. Food Technol Biotech 49:359–363

Dlugokenski REF, Sella SRBR, Guizelini BP, Vandenberghe LPS, Woiciechowski AL, Soccol CR, Minozzo JC (2011) Use of soybean vinasses as a germinant medium for a *Geobacillus stearothermophilus* ATCC 7953 sterilization biological indicator. Appl Microbiol Biot 90:713–719

Endres JG (2001) Soy protein products: characteristics, nutritional aspects and utilization. American Oil Chemists' Society Press, Champaign

Hettiarachchy N, Kalapathy U (1997) Soybean protein products. In: Liu K (ed) Soybeans: chemistry, technology and utilization. Aspen Publishers, Gaithersburg, pp 379–411

IBGE (2015) Levantamento Sistemático da Produção Agrícola. ftp://ibge.gov.br/Producao_Agricola/Levantamento_Sistematico_da_Producao_Agricola_[mensal]/Fasciculo/lspa_201508.pdf Accessed 08 Oct 2015

Karp SG, Igashiyama AH, Siqueira PF, Carvalho JC, Vandenberghe LPS, Thomaz-Soccol V, Coral J, Tholozan JL, Pandey A, Soccol CR (2011) Application of the biorefinery concept to produce L-lactic acid from the soybean vinasse at laboratory and pilot scale. Bioresource Technol 102:1765–1772

Letti LAJ (2007) Production of bioethanol by soybean molasses fermentation by *Zymomonas mobilis*. Dissertation, Federal University of Paraná/University of Provence

Letti LAJ, Karp SG, Woiciechowski AL, Soccol CR (2012) Ethanol production from soybean molasses by *Zymomonas mobilis*. Biomass Bioenergy 44:80–86

Long CC, Gibbons WR (2013) Conversion of soy molasses, soy solubles, and dried soybean carbohydrates into ethanol. IJABE 6:62–68

Lusas EW, Riaz MN (1995) Soy protein products: processing and use. J Nutr 125:573–580

Melo AL, Soccol VT, Soccol CR, Nogueira Júnior M (2009) Evaluation of *Bacillus sphaericus* bioinsecticide produced with white soybean meal as culture medium for the control of *Culex* (Culex) *quinquefasciatus*. Cad Saúde Pública 25:563–569

Navarro AR, Sepúlveda MC, Rubio MC (2000) Bio-concentration of vinasse from the alcoholic fermentation of sugar cane molasses. Waste Manage 20:581–585

Peisker M (2001) Manufacturing of soy protein concentrate for animal nutrition. Cahiers Options Mediterraneennes 54:103–107

Prasad S, Singh A, Joshi HC (2007) Ethanol as an alternative fuel from agricultural, industrial and urban residues. Resour Conserv Recy 50:1–39

Rojas MJ, Siqueira PF, Miranda LC, Tardioli PW, Giordano RLC (2014) Sequential proteolysis and cellulolytic hydrolysis of soybean hulls for oligopeptides and ethanol production. Ind Crop Prod 61:201–210

Romão BB, Silva FB, Resende MM, Cardoso VL (2012) Ethanol production from hydrolyzed soybean molasses. Energy Fuel 26:2310–2316

Roukas T (1996) Ethanol production from non-sterilized beet molasses by free and immobilized *Saccharomyces cerevisiae* cells using fed-batch culture. J Food Eng 27:87–96

Sanada CTN, Karp SG, Spier MR, Portella AC, Gouvêa PM, Yamaguishi CT, Vandenberghe LPS, Pandey A, Soccol CR (2009) Utilization of soybean vinasse for α-galactosidase production. Food Res Int 42:476–483

Silva FB, Romão BB, Cardoso VL, Coutinho Filho U, Ribeiro EJ (2012) Production of ethanol from enzymatically hydrolyzed soybean molasses. Biochem Eng J 69:61–68

Siqueira PF (2007) Production of bio-ethanol from soybean molasses by *Saccharomyces cerevisiae*. Dissertation, Federal University of Paraná

Siqueira PF, Karp SG, Carvalho JC, Sturm W, Rodríguez-León JA, Tholozan JL, Singhania RR, Pandey A, Soccol CR (2008) Production of bio-ethanol from soybean molasses by *Saccharomyces cerevisiae* at laboratory, pilot and industrial scales. Bioresource Technol 99:8156–8163

Soccol CR, Vandenberghe LPS, Costa B, Woiciechowski AL, Carvalho JC, Medeiros ABP, Francisco AM, Bonomi LJ (2005) Brazilian biofuel program: an overview. J Sci Ind Res 64:897–904

Soccol CR, Karp SG, Noll MA, Soccol VT, Suarez CB, Pandey A (2009) Application of biorefinery concept for soybean industrialization in Brazil. In: Pandey A, Larroche C, Soccol CR, Dussap CG (eds) New horizons in biotechnology. Asiatech Publishers, New Delhi, pp 186–207

Soccol CR, Karp SG, Siqueira PF, Sanada CTN, Soccol VT, Pandey A (2013) Biorefinery concept applied to valorization of agro-food coproducts and wastes: integrated process for waste recycling. In: Soccol CR, Pandey A, Larroche C (eds) Fermentation processes engineering in the food industry, 1st edn. CRC Press, New York, pp 430–464

Togrul H, Arslan N (2004) Mathematical model for prediction of apparent viscosity of molasses. J Food Eng 62:281–289

USDA (2015) World agricultural supply and demand estimates. http://www.usda.gov/oce/commodity/wasde/latest.pdf. Accessed 08 Oct 2015

Weingartner V (2010) Produção, purificação e identificação de mananase, obtida por fermentação no estado sólido utilizando cascas de soja e *Aspergillus niger*. Dissertation, Federal University of Paraná

Zech M, Görisch H (1995) Invertase from *Saccharomyces cerevisiae*: reversible inactivation by components of industrial molasses media. Enzyme Microb Tech 17:41–46

Bioethanol Wastes: Economic Valorization

Eduardo Bittencourt Sydney, Carlos José Dalmas Neto, Alessandra Cristine Novak, Adriane Bianchi Pedroni Medeiros, Régis Nouaille, Christian Larroche and Carlos Ricardo Soccol

Abstract The valorization of solid, liquid, and gaseous wastes from industrial activities is important to achieve positive economic and environmental balances. The ethanol industry is one of the biggest examples of biorefineries around the world, generating enormous amounts of wastes and reusing them within the industry. New technologies to add value to these residues are underdevelopment and will chance the panorama of the sector. This chapter describes technologies to reuse and valorize the solids, liquids, and gaseous wastes generated during the production of ethanol through the fermentation of sugarcane. Special attention is given to the liquid and gaseous wastes, more specifically vinasse and CO_2, which can be used in microalgal and cyanobacterial cultures for the production of biomolecules such as lipids (to biodiesel), proteins, and pigments. Technologies to produce biogas and biohydrogen from vinasse are also presented and discussed.

E.B. Sydney (✉) · A.C. Novak
Federal University of Technology—Paraná (UTFPR), Campus Toledo Rua Cristo Rei 19, Toledo, PR 85902-490, Brazil
e-mail: eduardosydney@utfpr.edu.br

A.C. Novak
e-mail: alessandrac@utfpr.edu.br

C.J.D. Neto · A.B.P. Medeiros · C.R. Soccol
Federal University of Paraná (UFPR), Curitiba, PR 81531-970, Brazil
e-mail: carlos.dalmas@ufpr.br

A.B.P. Medeiros
e-mail: adrianebpm@ufpr.br

C.R. Soccol
e-mail: soccol@ufpr.br

R. Nouaille
AFYREN SAS Biopôle Clermont Limagne, 63360 Saint Beauzire, France
e-mail: regis.nouaille@afyren.com

C. Larroche
GePEB - Institut Pascal, UMR 6602 Univ Blaise Pascal/CNRS/IFMALaboratoire d'Excellence IMobS3, 24, avenue des Landais, BP 20206 63174 Aubière Cedex, France
e-mail: christian.larroche@univ-bpclermont.fr

C.R. Soccol et al. (eds.), *Green Fuels Technology*,
Green Energy and Technology, DOI 10.1007/978-3-319-30205-8_11

Finally, a chemical process that promotes vinasse treatment (COD, BOD, and turbidity reduction) coupled to the mitigation of CO_2 are presented. All these technologies promote the valorization of the ethanol industry wastes, bringing economic advantages to the segment and immensurable environmental benefits.

Keywords Ethanol · Biorefineries · Microalgae · Biohydrogen · CO_2 fixation

1 Introduction

The ethanol industry is probably the most known and studied industrial activity in Brazil. Historically, it dates from the 1970s when the increase of petroleum prices became unaffordable to the country. This opened the opportunity to encourage the production of ethanol based on favorable climate and soil characteristics and the impressive adaptation and productivity of sugarcane (*Saccharum* spp) in Brazilian soil (especially in the southeast region). Brazil is the world′s largest sugarcane producer. The volume of sugarcane processed in Brazil will be 663 million tons in the 2015/2016 crop. Currently, Brazil accounts for a third of world production of sugarcane, 20 % of production and 40 % of global sugar exports, and 30 % of production and 60 % of global ethanol exports (CONAB 2015).

The Brazilian ethanol industry experienced years of economic prosperity. In 1971 a Federal Law determined the blend of 5 % ethanol in gasoline, but it was only after 1973 (first petroleum crisis), when Brazilian imports of petroleum quadrupled (reaching 2.5 billion dollars) that the government introduced the National Alcohol Program (Proálcool—1975), which subsidized and encouraged the production of ethanol as a substituent for petroleum-based fuels (Soccol et al. 2010). This resulted in the development of ethanol-fueled cars by big companies, especially after 1978, such as Ford, Volkswagen, Fiat, and General Motors at that time. In 1985, sales of ethanol-fueled cars peaked and accounted for 96 % of all new cars sold (UNICA 2015).

However, an economic crisis that began in 1986 followed by inflation of 1764.86 % in 1989, in addition with dropping petroleum prices, resulted in cuts of ethanol incentives, which collapsed the ethanol industry. During this process, the industry adapted its process to recover and transform all types of residues (solid, liquid, and gaseous) into commercial products. Today, Brazilian ethanol industry is one of the biggest examples of biorefineries in the world. Total production of ethanol in the 2015/16 crop is estimated at 28.82 billion liters. The anhydrous ethanol used in mixture with gasoline, it is expected 12.08 billion liters. For hydrous ethanol, used in flex fuel vehicles, the expectation is that there will be a reduction of 1.2 % compared with production in the previous harvest, which is equivalent to less 198.23 million liters (CONAB 2015).

Traditionally, ethanol is produced in Brazil and many other tropical countries through the fermentation of the juice extracted from sugarcane. During this process, many liquid, solid, and gaseous wastes are generated and can be reused. In this chapter, the classic and modern technologies for the valorization of solid, liquid, and gaseous wastes from the fermentation ethanol industry are described and discussed, reinforcing the important role played by biotechnology in this theme.

2 Ethanol and Sugar Production from Sugarcane

The world ethanol production in 2014 reached around 93 billion liters (Renewable Fuel Association 2015). The USA production represented 58 % and Brazil 25 %. Other countries with representative production were China, Canada, Thailand, Argentina, and India (Renewable Fuels Association 2015).

In Brazil and in many tropical countries ethanol is produced through the fermentation of sugarcane juice by yeast. This process generates a great amount of solid, liquid, and gaseous wastes that can be reused by the industry. The ethanol and sugar production process is presented in a simplified flowchart (Fig. 1), highlighting the key residues.

Sugarcane is harvested mechanically or manually and transported by special designed trucks to the industry. When arrived, samples of sugarcane are withdrawn in order to determine the concentration of total reducing sugars (TRS), which is the main parameter and governs the whole industrial process.

Sucrose content in sugarcane varies between 10 and 15 %. The payment system adopted by the ethanol industry is based on TRS on sugarcane. This resulted in big economic benefits for the industrials and favored the adoption of better agronomic techniques, including genetically improvement (increased TRS, climate and soil adaptation, plague control—there are more than 600 varieties in Brazil). Besides TRS analysis to the determination of sugarcane quality, many other analyses are carried within the ethanol industry (Oliveira 2012).

The sugarcane is washed to remove impurities and it is chopped, if needed (mechanical harvest already chops sugarcane). After grinding, the juice is removed from sugarcane through pressure systems (pressing, for example) and used as raw material for sugar and ethanol production.

The choice between sugar or ethanol production depends on the market and technical aspects. The sugarcane juice directed to sugar production passes through evaporation. At this point sugar is already in crystallized form and is centrifuged. The sucrose crystals are separated and the very viscous liquid that is left is called molasses. Molasses is generally sent to the fermentation unit and mixed to sugarcane juice for fermentation. Brazil is the biggest sugar producer in the world, representing 25 % of the total production (UNICA 2015).

Ethanol is produced from the sugarcane juice, through a classic fermentation process, in which yeasts transform sugarcane juice, molasses, or a molasses–juice mixture into ethanol. Fermentation is carried in large open reactors (200–500 m^3) generally during 8 h. This is a biological process that can be represented by the stoichiometric equation

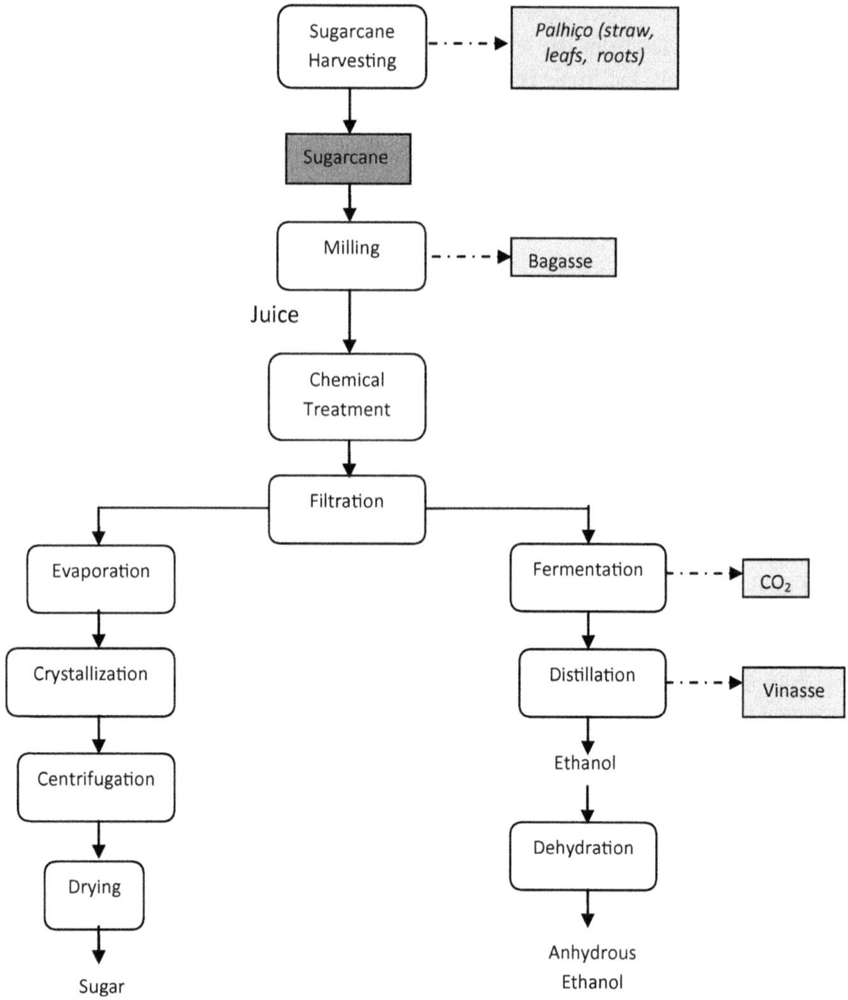

Fig. 1 Flowchart of ethanol and sugar production process showing the main residues

$$C_{12}H_{22}O_{11} + H_2O \rightarrow C_6H_{12}O_6 + C_6H_{12}O_6 \quad (1)$$

$$C_6H_{12}O_6 \rightarrow 2CH_3CH_2OH + 2CO_2 + 23,5 \quad \text{kcal} \quad (2)$$

Ethanol is recovered by distillation (hydrous ethanol). Anhydrous ethanol is produced through a dehydration process, such as molecular sieve. Each ton of sugarcane produces 85 L of ethanol, representing a process yield of approximately 83 %. In 2014/2015, 28.394 million m^3 of ethanol (59 % of hydrated ethanol) and 35.548 million tons of sugar were produced in Brazil. Sugarcane productivity in 2014/2015 was approximately 74 t/ha (UNICA 2015). According to a study carried

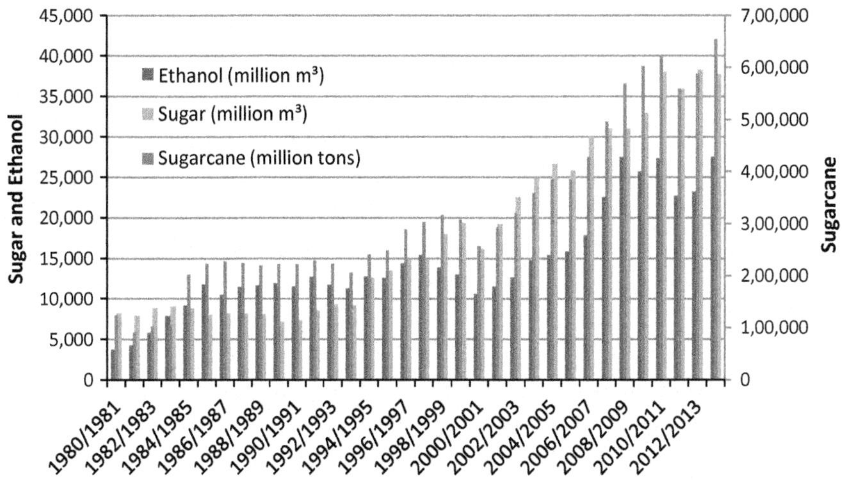

Fig. 2 Evolution of sugarcane harvest and ethanol and sugar production between 1980 and 2013, in Brazil. Adapted from UNICA (2015)

by the Federation of the Industries of São Paulo State—Brazil (FIESP 2013), sugarcane productivity in 2023/2024 must be at 82 t/ha to meet the country's needs.

The evolution of sugarcane harvest and sugar and ethanol production in Brazil is presented in Fig. 2.

2.1 *Potential of Sugarcane in Energy Production*

In 2013 the Brazil had more than 400 plants of sugar and alcohol, all of them are self-reliant in energy, thanks to the production of steam through the burning of bagasse in boilers. However, only just over 20 % of the plants traded their electric power surpluses on the market due to the cost of cogeneration implementation. The new industries already have such adjustments as are necessary for the production and provision of energy (Novacana 2015).

The National Energy Balance presents the sugarcane and its derivatives as the second largest energy source in the country, in ton of oil equivalent ahead of hydroelectricity and behind only oil. In large part this is due to the burning of bagasse for power consumption in plants and to supply the public network (Novacana 2015).

Figure 3 shows clearly that the current model of utilization of the sugarcane is outdated, where only a part of its energy potential is actually recovered, in the form of alcohol (24 %). The other fraction (76 %) of energy is present in the palhiço and bagasse. Much of the "palhiço", the waste of sugarcane harvest, is currently left in the soil. Sugarcane bagasse is also little reused, only a portion is burned for power generation, or more recently harnessed to produce cellulosic ethanol.

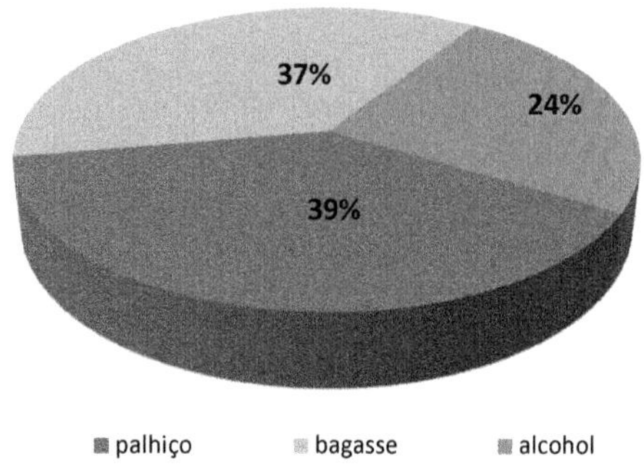

Fig. 3 Energy contained in fractions of cane per hectare. *Source* Adapted from Ripoli and Molina (1991)

It is estimated that the sugar–ethanol sector can meet up to 15 % of the energy demand of the country, surpassing the equivalent to the double generation of the Itaipu hydroelectric (Galbiati et al. 2010). If all the biomass of cane available in the country was fully exploited, it would be possible to add a power supply volume about 11,000 MW until the harvest of 2018–2019 (Novacana 2015).

The sugarcane harvest period, from March to November, coincide with times when the offer of hydroelectricity is normally smaller, because of the decrease of rainfall. As well as the system working closer to the limit, the surplus in electricity from sugar cane can be an interesting alternative (EMBRAPA 2015).

In addition to generating clean energy, the plants are enabled to request the issuance of certifications carbon credit projects. However, this is not about a simple process, since the credits are issued directly by the United Nations (UN), which makes the action something expensive and time-consuming, of the order of 2–3 years.

On the other hand, the initial investment for energy production is quite high. Some surveys indicate around U$400.000 per Megawatt (MW) produced. Yet it is a very viable investment since the time of return of capital employed is between 5 and 7 years. Some industrial options, for example, are on the order of 1–13 years to return to business. Another advantage in the deployment of this system of energy production is the sale of the surplus to the concessionaires. These are long-term contracts, on the order of 20 years, which guarantee a source of income, much less linked to oscillations of market (Martins 2009).

3 Solid, Liquid, and Gaseous Wastes from Sugar/Ethanol Industry

As presented in Fig. 2, the ethanol and sugar volume of production in Brazil is enormous. This means that a lot of raw material is processed and big volumes of wastes are generated. These wastes, if not properly discarded, pose a huge environmental impact.

Since ethanol and sugar are relatively cheap products, the valorization and reuse of wastes within the industry are of great economic impact. In this context, the sector developed interesting strategies. The wastes generated during the whole process are sugarcane bagasse, vinasse, carbon dioxide, filter cake, and yeast. Except for carbon dioxide, the others are currently reused in Brazilian industries of ethanol and sugar. The origin, characteristics and technologies covering these wastes are described next.

3.1 Sugarcane Bagasse

The bagasse is the fibrous residue resulting from sugarcane juice extraction. The proportion of this waste depends on the amount of fiber that the sugarcane cultivars present, in addition to the milling conditions to which the cane is submitted. Generally, the genetic material in use in the sugar cane mills presents a content of approximately 280 kg of bagasse (with 50 % of humidity) on every ton of cane processed (Rodrigues et al. 2003). Considering the Brazilian production of sugarcane, the generation of bagasse could be currently estimated at 187 million tons/year. A portion of this material is burned for energy production; however, the remainder is not yet used and generates environmental problems. The world processed bagasse reaches only about 54 million tons (Peng et al. 2010).

The bagasse of sugarcane is being harnessed in Brazil for several years on a large scale. It is mainly used as a fuel in boilers, cogeneration of heat and electricity. In 2012, 11.2 % of the energy consumed by the country came from the use of sugarcane bagasse (EPE 2013).

Morphologically, the sugarcane bagasse is very heterogeneous material composed by 32–44 % of cellulose, 27–32 % of hemicellulose 19–24 % of lignin, and 4.5–9.0 % of ashes (Jacobsen and Wyman 2002).

Several studies have been carried out and focus on the use of the sugarcane bagasse fractions. This material rich in cellulose, hemicellulose, and lignin can be the raw material for the production of fuels and other chemical derivatives.

According to Soccol et al. (2010) the development of technologies that enable the production of ethanol from sugarcane bagasse in Brazil has a promising scenario. This is because the production process can be attached to the units of sugar/ethanol already in operation, requiring lower investments, infrastructure, logistics, and energy supply. Furthermore, there are no transportation costs as the bagasse is generated in the own industrial units. From 10 million tons of dry biomass, 600 million liters of ethanol could be produced, considering only the cellulosic fraction.

The production of ethanol from sugarcane bagasse requires steps of pretreatment of the material. Due to its lignocellulosic composition, the bagasse must be subjected to physical or chemical treatment, or a combination of both to facilitate the subsequent enzymatic hydrolysis of cellulose and hemicellulose. The pretreatment will remove structural and compositional barriers of lignocellulosic materials, promoting an improvement in the rate of hydrolysis and increased income fermentable sugars from cellulose and hemicelluloses. The sugars released in the hydrolysis will be used in the fermentation process. The cost of pretreatment of the bagasse can reach 20 % of the total of cellulosic ethanol production (Yang and Wyman 2008). The pretreatment step should be well established to obtain the maximum yield of fermentable sugars. In addition to improving the digestibility of cellulose, the formation of inhibitory compounds during the pretreatment should be minimized. These compounds derived from sugars and lignin may become an obstacle to the fermentation process. And also an ideal pretreatment step has a low energy demand (Galbe and Zacchi 2007).

The bagasse is an energy source for cogeneration plants, which when burned, generates thermal energy in the form of steam and electricity. The operation occurs as follows: in a furnace the bagasse is burned, providing heat for vaporization of water in the boiler. The superheated steam proceeds to the turbine connected to a generator, the movement of the turbine blades produce electricity. The exhausted steam turbine can also be harnessed as heating fluid in the plant.

The Steam Cycle with back pressure steam turbine is the most common kind of cogeneration system in the Brazilian sugarcane industries. In this case the process determines the quantity of steam that can be produced by the boiler once there is not a condensation system. This kind of cogeneration system can operate just during the crush season when the factory is in operation and the steam demand exists (Ensinas et al. 2006).

3.2 *Palhiço*

The palhiço (or palhito) is the waste material from cane harvest composed of straw, green leaves, grinding wheels, roots, and weeds. It is disposable on the soil surface after cane harvest mechanize when burning is not performed. It has high potential for the production of electrical energy. The palhiço is released by the gatherer machine in the form of a layer spread on the soil surface with a thickness quite uneven. It is left in the ground for drying for about 10 days, which results in a moisture slightly greater than 20 % on average (varying from about 15–30 %) (Ripoli and Gamero 2007). A single pass of the windrow machine reduces contamination with earth. The material is subsequently collected, according to the system being used (Ripoli and Gamero 2007). With these moisture conditions the lower calorific value (LCV) of this residue is approximately 2750 kcal/kg. Assuming a bagasse with 50 % humidity and 1.8 % sugar content, the LCV is 1803 kcal/kg. Comparing in terms of energy production, the burning of the palhiço has advantages in relation to bagasse.

Only 40–50 % of the palhiço is taken from the soil due to agronomic advantages that it offers to the land, such as: maintenance of soil moisture, organic matter, and increase emergency control of weed plants (EMBRAPA 2015).

The use of palhiço as energy source on a large scale is still hampered by high recovery costs, which involves gathering, baling, transportation, size reduction, and residue utilization technology. There are also difficulties related to the heterogeneity and low-density material, in addition to using currently equipment low operational capacity and high costs, since the equipment is specific to collect forage (Michelazzo and Braunbeck 2008).

3.3 *Vinasse*

At the end of the fermentation, practically 100 % of the sugar (sucrose) present in the culture media is consumed by the yeast (usually a *Saccharomyces*), resulting in a liquid called wine. The wine has a concentration of ethanol (% in volume) between 6 and 10 GL, which is recovered by distillation in the top part of distillation columns. The present volatile substances are separated based on their different boiling points.

Vinasse is removed at the bottom of the distillation columns. It consists of ethanol-free fermented broth and contains some organic solids in suspension as well as minerals, residual sugar, and some volatile compounds. Considering the ethanol concentration in the wine, vinasse is generated in an average proportion of 12–15 L for each liter of alcohol produced. It means that more than 330 million cubicmeter of vinasse were produced in 2013/2014, which is astounding.

The physicochemical characteristics of vinasse are depicted in Table 1. Because of its production rate and its chemical characteristics vinasse constitutes the largest pollution source of the Brazilian ethanol industry.

Currently, the destination given to vinasse is its aspersion over sugarcane plantations. This practice has totally or partially replaced the use of chemical fertilizers. Vinasse is usually stored in depuration lagoons (Fig. 4) prior use. Channels are built through sugarcane plantations where vinasse drains and a motor pump truck is responsible to sprinkle the liquid (Fig. 5). Its application as fertilizer has some advantages, especially in terms of productivity, but the amount used might be well determined, respecting the soil's ion retention capacity.

There is a maximum rate of vinasse application in the field based on vinasse and soil compositions. At the state of São Paulo (Brazil), the regulamentation that governs the use of vinasse as fertilizer is the technical guide P4.231—Stillage: criteria and procedures for agricultural soil application (CETESB—Technological Environmental Sanitation Company of São Paulo), which is used by other states of the country. However, in most places inspections are very difficult to be carried and controlled. The problem is that enormous volumes of vinasse is daily produced and it has no destination besides its use in soil, which results, usually, in indiscriminate use.

The direct application of vinasse in the soil in high rates can cause salinization, leaching of metals present in the soil to groundwater, changes in soil quality due to unbalance of nutrients, alkalinity reduction, crop losses, increase of phytotoxicity and unpleasant odor (Christofoletti et al. 2013). The most easily observed effects are productivity reduction, late maturation, and low sucrose content in sugarcane plants

Table 1 Physico-Chemical characterizations of Vinasse (media of 64 samples from 28 ethanol industries from São Paulo–Brazil) (Hassuda et al. 1991)

Parameter	Unit	Medium value
pH		4.15
Soluble solids	Brix	18.65
DBO_5	mg/L O_2	16,494.76
DQO	mg/L O_2	28,450.00
Calcium	mg/L CaO	515.25
Chloride	mg/L Cl^-	1218.91
Cooper	mg/L CuO	1.20
Iron	mg/L Fe_2O_3	25.17
Phosphorus	mg/L P_2O_4	60.41
Magnesium	mg/L MgO	225.64
Manganese	mg/L MnO	4.82
Nitrogen	mg/L N	356.63
Ammonia nitrogen	mg/L N	10.94
Potassium	mg/L K_2O	2034.89
Sodium	mg/L Na^+	51.55
Sulfate	mg/L $SO_4{}^{-2}$	1537.66
Sulfite	mg/L $SO_3{}^{-2}$	35.90
Zinc	mg/L ZnO	1.70
Ethanol-CG	mL/L	0.88
Glycerol	mL/L	5.89

Fig. 4 Depuration lagoon where vinasse is stored at an ethanol and sugar industry (São Paulo, Brazil). Vinasse flow at the moment the pictures was taken was 350 $m^3\ h^{-1}$. *Source* The authors

(Pinto 1999). Moreover, Christofoletti et al. (2013) reviewed and described that vinasse can contribute to disseminated endemic diseases in water bodies, are toxic to the aquatic (surface and groundwater) and terrestrial environments (greenhouse gas emissions, soil contamination, genotoxicity).

When vinasse is produced in excess and cannot be used as fertilizer, which is common, industries throw it in areas called "sacrifice zones." In this area the soil

Fig. 5 Channels for the distribution of vinasse along the field (*left*) and coated channel in an ethanol and sugar industry located in São Paulo, Brazil. *Source* The authors

becomes very salty and acid causing desertification and rendering it unusable for any other purpose. In long term these characteristics are also noted in productive land, causing productivity decrease, late maturing, and decrease in sucrose content (Pinto 1999). In 1986 40 % of the vinasse produced in Brazil was not used as fertilizer and was thrown in sacrifice zones (Angenent and Wrenn 2008). Unfortunately, no updated data collection is available. Informal conversations with the environmental manager of an industry in São Paulo indicated that this number is approximately 25 % (which means that 82.629 thousand cubicmeter of vinasse are simply discarded).

The biggest concern around vinasse disposal is that the biggest ethanol production area in Brazil is located above the Guarani Aquifer—the second largest aquifer systems in Brazil and an important source of fresh water; it covers 1.200,000 km^2 (460.000 m^2) (Brazil, Argentina, Paraguay, and Uruguay) with a volume of about 40.000 km^3 (9.600 m^3) (Iritani and Ezaki 2012). The Guarani Aquifer is in an important strategic reserve for supplying the population and for the development of economic activities. If the indiscriminate use of vinasse in the soil is not controlled, serious problems of infiltration and consequent pollution of such underground water may occur.

The composition of vinasse is not standardized according to soil use, which is an issue of environmental concern because vinasse characterization varies significantly according to each sugarcane processing plant. For example, the chemical oxygen demand (COD) of organic matter content can vary between 10 and 65 g L^{-1}.

Consequently, the impacts caused by potential GHG emissions resulting from vinasse organic matter degradation on soil, in addition to unpleasant odors generated and possibly attracting insects, are not considered but are likely to occur (Moraes et al. 2014).

Since 1990s, Hassuda et al. (1991) identified infiltration problems due to vinasse aspersion in Bauru Aquifer (SP-Brazil), another Brazilian aquifer, increasing the fear of similar problems in the Guarani Aquifer. This problem is not only related to the sacrifice zones cited before, but also due to the lack of protection given to avoid vinasse infiltration in the soil (uncoated channels in Fig. 5).

New government regulations are now forcing the industries to coat the channels (Fig. 5), but inspection is very limited.

During the last decades, ethanol production has increased very rapidly. Brazil is, nowadays, the second higher ethanol producer in the world. Recent international incentive and demand for biofuels production influenced Brazilian ethanol industries, increasing production. Thus, the problem of vinasse disposal will worsen. Indeed, its continuous discharge onto land can endanger the chemical and physical structure of the soil, reduce yields and lead to serious groundwater pollution problems. In this context, it is of great importance to give a more rational destination to vinasse or at least reduce its toxicity.

In this section, technologies of vinasse usage for biogas, biomethane, and biohydrogen from anaerobic fermentation, microalgal biodiesel, proteins, and pigment productions will be presented. Figure 6 presents an overview of all the technologies based on vinasse utilization that will be described in this chapter.

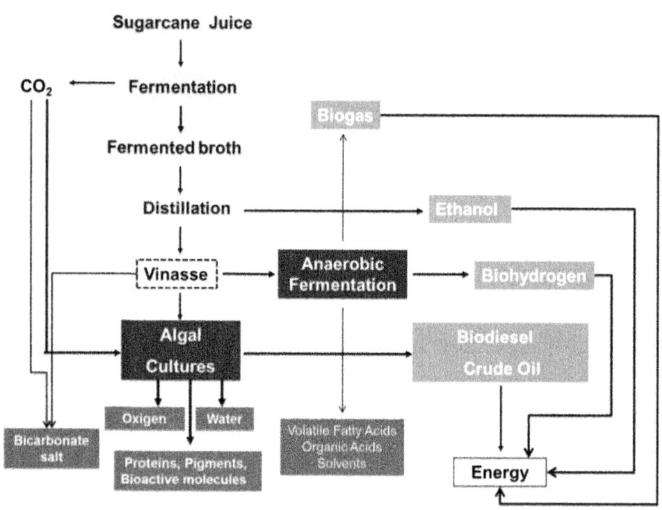

Fig. 6 Technologies of reuse of the vinasse generated from the ethanol production process to produce biomolecules, chemicals, and energy (biogas, biodiesel, crude oil, and hydrogen)

3.3.1 Vinasse for Biogas Production

Regarding the composition depicted in Table 1, vinasse is an interesting substrate for microorganism growth because it presents a great amount of micronutrients. Iron, magnesium, phosphorus, and nitrogen content are interesting for the development of biogas production.

Because low amounts of fermentable carbon are present in its composition, vinasse might be enriched with a carbohydrate source to allow the production of great quantities of biogas. Some cheap fermentable carbon sources are available in Brazil, especially in the ethanol/sugar industries, where vinasse is generated: sugarcane molasses, sugarcane juice, and sugarcane bagasse.

The first pilot plant for the production of biogas in Brazil was installed at Usina São Martino (São Paulo—Brazil). Nowadays, other industries use the technology (pilot plants) to produce energy from biogas. Biogas is produced from vinasse at a rate of 9.5 m^3 m^{-3} (Elia Neto et al. 2009), with a methane content of approximately 70 % (Rodrigues et al. 2012). According to Moraes et al. (2014) computational simulations, 0.29 m^3 kg_{COD}^{-1} of biogas can be produced from vinasse, but this scenario considers the fermentation of pure vinasse (without supplementations with carbon sources).

Biogas from vinasse supplemented with sugarcane bagasse, molasses and/or sugarcane juice is an interesting alternative to increase biogas production, but competes with ethanol production (first and second ethanol generation), requiring careful technical and economic evaluation. On the other hand, the technology is simple and easy to be adapted to the existing vinasse ponds.

The main point of biogas is that it can be produced in a decentralized manner, facilitating the distribution energy and, thus, reducing costs. Considering the economic characteristics of the country, which is based on primary and secondary sector of production, there are large quantities of wastes (liquid and solid) capable of being transformed through anaerobic fermentation to biogas or biomethane.

Moreover, the liquid waste from the biodigesters presents characteristics of fertilizer that can be still used in fertigation of sugarcane plantations, but with less toxicity to soil. The biodigestion of vinasse functions as an intermediary step to recover part of its energy in form of biogas, maintaining its characteristics as fertilizer.

Besides biogas, biomethane (purified methane form biogas) for industrial use and/or as fuel shows up as an interesting alternative to Brazil economy. It is important to consider that biogas production and regulation is developing quickly in Brazil these days. In January 2015 the National Agency of Petroleum, Gas and Biofuels (ANP) published a resolution that determines which characterizes the quality parameters for biomethane. It is expected until the end of 2015 another by ANP to regulates the safety of biogas producing facilities. With these regulations the biogas sector should have a very fast growth in the coming years.

Nowadays, biogas producing facilities in Brazil are predominantly rural, carried in covered lagoons without agitation or temperature control. Since the anaerobic fermentation of vinasse to biogas is possible, it is also possible to couple it to hydrogen production facilities. In this context hydrogen can be produced from vinasse in two ways: (i) reformation of biogas and (ii) dark fermentation of organic matter. In this chapter only the direct H_2 production from fermentation will be discussed.

3.3.2 Hydrogen Production in Vinasse-Based Medium Through Dark Fermentation

The advantages of the fermentative hydrogen production are the broad spectrum of applicable substrates as well as high hydrogen production yields (Nandi and Sengupta 1998). The possibility of coupling the energetic hydrogen production from biomass with the simultaneous treatment of waste materials is an addition crucial advantage. Both biohydrogen production and methane from anaerobic digestion are CO_2 neutral since the carbon released by their combustion is derived, directly or indirectly, from recently fixed atmospheric CO_2 (Reith 2001). Moreover, the emitted carbon associated with hydrogen produced by microbial fermentation is released during the fermentation rather than during its utilization, thus potentially allowing easy capture of CO_2.

Among the wide range of by-products of diverse microbial metabolism, the two pathways producing hydrogen from carbohydrates are associated with acetate and butyrate. The theoretical yield of H_2 per mole of glucose associated to the production of acetate and butyrate is described as the following reactions:

$$C_6H_{12}O_6 + 2H_2O \rightarrow 2CH_3COO^- + 2CO_2 + 4H_2 \qquad \Delta G'0 = -206\ \text{kJ mol}^{-1}$$
$$C_6H_{12}O_6 + 4H_2O \rightarrow 2CH_3CH_2CH_2COO^- + 2CO_2 + 2H_2 \qquad \Delta G'0 = -264\ \text{kJ mol}^{-1}$$

Biological hydrogen production from vinasse emerges as an interesting alternative since biohydrogen production by anaerobic bacteria is half the metabolic path to produce biogas. This means that if biogas from vinasse is possible, biohydrogen is also.

Because vinasse does not contain fermentable sugars in its composition (or in too low concentrations) one possibility is to produce biohydrogen through photofermentation. Physiological studies of cyanobacteria have identified many producing strains, such as *Spirulina platensis* (Ayoama et al. 1997), *Anabaena cylindrica* (Jeffries et al. 1978), *Cycas revoluta* (Margheri et al. 1990), and others. One of the greatest drawbacks of this technology, besides dependence of light (which influences in bioreactors design, difficulties in large-scale production, among others) is that hydrogen production by cyanobacteria occurs in such a limiting environment that that cell death is a natural consequence. Moreover, in this specific case, vinasse is a dark turbid liquid, affecting the passage of light and, therefore, decreasing productivity.

On the other hand, the addition of cheap and available carbon sources to vinasse allows driving of dark anaerobic fermentation processes, which are described to be much cheaper than the photosynthetic and present higher yields. Considering the usage of molasses or sugarcane juice as carbon sources, which are highly available in sugar/ethanol industries, they do not burden on the cost of the medium for biohydrogen production. At this point, promotion and maintenance of anaerobic environment are the processes that will probably impact most significantly the price of the final product. The effluent from dark fermentation can be used as substrate for photofermentation process.

Two possibilities arise regarding biological hydrogen production through dark fermentation,

Production of biohydrogen and purification and reuse of the coproducts (VFAs —volatile fatty acids): Organic acids are some of the end products of anaerobic metabolism to produce biohydrogen, especially C2 and C4 acids. These organic acids can be produced and sold as commodity chemicals or further processed into higher value chemicals, biofuels, or bioproducts. Among the acids produced during biohydrogen production are acetic, butyric, succinic, lactic, formic and propionic acids. Usually, in biohydrogen processes it is observed a preferential production of acetic and butyric acids. Considering the economic issues associated to biohydrogen production systems, the recovery or reuse of such VFAs are of great interest since H_2 production is high.

Coupled biohydrogen and biogas production: Generally the volatile fatty acids produced during bioH_2 production are not recovered, but used in sequential processes as substrate for microbial methane or solvent production. The process is composed, then, of two bioreactors: the first one to biohydrogen production, where the raw materials are fermented by anaerobic microorganisms and produces biohydrogen and VFAs, and the other to biogas production, where the VFAs from the previous reactor work as substrate for methane production by methanogenics. At the end, both biohydrogen and biogas are recovered.

The biohydrogen can be used in chemical industry or in fuel cells for the production of electricity. Otherwise, the hydrogen-rich biogas can be used for heat generation through direct combustion or in boilers (specific heat of hydrogen is 14304 J kg^{-1} K^{-1}).

3.3.3 Evaluation of Vinasse for Biohydrogen Production by Strict Aanaerobic Consortia—A Case Study

The gas produced in anaerobic dark fermentation of supplemented vinasse varies in relation to the strain or consortia used and culture conditions (partial pressure of hydrogen, temperature, pH, etc.). Preliminary studies on the evaluation of using vinasse as culture medium for biohydrogen and VFAs production by anaerobic bacteria were carried in the Bioprocess Engineer Laboratory of Federal University of Paraná (Brazil) and are described in this section.

Two known *Clostridium* strains, *Clostridium saccharoperbutylacetonicum* and *Clostridium beijerinckii* purchased from ATCC (ATCC 27021 and 8260, respectively) and one natural vinasse consortium (VINA) were evaluated in a first step of the work. Gas production in Hungate tubes cultures was measured and analyzed twice in a week or daily according to the experiment. Those cultures degassed daily were considered free of H_2 partial pressure. The gas sampled from the headspace was analyzed using a MicroGC Agilent 300A.

During 20 generations the gas produced during fermentation was measured and analyzed twice a week (fourth and seventh days of fermentation). Results of average hydrogen production rate (in mL L^{-1} day^{-1}) and average total production (in mL_{H2} L^{-1}) of each strain are showed in Table 2. It can be observed that, besides H_2 concentration in gaseous phase did not present considerable differences, the total H_2 production (mL L^{-1}) was much higher in those cultures degassed daily (considered free of H_2 partial pressure). Table 3 presents the complete analysis of gaseous phase of the culture from Table 2.

Since gas production was greatly increased by minimizing H_2 partial pressure, analysis of the liquid phase was also carried. The VFAs analysis at the last day of cultivation (seventh) is showed in Table 4. It can be noticed that in those cultures under H2 partial pressure there was production of more oxidized VFAs (propionate and lactate), which contains more hydrogen in its composition. This is compatible with the results in hydrogen production presented in Table 2. Moreover, in cultures free of H2 pressure acetate and butyrate were the main products of strains ATCC 27021 and ATCC 8260.

Based on VFAs and gas analysis it was noted that high butyrate/acetate ratio is related to higher hydrogen content in the gas phase (Table 5). At the same time there is a relation between the amount of gas produced and the hydrogen concentration in the gaseous phase.

These studies showed that vinasse is an interesting source of nutrients and promotes the biohydrogen and volatile fatty acids. Very high hydrogen content in the gaseous phase (33 and 36 %) was achieved by pure cultures, which also presented higher susceptibility to hydrogen partial pressure. It was also observed the effect of the profile of volatile fatty acids and the hydrogen production, resulting in the determination of a relation between the Butyrate/Acetate content and the concentration of hydrogen in the exiting gas.

Table 2 Hydrogen production by 3 strains grown in vinasse-based medium

Strain	Carbon source (10 g L^{-1})	H_2 (ml/L/day)	Total H_2 (ml/L)	Hydrogen in gas phase (%)
Cultures with H_2 partial pressure				
VINA	Sucrose	262.6	1838.2	12
ATCC 27021	Sucrose	587.8	4114.6	36
ATCC 8260	Sucrose	635.3	4447.1	34
Cultures free of H_2 partial pressure				
VINA	Sucrose	403.9	2827.3	13
ATCC 27021	Sucrose	2526.3	17684.1	33
ATCC 8260	Sucrose	1895.8	13270.6	24

Results represent an average of 20 generation measurements

Table 3 Typical gas analysis carried in a MicroGC equipped with a MoleSieve 5A and PLOT U columns

Strain	Medium	H_2	O_2	N_2	CH_4	CO_2	H_2S	H_2O
Cultures under H_2 partial pressure								
VINA + vinSa	Vinasse + 10 g L^{-1} sucrose	11.89	1.57	8.46	0.00	80.00	0.07	1.34
ATCC 8260		33.66	0.47	2.55	0.00	68.69	0.01	1.12
ATCC 27021		36.09	1.05	4.12	0.00	61.85	0.03	1.15
Cultures avoiding H_2 partial pressure								
VINA + vinSa	Vinasse + 10 g L^{-1} sucrose	13.06	1.30	5.23	0.00	80.76	0.42	1.37
ATCC 8260		24.47	1.27	5.43	0.00	71.12	0.04	1.49
ATCC 27021		33.54	0.81	3.94	0.00	64.62	0.02	1.41

Results are presented in %

Table 4 VFAs concentration (g L^{-1}) at the 7th day of fermentation in vinasse-based medium

Strain	Acetate	Formate	Butyrate	Ethanol	Propionate	Lactate	Succinate
Cultures with H_2 partial pressure							
ATCC 27021	1.79	0	3.53	0	1.42	0.79	0
ATCC 8260	1.64	0	4.28	0	1.39	0.59	0
VINA	1.7	0.6	2.30	1.80	0.90	0.25	0
Cultures free of H_2 partial pressure							
ATCC 27021	1.14	0	3.61	0	−1.10	−0.37	0.16
ATCC 8260	1.94	0.08	2.55	0.16	−1.10	−0.37	0.24
VINA	0.98	0.40	0.41	2.48	−0.70	−0.37	0

The negative concentrations indicate the consumption of these metabolites in comparison to the non-fermented medium

Table 5 Butyrate/acetate ratio, gas produced and hydrogen content in the gas phase in the experiments carried with each strain

	VINA	ATCC 8260	ATCC 27021
Butyrate/Acetate	0.42	1.31	3.17
H_2 (%)	13.40	24.00	33.00
Gas (L_{gas}/L_{medium})	21.10	55.29	53.59

Butyrate/Acetate ratio was considered based on VFAs analysis of the seventh day of fermentation

3.3.4 Vinasse for Microalgae Biomass Production

Vinasse has considerable amounts of nitrogen, phosphorous , especially potassium in its composition, besides some micronutrients. Hence, it works as fertilizer to plants, as described earlier in this chapter.

Microalgae and cyanobacteria are microscopic organisms that typically grow suspended in a liquid medium and are able to use the solar energy to combine water with carbon dioxide to create biomass. Since the carbon metabolism of plants and

Table 6 Microalgae and cyanobacteria strains evaluated for biomass production in the BioOil Project

Microalgae/Cyanobacteria Strains	
Spirulina platensis	*Spirulina maxima*
Dunaliella tertiolecta	*Dunaliella salina*
Botryococcus braunii	*Neochloris oleoabundans*
Scenedesmus obliquus	*Phaeodactylum tricornutum*
Tetraselmis chuii	*Chlorella minutissima*
Chlorella kessleiri	*Chlorella vulgaris*
Synechococcus elongatus	*Synecococcus nidulans*

microalgae are essentially the same (dependence of light, CO_2, and micronutrients to produce biomass through photosynthesis), those materials that work as fertilizer for plants probably work as fertilizer for microalgae.

In this context the Bioprocess Engineer and Biotechnology Department of the Federal University of Paraná (UFPR)—Brazil developed, in partnership with Ouro Fino Agronegócio Ltda, the Ourofino BioOil—Brazil project. This project is being carried since 2004 and resulted in 6 patents (BR 13 2012 025497 1, PI 1106809-4, PI 0804115-6, PI 0706144-7, PI 0706170-6, and PI 0705520-0). Some of the most important results of the microalgae subproject obtained during these years are described in this chapter.

The BioOil project constitutes a set of innovative technologies, environmentally sustainable, based on the reuse of liquid and gaseous wastes from the ethanol industry. It was carried between 2004 and 2014. More specifically, the BioOil project involves the use of vinasse from the ethanol distillation columns and the CO_2 produced in ethanol fermentation to produce microalgae biomass for use in animal feed, food production, medicine, biofertilizers, bioenergy (biodiesel, hidrocarbons, etc.) and/or extraction of other bioactive compounds (pigments, bioactive proteins, etc.). Moreover, the microalgal process objective the reduction of BOD and/or COD (biological and chemical oxygen demand, respectively) of vinasse, the production of oxygen (O_2) to reduce the greenhouse effect.

Fourteen different strains of microalgae/cyanobacteria were evaluated within the scope of this project (Table 6). Between them, the microalgae/cyanobacteria from the genre *Chlorella* and *Spirulina*, *Dunaliella*, and *Botryococcus* have special commercial interest, producing protein-rich biomass, pigments and biofuels, respectively. Other genre and strains were tested considering their potential and specific characteristics described in the scientific literature.

Biodiesel and BioOil Production from Oleaginous Microalgae

This project is the first to be developed in Brazil that presents an alternative for the production of biodiesel and/or fuel oil than the conventional technologies (from vegetal and animal fats). In this project, biodiesel was produced from the lipids accumulated in microalgae cells, which are produced in vinasse medium. The main advantage of this process is productivity, while plants and animals take long periods to double its biomass, microalgae duplicate each 2–6 h. It is, then, possible to produce one or more bioreactors for the same amount of lipids/oils in area hundreds of times smaller than that used for plants. The process control in photobioreactors is much simpler than the controls of an agricultural crop and is independent of climatic variations and phytosanitary questions. Another important advantage of this process is the reduction of some pollutant characteristics of vinasse (but maintaining its fertilizer potential), reducing risks of environmental problems caused by its disposal in soil (data not presented).

At a first step, microalgae were screened according to higher lipid productivity (biomass production × lipid content × time of cultivation) at media with different vinasse concentration (1–100 %). The dried biomass obtained at the end of culture had their lipid content quantified by extraction methanol:chloroform 1:1 followed by an hexane liquid–liquid extraction (Sydney et al. 2010).

To complete remove the lipids within the cells of *Chlorella vulgaris*, which has a thick cell wall that has difficult solvent extraction, hydrolysis was carried out. To 1 g of biomass it was added 20 mL of methanol anhydrous and 1 mL of concentrated HCl and placed in a water bath at 90 °C for 30 min. Then, a liquid–liquid extraction was carried out with hexane. This process increased lipid extraction from *C. vulgaris* from 3.36 to 17.9 % (5 folds) (Table 7).

Those microalgae with higher potential of lipid production in vinasse medium were scaled up to a 12 L (working volume) photobioreactor. Table 7 presents the results achieved at bioreactor scale. The highest lipid production in vinasse medium was observed for *Botryococcus braunii*, followed by *Synecococcus nidulans* and *C. vulgaris*. The higher lipid content was achieved with B. braunii (19.5 %) and *C. vulgaris* (17.9 %), while higher biomass production was achieved with *S. platensis*.

Table 7 Biomass, lipid content, and lipid production by different microalgae and cyanobacteria cultured in vinasse-based medium

Strain	Vinasse in medium (%)	Biomass production (g L^{-1})	Lipid content in cells (%)	Lipids (mg L^{-1})
Spirulina platensis	40	3.13	6.3	197.19
Spirulina platensis	30	2.23	5.8	129.34
Botryococcus braunii	10	2.19	16.5	361.35
Botryococcus braunii	20	1.88	15.8	297.04
Botryococcus braunii	30	1.65	19.5	321.75
Scenedesmus obliquus	30	2.15	10.1	217.15
Scenedesmus obliquus	20	1.68	9.6	161.28
Chlorella vulgaris	30	1.5	3.36	50.40
Chlorella vulgaris (hydrolyzed)	30	1.5	17.9	267.90
Synecococcus nidulans	30	2.23	13.0	289.90

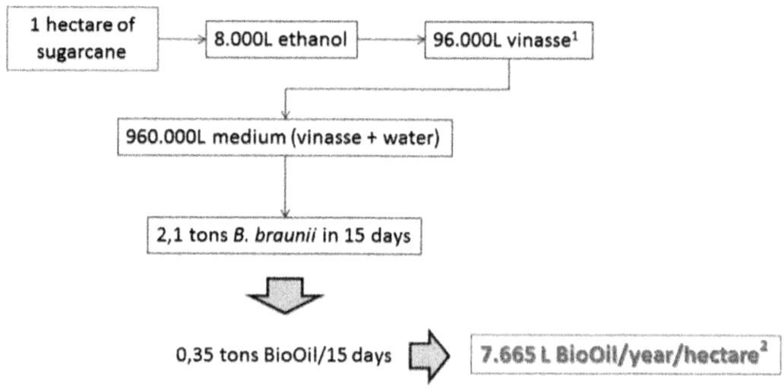

Fig. 7 Calculation of biooil production by the microalgae *Botryococcus braunii* in vinasse medium

According to the results (Table 7) we can estimate the amount of microalgal oil that can be produced in vinasse medium. This calculation, using the biomass production and lipid content of *B. braunii* cultivated in 10 % vinasse medium (Table 7) is presented in Fig. 7. For each liter of ethanol produced it was considered that 12 L of vinasse is generated. The BioOil density was around 0.9 kg/L.

The biomass obtained from the cultivation of different microalgae were analyzed in a CHNS Analyzer (Flash 2000 series, Thermo Scientific) in order to evaluate the correlation between the Carbon/Nitrogen proportion of cells and the lipid content. As expected, the higher the relation C/N (carbon/nitrogen), higher the lipid content within cells. The determination of a direct relation between C/N and lipid content configures a very powerful tool in the bioprospection of oleaginous microalgae.

This lipids obtained can be directly burned to produce heat or transformed to Biodiesel by transesterification. Moreover, the lipid-rich biomass can be transformed into BioOil by pyrolysis; this process is described in Dalmas et al. 2013.

The fatty acid composition of the biodiesel obtained from *B. braunii*, *S. obliquus,* and *C. vulgaris* were performed in a Shimadzu GC-2010 gas chromatograph (GC) equipped with a hydrogen flame ionization detector (Table 8). Fatty acid methyl esters (FAME) for analysis were prepared by the method proposed by Hartmann and Lago (1973).

The fatty acids methyl ester profile presented in Table 8 shows a predominance of palmitic (C16:0) and oleic (C18:1) acids along with considerable amounts of estearic (C18:0), linolenic (C18:2), and alpha-linolenic (C18:3) acids. These fatty acids are common in traditional vegetable oils, such as palm and soybean, and are adequate for biodiesel production.

Microalgae cultures were then scaled up to 10 m^3 cultures in open pond reactors (raceways) (Fig. 8). At this stage, a control of CO_2 injection based on pH was developed. During microalgal growth the pH tends to increase, a consequence of its metabolism (CO_2 consumption through photosynthesis). The use of CO_2 to control this pH also works as carbon source to support biomass growth and lipid production. CO_2 was injected at 5 % concentration.

Table 8 Fatty Acids Methyl Esters (FAME) obtained by the transesterification of microalgal lipids produced in vinasse medium

	Botryococcus braunii	*Botryococcus braunii*	*Scenedesmus obliquus*	*Chlorella vulgaris*[a]
Medium	Synthetic (MBM)	10 % vinasse	30 % vinasse	30 % vinasse
FAME	% FAME	% FAME	% FAME	% FAME
C6:0	0.6112	3.5961	0.8031	0.0000
C8:0	0.2194	0.2338	0.3952	0.0000
C10:0	0.4100	0.3647	1.0449	0.0000
C11:0	0.0000	0.0000	0.0000	0.0000
C12:0	0.1473	1.2305	0.1400	0.0000
C13:0	1.2952	0.5641	0.8409	0.0000
C14:0	1.5454	1.7607	1.4905	0.0000
C14:1	0.6749	2.8270	0.5321	0.0000
C15:0	2.0518	0.9836	1.9298	0.0000
C15:1	0.9244	0.8710	1.1489	0.0000
C16:0	29.9777	20.0218	26.2576	25.5286
C16:1	7.5951	13.9675	11.7846	0.0000
C17:0	0.1093	2.2754	2.2183	0.0000
C17:1c	4.9775	6.2693	2.9434	0.0000
C18:0	4.6811	4.8371	5.7775	0.0000
C18:1	19.7419	20.1950	11.1342	26.7700
C18:2n6	15.1572	6.0620	25.7606	47.7015
C20:0	0.2207	1.0128	0.2198	0.0000
C18:3n6	4.0591	6.9451	2.8077	0.0000
C20:1	1.2284	0.1040	0.5109	0.0000
C18:3n3	2.6007	0.7447	0.7425	0.0000
C21:0	0.2966	0.1998	0.2133	0.0000
C20:2	0.0000	0.0000	0.0000	0.0000
C22:0	0.0000	0.0000	0.0000	0.0000
C20:3n6	0.0000	0.5572	0.0000	0.0000
C22:1n9	0.1473	0.0000	0.0000	0.0000
C20:3n3	0.0121	0.2390	0.0516	0.0000
C20:4n6	0.4902	0.1051	0.7805	0.0000
C23:0	0.0000	0.2408	0.0000	0.0000
C22:2	0.0462	0.4320	0.0643	0.0000
C24:0	0.0000	0.0000	0.0000	0.0000
C20:5n3	0.1923	0.2340	0.1901	0.0000
C24:1	0.0000	0.5382	0.0000	0.0000
C22:6n3	0.5870	2.5876	0.2177	0.0000

[a]Hydrolyzed biomass

Fig. 8 Carbon dioxide injection system in the raceways for microalgal biomass production. *Source* The authors

The broth resultant from microalgae growth can be recirculated and reused in new cycles of biomass production. Tests carried in large-scale cultures indicated that up to three recirculation cycles are viable. The effluent from each cycle were analyzed for COD and BOD in order to determine if the microalgae growth promotes remediation of vinasse (Table 9). It was observed that a great reduction in COD and BOD was achieved since the first culture (first cycle) (98 % and 89 %, respectively). At the last cycle (third) BOD and COD were virtually removed, which reinforce the environmental importance of this technology.

Besides the lipids (and biodiesel), the residual microalgal biomass contains great amounts of proteins and can be used for feed purposes. Considering the calculations of Fig. 9, after lipid extraction 42.6 tons of microalgal meal are generated per hectare per year. According to analysis carried in the lab, this microalgal meal free of lipids contains approximately 50 % of proteins, which opens the possibility to be used in food and feed.

The BioOil project yields are resumed in Fig. 9, showing the use and recirculation of vinasse to produce lipids that can be converted to biodiesel or burned and a proteic microalgal biomass free of lipids that can be used in food and/or feed.

Table 9 Biomass production and COD and BOD removal when the vinasse from a microalgal culture was recirculated (cycles) for further microalgal cultures

	Biomass (g/L)	Lipids (%)	COD (mg/L O_2)	BOD (mg/L O_2)
Start	0.21	–	15,075	9870
1st cycle	2.67	19.79	3180	1058
2nd cycle	1.83	20.71	335	117
3rd cycle	1.42	20.05	16	4.99

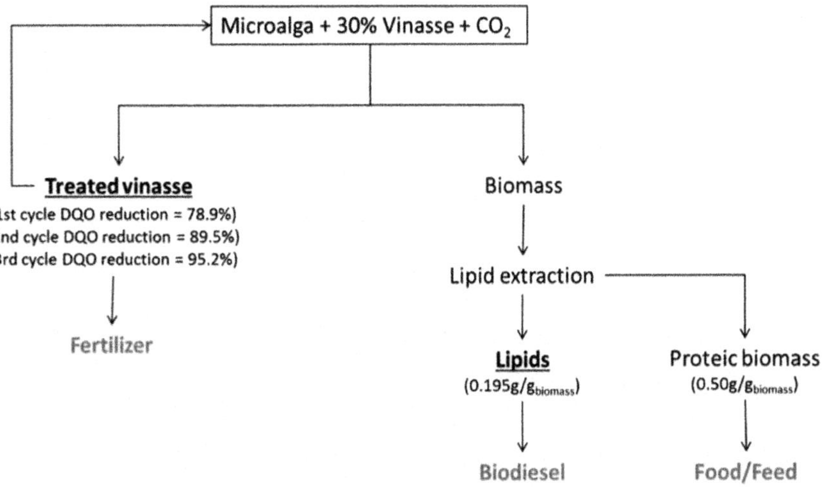

Fig. 9 Summary of the Biooil project, showing the production of lipids and protein-rich biomass and the reduction of the pollutant potential of vinasse

Microalgal Biomass for Food and Feed

Some cyanobacteria constituted of high concentration of proteins and are used for years in animal and human nutrition. The biggest example is *Spirulina*, a filamentous blue-green alga began to be cultivated in US and used as food supplement, aquaculture feed, and food coloring.

Most Spirulina can be harvested and processed easily because of the big size of cells. This, together with the best pH of growth (between 8.5 and 9.5, making it difficult to contaminate) and its protein-rich composition (55–70 %, according to Phang et al. 2000), made *Spirulina* the most known and large-scale produced cyanobacteria around the world.

During the studies carried in BioOil project, Spirulina presented a great capacity to grow in vinasse medium. Among the microalgae and cyanobacteria evaluated (Table 7), it presented the higher biomass production (3.13 and 2.23 in 40 % and 30 % vinasse medium, respectively). This potential to grow on vinasse, in addition to the ease on recovering cells (by simple filtration) and the existence of established industrial process of Spirulina biomass production called the attention to an opportunity of producing high protein content biomass to food and feed.

A series of experiments were carried with Spirulina in media with a higher range of vinasse concentration (25, 50, 75, and 100 %). The results are presented in Table 10.

A process of adaptation of *Spirulina* to higher concentration of vinasse was carried. The culture adapted to the medium containing 25 % vinasse was inoculated in a medium containing 30 % vinasse. The culture was maintained during 1 month, then being inoculated to a medium containing 35 % vinasse. Surprisingly biomass

Table 10 *Spirulina platensis* biomass production in different vinasse concentration media

Vinasse (%)	Water (%)	Biomass (t = 0)	Biomass (t = 14 days)	Biomass production (g L^{-1})
25	75	0.205	3.374	3.169
50	50	0.195	3.162	2.967
75	25	0.199	1.435	1.236
100	0	0.197	0.974	0.777

Fig. 10 10 m^3 raceway of *Spirulina platensis* biomass production in 50 % vinasse medium (*left*) and the filtered biomass drying in the sun (*right*). *Source* The authors

production increased, reaching a maximum of 4.47 g L^{-1} in 15 days in medium composed of vinasse and water at same proportions (50 % vinasse) and injection of CO_2 at 5 % (v/v air). This is 50 % more biomass than the first experiments (Table 10). This culture was than scaled up to 10 m^3 open pond reactors (raceways) (Fig. 10).

Biopigment Production from Microalgae in Vinasse Medium

Astaxanthin is carotenoid traditionally used in the aquaculture feed industry as a source of pigmentation for fish in aquaculture (especially salmonids) and for eggs in the poultry industry (Cifuentes et al. 2003). But because of its high antioxidant potential, greater than β-carotene and tocopherol (Naguib 2000), this ketocarotenoid has also received an increased interest for clinical applications.

Two microorganisms can naturally produce higher amounts of Astaxanthin: the microalgae *Haematococcus pluvialis* and the yeast *Phaffia rhodozym*. The most advantage in using microalgae to produce Astaxanthin is the independence of organic carbon sources for growth, which allows the use of a variety of agroindustrial liquid wastes.

H. pluvialis is a bi-flagellated microalgae that when in stress conditions enters a resting stage and transforms into cysts, which are enlarged cells with a thick and resistant cell wall and accumulate large amounts of carotenoids, especially Astaxanthin. It is the microorganism capable to accumulate up to 4 % of Astaxanthin (Aflalo et al. 2007). The production of Astaxanthin is generally separated in two steps: biomass production phase (vegetative cells) and carotenoid accumulation phase (cell encystent and Astaxanthin production). Carotenoid accumulation occurs in stress conditions which can be caused by salinity, light exposure, addition of acetate, C: N ratio, among others. Biomass production phase is, however, the limiting step in this technology and deserves special attention in process development. According to Li et al. (2011) Cyanotech Inc., Algatechnologies, Ltd., Mera Pharmaceuticals Inc., and Biogenic Inc. produces Astaxanthin from *Haematococcus* with costs up to U$3,000.00/kg. One alternative to reduce this is to use alternative medium for vegetative grow.

In this context, within the Biooil Project, vinasse was tested as medium for the production of Astaxanthin from *H. pluvialis*. The study started by cultivating *H. pluvialis* in different vinasse concentrations (1, 3, 5 %, and from 10 to 50 %) diluted in distilled water. Cultures were carried during 15 days in 2500 lx and at 25 °C. The results of biomass production are presented in Table 11.

According to the results in Table 11, vinasse concentration for *H. pluvialis* biomass production was fixed in 3 % (10 % vinasse medium presented good growth, but presented problems in pigment accumulation phase due to its turbidity —data not shown). Another set of experiments with supplementation of salts to vinasse were carried in order to optimize biomass growth. According to vinasse composition and the BBM composition, nitrate (250 mg L^{-1}) and magnesium (75 mg L^{-1}) was added to vinasse medium.

With the addition of nitrate as nitrogen source and magnesium the biomass production was much higher than on pure 3 % vinasse (1.77 g L^{-1} versus 0.7 g L^{-1}). On day 14 NaCl (0.7 % m/v) was added to the culture to stimulate carotenoid production. It was noted a biomass decrease of approximately 8 %, which is high considering large-scale systems.

Then other strategies for pigment production were evaluated. Four experiments were carried testing different conditions for pigment production and are presented in Table 12. Biomass production phase was carried separately at 2500 lx, 12:12 h photoperiod and at 25 °C. Astaxanthin accumulation phase was carried by 10 days.

Table 11 Biomass production of *Haematococcus pluvialis* in different vinasse concentrations

Medium	Biomass production (g L^{-1})	Medium	Biomass production (g L^{-1})
BBM[1]	0.82		
1 % vinasse	0.15	20 % vinasse	0.50
3 % vinasse	0.70	30 % vinasse	0.05
5 % vinasse	0.66	40 % vinasse	0.02
10 % vinasse	0.69	50 % vinasse	0.08

[1]Bold's Basal Medium

Table 12 Astaxanthin content in the final biomass of *H. pluvialis* cultured in different stress conditions

Experiment	Stress condition			Astaxanthin (%)
	Illumination/photoperiod	Salinity	Nitrate (mg/L)	
1	Artificial light: 2500 lx/12 h:12 h	–	–	2.18
2	Artificial light: 2500 lx/12 h:12 h	0.7 % NaCl	–	1.84
3	Artificial light: 2500 lx/12 h:12 h	0.7 % NaCl	200	1.51
4	Natural light: 6500–11,000 lx/ approx. 10 h:14 h	0.7 % NaCl	–	2.30
5	Natural light: 6500–11,000 lx/ approx. 10 h:14 h	–	–	2.99

Nitrogen content at the end of growth phase was quantified by the method of Cataldo et al. (1975) to guarantee nitrogen exhaustion prior Astaxanthin production phase. To the experiment three nitrates were added in order to evaluate the effect of nitrogen in pigmentation. Experiments 4 and 5 were carried in extern area and were, thus, influenced by natural light regime (duration and incidence) and temperature (18–32 °C). Astaxanthin was extracted and quantified according to the methodology described by Aquasearch (1999).

In the experiments presented in Table 12 biomass loss after NaCl addition was close to 12 %. The presence of nitrogen in pigment accumulation phase showed negative effect, as already described in the literature. Surprisingly, NaCl addition did not present the expected positive effect in Astaxanthin accumulation by *H. pluvialis* cultured in vinasse. On the other hand, light incidence significantly increased pigment production (37 % higher).

3.4 Carbon Dioxide

The Framework Convention on Climate Change, signed in Rio de Janeiro in 1992, made global warming a major focus and the development of technologies for reducing/absorbing greenhouse gases (GhG) gained importance.

Nowadays the International Energy Agency, an autonomous organization constituted by 29 member countries, works with a scenario of lowering carbon dioxide emissions to 2050 in order to have a 50 % chance of limiting average global temperature increase to 2 °C. To become a reality, it is necessary to "cut energy- and process-related CO_2 emissions by almost 60 % by 2050 (compared with 2012) and ensuring they continue to decline thereafter" (IEA 2015). This reduction includes mainly the power, transportation and industry sectors, each contributing with 52 %, 19 %, and 15 %to the reduction in emissions. In this context the technologies that will contribute most are end-use energy efficiency (36 %), the use of renewable in power generation (31 %) and carbon capture and storage (16 %) (analysis from the data presented by IEA 2015).

The ethanol industry produces a lot of carbon dioxide, but this is a clean CO_2 because the carbon comes from the sugarcane plants, which absorb atmospheric CO_2 for photosynthesis. On the other hand, the ethanol industry has great potential to reuse the CO_2 generated from fermentation of sugarcane juice and molasses. From Eq. 1 (production of ethanol from sucrose), considering the density of ethanol equal to 789 kg m^{-3}, it could be state that 1 m^3 of ethanol produces approx. 754.7 kg of CO_2. In 2013/2014, then, the 27.543 million cubicmeter of ethanol was produced accompanied of 2.07×10^7 tons of CO_2. An average ethanol industry (1500 m^3/day ethanol) produces, alone, 1132 tons CO_2 each day.

The ethanol industry must develop technologies to reuse all this CO_2, which today is simply liberated to the atmosphere, generating added value products.

Various CO_2 mitigation strategies have been thus investigated, which can be generally classified into two categories: (1) chemical reaction-based and (2) biological CO_2 mitigation.

3.4.1 CO_2 Mitigation Through Microalgal Cultures

Microalgae and cyanobacteria are interesting alternatives in the development of process capable to promote CO_2 mitigation since they naturally consume CO_2 through photosynthesis and did not depend on organic carbon availability to be cultured, allowing the development of technologies in liquid wastes from a variety of activities.

Considering the scenario presented by the International Energy Agency, the technologies related to the production of biodiesel from microalgae in vinasse medium, presented in this chapter, promotes advances in two of the cited areas: renewable energy and carbon dioxide capture. The evaluation of the capacity of microalgal/cyanobacterial cultures to capture carbon dioxide plays, thus, an essential role in the subject. Carbon dioxide and microalgal cultivation balances were carried in order to evaluate the potential of carbon dioxide uptake of the technology.

As an example, the mass balance of the culture of *B. braunii* in 10 % vinasse, which produced 2.19 g/L of biomass in 15 days and 16.5 % of lipids within cells (Table 7) is presented in Tables 13 and 14. Lipids elemental formula was calculated considering the FAME composition presented in Table 8, while biomass elemental formula was determined from a lipid extracted microalgal biomass in a CHNS analyser (Flash 2000 series, Thermo Scientific) (oxygen content in biomass was considered as the amount to reach 100 %).

It can be observed that the production of biomass of *B. braunii* in 10 % vinasse medium resulted in the uptake of 5.44 $g_{CO_2}\ g_{biomass}{}^{-1}\ L^{-1}$. Comparing to the culture respirometric analysis of the same strain of *B. braunii* cultivated in synthetic

Table 13 Mass balance of biomass production of *B. braunii* cultured in 10 % vinasse and produced 2.19 g L^{-1} biomass with 16.5 % of lipids

	Water +	Carbon dioxide +	Nitrogen source =	Biomass +	Oxygen +	Lipids
	H_2O	CO_2	HNO_3	$CH_{1,78}N_{0,15}O_{0,52}$	O_2	$CH_{1,85}O_{0,11}$
mol(s)	1.74	2.00	0.15	1.00	1.00	1.00
g	31.32	88.00	9.45	24.20	32.00	15.61

Table 14 Calculations considered the data in Table 13 to estimate the carbon dioxide uptake by *B. braunii* culture

Biomass (g/L)	Lipid (%)	Lipid ($g/g_{biomass}/L_{medium}$)	CO_2 uptake (g/L/15 days)	O_2 production (g/L)	Nitrogen uptake (mg/L)
2.19	16.5	0.52	5.44	7.91	190.04

medium carried by Sydney et al. (2010), which resulted in the uptake of 7.5 $g_{CO_2}\ g_{biomass}{}^{-1}\ L^{-1}$ in a biomass in with 33 % of lipids (88 % of the CO_2 was directed to biomass production), and considering that *B. braunii* in 10 % vinasse medium could accumulate this same 33 % of lipids, the results would be similar (carbon uptake in vinasse medium would reach 6.89 $g_{CO_2}\ g_{biomass}{}^{-1}\ L^{-1}$).

Furthermore, Table 14 gives important information about oxygen production and nitrogen demand of the culture, which is crucial in industrial process development.

3.4.2 Bicarbonate Production

The growth of the microalgal/cyanobacterial biomass in liquid media is usually limited by the low solubility of CO_2, which depends on factors such as temperature, pressure, partial pressure of other gases. The biomass productivity, and consequently the CO_2 fixation rate of microalgae and cyanobacteria, directly depends on this solubility as well as chemical and biochemical factors linked to photosynthesis.

Nutritionally favorable, vinasse has nonideal physicochemical characteristics for microalgae and cyanobacteria, such as high turbidity and acidity. The turbidity affects the passage of light through the medium, which is a limiting factor since photosynthetic organisms depend on it for photosynthesis. Even in synthetic medium, high biomass concentrations cause self-shadowing effect of the cells, hampering above 4 grams per liter production. In this aspect, the color and the turbidity of vinasse represent further difficulty.

In order to overcome these problems and to bring productivity improvements to algal biomass production processes and their products (proteins, lipids, pigments, among others), a vinasse treatment process was developed within the BioOil project.

When carbon dioxide (CO_2) dissolves in water, it forms carboxylic acid. At an alkaline pH, it dissociates into bicarbonate ions carrying one negative charge (HCO_3^-), while at an even higher pH it further dissociates into carbonate ions (CO_3^{2-}) carrying two negative charge. In very high pH solutions carbon dioxide reacts readily with bases, generating bicarbonate and carbonate salts. These bicarbonate and carbonate ions can be used by microalgae and cyanobacteria as carbon sources for growth. Depending on the amount of base added, bicarbonate and carbonate salts precipitate during the carbonation process and can be removed by a simple precipitation or centrifugation to be used as a fertilizer, in feed composition, in soil pH correction or even sold in purified form.

In the developed process, vinasse was used instead of water to fixate CO_2 from the fermentation vats. During the performed tests it was noticed that when the pH of the vinasse is raised above 10 precipitation of solids occurs. This alters the original composition and can directly influence the algal growing process. It is important, thus, to use a base capable of increase vinasse pH without causing significant composition changes but helping chemical carbon dioxide fixation.

Thus, different treatments of vinasse were carried by the addition of different bases to it. After each treatment the total solids, ash, and organic matter (Table 15) were determined in vinasse. Analyses were performed according to methods described by the Association of Analytical Communities (AOAC).

Among the bases analyzed, it can be seen that the addition of kelp caused the removal of almost 100 % of the ash in vinasse, which will certainly harm the subsequent cultivation of microalgae and/or cyanobacteria (lack of micronutrients). Moreover, the addition of soda (NaOH) gave the most interesting results have not caused significant changes in the composition of vinasse. Besides the use of hydrated lime and sodium carbonate also showed interesting results.

The vinasse treated with lime was analyzed as to its composition and compared with standard vinasse (Table 16). It can be seen that the addition of lime did not

Table 15 Analysis of precipitated mass, ash and organic matter from different treatments of vinasse to increase vinasse pH aiming bicarbonate/carbonate production

Treatment	Total mass precipitate (g/L)	Ash (g/L)	Organic Matter (g/L)	% of removal (in relation to standard vinasse)		
				Total precipited	Ash	Organic matter
Kelp	24.9	12.4	12.5	41.5	99.5	26.3
Hydrated lime	23.3	7.2	16.1	38.8	57.4	33.9
$CaCO_3$	19.4	6.7	12.8	32.4	53.5	26.8
NaOH	9.5	4.2	5.2	15.8	33.9	11.0
K_2CO_3	27.0	10.8	16.2	45.0	86.5	34.1
Standard vinasse						
Pure vinasse	60.1	12.5	47.6	0	0	0

Table 16 Comparison between the ion composition (nitrogen, sodium, calcium, potassium, magnesium, and phosphorous) of natural vinasse and vinasse treated with lime

	Total Nitrogen (mg L^{-1})	Sodium (mg L^{-1})	Calcium (mg L^{-1})	Potassium (mg L^{-1})	Magnesium (mg L^{-1})	Total phoshorous (mg L^{-1})
Standard vinasse	137	19.5	165	646	64.3	38.14
Vinasse treated with lime	116	20.2	1323	697	14.3	30.96

Table 17 Analysis of the variation of the NaOH addition (1 M or 2 M) to vinasse in the bicarbonate purity and process yield

	Reaction time (h)	Final pH	CO_2 fixation (kg/m^3/h)	Precipitated salt (g $L^{-1}_{vinasse}$)	Bicarbonate Purity (%)	Process yield (%)
Vinasse + 1 M NaOH	6	7.8	5.79	50.10	62	79
Vinasse + 2 M NaOH	5	8.1	6.25	31.25	70.73	50

cause major changes in the main components of vinasse except for magnesium concentration. The BOD and COD analyses of both stillage were also performed and showed that the treatment with lime resulted in no significant change (decrease of 8.1 % in 1.15 % BOD and COD).

Two experiments were conducted to determine the carbon-fixing potential vinasse basified with sodium hydroxide at different NaOH concentrations (1 and 2 M) (Table 17). It used conventional agitated reactor type with total capacity of 5L (working 4 L) and height/diameter equal to 0.61. Agitation was maintained at 300 rpm and injecting pure CO_2 in the flow of 3 L/min (0.75 vvm). Temperature was kept at 40 °C and pH was monitored. Samples were taken hourly and dried to quantify the production of bicarbonate salt. The end of the reaction was considered when the pH approached neutrality or when pH did not decrease after 5 min of continuous CO_2 injection. At the end of the experiment the treated vinasse was titrated to determine the amount of CO_2 fixed. The agitation was turned off and the solution left resting. The precipitated salt was recovered and dried at 50 °C and weight. Bicarbonate content on the precipitated salt was determined by titration. A solution of 1.5 g L^{-1} salt in distilled water was titrated using phenolphthalein (for quantification of carbonates and hydroxides) and methyl orange (quantification bicarbonates). The salt was analyzed using Hach® kits for the presence of phosphates, nitrates, and nitrites.

From Table 17 we can state that sodium bicarbonate production in vinasse reached better results when 1 M NaOH was used. Figure 11 presents the bioreactor used and the precipitated salt at the end of the process.

These are initial experiments that are being optimized. To optimize the reaction yield, vinasse entry should be preferably in countercurrent with CO_2 (vinasse enters

Fig. 11 From *left* to *right* Reactor used in bicarbonate production experiments, detail of the precipitated salt after the reaction is complete, and dry produced salt. *Source* The authors

Table 18 Evaluation of BOD, COD, and turbidity of vinasse treated with 2 M NaOH used for bicarbonate production

	Standard vinasse	Vinasse treated with 2 M NaOH
BOD (mg L^{-1})	10,151	6126
COD (mg L^{-1})	15,371	11,415
Turbidity (NUT)	1148.1	458

from the top of the reactor and CO_2 enters from the bottom). The reactor should preferably be of the type Bubble Column or Air-Lift because the gas–liquid contact time is higher. The relationship between height and diameter of the tank (Height/Diameter) plays a key role by allowing greater mass transfer gas–liquid. The injected CO_2 can be captured at the top of the tank and recirculated.

The vinasse treated with 1 M NaOH was analyzed for turbidity, BOD and COD according to the methods described in Standard Methods for the Examination of Water and Wastewater (21th ed. 2005) (Table 18). When compared to standard vinasse it was observed a 39.6 % decreases in BOD, 25.7 % in COD, and 60.1 % in turbidity. These results show the potential of this new technology to reduce the polluting nature of vinasse and also to reduce it turbidity, interesting for algal cultivation.

Experiments to evaluate the effect of treated vinasse in cyanobacteria growing were then carried. Cyanobacteria from the *Spirulina* genre were used. The cyanobacterial cultures were grown on different concentrations of vinasse (diluted with water) treated with 1 M NaOH. The results are shown in Table 19.

The treatment of vinasse with NaOH allowed the growth of the cyanobacteria in high concentrations of vinasse. On the previous studies described in this chapter, it was never achieved such a great biomass production in vinasse concentrations of 70 %. Among the *Spirulina* strains tested, *S. maxima* presented the better adaptation to treated vinasse and shows up as a potential strain for process scale up.

Table 19 *Spirulina* biomass produced in different treated vinasse concentrations

Treated vinasse		
Strain	Vinasse (%)	Biomass (g L^{-1})
S. platensis	30	0.9
	50	0.94
	70	1.02
	100	0.12
S. maxima	30	1.54
	50	1.37
	70	2.22
	100	1.57
S. laxissima	30	0.45
	50	1.39
	70	1.21
	100	1.07

Another experiment with treated vinasse was carried out with the microalgae *Botryococcus braunii*. For the preparation of the medium 1 M L^{-1} of NaOH was added to pure vinasse and CO2 was injected at 0.75 vvm. The carbonation reaction took 3:30 h to reach pH 7.5 and resulted in the production of 45.15 g L^{-1} $NaHCO_3$ (60 % yield). The treated vinasse was diluted with distilled water to 30 %; 1 g L^{-1} of $NaNO_3$ was added to the medium.

At the end of 15 days, 2.22 g L^{-1} of biomass was produced, which represents 34 % more than in nontreated vinasse (Table 7). Further analysis showed that the biomass contained 13.25 % of lipids (against 19.5 % in nontreated vinasse).

4 Conclusions

The ethanol industry is one of the biggest bioprocess in terms of volume of production in the world. Together with the production of ethanol and sugar, solid, liquid, and gaseous wastes are generated in great amounts. Most of them are reused within the industry, but new promising technologies are being developed and existing technologies being improved.

Sugarcane bagasse is the solid residue produced in larger quantities. Today burned in boilers to produce heat, vapor, and bioelectricity. In the Brazilian energy context, this biomass is of great importance and is receiving special attention as an alternative to the diversification of the energy matrix. New technologies of production of ethanol from bagasse are under development all around the world and will soon play a central role in the energy sector.

The liquid residue generated in the ethanol production in those countries that use fermentation of carbon compounds to produce ethanol is a liquid dark residue called vinasse. Vinasse is produced at a rate of 12:1 (vinasse: ethanol) and is generally used as fertilizer. But its physicochemical characteristics (COD, BOD, high potassium concentration, low pH) can cause environmental problems such as underground water contamination and lixiviation if usage is not controlled. In this context the development of technologies to promote a more rational destination to vinasse, reducing its pollutant characteristics and coupled to the production of value added bioproducts is of great importance.

Considering this, and because of its fertilizer characteristics, vinasse was evaluated for the cultivation of microalgae and cyanobacteria within the scope of a Project called Biooil. Besides the use of vinasse, the project proposes the use of the CO_2 from the fermentation vats in the algal cultures. These studies started aiming the production of oleaginous microalgae to the production of biodiesel. Best results were achieved with *B. braunii*, *C. vulgaris* and *S. nidulans*. In the context of this project, mass balances were carried to determine the carbon dioxide uptake in microalgal cultures. This is of great interest considering the necessity to reduce carbon dioxide emissions.

The success in the preliminary works opened new possibilities that also involved the production of protein-rich biomass (*Spirulina* strains) and pigments (*H. pluvialis*) through microalgae/cyanobacteria. Further treatments of vinasse to reduce turbidity and optimize microalgal and cyanobacteria resulted in the development of a technology capable of promoting growth and the carbon dioxide capture. This consisted of the addition of bases to vinasse and the injecton of CO_2 (which can be from the ethanol fermentation vats), resulting in the production of bicarbonate salt (that can be recovered) and the significant reduction of vinasse turbidity, COD and BOD.

Moreover, vinasse was evaluated as a nutrient source for the production of biohydrogen through anaerobic fermentation. Interesting hydrogen content (>30 %) and productivity ($> 2.5\ L_{H_2}/L/day$) were achieved in preliminary experiments. This opens the possibility to transform existing pilot scale technologies of biogas production from vinasse in a two-step process that generates two gaseous sources of energy: biohydrogen and biogas, which can be used within the ethanol industry in a biorefinery system.

All the technologies presented and discussed in this chapter allows the valorization of the liquid, solid, and gaseous ethanol industry, which is today one of the biggest examples of biorefinery. These new technologies foment the production of higher value added products and are the future of the segment.

References

Aflalo C, Meshulam Y, Zarka A, Boussiba S (2007) On the relative efficiency of two- vs. one-stage production of astaxanthin by the green alga *Haematococcus pluvialis*. Biotechnol Bioeng 98:300–305

Angenent LT, Wrenn BA (2008) Optimizing mixed-culture bioprocesses to convert wastes into biofuels. In: Wall JD, Harwood CS, Demain A (eds) Bioenergy. ASM Press, Washington, pp 179–194

Aoyama K, Uemura I, Miyake J, Asada Y (1997) Fermentative metabolism to produce hydrogen gas and organic compounds in a Cyanobacterium, *Spirulina platensis*. J Ferment Bioeng 83:17–20

Aquasearch Inc. (1999) Analysis of total astaxanthin in algae meal prepared from *Haematococcus pluvialis*. Technical report available at http://www.fda.gov/ohrms/dockets/dailys/00/jun00/061900/rpt0065_tab8.pdf. Acessed on 27 Aug 2015

Cataldo DA, Haroon M, Schrader LE, Youngs VL (1975) Rapid colorimetric determination of nitrate in plant tissue by nitration of salicylic acid. Commun Soil Sci Plan 6:71–80

Christofoletti CA, Escher JP, Correia JE, Marinho JFU, Fontanetti CS (2013) Sugarcane vinasse: environmental implications of its use. Waste Manage 33:2752–2761

Cifuents AS, González MA, Vargas S, Hoeneisen M, González N (2003) Optimization of biomass, total carotenoids and astaxanthin production in *Haematococcus pluvialis* Flotow strain Steptoe (Nevada, USA) under laboratory conditions. Biol Res 36:343–357

CONAB (2015) Companhia Nacional de Abastecimento. Acompanhamento da safra brasileira de cana-de-açúcar. – v. 1 – Brasília. http://www.conab.gov.br. Accessed 30 Sept 2015

Dalmas Neto CJ, Sydney EB, Assmann R, Coraucci Neto D, Soccol CR (2013) Production of biofuels from algal biomass by fast pyrolysis. In: Pandey A, Lee D-J, Chisti Y, Soccol CR (eds) Biofuels from algae, 1st ed. Elsevier, London, pp 143–153

Elia Neto A, Shitanku A, Pio AAB, Conde AJ, Gianetti F, Donzelli JL (2009) Manual de Conservação e Reúso de Água na Agroindústria Sucroenergética. ANA - Agência Nacional de Águas, FIESP - Federação das Indústrias do Estado de São Paulo, UNICA - União da Indústria da Cana-de-açúcar, CTC - Centro de Tecnologia Canavieira, Brasília, p 239

EMBRAPA (2015) Empresa Brasileira de Pesquisa Agropecuária (http://www.agencia.cnptia.embrapa.br/gestor/cana-de-acucar/arvore/CONTAG01_131_22122006154842.html). Accessed 28 Sept 2015

Ensinas AV, Nebra SA, Lozano MA, Serra L (2006) Analysis of cogeneration systems in sugarcane factories – Alternatives of steam and combined cycle power plants. In: Proceedings of ECOS 2006 Aghia Pelagia, Crete, Greece 12–14 July 2006

EPE – Empresa de Pesquisa Energética (2013) BRAZILIAN ENERGY BALANCE 2013|year 2012. (https://ben.epe.gov.br/downloads/Relatorio_Final_BEN_2013.pdf). Acessed 15 Sept 2015

FIESP – Federation of Industries of São Paulo (2013) Outlook Fiesp 2023: Projeções para o agronegócio brasileiro. São Paulo: FIESP. ISBN 978-85-7201-014-6

Galbe M, Zacchi G (2007) Pretreatment of lignocellulosic materials for efficient bioethanol production. Adv Biochem Eng Biotechnol 108:41–67

Galbiati JK, Lavanholi MGDP, Gallo CA (2010) Produção de energia elétrica a partir da queima do bagaço de cana-de-açúcar. Nucleus-Ituverava 7:127–138

Hartman L, Lago RCA (1973) Rapid preparation of fatty acids methyl esters. Lab Pract Lond 22:475–476

Hassuda S, Rebouças AC, Cunha RCA (1991) Impactos da infiltração da vinhaça de cana no aqüífero Bauru. Bol IG-USP Publ Espec 9:169–171

Iritani MA, Ezaki S (2012) As águas subterrâneas do Estado de São Paulo, 3rd ed. Secretaria de Estado do Meio Ambiente – SMA, São Paulo

Jacobsen SE, Wyman CE (2002) Xylose monomer and oligomer yields for uncatalyzed hydrolysis of sugarcane bagasse hemicelluloses at varying solids concentration. Ind Eng Chem Res 41:1454–1461

Jeffries TW, Timourien H, Ward RL (1978) Hydrogen production by *Anabaena cylindrica*: effect of varying ammonium and ferric ions, pH and light. Appl Environ Microbiol 35:704–710

Li J, Zhu D, Niu J, Shen S, Wang G (2011) An economic assessment of astaxanthin production by large scale cultivation of *Haematococcus pluvialis*. Biotechnol Adv 29:568–574

Margheri MC, Tredici MR, Allotta G, Vagnoli L (1990) Heterotrophic metabolism and regulation of uptake hydrogenase activity in symbiotic cyanobacteria. In: Polsinelli M, Materassi R, Vincenzini M. Dordrecht (eds) Developments in plant and soil sciences—biological nitrogen fixation. Kluwer Academic Publisher, Boston, pp 481–486

Martins R (2009) Energia produzida a partir do bagaço da cana é economicamente viável. Available at: http://www.inovacaotecnologica.com.br/noticias/noticia.php?artigo=energia-produzida-partir-bagaco-cana-economicamente-viavel&id=010175090810#.VdMQ9_lVhHw. Accessed 23 Sept 2015

Michelazzo MB, Braunbeck OA (2008) Análise de seis sistemas de recolhimento do palhiço na colheita mecânica da cana-de-açúcar R Bras Eng Agríc Ambiental 12:546–552

Moraes BS, Junqueira TL, Pavanelloa LG, Cavaletta O, Mantelattoa PE, Bonomia A, Zaiat M (2014) Anaerobic digestion of vinasse from sugarcane biorefineries in Brazil from energy, environmental, and economic perspectives: profit or expense? Appl Energy 113:825–835

IEA—International Energy Agency (2015) Energy technologies perspectives. Available at http://www.iea.org/etp/explore/. Accessed 2 Oct 2015

Naguib YM (2000) Antioxidant activities of astaxanthin and related carotenoids. J Agric Food Chem 48:1150–1154

Nandi R, Sengupta S (1998) Microbial production of hydrogen: an overview. Crit Rev Microbiol 24:61–84

NOVACANA (2015) Vantagens da bioeletricidade do bagaço de cana para o Brasil http://www.novacana.com/estudos/vantagens-da-bioeletricidade-do-bagaco-de-cana-para-o-brasil-120913/

Oliveira ER (2012) Procedimentos e normas para o acompanhamento de análise da qualidade da cana-de-açúcar. Technical Rapport. ORPLANA – Organization of Sugarcane Growers from the South-Central Region of Brazil

Peng F, Ren JL, Xu F, Bian J, Peng P, Sun RC (2010) Fractional study of alkali-soluble hemicelluloses obtained by graded ethanol precipitation from sugarcane bagasse. J Agric Food Chem 58:1768–1776

Phang SM, Miah MS, Chu WL, Hashim M (2000) Spirulina culture in digested sago starch factory waste water. J Appl Phycol 12:395–400

Pinto CP (1999) Tecnologia da Digestão Anaeróbia da Vinhaça e Desenvolvimento Sustentável. Dissertation Universidade Estadual de Campinas

Reith JH (2001) Report of the workshop biological hydrogen production, 4 Oct, Utrecht, The Netherlands (In Dutch)

Renewable Fuels Association (2015) http://www.ethanolrfa.org/resources/industry/statistics/. Accessed 30 Sept 2015

Ripoli MRC, Gamero CA (2007) Palhiço de cana-de-açúcar: Ensaio padronizado de recolhimento por enfardamento cilíndrico. Energ Agric Botucatu 22:75–93

Ripoli TC, Molina WF Jr (1991) Cultura canavieira: um desperdício energético. Maquinaria Agrícola 3:2–3

Rodrigues A, Santos RF, Avaci AB, Rosa HA, Chaves LI, Gasparin E (2012) Estimativa do potencial de geração de energia elétrica a partir da vinhaça. Acta Iguazu 1:80–93

Rodrigues RCLB, Felipe MGA, Sil JBA, Vitolo M (2003) Response surface methodology for xylitol production from sugarcane bagasse hemicellulosic hydrolyzate using controlled vacuum evaporation process variables. Process Biochem 38:1231–1237

Soccol CR, Vandenberghe LPS, Medeiros ABP, Karp SG, Buckeridge M, Ramos LP, Pitarelo AP, Ferreira-Leitão V, Gottschalk LMF, Ferrara MA, Bon EPS, Moraes LMP, Araújo JA, Torres FAG (2010) Bioethanol from lignocelluloses: Status and perspectives in Brazil. Bioresour Technol 101:4820–4825

Sydney EB, Sturm W, de Carvalho JC, Soccol VT, Larroche C, Pandey A, Soccol CR (2010) Potential carbono dioxide fixation by industrially important microalgae. Bioresour Technol 101:5892–5896

UNICA (2015) Brazilian Sugarcane Industry Association. http://www.unica.com.br/. Accessed 22 Aug 2015

Yang B, Wyman CE (2008) Pre-treatment: the key to unlocking low-cost cellulosic ethanol. Biofuels Bioprod Bioref 2:26–40

General Assessment of the Currently Available Biodiesel Production Technologies

Eduardo J.M. de Paiva, Vinícius Kothe, Marcos Lúcio Corazza, Ângela Silva, Shirley Nakagaki, Fernando Wypych and Luiz Pereira Ramos

Abstract With the predicted depletion of crude oil reservoirs in the next decades as well the need for developing renewable energy and sustainable production processes, low-carbon technologies will have a crucial role in securing energy supplies for future generations and offsetting the environmental impact of fossil fuels production and use. In this scenario, biodiesel offers well known advantages over petrodiesel because it is an environmental friendly fuel whose fuel properties are attractive and its overall emission profile of greenhouse gases and sulfur is low. This work provides a broad overview about biodiesel production technologies after more than two decades of intensive R&D, highlighting the main aspects related to feedstock availability, conversion techniques, and process economics. In the vanguard of biodiesel production technologies are the advent of novel reactor concepts, the use of process intensification, and the development of novel catalytic systems. These technologies, combined with alternatives feedstocks such as algae, non-edible vegetable oils, industrial soapstocks, and waste oils and greases, are likely to pave the road for the establishment of more environmental friendly and economically affordable processes.

Keywords Biodiesel · Feedstock availability · Conversion technologies · Catalysis · Process economics

E.J.M. de Paiva · M.L. Corazza
Department of Chemical Engineering, Federal University of Paraná, Curitiba, PR, Brazil

Â. Silva · S. Nakagaki
Catalysis and Bioinorganic Chemistry Laboratory, Department of Chemistry, Federal University of Paraná, Curitiba, PR, Brazil

V. Kothe · F. Wypych · L.P. Ramos (✉)
Research Center of Applied Chemistry/INCT Energy and Environment (INCT E&A), Department of Chemistry, Federal University of Paraná, Curitiba, PR, Brazil
e-mail: luiz.ramos@ufpr.br

C.R. Soccol et al. (eds.), *Green Fuels Technology*,
Green Energy and Technology, DOI 10.1007/978-3-319-30205-8_12

1 Introduction

Since the advent of the Industrial Revolution in the late eighteenth century, the control and management of energy sources became a mandatory requirement for modern society development. In 2013, the most consumed energy came from fossil fuels, which accounted for 82.67 % among other energy sources in which 30.92 % consisted of crude oil, 28.95 % of coal and 22.81 % of natural gas. The petroleum-derived fuels play a vital role in the development of industrial growth, agricultural technologies, domestic needs, transportation, and many other basic human needs. Globally, 11 billion tons of fossil fuel has been consumed annually. At this rate, these sources will soon be exhausted and this may contribute to soaring fossil fuels (Wan Ghazali et al. 2015). Due to a continuous growth in human population, most of the world's total energy production is used for industrial applications, transportation, and power generation (Avhad and Marchetti 2015). In an overview about the energy world demand set by the U.S Energy Information Administration (EIA 2013), the total energy consumed in 2010 was 5.53×10^{20} J and is predicted to rise to 8.65×10^{20} J by 2040. Accordingly, the total world energy consumption will grow by 56 % between 2010 and 2040. More than 85 % of the increase in global energy demand from 2010 to 2040 is expected to occur among developing nations, led by China and India, which have been driven by strong economic growth and expanding populations (together they account for 10 % of world energy consumption in 1990 and nearly 24 % in 2010).

Transportation is currently the second largest energy consuming sector and is increasing by an average of 1.1 % per year (EIA 2013). In the current situation, the foremost amount of energy is supplied by conventional fossil fuel resources, such as gasoline, liquefied petroleum gas, diesel fuel, and natural gas. The use of fossil fuels is directly associated with the increase in greenhouse gases and global warming. According to United Nations Organization (UNO), the average Earth's temperature will rise 1.8–4 °C until 2100, thus accelerating the glaciers melting, raising the oceans levels, and causing intense hurricanes. Besides, the hydric balance on Earth should be shifted causing severe dry or flooding seasons in some areas (UNO 2010).

In addition to serious environmental issues, dwindling reserves of crude oil, oscillating petroleum fuel prices, and the overconsumption of liquid fuels, especially for the transportation sector, have increased the need to find alternative "green" sources of energy that are sustainable, environmentally tolerable, economically competitive, and easily available. The numerous modes of renewable energy resources are anticipated to play a significant role in resolving the world's future power situation; therefore, over the past few years, researchers have driven their attention towards finding an appropriate replacement for fossil fuels. Renewable energy resources such as solar energy, wind energy, hydro-energy, and biofuels (biodiesel, bioethanol, biogas, and biomass) have been considered as potential alternatives to reduce the worldwide dependency on the use of fossil fuels (Panwar et al. 2011; Avhad and Marchetti 2015).

According to the International Energy Agency (IEA 2014), low-carbon technologies will have a crucial role to secure energy supplies for the future and to offset the environmental impact of fossil fuels. In addition to the energy efficiency, many kinds of renewable energy, carbon capture and storage, nuclear power and new transport technologies must be widely deployed to reach the emission goals. In a broad perspective, the biodiesel production chain uses the primary source of energy (solar energy) to generate live feedstock; thus, the carbon cycle is favored and the benefits related to its use are obvious.

The American Society for Testing and Materials (ASTM) defines biodiesel as a mono-alkyl esters derived from lipid feedstock such as vegetable oils or animal fats (Marchetti 2010). The major components of plant oils and animal fats are triacylglycerols (TAGs), which are esters of fatty acids and glycerol. TAGs consist of different fatty acid composition that influences both physical and chemical properties of vegetable oils and animal fats, being also critical to determine the quality of their corresponding biodiesel. Two kinds of fatty acids are found in natural lipid sources: saturated and unsaturated fatty acids, which may include one or more carbon-carbon double bonds in their hydrocarbon chain. The most common fatty acids found in nature are palmitic acid (16:0), stearic acid (18:0), oleic acid (18:1), linoleic acid (18:2), and linolenic acid (18:3). Besides fatty acids and their corresponding acylglycerols, additional components such as phospholipids, carotenes, tocopherols, sulphur compounds, and water are often present in such plant oils (Demirbas 2007; Knothe 2009; Marchetti 2010, 2012; Avhad and Marchetti 2015).

Biodiesel has known advantages compared to regular diesel since it is an environmental friendly fuel, with better lubricity and combustion performance and reduced emissions of greenhouse gases and sulfur, which is practically absent (Bowman et al. 2006; Kiss 2010). Besides, its use engenders numerous societal benefits: rural revitalization, creation of new jobs, and reduced global warming (Kiss et al. 2008). Its physical properties have been reviewed widely (Aransiola et al. 2010, 2012; Jain and Sharma 2010; Juan et al. 2011) and some of which are dependent on the feedstock employed for its production. The flash point of biodiesel is significantly higher than that of petroleum diesel or gasoline, thus making it one of the safest fuels available. However, the calorific value of biodiesel (37.27 MJ/L) is about 9 % lower than that of the regular petroleum diesel.

The environmental and physical–chemical biodiesel advantages over petroleum diesel cannot be overemphasized: it is safe, renewable, nontoxic and biodegradable, it contains no sulphur and is a better lubricant, and it has higher cetane number, combustion efficiency, and flash point, among others (Knothe 2009; Marchetti 2010; Aransiola et al. 2014). Additionally, biodiesel production could also provide an opportunity to improve the domestic oil market in developing countries, and enhance the net income of farms and agro-industrial activities (Demirbas 2007; Knothe 2009). However, after more than three decades of intense biodiesel research, some questions still remain and others arose as a result of its production in large scale.

The main questions to be answered are bound to the economic feasibility of biodiesel production and to its quality and storage stability. The answer is not straightforward and depends on many different factors such as the vegetable oils market, petroleum diesel prices, and feedstock type and availability (edible or non-edible vegetable oils, animal fats, waste cooking oils, and algae lipids), as well as the routes employed for its production (catalytic, non-catalytic, physical methods, supercritical, pyrolysis, etc.).

2 Biodiesel Production (Economics vs. Feedstock)

Most of the biodiesel produced worldwide comes from edible sources of vegetable oils. However, the use of edible oils is controversial because it competes with food applications and issues related to land usage. Biodiesel can also be produced from other feedstocks such as non-edible vegetable oils, soapstocks, waste cooking oil, rendered materials and greases. Although not available inasmuch, these alternative sources are appealing due to their low cost and favorable impact on waste management.

According to Rico and Sauer (2015), the world biodiesel production in 2014 was led by the United States of America (USA), followed by Indonesia and Brazil. By contrast, Germany and Argentina decreased their annual production since 2010 and 2012, respectively. Other countries with a high annual production of vegetable oils, especially palm oil in Malaysia, produce small biodiesel quantities but they are also important players in this market because most of their oil production is exported abroad, a large share of it being used for biodiesel production elsewhere.

In USA, the government has encouraged the use of alternative fuels since 1992. Most of the biodiesel produced in this country comes from soybean oil, but the contribution of other sources in increasing. In 2013, the Environmental Protection Agency (EPA) proposed that 5 billion liters of biomass-based diesel should be used annually under the so-called Renewable Fuel Standard Program, 2.3 billion liters of which coming from soybeans, 1.17 billion liters from corn, and 1.48 billion liters from waste materials (Reitze Jr 1993). In this growing market, critical variables were soon identified such as the dependency on soybean oil price, shifts in the international petroleum market, and federal and state subsidies provided to industry (Hay 2014; Rico and Sauer 2015).

In the European Union, especially in Germany, sales and tax incentives provided a cost reduction in the production process such that biodiesel could access the market and compete with crude oil and its derivatives. The Common Agricultural Policy (CAP) was first to implement incentives for biofuel production and, after releasing these guidelines, Europe increased its biodiesel production and consumption, especially in Germany. Biofuels were granted lower tax rates and farmers of EU countries were encouraged to plant oil seeds. With this, the EU biodiesel production experienced a rapid growth and this has been a matter of discussion not only in terms of land usage locally and globally (e.g., Malaysia and Indonesia) but also in relation to its impact in the vegetable oil market (ICCT 2013). The incentives have included tax exemptions and increasing taxes on diesel (Rico and Sauer

2015). After several years of growth, there has been a decrease in biodiesel consumption and this was partly due to a significant reduction in these tax exemptions, which was motivated by concerns over its impact on food prices and the emission of greenhouse gases that could be related to the production of biofuels (Tyner 2009).

The world industrial production of biodiesel is still dependent of edible vegetable oils feedstock mostly due to several issues, such as infrastructure and logistics to grow and harvest industrial crops, regional climate, technology to convert biomass into biodiesel, and the economics. Rico and Sauer (2015) argued that under normal conditions, without subsidy policies and incentives, biodiesel is only viable if its market price, corrected by the heating value, is equal or lower than diesel and/or higher than that of vegetable oils. Undoubtedly, the natural choice for an edible oil producer is to sell their products in the food market rather than in the energy market.

The major cost factor in the production of biodiesel is the price of the raw material. Conversion costs usually account for about 10 % in large facilities, and between 25 and 40 % in small production units (Tomei and Upham 2009). In recent years, there has been a search for cheap raw materials together with a quest for the most economic processing method (Aransiola et al. 2014). A lot of research have been focused to produce biodiesel from non-edible resources (Achten et al. (2008); Cynthia and Lee 2011; Raja et al. 2011; Banković-Ilić et al. 2012; Paiva et al. 2013; Atabani et al. 2013). According to Sajid et al. (2016), many potential non-edible biomass feedstocks have been investigated, mainly oils from non-edible vegetables such as *Jatropha curcas* (Jatropha) (Sahoo and Das 2009), *Linum usitatissimum* (Linseed) (Borugadda and Goud 2012), *Simmondsia chinesnsis* (Jojoba) (Shah et al. 2014), *Hevea brasiliensis* (rubber seed) (Bharathiraja et al. 2014), *Azadirachta indica* (Neem) (Ali et al. 2013), Cotton seed (Royon et al. 2007), *Nicotianna tabacum* (tobacco) (Usta et al. 2011), and other interesting potential sources such as Orbinya and Attalea (babassu palm) (Paiva et al. 2013), among others. The use of non-edible sources has the potential to eliminate the competition between food and energy. However, non-edible feedstocks still require the use of land for growth but this is much less compared to those for edible biomass feedstocks (Sahoo and Das 2009; Sajid et al. 2016).

Some authors developed indicators with arguments for and against the production of liquid biofuels and their impact on land use. Rathmann et al. (2010) found that liquid biofuels have no significant impact on the land use, although they influenced food prices, but this was not a determining factor in the modification of price trends.

A potential alternative to non-edible vegetable oils is the use of waste cooking oil (WCO) as a feedstock to produce biodiesel (Naima and Liazid 2013; Gopal et al. 2014). WCO may come from different sources including commercial, industrial, and domestic establishments. There are some advantages of using WCO to produce biodiesel: it can help to decrease the amount of farmland, which is necessary for biodiesel producing crops; the usage of WCO helps to reduce biodiesel production costs (Zhang et al. 2003; Kulkarni and Dalai 2006); and its use as a feedstock reduces the cost of waste product removal and treatment because the WCO waste management is normally problematic. However, there are also few disadvantages associated with the use of WCO. During the cooking process, free fatty acid and other products such as acylglycerol dimers and trimers can be formed in the oil and these products may affect transesterification and the final biodiesel properties (Kulkarni and Dalai 2006). Foremost, one of the main complications is the logistics associated with the collection and supply of WCO (Ramos et al. 2013; Cho et al. 2015).

Sajid et al. (2016) carried out a comparative study between the production chain of non-edible *Jatropha curcas* and WCO using alkali-catalyzed transesterification. For this, the Life Cycle Analysis (LCA) methodology was used to assess the environmental impacts (greenhouse gas emissions, resource consumption and depletion, human health impacts, etc.) of the entire biodiesel production chain. Process simulation was performed on the basis of biomass primary production, harvesting, industrial processing, and the biodiesel end use. In this study, the use of WCO required no special energy input other than collecting it from various sources while the production of non-edible jatropha oil involved the use of land for cultivation, even though with much lower agronomic requirements compared to edible oil sources. However, the sources of WCO are scattered and its collection requires high transportation coverage and costs. From the standpoint of production, the use of WCO demanded more energy and more chemicals in the production process while the use of jatropha oil produced a lower environmental impact. These authors concluded that the decision of using either raw material should also be based on the process economic analysis.

The use of animal fats and rendered materials poses as a good alternative to complement or even supply biodiesel production demands. Animal fats are primarily derived from animal processing facilities and by the rendering process. The main animal fats include tallow from processing cattle, lard and white grease from swine processing, and poultry fat from the processing of chicken, turkey, or other birds. The fats/oils generated by fish processing plant and leather industry fleshing wastes have also been found to be viable biodiesel feedstocks (Alptekin et al. 2012; Jayasinghe and Hawboldt 2012; Ong et al. 2013).

Some animal fat is already in used at the industrial scale (e.g., chicken, tallow, lard fats) for biodiesel production (da Cunha et al. 2009; Schörken and Kempers 2009; Kondamudi et al. 2009). Tallow is the second source employed for biodiesel production in Brazil since 2008 and its use accounted for 22 % of biodiesel production in 2014 (Rico and Sauer 2015). The use of animal fat wastes offers economic, environmental, and food security advantages over the more commonly used edible vegetable oils. Moreover, animal fat and waste materials (AFWs) contain high levels of saturated fatty acids and FFA requiring more complex production techniques; however, the low degree of unsaturation (or iodine number) of AFWs fatty acids has several advantages such as high calorific value, high cetane number and good oxidation stability (Tong et al. 2010; Alptekin et al. 2012).

Adewale et al. (2015) reviewed the current technologies employed in the conversion of AFWs to biodiesel as well the conversion of important AFW sources such as chicken fat, lard, tallow, fish oil, and leather industrial wastes by methods such as homogeneous and heterogeneous catalysis (including enzymatic conversion), supercritical conditions, ultrasound and microwave-assisted transesterification, as well as pyrolysis. In such review, the authors present the pros and cons associated with each technique. For example, homogeneous transesterification of AFW requires some pre-treatment due the high FFA concentration as well pre-heating due the high level of saturated compounds and problems related to side reactions such as soap formation and the elevated number of downstream purification process, which constitutes a real problem in this route. Among all the techniques covered by these authors, the assisted techniques combined with enzymatic catalysis was shown to have some advantages, such as reductions in the overall energy consumption and in the production cost of biodiesel synthesis given their capability to reduce reaction time, temperature and pressure required for optimal conversion.

Da Rós et al. (2012) reported the optimal conditions for the microwave-assisted enzymatic synthesis of biodiesel from beef tallow. An almost total conversion of the original beef tallow's fatty acids was obtained at the following experimental conditions: beef tallow to ethanol molar ratio of 1:6, 50 °C and 8 h of reaction time. Microwave processing was shown to be energy efficient, causing no destructive effects on the enzymes and reaching higher biodiesel yields compared to conventional heating.

Armenta et al. (2007) studied the ethanolysis of fish oil using ultrasound irradiation by examining the effect of different parameters on the production process, such as ultrasonic device (bath vs. probe), catalyst type (KOH vs. sodium methoxide), catalyst concentration (0.5 vs. 1 %), temperature (20 vs. 60 °C), and duration of exposure (10 vs. 90 min). Sodium ethoxide was a more efficient catalyst than KOH. Conversions of 98 % were achieved after 30 min at 60 °C in the presence of 0.8 % of sodium methoxide.

Depending of the characteristics of alternative raw materials such as AFW and fish oils, their blend with vegetable oils seems to be a natural choice for biodiesel production, therefore providing a dilution factor for unfavorable properties such as low oxidation stability. Of course, the choice for a raw material and the selection of a conversion process would always demand a careful economical evaluation because low cost feedstocks may involve much more costly production processes. However, production of biodiesel from AFW and other low cost oils and greases eliminates the problems associated with waste management and handling of undesirable by-products of the food industry.

As mentioned before, biodiesel from microalgae has raised as a promising alternative. The use of microalgae for fuels and chemicals has many advantages over traditional sources including soybeans (Lee 2011). It is grown in open ponds, exerting zero demand on arable land. It has the potential for up to 100 times greater biodiesel yield compared to soybeans (Lee 2011) and has the advantages of a high growth rate, short maturity, high biomass productivity, and low environmental impact (Gonçalves et al. 2013). However, developing the technology for converting microalgae lipids to biodiesel is more challenging when compared to the use of plant oils mainly due to the high water content of microalgae slurries and the difficulty in extracting the oil from the microalgae biomass leading to relatively high production costs (Milledge and Heaven 2012). The common process for obtaining oil from microalgae consists of microalgae harvesting, biomass drying and oil extraction (Cooney et al. 2009), in which the drying step consumes a large amount of energy. In addition, some of the downstream technologies still suffer the drawbacks of high energy consumption and low treatment capacity (as seen, for example, with centrifugation and supercritical fluid extraction), limiting the commercialization of these methods (Chen et al. 2015). It therefore presents an attractive option as a raw material source if the cost of microalgae oil production becomes favorable. Strategies for reducing production costs by employing genetic engineering to improve the microalgae oil production capacity and the modern concepts of integrated bio-refinery could lead the way to reduce the overall production costs.

Many researches have been focused on alternatives to avoid the energy consumption and reduce the costs associated with microalgae oil extraction. Lardon et al. (2009) indicated that 90 % of the energy consumption for biodiesel production resides in the lipid extraction from the microalgae biomass. Hence, extraction is considered one of the most limiting steps for the development of a viable microalgae-based biodiesel production process. Direct transesterification (also called in situ transesterification) and both microwave- and ultrasound-assisted techniques are the most commonly routes investigated in the last decade (Carvalho Júnior et al. 2011; Cao et al. 2013; Guldhe et al. 2014). Chen et al. (2015) evaluated the potential scale up and commercial applications of a multi-step process using *Chlamydomonas* sp., which was reported to have a dry biomass content of 31.3 % and an oil content of 26.3 % dry biomass. The conversion strategies consisted of microwave disruption, partial water removal, wet oil extraction using solvents and co-solvents (i.e., hexane and methanol) and homogeneous and heterogeneous transesterification, thus skipping the wet extraction step. Conversions around 90 % were achieved after a two-step oil extraction and transesterification.

Zhang et al. (2015) studied an innovative process using direct transesterification of dry microalgae powder (*Chlorella* sp.) that was assisted by a mixture of solvents. The use of 75 % ethanol (in this case used as a solvent, transesterification reagent and bactericide) in hexane, acetone, petroleum ether, carbon tetrachloride, ethyl ether, *n*-butanol, and chloroform was tested for in situ acid transesterification at following conditions: 0.3 g of dried microalgae biomass, 6–9 mL of mixed solvents, 0.6–1.0 mL of sulfuric acid, 90–100 °C and 2 h. The best yields (around 90 %) were achieved with ethanol in hexane and ethanol in petroleum ether, but the use of acetone also resulted in good yields (around 85 %).

These research cases show the enormous potential to be unlocked using microalgae biomass. It is by far the most environmentally friendly feedstock and its requirements for growth do not demand the use of arable land. Besides, from the energy standpoint, microalgae directly entrap carbon dioxide and convert solar energy into useful biomass.

3 Conversion Techniques

The cost of vegetable oils may account for 60–80 % of the total production cost of biodiesel (Helwani et al. 2013; Avhad and Marchetti 2015), depending on the raw material chosen for conversion. Hence, the remaining 20–40 % must be minimized during the conversion processes. As reviewed, several researches have tried to find the best, cheapest and environmentally friendly route to convert the potential feedstock into a biodiesel fuel that is suitable for replacing petroleum diesel in the transportation sector. The best route to be set is not a simple task and it is a direct function of the type of biomass available locally (vegetable oils, algae oils, WCO, or AFWs).

The most widely used route in industrial scale as well in laboratory and pilot plants is the transesterification by homogenous basic catalysis, commonly using NaOH, KOH, sodium and potassium methoxides. Its spread use around the world may be attributed to its quick conversion times, high yields of esters and mild process conditions. However, this route is not suitable for high free fatty acid (FFA) feedstocks due to the undesired side reaction of saponification. Strategies to recover glycerol in high yields and to reduce the energy consumption in the biofuel downstream processing has been reported elsewhere (Fukuda et al. 2001; Noureddini et al. 2005).

A review in the early literature (Ma and Hanna 1999) shows that the direct use of vegetable oils and its blends into diesel engines has failed due to the long-term problems caused by the accumulation of carbon deposits, atomization problems, gelling of lubricant and polymerization during storage and combustion (Fukuda et al. 2001; Aransiola et al. 2014). Microemulsions could overcome the viscosity problem associated with direct use of vegetable oils but the long-term engine problems remained unchanged. The use of pyrolysis was also assessed but uncontrolled side reactions complicate the achievement of a good product selectivity and stability during storage and handling. The presence of glycerol in TAGs directs to the formation of intermediates, particularly unsaturated aldehydes, ketenes, ethyl radicals, propylene, and ethylene oxides that are able to initiate condensation and polymerization reactions that are detrimental to the fuel performance (Santos et al. 2010; Avhad and Marchetti 2015).

Aransiola et al. (2014) made an extensive review on recent trends in biodiesel production focusing on non-conventional processes such as reactive distillation and the use of supercritical conditions. Avhad and Marchetti (2015) presented a survey on the available technologies to convert biomass into biodiesel, including details about heterogeneous, homogeneous and enzymatic catalysts. Adewale et al. (2015) presented an extensive review on the current technologies for the conversion of AFW. Qiu et al. (2010) presented a complete review about the use of novel reactor designs and non-conventional technologies to convert biomass into biodiesel. In the light of these works, a brief discussion about the currently available technologies will be presented in the following section. Also, a detailed discussion about heterogeneous catalysts will be presented due to its potential to reduce costs and deliver suitable processes of low environmental impact.

3.1 Conventional Techniques

Transesterification of vegetable oil using homogeneous catalysts involves the use of catalyst in liquid form, mainly acid and alkaline catalysts. The basic factor in the acid catalysis is the protonation of the carboxyl group in TAGs and the alcohol attacking the protonated carbon to create a tetrahedral intermediate. However, in a homogeneous-base catalyzed reaction, the important factor is to promote the nucleophilic attack of the alkoxide on the electrophilic carbon atom of the TAG ester group (Schuchardt et al. 1998). In a stoichiometric transesterification reaction, one TAGs molecule reacts with three alcohol molecules to produce three moles of fatty acid alkyl esters (FAAE) and one molecule of glycerol (Fig. 1) (Marchetti et al. 2007; Marchetti 2013). Generally, an excess of alcohol is used to shift the reaction equilibrium to the product formation. However, other factors may be influential such as the reaction temperature (should remain in the range of 60–80 °C), effective agitation, and the concentration of the reaction catalyst (Schuchardt et al. 1998; Ma and Hanna 1999).

In general, the alcohol used in this reaction is methanol because of its great reactivity, availability, and cost. Besides, methanol can be obtained relatively free of water. In Brazil, the transesterification employing ethanol has been intensively studied because this reagent is produced from renewable resources (sugarcane) in large scale, therefore contributing in environmental issues and offering good opportunities to the entire agro-industrial production chain (Cordeiro et al. 2011).

Glycerol is a reaction co-product that has a high commercial value in several industrial segments, such as in the production of pharmaceuticals, food, additives, and polymers (Ma and Hanna 1999). There is a recurring concern about ways to use glycerol as its production should increase with the increasing demand for biodiesel, requiring its utilization by the global industry in some way (Mota et al. 2009).

The base-catalysed transesterification reaction mechanism is given in Fig. 2 (Ma and Hanna 1999; Thanh et al. 2012). The formation of alkoxide catalytic species is observed in step 1. However, this species can also be added directly to the reaction

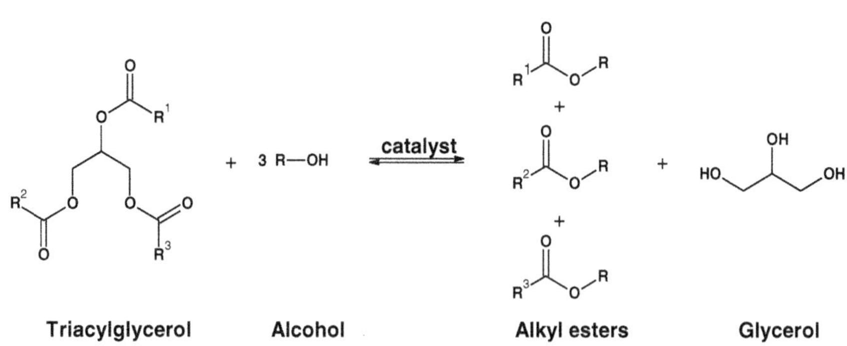

Fig. 1 Schematic representation of the catalyzed transesterification reaction

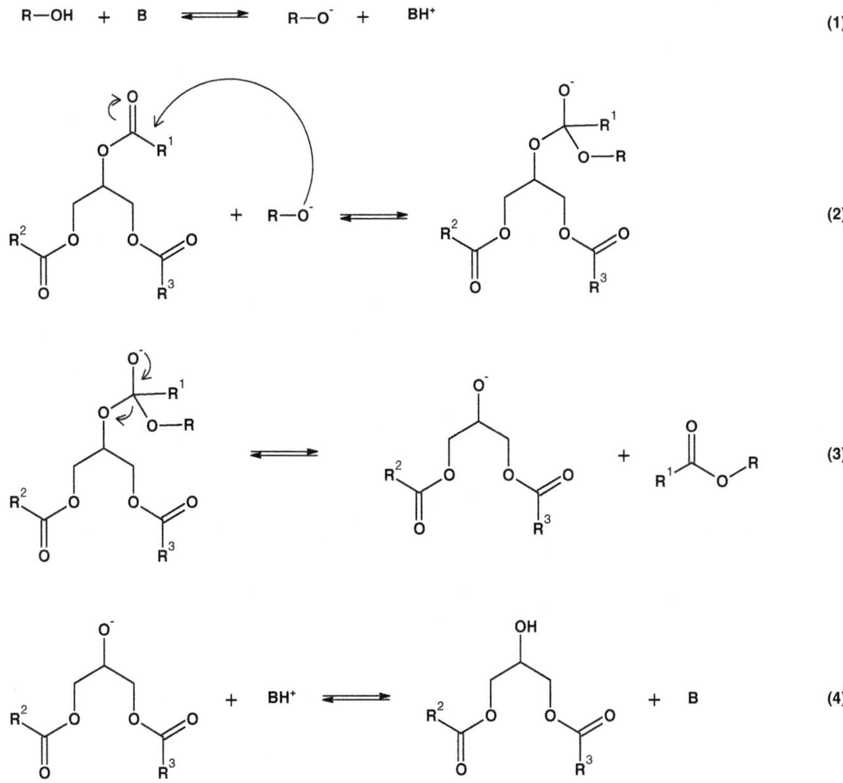

Fig. 2 Schematic representations of the sequential steps of the ester formation in the transesterification reaction catalyzed by bases

medium in the form of an alcoholic solution of a metal alkoxide. Further, alkoxide species promote a nucleophilic attack to the carbon atom of TAG ester group, leading to the formation of a tetrahedral intermediate (step 2). This species eliminates an intermediate methyl ester molecule leading to the formation of another alkoxide ion in the form of a diacylglycerol (DAG) molecule. After two more cycles, two new molecules of fatty acid alkyl esters (FAAE) are produced along with one glycerol molecule and, at the end of the process, the catalytic species is regenerated (Ma and Hanna 1999; Thanh et al. 2012). It is worth mentioning that, when NaOH, KOH, K_2CO_3, or any other catalytic species alike are used, one mole of water is formed for every mole of catalytic species that is release in the reaction medium and this may lead to a partial saponification of the feedstock, depending on the reaction conditions (Ma and Hanna 1999).

The formation of soap due to the reaction between FFA and the homogeneous base catalyst is considered a severe drawback for biodiesel production due to the fact that it reduces the final ester yield and complicates the subsequent separation and purification steps. Alternatively, a two-step process has been proposed to circumvent this problem, which consists of an acid-catalyzed esterification of FFA followed by basic homogeneous transesterification (Canakci and Van Gerpen 2001). When acid catalysis is used to convert acid oils, the acid-catalyzed transesterification is the rate limiting step because the esterification kinetics is much faster (Sun et al. 2010; Avhad and Marchetti 2015).

The fatty acid esterification is a reversible reaction involving a fatty acid and an alcohol to produce a FAAE that is usually assisted by an acid catalyst (Fig. 3) (Solomons and Fryhle 2002). Being reversible, several factors are able to influence the reaction equilibrium such as the reactants molar ratio, temperature and catalyst concentration (Ma and Hanna 1999). The reaction mechanism can be described in four steps (Fig. 4) (Solomons and Fryhle 2002): (a) the protonation of the carboxylic acid with a Brønsted-Lowry acid, thus facilitating the nucleophilic attack of the alcohol to the carboxyl carbon atom; (b) the formation of a tetrahedral intermediate, which subsequently undergoes a rearrangement, releasing one molecule of water; (c) the formation of the ester molecule; and (d) the regeneration of the reaction catalyst.

The formation of the tetrahedral intermediate species is fundamental in the esterification mechanism catalyzed by a Brønsted-Lowry acid. However, likewise for the transesterification mechanism, the tetrahedral intermediate cannot be isolated because it is unstable under the reaction conditions, therefore undergoing a fast dehydration that results in the formation of the desired FAAE (Carey 2011).

Fig. 3 Schematic representations of the esterification reaction

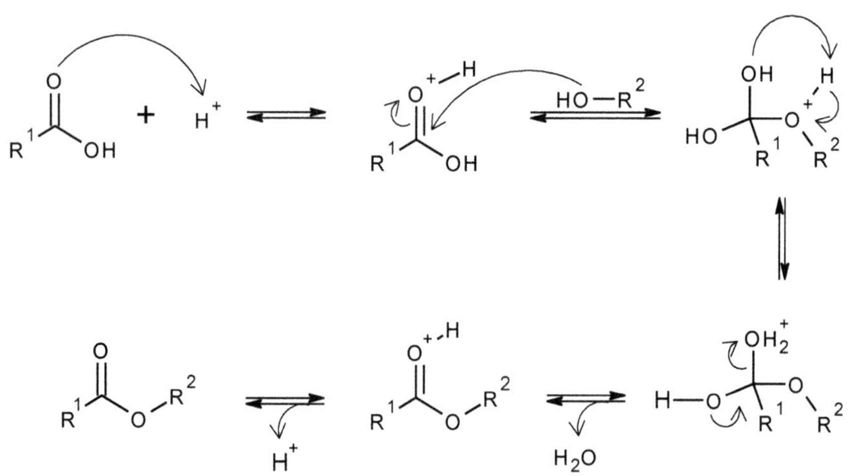

Fig. 4 Schematic representation of the reaction mechanism of the esterification reaction catalyzed by a Brönsted-Lorry acid species

Camacho et al. (2005) reported a study involving the production of biodiesel by esterification of palm oil fatty acids. The reaction was conducted at 130 °C in a Parr 4842 reactor using different homogeneous acid catalysts and different alcohol-to-acid molar ratios. Short chain alcohols were employed such as methanol and ethanol, both in their anhydrous and hydrated forms, in order to optimize the experimental conditions. The best reaction conversions were observed when anhydrous ethanol and methanol were used for the reactions catalyzed by sulfuric or methanesulfonic acids, compared to phosphoric and trichloroacetic acids. These results were attributed to the small percentage of water present in the first two acids, since the latter two had about 15 wt.% of water.

Su (2013a, b) suggested that, unlike sulfuric acid (H_2SO_4) and nitric acid (HNO_3), hydrochloric acid (HCl) can be recovered from the post-reaction mixture and be reutilized as catalysis for the next esterification reaction cycle. A lipase hydrolyzed feedstock of low acid value was subjected to the HCl-catalyzed esterification process. HCl did not mix with the formed ester phase and remained in the alcohol phase probably due to the presence of 63 % of water in it. This study suggested that 99.75 % of HCl can be recovered after the esterification process compared to 57.75 and 69.25 % for HNO_3 and H_2SO_4, respectively. Furthermore, higher than 97 % FFA conversions were obtained in 120 min of methanolysis and the presence of water seemed to facilitate the separation process. The above mentioned report suggest that a combination of both acid-catalyzed pre-treatment, adjusted only to convert FFA into esters, followed by a separation process, acid neutralization and basic homogeneous transesterification is a viable route for industrial applications.

Most if not all of the currently used biodiesel production routes are performed in the presence of catalytic species. Emphasis has been given to the use of basic and acid homogeneous catalysts; however, the processes can also be mediated by enzymes. Enzymes are biological catalysts that present high specificity, regioselectivity, and enantioselectivity (Akoh et al. 2007; Guldhe et al. 2015). For the synthesis of biodiesel, studies employing lipases (glycerol ester hydrolases, EC 3.1.1.3) are increasing steadily, which are the enzymes frequently employed to catalyze the hydrolysis of fats and oils to produce free fatty acids, acylglycerols, and free glycerol (Jaeger et al. 1999; Ros et al. 2012).

Lipases have also been used to allow the direct use of residual oils and greases in the biodiesel production process. In general, enzymes are less sensitive to the high water and free fatty acid contents that are often seen in residual oils and, depending on the reaction conditions, such biocatalysts may be applied in both esterification and transesterification of fatty materials. However, the use of enzymes is still hindered by several factors including their high production cost, the use for long reaction times to achieve high yields, the need of an organic solvent to improve their catalytic performance and, in some cases, the need for a strict control of water during the entire conversion process. In general, the immiscibility of lipid substrates in aqueous environment makes the catalytic reaction a heterogeneous process, forming a liquid–liquid interface in which the lipases interact with the substrates and exert their catalytic activity the reaction (Akoh et al. 2007).

Major advances have been observed in recent years in processes optimization involving enzymatic catalysis, either through immobilization, regeneration, protein engineering, and production of recombinant lipase in large scale, bring the hope of reducing the overall cost of biodiesel production using this catalytic route (Akoh et al. 2007; Tran and Balkus Jr 2011). Enzyme immobilization plays an important role in this process because it facilitates the catalyst recovery and reuse. The immobilization also provides a better stability to lipases, avoiding their denaturation and increasing their catalytic activity when compared to free lipases (Jaeger et al. 1999; Shah et al. 2004; Robles-Medina et al. 2009; Zhao et al. 2015).

Table 1 shows some examples in which enzymes were used for the transesterification of oils and fats. In these reports, enzymes are either in free form or immobilized in various supports. As in other catalytic systems, the use of biocatalysis to produce biodiesel has advantages and disadvantages, but this option has been configured as a suitable alternative for the future (Table 2) (Royon et al. 2007).

Table 1 Biodiesel production resulted via enzymatic catalysis of transesterification reaction of vegetable oils

Oil	Catalyst	Conditions (alcohol/solvent)	Yield (%)
Soybean oil[a]	*Burkholderia cenocepacia* immobilized lipase	Methanol/isooctane	98
Jatropha curcas oil[a]	*Pseudomonas cepacia* immobilized lipase	Ethanol alone	98
Jatropha curcas oil[b]	*Enterobacter aerognes* immobilized lipase	Methanol/t-butanol	94
Sunflower oil[a]	*Candida antarctica* immobilized lipase	Methanol/water	99
Castor oil[c]	Lipozyme IM	Ethanol/hexane	99
Jatropha curcas oil[d]	*Candida antarctica* immobilized lipase B (Novozym-435)	Ethyl acetate alone	91
Sunflower oil[e]	*Candida antarctica* immobilized lipase B (Novozym-435)	Ethyl acetate alone	93
Soybean oil[f]	LipB68 (*Pseudomonas fluorescens*) recombinant	Methanol/heptane	92
Cottonseed oil[g]	*Candida antarctica* lipase	Methanol/t-butanol	97
Cottonseed oil[g]	*Candida antarctica* immobilized lipase B (Novozym-435)	Methanol/t-butanol	97
Palm oil[h]	*Rhizopus oryzae* lipase	Methanol alone	55

[a]Lotti et al. (2015); [b]Kumari et al. (2009); [c]Akoh et al. (2007); [d]Modi et al. (2007); [e]Modi et al. (2006); [f]Luo et al. (2006); [g]Royon et al. (2007); [h]Pizarro and Park (2003)

Table 2 Main advantages and disadvantages of chemical and enzymatic processes for biodiesel production

Process	Advantages	Disadvantages
Homogeneous catalysis (basic or acid)	Experimental simplicity High yield Short reaction time	Recover and reuse of the catalyst species is not possible Very sensitive to water precluding the use of hydrated alcohol Low purity of the obtained products
Biocatalysis	Easy separation of the catalyst (when immobilized) High purity of the products Little sensitive to water enabling the use of hydrated alcohol	Long reaction time Enzyme high cost Low yield percentage when free enzyme is used

3.2 Novel Reactor Technologies (Catalytic and Non-catalytic Process)

Esterification and transesterification under supercritical conditions are non-catalyzed processes in which the mixture becomes homogeneous where the reaction occurs. Hence, this method is suitable for any type of raw material, especially those that are otherwise difficult to treat using conventional methods, such as animal fats and acid oils (Al-Zuhair et al. 2012). Another advantage of this process is that the alcohol recovery is facilitated due to the high temperature of the stream leaving the reactor chamber (Bertoldi et al. 2009). A sequence of flash separation units might be enough to recovery the alcohol used in excess.

A supercritical fluid (SCF) is obtained when any substance (compound, mixture or element) is brought above its critical pressure (Pc) and critical temperature (Tc), where distinct liquid and gas phases do not exist. SCFs can diffuse through solids like a gas, and dissolve materials like a liquid (Bernal et al. 2012; Adewale et al. 2015).

The typical ranges of operational conditions for non-catalyzed transesterification reactions have been reported as 280–400 °C and 10–30 MPa (Marulanda et al. 2010). Supercritical processes offer several advantages over conventional transesterification because these can be performed in the absence of a catalyst, enable the use of a wide range of feedstocks (vegetable oils, algae lipids, AFWs and WCO), can be coupled to an oil extraction unit and be combined with enzymatic catalysis, thus maximizing yields and selectivity in short reaction times. However, some important drawbacks are the maintenance of high operational conditions that can lead to isomerization of methyl esters (Imahara et al. 2008) and the need of high molar ratios between alcohols and feedstock. Several studies have shown that the thermal stability of unsaturated methyl esters may be compromised in this process (Quesada-Medina and Olivares-Carrillo 2011; Olivares-Carrillo and Quesada-Medina 2012).

In recent years, some process intensification techniques have been proposed and studied in order to optimize and improve biodiesel production. These intensification process units generally intend to promote the reaction and separation steps into a single unit operation, or accelerate both the reaction rate and the mass transfer steps. Some of these techniques are reactive distillation, extractive reaction, ultrasound and microwave irradiation.

Reactive distillation columns (RDC) consist of a core reactive zone completed by rectifying and stripping separation sections whose extent depends on the separation behavior of the reaction mixture. Its concept is based on the volatility difference between esters and fatty acids allowing continuous stripping on the components at the top and at the bottom of the column (Kiss 2011). Kiss et al. (2008) concluded using computational simulators (ASPEN) that the process is feasible for the esterification of high FFA feedstocks. This process offers some advantages over conventional biodiesel production processes, such as high productivities in short reaction times, no need for alcohol in excess due to the continuous removal of water and products, lower capital investment because further separation units are not needed, and no need of neutralization and separation of catalysts when fixed bed catalysts are employed. Despite these clear advantages, most studies carried out so far have been focused on process simulation only, and to explore the true potential of this technology, detailed economics coupled to pilot plant data are still required to overcome the challenge of scaling it up to industrial scale.

Ultrasound is the process of propagating an oscillating sound pressure wave with a frequency greater than the upper limit of the human hearing range. Ultrasound frequencies range between 20 kHz and 10 MHz, with associated acoustic wavelengths in liquids of roughly 100–0.15 mm. Application of ultrasound in the chemical processing enhances both mass transfer and reaction rates, leading to shorter reaction times and significant savings in reagents. Acoustic cavitation, which is the most important element of sonication, is the formation, growth, and implosive collapse of bubbles in a liquid irradiated with sound or ultrasound. The propagation of sound through the liquid causes expansion and compression waves, which forms an aerosol of solvent, solute vapors and previously dissolved gases. This phenomenon causes bubbles to grow and recompress due to the high mass transfer rates arising from the formation of a micro-emulsion through ultrasonic cavitation (Singh et al. 2007). The application of ultrasound technology to the transesterification of different vegetable oils has been widely investigated and reported (Singh et al. 2007; Kumar et al. 2010, 2012; Paiva et al. 2013).

This technique has proven to enhance reaction rates by overcoming the inherent issues of slow batch reaction rates and time-consuming phase separation of conventional transesterification processes (Veljković et al. 2012; Paiva et al. 2013). Similarly to other novel techniques, the use of ultrasound has been assessed mainly in bench and small-scale pilot plants. However, ultrasound reactors are commercially available with a capacity to deliver hundreds of cubic meters of biodiesel per hour. Anyway, the decision about using this technology in large scale would depend on a survey concerning investment and maintenance costs.

Transesterification is a mass transfer limited and reversible reaction commonly carried out in a batch type process. Hence, other types of process intensification technologies have been developed in recent years and applied to improve mixing and mass/heat transfer between the two liquid phases. These technologies either utilize novel reactors or coupled reaction/separation processes such as static mixers, micro-channel reactors, and cavitational reactors among others techniques (Qiu et al. 2010).

Static mixers consist of specially designed motionless geometric elements enclosed within a pipe or a column that creates effective radial mixing of two immiscible liquids as they flow through. Recently, they have been used in continuous biodiesel synthesis in combination with other equipments (Noureddini et al. 1998; Peterson et al. 2002; Qiu et al. 2010). Thompson and He (2007) used a stand-alone closed-loop static mixer system as a continuous-flow reactor to produce biodiesel from canola oil using methanol and sodium hydroxide. The system is composed of two stainless steel static reactors (4.9 mm ID × 300 mm long) including 34 fixed right- and left-hand helical mixing elements. High quality ASTM D6584 biodiesel was obtained after optimization of the experimental conditions.

Micro-channel reactors achieve rapid reaction rates by improving the efficiency of heat and mass transfer and utilizing high surface area/volume ratio and short diffusion distances (Kobayashi et al. 2006; Qiu et al. 2010). Sun et al. (2008) studied KOH-catalyzed transesterification of unrefined rapeseed oil and cottonseed oil with methanol in capillary microreactors with inner diameters of 0.25 mm. With a residence time of 5.89 min, a 99.4 % yield of methyl esters was obtained at 60 °C with 1 wt% KOH and a methanol to oil molar ratio of 6:1.

Cavitational reactors use acoustic energy or flow energy to generate the cavitation phenomena that results in process intensification. During cavitation, the violent collapse of the cavities produced by the pressure changes releases large magnitude of energy over a small location, and brings about very high temperatures and pressures. Cavitation also intensifies the mass transfer rate by generating local turbulence and liquid micro-circulation in the reactor (Kumar et al. 2010, 2012). Kelkar et al. (2007) investigated two different reactors based on acoustic and hydrodynamic cavitation and compared their performance in biodiesel synthesis from vegetable oils. At a reaction time of 15 min, more than 90 % in biodiesel yield was obtained during transesterification of vegetable oils with methanol in the presence of sodium hydroxide using both reactor designs.

3.3 Heterogeneous Catalysis

To circumvent some of the drawbacks of homogeneous (acid/base) catalytic systems for biodiesel production, many heterogeneous catalysts have been studied intensively in both lab and pilot scales, and some of these have already been scaled-up to industrial applications. Heterogeneous catalysis is an attractive route for biodiesel production because the sustainability of the process and the quality of the final products are more satisfying than what is usually observed in homogeneous processes (Melero et al. 2009; dos Santos et al. 2011; Gole and Gogate 2012; Bail et al. 2013).

Solids employed in heterogeneous processes greatly simplify the separation and purification of the reaction co-products. This is mainly because such catalysts, when truly heterogeneous, do not leach to the reaction products and do not produce soaps (Melero et al. 2009; dos Santos et al. 2011; Gole and Gogate 2012; Bail et al. 2013). In addition, the heterogeneous processes allow the catalyst recovery and reuse in several consecutive reaction cycles, no generation of waste waters or waste solvents, and easy separation of reaction products. However, most solid catalysts do not compete with homogeneous alkaline or acid catalysts because, in general, longer times and higher temperatures are required for optimal performance. There are only a few reports describing the development of fast and efficient heterogeneous catalytic processes working at low reaction temperatures (Dabdoub et al. 2009).

The industrial use of heterogeneous catalysis is still at an early stage. Indeed, in a search for technologies available at industrial scale, most industries offer variations of the traditional technology based on alkaline transesterification in homogeneous media. However, these factors do not invalidate the use of solid catalysts because the possibility of recycling and reuse is expected to offer an economic compensation in long term. Therefore, the current technological challenge is to develop suitable heterogeneous catalysts that exhibit high stability, selectivity, and catalytic efficiency while promoting a better process sustainability (dos Santos et al. 2011, 2015; Bail et al. 2013).

Ideal solid catalysts must have high porosity, high stability, high concentration of acidic or basic sites (Lewis and Brønsted-Lowry acidity-basicity), and high surface area, among others. Furthermore, the presence of a double acid character in the solid (Lewis and Brønsted-Lowry acid sites) is often desired and idealized for situations in which the feedstock has a high acid value (Melero et al. 2009; Wilson and Lee 2012; Sánchez-Vázquez et al. 2013; Pirez et al. 2014).

According to Wilson et al. (2008), the preparation of heterogeneous catalysts must optimize the effect of its acid/basic strength of the solid, the solids surface hydrophobicity, and the pore architecture. These parameters are essential for studying and designing of new solids with high catalytic interest (Wilson and Lee 2012; Sánchez-Vázquez et al. 2013; Pirez et al. 2014).

From a chemical standpoint, many solids are able to catalyze both esterification and transesterification in a single process. Examples of these bifunctional catalytic systems are zeolites, inorganic oxides, nanometric mesoporous materials, coordination compounds, organic polymers, and ion exchange resins, among others (Dabdoub et al. 2009; dos Santos et al. 2011, 2015; Bail et al. 2013; Sánchez-Vázquez et al. 2013). Table 3 shows some of the most relevant examples of solids already used as heterogeneous catalysts for biodiesel production.

Some reports describe different methods for replacing inorganic acids such as H_2SO_4 by solids with high acid character such as niobium or tungsten oxides, Amberlyst® resins, Nafion®, mesoporous aluminosilicates, heteropolyacids, silica mesoporous functionalized, and modified zirconia, among others. The literature also features several examples of basic solids with high potential for use in transesterification reactions, similarly to NaOH or KOH (Okuhara 2002; Corma and Garcia 2003). Numerous alkali and alkaline earth based metal oxide catalysts have been tested for transesterification, such as magnesium oxide (MgO), calcium oxide (CaO), strontium oxide (SrO), alkali and alkaline earth metal supported oxides, and mixed metal oxides, as well as hydrotalcites and anionic resins (Kawashima et al. 2008, 2009; Thanh et al. 2012). CaO has been the most intensively used heterogeneous catalyst because of its higher basicity, lower solubility, lower price, and easy and safe handling compared to chemicals such as KOH (Kawashima et al. 2008, 2009; Reyero et al. 2014).

By treating CaO at 700 °C, the resulting solids were able to convert vegetable oils and animal fats into FAAE by transesterification (Di Serio et al. 2008; Boey et al. 2011). Granados et al. (2007) reported that the improvement of CaO catalytic activity by calcination is associated to the removal the surface carbonate and hydroxyl groups. Yields higher than 90 % were obtained after 90 min using 1 wt% of catalyst, 50 g of oil (refined sunflower oil, food grade), 60 °C, 1000 rpm, and a methanol to oil molar ratio of 13:1. CaO could also be re-used for several runs without significant deactivation. The yield of fatty acid methyl esters (FAME) decreased from 90 to 80 % after the second reaction run and this catalytic performance was maintained even after eight cycles of reuse.

Calcined CaO (900 °C for 1.5 h) was also reported by Kouzu et al. (2008a) as a very active catalyst for the transesterification of soybean oil with methanol. Reactions were carried out under the following conditions: 100 mL of soybean oil, 0.8 g of catalyst (14 mmol) and 50 mL of methanol under reflux. FAME yields over 99 % were obtained in 2 h but a lower catalytic activity was observed when a new reaction cycle was carried out using the recovered solids after reactivation.

Catalysts with sulfonic acid groups have been reported to display an excellent catalytic activity in esterification reactions (Tesser et al. 2005). These catalysts are usually polymeric such as in the case of sulfonated cross-linked polystyrene-divinylbenzene copolymers and their acidity is usually analogous to that of *p*-toluene-sulfonic acid. Although more expensive than inorganic acids, these solids are less corrosive and minimize the environmental impact of the conversion process, possibly compensating for their relatively high cost (Tesser et al. 2005).

One important work about the use of strong acid ion-exchange resins (Relite CFS) for biodiesel production was reported by Tesser et al. (2005). These authors studied the esterification kinetics of oleic acid in the presence of TAGs. The

Table 3 Some examples of solids catalyst prepared and used in esterification/transesterification reaction for biodiesel production

Solid catalyst	Reaction	Reaction time (h)	Oil:alcohol molar ratio	Temperature (° C)	Conversion (%)
MoO_3/SiO_2[a]	Esterification: lauric acid/ethanol	3	1:12	120	95.0
WOx/ZrO_2[b]	Esterification: palmitic acid/methanol	6	1:12	120	97.5
Na_2MoO_4[c]	Esterification: oleic acid/ethanol	6	1:12	120	95.1
Na_2MoO_4[c]	Esterification: lauric acid/methanol	6	1:12	120	97.1
CaO/MgO[d]	Transesterification: rapeseed oil/methanol	2	1:13	64,5	92.0
PDVB-SO_3H-SO_2CF_3[e]	Transesterification: tripalmitin/methanol	16	1:90	65	91.4
VOP[f]	Transesterification: soybean oil/methanol	1	–	150–180	80.0
K_2CO_3, Na_2CO_3 and $CaCO_3$[g]	Transesterification: castor oil/methanol	10	1:6	–	–
MgO/Al_2O_3[h]	Transesterification: soybean oil/methanol	1	1:12	180	94.0
Nb_2O_5/MCM-41[i]	Transesterification: sunflower oil/methanol	4	1:12	200	95.0
Zr-SBA-15[j]	Transesterification of low-grade feedstock	6	1:50	209	96.0
MCM-41-guanidine[k]	Transesterification: soybean oil/methanol	3	–	70	99.0
ZnO-nCaO[l]	Transesterification: butyrate/methanol	2	1:12	60	90.0
FA/Na-X[m]	Transesterification: sunflower oil/methanol	8	1:6	65	83.5
$PrSO_3H$/SBA-BTSB100 %[n]	Esterification: palmitic acid/methanol	–	1:30	60	94.0
CaO, SrO, and K_3PO_4[o]	Transesterification: used frying oil/methanol	3	1:6	65	92.0

[a]Gole and Gogate (2012); [b]dos Santos et al. (2011); [c]Bail et al. (2013); [d]Sánchez-Vázquez et al. (2013); [e]Watanabe et al. (2001); [f]Liu et al. (2013); [g]Wilson et al. (2008); [h]Gallo et al. (2008); [i]García-Sancho et al. (2011); [j]Iglesias et al. (2014); [k]de Lima et al. (2014); [l]Alba-Rubio et al. (2010); [m]Babajide et al. (2012); [n]Pirez et al. (2014); [o]Viola et al. (2012)

reactions were carried out at 85 °C using methanol to oleic oil molar ratios of 8:1 and 10:1 and 2.5 wt% of catalyst. Relite CFS had a good catalytic activity under these reaction conditions, reaching 80 % conversion of free fatty acids after 2 h.

In summary, four major groups of catalysts have been employed in the biodiesel synthesis: basic and acid catalysts (homogeneous system) such as NaOH, KOH, CH_3NaO, CH_3KO, H_2SO_4 and H_3PO_4; heterogeneous solid acid or basic catalysts such as $CaCO_3$ and Nb_2O_5; polymeric materials such as ion-exchange resins; and free and immobilized enzymes. In this section, some of the most promising catalytic systems were reviewed in the light of their potential for industrial application. However, many other catalytic systems have been studied so far and the readers are referred to several reviews listed in this chapter to obtain more information about their principles and properties. In this scenario, there is a continuous search for stable heterogeneous catalytic systems that are more tolerant to water and free fatty acids. This would allow the more extensive use of low cost feedstocks in the biodiesel production chain.

3.4 Layered Materials as Catalysts for Biodiesel Production

Layered materials are also very attractive for biodiesel production because they are able to perform a double role in the reaction medium, acting as catalyst as well as surfactants. As a result, their catalytic performance approaches homogeneous during the reaction course but the solids can be recovered like any heterogeneous catalyst by simple method of precipitation (de Paiva et al. 2015b). These catalysts can be divided in metallic carboxylates and layered glycerolates. Although many attempts have been made to synthesize and characterize metal/glycerol complexes (Radoslovich et al. 1970; Rodrique et al. 1978; Bruylants et al. 1980; Taylor et al. 1992; Mendelovici et al. 1990, 2014), the first reports describing the synthesis and structure resolution of zinc monoglycerolates was published in 1983, with the crystals being produced by heating ZnO with glycerol at 220 °C (Hambley and Snow 1983). Figure 5a show the structure of layered zinc monoglycerolate ($C_3H_6O_3Zn$) along its b-axis. The crystals have a monoclinic structure with the space group symmetry $P2_1/c$ with the following cell parameters: $a = 8.110$ Å; $\alpha = 6.404°$; $\beta = 8.714°$ and $\gamma = 93.44°$. Zinc atoms occupy a distorted trigonal bi-pyramidal geometry, where one glycerol molecule is bonded to three different zinc atoms as shown in Fig. 5b.

Figure 6 shows the structure of layered calcium diglycerolate ($Ca(C_3H_7O_3)_2$) (León-Reina et al. 2013), which was determined using the Rietveld structure refinement method. Calcium diglycerolate crystals ranging from 1 to 5 mm were prepared by calcining high purity $CaCO_3$ to obtain fresh CaO, which was reacted with a mixture of 100 mL of methanol and 35 mL of glycerol at 50 °C overnight. To avoid carbonation or hydration, the solid was stored under protective nitrogen atmosphere. The structure is orthorhombic with *F222* space group symmetry and the following cell parameters: $a = 21.3356$ Å; $b = 13.53610$ Å; $c = 13.35806$ Å.

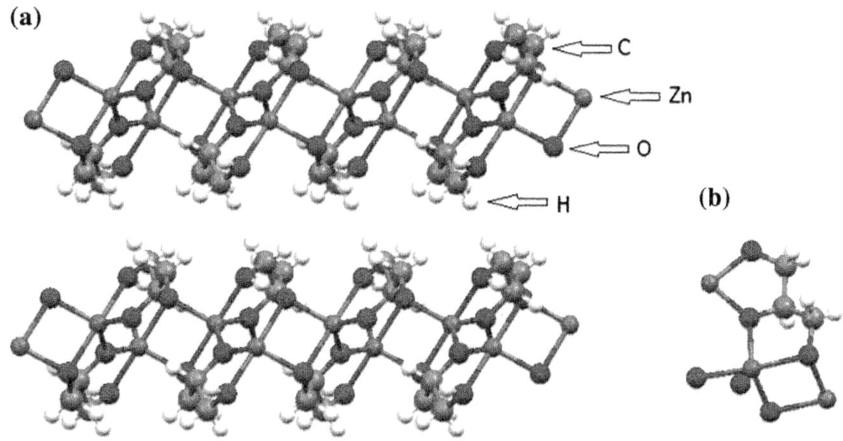

Fig. 5 Structure of layered zinc monoglycerolate ($C_3H_6O_3Zn$): **a** view along the b-axis of the structure and **b** detail of the structure according to Hambley and Snow (1983)

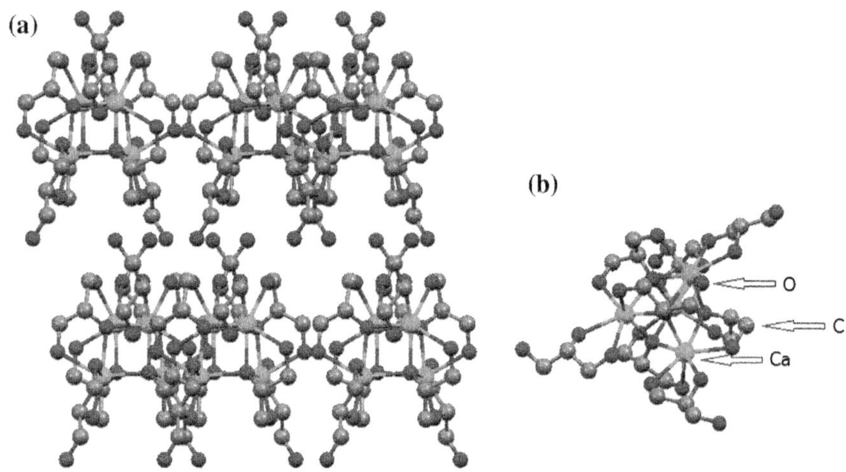

Fig. 6 **a** Structure of layered calcium diglycerolate ($Ca(C_3H_7O_3)_2$) and **b** detail of the structure; hydrogen atoms positions were not refined by León-Reina et al. (2013)

The structure can be explained by a combination of ($Ca_4(C_3H_7O_3)_8$) tetramer connected through H bonds. In the tetramer, Ca has pentagonal bi-pyramid geometry where one glycerol molecule is bonded to three different calcium atoms as shown in Fig. 6b.

The catalytic activity of calcium monoglycerolate ($C_3H_6O_3Ca$) and calcium diglycerolate ($Ca(C_3H_7O_3)_2$) for the transesterification of vegetable oils was probably first described by Kouzu et al. (2008a, b, 2009a, b, 2010a, b) and Kouzu and Hidaka (2012). In general, metal glycerolates are emerging as potential catalysts for biodiesel production. However, the mechanism involved in the conversion of TAG to fatty acid alkyl esters is still unknown.

In their earlier work, Kouzu et al. (2008b) obtained calcium oxide by calcining calcium carbonate at 900 °C for 1.5 h and used in the methanolysis of soybean oil under reflux in a 500 mL four-neck glass reactor. After recovery of the catalyst, the authors observed that CaO was transformed "in situ" in calcium diglyceroxide or calcium diglycerolate $Ca(C_3H_7O_3)_2$. The recovered catalyst was also synthesized by reacting glycerol (50 vol%) and methanol for 2 h under reflux in the presence of CaO. The catalytic activity of the solids was investigated by reacting soybean oil with methanol and 14 mmol of the catalyst at 60 °C. High FAME yields were obtained in both cases but the activity of calcium diglycerolate seemed to be more resistant to moisture and CO_2 than CaO. Also, the structure of calcium diglycerolate was preserved and had less instability than CaO when in contact with air (Kouzu et al. 2010b). In another work, Kouzu et al. (2010a) observed that around 10 % of the catalyst was leached out to the solution and that calcium glyceroxide was not the active catalyst but acted as a precursor of "calcium-X", the given name to CH_3O–Ca–$O(OH)_2C_3H_5$ compounds that are prepared by immersing calcium diglycerolate in methanol.

After these pioneering studies, the synthesis and characterization of different glycerolates and their structure, properties, and catalytic activity in esterification reactions have provoked a considerable interest in the literature (León-Reina et al. 2013; Lisboa et al. 2014; Reinoso et al. 2014; Reyero et al. 2014). Some of these results are described below.

Reinoso et al. (2014) synthesized zinc glycerolate by heating 0.05 mol of zinc acetate dehydrate and 3.4 mol of glycerol (containing 2 % of water) at 160 °C for 1 h under 500 rpm. Afterwards, the solids were washed with ethanol and dried at 40 °C for 1 h. The catalytic performance of these solids was tested in transesterification reactions at different temperatures using a methanol:oil molar ratio of 30:1 and 3 wt% of catalyst in relation to the oil mass. After 6 h at 100 °C, a TAG conversion of 95.0 % and a FAME yield of 65.9 % were obtained but, at 140 °C, both TAG conversion and FAME yield were slightly higher. These authors also checked the reuse of the catalyst after filtration and washing with a 1:1 volumetric mixture of ethyl ether and ethyl alcohol. The catalyst structure remained unchanged after five consecutive reaction cycles and no evidence of calcium leaching was observed even at 140 °C. However, the presence of 0.5 % water or 10 % stearic acid in the reaction medium at 140 °C led to a small decrease in FAME yield and a partial conversion of zinc glycerolate into zinc carboxylates such as zinc stearate.

Reyero et al. (2014) synthesized calcium diglycerolate ($Ca(C_3H_7O_3)_2$) by reacting 3 g of CaO with 10 g of glycerol and 44 g of methanol at 60 °C under mechanical stirring for 3 h, followed by filtration, washing with THF and drying overnight at 60 °C under vacuum. Calcium monoglycerolate ($CaC_3H_6O_3$) was synthesized by reacting $Ca(OH)_2$ and glycerol (mass ratio of 1:10) at 180 °C for 2 h in a autoclave reactor, followed by washing twice with ethanol and drying overnight at 60 °C under vacuum. Both solid catalysts were tested in the methanolysis of sunflower oil at 60 °C, with calcium monoglycerolate being less active then calcium diglycerolate. This difference was attributed to the higher basicity and surface area of the most active catalyst. One important finding is that both calcium glycerolates are sensitive to atmospheric air, being probably deactivated by CO_2. Exposure to the atmosphere at room temperature for 2 h reduced the catalytic activities of calcium diglycerolate and calcium monoglycerolate from 83.4 and 29.9 % to 2.9 and 8.9 %, respectively. This information suggests that CaO, calcium glycerolates and probably other types of metal glycerolates must be stored under protective inert atmosphere.

Calcium monoglycerolate, calcium diglycerolate without and with an excess of glycerolate anions, and other metals glycerolates (strontium, barium and zinc) have also been synthesized and evaluated as active materials for the methanolysis of refined soybean oil using a methanol:oil molar ratio of 50:1 and 2 wt% of catalyst for 2 h at 500 rpm under methanol reflux (Lisboa et al. 2014). Calcium, strontium, barium and zinc monoglycerolates converted 62.79, 99.20, 99.00 and 0.79 % of soybean oil TAGs into methyl esters and both calcium glycerolates presented conversions of 98.63 and 99.90 %, respectively. Only calcium monoglycerolate, calcium diglycerolate and zinc monoglycerolate had the structure preserved after the first reaction cycle. Calcium diglycerolate rich in glycerolate anions and strontium monoglycerolate were completely decomposed to a mixture of carbonate and hydroxide while barium monoglycerolate was completely dissolved in the reaction medium. After three consecutive reaction steps, all the evaluated materials lost their catalytic activity. However, no measures were taken to prevent the contact the solid catalysts with atmospheric CO_2 and this was probably the reason for deactivation.

Methanolysis of refined soybean oil with zinc monoglycerolates was also evaluated at high temperatures under variable molar ratios using 2 wt% of the dried (not anhydrous) catalyst in relation to the oil mass (da Silva Lisboa et al. 2013). Reactions were carried out for 2 h under an agitation of 500 rpm. The best conversion of 96.9 % was obtained at 120 °C when a methanol:oil molar ratio of 50:1 was used. The catalyst preserved its activity and structure after four consecutive reaction cycles (or three cycles of reuse). A small loss in catalytic activity was observed, with 98.4 % in the first, 95.1 % in the second, 94.9 % in the third and 89.1 % in the fourth reaction cycle. This was attributed to physical losses that might have occurred during this reaction sequence, even though the amounts of catalyst and reagents were tentatively adjusted after each reaction cycle.

Layered materials such as layered double hydroxides (LDHs) and layered hydroxide salts (LHSs) have also been used in many catalytic applications including both esterification and transesterification of several lipid sources for biodiesel production (Cordeiro et al. 2011). Cordeiro et al. (2008) used zinc hydroxide nitrate ($Zn_5(OH)_8(NO_3)_2.2H_2O$) in the catalytic conversion of lauric acid to methyl laurate. High yields were obtained but the solids were rearranged to a different layered structure after the first reaction cycle. Later, it was demonstrated that zinc hydroxide nitrate reacted in situ with lauric acid to produce zinc laurate, which was identified as the actual catalytic active phase for esterification. This material showed good stability and catalytic activity in several consecutive reaction cycles.

Di Serio et al. (2005) was one of the first to report the catalytic esterification of fatty acids employing layered metal carboxylates or layered metal soaps such as zinc acetate and zinc stearate. These authors revealed that these catalysts have surfactant properties, promoting a better contact between both oil and alcohol phases. In this study, the best results were achieved with zinc stearate.

Figure 7 shows the structure of zinc octanoate (Lacouture et al. 2000), which is supposed to represent the structure of other long chain zinc carboxylates. In this structure, two zinc cations are connected by a bidentade carboxylate bridge. This fully extended all trans conformation generate a two-dimensional layers that are packed along the longer axis (axis a) of the unit cell and held together by Van der Waals forces. Each zinc cation is coordinated tetrahedrically to four different oxygen atoms belonging to different carboxylate groups. The crystallinity is monoclinic and the cell parameters are: a = 21.093 Å; b = 4.6905 Å; c = 9.2544 Å; α = 90.00°; β = 101.323°; and γ = 90.00°. With this, the structure resembles a two-dimensional micelle where the interior is hydrophilic and the outside is hydrophobic.

Lisboa et al. (2012) synthesized several metal laurates and investigated their catalytic use in the methyl esterification of lauric acid. The laurates were synthesized by metathesis, where 4.33 g (21.6 mmol) of lauric acid were dissolved in 20 mL of methanol and neutralized by sodium hydroxide. The obtained sodium laurate was dissolved in 30 mL of distilled water at room temperature and the metal

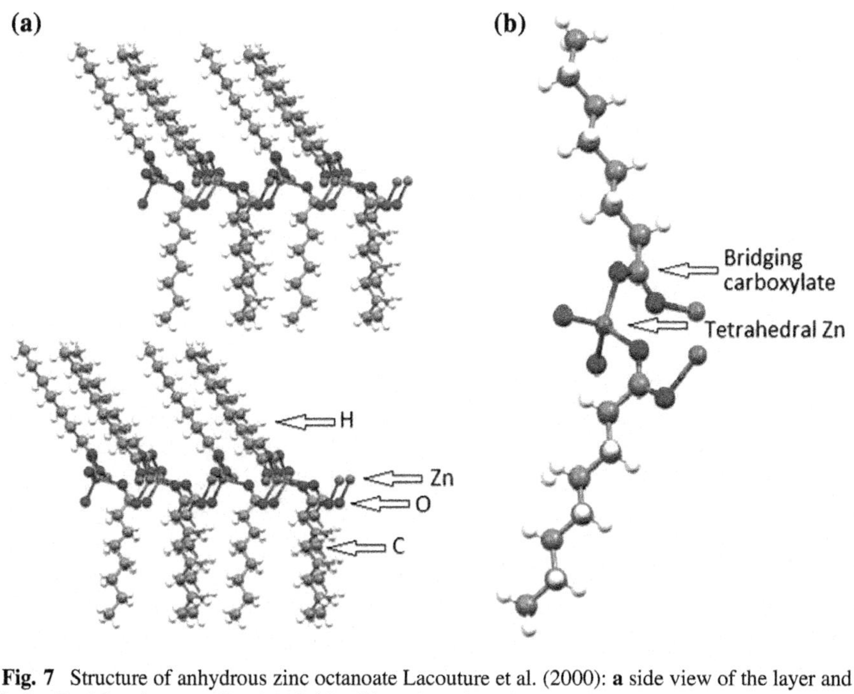

Fig. 7 Structure of anhydrous zinc octanoate Lacouture et al. (2000): **a** side view of the layer and **b** detail of the zinc coordination, bridged by one carboxylate group

salt solution was added under vigorous stirring to the aqueous/alcoholic sodium laurate solution. After addition, each system was submitted to magnetic stirring for 1 h. With this procedure, manganese (II), copper (II), lantanum (III) and nickel (II) were prepared. To evaluate the performance of these catalysts, eleven experiments were performed using a factorial design. Compared to the thermal control, catalytic conversions between 80 and 90 % were observed in all cases but only manganese laurate demonstrated the ability to catalyze the esterification of lauric acid with ethanol; besides, only copper and lanthanum laurates could be recovered intact after one reaction cycle.

The use of zinc carboxylates for the esterification of a commercial FFA mixture (90 % oleic acid) with methanol was also investigated (de Paiva et al. 2015b). Synthetic zinc laurate was obtained by neutralizing lauric acid with sodium hydroxide and the resultant salt was reacted with $ZnCl_2$ in aqueous solution. Also, a commercial carboxylate preparation called zinc stearate (SIM Estearina, Curitiba, PR, Brazil) was also tested under similar reaction conditions. This sample consisted of a single crystallographic phase where stearate and palmitate anions (48.06 % and 45.36 %, respectively) are simultaneously intercalated in-between the layers. Both catalysts had similar catalytic performances. Conversions around 92 % were obtained at 165 °C after 120 min using an ethanol:FFA molar ratio of 8:1 and a catalyst loading of 5 wt%. The reaction rate and equilibrium constants were described and the apparent activation energy was estimated from the slope of the Arrhenius plot as being around 80 kJ/mol, which is close to the 68 kJ/mol reported by Zhang et al. (2012).

Although the catalysts could be recovered after the reaction, the intercalated carboxylate anions changed during the reaction course, being preferably replaced by the most abundant species of the FFA mixture (oleate anion) (de Paiva et al. 2015b). After 24 h at 150 °C and with a molar ratio of 3:1, zinc laurate (95.79 % pure; 5 wt%) changed to 13.94 % laurate, 24.92 % stearate, 41.42 % oleate and 4.97 % palmitate. Following the same procedure, commercial zinc stearate was characterized after 24 h at 165 °C under a slightly higher molar ratio of 8:1 and its chemical composition changed to 10.91 % stearate, 84.35 % oleate and 2.89 % laurate. These materials showed a unique behavior by changing the composition according to the main FFA present in the medium while the catalytic performance remained the same (Ramos et al. 2015).

Paiva et al. (2015a) also studied the influence of long chain primary alcohols (1-butanol and 1-hexanol) in the esterification of high oleic FFA mixture using commercial zinc stearate as the solid catalyst. The investigated alcohol:FFA molar ratios (MR) were 3:1, 8:1 and 12:1 for 1-butanol and 3:1 and 12:1 for 1-hexanol while the amount of zinc stearate was fixed at 5 wt% in relation to the mass of fatty acids. The investigated temperatures were of 135, 150, and 165 °C for butanolysis and 150 and 165 °C for hexanolysis, respectively. The reaction conversion increased with the increasing alcohol chain length. Conversions of oleic acid reached 92 % with gains of 42 % in relation to the reaction control, highlighting the high catalytic activity of this material. Another observation is that a quasi-homogenous kinetic model was able to predict conversions with both alcohols, indicating that the adsorption and diffusion phenomena are negligible. This also evidenced the emulsifying role of this catalyst, which promotes a better dispersion of the alcohol into the oleic phase. Moreover, this catalyst system was highly resistant to water and could be easily and almost completely recovered after reaction completion by adding a small amount of acetone. Beyond the kinetic modeling, the apparent activation energy estimated from the slope of the Arrhenius plot was about 66 kJ/mol, once more being similar to values reported elsewhere (Brahmkhatri and Patel 2012; de Jong et al. 2009).

4 Future and Perspectives

Despite the current growth, sustainability of the biodiesel industries may be limited due to the industry's inability to secure cheap feedstock. However, the economic scenario should change in the near future and biodiesel production may become favorable considering the depletion and scarcity of fossil fuels. The current subsidy practices adopted by some countries (e.g., United States, UE, Brazil and others) have stimulated biodiesel production and assured competitive prices toward petroleum diesel. This type of policies can generate benefits in a long-term period. In the Brazilian case, an economy of 11.3 billion dollars was obtained with diesel imports in 2005–2014 (Rico and Sauer 2015).

Strategies focused in the reduction of biodiesel production cost will be responsible to ensure competitiveness with research focused in WCO, AFWs, algae lipids and non-edible vegetable oils. Haas (2005) showed that the quality of biodiesel obtained from a low value lipid soapstock was similar to any other. Also, the combination of different feedstocks (blends of edible, non-edible with WCO and AFWs) can assure desired properties of biodiesel (cetane number, cloud point, particulate emissions etc.). In this sense, various integration strategies have been studied and these are provided in the review by Hasheminejad et al. (2011).

Comparative studies have been carried out with different processes. West et al. (2008) studied the economics aspect using HYSYS/ASPEN simulator with four different processes: two homogeneous, a heterogeneous and a supercritical process. The heterogeneous and supercritical processes were both superior economically, even though only the heterogeneous proved to have a positive net after tax rate of return. Marchetti and Errazu (2008) reached the same favorable conclusion with respect to use of heterogeneous catalysts.

Novel reactor technology (RDC, cavitational reactors, assisted ultrasounds techniques and others) combined with heterogeneous catalysis can be the threshold point in which biodiesel production becomes economically attractive and environmental friendly helping to reach the goals of obtaining self-sustainable and greener process. These goals will be accomplished when the current vanguard research in this field starts to be assessed from an economic standpoint point and successful bench technologies are transferred to pilot and industrial scales.

Acknowledgments The authors are grateful to CNPq and CAPES (Brazil) for the financial support to carry out this study in the form of research grants (Grants 575812/2008-7, 558836/2010-0, 306920/2013-1, and 406737/2013-4) and Ph.D. scholarships for our graduate students A.S. and E.J.M.P.), respectively. The financial support of the National Institute of Science and Technology (INCT) in Energy and Environment is also greatly acknowledged.

References

Achten WMJ, Verchot L, Franken YJ, Mathijs E, Singh VP, Aerts R, Muys B (2008) Jatropha bio-diesel production and use. Biomass Bioenerg 32(12):1063–1084. doi:10.1016/j.biombioe.2008.03.003

Adewale P, Dumont MJ, Ngadi M (2015) Recent trends of biodiesel production from animal fat wastes and associated production techniques. Renew Sust Energ Rev 45:574–588. doi:10.1016/j.rser.2015.02.039

Akoh CC, Chang SW, Lee GC, Shaw JF (2007) Enzymatic approach to biodiesel production. J Agric Food Chem 55(22):8995–9005. doi:10.1021/jf071724y

Al-Zuhair S, Hussein A, Al-Marzouqi AH, Hashim I (2012) Continuous production of biodiesel from fat extracted from lamb meat in supercritical CO2 media. Biochem Eng J 60:106–110. doi:10.1016/j.bej.2011.10.010

Alba-Rubio AC, Santamaria-Gonzalez J, Merida-Robles JM, Moreno-Tost R, Martin-Alonso D, Jimenez-Lopez A, Maireles-Torres P (2010) Heterogeneous transesterification processes by using CaO supported on zinc oxide as basic catalysts. Catal Today 149(3–4):281–287. doi:10.1016/j.cattod.2009.06.024

Ali MH, Mashud M, Rubel MR, Ahmad RH (2013) Biodiesel from neem oil as an alternative fuel for diesel engine. Procedia Eng 56:625–630. doi:10.1016/j.proeng.2013.03.169

Alptekin E, Canakci M, Sanli H (2012) Evaluation of leather industry wastes as a feedstock for biodiesel production. Fuel 95:214–220. doi:10.1016/j.fuel.2011.08.055

Aransiola EF, Betiku E, Ikhuomoregbe DIO, Ojumu TV (2012) Production of biodiesel from crude neem oil feedstock and its emissions from internal combustion engines. Afr J Biotechnol 11:6178–6186

Aransiola EF, Betiku E, Layokun S, Solomon B (2010) Production of biodiesel by transesterification of refined soybean oil. IJBCS 4(2):391–399

Aransiola EF, Ojumu TV, Oyekola OO, Madzimbamuto TF, Ikhu-Omoregbe DIO (2014) A review of current technology for biodiesel production: State of the art. Biomass Bioenerg 61:276–297. doi:10.1016/j.biombioe.2013.11.014

Armenta RE, Vinatoru M, Burja AM, Kralovec JA, Barrow CJ (2007) Transesterification of fish oil to produce fatty acid ethyl esters using ultrasonic energy. J Am Oil Chem Soc 84(11):1045–1052. doi:10.1007/s11746-007-1129-2

Atabani AE, Silitonga AS, Ong HC, Mahlia TMI, Masjuki HH, Badruddin IA, Fayaz H (2013) Non-edible vegetable oils: a critical evaluation of oil extraction, fatty acid compositions, biodiesel production, characteristics, engine performance and emissions production. Renew Sust Energ Rev 18:211–245. doi:10.1016/j.rser.2012.10.013

Avhad MR, Marchetti JM (2015) A review on recent advancement in catalytic materials for biodiesel production. Renew Sust Energ Rev 50:696–718. doi:10.1016/j.rser.2015.05.038

Babajide O, Musyoka N, Petrik L, Ameer F (2012) Novel zeolite Na-X synthesized from fly ash as a heterogeneous catalyst in biodiesel production. Catal Today 190(1):54–60. doi:10.1016/j.cattod.2012.04.044

Bail A, dos Santos VC, de Freitas MR, Ramos LP, Schreiner WH, Ricci GP, Ciuffi KJ, Nakagaki S (2013) Investigation of a molybdenum-containing silica catalyst synthesized by the sol-gel process in heterogeneous catalytic esterification reactions using methanol and ethanol. Appl Catal B-Environ 130:314–324. doi:10.1016/j.apcatb.2012.11.009

Banković-Ilić IB, Stamenković OS, Veljković VB (2012) Biodiesel production from non-edible plant oils. Renew Sust Energ Rev 16(6):3621–3647. doi:10.1016/j.rser.2012.03.002

Bernal JM, Lozano P, Garcia-Verdugo E, Burguete MI, Sanchez-Gomez G, Lopez-Lopez G, Pucheault M, Vaultier M, Luis SV (2012) Supercritical synthesis of biodiesel. Molecules 17 (7):8696–8719. doi:10.3390/molecules17078696

Bertoldi C, da Silva C, Bernardon JP, Corazza ML, Cardozo FL, Oliveira JV, Corazza FC (2009) Continuous production of biodiesel from soybean oil in supercritical ethanol and carbon dioxide as cosolvent. Energ Fuel 23(10):5165–5172. doi:10.1021/ef900402r

Bharathiraja B, Chakravarthy M, Kumar RR, Yuvaraj D, Jayamuthunagai J, Kumar RP, Palani S (2014) Biodiesel production using chemical and biological methods—a review of process, catalyst, acyl acceptor, source and process variables. Renew Sust Energ Rev 38:368–382. doi:10.1016/j.rser.2014.05.084

Boey PL, Maniam GP, Abd Hamid S (2011) Performance of calcium oxide as a heterogeneous catalyst in biodiesel production: a review. Chem Eng J 168(1):15–22. doi:10.1016/j.cej.2011.01.009

Borugadda VB, Goud VV (2012) Biodiesel production from renewable feedstocks: status and opportunities. Renew Sust Energ Rev 16(7):4763–4784. doi:10.1016/j.rser.2012.04.010

Bowman M, Hilligoss D, Rasmussen S, Thomas R (2006) Biodiesel: a renewable and biodegradable fuel. Hydrocarb Process 85(2):103–106

Brahmkhatri V, Patel A (2012) Esterification of lauric acid with butanol-1 over H3PW12O40 supported on MCM-41. Fuel 102:72–77. doi:10.1016/j.fuel.2012.05.053

Bruylants P, Munaut A, Poncelet G, Ladriere J, Meyers J, Fripiat J (1980) Ir and Mossbauer study of iron glycerolates. J Inorg Nucl Chem 42(11):1603–1611. doi:10.1016/0022-1902(80)80324-1

Camacho L, Carvalho LG, Britto PP, Santos RTP, Aranda DAG (2005) Efeito da Natureza e Concentração de Ácidos Homogêneos na Esterificação de Ácidos Graxos. Paper presented at the 3th Congresso Brasileiro de P&D em Petróleo e Gás, Salvador

Canakci M, Van Gerpen J (2001) Biodiesel production from oils and fats with high free fatty acids. T ASAE 44(6):1429–1436

Cao H, Zhang Z, Wu X, Miao X (2013) Direct biodiesel production from wet microalgae biomass of Chlorella pyrenoidosa through in situ transesterification. Biomed Res Int 2013:930686. doi:10.1155/2013/930686

Carey FA (2011) Química Inorgânica, vol 2. 7 th edn. AMGH, Porto Alegre

Carvalho Júnior RM, Vargas JVC, Ramos LP, Marino CEB, Torres JCL (2011) Microalgae biodiesel via in situ methanolysis. J Chem Technol Biot 86(11):1418–1427. doi:10.1002/jctb.2652

Chen CL, Huang CC, Ho KC, Hsiao PX, Wu MS, Chang JS (2015) Biodiesel production from wet microalgae feedstock using sequential wet extraction/transesterification and direct transesterification processes. Bioresour Technol 194:179–186. doi:10.1016/j.biortech.2015.07.021

Cho S, Kim J, Park H-C, Heo E (2015) Incentives for waste cooking oil collection in South Korea: a contingent valuation approach. Resour Conserv Recy 99:63–71. doi:10.1016/j.resconrec.2015.04.003

Cooney M, Young G, Nagle N (2009) Extraction of bio-oils from microalgae. Sep Purif Technol 38(4):291–325. doi:10.1080/15422110903327919

Cordeiro CS, Arizaga GGC, Ramos LP, Wypych F (2008) A new zinc hydroxide nitrate heterogeneous catalyst for the esterification of free fatty acids and the transesterification of vegetable oils. Catal Commun 9(11–12):2140–2143. doi:10.1016/j.catcom.2008.04.015

Cordeiro CS, da Silva FR, Wypych F, Ramos LP (2011) Catalisadores heterogêneos para a produção de monoésteres graxos (biodiesel). Quim Nova 34(3):477–486

Corma A, Garcia H (2003) Lewis acids: from conventional homogeneous to green homogeneous and heterogeneous catalysis. Chem Rev 103(11):4307–4365. doi:10.1021/cr030680z

Cynthia OB, Lee KT (2011) Feasibility of Jatropha oil for biodiesel: economic analysis. Paper presented at the World Renewable Energy Congress, Linköping—Sweden da Cunha ME, Krause LC, Moraes MSA, Faccini CS, Jacques RA, Almeida SR, Rodrigues MRA, Caramão EB (2009) Beef tallow biodiesel produced in a pilot scale. Fuel Process Technol 90 (4):570–575. doi:10.1016/j.fuproc.2009.01.001

da Rós PCM, de Castro HF, Carvalho AKF, Soares CMF, de Moraes FF, Zanin GM (2012) Microwave-assisted enzymatic synthesis of beef tallow biodiesel. J Ind Microbiol Biotechnol 39(4):529–536. doi:10.1007/s10295-011-1059-8

da Silva Lisboa F, da Silva FR, Ramos LP, Wypych F (2013) Zinc monoglycerolate as highly active and reusable catalyst in the methyl transesterification of refined soybean oil. Catal Lett 143(11):1235–1239

Dabdoub MJ, Bronzel JL, Rampin MA (2009) Biodiesel: visão crítica do status atual e perspectivas na academia e na indústria. Quim Nova 32(2):776–792

de Jong MC, Feijt R, Zondervan E, Nijhuis TA, de Haan AB (2009) Reaction kinetics of the esterification of myristic acid with isopropanol and n-propanol using p-toluene sulphonic acid as catalyst. Appl Catal A 365(1):141–147. doi:10.1016/j.apcata.2009.06.009

de Lima AL, Mbengue A, San Gil RA, Ronconi CM, Mota CJ (2014) Synthesis of amine-functionalized mesoporous silica basic catalysts for biodiesel production. Catal Today 226:210–216

de Paiva EJ, Corazza ML, Sierakowski MR, Warna J, Murzin DY, Wypych F, Salmi T (2015a) Influence of two different alcohols in the esterification of fatty acids over layered zinc stearate/palmitate. Bioresour Technol 193:337–344. doi:10.1016/j.biortech.2015.06.079

de Paiva EJM, Sterchele S, Corazza ML, Murzin DY, Wypych F, Salmi T (2015b) Esterification of fatty acids with ethanol over layered zinc laurate and zinc stearate—kinetic modeling. Fuel 153:445–454. doi:10.1016/j.fuel.2015.03.021

Demirbas A (2007) Importance of biodiesel as transportation fuel. Energ Policy 35(9):4661–4670. doi:10.1016/j.enpol.2007.04.003

Di Serio M, Tesser R, Dimiccoli M, Cammarota F, Nastasi M, Santacesaria E (2005) Synthesis of biodiesel via homogeneous Lewis acid catalyst. J Mol Catal A-Chem 239(1–2):111–115. doi:10.1016/j.molcata.2005.05.041

Di Serio M, Tesser R, Pengmei L, Santacesaria E (2008) Heterogeneous catalysts for biodiesel production. Energ Fuel 22(1):207–217. doi:10.1021/ef700250g

dos Santos VC, Bail A, Okada HdO, Ramos LP, Ciuffi KJ, Lima OJ, Nakagaki S (2011) Methanolysis of soybean oil using tungsten-containing heterogeneous catalysts. Energ Fuel 25 (7):2794–2802. doi:10.1021/ef200055j

dos Santos VC, Wilson K, Lee AF, Nakagaki S (2015) Physicochemical properties of WOx/ZrO2 catalysts for palmitic acid esterification. Appl Catal B-Environ 162:75–84. doi:10.1016/j.apcatb.2014.06.036

EIA (2013) International energy outlook. U.S Energy Administration. http://www.eia.gov/forecasts/ieo/. Accessed 26 Aug 2015

Fukuda H, Kondo A, Noda H (2001) Biodiesel fuel production by transesterification of oils. J Biosci Bioeng 92(5):405–416. doi:10.1016/s1389-1723(01)80288-7

Gallo JMR, Pastore HO, Schuchardt U (2008) Study of the effect of the base, the silica and the niobium sources on the [Nb]-MCM-41 synthesized at room temperature. J Non-Cryst Solids 354(15–16):1648–1653. doi:10.1016/j.jnoncrysol.2007.10.010

García-Sancho C, Moreno-Tost R, Mérida-Robles JM, Santamaría-González J, Jiménez-López A, Maireles-Torres P (2011) Niobium-containing MCM-41 silica catalysts for biodiesel production. Appl Catal B-Environ 108–109:161–167. doi:10.1016/j.apcatb.2011.08.025

Gole VL, Gogate PR (2012) Intensification of synthesis of biodiesel from nonedible oils using sonochemical reactors. Ind Eng Chem Res 51(37):11866–11874. doi:10.1021/ie2029442

Gonçalves AL, Pires JCM, Simões M (2013) Green fuel production: processes applied to microalgae. Environ Chem Lett 11(4):315–324. doi:10.1007/s10311-013-0425-3

Gopal KN, Pal A et al (2014) Investigation of emissions and combustion characteristics of a CI engine fueled with waste cooking oil methyl ester and diesel blends. Alexandria Eng J 53 (2):281–287

Granados ML, Poves MDZ, Alonso DM, Mariscal R, Galisteo FC, Moreno-Tost R, Santamaria J, Fierro JLG (2007) Biodiesel from sunflower oil by using activated calcium oxide. Appl Catal B-Environ 73(3–4):317–326. doi:10.1016/j.apcatb.2006.12.017

Guldhe A, Singh B, Mutanda T, Permaul K, Bux F (2015) Advances in synthesis of biodiesel via enzyme catalysis: novel and sustainable approaches. Renew Sust Energ Rev 41:1447–1464. doi:10.1016/j.rser.2014.09.035

Guldhe A, Singh B, Rawat I, Ramluckan K, Bux F (2014) Efficacy of drying and cell disruption techniques on lipid recovery from microalgae for biodiesel production. Fuel 128:46–52. doi:10.1016/j.fuel.2014.02.059

Haas MJ (2005) Improving the economics of biodiesel production through the use of low value lipids as feedstocks: vegetable oil soapstock. Fuel Process Technol 86(10):1087–1096. doi:10.1016/j.fuproc.2004.11.004

Hambley TW, Snow MR (1983) The crystal and molecular-structure of Zinc(Ii) monoglycerolate. Aust J Chem 36(6):1249–1253. doi:10.1071/CH9831249

Hasheminejad M, Tabatabaei M, Mansourpanah Y, Khatami far M, Javani A (2011) Upstream and downstream strategies to economize biodiesel production. Bioresour Technol 102(2):461–468. doi:10.1016/j.biortech.2010.09.094

Hay JF (2014) Soybean as a biodiesel feedstock. Nebraska University, Lincon

Helwani Z, Aziz N, Bakar MZA, Mukhtar H, Kim J, Othman MR (2013) Conversion of Jatropha curcas oil into biodiesel using re-crystallized hydrotalcite. Energ Convers Manage 73:128–134. doi:10.1016/j.enconman.2013.04.004

ICCT (2013) Vegetable oil markets and the EU biofuel. International Council on Clean Transportation

IEA (2014) Worldwide engagement for sustainable energy strategies. vol 2014. International Energy Agency

Iglesias J, Melero JA, Bautista LF, Morales G, Sanchez-Vazquez R (2014) Continuous production of biodiesel from low grade feedstock in presence of Zr-SBA-15: catalyst performance and resistance against deactivation. Catal Today 234:174–181. doi:10.1016/j.cattod.2014.01.004

Imahara H, Minami E, Hari S, Saka S (2008) Thermal stability of biodiesel in supercritical methanol. Fuel 87(1):1–6. doi:10.1016/j.fuel.2007.04.003

Jaeger KE, Dijkstra BW, Reetz MT (1999) Bacterial biocatalysts: molecular biology, three-dimensional structures, and biotechnological applications of lipases. Annu Rev Microbiol 53:315–351. doi:10.1146/annurev.micro.53.1.315

Jain S, Sharma MP (2010) Prospects of biodiesel from Jatropha in India: a review. Renew Sust Energ Rev 14(2):763–771. doi:10.1016/j.rser.2009.10.005

Jayasinghe P, Hawboldt K (2012) A review of bio-oils from waste biomass: focus on fish processing waste. Renew Sust Energ Rev 16(1):798–821. doi:10.1016/j.rser.2011.09.005

Juan JC, Kartika DA, Wu TY, Hin TY (2011) Biodiesel production from Jatropha oil by catalytic and non-catalytic approaches: an overview. Bioresour Technol 102(2):452–460. doi:10.1016/j.biortech.2010.09.093

Kawashima A, Matsubara K, Honda K (2008) Development of heterogeneous base catalysts for biodiesel production. Bioresour Technol 99(9):3439–3443. doi:10.1016/j.biortech.2007.08.009

Kawashima A, Matsubara K, Honda K (2009) Acceleration of catalytic activity of calcium oxide for biodiesel production. Bioresour Technol 100(2):696–700. doi:10.1016/j.biortech.2008.06.049

Kelkar MA, Gogate PR, Pandit AB (2007) Cavitation as a novel tool for process intensification of biodiesel synthesis. Paper presented at the Proceedings of the 6th International Symposium on Catalysis in Multiphase Reactors, Pune, India

Kiss AA (2010) Separative reactors for integrated production of bioethanol and biodiesel. Comput Chem Eng 34(5):812–820. doi:10.1016/j.compchemeng.2009.09.005

Kiss AA (2011) Heat-integrated reactive distillation process for synthesis of fatty esters. Fuel Process Technol 92(7):1288–1296. doi:10.1016/j.fuproc.2011.02.003

Kiss AA, Dimian AC, Rothenberg G (2008) Biodiesel by catalytic reactive distillation powered by metal oxides. Energ Fuel 22(1):598–604. doi:10.1021/ef700265y

Knothe G (2009) Improving biodiesel fuel properties by modifying fatty ester composition. Energ Environ Sci 2(7):759. doi:10.1039/b903941d

Kobayashi J, Mori Y, Kobayashi S (2006) Multiphase organic synthesis in microchannel reactors. Chem Asian J 1(1–2):22–35. doi:10.1002/asia.200600058

Kondamudi N, Strull J, Misra M, Mohapatra SK (2009) A green process for producing biodiesel from feather meal. J Agric Food Chem 57(14):6163–6166. doi:10.1021/jf900140e

Kouzu M, J-S Hidaka (2012) Transesterification of vegetable oil into biodiesel catalyzed by CaO: a review. Fuel 93:1–12. doi:10.1016/j.fuel.2011.09.015

Kouzu M, J-S Hidaka, Komichi Y, Nakano H, Yamamoto M (2009a) A process to transesterify vegetable oil with methanol in the presence of quick lime bit functioning as solid base catalyst. Fuel 88(10):1983–1990. doi:10.1016/j.fuel.2009.03.013

Kouzu M, Hidaka J, Wakabayashi K, Tsunomori M (2010a) Solid base catalysis of calcium glyceroxide for a reaction to convert vegetable oil into its methyl esters. Appl Catal a-Gen 390 (1–2):11–18. doi:10.1016/j.apcata.2010.09.029

Kouzu M, Kasuno T, Tajika M, Sugimoto Y, Yamanaka S, Hidaka J (2008a) Calcium oxide as a solid base catalyst for transesterification of soybean oil and its application to biodiesel production. Fuel 87(12):2798–2806. doi:10.1016/j.fuel.2007.10.019

Kouzu M, Kasuno T, Tajika M, Yamanaka S, Hidaka J (2008b) Active phase of calcium oxide used as solid base catalyst for transesterification of soybean oil with refluxing methanol. Appl Catal a-Gen 334(1–2):357–365. doi:10.1016/j.apcata.2007.10.023

Kouzu M, Tsunomori M, Yamanaka S, Hidaka J (2010b) Solid base catalysis of calcium oxide for a reaction to convert vegetable oil into biodiesel. Adv Powder Technol 21(4):488–494. doi:10.1016/j.apt.2010.04.007

Kouzu M, Yamanaka S, Hidaka J, Tsunomori M (2009b) Heterogeneous catalysis of calcium oxide used for transesterification of soybean oil with refluxing methanol. Appl Catal a-Gen 355 (1–2):94–99. doi:10.1016/j.apcata.2008.12.003

Kulkarni MG, Dalai AK (2006) Waste cooking oil an economical source for biodiesel: a review. Ind Eng Chem Res 45(9):2901–2913. doi:10.1021/ie0510526

Kumar D, Kumar G, Johari R, Kumar P (2012) Fast, easy ethanomethanolysis of Jatropha curcus oil for biodiesel production due to the better solubility of oil with ethanol in reaction mixture assisted by ultrasonication. Ultrason Sonochem 19(4):816–822. doi:10.1016/j.ultsonch.2011.11.004

Kumar D, Kumar G, Poonam Singh CP (2010) Fast, easy ethanolysis of coconut oil for biodiesel production assisted by ultrasonication. Ultrason Sonochem 17(3):555–559. doi:10.1016/j.ultsonch.2009.10.018

Kumari A, Mahapatra P, Garlapati VK, Banerjee R (2009) Enzymatic transesterification of Jatropha oil. Biotechnol Biofuels 2(1):1. doi:10.1186/1754-6834-2-1

Lacouture F, Peultier J, Francois M, Steinmetz J (2000) Anhydrous polymeric zinc(II) octanoate. Acta Crystallogr C 56(Pt 5):556–557. doi:10.1107/S0108270100001876

Lardon L, Helias A, Sialve B, Steyer JP, Bernard O (2009) Life-cycle assessment of biodiesel production from microalgae. Environ Sci Technol 43(17):6475–6481. doi:10.1021/es900705j

Lee DH (2011) Algal biodiesel economy and competition among bio-fuels. Bioresour Technol 102 (1):43–49. doi:10.1016/j.biortech.2010.06.034

León-Reina L, Cabeza A, Rius J, Maireles-Torres P, Alba-Rubio AC, López Granados M (2013) Structural and surface study of calcium glyceroxide, an active phase for biodiesel production under heterogeneous catalysis. J Catal 300:30–36. doi:10.1016/j.jcat.2012.12.016

Lisboa FD, Gardolinski JEFD, Cordeiro CS, Wypych F (2012) Layered metal laurates as active catalysts in the methyl/ethyl esterification reactions of lauric acid. J Brazil Chem Soc 23(1):46–U379. doi:10.1590/s0103-50532012000100008

Lisboa FdS, Silva FRd, Cordeiro CS, Ramos LP, Wypych F (2014) Metal glycerolates as catalysts in the transesterification of refined soybean oil with methanol under reflux conditions. J Brazil Chem Soc. doi:10.5935/0103-5053.20140144

Liu F, Zheng A, Noshadi I, Xiao F-S (2013) Design and synthesis of hydrophobic and stable mesoporous polymeric solid acid with ultra strong acid strength and excellent catalytic activities for biomass transformation. Appl Catal B-Environ 136–137:193–201. doi:10.1016/j.apcatb.2013.01.063

Lotti M, Pleiss J, Valero F, Ferrer P (2015) Effects of methanol on lipases: molecular, kinetic and process issues in the production of biodiesel. Biotechnol J 10(1):22–30. doi:10.1002/biot.201400158

Luo Y, Zheng Y, Jiang Z, Ma Y, Wei D (2006) A novel psychrophilic lipase from Pseudomonas fluorescens with unique property in chiral resolution and biodiesel production via transesterification. Appl Microbiol Biotechnol 73(2):349–355. doi:10.1007/s00253-006-0478-3

Ma F, Hanna MA (1999) Biodiesel production: a review1 journal series #12109, Agricultural Research Division, Institute of Agriculture and Natural Resources, University of Nebraska-Lincoln. 1. Bioresour Technol 70(1):1–15. doi:10.1016/s0960-8524(99)00025-5

Marchetti JM (2010) Biodiesel production technologies, 1st edn. Nova Science Publisher Inc., New York

Marchetti JM (2012) A summary of the available technologies for biodiesel production based on a comparison of different feedstock's properties. J Therm Anal 90(3):157–163. doi:10.1016/j.psep.2011.06.010

Marchetti JM (2013) Influence of economical variables on a supercritical biodiesel production process. Energ Convers Manage 75:658–663. doi:10.1016/j.enconman.2013.07.039

Marchetti JM, Errazu AF (2008) Technoeconomic study of supercritical biodiesel production plant. Energ Convers Manage 49(8):2160–2164. doi:10.1016/j.enconman.2008.02.002

Marchetti JM, Miguel VU, Errazu AF (2007) Possible methods for biodiesel production. Renew Sust Energ Rev 11(6):1300–1311. doi:10.1016/j.rser.2005.08.006

Marulanda VF, Anitescu G, Tavlarides LL (2010) Investigations on supercritical transesterification of chicken fat for biodiesel production from low-cost lipid feedstocks. J Supercrit Fluid 54 (1):53–60. doi:10.1016/j.supflu.2010.04.001

Melero JA, Iglesias J, Morales G (2009) Heterogeneous acid catalysts for biodiesel production: current status and future challenges. Green Chem 11(9):1285–1308. doi:10.1039/b902086a

Mendelovici E, Sagarzazu A, Villalba R (2014) Comparative thermal transformations of synthetic Fe–Mn glycerate (alkoxides). J Therm Anal 40(3):1115–1122. doi:10.1007/bf02546872

Mendelovici E, Villalba R, Sagarzazu A (1990) Synthesis and characteristics of Fe–Mn glycerate (alkoxides), new precursors of ferrite structures. J Mater Sci Lett 9(1):28–31. doi:10.1007/Bf00722859

Milledge JJ, Heaven S (2012) A review of the harvesting of micro-algae for biofuel production. Rev Environ Sci Bio/Technol 12(2):165–178. doi:10.1007/s11157-012-9301-z

Modi MK, Reddy JR, Rao BV, Prasad RB (2007) Lipase-mediated conversion of vegetable oils into biodiesel using ethyl acetate as acyl acceptor. Bioresour Technol 98(6):1260–1264. doi:10.1016/j.biortech.2006.05.006

Modi MK, Reddy JRC, Rao BVSK, Prasad RBN (2006) Lipase-mediated transformation of vegetable oils into biodiesel using propan-2-ol as acyl acceptor. Biotechnol Lett 28(9):637–640. doi:10.1007/s10529-006-0027-2

Mota CJA, Da Silva CXA, Gonçalves VLC (2009) Gliceroquímica: novos produtos e processos a partir da glicerina de produção de biodiesel. Quim Nova 32(3):639–648

Naima K, Liazid A (2013) Waste oils as alternative fuel for diesel engine: a review. J Petroleum Technol Altern Fuels 4(3):30–43

Noureddini H, Gao X, Philkana RS (2005) Immobilized Pseudomonas cepacia lipase for biodiesel fuel production from soybean oil. Bioresour Technol 96(7):769–777. doi:10.1016/j.biortech.2004.05.029

Noureddini H, Harkey D, Medikonduru V (1998) A continuous process for the conversion of vegetable oils into methyl esters of fatty acids. J Am Oil Chem Soc 75(12):1775–1783. doi:10.1007/s11746-998-0331-1

Okuhara T (2002) Water-tolerant solid acid catalysts. Chem Rev 102(10):3641–3666. doi:10.1021/cr0103569

Olivares-Carrillo P, Quesada-Medina J (2012) Thermal decomposition of fatty acid chains during the supercritical methanol transesterification of soybean oil to biodiesel. J Supercrit Fluid 72:52–58. doi:10.1016/j.supflu.2012.08.012
Ong LK, Kurniawan A, Suwandi AC, Lin CX, Zhao XS, Ismadji S (2013) Transesterification of leather tanning waste to biodiesel at supercritical condition: Kinetics and thermodynamics studies. J Supercrit Fluid 75:11–20. doi:10.1016/j.supflu.2012.12.018
Paiva EJ, da Silva ML, Barboza JC, de Oliveira PC, de Castro HF, Giordani DS (2013) Non-edible babassu oil as a new source for energy production-a feasibility transesterification survey assisted by ultrasound. Ultrason Sonochem 20(3):833–838. doi:10.1016/j.ultsonch.2012.11.003
Panwar NL, Kaushik SC, Kothari S (2011) Role of renewable energy sources in environmental protection: a review. Renew Sust Energ Rev 15(3):1513–1524. doi:10.1016/j.rser.2010.11.037
Peterson CL, Cook JL, Thompson JC, Taberski JS (2002) Continuous flow biodiesel production. Appl Eng Agric 18(1):5–11
Pirez C, Lee AF, Jones C, Wilson K (2014) Can surface energy measurements predict the impact of catalyst hydrophobicity upon fatty acid esterification over sulfonic acid functionalised periodic mesoporous organosilicas? Catal Today 234:167–173. doi:10.1016/j.cattod.2014.01.042
Pizarro AVL, Park EY (2003) Lipase-catalyzed production of biodiesel fuel from vegetable oils contained in waste activated bleaching earth. Process Biochem 38(7):1077–1082
Qiu ZY, Zhao LN, Weather L (2010) Process intensification technologies in continuous biodiesel production. Chem Eng Process 49(4):323–330. doi:10.1016/j.cep.2010.03.005
Quesada-Medina J, Olivares-Carrillo P (2011) Evidence of thermal decomposition of fatty acid methyl esters during the synthesis of biodiesel with supercritical methanol. J Supercrit Fluid 56 (1):56–63. doi:10.1016/j.supflu.2010.11.016
Radoslovich EW, Raupach MR, Slade PG, Taylor RM (1970) Crystalline cobalt, zinc, manganese, and iron alkoxides of glycerol. Aust J Chem 23(10):1963. doi:10.1071/ch9701963
Raja SA, Smart DSR, Lee CLR (2011) Biodiesel production from Jatropha oil and its characterization. Res J Chem Sci 1(1):81–87
Ramos TR, Gomes MI, Barbosa-Povoa AP (2013) Planning waste cooking oil collection systems. Waste Manag 33(8):1691–1703. doi:10.1016/j.wasman.2013.04.005
Ramos LP, Brugnago RJ, Silva FR, Cordeiro CS, Wypych F (2015) Simultaneous esterification and transesterification of acid oils using layered zinc carboxylates as bifunctional catalysts. Quim Nova 38(1):46–54. doi:10.5935/0100-4042.20140274
Rathmann R, Szklo A, Schaeffer R (2010) Land use competition for production of food and liquid biofuels: an analysis of the arguments in the current debate. Renew Energ 35(1):14–22. doi:10.1016/j.renene.2009.02.025
Reinoso DM, Damiani DE, Tonetto GM (2014) Zinc glycerolate as a novel heterogeneous catalyst for the synthesis of fatty acid methyl esters. Appl Catal B-Environ 144:308–316. doi:10.1016/j.apcatb.2013.07.026
Reitze AW Jr (1993) Regulation on fuels and fuel additives under section 211 of the clean air act. TulSa lJ 29:485
Reyero I, Arzamendi G, Gandia LM (2014) Heterogenization of the biodiesel synthesis catalysis: CaO and novel calcium compounds as transesterification catalysts. Chem Eng Res Des 92 (8):1519–1530. doi:10.1016/j.cherd.2013.11.017
Rico JAP, Sauer IL (2015) A review of Brazilian biodiesel experiences. Renew Sust Energ Rev 45:513–529. doi:10.1016/j.rser.2015.01.028
Robles-Medina A, Gonzalez-Moreno PA, Esteban-Cerdan L, Molina-Grima E (2009) Biocatalysis: towards ever greener biodiesel production. Biotechnol Adv 27(4):398–408. doi:10.1016/j.biotechadv.2008.10.008
Rodrique L, Delvaux G, Dardenne A (1978) Synthèse à haute température d'un glycérolate de fer au départ d'oxalate ferreux et ferrique. Powder Technol 19(1):93–101. doi:10.1016/0032-5910(78)80076-x

Ros PC, Castro HF, Carvalho AK, Soares CM, Moraes FF, Zanin GM (2012) Microwave-assisted enzymatic synthesis of beef tallow biodiesel. J Ind Microbiol Biotechnol 39(4):529–536. doi:10.1007/s10295-011-1059-8

Royon D, Daz M, Ellenrieder G, Locatelli S (2007) Enzymatic production of biodiesel from cotton seed oil using t-butanol as a solvent. Bioresour Technol 98(3):648–653. doi:10.1016/j.biortech.2006.02.021

Sahoo PK, Das LM (2009) Process optimization for biodiesel production from Jatropha, Karanja and Polanga oils. Fuel 88(9):1588–1594. doi:10.1016/j.fuel.2009.02.016

Sajid Z, Khan F, Zhang Y (2016) Process simulation and life cycle analysis of biodiesel production. Renew Energ 85:945–952. doi:10.1016/j.renene.2015.07.046

Sánchez-Vázquez R, Pirez C, Iglesias J, Wilson K, Lee AF, Melero JA (2013) Zr-containing hybrid organic-inorganic mesoporous materials: hydrophobic acid catalysts for biodiesel production. Chem Cat Chem 5(4):994–1001. doi:10.1002/cctc.201200527

Santos AL, Martins DU, Iha OK, Ribeiro RA, Quirino RL, Suarez PA (2010) Agro-industrial residues as low-price feedstock for diesel-like fuel production by thermal cracking. Bioresour Technol 101(15):6157–6162. doi:10.1016/j.biortech.2010.02.100

Schörken U, Kempers P (2009) Lipid biotechnology: industrially relevant production processes. Eur J Lipid Sci Tech 111(7):627–645. doi:10.1002/ejlt.200900057

Schuchardt U, Sercheli R, Vargas RM (1998) Transesterification of vegetable oils: a review. J Brazil Chem Soc 9(3):199–210

Shah M, Ali S, Tariq M, Khalid N, Ahmad F, Khan MA (2014) Catalytic conversion of jojoba oil into biodiesel by organotin catalysts, spectroscopic and chromatographic characterization. Fuel 118:392–397. doi:10.1016/j.fuel.2013.11.010

Shah S, Sharma S, Gupta MN (2004) Biodiesel preparation by lipase-catalyzed transesterification of Jatropha oil. Energ Fuel 18(1):154–159. doi:10.1021/ef030075z

Singh AK, Fernando SD, Hernandez R (2007) Base-catalyzed fast transesterification of Soybean oil using ultrasonication. Energ Fuel 21(2):1161–1164. doi:10.1021/ef060507g

Solomons TWG, Fryhle CB (2002) Química Orgânica 2, 7th edn. LTC, Rio de Janeiro

Su CH (2013a) Kinetic study of free fatty acid esterification reaction catalyzed by recoverable and reusable hydrochloric acid. Bioresour Technol 130:522–528. doi:10.1016/j.biortech.2012.12.090

Su CH (2013b) Recoverable and reusable hydrochloric acid used as a homogeneous catalyst for biodiesel production. Appl Energ 104:503–509. doi:10.1016/j.apenergy.2012.11.026

Sun H, Ding Y, Duan J, Zhang Q, Wang Z, Lou H, Zheng X (2010) Transesterification of sunflower oil to biodiesel on ZrO2 supported La2O3 catalyst. Bioresour Technol 101(3):953–958. doi:10.1016/j.biortech.2009.08.089

Sun J, Ju J, Ji L, Zhang L, Xu N (2008) Synthesis of biodiesel in capillary microreactors. Ind Eng Chem Res 47(5):1398–1403. doi:10.1021/ie070295q

Taylor RM, Slade PG, Aldous GL, Wilding IR, Siddiqui O, Whitehouse MW (1992) Preparation and properties of a glycerolatocalcium complex. Aust J Chem 45(7):1179–1185. doi:10.1071/CH9921179

Tesser R, Di Serio M, Guida M, Nastasi M, Santacesaria E (2005) Kinetics of oleic acid esterification with methanol in the presence of triglycerides. Ind Eng Chem Res 44(21):7978–7982. doi:10.1021/ie050588o

Thanh LT, Okitsu K, Boi LV, Maeda Y (2012) Catalytic technologies for biodiesel fuel production and utilization of glycerol: a review. Catalysts 2(1):191–222. doi:10.3390/catal2010191

Thompson JC, He BB (2007) Biodiesel production using static mixers. T Asabe 50(1):161–165

Tomei J, Upham P (2009) Argentinean soy-based biodiesel: an introduction to production and impacts. Energ Policy 37(10):3890–3898. doi:10.1016/j.enpol.2009.05.031

Tong D, Hu C, Jiang K, Li Y (2010) Cetane number prediction of biodiesel from the composition of the fatty acid methyl esters. J Am Oil Chem Soc 88(3):415–423. doi:10.1007/s11746-010-1672-0

Tran DN, Balkus Jr KJ (2011) A perspective of recent progress in immobilization of enzymes. ACS Catal 1

Tyner WE (2009) The integration of energy and agricultural markets. In: Proceedings of the 27th international association of agricultural economists conference. Beijing, China, 08/16/2009
UNO (2010) United nations organization. www.un.org/english/. Accessed 26 Aug 2015
Usta N, Aydogan B, Con AH, Uguzdogan E, Ozkal SG (2011) Properties and quality verification of biodiesel produced from tobacco seed oil. Energ Convers Manage 52(5):2031–2039. doi:10.1016/j.enconman.2010.12.021
Veljković VB, Avramović JM, Stamenković OS (2012) Biodiesel production by ultrasound-assisted transesterification: State of the art and the perspectives. Renew Sust Energ Rev 16(2):1193–1209. doi:10.1016/j.rser.2011.11.022
Viola E, Blasi A, Valerio V, Guidi I, Zimbardi F, Braccio G, Giordano G (2012) Biodiesel from fried vegetable oils via transesterification by heterogeneous catalysis. Catal Today 179(1):185–190. doi:10.1016/j.cattod.2011.08.050
Wan Ghazali WNM, Mamat R, Masjuki HH, Najafi G (2015) Effects of biodiesel from different feedstocks on engine performance and emissions: a review. Renew Sust Energ Rev 51:585–602. doi:10.1016/j.rser.2015.06.031
Watanabe Y, Shimada Y, Sugihara A, Tominaga Y (2001) Enzymatic conversion of waste edible oil to biodiesel fuel in a fixed-bed bioreactor. J Am Oil Chem Soc 78(7):703–707. doi:10.1007/s11746-001-0329-5
West AH, Posarac D, Ellis N (2008) Assessment of four biodiesel production processes using HYSYS. Plant. Bioresour Technol 99(14):6587–6601. doi:10.1016/j.biortech.2007.11.046
Wilson K, Hardacre C, Lee AF, Montero JM, Shellard L (2008) The application of calcined natural dolomitic rock as a solid base catalyst in triglyceride transesterification for biodiesel synthesis. Green Chem 10(6):654–659. doi:10.1039/b800455b
Wilson K, Lee AF (2012) Rational design of heterogeneous catalysts for biodiesel synthesis. Cat Sci Tec 2(5):884. doi:10.1039/c2cy20038d
Zhang Y, Dube MA, McLean DD, Kates M (2003) Biodiesel production from waste cooking oil: 2. Economic assessment and sensitivity analysis. Bioresour Technol 90(3):229–240. doi:10.1016/s0960-8524(03)00150-0
Zhang H, Ding J, Qiu Y, Zhao Z (2012) Kinetics of esterification of acidified oil with different alcohols by a cation ion-exchange resin/polyethersulfone hybrid catalytic membrane. Bioresour Technol 112:28–33. doi:10.1016/j.biortech.2012.02.104
Zhang Y, Li Y, Zhang X, Tan T (2015) Biodiesel production by direct transesterification of microalgal biomass with co-solvent. Bioresour Technol. doi:10.1016/j.biortech.2015.07.052
Zhao X, Qi F, Yuan C, Du W, Liu D (2015) Lipase-catalyzed process for biodiesel production: Enzyme immobilization, process simulation and optimization. Renew Sust Energ Rev 44:182–197. doi:10.1016/j.rser.2014.12.021

Biodiesel Production by Hydroesterification: Simulation Studies

Donato Alexandre Gomes Aranda and Guilherme Duenhas Machado

Abstract Due to environmental and economic issues, alternative fuels have been developed as a way to achieve sustainability. Biodiesel is a biofuel substitute for petroleum diesel and has been produced mainly by the transesterification reaction, a process performed in two stages, chemical reaction, and product separation, which has several disadvantages. Biodiesel production considering an alternative method, by hydroesterification from fatty matter has many advantages, but available technology still requires distinct steps for reaction and separation. This chapter presents computational simulations of a multistep process of soybean oil biodiesel production by hydroesterification. The hydrolysis reaction (first step) uses a packed bed reactor and the esterification reaction (second step) utilizes a reactive distillation column, which combines the reaction and separation steps in the same piece of equipment. Several chemical processes employ this intensification technique but its application to esterification processes is recent. The simulation results show that the methodology used is consistent and sensitivity analysis suggest that conversions close to 99 % are possible.

Keywords Biodiesel · Reactive distillation · Hydroesterification · Computer simulation

---

D.A.G. Aranda (✉)
Departamento de Engenharia Química, Escola de Química – Universidade Federal do Rio de Janeiro, Rio de Janeiro, RJ, Brazil
e-mail: donato@eq.ufrj.br

G.D. Machado
Departamento de Engenharia Química – Universidade Tecnológica Federal do Paraná, Londrina, PR, Brazil

C.R. Soccol et al. (eds.), *Green Fuels Technology*,
Green Energy and Technology, DOI 10.1007/978-3-319-30205-8_13

1 Environmental Aspects

Due to economic subjects and the problem of global warming, in the last years an increasingly important has been devoted to renewable energy sources, defined as energy that comes from resources which are naturally replenished on a human timescale (Madras et al. 2004; Valliyappan et al. 2008). Important studies show that biodiesel (fatty acid alkyl esters) is a potential and viable replacement for petroleum diesel fuel (Altin et al. 2001; Ma and Hanna 1990). Started in the 1990s, commercial biodiesel production has spread around the world reaching significant production in Europe, USA, Southeast Asia, and becoming a promising opportunity in Latin America and some African countries.

This biodegradable fuel can be obtained mainly by transesterification reaction, esterification reaction, or by hydroesterification reaction using homogeneous or heterogeneous reaction systems (Knothe et al. 2005; Oliveira et al. 2005). The main route used widely around the world for the industrial production of biodiesel apply the transesterification reaction of refined oils by alkaline catalysts, like potassium and sodium alkoxides, since the process is relatively simple and reaches conversions above 95 % (Fukuda et al. 2001).

2 The Conventional Biodiesel Process

In the transesterification reaction, a triglyceride molecule reacts with three molecules of alcohol to produce three monoester molecules (biodiesel) and a glycerol molecule as coproduct (Fig. 1).

On the other hand, the biodiesel production by transesterification has some drawbacks, essentially in the step of purification, causing waste generation and loss of resources. Feedstock specifications like free fatty acids (FFA) and humidity content in this process must assume minimal values, once are key parameters to accelerate an undesired saponification reaction, which is not desired. The existence of FFA and water content in the reaction bulk favor soap generation, which inhibits

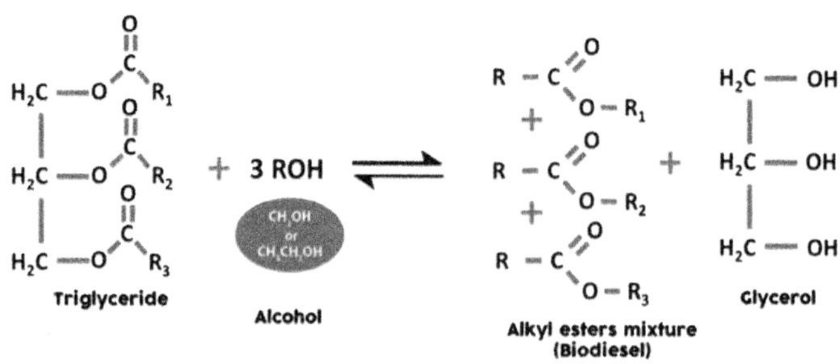

Fig. 1 The transesterification reaction

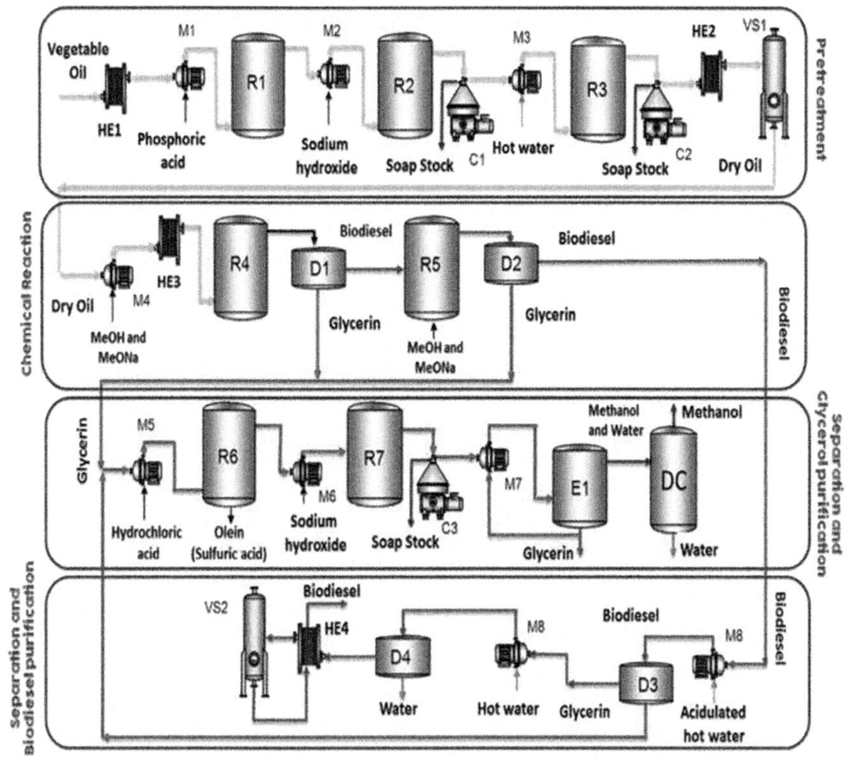

Fig. 2 Detailed flowchart of conventional biodiesel production by transesterification with alkaline homogeneous catalysis—*R* reactor, *TC* heat exchanger, *D* decanter, *M* mixer, *VS* vacuum dryer, *DC* distillation column and *C* centrifuge (Dedini 2015; Crown Iron 2015; Rocha et al. 2010)

the separation of the alkyl esters and glycerol, and contributes to emulsion formation during the washing step (Van Gerpen 2005; Van Gerpen et al. 2004). A conventional flowchart process for biodiesel production can be seen in Fig. 2.

Raw materials with higher FFA content need to be pretreated. Usually, is applied a pre-esterification step with corrosive acid catalysts like sulfuric acid. With this strategy, FFA is reduced and non-desirable soap production decreases in the transesterification step. Nevertheless, besides to chemicals extra costs, sulfate ions can form incrustations that increase operating costs for maintenance. Palm oil biodiesel is a typical case for pre-esterification process since FFA in crude palm oil is usually about 5 %. The same treatment is necessary for other palm crops like "macauba" (*Acrocomia aculeata*) and wet environment oils like algae oils.

Despite these economic and process disadvantages, few technical changes have occurred in the transesterification process in the last decades. In fact, yields higher than 99 % are usually obtained mainly using low acid feedstocks like rapeseed and soybean oils.

3 Alternatives to the Conventional Process

3.1 *Heterogeneous Catalysis*

Others catalysts and chemical reactions can be evaluated as techniques of overcoming the limitations of conventional process by homogeneous transesterification. Regarding the type of catalyst, the use of heterogeneous catalysis can be an alternative to conventional homogeneous catalysis. Production processes that use heterogeneous catalysis have the following benefits (Aranda et al. 2009; Di Serio et al. 2008; Ondrey 2004):

1. Easier separation of products and better removal of the catalyst;
2. High purity of glycerol and biodiesel;
3. Salt-free glycerin;
4. Elimination of the alkaline catalyst neutralization process;
5. Reduced problems with the saponification reaction;

In this context, inorganic heterogenous catalysts, which are insoluble in the reactive medium and so can be separated more easily, can be mentioned (Aransiola et al. 2014). Biodiesel production cost can be potentially reduced bay heterogeneous catalysts use, making it competitive against fossil diesel. It can minimize waste generation, facilitate the purification of monoesters and reuse of the solid catalyst, decrease reaction time, yield higher conversions, and allow the use of low quality raw materials, thereby reducing costs (Vicente et al. 2004).

Still, heterogeneous catalysis processes could require more severe reaction conditions, such as high temperatures, so that the reaction rate can be similar with that of homogeneous processes (Bournay et al. 2005).

3.2 *The Esterification Reaction*

About the type of chemical reaction, the esterification of FFA with short-chain alcohols (i.e., methanol or ethanol) is another way to produce biodiesel. In this process, a previous step of hydrolysis of oils and fats is required as a pretreatment to increase the FFA concentration producing a more complete conversion. Such step increases the variety of raw materials usable for biodiesel production (Warabi et al. 2004). Studies show that this reaction of fatty acid esterification is faster and occurs in a single step, different three stages of the transesterification of triglycerides (Kusdiana and Saka 2001). Industrially, a catalytic process for the heterogeneous esterification of fatty acids was developed and the technology was applied in a 12,000 ton/year biodiesel plant in Belém-PA-Brazil that was started up in April, 2005. An esterification reaction scheme can be found in Fig. 3.

This procedure of esterification reaction, however, still requires distinct steps of chemical reaction and products separation. This process could be even more

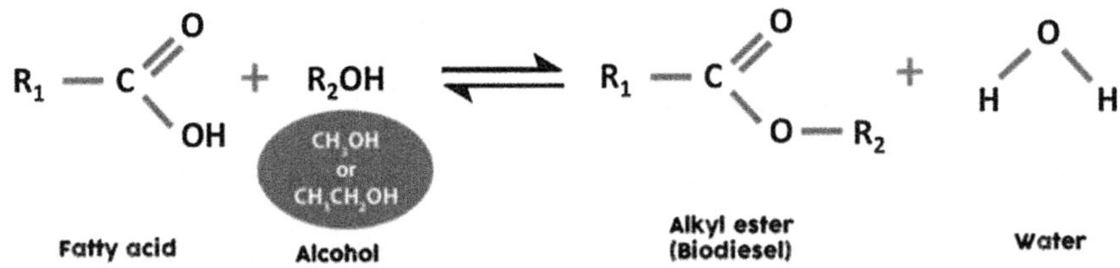

Fig. 3 The esterification reaction

attractive if it could be run in a single integrated step, in which reaction and separation pass in a single device, as in a distillation column with a reactive sector.

3.3 *The Reactive Distillation Column (RDC)*

The reactive distillation is a hybrid operation that combines two of the main tasks in chemical engineering: chemical reaction and physical separation. The first patents for this process emerged in the 1920s. However, little is developed before the 1980s (Chen et al. 2000; Agreda and Partin 1984) when the reactive distillation has gained increased attention as an alternative method that could be used instead of conventional distillation and chemical reaction consecutive process. A sketch of a reactive distillation column (RDC) applied to esterification reaction is seen in Fig. 4.

Fig. 4 Schematic of a reactive distillation column for the esterification reaction

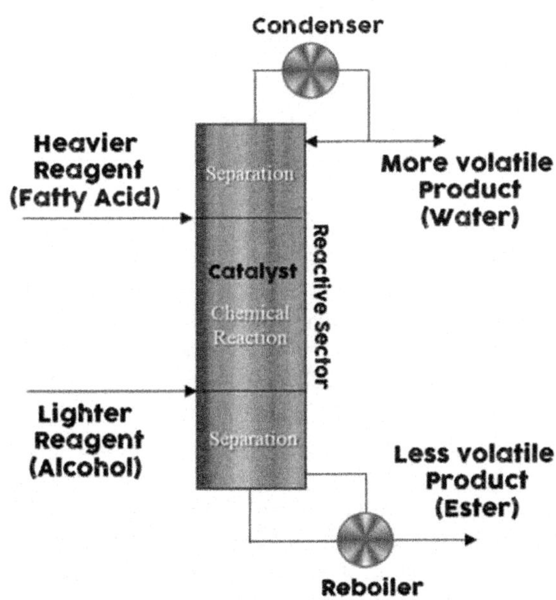

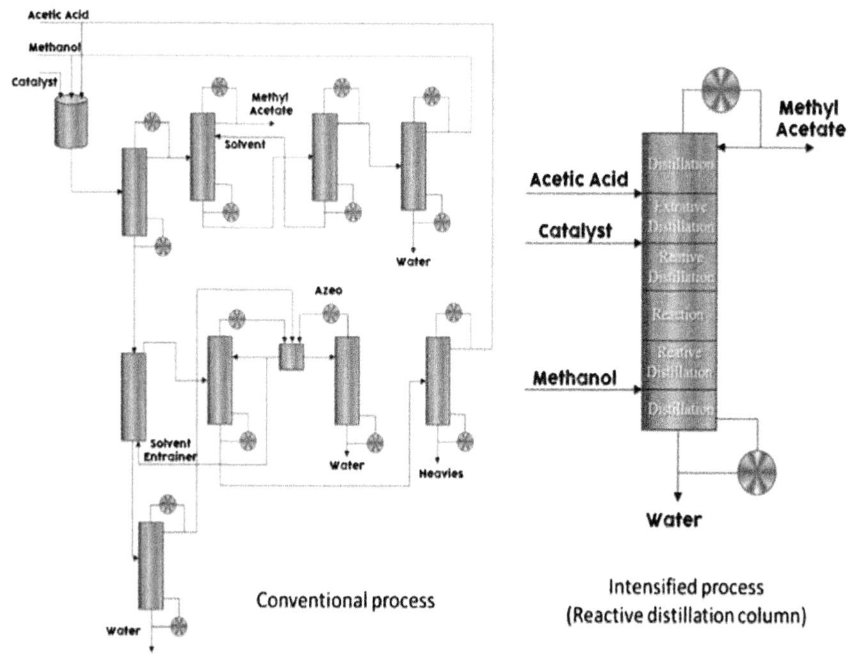

Fig. 5 Methyl acetate synthesis process modification by the reactive distillation technology use (Eastman Chemicals—adapted from Stankiewicz 2003)

Industrial Eastman Chemicals case study process for methyl acetate synthesis is an example of the benefits of this combined process. The processing costs have been substantially reduced (around 80 %) by elimination of units as well as the possibility of heat integration. Using this integration procedure, the conventional process, including 11 different steps and involving 28 equipments was replaced by only a highly integrated RDC. (Stankiewicz and Moulinjin 2002; Krafczyk and Gmehling 1994). Figure 5 shows a schematic of this process intensification improvement.

Besides the industrial application of the methyl acetate production and other esterification reactions cases using reactive distillation, several studies suggest the use of this technology directed to the production of biodiesel. The potential applicability of reactive distillation for biodiesel production has motivated several publications such as those of Gomez-Castro et al. (2011), Hernandez et al. (2010) and especially those of Dimian and Bildea (2008), Dimian et al. (2009), Machado et al. (2011, 2013, 2015), Kiss (2010, 2011, 2012).

RDC can contribute to the search for an advantageous alternative technology for biodiesel production by esterification reaction (Krafczyk and Gmehling 1994; Steinigeweg and Gmehling 2002). However, verifying the fat materials available and currently used in the production of biodiesel, an additional hydrolysis is still required since for the esterification reaction, such fatty acids must be available.

3.4 The Hydroesterification Process

From this necessity, considering alternative types of chemical reactions, hydroesterification—hydrolysis followed by esterification—is, therefore, very promising. It has several advantages, as the ability to occur with to any fatty materials with high acidity (FFA) and water content. In the first stage, all triacylglycerides are converted to fatty acids on the action of water. The hydrolysis reaction is shown in Fig. 6.

In the second stage, the formed fatty acids are esterified using methanol or ethanol. Commercial biodiesel plants successfully use this process (Biobrax, in the Brazilian state of Bahia). The biodiesel and glycerin produced in this process have very high purity when compared with the current method used to produce biodiesel. Niobium oxide (Nb_2O_5) can be used as catalyst for both hydrolysis and esterification steps in biodiesel production (Aranda et al. 2009; Rocha et al. 2010).

Both steps can be accelerated by acid catalysts and become more attractive when coupled to the use of heterogeneous catalysts, eliminating soap formation, reducing the number of separation units, enabling catalyst reuse, and producing high purity, salt-free glycerin (Aranda et al. 2009).

The great advantage of hydroesterification compared to transesterification is that the free fatty acid is reactant in the same reaction, thus not being a limitation in terms of specification of raw material (Lotero et al. 2005; Ma and Hanna 1990; Di Serio et al. 2008). This makes it possible to use raw materials of high acidity, such as macaúba (*A. aculeata*) oil, with difficulties associated to harvesting and fruit processing, and of very high acidity (above 30 mg KOH/g), without the need for pretreatment of the feeding oil via a neutralization reaction. Similar constraints of algal biodiesel include humidity and high acidity of algae oil, as well (Chisti 2013).

On the other hand, hydrolysis processes of raw materials are known and well established, with current operation on industrial scale. Moreover, the esterification step of fatty acids still has a number of technical challenges regarding the mode of operation (batch or continuous). Thus, it opens up opportunity to use RDC

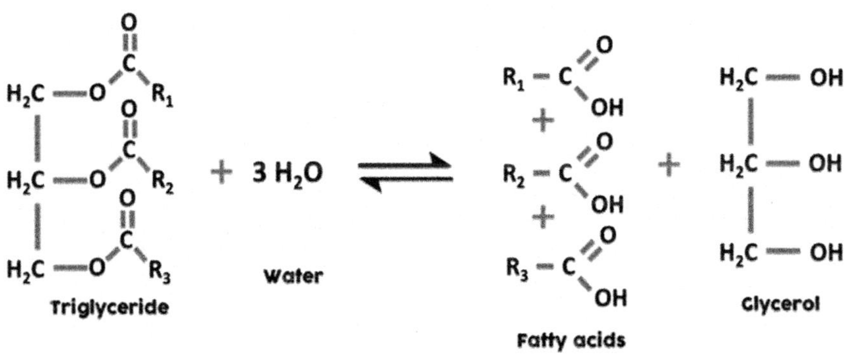

Fig. 6 The hydrolysis reaction

technology in the esterification step. Reactive distillation is a hybrid operation that combines physical separation and chemical reaction in the same piece of equipment (Chen et al. 2000). Thus, the use of RDC increases productivity and selectivity, reduces energy use, eliminates the need for solvents; in other words, it leads to process intensification.

The following will be presented computer simulations of a continuous multistage process for the production of biodiesel by hydroesterification. The hydrolysis reaction is performed in a packed bed reactor (PBR), while the esterification step is conducted in a RDC. In both cases, sensitivity analysis of design parameters and operating conditions are performed in order to optimize the operation of each step.

4 Mathematic Modeling

4.1 Hydrolysis Reaction

The choice of the tubular reactor for the hydrolysis step is based on the fact that this reaction is conducted at high pressure and temperature, which can cause thermal degradation of the components if the residence time in the reactor is large. A mass balance is performed in a differential catalyst mass element dW, for component A, involved in one single reaction over the total mass of catalyst, W. Isothermal conditions and constant pressure along the PBR reactor are considered in this work. The tubular reactor can be described as in Fig. 7.

Based on Fig. 7, a mass balance for component A can be performed by the Eq. 1:

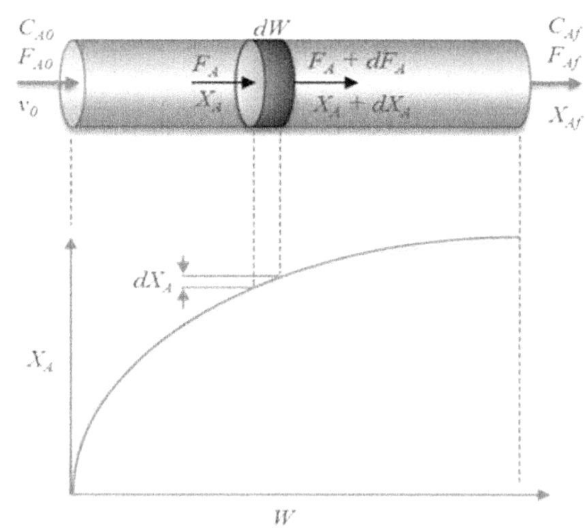

Fig. 7 Schematic of a differential mass element dW in a PBR (adapted from Machado et al. 2013)

$$F_A - (F_A + \mathrm{d}F_A) = r'_A \mathrm{d}W, \tag{1}$$

where F_A is the input molar stream of A, F_A + $\mathrm{d}F_A$ is output molar stream of A and $r'_A \mathrm{d}W$ is the consumption/generation rate of A by reaction.

Equation 2 shows that the mass balance can be rewritten in terms of the conversion of component A, X_A:

$$F_A = F_{A0}(1 - X_A), \tag{2}$$

where F_{A0} is the molar inlet stream of compound A in the PBR reactor and X_A is the conversion of component A. Written in terms of conversion, the differential mass balance for component A is seen in Eq. 3:

$$\frac{\mathrm{d}X_A}{\mathrm{d}\left(\frac{W}{F_{A0}}\right)} = r'_A \tag{3}$$

Its integrated form is the design equation of a tubular reactor, Eq. 4:

$$\frac{W}{F_{A0}} = \int_{o}^{X_A} \frac{\mathrm{d}X_A}{r'_A} \tag{4}$$

In this study, a pseudo-homogeneous model is used to express the chemical reaction rate as a function of reactants concentrations. Equation 5 shows this model, which considers a chemical reaction ($aA + bB \Leftrightarrow cC + dD$) between two reactants forming two products.

$$r'_A = k_1 C_A C_B - k_{-1} C_C C_D, \tag{5}$$

where C_A and C_B are the molar concentrations of components A and B, respectively, C_C and C_D are the molar concentrations of components C and D, respectively, k_1 and k_{-1} are reaction constants of the direct and reverse reactions, given by the Arrhenius Eq. 6:

$$k_i = k_{0i} \exp\left(\frac{-E_{ai}}{RT}\right), \tag{6}$$

where T is the temperature, R is the universal gas constant, k_{0i} is the pre-exponential factor and E_{ai} is the activation energy of reaction i.

The concentration of all species in the reaction can be written as a function of volumetric flow rate (v_0), molar stream of the species (F_{i0}), conversion of component A (X_A) and their stoichiometric coefficients as in Eq. 7:

$$C_A = \frac{F_{A0}}{v_0}(1 - X_A); \quad C_B = \frac{F_{A0}}{v_0}\left(\theta_b - \frac{b}{a}X_A\right);$$
$$C_C = \frac{F_{A0}}{v_0}\left(\theta_c - \frac{c}{a}X_A\right); \quad C_D = \frac{F_{A0}}{v_0}\left(\theta_d - \frac{d}{a}X_A\right), \tag{7}$$

where $\theta_i = \frac{F_{A0}}{F_{i0}}$.

The design equation of PBR, for the particular reaction considered, is obtained by combining Eqs. 5–7 into Eq. 4, resulting in Eq. 8

$$W = \frac{v_0^2}{F_{A0}} \int_o^{X_A} \frac{dX_A}{k_{01} \exp\left(\frac{-E_{a1}}{RT}\right)(1 - X_A)\left(\theta_B - \frac{b}{a}X_A\right) - k_{0-1} \exp\left(\frac{-E_{a-1}}{RT}\right)\left(\theta_C - \frac{c}{a}X_A\right)\left(\theta_D - \frac{d}{a}X_A\right)} \tag{8}$$

In order to obtain the catalyst mass, Eq. 8 is numerically integrated, resulting in a composition profile of each component throughout the equipment by specifying the conversion of component A, X_A, and the reactor feed flow, for known kinetic constants in function of temperature. The integration step (dX_A) used was 0.01.

4.2 Esterification Reaction

As seen above, the hydrolysis reaction is conducted in a reactor PBR. In esterification reaction, a RDC takes place. The modeling of this step is based on concerns of Chen et al. (2000), Alfradique and Castier (2005), Machado et al. (2011, 2013, 2015) for the simulation of RDC.

The methodology takes some considerations into account, which are detailed below. The heat of reaction is considered insignificant when compared to the value of heat of vaporization in the energy balance. Heat transfer occurs in the reboiler (bottom) and in the condenser (top) but the interior stages of the column are assumed adiabatic. The liquid phase governs the chemical reactions occurrence and is controlled by chemical kinetics. Thermodynamic modeling does not consider the possibility of vapor–liquid–liquid equilibrium (VLLE) formation. The VLLE occurrence has a vigorous dependence on the temperature and, and in the cases in this study considerably high temperatures are used to prevent the formation of two liquid phases. Besides, the kinetic models used in the validation case (Steinigeweg and Gmehling 2003) and the case proposed (Rocha et al. 2010) are obtained under temperatures ranging from 150 to 200 °C and, in this range, the occurrence of two different liquid phases was not observed.

We consider pseudo-homogeneous kinetics that does not account for the influence of adsorption as a possible limiting step to reaction rate. Each stage is considered as a continuous stirred-tank reactor (CSTR). The stream of liquid and vapor leaving the stages are in phase equilibrium, the vapor phase has ideal gas behavior,

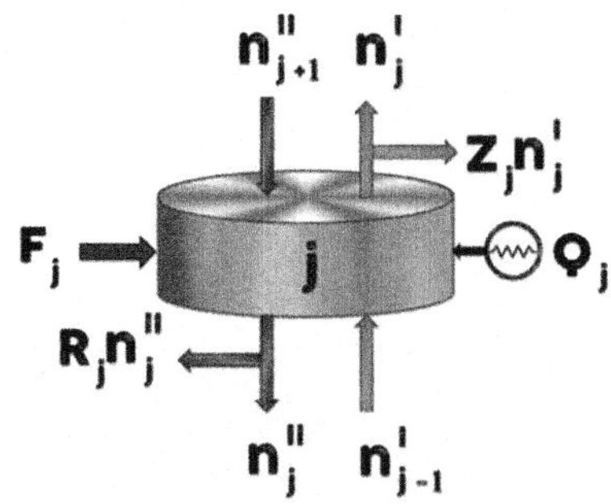

Fig. 8 Theoretical stage j configuration for a reactive distillation column (RDC) (adapted from Alfradique and Castier 2005)

and the liquid phase is considered as a nonideal solution. Models of excess Gibbs free energy describe the liquid phase behavior. The generic plate scheme adopted is shown in Fig. 8.

For this step, the model equations for the distillation column are formulated as follows. Equation 9 shows the mass balance of i component in a theoretical stage j as a residual function, $f_{i,j}^{m}$.

$$f_{i,j}^{m} = (R_j + 1)n_{i,j}^{\mathrm{II}} + (Z_j + 1)n_{i,j}^{\mathrm{I}} - (n_{i,j+1}^{\mathrm{II}} + n_{i,j-1}^{\mathrm{I}} + F_{i,j} + \sum_{k=1}^{nr} \nu_{i,k}\xi_{k,j}) = 0, \quad (9)$$

where the molar flow of the compound i in the liquid leaving the stage j is $(R_j + 1)n_{i,j}^{\mathrm{II}}$, $R_j n_{i,j}^{\mathrm{II}}$, is the flow rate of the compound i in the liquid side stream, $n_{i,j}^{\mathrm{II}}$ is the flow rate of the i compound in the liquid getting the next stage. $(Z_j + 1)n_{i,j}^{\mathrm{I}}$ is the molar flow of the compound i in the vapor exiting the stage j, $Z_j n_{i,j}^{\mathrm{I}}$ is the molar flow of the compound i in the vapor side stream, and $n_{i,j}^{\mathrm{I}}$ is the molar flow of the compound i in the vapor reaching the subsequent stage. In same equation, $\nu_{i,k}$ is the stoichiometric coefficient of component i in reaction k, $\xi_{k,j}$ is the extent of reaction and nr denotes the number of independent chemical reactions.

Considering that the upward (vapor) and downward (liquid) streams leaving each stage are in equilibrium, the equations describing this condition are defined by another residue function, $f_{i,j}^{\mathrm{eq}}$, based on the isofugacity condition:

$$f_{i,j}^{\mathrm{eq}} = \ln\left(x_{i,j}^{\mathrm{I}} P_j\right) - \ln\left(x_{i,j}^{\mathrm{II}} \gamma_{i,j}^{\mathrm{II}} P_{i,j}^{\mathrm{sat}}\right) = 0 \quad (10)$$

In Eq. 10, the fugacity coefficients of each component in the vapor phase are assumed to be equal to 1, and the Poynting factor correction is neglected. $x_{i,j}^{\mathrm{I}}$ and $x_{i,j}^{\mathrm{II}}$ are the mole fractions of compound i in the vapor and liquid streams, respectively, leaving the stage j, P_j is the pressure of stage j, $P_{i,j}^{\mathrm{sat}}$ is the vapor pressure of component i in stage j, $\gamma_{i,j}^{\mathrm{II}}$ is the activity coefficient of component i in the liquid stream leaving the stage j.

The reaction rate expression is similarly expressed by a residue ($f^r_{k,j}$) function given by Eq. 11:

$$f^r_{k,j} = \ln k_{k,j} + \sum_i^{nc} \alpha_{i,k} \ln\left(\frac{X^{\mathrm{II}}_{i,j}}{X^{\mathrm{II}}_j}\right) - \ln \xi_{k,j} = 0, \quad (11)$$

where V^{II}_j represents the molar volume of an ideal liquid solution in stage j, kinetic constant of reaction k in stage j is $k_{k,j}$ and the kinetic order of component i in reaction k is $\alpha_{i,k}$. For kinetic data correlated as function of activities, as applied in the validation case, Eq. 12 is generated

$$f^r_{k,j} = \ln k_{k,j} + \sum_i^{nc} \alpha_{i.k} \ln\left(x^{\mathrm{II}}_{i,j}\gamma^{\mathrm{II}}_{i,j}\right) - \ln \xi_{k,j} = 0 \quad (12)$$

Equation 13 shows the residue function for the energy balance, expressed by f^h_j

$$f^h_j = (R_j + 1)H^{\mathrm{II}}_j + (Z_j + 1)H^{\mathrm{I}}_j - \left(H^{\mathrm{II}}_{j+1} + H^{\mathrm{I}}_{j-1} + H_{F_j} + Q_j\right) = 0 \quad (13)$$

being $(Z_j + 1)H^{\mathrm{I}}_j$ and $(R_j + 1)H^{\mathrm{II}}_j$ are the total enthalpies of the liquid and vapor streams leaving the stage j, H^{I}_{j-1} and H^{II}_{j+1} are the total enthalpies of the liquid and vapor streams entering stage j, total feed stream enthalpy is given by H_{F_j} and Q_j is the heat added or removed of a theoretical stage j.

To specify the condenser and reboiler behavior an extra equation is used, variable E_j, which is set by the ratio between the total molar streams of vapor and liquid leaving a given stage of the reactive distillation column (stage j). a The residue function for this E_j definition, f^{vl}_j, can be written as detailed by Eq. 14

$$f^{vl}_j = (Z_j + 1)\sum_{i=1}^{nc} n^{\mathrm{I}}_{i,j} - E_j(R_j + 1)\sum_{i=1}^{nc} n^{\mathrm{II}}_{i,j} = 0 \quad (14)$$

Values of $E_1 \neq 0$ and $E_N = 0$ are used in the extremes of the column (bottom and top, respectively). Thus, total condenser and partial reboiler are considered in all cases. The value of E_j is calculated for the internal stages of the column (from 2 to $N - 1$).

Additional details about the solution method are available in Alfradique and Castier (2005) study.

4.2.1 Thermodynamic Modeling and Computational Implementation

The nonideal liquid phase behavior is described by the UNIFAC Dortmund excess Gibbs free energy model (Gmehling et al. 1993). For the proposed case we took the weighted average of the three components (fatty acids or fatty acid esters) that constitute the pseudo-components to compute the matrix of subgroups in the UNIFAC method.

The liquid (h^L) and vapor (h^V) molar enthalpies were calculated using the Eqs. 15 and 16, respectively

$$h^L = \sum_{i=1}^{nc} x_i \int_{T_{\text{ref}}}^{T} c_{p,i}^L \mathrm{d}T + h^E \tag{15}$$

$$h^V = \sum_{i=1}^{nc} y_i \left(\Delta h_i^{\text{vap}} + \int_{T_{\text{ref}}}^{T} c_{p,i}^L \mathrm{d}T \right), \tag{16}$$

where the molar enthalpy of vaporization of component i in the system is Δh_i^{vap}, h^E is the molar excess enthalpy and $c_{p,i}^L$ the molar specific heat of component i in the liquid phase. The temperature of reference (T_{ref}) considered was 298.15 K.

In the proposed case, the pseudo-component vapor pressure was not evaluated by the weighted average of the parameters A, B, and C of the Antoine equation. A pseudo-component data of vapor pressure at certain temperatures (between 340 and 500 K) was made, using the Eq. 17:

$$P_i^{\text{sat}} = \sum_{1}^{3} f_i \cdot P_i = f_1 \cdot P_1 + f_2 \cdot f_3 \cdot P_3, \tag{17}$$

where f_i is the mole fraction of component i in the pseudo-component and P_i is the vapor pressure of component i calculated at a given temperature. So, the data generated were used to obtain the Antoine parameters of the pseudo-components, soybean oil and hydrolyzed soybean oil (HSO).

Here, HSO is defined as a pseudo-component that represents a mixture of the following fatty acids with their respective molar compositions: 58.2 % of linoleic acid, 25.1 % of oleic acid and 16.7 % of palmitic acid. In the same way, biodiesel is defined as a pseudo-component that represents a mixture of fatty acid esters with the following composition: 58.2 % of ethyl linoleate, 25.1 % ethyl oleate and 16.7 % of ethyl palmitate.

The Clausius–Clapeyron equation was used to estimate the molar enthalpy of vaporization, Eq. 18

$$\Delta h_i^{\text{vap}} = RT^2 \frac{\mathrm{d} \ln P^{\text{sat}}}{\mathrm{d}T} \tag{18}$$

The mathematical formulation was implemented using the programming language FORTRAN 77. Residue functions expressions, the Jacobian matrix and the activity coefficient models were generated using the Thermath program (Castier

1999). The technique applied is analogous to that used by Alfradique and Castier (2005), Machado et al. (2011, 2013, 2015). Property model parameters from National Institute of Standards and Technology (NIST) and Design Institute for Physical Properties (DIPPR) Project 801 databases for pure compounds were used in the simulations (DIPPR 2013).

4.2.2 Reaction Conditions and Alcohol Used

The kinetic data used took into account the conditions under which they were obtained experimentally. Considering the validation case, temperature ranges were the same adopted by authors (Steinigeweg and Gmehling 2003). For the proposed case, the hydrolysis reaction was conducted experimentally at 260 °C while the esterification reaction under 200 °C (Rocha et al. 2010).

Thus, we consider such temperatures in the first step (hydrolysis reaction) and in the second step (esterification reaction) taking into account the reactive sector of a RDC with an average temperature in very close to the value of the experimental data.

About the alcohol used, in validated case methanol was used in simulations according the literature. On the other hand, the choice of ethanol in the transesterification reaction has shown disadvantages in that it has higher amount of humidity as compared to methanol. Such water composition can accelerate unwanted reactions in the conventional process.

For proposed case, ethanol was chosen in the hydroesterification process such a disadvantage does not occur, since water is produced therein. On the other hand, taking into account the clean biodiesel proposal also thought up the use of a renewable reagent such as vegetable oil.

5 Results

5.1 Hydrolysis Reaction

The hydrolysis of soybean oil with water, producing HSO and glycerol takes place according to the following stoichiometry, relationship shown in the Eq. 19:

$$\text{Soybean Oil}\,(1) + 3.\text{Water}(2) \rightleftharpoons 3.\,\text{HSO}(3) + \text{Glycerol}(4) \qquad (19)$$

Due to the difficulty in representing the main components that comprise the soybean oil mixture and the lack of experimental data for these molecules, the physical properties of soybean oil properties are assumed to be identical to those of triolein in this hydrolysis step.

As said before, the molar composition of fatty acids in the HSO is 58.2 % linoleic acid, 25.1 % oleic acid, and 16.7 % palmitic acid. To account for that, the

Table 1 Specification for the packed bed reactor (PBR) and hydrolysis reaction

Variables	Specifications	
Pressure	Constant	56 bar
Temperature	Constant	260 °C
	v_0	0.13 m^3/min
Feed	Molar ratio (water/soybean oil)	5.7
	F_0	30.0 mol/min
Arrhenius equation	Pre-exponential factor	156.90 (direct)/2.42 (reverse)
Parameters	Activation energy	−40,421.9 J/mol

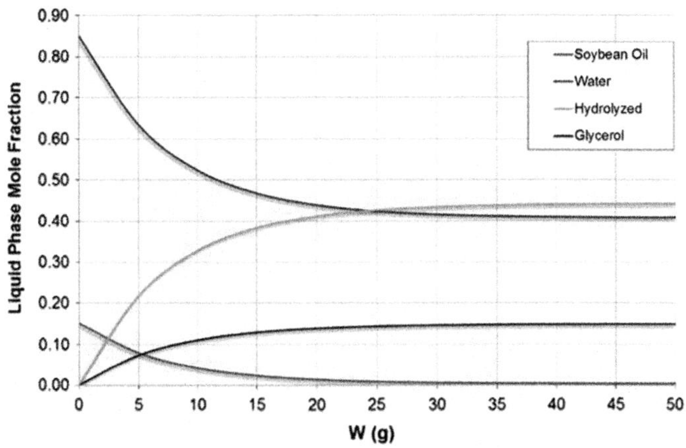

Fig. 9 Liquid phase mole fraction of components to hydrolysis reaction at PBR output (adapted from Machado et al. 2015)

HSO is modeled as a pseudo-component that represents a mixture with that composition.

The hydrolysis reaction is considered reversible. The rate model used is the pseudo-homogeneous with dependence of the molar concentrations of the reactants in the same way of Eqs. 5 and 6.

The kinetic data of the hydrolysis reaction are from Rocha et al. (2010) who used niobium oxide (Ni_2O_5) as catalyst. The specifications the process and the reaction rate constants for this catalyst are given in Table 1.

Figure 9 displays the liquid phase mole fractions along the simulated PBR. The soybean oil fed to the reactor is almost completely consumed with 50 g mass catalyst, while excess water in the feed stream leaves the reactor with 40 % molar composition in the output stream. This excess humidity can be removed by flashing in subsequent operations.

The HSO and glycerol generated in the chemical reaction form two phases that are essentially immiscible at mild temperatures. Thus, these products can be separated by decantation or also by centrifugation.

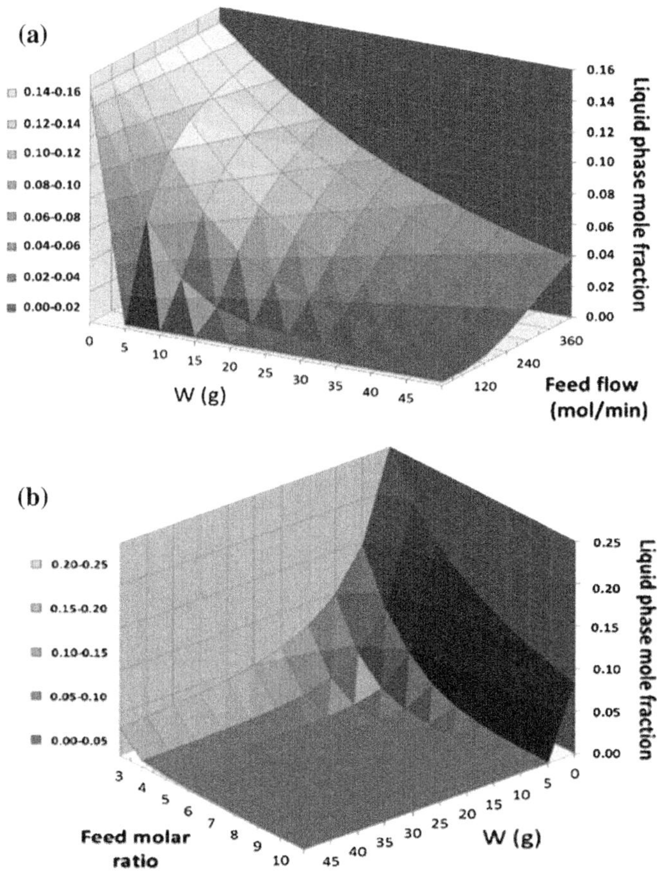

Fig. 10 Surface composition of soybean oil (1) in the liquid phase as a function of feed flow rate by the mass of catalyst (*W*) (**a**) and surface composition as a function of feed molar ratio (**b**)

Figure 10 shows the liquid phase mole composition surface of soybean oil at reactor output for several values of catalyst mass (from 0 to 50 g) as function of the feed flow (a) and feed molar ratio of water/soybean oil (b), respectively.

From Fig. 10a, the minimum flow provides the best conversion results. From 100 mol/min, there is a sharp drop of the soybean oil conversion with increasing feed flow rate. From the feeding molar ratio, Fig. 10b, the smaller value of this variable represents to the process a lower processing cost in terms of raw materials. The stoichiometry of the hydrolysis reaction analyzed between water and soybean oil is 3:1. For a value twice of this ratio, the conversion curve has almost stable value. As a way to avoid costly excess water recirculation, it was decided to consider a molar ratio close to 6:1, more specifically 5.7. Industrial conditions of

SGS (Ponta Grossa/PR, Brazil) reported 16:1 molar ratio which, in this work, generated above the 99.6 % conversion.

5.2 *Esterification Reaction*

5.2.1 Validation Case

Esterification of Decanoic Acid with Methanol

The methodology used in the esterification step using RDC was validated in this case. The experimental data obtained from experimental studies of Steinigeweg and Gmehling (2003) will be used to validate the mathematical modeling applied here.

The esterification of decanoic acid (1) with methanol (2) generating methyl decanoate (3) and water (4) is set by a stoichiometric association relationship better shown in the Eq. 20

$$\underset{(1)}{C_9H_{19}COOH} + \underset{(2)}{CH_3OH} \Leftrightarrow \underset{(3)}{C_9H_{19}COOCH_3} + \underset{(4)}{H_2O} \tag{20}$$

The chemical reaction of esterification is considered to be of first order with respect to decanoic acid and methanol. The inverse reaction (hydrolysis) is considered to be of first order with respect to methyl decanoate and water. These assumptions are the same employed by Steinigeweg and Gmehling (2003) to develop a pseudo-homogeneous reaction rate model dependent on the activity of reagents, as seen in Eq. 21

$$r_1' = k_1 \,.\, a_1 \,.\, a_2 - k_{-1} \,.\, a_3 \,.\, a_4 \tag{21}$$

The catalyst used was a strongly acid ion exchange resin commercially called Amberlyst 15.

The simulated column had 20 stages (reboiler, 18 adiabatic plates, and condenser). The specifications of the feed are presented in Table 2.

Table 3 shows the results obtained in the simulations of this work.

Figure 11 shows the mole fraction profiles in the liquid phase. In general, the profiles obtained in this work show the same tendency of the experimental and simulation data available. The largest deviations occur between stages 2 and 15. Figure 12 shows the temperature and the extents of the esterification (direct) and hydrolysis (reverse) reactions along the column.

In the extremes (top and bottom of the column), the results are in excellent agreement with the experimental values. More pronounced deviations occur in the intermediate stages. These differences between the simulation results of this study and of the literature can be attributed to some modeling issues. Here, Eq. (21), a pseudo-homogeneous model, is used to model the reaction rate, while the cited

Table 2 Specifications of the reactive distillation column for validation process

Variables	Specifications	Value/location
Pressure	All stages	1.0132 bar
Stages	n	20
Condenser	Total	Stage 20
Reboiler	Partial	Stage 1
Reflux ratio	Condenser	0.5
Reactive zone		Stages 7–14
Catalyst	Katapak-SP packing filled with Amberlyst 15 resin	189.6 g
Feed 1	0.25 mol/min	
	1.0132 bar, 331.19 K	Stage 14
	Decanoic acid	
Feed 2	0.483 mol/min	
	1.0132 bar, 337.65 K	Stage 06
	Methanol	
Reaction equation	Pre-exponential factor	$9.1164.10^5$ (direct)/ $1.4998.10^4$ (reverse)
Parameters	Activation energy	−68,710 J/mol

Table 3 Comparison between simulation results and data from literature

Column location/properties		Steinigeweg and Gmehling (2003)	Experimental	Machado et al. (2011)
Liquid phase mole fraction				
Column top	1	0	0	0
	2	0.716	0.76	0.763
	3	0	0	0
	4	0.277	0.24	0.237
Column bottom	1	0.366	0.428	0.511
	2	0.303	0.22	0.103
	3	0.303	0.332	0.386
	4	0.001	0	0
Temperature (K)				
	Stage 1	363.53	–	363.49
	Stage 11	347.28	–	338.83
	Stage 20	341.27	–	341.39
Conversion	(% decanoic acid)	42.99	–	42.99

literature results are based on a heterogeneous model that considers adsorption as a limiting step. Figure 12 shows that the esterification reaction is favored close to the feed location of fatty acid. This region has the highest temperature of the reactive zone.

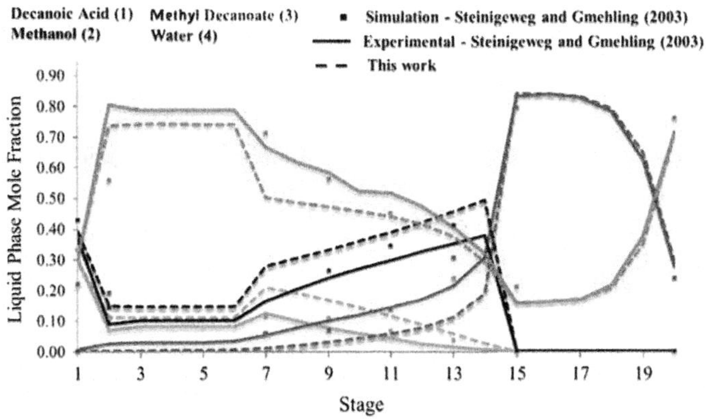

Fig. 11 Comparison of liquid phase composition along the reactive distillation column of validation case with literature (adapted from Machado et al. 2011)

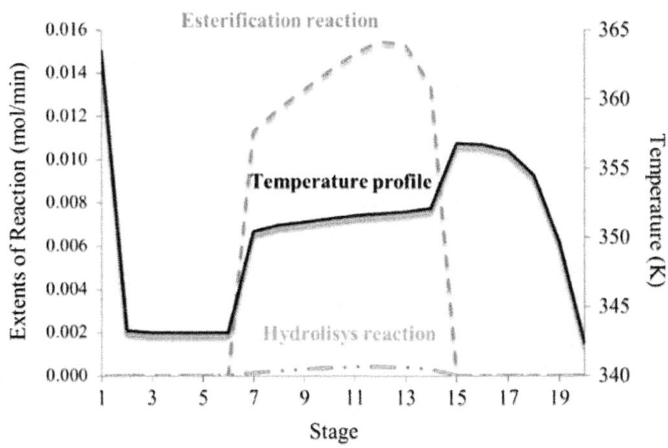

Fig. 12 Temperature profile and reaction rate along the reactive distillation column of validation case

The good agreement between the simulation results and the literature data suggests that the methodology adopted here is valid.

5.2.2 Proposed Case

Esterification of Hydrolyzed Soybean Oil (HSO) with Ethanol

The esterification of HSO with ethanol forming a mixture of esters of biodiesel and water takes place according to stoichiometry relationship expressed in Eq. 22

$$\text{Hydrolyzed (1)} + \text{Ethanol (2)} \rightleftharpoons \text{Biodiesel (3)} + \text{Water (4)} \tag{22}$$

Here, "biodiesel" refers to the group of esters formed regardless of purity specifications for commercialization. In this work, as in the case of HSO, biodiesel is considered a pseudo-component that represents a mixture of fatty acid esters with the following molar composition: 58.2 % of ethyl linoleate, 25.1 % ethyl oleate and 16.7 % of ethyl palmitate. We follow the same considerations used by Machado et al. (2013, 2015) to calculate the thermodynamic properties of this pseudo-component.

The Antoine equation was used to calculate the vapor pressure. In the simulations performed here, Antoine equation (Hala et al. 1984) was used to calculate the vapor pressure, as displayed in Eq. 23:

$$\ln P^{sat}(\text{bar}) = A - \frac{B}{T(\text{K}) + C} \tag{23}$$

To obtain, the A, B, and C parameters of HSO and biodiesel, we generated pseudo vapor pressure data at selected temperatures using the Eq. 24.

$$P_i^{sat} = \sum_{1}^{nc} f_i \cdot P_i \tag{24}$$

where f_i is the mole fraction of component i in the pseudo-component, P_i is the vapor pressure of component i calculated at a given temperature, and nc is the number of components considered in each pseudo-component. In these calculations, we used temperatures between 340 and 500 K. The vapor pressure of each pure component (P_i) was calculated using an extended Antoine equation, as noted in Eq. 25:

$$\ln P^{sat}(\text{Pa}) = A + \frac{B}{T(\text{K})} + C \cdot \ln[T(\text{K})] + D \cdot T^E(\text{K}) \tag{25}$$

Table 4 presents the parameters used in Eq. 25 for each pure component considered in the pseudo-components (HSO and biodiesel). Those parameters for pure components were obtained from National Institute of Standards and Technology (NIST) database (NIST 2008) and Design Institute for Physical Properties (DIPPR) database (DIPPR 2013).

Table 5 presents the Antoine parameters obtained for the HSO and biodiesel. It also presents parameters used for water and ethanol. The parameters for ethanol and water can be used for temperature ranges of (364.80–513.91) and (379.00–573.00) K, respectively.

The molar volume and heat capacity in liquid phase of HSO and biodiesel were also calculated by the weighted average of the properties of the pure components considered in each pseudo-component. Table 5 shows the values of these properties

Table 4 Parameters for pure components used in the extended Antoine equation

	Pure component	Extended Antoine equation parameters					*T* (K) range
		A	*B*	*C*	*D*	*E*	
HSO	Linoleic acid	141.470	−18,229.000	−15.692	2.78E−18	6	387.5–581.25
	Oleic acid	162.450	−19,053.000	−18.997	4.27E−06	2	286.53–550.00
	Palmitic acid	222.960	−23,415.000	−27.085	7.02E−18	6	335.66–624.15
Biodiesel	Ethyl linoleate	106.867	−15,510.000	−12.778	2.50E−18	6	218.20–777.8
	Ethyl oleate	106.085	−14,392.300	−11.137	2.44E−18	6	253.67–772.10
	Ethyl palmitate	118.986	−15,352.600	−12.820	1.19E−18	6	297.15–759.40

Table 5 Physical parameters and Antoine equation parameters used in the simulations of this work (Machado et al. 2015)

	Liquid molar volume (cm^3/mol)	Cp liquid [J/(mol K)]	Antoine equation parameters		
			A	*B*	*C*
HSO	312.025	705.853	11.293	5783.221	−122.287
Biodiesel	347.508	801.960	11.390	6504.020	−78.000
Ethanol	40.58	124.8	11.340	3299.110	−61.820
Water	18.07	78.11	8.200	1482.550	−198.040

used in the simulations. The value in the liquid phase of the heat capacity and molar volume were obtained for the temperatures of 450 and 298 K, respectively, from the National Institute of Standards and Technology (NIST) (NIST 2015) and Design Institute for Physical Properties (DIPPR) databases (DIPPR 2013).

The esterification reaction is considered to be of first order with respect to concentrations of HSO and ethanol, while the inverse reaction (hydrolysis) follows a first order kinetic with respect to the concentrations of biodiesel and water. From these considerations, we propose the following pseudo-homogeneous model as seem in Eq. 26.

$$r_1' = k_1 \,.\, C_1 \,.\, C_2 - k_{-1} \,.\, C_3 \,.\, C_4 \tag{26}$$

The rate constants of that equation were fitted from the experimental kinetic data of Rocha et al. (2010) who studied esterification of HSO using the same niobium oxide catalyst employed for the soybean oil hydrolysis.

The simulated column has 20 stages: 1 partial reboiler, 18 adiabatic plates (12 of them reactive), and 1 total condenser. All specifications of the RDC are shown in Table 6.

Figure 13 shows the liquid phase mole fraction profile to the components of the esterification reaction along the RDC.

From this figure, the desired product, biodiesel, is removed from the bottom of the column with considerable purity (95.89 %). Excluding the amount of ethanol in the bottom stream (which may be separated by a flash process), the purity of the biodiesel produced reaches 99 %. This is due to the use of high heat duty on the reboiler. According to Machado et al. (2013), the reboiler temperature can exceed

Table 6 Specifications for reactive distillation column (RDC) and esterification reaction

Variables	Specifications	Value/location
Pressure	All stages	1.0132 bar
Stage numbers		20
Condenser	Total	Stage 20
Reboiler	Partial	Stage 1
Reflux ratio	Condenser	0.001
	Reboiler	0.17
Reactive sector	Niobium oxide (Nb_2O_5)	Stages 6–17
Catalyst		500 g per stage
Feed	1.0132 bar	
1 (HSO)	13.19 mol/min	Stage 18
	480.15 K	
2 (Ethanol)	15.76 mol/min	Stage 6
	351.15 K	
Reaction Arrhenius	Pre-exponential factor	16.13 (direct)/0.72 (reverse)
Parameters	Activation energy	−24,117 J/mol

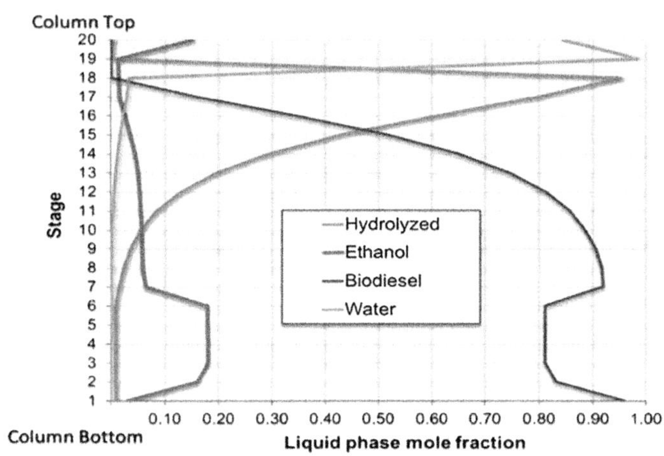

Fig. 13 Liquid phase composition along the reactive distillation column for proposed case (adapted from Machado et al. 2015)

260 °C. This high temperature could degrade the biodiesel formed and removed in the bottom stream.

Reboiler Heat Duty Sensitivity Analysis

In order to obtain the optimum operating parameters, the influence of the heat duty on the reboiler was studied. Figure 14a displays a temperature surface and Fig. 14b

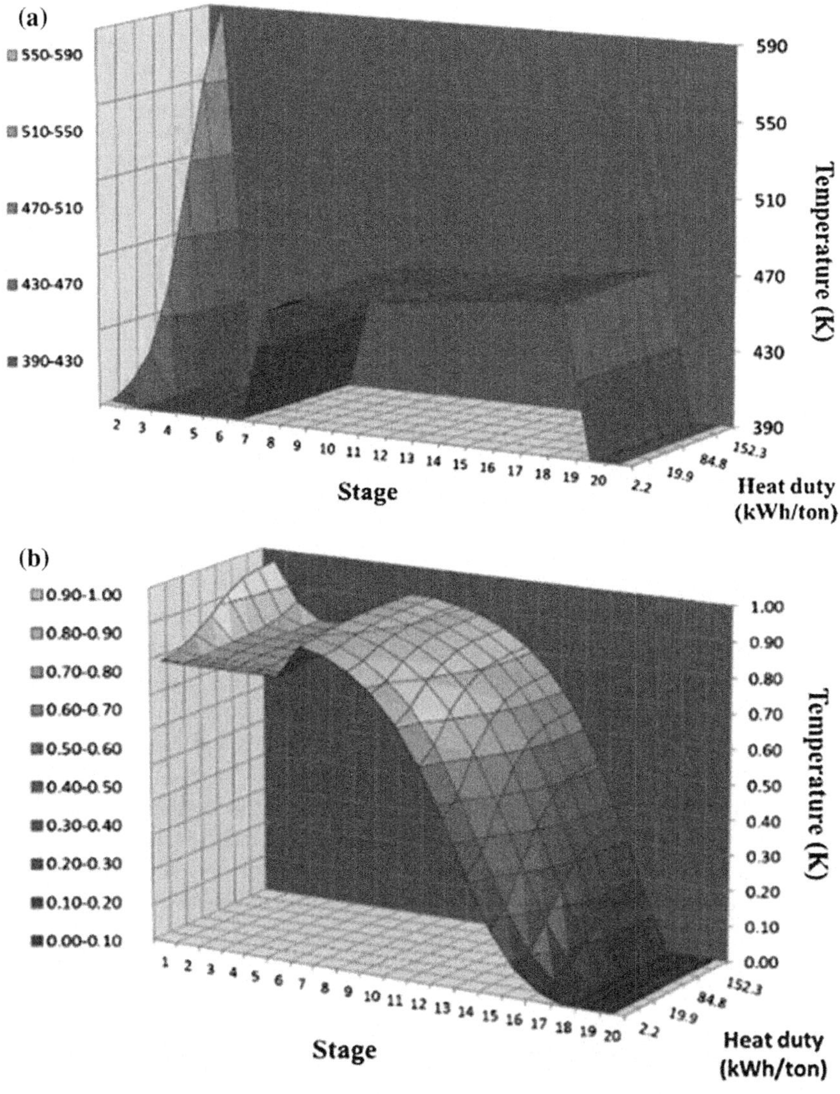

Fig. 14 Temperature surface, and **a** biodiesel molar composition **b** along the reactive distillation column. Reboiler heat duty sensitivity analysis (adapted from Machado et al. 2015)

exhibits biodiesel surface molar composition along the RDC as function of the reboiler heat duty. Figure 15 shows the conversion of HSO in this sensitivity analysis.

Figure 14a shows a remarkable change in temperature. At the lower limit with minimum heat duty on reboiler, the temperature in the column bottom is approximately 390.9 K, when the upper limit, 153.3 kWh/ton fatty esters (biodiesel) is

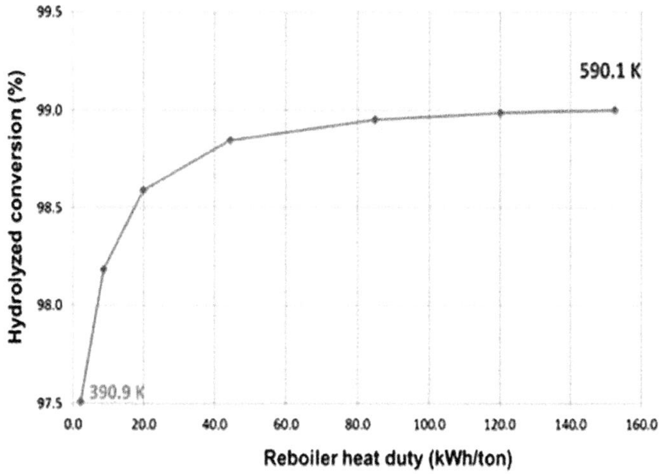

Fig. 15 Hydrolyzed conversion curve generated by reboiler heat duty sensitivity analysis (adapted from Machado et al. 2015)

used, the temperature reaches around 590.1 K. Such temperature can promote thermal degradation of biodiesel and also increase the cost of operation of the RDC, as more energy is required. Moreover, the increased heat duty in the reboiler increased conversion by only about 1.88 %, as can be noted on Fig. 14b which exhibits a small increase in the biodiesel generation with a and consequently better evaluated on Fig. 15.

Therefore, from the viewpoint of conversion, it is not recommend to try to gain efficiency by increasing the heat load. According to Machado et al. (2013), setting the reboiler off and using thermal energy in feed stream can reduce energy consumption and could avoid exposure of the bottom product to elevated temperatures.

HSO Feed Temperature Sensitivity Analysis

As stated previously, the thermal effect on the reboiler can be unfavorable to the stability of the produced biodiesel. In order to minimize this effect, a sensitivity analysis using HSO (fatty acids) feed temperature as an independent variable was made, with values in the range between 380 and 533 K. In such analysis, the energy supplied to the process by the reboiler was kept minimal (about 2.2 kWh/ton biodiesel produced), not zero for computer code convergence issues. Figures 16 and 17 show the results for this sensitivity analysis.

Considering Fig. 16a, it can be seen that the temperature of the HSO feed stream has a strong influence on the temperature profile along the column. The biodiesel liquid phase mole fraction profile of Fig. 16b differs substantially from the profile is shown in Fig. 14b. Figure 17 shows considerable increase in the HSO conversion, showing that this variable is very important for the performance of the RDC. As the

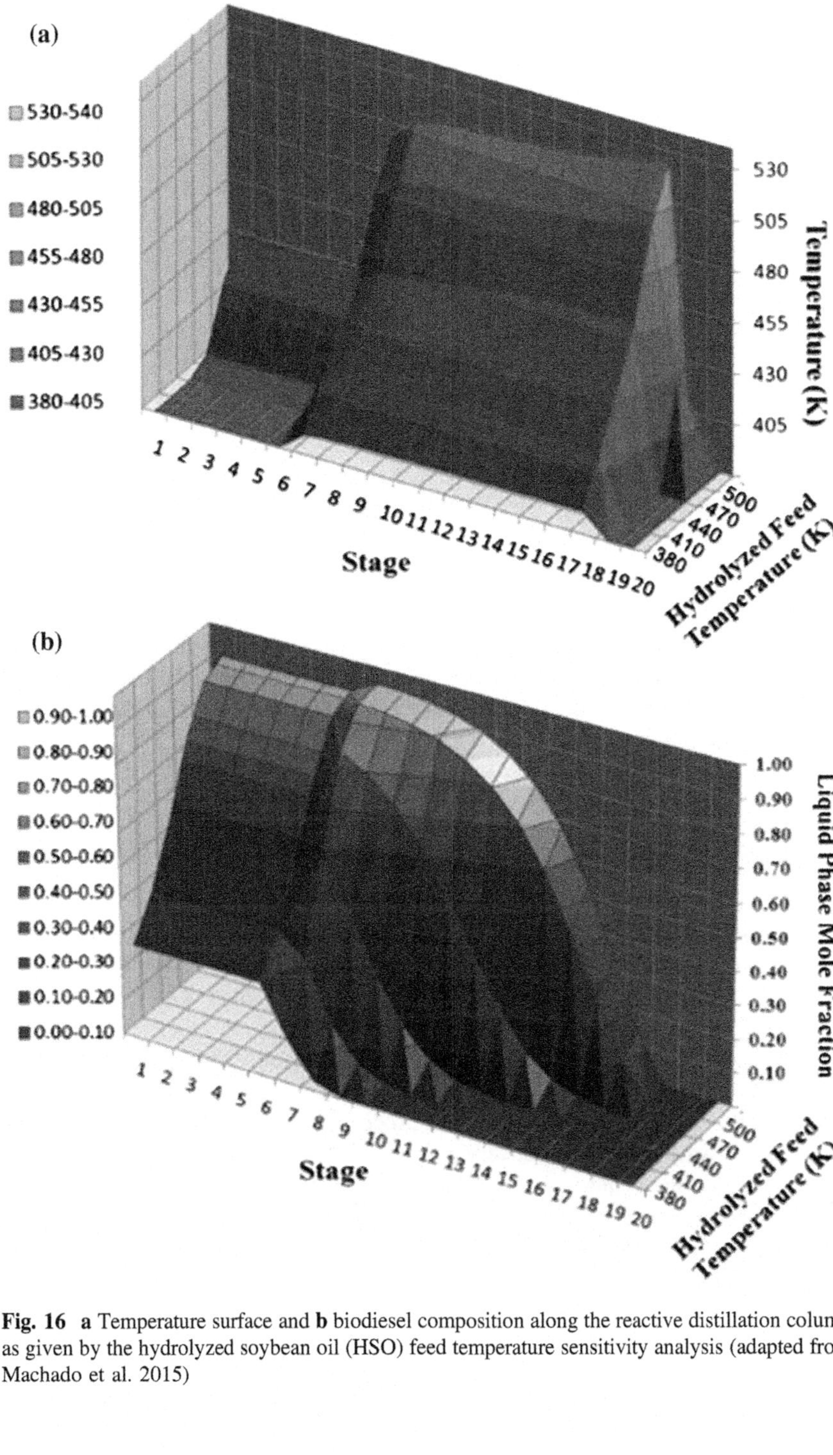

Fig. 16 **a** Temperature surface and **b** biodiesel composition along the reactive distillation column as given by the hydrolyzed soybean oil (HSO) feed temperature sensitivity analysis (adapted from Machado et al. 2015)

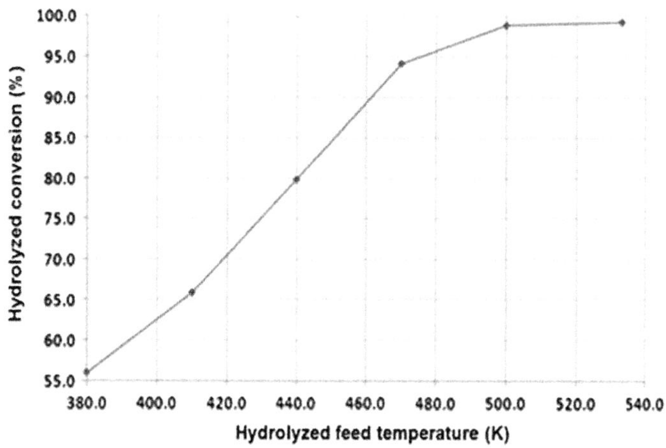

Fig. 17 Hydrolyzed conversion curve generated by hydrolyzed feed temperature sensitivity analysis of proposed case (adapted from Machado et al. 2015)

HSO conversion exhibits a wider range in Fig. 17 (55–99 %), increasing the feed temperature of HSO seems to be a better optimization strategy than increasing the reboiler heat transfer rate. Besides, in this procedure, with minimum thermal load in the reboiler, temperatures in the range 364–424 K are found in the bottom product. Such condition is not so extreme when compared with the full use of the reboiler, as shown in the previous analysis.

6 The Combined Hydroesterification Process

Both hydrolysis and esterification reactions were combined in a global process. A flowchart proposed for the combined hydrolysis process executed in a PBR reactor, followed by esterification reaction performed in a RDC, it is seen in Fig. 18.

In Fig. 18, the term "MIX" represents the mixing of reactants before the PBR. The term "SEP" identifies separation processes. Table 7 shows the main and recovery flows of the hydroesterification process. It also shows that the recycles of ethanol and water allow the recovery of up to 20 % of the ethanol fed into the RDC and 98.5 % of the water required in the PBR.

If we use the benefit of hydroesterification process, i.e., leave a grease field of low quality and additionally imagine generic separation equipment for the process as a whole we can suggest the Fig. 19.

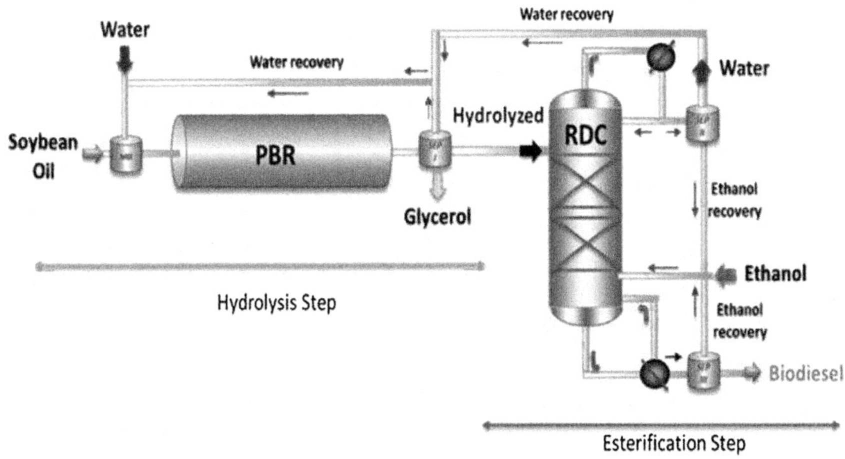

Fig. 18 Schematic flowchart for the combined process of hydrolysis followed by esterification to biodiesel production from soybean oil (adapted from Machado et al. 2015)

Table 7 Main and recovery streams of the hydroesterification process

	PBR in	PBR out	RDC in	RDC out	
Main streams (kg/min)					
Soybean oil	3.8941	0.0709	–	–	
Water	0.4572	0.2195	0.0014	0.2374	
Hydrolyzed	–	3.7264	3.7264	0.026	
Glycerin	–	0.405	–	–	
Ethanol	–	–	0.727	0.1109	
Biodiesel	–	–	–	4.0684	
	SEP I out	SEP II out	SEP III out	PBR req	RDC req
Recovery streams (kg/min)					
Ethanol	–	0.0554	0.0554	–	0.727
Water	0.2195	0.2374	–	0.4572	–

For Table 7, it is noted that the recycles of ethanol and water are beneficial to the process and can be recovered up to 20 % of the ethanol fed into the RDC to the esterification reaction and even 98.5 % of the required water in the hydrolysis reaction in the PBR reactor, which are added in excess of the respective processes. Thus, almost all of the water necessary for the hydrolysis process can be recovered therein.

Moreover, although soybean oil is used as raw material, it is noteworthy that this combined process of hydroesterification can use any feedstock with low cost having considerable humidity and fatty acids amounts, since they are both usable in the process and do not hinder the chemical reactions.

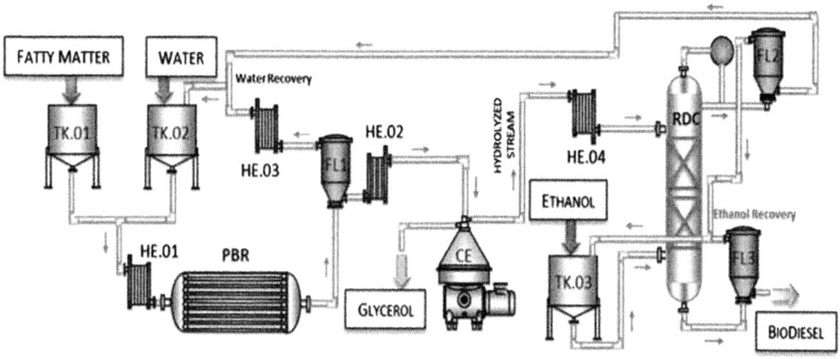

Fig. 19 Schematic flowchart for the combined process of hydrolysis followed by esterification to biodiesel production from any fatty matter (*FL* Flash, *TK* Tank, *CE* Centrifuge, *HE* Heat exchanger)

7 Considerations

In this chapter, computer simulations were presented for the steps of hydrolysis of soybean oil in a tubular reactor (PBR) and esterification of fatty acids (HSO) in a RDC. Such steps were combined in sequence to produce biodiesel by hydroesterification.

This is the first study considers a mixture of fatty acids using a pseudo-homogeneous kinetic model based on experimental solid catalysis data. The simulations have been performed with a validated non-commercial computational code, used in previous works of the group.

In the hydrolysis step, sensitivity analysis showed that a feed ratio of 5.7 (mol of water per mole of soybean oil) can lead to high conversions at the 30 mol/min feed flow.

An observed benefit of this combined process is that the water produced in the esterification step in the RDC can be recovered in the hydrolysis reaction in the reactor PBR, in previous step, together with the water fed in excess therein.

The results obtained in validated case showed good agreement with experimental and simulated data available in literature, validating the simulation procedures for the RDC step.

For this step of esterification, RDC sensitivity analysis shows that the increase of the HSO feed temperature with the use of minimum energy on reboiler can be a more suitable alternative optimization, also avoiding biodiesel exposure to high temperatures in bottom column, since it could be exploit the high temperature (533 K) of the HSO leaving the hydrolysis reaction.

The methodology used was consistent and the results of sensitivity analysis showed operating conditions optimized for conversions above 99 %. Regarding the specifications, even ignoring the SEP III separation process, the produced biodiesel has purity of 98.3 % on mass ester content, higher than current international specs requirements of 96.5 %.

References

Agreda V, Partin L (1984) Reactive distillation process for the production of methyl acetate. US Patent 4,435,595

Alfradique MF, Castier M (2005) Modeling and simulation of reactive distillation columns using computer algebra. Comput Chem Eng 29:1875–1884. doi:10.1016/j.compchemeng.2005.04.002

Altin R, Çetinkaya S, Yucesu HS (2001) The potential of using vegetable oil fuels as fuel for diesel engines. Energy Conv Man 42:529. doi:10.1016/S0196-8904(00)00080-7

Aranda DAG, Gonçalves JA, Peres JS et al (2009) The use of acids, niobium oxide, and zeolite catalysts for esterification reactions. J Phys Org Chem 22:709. doi:10.1002/poc.1520

Aransiola EF, Ojumu TV, Oyekola OO et al (2014) A review of current technology for biodiesel production: state of the art. Biomass Bioenergy 61:276–297. doi:10.1016/j.biombioe.2013.11.014

Bournay L, Casanave D, Delfort B et al (2005) New heterogeneous process for biodiesel production: a way to improve the quality and the value of the crude glycerin produced by biodiesel plants. Catal Today 106:190–192. doi:10.1016/j.cattod.2005.07.181

Castier M (1999) Automatic implementation of thermodynamic models using computer algebra. Comput Chem Eng 23:1229–1245. doi:10.1016/S0098-1354(99)00286-0

Chen F, Huss RS, Malone MF et al (2000) Multiple steady states in reactive distillation: kinetic effects. Comput Chem Eng 24:2457. doi:10.1016/S0098-1354(01)00750-5

Chisti Y (2013) Constraints to commercialization of algal fuels. J Biotechnol 167:201–214. doi:10.1016/j.jbiotec.2013.07.020

Crown Iron (2015) http://www.crowniron.com/technologies/biodiesel.cfm. Accessed 23 May 2015

Dedini (2015) http://www.dedini.com.br/pt/biodiesel.html. Accessed 13 Aug 2015

Di Serio M, Tesser R, Pengmei L et al (2008) Heterogeneous catalysts for biodiesel production. Energy Fuels 22:207. doi:10.1021/ef700250g

Dimian AC, Bildea CS (2008) Chemical process design—computer-aided case studies. Wiley, Weinheim

Dimian AC, Bildea CS, F Omota et al (2009) Innovative process for fatty acid esters by dual reactive distillation. Comput Chem Eng 33:743–750. doi:10.1016/j.compchemeng.2008.09.020

DIPPR—Design Institute for Physical Properties—Database Project 801 (2013) Full version

Fukuda H, Kondo A, Noda H (2001) Biodiesel fuel production by transesterification of oils. J Biosci Bioeng 92:405. doi:10.1016/S1389-1723(01)80288-7

Gmehling J, Li J, Schiller M (1993) A modified UNIFAC model. 2. Present parameter matrix and results for different thermodynamic properties. Ind Eng Chem Res 32:178–193. doi:10.1021/ie00013a024

Gomez-Castro FI, Rico-Ramirez V, Segovia-Hernandez JG et al (2011) Esterification of fatty acids in a thermally coupled reactive distillation column by the two-step supercritical methanol method. Chem Eng Res Des 89:480–490. doi:10.1016/j.cherd.2010.08.009

Hala E, Boulblik T, Fried V (1984) Vapour pressure of pure substances, vol 17. Elsevier, Amsterdam, p 972

Hernandez S, Segovia-Hernandez JG, Juarez-Trujillo L et al (2010) Design study of the control of a reactive thermally coupled distillation sequence for the esterification of fatty organic acids. Chem Eng Commun 198:1–18. doi:10.1080/00986445.2010.493102

Kiss AA (2010) Separative reactors for integrated production of bioethanol and biodiesel. Comp Chem Eng 34:812–820. doi:10.1016/j.compchemeng.2009.09.005
Kiss AA (2011) Heat-integrated reactive distillation process for synthesis of fatty esters. Fuel Process Technol 92:1288–1296. doi:10.1016/j.fuproc.2011.02.003
Kiss AA, Bildea CS (2011) Integrated reactive absorption process for synthesis of fatty esters. Bioresour Technol 102:490–498. doi:10.1016/j.biortech.2010.08.066
Kiss AA, Segovia-Hernández JG, Bildea CS (2012) Reactive DWC leading the way to FAME and fortune. Fuel 95:352–359. doi:10.1016/j.fuel.2011.12.064
Knothe G, Van Gerpen J, Krahl J (2005) The biodiesel handbook. AOCS Press, Champaign
Krafczyk J, Gmehling J (1994) Use of catalyst packages for the production of methyl acetate by reactive rectification. Chem Ing Tech 66:1372–1375. doi:10.1002/cite.330661014
Kusdiana D, Saka S (2001) Kinetics of transesterification in rapeseed oil to biodiesel fuels as treated in supercritical metanol. Fuel 80:693–698. doi:10.1016/S0016-2361(00)00140-X
Lotero E, Liu YJ, Lopez DE (2005) Synthesis of biodiesel via acid catalysis. Ind Eng Chem Res 44:5353–5363. doi:10.1021/ie049157g
Ma F, Hanna MA (1990) Biodiesel production: a review. Bioresour Technol 70:1–15. doi:10.1016/S0960-8524(99)00025-5
Machado GD, Aranda DAG, Castier M et al (2011) Computer simulation of fatty acid esterification in reactive distillation columns. Ind Eng Chem Res 50:10176–10184. doi:10.1021/ie102327y
Machado GD, Pessoa FLP, Castier M et al (2013) Biodiesel production by esterification of hydrolyzed soybean oil with ethanol in reactive distillation columns: simulation studies. Ind Eng Chem Res 52:9461–9469. doi:10.1021/ie400806q
Machado GD, Souza TL, Aranda DAG et al (2015) Computer simulation of biodiesel production by hydroesterification. Chem Eng Process. doi:10.1016/j.cep.2015.10.015
Madras G, Kolloru C, Kumar R (2004) Synthesis of biodiesel in supercritical fluids. Fuel 83:2029–2033. doi:10.1016/j.fuel.2004.03.014
NIST—National Institute of Standards and Technology—Chemistry WebBook (2015) Standard Reference Database Number 69
Oliveira D, Luccio M, Faccio C et al (2005) Optimization of alkaline transesterification of soybean oil and castor oil for biodiesel production. Appl Biochem Biotechnol 121:231. doi:10.1385/ABAB:122:1-3:0553
Omota F, Dimian AC, Bliek A (2003a) Fatty acid esterification by reactive distillation. Part 1: equilibrium-based design. Chem Eng Sci 58:3159–3174. doi:10.1016/S0009-2509(03)00165-9
Omota F, Dimian AC, Bliek A (2003b) Fatty acid esterification by reactive distillation. Part 2: kinetics-based design for sulphated zirconia catalysts. Chem Eng Sci 58:3175–3185. doi:10.1016/S0009-2509(03)00154-4
Ondrey G (2004) Biodiesel production using a heterogeneous catalysts. Ind Eng Chem Res 10:13. doi:10.1016/j.biortech.2010.10.080
Rocha LLL, Ramos ALD, Antoniosi Filho NR (2010) Production of biodiesel by a two-step niobium oxide catalyzed hydrolysis and esterification. Lett Org Chem 7:571. doi:10.2174/157017810793362343
Stankiewicz A (2003) Reactive separations for process intensification: an industrial perspective. Chem Eng Process 42:137–144. doi:10.1016/S0255-2701(02)00084-3
Stankiewicz A, Moulinjin J (2002) Process intensification. Ind Eng Chem Res 4:1920–1924. doi:10.1021/ie011025p
Steinigeweg S, Gmehling J (2002) n-Butyl acetate synthesis via reactive distillation: thermodynamic aspects, reaction kinetics, pilot-plant experiments, and simulation studies. Ind Eng Chem Res 41:5483. doi:10.1021/ie020179h
Steinigeweg S, Gmehling J (2003) Esterification of a fatty acid by reactive distillation. Ind Eng Chem Res 42:3612. doi:10.1021/ie020925i
Valliyappan T, Bakhshi N, Dalai AK (2008) Pyrolysis of glycerol for the production of hydrogen or syn gas. Bioresour Technol 99:4476. doi:10.1016/j.biortech.2007.08.069

Van Gerpen J (2005) Biodiesel processing and production. Fuel Process Technol 86:1097–1107. doi:10.1016/j.fuproc.2004.11.005

Van Gerpen J, Shanks B, Pruszko R et al (2004) Biodiesel. Production Technology, National Renewable Energy Laboratory—NREL, Colorado

Vicente G, Martinez M, Aracil J (2004) Integrated biodiesel production: a comparison of different homogeneous catalysts systems. Bioresour Technol 92:297–305. doi:10.1016/j.biortech.2003.08.014

Warabi Y, Kusdiana D, Saka S (2004) Reactivity of triglycerides and fatty acids of rapeseed oil in supercritical alcohols. Bioresour Technol 91:283–287. doi:10.1016/S0960-8524(03)00202-5

Biodiesel and Bioethanol from Microalgae

Etiele Greque de Morais, Luiza Moraes, Michele Greque de Morais and Jorge Alberto Vieira Costa

Abstract Microalgae are a group of unicellular or multicellular photosynthetic microorganisms that have the ability to use sources of organic and inorganic carbon for their development. Microalgal biomass has been proposed as a raw material for the production of energy and other products due to its high productivity, flexibility, capacity for using wastewater, and the fact that nonfertile land is required for cultivation. These microorganisms can synthesize and store lipids in the form of triacylglycerols and fermentable carbohydrates, which are used for the production of biodiesel and bioethanol, respectively. The application of nutrient-rich industrial effluents for the cultivation and the application of microalgal biomass within the photobiorefinery concept are matters that research groups around the world have considered to reduce production costs and to make sustainable energy production using these microorganisms feasible.

Keywords Biofuels · Carbohydrates · Energy · Lipids

1 Introduction

Global warming coupled with the reduction of fossil resources and energy insecurity are the main reasons for the search for renewable energy sources to meet the growing demand and to minimize their environmental impact (Zhao and Su 2014). According to John et al. (2011), biofuels generated from biomass are promising energy sources that can contribute to the reduction of climate changes, improve the depletion of fossil fuels, and enhance energy security.

E.G. de Morais · L. Moraes · J.A.V. Costa (✉)
Laboratory of Biochemical Engineering, College of Chemistry and Food Engineering, Federal University of Rio Grande, Rio Grande-RS, Brazil
e-mail: jorgealbertovc@terra.com.br

M.G. de Morais
Laboratory of Microbiology and Biochemical, College of Chemistry and Food Engineering, Federal University of Rio Grande, Rio Grande-RS, Brazil

C.R. Soccol et al. (eds.), *Green Fuels Technology*,
Green Energy and Technology, DOI 10.1007/978-3-319-30205-8_14

Microalgae are photosynthetic microorganisms with relatively simple requirements for growth when compared with other biomass sources. During photosynthesis, microalgae convert sunlight, water, and CO_2 into O_2 and biomass. The biomass that is produced may exhibit a wide range of biocompounds that can be converted to biofuels, such as biogas, biodiesel, biohydrogen, bioethanol, and biogasoline. Additionally, the direct burning of biomass for energy production can be performed through pyrolysis (Medipally et al. 2015; Costa and Morais 2011).

Microalgae can accumulate high levels of carbohydrates in the biomass, and these carbohydrates may be potentially used as a substrate in fermentation to produce bioethanol. Furthermore, some species are known to accumulate lipids under stress conditions and can thus be used as raw material for biodiesel production (Costa and Morais 2011; Chen et al. 2013). The microalgal biomass belong to the third generation of biofuels and is considered an alternative energy source to fossil fuels without having the drawbacks associated with the first and second generations, such as food security, land use, and water scarcity. The production of microalgal biomass to obtain biofuels has the following advantages over traditional energy crops: increased biomass productivity in terms of the land area required for cultivation; the ability to use liquid effluents (wastewater), gaseous (combustion gas) as a source of nutrients; the use of nonarable land, not competing with food production; a short harvest time; and the possibility of manipulating culture conditions to produce the compounds of interest, which provide the supplies needed to meet the demand for biofuels (Brennan and Owede 2010). Therefore, many benefits are verified to make feasible the application microalgae as a source of biomass clean, efficient, and sustainable to produce biofuels and coproducts of high added value in photobiorefineries (Vanthoor-Koopmans et al. 2013).

This chapter discusses the main aspects with respect to the production of biofuels from microalgae, with a primary focus on biodiesel and bioethanol and on the viability of the process by applying the photobiorefinery concept.

2 Microalgae

Microalgae are organisms that play a key role in aquatic ecosystems. It is estimated that approximately 40 % of the total photosynthesis is performed by these microorganisms. Microalgae are responsible for the base of most aquatic food chains. In some coastal environments, the concentration of microalgae biomass can equal or even exceed bacteria (Moreno-Garrido 2008). These microorganisms react to changes in their external environment with intracellular alterations. Therefore, manipulating the culture conditions and the presence or absence of nutrients stimulates the biosynthesis of compounds.

Prokaryotic microalgae (cyanobacteria) have no membrane-bound organelles (plastids, mitochondria, nucleus, Golgi complex, and flagella) and are more similar to bacteria than to algae. Eukaryotic strains, which cover a large number of species, have organelles that control the cellular functions that allow survival and reproduction. Eukaryotic microalgae are categorized into a variety of classes that are primarily defined by their pigmentation, cellular structure, and life cycle. The most important classes are green algae (Chlorophyta), red algae (Rhodophyta), and diatoms (Bacillariophyta) (Khan et al. 2009).

Microalgae metabolism can be autotrophic or heterotrophic. Autotrophic metabolism requires only inorganic compounds, such as CO_2, salts and solar energy, for development. Heterotrophic metabolism does not involve photosynthesis; therefore, it requires an external source of organic compounds that can be used as a nutrient and energy source. Some photosynthetic species are mixotrophic and have the ability to perform photosynthesis and use exogenous organic sources (Lee 1980). The involved metabolism can be varied according to the pH changes that depend on the development stoichiometry of the microalgae. *Chlorella vulgaris* and *Spirulina platensis* are examples of species that grow under autotrophic, heterotrophic, and mixotrophic conditions (Chojnacka and Marques-Rocha 2004).

In addition to organic carbon substrates, vitamins, salts, and other nutrients (nitrogen and phosphorus) are vital for microalgae growth, and equilibrium between operational parameters (pH, temperature and light intensity) is also required (Williams 2002). Therefore, it is important to define the influence of these parameters and their correlation prior to attempting to manipulate them. In this way, it is possible to obtain control over the composition of microalgal biomass even in large-scale crops (Chisti 2008).

The microalgae growth phases are defined as: (1) lag phase; (2) transitional phase (3) exponential growth phase; (4) linear growth phase; (5) slowing down; (6) stationary phase; and (7) decline or cell death. In the first phase (lag or adaptation), the microorganism synthesizes the enzymes that are required for metabolism of the compounds present in the medium, and this phase occurs immediately after inoculation. This step is followed by phase 2, which starts cell reproduction, and the specific growth rate increases until the exponential phase (phase 3), in which the speed is constant and maximum. In stage 4, the reproduction speed is constant until the deceleration phase (step 5), which occurs due to the exhaustion of one or more nutrients in the medium. The stationary phase (6) is where the microorganism reaches the maximum cell concentration and cell reproduction speed is equal to death until phase 7, where the cell concentration decreases at a rate that exceeds the reproduction (Schmidell et al. 2001).

Generally, microalgal cultures in the exponential growth phase contain more protein in their biomass due to cell multiplication. The production of lipids and carbohydrates occurs in the stationary phase as a form of energy reserve. The microalgal biomass is rich in polar lipids during the exponential phase and accumulates triglycerides (reserve) in the stationary phase (Dunstan 1993).

Microalgae have been the focus of numerous investigations at research centers in several countries. Culturing microalgae is conducted for various purposes, such as obtaining pharmaceutical, food and fertilizers, and have more recently been proposed as a source of energy (Scapin 2005; Scragg et al. 2003). Studies have shown their beneficial application in biofuel production (Gao et al. 2010; Ho et al. 2011) and in the mitigation of gases that cause global warming (Morais and Costa 2007).

3 Microalgae Biofuels Versus Other Biomass

Biofuels are produced from several biomasses and can be classified into liquid, gaseous, or solid forms (Demirbas 2008). Among the varieties produced, the standouts are bioethanol, biomethanol, biodiesel, biohydrogen, and biogas (Nigam and Singh 2011). These biofuels have emerged as strategically important sources of sustainable fuels and are considered promising in reducing greenhouse gas emissions and the production of new energy resources (Jaecker-Voirol et al. 2008). Biofuels are generally classified into two groups: (i) raw rivers, such as firewood, wood chips, pellets, animal waste, forestry, and cultures, (ii) and side comprising, such as bioethanol, butanol, biodiesel, and biohydrogen (Russo et al. 2012; Dragone et al. 2010).

The bioethanol and butanol produced via the fermentation of starch from wheat, barley, corn, potato, sugar cane, beet sugar, and biodiesel produced chemically from rapeseed, soybean, sunflower, palm, coconut, and animal fats are considered first-generation biofuels (Dragone et al. 2010). The production of bioethanol and biodiesel from *Jatropha*, cassava, *Miscanthus* and bioethanol, and butanol from lignocellulosic raw materials are considered second-generation biofuels (Dragone et al. 2010; Sims et al. 2010). However, despite the high diffusion of these two generations of biofuels, these have negative impacts on food security and cause water scarcity and deforestation (FAO 2007, 2008; Naik et al. 2010). Therefore, it is necessary to search for more viable biomass sources for energy production to replace or supplement fossil fuels (Medipally et al. 2015).

Microalgae can provide a high productivity of biocompounds, such as lipids and carbohydrates (Wijffels and Barbosa 2010; Costa and Morais 2011). These microorganisms are considered promising raw materials for the production of third-generation biofuels (Maity et al. 2014) such as biodiesel, bioethanol, biohydrogen, and biogas (Costa and Morais 2011). Some advantages of microalgae use for biofuel production are that it has a high rate of cell growth, a lower water demand compared with terrestrial cultures, and that its cultivation can be performed on nonarable land, which minimize environmental impacts and does not compromise food production (Demirbas 2011; Medipally et al. 2015). In addition, microalgae have the ability to absorb nutrients from wastewater (sewage) (Chiu et al. 2015), solids (Vaz et al. 2016), and gas (combustion gas) (Radmann et al. 2011), thus, they produce potentially more profitable biomass than does conventional farming (Demirbas 2011). The fusion of all of these features enables the application of biomass in both renewable energy production and biocompound extraction, with applications in nutrition, cosmetics, medicine, and chemicals (Zhu 2015).

The use of microalgae for biofuel production may also have some limitations, such as a small cell size, expensive biomass harvest, and high costs. However, these limitations can be overcome by improving cultivation technologies with respect to harvesting, drying (Medipally et al. 2015), and the conversion of biomass into biofuels (Brennan and Owende 2010; Costa and Morais 2011). In addition to these modifications, the application of genetic engineering and the development of photobioreactors could potentially contribute to an increase in cell growth rates and an accumulation of reserve compounds, such as lipids and carbohydrates (Medipally et al. 2015; Brennan and Owende 2010).

In the context of clean and sustainable energy development, microalgal biomass has favorable characteristics compared with other raw materials for use in the production of biofuels and high value-added coproducts (Subhadra 2010). Obtaining coproducts along with the production of microalgal biofuels could facilitate the marketing of these and solidify the concept of microalgae photo-biorefineries (Zhu 2015).

4 Microalgal Biodiesel

Biodiesel is a renewable and biodegradable fuel that consists of a mixture of methyl and ethyl esters of fatty acids that are obtained through the transesterification reaction with any triglyceride with a short chain alcohol. Compared with conventional diesel, biodiesel stands out because of its ability to equilibrate the negative balance generated by the emission into the atmosphere. Compared with diesel, biodiesel reduces the emission of sulfur compounds (SO_x), carbon monoxide (CO), and particulate matter (PM) but increases the emission of nitrogen oxide (NO_x) (Hoekman and Robbins 2012; Visentainer and Santos 2013)

The raw material is one of the most important points in biodiesel production and represents 70 % of this biofuel's production cost. For Brazilian biodiesel, vegetable oils and animal fats are the main raw materials used. The feasibility of using oil depends on various aspects, such as the oil content, yield per unit area, adaptation for different production systems, seasonality, and plant life cycle (Visentainer and Santos 2013). Biological systems are commonly characterized by transforming energy-poor compounds into high value-added products. The transformation of solar energy, which is the most abundant energy source on our planet, into organic compounds occurs during photosynthetic metabolism, and microalgae are the most efficient at doing this (Henrikson 1994). The use of microalgal biomass for biodiesel production draws attention due to the high productivity in biomass and the crop capacity in inhospitable places (Demirbas 2006).

Many species of microalgae can be induced to accumulate considerable amounts of lipids, which contribute to obtaining a higher yield of oil. The average lipid content varies between 1 and 70 % by dry weight according to changes in the culture medium. The microalgae *Chlorella*, *Dunaliella*, *Nannochloropsis*, *Scenedesmus*, and *Tetraselmis* present lipid concentrations between 20 and 50 %.

In the selection of an appropriate species for biofuel production, one should consider other factors, such as the composition of existing fatty acids from different species of microalgae. The fatty acid composition has a significant effect on the qualitative and quantitative characteristics of the produced biodiesel. The microalgal biomass may be composed of saturated and unsaturated fatty acids with 12–22 carbon atoms and some of families ω3 and ω6 (Spolaore et al. 2006; Chisti 2008; Li et al. 2008; Wang et al. 2008).

Nautiyal et al. (2014) studied the characteristics of biodiesel obtained from *S. platensis*. The viscosity and oil density are directly related to the atomization process during combustion, with lower values of these parameters indicating better fuel ignition performance (Ramírez-Verduzco et al. 2012). The density of the oil obtained from the microalgae fits European standards, is equivalent to sunflower oil, and approaches the values for diesel and palm. Microalgal oil has a viscosity within the American standards and is greater than those of sunflower, palm, and diesel oils (Table 1).

The acidity is a measure of free fatty acids in the fuel and must be the lowest possible to avoid forming soaps during transesterification, which may cause problems in the engine especially in the nozzle (Naureen et al. 2015). The acidity of the oil obtained using *S. platensis* is 0.45 mg KOH g^{-1}, which is appropriate according to both European and American standards. Microalgal oil has a lower calorific value than diesel does. A greater calorific value results in less fuel consumed because less fuel is required to maintain the potency of the specific ignition. The gross calorific value of microalgae is due to a greater amount of oxygen in biofuel (Nautiyal et al. 2014).

The pour point is the lowest temperature at which a fuel can flow without gelling. The microalgal biodiesel has a pour point of −18 °C, which is close to that of diesel. With respect to the flash point, the oil obtained from *S. platensis* is within the American and European standards and is lower than the palm and sunflower oil.

Table 1 Characteristics of *Spirulina platensis* (Nautiyal et al. 2014), sunflower (Naureen et al. 2015), palm (Sarin et al. 2007), diesel oils (Singh and Singh 2010), American standards (ASTM D6751 2015) and European standards (EN 14214 2003)

Properties	Oils					
	S. platensis	Sunflower	Palm	Diesel	D6751	EN14214
Density (kg m^{-3})	860	860	876	855	870–900	860–900
Viscosity at 40 °C (mm^2 s^{-1})	5.7	4.72	4.76	3.06	1.9–6.0	3.5–5.0
Acidity (mg KOH g^{-1})	0.4	0.07	–	–	<0.8	<0.5
Calorific power (MJ kg^{-1})	41.4	–	–	43.8	–	32.9
Flash point (°C)	130	183	170	76	100–170	>101
Pour point (°C)	−18	−5	−17	−16	−31	–

4.1 Lipid Composition in Microalgae

Lipids are essential for the maintenance of various structures of living organisms and act in several metabolic processes involved in biological functions, such as membrane components, thermal insulation and energy reserves. The microalgae lipid compounds are glycerol, sugars, bases, or esterified saturated fatty acids or unsaturated fatty acids. When they are hydrolyzed, lipids release fatty acids and glycerol. These nutrients are mainly stored in microalgae vacuoles as a reserve material, and the lipid content stored by microalgae can be high in relation to the total remaining compounds (Lourenço 2006).

Responses to stimulation or environmental changes are inherent in all living organisms. In microalgae, the response of cells to environmental conditions defines the factors as limiting when there is a reduction in the growth rate and/or a biochemical reaction without the need for cell acclimation or stress that involves a metabolic imbalance that demands biochemical adjustments before the cells may establish a new state of growth or biosynthesis (Richmond 2004).

Changes in nitrogen content, light intensity, temperature, and salinity or CO_2 concentration during microalgae cultivation are capable of increasing the lipid concentration. The limitation of nitrogen in the culture medium is one of the most effective methods for lipid accumulation and results not only in accumulation but also in a change in the free fatty acid composition. However, improvements in the lipid concentration do not result in a biomass increase (Brennan and Owende 2010).

Some microalgae have demonstrated the ability to accumulate large amounts of total lipids, but lipid accumulation is slow in species such as *Chlorella*, *Scenedesmus*, and *Spirulina*. The accumulation of fatty acids is closely linked to the stages of microalgal growth. Lipids act as an energy source for the microorganism when they accumulate during adverse conditions and in the stationary phase of growth, where growth stagnation occurs due to the depletion of nutrients in the medium (Adarme-Veja et al. 2014).

4.2 Lipid Metabolism in Microalgae

Based on the homologous sequence and some similar biochemical characteristics of the genes and/or enzymes isolated from microalgae and higher plants that are involved in lipid metabolism, it is believed that the basic pathways of biosynthesis of fatty acids and triglycerides acids (TAGs) in microalgae are similar to those of higher plants. The fatty acid synthesis in microalgae occurs mostly in the thylakoid membrane and the stroma region of the chloroplast. The biosynthetic pathway of lipids in microalgae occurs in four steps: accumulation of carbohydrates in the cell, acetyl-CoA formation followed by malonyl-CoA formation, palmitic acid synthesis, and the synthesis of long chain fatty acids (Pandey et al. 2014). The accumulation of energy-rich compounds is the first step in the biosynthesis of lipids by

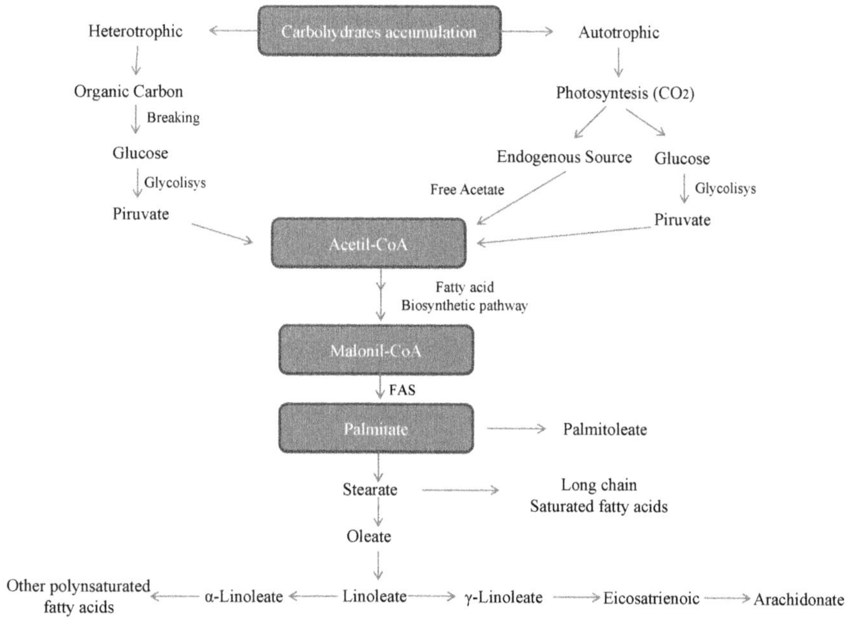

Fig. 1 Summarized scheme of microalgae lipids' metabolism

microalgae. This accumulation of carbon varies in autotrophic and heterotrophic conditions (Fig. 1).

Autotrophic organisms synthesize their own carbohydrates through photosynthesis, whereas heterotrophic organisms assimilate carbohydrates from the cell's environment (Madigan et al. 2010). In the second step, photosynthesis provides an endogenous source of acetyl-CoA, thereby activating synthesis in the stroma through free acetate or from the glucose conversion to pyruvate in the cytosol during glycolysis. This acetyl-CoA is preferentially transported from the plastid to the cytosol, where it is converted to fatty acid and then to TAG, which is, again, transported to the cytosol to form lipids. The acetyl-CoA is maintained by the Calvin cycle and glycolysis, and pyruvate kinase (PK) is maintained by pyruvate synthesis, which occurs in the chloroplast. The first reaction of the biosynthetic pathway of fatty acids is the formation of malonyl-CoA from acetyl-CoA and CO_2 and is catalyzed by the enzyme acetyl-CoA carboxylase (Pandey et al. 2014).

After malonyl-CoA molecules form through a repetitive sequence of reactions, which is catalyzed in 4 steps by a system known as fatty acid synthase (FAS), long carbon chain fatty acids are built. A saturated acyl grouping, which is produced in each series of reactions in steps 4, becomes the substrate for subsequent condensation with an activated malonila group. In each passage through the cycle, the chain of the fatty acyl group is increased by 2 carbons.

With the FAS system, the synthesis of fatty acids leads to a single product, and no intermediaries are released. When the chain length reaches 16 carbon atoms (palmitic acid), the product leaves the cycle. Palmitate is the precursor of stearate and the long chain saturated fatty acids, palmitoleic, and oleic. Palmitic acid is modified and elongated to stearate (18:0), and longer saturated fatty acids are formed from the addition of acetyl groups through the stretching system action of the fatty acids present in the endoplasmic reticulum (ER) and the smooth mitochondria. The stretching mechanism ER is identical to palmitate synthesis involving the 2 carbons donated by malonyl-CoA, which is followed by reduction, dehydration, and reduction to produce saturated 18-carbon stearoyl-CoA (Nelson and Cox 2011).

In microalgae, oleate (the stearoyl-CoA) is converted to α- and γ-linoleate. The α-linoleate is further converted to another polyunsaturated fatty acid as γ-linoleate becomes eicosatrienoato or even arachidonate (Thelen and Ohlogge 2002). Mammals are not able to convert oleate to linoleate or α-linoleate because they lack the enzymes that introduce double bonds at carbon 9 (C9). In this way, all of the fatty acids that contain double bonds at C9 must be added to the diet and are called essential fatty acids (Nelson and Cox 2011).

4.3 Microalgae Lipid Extraction for Biodiesel Production

During the extraction, the microalgal biomass is exposed to solvents that extract the lipid matrix. Once it is separated from the cell and after the solvent and water evaporate, the lipid mass can be measured gravimetrically. The technology used to extract lipids for biodiesel production should have high specificity to minimize the extraction of non-lipid materials, such as proteins and carbohydrates. To reduce the downstream steps (fractionation/purification), the extraction technology must be selective with respect to glycerides and should not extract nonconvertible lipid fractions form the biodiesel, such as polar and neutral lipids (free fatty acids, hydrocarbons, sterols, ketones, carotenoids, and chlorophylls). Moreover, the selected technique must be efficient in terms of energy and time, should not be reactive with lipids and should be secured to the analyst (Kates 1986; Medina et al. 1998).

The principle of lipid extraction through organic solvents can be summarized using the basic chemical principle that "like dissolves like." Due to interactions between the long hydrophobic fatty acid chains, neutral lipids participate in a van der Waals bonding to form globules in the cytoplasm (Kates 1986; Medina et al. 1998). The mechanism for solvent extraction can be divided into five steps. When the microalgal cell is exposed to a nonpolar organic solvent, such as hexane or chloroform, the solvent penetrates through the cell membrane to the cytoplasm (step 1) and interacts with neutral lipids through van der Waals forces (step 2) to form an organic solvent-lipid complex (step 3). This complex organic solvent-lipid, through a concentration gradient, diffuses through the cell membrane (step 4), and the organic solvent forms a film around the cell (step 5). As a result, the neutral lipids are extracted from the cells and remain dissolved in the polar organic solvent. This static organic solvent film is formed due to the interaction between the organic solvent and the cell wall. This film involves microalgal cell and remains unaltered by any solvent flow or agitation.

Some neutral lipids can be found in the cytoplasm in a complex with polar lipids. This complex is strongly bonded through hydrogen bonds to the cell membrane proteins. The van der Waals interactions that are formed between the nonpolar solvent and neutral lipids in the lipid-solvent complex are not sufficient to break these associations between the lipids and proteins. Conversely, polar solvents (such as methanol or isopropanol) are able to break these associations forming hydrogen bonds with the polar lipids (Kates 1986; Medina et al. 1998). The mechanism whereby polar and nonpolar organic solvents in a mixture are able to extract membrane-associated lipids can be divided into the same five steps described above.

Polar and apolar solvents penetrate through the cell membrane into the cytoplasm and interact with the lipid complex. During this interaction, the nonpolar organic solvent involves the lipid complex that forms van der Waals associations with the complex neutral lipids. At the same time, the polar organic solvent involves the lipid complex and forms hydrogen bonds with the polar lipid complex. The hydrogen bonds are sufficiently strong to displace the lipid–protein associations that link the complex to the cell membrane, and diffusion of the organic solvent-lipid complex through the cell membrane's organic solvent film occurs. Several authors have described different lipid extraction methods using organic solvents with small differences; the most used are those of Bligh and Dyer (1959) and Folch et al. (1957).

4.4 *Transesterification of Lipids Extracted from Microalgae for Biodiesel Production*

The transesterification is the chemical conversion of triacylglycerides and methyl esters using alcohol and a catalyst. The reaction takes 3 mol of alcohol and one mole of triacylglycerides for every mole glycerol and 3 mol of methyl esters. Alcohols that are commonly used in transesterification include methanol, ethanol, propanol, butanol, and amyl alcohol. Methanol is the most widely used alcohol due to its low cost and physical and chemical advantages (Bahadar and Khan 2013).

The fatty acid methyl esters resulting from the transesterification of lipids are used as biofuel. The glycerol obtained from the transesterification is a polyol that, in general, is combined with a triglyceride. The crude glycerol obtained from the biodiesel transesterification contains approximately 20 % impurities, such as catalyst, alcohol, fatty acids, salts, and water. The content of the impurities depends on the feedstock and the type of catalysts used to produce the biodiesel. The crude glycerol has little direct application; however, once it is purified, the glycerol can be used in the chemical industry (production of paints, cellophane and paper), food (emulsifier), and cosmetics (moisturizing). However, the demand for glycerol is less than the amount produced by transesterification (Beatriz et al. 2011). Glycerol has also been studied as a carbon source for the cultivation of various microorganisms, such as microalgae (Abad and Turon 2012).

As the fatty acid composition, water content and free fatty acids can vary in the raw material for the production of biodiesel, some care is necessary in choosing the catalyst for the process of obtaining biodiesel to obtain a higher yield, higher quality and fewer effluents. Catalysts include acids, bases, and enzymes, such as lipases (Visentainer and Santos 2013).

The basic homogeneous catalysts, which are most frequent, occur and involve the presence of catalytic base. Experimentally, base-catalyzed reactions are faster than acid-catalyzed reactions. The reactions catalyzed by bases may be optimized at 60 °C and atmospheric pressure for 90 min (Bahadar and Khan 2013). In a basic homogeneous catalysis application for microalgal biomass, the presence of free fatty acids (FFA) in the raw material (more than 0.5 % $v\ v^{-1}$) hinders the application of bases for the transesterification reaction. This occurs because the FFA react with the basic catalyst and form soap. The saponification reaction decreases the availability of triglycerides to form the corresponding esters and increases the solubility of the esters formed, and glycerol esters are lost when separating the biofuel from the glycerol (Visentainer and Santos 2013). Therefore, acid catalysis may be an alternative and is not sensitive to the presence of FFA in the oil.

Acid-catalyzed transesterification involves catalytic acid and compared with alkaline catalysis, has a long reaction. When using acidic homogeneous catalysis, there is an esterification of the FFA (converted to alkyl ester), and transesterification occurs simultaneously. The optimum condition for 100 % catalyzed acidic transesterifications is a ratio 56:1 methanol: oil and a temperature of 30 °C (Bahadar and Khan 2013). However, another alternative would be the combination of acidic and basic catalysis. In this method of catalysis, lipids undergo pretreatment with acid to reduce the FFA content prior to performing the base-catalyzed transesterification reaction.

The enzyme catalyst stands out because it can be used with any source of triglycerides and is independent of the amount of FFA. With the application of enzymes, it is not necessary to carry out neutralization and washing of the biodiesel obtained because this is produced in the neutral range. This generates a much lower amount of waste than does using chemical catalysts. Moreover, this type of catalysis facilitates the recovery of glycerin because the enzymes are more specific than the chemical source facilitating the purification steps (Visentainer and Santos 2013). The transesterification using lipase at low temperatures is able to produce greater amounts of esters than other methods of transesterification. Using immobilized lipase at 75 % with 10 % water, a biodiesel conversion efficiency of 98.2 % is obtained in 12 h (Bahadar and Khan 2013).

Another method for lipid esterification in the formation of biodiesel is in situ transesterification. This reaction involves the extraction and esterification steps together in a single step. The use of this method to obtain biodiesel from microalgal biomass can be a viable option because it reduces the extraction expenses and reaction time (Pandey et al. 2014). Johnson and Wen (2009) demonstrated that the application of a direct method for obtaining biodiesel from *Schizochytrium limacinum* produced larger quantities of methyl esters than did conventional extraction methods.

5 Microalgae Bioethanol

Among the existing biofuels, it is assumed that ethanol will become more globally widespread (John et al. 2011). Approximately 9 % of the world production of ethanol is obtained synthetically from ethene, whereas the remaining 91 % is produced using a biochemical pathway known as bioethanol (Demirbas 2005). This biofuel is considered one of the most promising markets because it has emerged as the main alternative to supplement and replace gasoline in the transportation sector (Balat and Balat 2009). Bioethanol has biodegradable characteristics (John et al. 2011) and produces cleaner combustion than fossil fuels due to the presence of oxygen in their composition (Mussatto et al. 2010). The use of bioethanol also contributes to reducing greenhouse gas emissions (CO_2, NO_x, and SOx) and particulate matter into the atmosphere (Demirbas 2005); these are considered enhancers of global warming.

Bioethanol is derived from renewable sources, such as vegetable raw materials (wheat, beet, corn, straw, and wood), and has the general chemical formula $(CH_2O)_n$. These raw materials are composed of carbohydrates (cellulose and hemicellulose) that can be hydrolyzed and then converted in the fermentation step to bioethanol by the action of microorganisms, which are typically yeast (Demirbas 2008).

In the contemporary period, bioethanol is produced on a commercial scale from raw materials, such as corn, sugar cane, and beet (Scholz et al. 2013). The global production of biofuels has increased considerably since the oil crisis in 1970, and between 1975 and 2006, the market grew from fewer than one billion liters to more than 39 billion liters (Licht 2006). In this scenario, the United States and Brazil stand out being the main bioethanol producers using corn and sugarcane, respectively (Mussatto et al. 2010). In 1976, Brazil created a National Alcohol Program (Proálcool) and today is one of the most developed nations in the sector (Costa and Morais 2011), producing in harvest year 2013 and 2014 more than 28 million m^3 of ethanol (Brasil 2015).

The concerns about food security and the demand for food are the main limiting for the use of agricultural crops for the production of renewable fuels. Therefore, the use of alternative feedstocks is necessary for the sustainable development of biofuel production (Medipally et al. 2015).

In the energy panorama, the microalgal biomass has been considered interesting alternative to traditional crops for bioethanol production because it does not have the disadvantages inherent to first- or second-generation biofuels (Chen et al. 2013). Microalgae are potential substrates for fermentation because they have a composition that is high in carbohydrates. These compounds may be directly available for fermentation or available after a pretreatment of the biomass (Costa and Morais 2011).

The carbohydrate concentration in microalgae can vary between 4 and 64 % (Becker 1994), and this amount can be influenced by the species and growing conditions used (Chen et al. 2013). Microalgae, such as *Dunaliella*, *Spirogyra*, *Chlorella*, *Porphyrydium*, *Anabaena* and *Scenedesmus* have a high content of carbohydrates in their biomass (Becker 1994) and may be considered potential raw materials for bioethanol production.

The microalgae bioethanol production process has been investigated, but this biofuel is not yet produced on a commercial scale (Harun et al. 2010a). According to Cheryl (2008), the production of bioethanol from microalgal biomass is promising, with an estimated production between 46,760 and 140,290 L ha^{-1} $year^{-1}$, which is greater than the contribution from any other raw material.

In this direction, many studies have been developed to facilitate bioethanol production process from microalgal biomass. Choi et al. (2010) evaluated the bioethanol production potential from *Chlamydomonas reinhardtii*. The authors obtained 57 % efficiency in the bioethanol production process by using an enzymatic hydrolysis as the biomass pretreatment. Harun et al. (2010b) reported the production of 38 % w w^{-1} of ethanol by using biomass from the microalga *Chlorococcum* sp. as feedstock for fermentation.

The production of bioethanol from microalgal biomass consists of a process with several steps (Fig. 2), which begins with the cultivation of these microorganisms followed by the recovery of the biomass from the liquid medium, cell disruption for the release of the metabolite of interest, saccharification (hydrolysis), fermentation, and finally separation by distillation (Chen et al. 2013; Mussatto et al. 2010).

Bioethanol can be purified and used as fuel. The CO_2 produced during the fermentation of the microalgal biomass may be used as a nutrient in the culture itself, which closes the carbon cycle (Ferreira et al. 2012). The residual biomass from the pretreatment step can also be used as a substrate for the production of other

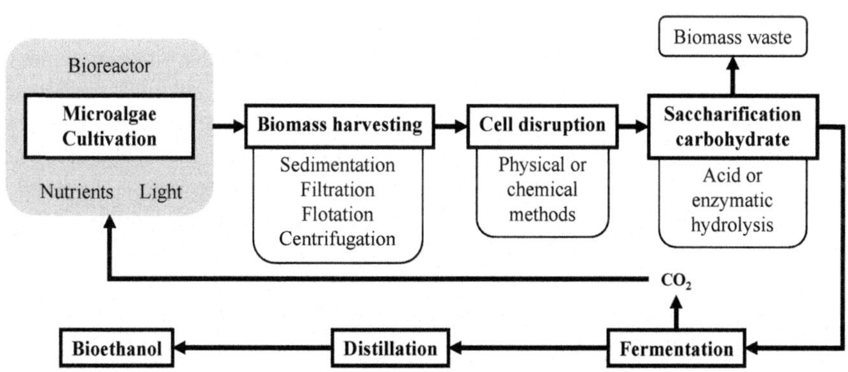

Fig. 2 Bioethanol production process from microalgal (adapted from Mussatto et al. 2010)

biofuels, such as biogas production from an anaerobic digestion process (Ueda et al. 1996; Harun et al. 2010a) (Fig. 2).

5.1 Composition of Carbohydrates in Microalgae

Carbohydrates present in microalgal biomass vary in both quantity and quality. In some strains of microalgae, 70 % of such compounds are found (Chu et al. 1982). In cells, carbohydrates can be accumulated in plastids, such as for storage (starch), or can become major cell wall components (cellulose, pectin and sulfated polysaccharides) (Rismani-Yazdi et al. 2011; Rangel-Yagui et al. 2004).

The walls of microalgae cells consist essentially of an inner layer and an outer layer. The composition of the outer cell wall varies according to species but usually contains specific polysaccharides, such as pectin, agar, and alginate. The inner layer of the cell wall of microalgae is mainly composed of cellulose and other cellulose materials (such as hemicellulose and glycoprotein) (Yamada and Sakaguchi 1982). As storage material, starch is the predominant carbohydrate in microalgae groups. Starch is composed of amylose and amylopectin and has a structure similar to that found in higher plants (Ball 2002). However, only green algae have the same starch accumulation mechanism inside the chloroplast (Dauvillée et al. 2009). Doucha and Lívanský (2009) confirmed that the microalgae *Chlorella* sp., under the culture conditions used to reduce the protein content, was able to accumulate starch as 70 % of the dry matter.

Among the microalgae monosaccharides, which include glucose, mannose, ribose/xylose, rhamnose, and fructose (Chu et al. 1982). For some microalgae, polymers of glucose produced by cellulose and starch are predominant components in the cell walls and in reservation products (Metting 1996). Therefore, the composition and the carbohydrate content in the microalgal biomass become a promising feedstock for bioethanol production.

In addition to the composition and content of carbohydrates in microalgae varying between species, these parameters can be influenced by nutritional factors (nitrogen, phosphorus and carbon), pH, light, temperature, and salinity (Chen et al. 2013). The starch in cyanobacteria is in general, accumulated by nitrogen depletion in the cyanobacteria *Synechococcus* sp. during cultivation and in *Synechocystis* sp. PCC 7002, *Arthrospira platensis* PCC 6803, *Arthrospira maxima*, *Anabaena variabilis*, and *Anacystis nidulans* (Xu et al. 2013; Lehmann and Wöber 1976).

In crops of *Chlorella* genus with nitrogen scarcity, an increase was observed in the carbohydrate content in the cells, and some strains had large amounts of stored starch (Richmond 1986). Studies by Margarites and Costa (2014) showed that greater levels of carbohydrates (69.2 % w w^{-1}) were verified when the microalgae *Chlorella minutissima* was cultivated with a lower concentration of nitrogen and without the addition of the phosphorus source.

The high salinity induced by adding 1.2 mol L^{-1} of NaCl in the cultivation of *Spirulina maxima* Sosa4 and *S. platensis* M2 reduced protein synthesis and carbohydrate accumulation (Tomaselli et al. 1993). Ho et al. (2012) evaluated the effect of light intensity on the carbohydrate content of *Scenedesmus obliquus* CNW-N. The authors found an increase in the glucose content in the biomass (53.4–73.1 %) when the light intensity increased from 60 to 180 μmol m^{-2} s^{-1}. Glucose represents approximately 73–80 % of the total carbohydrates produced from *S. obliquus* CNW-N, which confirms the potential of biomass for bioethanol production.

5.2 Carbohydrate Metabolism in Microalgae

Carbohydrates are the primary products derived from photosynthesis and metabolism of carbon fixation, which is known as the Calvin Benson cycle. Photosynthesis is a biological process that uses NADPH and ATP to convert the captured CO_2 from the air and produce sugars, such as glucose, and other metabolites. In the Calvin cycle, the first stage in the assimilation of CO_2 in biomolecules is the carbon fixation reaction: CO_2 condensation with a five-carbon acceptor and ribulose-1,5-bisphosphate forms two molecules of 3-phosphoglycerate. In the second stage, 3-phosphoglycerate is reduced to triose-phosphate. A total of three molecules of CO_2 are fixed to three molecules of ribulose 1,5-bisphosphate to form six molecules of glyceraldehyde-3-phosphate (18 carbon) that is in equilibrium with dihydroxyacetone phosphate. In the third stage, five of the six triose-phosphate molecules (15 carbons) are used to regenerate three molecules of ribulose-1,5-biphosphate (15 carbons), which is the starting material. The sixth triose-phosphate molecule remains in the liquid product of the photosynthesis and may be used to produce hexoses for fuel and building blocks or sucrose or starch for storage (Nelson and Cox 2011).

Among the fermentable carbohydrates, cellulose is a main compound present in green algae (Radakovits et al. 2010; Chen et al. 2013). The synthesis of the cellulose is a complex process that involves many enzymatic reactions. The starting substrate for the synthesis is UDP-glucose, which is formed from UDP and fructose reaction catalyzed for the synthesis of sucrose (Nelson and Cox 2011).

The metabolic pathways of energy-rich molecules (e.g., carbohydrates and lipids) are closely linked. Studies have shown that there is competition between the synthesis of lipids and carbohydrates (starch) as the main precursor for the synthesis of triacylglycerols (TAG) is glycerol-3-phosphate (G3P), which is produced by the catabolism of glucose (glycolysis). Accordingly, to increase the production of biofuels from microalgae, it is vital to understand and manipulate the metabolisms related to the achievement of greater carbohydrate accumulation, such as an increase of glucan and the storage and reduction of starch degradation (Ho et al. 2012; Rismani-Yazdi et al. 2011).

With the development of genetic engineering and a better understanding of carbohydrate metabolism, microalgal strains that accumulate higher concentrations of this compound can be developed. Therefore, the biofuel production process from microalgal biomass becomes more efficient and feasible than when using other raw materials (Chen et al. 2013).

5.3 Bioethanol Production Through Carbohydrates Extracted from Microalgal Biomass

The bioethanol production process mainly includes upstream and downstream operations that involve the pretreatment of biomass, saccharification, fermentation, and product recovery. The pretreatment of biomass is considered an important step in the process because it is essential that fermentable sugars are released and made available to the fermentation process (Rabelo et al. 2009; Hernández et al. 2015). The cell wall rupture should be performed because the majority of carbohydrates are located either inside the cell wall (such as cellulose and hemicellulose) or intracellularly in the form of starch (Domozych et al. 2012; Richmond 2004).

The pretreatment of biomass also contributes to an increase in the surface area and the solubility of the sugars and improves the digestibility of substrate (Harun et al. 2011). Different methods have been tested to promote cell rupture and release of cell wall carbohydrates into monosaccharides, such as physical methods including homogenized with high pressure, microwave, ultrasound, and heat (Halim et al. 2012; Hernández et al. 2014). Most carbohydrates in microalgal biomass are not easily fermentable to produce bioethanol. Therefore, before the fermentation step, the polysaccharide must be hydrolyzed into fermentable sugars (Hahn-Hägerdal et al. 2007) through the process called saccharification.

Saccharification is one of the most important steps because at this step conversion of complex polysaccharides occurs, which is mainly cellulose and starch into simple carbohydrates, through chemical, enzymatic and microbiological means (Anisha and John 2012). The saccharification process may be influenced by several factors, such as the crystallinity of the cellulose, the substrate surface area, the cell wall thickness, the porosity, the mass transfer, and the hemicellulose or lignin content (Alvira et al. 2010). Microalgae have a little or no lignin in their composition. Therefore, these microorganisms can be classified as cellulose-based materials. In this context, the cellulase enzymes obtained from fungi, bacteria, or protozoa may be applied to microalgal biomass hydrolysis (Harun and Danquah 2011b).

In general, hydrolysis may be carried out using acids or alkalis or by enzymatic means. Although acid hydrolysis is faster, easier, and less expensive, the acidic conditions may lead to decomposition of the sugars into unwanted compounds that inhibit the fermentation process (Gírio et al. 2010). In contrast to acid hydrolysis, enzymes are slower and more costly (Lynd et al. 2002). This less harmful environment can provide higher yields of glucose without creating inhibitory products (Ho et al. 2013).

After the saccharification process occurs, the hydrolyzed carbohydrates are fermented for bioethanol production using bacteria (*Zymomonas mobilis*, *Escherichia coli*), yeasts (*Saccharomyces cerevisiae*, *Candida* and *Pichia*), or filamentous fungi (Simas-Rodrigues et al. 2015; Anisha and John 2012). The fermentation process happens through the anaerobic degradation of organic substrates (carbohydrate or not), such as glucose, by microorganisms in the products to produce energy in the form of ATP. In this process, when the major products are ethyl alcohol and CO_2, the process is alcoholic fermentation (Lehninger et al. 2000), which has its overall reaction shown in Eq. (1) and a theoretical yield for ethanol formation of 0.511 $g_{ethanol}\ g_{glucose}^{-1}$.

$$C_6H_{12}O_6 \rightarrow 2C_2H_6O + 2CO_2 \quad (1)$$

Microalgal biomass fermentation can occur in two ways: separate hydrolysis and fermentation or simultaneous. When separated, the main advantages are providing the ideal conditions for hydrolysis and fermentation, the application of inexpensive chemicals, simple equipment, and less processing time (Thu et al. 2009). Moreover, when saccharification and fermentation occur simultaneously, greater hydrolysis rates and product yields are produced, and less sterile conditions, fewer enzymes, smaller operating volumes, shorter processing times, and lower costs are required (Balat et al. 2008).

Microalgal biomass for bioethanol has been produced using different strains and with varying yields, as shown in Table 2. Shirai et al. (1998) found that the use of hydrolysis from *Dunaliella* sp. biomass with commercial glucoamylase promoted the fermentation process and resulted in the conversion of 0.011 $g_{ethanol}\ g_{biomass}^{-1}$.

Table 2 Concentration of carbohydrates (% w w^{-1}) and microalgal biomass conversion to ethanol ($Y_{ethanol\ biomass}^{-1}\ g_{ethanol}\ g_{biomass}^{-1}$)

Microalgae	Carbohydrates concentration			$Y_{ethanol\ biomass}^{-1}$	References
	Total	*	**		
Chlamydomonas reinhardtii	59.7	43.6	44.7	0.235	Choi et al. (2010)
Chlorococcum humicola	32.5	11.3	15.2	0.520	Harun and Danquah (2011a)
Chlorococcum sp.	–	–	–	0.383	Harun et al. (2010b)
Chlorococcum infusionum	32.5	11.3	15.2	0.261	Harun et al. (2011)
Chlorella vulgaris	50.9	–	48.0	0.209	Ho et al. (2013)
Chlorella vulgaris	–	–	–	0.400	Lee et al. (2011)
Dunaliella sp.	–	–	–	0.011***	Shirai et al. (1998)

*% starch; **% glucose; *** theoretical conversion

Harun et al. (2010b) found that in *Chlorococcum* sp. bioethanol conversion, without biomass pretreatment, the lipid extraction process facilitated the release of fermentable sugars.

The yield differences observed in these studies (Table 2) can be attributed to fermentable carbohydrate concentrations in each microalgae and the ethanol conversion rate used in the process. The assortment of carbohydrates in microalgae makes it efficient direct use improbable especially when using *S. cerevisiae* is a necessary step of pretreatment and saccharification of the biomass to turn the biomass into fermentable carbohydrates. Furthermore, the microalgal biomass as substrate for fermentation can affect the mass balance with respect to ethanol because it does not take into account other macromolecules, such as lipids and proteins (Doan et al. 2012).

5.4 Direct Bioethanol Production in Microalgal Cultivation

In addition to the conventional production of ethanol through carbohydrates fermentation, other ways of producing this metabolite have been investigated. Direct production of ethanol in microalgal cultivation can occur when the microorganisms perform self-fermentation of intracellular carbohydrates. Another way in which direct production occurs is when genetic engineering is used by modifying the metabolic pathway of the carbon cycle.

Self-fermentation occurs when microalgae, in light and dark anaerobic environments, use intracellular starch to produce ethanol. However, several studies have reported that this process has low yield compared with ethanol production using conventional fermentation (Doan et al. 2012). Hirano et al. (1997) evaluated the production of ethanol from intracellular and conventional fermentation. The authors found that ethanol production through intracellular fermentation by *C. reinhardtii* (UTEX2247) showed a maximum concentration of 1 % ethanol (w w^{-1}). This study also indicated that the intracellular production of ethanol is simpler and less energy intensive than is the conventional process.

According to Puig et al. (2007), another factor that should be considered when auto fermentation occurs is the presence of contaminants if axenic cultures have not been employed. The secreted ethanol, even in low concentrations, can be used by contaminant microorganisms as nutrients and can limit the process efficiency.

Genetic engineering can also enable the production of compounds by microalgae by changing the metabolic pathways. The photosynthesis of microalgae is mainly based on the Calvin cycle, in which ribulose-1,5-bisphosphate in combination with CO_2 produces two molecules of 3-phosphoglycerate (3-PGA) that is used for the synthesis of glucose and other metabolites. Accordingly, attempts have been made to redirect the 3-PGA in ethanol by introducing genes that produce ethanol (alcohol dehydrogenase and pyruvate decarboxylase) (Wahlund et al. 1996).

Deng and Coleman (1999) reported the production of ethanol by genetic engineering methods in microalgae *Synechococcus* sp. strain PCC 7942. In this investigation, new genes were introduced to create a new route for the use of fixed carbon and resulted in ethanol synthesis. The encoding sequences for pyruvate decarboxylase (pdc) and alcohol dehydrogenase II (adh) from *Z. mobilis* bacterium were cloned into the vector pCB4 transport and were then used to transform *Synechococcus* sp. PCC 7942. The modified microalgae were able to synthesize ethanol, which diffused from the cells into the culture medium.

6 Industrial Effluents Use as Nutrients in Microalgal Cultivation

Microalgae are a promising alternative for sustainable biofuel production; however, methods to reduce production costs must be optimized (Leite et al. 2015). The use of industrial effluents is an alternative treatment of these wastes together with reduction of costs for cultivation for microalgal production of biofuel and other by-products (Table 3). Microalgae have the capacity to absorb different carbon sources for use in mixotrophic and heterotrophic cultures. In this way, microalgae are being studied for treating effluents, such as glycerol and gaseous effluents from flue gases (Chen et al. 2012; Leite et al. 2015). Furthermore, microalgae can effectively absorb nitrogen and phosphorus, which offers a possible alternative for the removal of inorganic nutrients for the treatment of wastewater and reducing eutrophication problems (Yang et al. 2011).

Table 3 Microalgae for effluent treatment

Microalga	Effluent	References
Chlorella ellipsoidea YJ1	Domestic secondary effluents	Yang et al. (2011)
Chlorella sp. ArM0029B	Residual nitrogen from biologically treated coke effluent	Chen et al. (2012)
Chlorella vulgaris		
Rhodosporidium toruloides	Distillery and domestic mixed wastewater	Ling et al. (2013)
Chlorella pyrenoidosa		
Scenedesmus sp. AMDD	Municipal wastewater	McGinn et al. (2012)
Chlorella vulgaris	Wastewater from ethanol and citric acid production	Valderrama et al. (2002)
Chorella sp.	Carbon dioxide capture from biogas	Chen et al. (2012)
Chlorella sp.	Biodiesel-derived glycerol	Leite et al. (2015)
Spirulina sp. LEB 18	Coal combustion flue gas	Morais and Costa (2008)
Spirulina sp. LEB 18	Effluent alcohol industry	Pandey et al. (2014)

Existing technologies to treat industrial and domestic effluents are expensive or only partially effective in removing pollutants. Microalgae growing in wastewater reduce the costs of treating this effluent, which can be a nutrient source for these microorganisms. Microalgae consume the waste for growth and for the production of lipids and carbohydrates that can be applied to obtaining biofuels. Inorganic nutrients are removed from wastewater, which reduces the risk of algae growth when wastewater is directly discharged to natural waters. The secondary effluent from domestic wastewater has low chemical oxygen demand (COD) and high concentrations of nitrogen and phosphorus, which is an appropriate culture medium for microalgal cultivation (McGinn et al. 2012).

The microalgae CO_2 biofixation mechanism is based on the ability of these microorganisms to perform photosynthesis but with greater attachment rates than terrestrial plants. Furthermore, combustion gases can be injected directly into the crops from the emitting source without cooling because many species are considered extremophiles and can withstand high temperatures.

Practical applications from waste effluents with microalgae are carried out in different parts of world. In Brazil, since 1996, the Biochemical Engineering Laboratory (LEB) at the Federal University of Rio Grande (FURG) has researched microalgae cultivation. The LEB has, since 2005, maintained a *Spirulina* production plant located in the President Medici Thermal Power Plant operated by the Central of Electricity Generation (CGTEE). The plant aims to conduct CO_2 biofixation from flue gases released by coal burning (Morais and Costa 2007). Ouro Fino Animal Health (Ribeirão Preto, SP, Brazil) uses waste from the ethanol industry for microalga biomass production and treats this waste (Pandey et al. 2014).

7 Microalgal Biorefinery for Production of Biofuels and Coproducts

The energetic planet needs are primarily met by the application of fossil fuels. However, these sources are finite and implementation and exploitation causes environmental problems, such as an increase of combustion gases generated from the burning of these materials. The search for alternative renewable sources has intensified to meet the rising demand for energy and raw materials. One of the potential sources to meet this challenge is a biomass that can be applied as a replacement or in conjunction with these fossil fuels (Yen et al. 2013). As a result, an industrial complex has been proposed similar to a petroleum refinery called a "biorefinery" to produce energy and chemicals from biomass.

Due to the premises of the energy crisis, global warming and climate change derived from the exploitation of fossil fuels, the microalgal biomass has been listed as a potential source of biofuels. Furthermore, microalgal cultures do not require large amounts of water; they have high productivity and have the possibility of

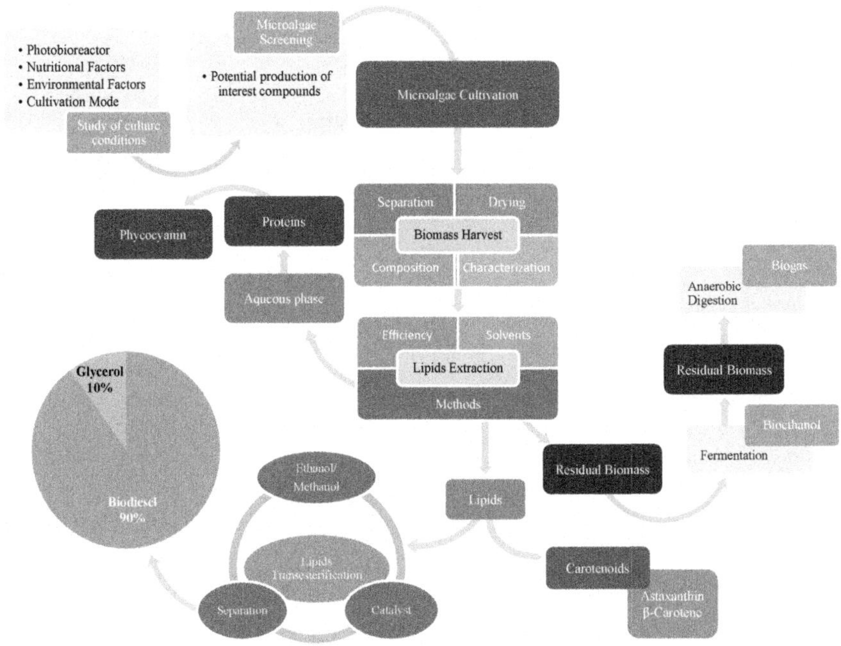

Fig. 3 Schematic diagram of a microalgal photobiorefinery

making this daily harvest even more attractive feedstock for sustainable power generation (Damiani et al. 2010). However, the application of these microorganisms for this purpose has a greater viability in combination with a biorefinery system (Zhu 2015). The microalgal photobiorefinery concept integrates the production of biofuels to coproducts with applications in food, pharmaceutical, and cosmetics with microalgae cultivation making the process of obtaining biofuels economically viable (Morais et al. 2015). High added value products, such as carotenoids and phycocyanin pigments can be produced, while also obtained biodiesel, bioethanol, and biomethane (Fig. 3).

Microalgae, such as *Spirulina* and *Chlorella* species, that have production capacity for a large amount of by-products have recommended to serve as the basis for microalgal photobiorefineries. These species are GRAS certified (generally recognized as safe), which indicates that their application in the food and cosmetic industries is allowed as long as the management and production comply with good manufacturing practices regulations (Andrade and Costa 2007). Biocompounds from the microalgal biomass may be obtained as proteins, carbohydrates, and lipids and can be converted into biofuels and high value-added compounds as natural dyes and fatty acids (Table 4).

Table 4 Microalgae, bioproducts and application

Microalgae	Bioproduct	Application	References
Spirulina fusiformis	C-phycocyanin	Pharmaceutical, cosmetic and food industry	Markou and Nerantzis (2013)
Nostoc muscorum	Phenolic		Mostafa (2012)
Chlorella protothecoides	Lutein		Mostafa (2012)
Dunaliella salina	β-carotene		Markou and Nrantzis (2013)
Chlorella vugaris	Fatty acids		Mitra et al. (2012)
Nannochloropsis gaditana	Biodiesel	Bioenergetic industry	Macías-Sánchez et al. (2009)
Spirulina maxima	Biodiesel		Baunillo et al. (2012)
Spirulina máxima	Biohydrogen		Ananyev et al. (2008)
Spirulina sp. LEB 18	Biomethane		Costa et al. (2008)
Chorella vulgaris FSP-E	Bioethanol		Shih-Hsin et al. (2013)
Chlamydomonas reinhardtii	Biohydrogen		Burgess et al. (2011)

8 Conclusion

The processes for the production of microalgae biofuels have been studied by many research centers in the biotechnology area. Biodiesel and bioethanol are the most studied due to their high demand, mostly in the transport area. Research has primarily involved the culture conditions focusing on obtaining greater concentrations of lipids and carbohydrates, more efficient processes and greater performance with respect to the extraction of biocompounds. The microalgae can be considered a more sustainable form of biofuel production. The application of industrial wastewater as a substrate in the culture medium and the introduction of the concept of the photobiorefinery are alternatives for enabling biofuel production from microalgal biomass.

References

Abad S, Turon X (2012) Valorization of biodiesel derived glycerol as a carbon source to obtain added-value metabolites: focus on polyunsaturated fatty acids. Biotechnol Adv 30:733–741

Adarme-Veja TC, Thomas-Hall SR, Schenk PM (2014) Towards sustainable sources for ômega-3 fatty acids production. Curr Opin Biotechnol 26:14–18

Alvira P, Tomás-Pejó E, Ballesteros M et al (2010) Pretreatment technologies for an efficient bioethanol production process based on enzymatic hydrolysis: a review. Bioresour Technol 101:4851–4861

ASTM D6751-15ce1 (2015) Standard specification for biodiesel fuel blend stock (B100) for middle distillate fuels. ASTM international, west conshohocken, PA

Ananyev G, Carrieri D, Dismukes GC (2008) Optimization of metabolic capacity and flux through environmental cues to maximize hydrogen production by the Cyanobacterium *Arthrospira* (*Spirulina*) *maxima*. Appl Environ Microbiol 74:6102–6113

Andrade MR, Costa JAV (2007) Mixotrophic cultivation of microalga *Spirulina platensis* using molasses as organic substrate. Aquaculture 264:130–134

Anisha GS, John RP (2012) Microalgae as an alternative feed stock for green biofuel technology. In: Johansen MN (ed) Microalgae biotechnology, microbiology and energy. Nova, New York, pp 277–294

Bahadar A, Khan MB (2013) Progress in energy from microalgae: a review. Renew Sustain Energy Rev 27:128–148

Balat M, Balat H (2009) Recent trends in global production and utilization of bio-ethanol fuel. Appl Energy 86:2273–2282

Balat M, Balat H, Öz C (2008) Progress in bioethanol processing. Prog Energy Combust Sci 34:551–573

Ball SG (2002) The intricate pathway of starch biosynthesis and degradation in the monocellular alga *Chlamydomonas reinhardtii*. Aust J Chem 55:49–59

Baunillo KE, Tan RS, Barros HR et al (2012) Investigations on microalgal oil production from *Arthrospira platensis*: towards more sustainable biodiesel production. RSC Adv 2:11267–11272

Beatriz A, Araújo YJK, Lima DP (2011) Glycerol: a brief history and their application in stereoselective syntheses. Quim Nova 34(2):306–319

Becker EW (1994) Microalgae: biotechnology and microbiology, 1st edn. Cambridge University Press, Cambridge, pp 178

Bligh EG, Dyer W (1959) A rapid method for total lipid extraction and purification. Can J Biochem Physiol 37:911–917

Brasil (2015) Ministério da Agricultura, Pecuária e Abastecimento. Secretaria de Produção e Agroenergia. Departamento da Cana-de-Açúcar e Agroenergia. Brazilian Sugarcane, Sugar and Ethanol Production Harvest Year 13/14 (in Portuguese)

Brennan L, Owende P (2010) Biofuels from microalgae—a review of technologies for production, processing, and extractions of biofuels and co-products. Renew Sustain Energy Rev 14:557–577

Burgess SJ, Tamburic B, Zemichael F et al (2011) Solar-driven hydrogen production in green algae. Adv Appl Microbiol 75:71–110

Chen C, Chang H, Kao P et al (2012) Biosorption of cadmium by CO_2-fixing microalga *Scenedesmus obliquus* CNW-N. Bioresour Technol 105:74–80

Chen CY, Zhao XQ, Yen HW et al (2013) Microalgae-based carbohydrates for biofuel production. Biochem Eng J 78:1–10

Cheryl (2008) Algae becoming the new biofuel of choice. Available via http://duelingfuels.com/biofuels/non–food–biofuels. Accessed 15 June 2015

Chisti Y (2008) Biodiesel from microalgae beats bioethanol. Trends Biotechnol 26:126–131

Chiu S-Y, Kao C-Y, Chen T-Y et al (2015) Cultivation of microalgal *Chlorella* for biomass and lipid production using wastewater as nutrient resource. Bioresour Technol 184:179–189

Choi SP, Nguyen MT, Sim SJ (2010) Enzymatic pretreatment of *Chlamydomonas reinhardtii* biomass for ethanol production. Bioresour Technol 101:5330–5336

Chojnacka K, Marques-Rocha FJ (2004) Kinetic and stoichiometric relationships of the energy and carbon metabolism in the culture of microalgae. Biotechnology 3:21–34

Chu F-LE, Dupuy JL, Webb KL (1982) Polysaccharide composition of five algal species used as food for larvae of the American oyster, *Crassostrea virginica*. Aquaculture 29(3–4):241–252

Costa JAV, Morais MG (2011) The role of biochemical engineering in the production of biofuels from microalgae. Bioresour Technol 102:2–9
Costa JAV, Santana FB, Andrade MR et al (2008) Microalga biomass and biomethane production in the south of Brazil. J Biotechnol 136:S430–S430
Damiani MC, Popovich CA, Constenla D et al (2010) Lipid analysis in *Haematococcus pluvialis* to assess its potential use as a biodiesel feedstock. Bioresour Technol 101:3801–3807
Dauvillée D, Deschamps P, Ral J-P et al (2009) Genetic dissection of floridean starch synthesis in the cytosol of the model dinoflagellate *Crypthecodinium cohnii*. Proc Natl Acad Sci USA 106 (50):21126–21130
Demirbas A (2005) Bioethanol from cellulosic materials: a renewable motor fuel from biomass. Energy Sources 27:327–337
Demirbas A (2006) Biodiesel production via non-catalytic SCF method and biodiesel fuel characteristics. Energy Convers Manage 47:2271–2282
Demirbas A (2008) Biofuels sources, biofuel policy, biofuel economy and global biofuel projections. Energy Convers Manage 49:2106–2116
Demirbas MF (2011) Biofuels from algae for sustainable development. Appl Energy 88:3473–3480
Deng MD, Coleman JR (1999) Ethanol synthesis by genetic engineering in cyanobacteria. Appl Environ Microbiol 65:523–528
Doan QC, Moheimani NR, Mastrangelo AJ et al (2012) Microalgal biomass for bioethanol fermentation: implications for hypersaline systems with an industrial focus. Biomass Bioenerg 46:79–88
Domozych DS, Ciancia M, Fangel JU et al (2012) The cell walls of green algae: a journey through evolution and diversity. Front Plant Sci 3(82):1–7
Doucha J, Lívanský K (2009) Outdoor open thin-layer microalgal photobioreactor: potential productivity. J Appl Phycol 21:111–117
Dragone G, Fernandes B, Vicente AA et al (2010) Third generation biofuels from microalgae. In: Méndez-Vilas A (ed) Current research, technology and education topics in applied microbiology microbial biotechnology. Formatex, pp 1355–1366
Dunstan GA (1993) Changes in the lipid composition and maximisation of the polyunsaturated fatty acid content of three microalgae grown in mass culture. J Appl Phycol 5(1):71–83
European Standards, European Norms, EM 14214:2003
FAO (2007) Sustainable bioenergy: a framework for decision makers. Food and Agriculture Organization, New York
FAO (2008) The state of food and agriculture. Food and Agriculture Organization, New York
Ferreira LS, Rodrigues MS, Converti A et al (2012) *Arthrospira* (*Spirulina*) *platensis* cultivation in tubular photobioreactor: use of no-cost CO_2 from ethanol fermentation. Appl Energy 92:379–385
Folch J, Lees M, Stanley GHS (1957) A simple method for the isolation and purification of total lipides from animal tissues. J Biol Chem 226(1):497–509
Gao C, Zhai Y, Ding Y et al (2010) Application of sweet sorghum for biodiesel production by heterotrophic microalga *Chlorella protothecoides*. Appl Energy 87:756–761
Gírio FM, Fonseca C, Carvalheiro F et al (2010) Hemicelluloses for fuel ethanol: a review. Bioresour Technol 101:4775–4800
Hahn-Hägerdal B, Karhumaa K, Fonseca C et al (2007) Towards industrial pentose-fermenting yeast strains. Appl Microbiol Biotechnol 74(5):937–953
Halim R, Danquah MK, Webley PA (2012) Extraction of oil from microalgae for biodiesel production: a review. Biotechnol Adv 30:709–732
Harun R, Danquah MK (2011a) Influence of acid pre-treatment on microalgal biomass for bioethanol production. Process Biochem 46:304–309
Harun R, Danquah MK (2011b) Enzymatic hydrolysis of microalgal biomass for bioethanol production. Chem Eng J 168:1079–1084
Harun R, Singh M, Forde GM, Danquah MK (2010a) Bioprocess engineering of microalgae to produce a variety of consumer products. Renew Sust Energ Rev 14:1037–1047

Harun R, Danquah MK, Forde GM (2010b) Microalgal biomass as a fermentation feedstock for bioethanol production. J Chem Technol Biotechnol 85:199–203

Harun R, Jason WSY, Cherrington T et al (2011) Exploring alkaline pre-treatment of microalgal biomass for bioethanol production. Appl Energy 88:3464–3467

Henrikson R (1994) Microalga *Spirulina*: superalimento del futuro. Urano DL, Barcelona (in Spanish)

Hernández D, Solana M, Riaño B et al (2014) Biofuels from microalgae: lipid extraction and methane production from the residual biomass in a biorefinery approach. Bioresour Technol 170:370–378

Hernández D, Riaño B, Coca M et al (2015) Saccharification of carbohydrates in microalgal biomass by physical, chemical and enzymatic pre-treatments as a previous step for bioethanol production. Chem Eng J 262:939–945

Hirano A, Ueda R, Hirayama S et al (1997) CO_2 fixation and ethanol production with microalgal photosynthesis and intracellular anaerobic fermentation. Pergamon 22:137–142

Ho SH, Chen CY, Lee DJ et al (2011) Perspectives on microalgal CO_2-emission mitigation systems—a review. Biotechnol Adv 29(2):189–198

Ho SH, Chen CY, Chang JS (2012) Effect of light intensity and nitrogen starvation on CO_2 fixation and lipid/carbohydrate production of an indigenous microalga *Scenedesmus obliquus* CNW-N. Bioresour Technol 113:244–252

Ho SH, Huang SW, Chen CY et al (2013) Bioethanol production using carbohydrate-rich microalgae biomass as feedstock. Bioresour Technol 135:191–198

Hoekman SK, Robbins C (2012) Review of the effects of biodiesel on NOx emissions. Fuel Process Technol 96:237–249

Jaecker-Voirol A, Durand I, Hillion G et al (2008) Glycerin for new biodiesel formulation. Oil Gas Sci Technol 63:395–404

John RP, Anisha GS, Nampoothiri KM et al (2011) Micro and macroalgal biomass: a renewable source for bioethanol. Bioresour Technol 102:186–193

Johnson MB, Wen ZY (2009) Production of biodiesel fuel from the microalga *Schizochytrium limacinum* by direct transesterification of algal biomass. Energ Fuels 23:5179–5183

Kates M (1986) Lipid extraction procedures techniques of lipidology isolation, analysis, and identification of lipids. Elsevier Science Publisher, Amsterdan

Khan SA, Hussain MZ, Prasad S et al (2009) Prospects of biodiesel production from microalgae in India. Renew Sustain Energy Rev 13:2361–2372

Lee RE (1980) Phycology, 1st edn. Cambridge University Press, Cambridge

Lee S, Oh Y, Kim D et al (2011) Converting carbohydrates extracted from marine algae into ethanol using various ethanolic *Escherichia coli* strains. Appl Biochem Biotechnol 164:878–888

Lehmann M, Wöber G (1976) Accumulation, mobilization and turn-over of glycogen in the blue-green bacterium *Anacystis nidulans*. Arch Microbiol 111:93–97

Lehninger AL, Nelson DL, Cox MM (2000) Lehninger principles of biochemistry, 3rd edn. Worth Publishers, New York

Leite GB, Paranjape K, Ahmed EM et al (2015) Utilization of biodiesel-derived glycerol or xylose for increased growth and lipid production by indigenous microalgae. Bioresour Technol 184:123–130

Li Y, Horsman M, Wu N et al (2008) Biofuels from microalgae. Biotechnol Prog 24:815–820

Licht FO (2006) World ethanol markets: the outlook to 2015. Agra Europe Special Report, Tunbridge Wells

Ling J, Nip S, Shim H (2013) Enhancement of lipid productivity of *Rhodosporidium toruloides* in distillery wastewater by increasing cell density. Bioresour Technol 146:301–309

Lourenço SO (2006) Cultivo de Microalgas Marinhas: Princípios e Aplicações. RiMa, São Paulo, Brasil (in Portuguese)

Lynd LR, Weimer PJ, van Zyl WH et al (2002) Microbial cellulose utilization: fundamentals and biotechnology. Microbiol Mol Biol Rev 66:506–577

Macías-Sánchez MD, Mantell C, Rodríguez M et al (2009) Comparison of supercritical fluid and ultrasound-assisted extraction of carotenoids and chlorophyll a from *Dunaliella salina*. Talanta 77:948–952
Madigan MT, Martinko JM, Dunlap PV et al (2010) Microbiologia de brock, 12th edn. Artmed
Maity JP, Bundschuh J, Chen CY et al (2014) Microalgae for third generation biofuel production, mitigation of greenhouse gas emissions and wastewater treatment: present and future perspectives—a mini review. Energy 78:104–113
Margarites ACF, Costa JAV (2014) Increment of carbohydrate concentration of *Chlorella minutissima* microalgae for bioethanol production. IJERA 4(11):80–86
Markou G, Nerantzis E (2013) Microalgae for high-value compounds and biofuels production: a review with focus on cultivation under stress conditions. Biotechnol Adv 31:1532–1542
Mcginn PJ, Dickinson KE, Park KC et al (2012) Assessment of the bioenergy and bioremediation potentials of the microalga *Scenedesmus* sp. AMDD cultivated in municipal wastewater effluent in batch and continuous mode. Algal Res 1(2):155–165
Medina AR, Grima EM, Gimenez AG et al (1998) Downstream processing of algal polyunsaturated fatty acids. Biotechnol Adv 16:517–580
Medipally SR, Yusoff F, Banerjee S et al (2015) Microalgae as sustainable renewable energy feedstock for biofuel production. BioMed Res Int 2015:1–13
Metting FB (1996) Biodiversity and application of microalgae. J Ind Microbiol Biotechnol 17:477–489
Mitra D, Van Leeuwen J, Lamsal B (2012) Heterotrophic/mixotrophic cultivation of oleaginous *Chlorella vulgaris* on industrial industrial co-products. Algal Res 1:40–48
Morais MG, Costa JAV (2007) Biofixation of carbon dioxide by *Spirulina* sp. and *Scenedesmus obliquus* cultivated in a three-stage serial tubular photobioreactor. J Biotechnol 129:439–445
Morais MG, Costa JAV (2008) Bioprocesses for removal of carbon dioxide and nitrogen oxide by microalgae for the utilization of gas generated during coal burning. Quim Nova 31(5):1038–1042
Morais MG, Vaz BS, Morais EG et al (2015) Biologically active metabolites synthesized by microalgae. BioMed Res Int 15:1–15
Moreno-Garrido I (2008) Microalgae immobilization: current techniques and uses. Bioresour Technol 99:3949–3964
Mostafa SM (2012) Microalgal biotechnology: prospects and applications soha additional information is available at the end of the chapter, Chapter 12, INTEC
Mussatto SI, Dragone G, Guimarães PMR et al (2010) Technological trends, global market, and challenges of bio-ethanol production. Biotechnol Adv 28(6):817–830
Naik SN, Goud VV, Rout PK et al (2010) Production of first and second-generation biofuels: a comprehensive review. Renew Sustain Energy Rev 14(2):578–597
Naureen R, Tariq M, Yusoff I et al (2015) Synthesis, spectroscopic and chromatographic studies of sunflower oil biodiesel using optimized base catalyzed methanolysis. Saudi J Biol Sci 22 (3):332–339
Nautiyal P, Subramanian KA, Datisdar MG (2014) Production and characterization of biodiesel from algae. Fuel Process Technol 120:79–88
Nelson DL, Cox MM (2011) Lehninger principles of biochemistry, 5th edn. W. H. Freeman
Nigam PS, Singh A (2011) Production of liquid biofuels from renewable resources. Prog Energy Combust Sci 37:52–68
Pandey A, Lee D-J, Chisti Y et al (eds) (2014) Biofuels from algae, 1st edn. Elsevier, Amsterdam
Puig S, Coma M, van Loosdrecht MCM et al (2007) Biological nutrient removal in a sequencing batch reactor using ethanol as carbon source. J Chem Technol Biotechnol 82:898–904
Rabelo SC, Maciel Filho R, Costa AC (2009) Lime pretreatment of sugarcane bagasse for bioethanol production. Appl Biochem Biotechnol 153(1–3):139–150
Radakovits R, Jinkerson RE, Darzins A et al (2010) Genetic engineering of algae for enhanced biofuel production. Eukaryot Cell 9:486–501

Radmann EM, Camerini FV, Santos TD et al (2011) Isolation and application of SO_X and NO_X resistant microalgae in biofixation of CO_2 from thermoelectricity plants. Energy Convers Manage 52:3132–3136
Ramírez-Verduzco LF, Rodríguez-Rodríguez JE, Jaramillo-Jacob ADR (2012) Predicting cetane number, kinematic viscosity, density and higher heating value of biodiesel from its fatty acid methyl ester composition. Fuel 91:102–111
Rangel-Yagui CDO, Danesi EDG, De Carvalho JCM et al (2004) Chlorophyll production from *Spirulina platensis*: cultivation with urea addition by fed-batch process. Bioresour Technol 92:133–141
Richmond A (1986) Microalgae of economic potential. In: Handbook of microalgal mass culture. CRC Press, Boston, pp 199–243
Richmond A (ed) (2004) Handbook of microalgae culture: biotechnology and applied phycology. Blackwell Science, Hoboken
Rismani-Yazdi H, Haznedaroglu BZ, Bibby K et al (2011) Transcriptome sequencing and annotation of the microalgae *Dunaliella tertiolecta*: pathway description and gene discovery for production of next-generation biofuels. BMC Genomics 12:148
Russo D, Dassisti M, Lawlor V et al (2012) State of the art of biofuels from pure plant oil. Renew Sustain Energy Rev 16(6):4056–4070
Sarin R, Sharma M, Sinhara YS et al (2007) Jatropha-palm biodiesel blends: an optimum mix for Asia. Fuel 86:1365–1371
Scapin AR (2005) Cultivo da microalga *Spirulina platensis* para obtenção de biomassa de alto valor nutricional. Argos, Chapecó (in Portuguese)
Schmidell W, Lima UA, Aquarone E et al (2001) Biotecnologia Industrial, 1st edn. Edgard Blücher LTDA, Brasil (in Portuguese)
Scholz MJ, Riley MR, Cuello JL (2013) Acid hydrolysis and fermentation of microalgal starches to ethanol by the yeast *Saccharomyces cerevisiae*. Biomass Bioenergy 48:59–65
Scragg AH, Morrison J, Shales SW (2003) The use of a fuel containing *Chlorella vulgaris* in a diesel engine. Enzyme Microb Technol 33:884–889
Shirai F, Kunii K, Sato C et al (1998) Cultivation of microalgae in the solution from the desalting process of soy sauce waste treatment and utilization of the algal biomass for ethanol fermentation. World J Microbiol Biotechnol 14:839–842
Simas-Rodrigues C, Villela HDM, Martins AP et al (2015) Microalgae for economic applications: advantages and perspectives for bioethanol. J Exp Bot 1–12
Sims REH, Mabee W, Saddler JN et al (2010) An overview of second-generation biofuel technologies. Bioresour Technol 101:1570–1580
Singh SP, Singh D (2010) Biodiesel production through the use of different sources and characterization of oils and their esters as the substitute of diesel: a review. Renew Sustain Energy Rev 14:200–216
Spolaore P, Joannis-Cassan C, Duran E et al (2006) Commercial applications of microalgae. J Biosci Bioeng 101:87–96
Subhadra BG (2010) Sustainability of algal biofuel production using integrated renewable energy park (IREP) and algal biorefinery approach. Energy Policy 38:5892–5901
Thelen JJ, Ohlrogge JB (2002) Metabolic engineering of fatty acid biosynthesis. Plant Metab Eng 4:12–21
Thu NM, Choi SP, Lee J et al (2009) Hydrothermal acid pretreatment of *Chlamydomonas reinhardtii* biomass for ethanol production. J Microbiol 19:161–166
Ueda R, Hirayama S, Sugata K, Nakayama H (1996) Process for the production of ethanol from microalgae. US Patent 5,578,472, 26 Nov 1996
Tomaselli L, Giovannetti L, Torzillo G (1993) Physiology of stress response in *Spirulina* spp. Bull Inst Océanogr Monaco 12:65–75
Valderrama LT, Del Campo CM, Rodriguez CM et al (2002) Treatment of recalcitrant wastewater from ethanol and citric acid production using the microalga *Chlorella vulgaris* and the macrophyte *Lemna minuscula*. Water Res 36:4185–4192

Vanthoor-Koopmans M, Wijffels RH, Barbosa MJ et al (2013) Biorefinery of microalgae for food and fuel. Bioresour Technol 135:142–149
Vaz BS, Costa JAV, Morais MG (2016) Use of solid waste from thermoelectric plants in the cultivation of microalgae. Braz Arch Biol Technol 59:1–8
Visentainer JV, Santos OO (2013) Aspectos reacionais e da química analítica na produção e controle de qualidade do biodiesel de óleos e gorduras. ISBN: 978-857:628:521-2 (in Portuguese)
Wahlund TM, Conway T, Tabita FR (1996) Bioconversion of CO_2 to ethanol and other compounds. ACS Div Fuel Chem Prepr 41:1403–1405
Wang B, Li Y, Wu N et al (2008) CO(2) bio-mitigation using microalgae. Appl Microbiol Biotechnol 79(5):707–718
Wijffels RH, Barbosa MJ (2010) An outlook on microalgal biofuels. Science 329:796–799
Williams JA (2002) Keys to bioreactor selection. CEP Mag 34–41
Xu Y, Guerra LT, Li Z et al (2013) Altered carbohydrate metabolism in glycogen synthase mutants of *Synechococcus* sp. strain PCC 7002: cell factories for soluble sugars. Metab Eng 16:56–67
Yamada T, Sakaguchi K (1982) Comparative studies on *Chlorella* cell walls: induction of protoplast formation. Arch Microbiol 132:10–13
Yang J, Xu M, Zhang X et al (2011) Life-cycle analysis on biodiesel production from microalgae: water footprint and nutrients balance. Bioresour Technol 102(1):159–165
Yen HW, Hu IC, Chen CY et al (2013) Microalgae-based biorefinery—from biofuels to natural products. Bioresour Technol 135:166–174
Zhao B, Su Y (2014) Process effect of microalgal-carbon dioxide fixation and biomass production: a review. Renew Sustain Energy Rev 31:121–132
Zhu L (2015) Biorefinery as a promising approach to promote microalgae industry: an innovative framework. Renew Sustain Energy Rev 41:1376–1384

Microbial Oil for Biodiesel Production

Carlos José Dalmas Neto, Eduardo Bittencourt Sydney, Luciana Porto de Souza Vandenberghe and Carlos Ricardo Soccol

Abstract Even if there is no consensus in the scientific community that climatic changes are related to the burning of fossil fuels, the extraction and use of fossil fuels is causing an imbalance of trade balances in several countries. Besides, there is the monopolization of economic power in the hands of a few countries. In this context, the diversification of energy matrixes around the world is of vital importance. Renewable energies, such as solar, wind, nuclear, and others appear as interesting alternatives, depending on the specific characteristics of the energy matrix of each country. Among the economy sectors that need improvements is transportation (i.e., freight and passenger transport). Only this sector consumes about one-third of all energy produced in the world, and is probably the most dependent of fossil fuels these days. The development of liquid biofuels with physico-chemical characteristics and similar prices to those derive from petroleum is the great challenge. This chapter presents a view of the biodiesel production and perspectives and the example of one Brazilian technology for biodiesel production from oleaginous yeast.

Keywords Microbial oil · Rhodosporidium toruloides · Biodiesel · Sugarcane

C.J.D. Neto · L.P. de Souza Vandenberghe · C.R. Soccol (✉)
Federal University of Paraná (UFPR), Curitiba, PR CEP 81531-980, Brazil
e-mail: soccol@ufpr.br

C.J.D. Neto
e-mail: carlos.dalmas@ufpr.br

L.P. de Souza Vandenberghe
e-mail: lvandenberghe@ufpr.br

E.B. Sydney
Federal University of Technology—Paraná (UTFPR)—Campus Toledo,
Rua Cristo Rei, 19, Toledo, PR 85902-490, Brazil
e-mail: eduardosydney@utfpr.edu.br

C.R. Soccol et al. (eds.), *Green Fuels Technology*,
Green Energy and Technology, DOI 10.1007/978-3-319-30205-8_15

1 Introduction

The registered world increase in total energy consumption in recent years is generated in the search of comfort and convenience. This continuous search brings increasing numbers of engines and machines, as well as their respective power consumption (Ministério da Agricultura 2015), resulting in the increase of fuel's consumption in the same proportion. In Brazil, between 2003 and 2012, it was noted an increase of 40 % in final energy consumption (Junior et al. 2014). In the next ten years, the country's total energy demand is expected to grow 5.3 % annually, reaching 372 million TOE (tons of oil equivalent) in 2020. The industrial and transportation sectors are and will continue to be the main responsible for energy consumption in the country, representing approximately 67 % of the total (Rodrigues and Batistela 2013). In worldwide scale, 30–35 % of all the energy consumed is used for passengers transport and goods (Dalmas Neto 2012). These modes of transport basically consume liquid and gaseous fuels such as diesel, aviation kerosene, gasoline, alcohol, and biodiesel. Among them, only the last two are renewable. There are also other renewable fuels, such as biobutanol and biohydrogen, however there are still no commercially producing units.

In 2011, the fossil fuels accounted for more than 80 % of the world's power generation, according to the International Energy Agency (World Energy Outlook 2011). The burning of these fuels generates, among other pollutants, carbon dioxide (CO_2).The scientific community continues to discuss whether global warming is caused by the excessive increase concentration of carbon dioxide in the atmosphere, but this idea is generally accepted. In the last 150 years, after the industrial revolution, the concentration of CO_2 in the atmosphere jumped from 280 to more than 400 ppm (parts per million), a record, reaching its peak in 2014 (Pandey et al. 2013). Before that, CO_2 concentration remained stable near 280 ppm for over a thousand years. This situation provoked a rush of the development of economically feasible and sustainable technologies for clean energy production and independent of fossil sources (Fischer et al. 2015).

In this context, the increase in world energy demand, the huge participation of fossil fuels (more than 80 %) in world energy matrix, and the highest concentration of CO_2 in the atmosphere in human history clearly indicates the need for improvements in renewable fuels. One of the most promising sources of renewable energy is the biodiesel. Biodiesel is a fuel derived from renewable biomass. It can be produced by transesterification and/or esterification of fats, greases materials obtained from vegetable, animal or microbial origin, such as algae and yeasts (Cassia et al. 2008).

The biodiesel produced by yeasts is sustainable, more efficient than biodiesel produced by oilseeds and approximately has the same production costs. The great advantages of the use of oleaginous yeast in the production of biodiesel, comparing with biodiesel from animal or plants, include the independence from whether conditions and need for large areas of lands, annual productivities of cultivars and possibility to couple the process with agroindustrial and domestic wastewater treatment (biorefinery concept), among others.

This chapter is addressed to the description of biodiesel production from oleaginous microorganisms' cultivation, alternative substrates for microbial oil production, extraction of lipids from cells and biodiesel production. Also, a promiscuous example of microbial oil production from sugarcane juice by yeasts and microalgae and its use as raw material for biodiesel production is presented as a case study. The project, called "Biooil Project", was developed in a partnership between the Ouro Fino Company and the Bioprocesses Engineering and Biotechnology Department of Federal University of Paraná-Brazil. The project aimed the use of sugarcane juice and molasses for oil-rich yeast and microalgae biomass production and biodiesel production. The technological developments resulted in several international patents.

2 Biodiesel

Chemically, biodiesel can be defined as fuel alkyl esters of long-chain carboxylic acids, produced from the transesterification and/or esterification of fats, greases materials of vegetable, animal, or microbial origin (Sitepu et al. 2014). According to the Brazillian Law number 11.097 from 13 January, 2005, biodiesel is defined as a fuel derived from renewable biomass that is used in internal combustion engines with compression ignition (Freitas et al. 2013).

Biodiesel can be obtained through different ways, varying the raw materials and the type of chemical reaction. Edible plant oils used for biodiesel worldwide are rapeseed (84 %), sunflower (13 %), palm oil (1 %), soybean, and others (2 %) (Atabani et al. 2012). These oils have primarily high oleic acid content and, thus, provide superior ignition quality, ideal melting point, kinematic viscosity as well as improved oxidative stability (Sitepu et al. 2014).

Brazilian soybean (*Glycine max*) production reached approximately 86 million tons (Freitas et al. 2013), from which approx. 21 % was used for oil extraction (Bonato et al. 2000) (corresponding to 18 million tons of oil and 16 million tons of biodiesel). Therefore, the volume of soybean produced in Brazil is adapted to support the national biodiesel production of 3.4 million tons. The other 12.6 million of tons of soybean oil that is produced is generally used for food applications.

Total world biodiesel production is rising significantly to meet the demand: production rose from 15,000 barrels per day in 2000 to 289,000 in 2008 (Atabani et al. 2012), with highest production in the European Union and USA. Plant oil comprises a large portion of the production cost of biodiesel (Miao and Wu 2006). The rising cost of edible plant oils, and public debate of the "food versus fuel" issue, have encouraged the development of so-called "second-generation" biodiesel from nonedible plant oils such as jatropha, jojoba, and waste oils such as cooking grease and animal fats (Sitepu et al. 2014).

The Brazilian biodiesel is produced mainly from soybean oil, 75 %, and animal fat (tallow), 22 % (Papanikolaou et al. 2004). The Brazilian biodiesel production jumped from 2.7 million m^3 for 3.4 million m^3, between 2012 and 2014, representing an increase of 26 % (Ministério da Agricultura 2015). Despite the high production of soybean in Brazil, the production of biodiesel from it faces some problems. The most alarming is that soybean is a commodity, therefore subjected to significant price variations, affecting its by-products. Another point concerns the increasing demand of energy, which appears as an opportunity for the diversification of the energy matrix. Moreover, soybean has lower biodiesel productivity (500 L/ha) when compared to other crops, such as palm (5000 L/ha). Since 80 % of the national biodiesel is produced from soybeans, soybean production must increase in this proportion in order not to cause a lack of this raw material, and consequently a demand-offer shock, unless there are other potential sources. Other alternative sources of biodiesel are under study, such as yeasts and microalgae, however, the economically feasibility of these process must be tested.

Biodiesel plays an important role in Brazilian energy matrix as an alternative fuel to replace fossil diesel. The Brazilian government created in 2004 the "National Programme of Biodiesel Production and Use", which aimed the inclusion of biodiesel in the Brazilian energy matrix, focusing the social inclusion and regional development. The strategy was to successively increase the blend of biodiesel in fossil diesel. According to the MME (Ministry of Mines and Energy/Brazil) the blend began in December 2004. In January 2008, the addition of 2 % of biodiesel in diesel (B2) became mandatory throughout the national territory. With the apparent maturation of the Brazilian market, this percentage was expanded by the National Council of Energy Policy [*Conselho Nacional de Política Energética* (CNPE)] to 5 % (B5) in January 2010, anticipating in three years the goal established by the law No. 11.097 from January 13, 2005. After many political disputes, law no 13.033 was published in September 24, 2014 DOU 25.9.2014, which establishes the following percentages of biodiesel, which is added to the diesel, for the final consumer (v/v) (Ministério de Minas e Energia 2015): (I) 6 % (B6), from 1 July, 2014; and (II) 7 % (B7), from 1 November, 2014. Other countries adopt more ambitious proportions of biodiesel added to fossil diesel, such as Argentina, where since December 2013 this percentage was established at 10 % (B10).

3 Single Cell Oil for Biodiesel Production

"Third-generation" biodiesel is under development using oil accumulating microbes such as microalgae, bacteria, yeasts, and other fungi. This is the Single Cell Oils (SCO), or microbial systems that produce and store oil. The SCO technology involves the production of lipids by microbial cells that convert substrates such as carbon dioxide, sugars, and organic acids to SCO. Cells are then lysed by solvent, mechanical, enzymatic, or other means for different industrial applications. The lipid is then separated from the cell fraction, and the neutral lipid undergoes chemical refining to produce an ester or other target molecule, by releasing the glycerol from the individual fatty acids. Typically, this is done by acid or base hydrolysis in the presence of an alcohol. While the methodology has been established, the current barrier in SCO production is developing a robust system that is cost competitive with petroleum-based fuels. This can be achieved by the development of strains able to convert low-cost substrates, grow quickly to high density, and produce larger quantities of neutral lipid, and development of improved harvesting and dewatering technologies (Sitepu et al. 2014).

It is important to note that the term SCO aims the production of fatty acids of nutritional value, especially to be incorporated to food products for children and infants (Papanikolaou et al. 2004). This project began in 1959 and was called the Torula Process, where the resulting cells were used as animal feed (Pometto et al. 2014). Lately new applications were developed, among which are the production of fatty acids, surfactants, and mainly biodiesel (Kalscheuer et al. 2006).

The production of microbial oils, or SCO, is now an economic reality. Microbial oils from oleaginous microorganisms, including bacteria, yeasts, and microalgae are promising feedstock for biodiesel. This product is very similar to vegetable oils in fatty acid composition and other properties (Meng et al. 2009). Microbial oil production has several advantages such as high productivities of biomass, abundance of raw and cheap materials, which may have low production costs in many situations. Comparing with the production of vegetable oils, the culture of oleaginous microorganisms is affected neither by seasons nor by climates, and what is more, oleaginous microorganisms can accumulate lipids within some hours or a few days (Xue et al. 2008) and it is easy to scale-up (Li et al. 2012).

Various microorganisms are commercially used as interesting sources for oil production (Ratledge 2004). Nearly a hundred microorganisms have already been recognized as producers of large amounts of oil. Some examples include *Cryptococcus albidus, Lipomyces starkeyi, Rhodotorula glutinis, Yarrowia lipolytica, Schizochytrium* spp., *Trichosporon fermentans, Rhodosporidium toruloides, Mortierella isabellina, Cunninghamella echinulata, Mucor* sp., *and others.*

Many authors have investigated the biodiesel production by microorganisms and some of these works are summarized in Table 1.

Most recently, genetically modified microorganisms have been developed for these purposes (Kalscheuer et al. 2006) by companies such as Solazyme, LS9, and Amyris.

4 Case Study: The Biooil Project

A case study of microbial oil production is described as an example of a very efficient process. The Biooil project, developed by OuroFino enterpris and the Bioprocess Engineering division of the Federal University of Paraná, UFPR, aimed the development of a technology capable of producing biodiesel from low-cost raw materials, through non-GMO microorganisms. The adopted carbon source was

Table 1 Microbial oil production from different microorganisms

Carbon source	Microorganism	Biomass (g/L)	Lipid (%)	Lipid production (g/L)	Time (h)	Reference
Hydrolyzed rice straw	*Trichosporon fermentans*	11.4	14.6	1.7	192	Huang et al. (2009)
Glucose	*Trichosporon fermentans*	ND	ND	13.6	120	Huang et al. (2009)
Xylose	*Trichosporon fermentans*	ND	ND	9.9	120	Huang et al. (2009)
Manose	*Trichosporon fermentans*	ND	ND	11.5	120	Huang et al. (2009)
Galactose	*Trichosporon fermentans*	ND	ND	13.9	120	Huang et al. (2009)
Cellobiose	*Trichosporon fermentans*	ND	ND	10.4	120	Huang et al. (2009)
Glycerol (100 g/L)	*Rhodosporidium toruloides*	35.3	46.0	16.2	120	Kiran et al. (2013)
Glycerol (200 g/L)	*Rhodosporidium toruloides*	16.2	53.1	8.6	120	Kiran et al. (2013)
Glycerol (300 g/L)	*Rhodosporidium toruloides*	4.0	50.6	2.0	120	Kiran et al. (2013)
Hydrolyzed corn cob	*Trichosporon coremiiforme*	20.4	37.8	7.7	192	Huang et al. (2013)
Glycerol	*Mortierella isabellina*	7.8	25.6	2.0	120	Huang et al. (2013)
Glycerol	*Cunninghamella echinulata*	6.2	53.2	3.3	120	Huang et al. (2013)
Glycerol	*Rhodotorula glutinis*	20.2	25.1	5.1	120	Huang et al. (2013)
Monosodiumglutamate wastewater	*Rhodotorula glutinis*	25.0	20.0	5.0	120	Huang et al. (2013)
Glucose and xylose	*Rhodotorula glutinis*	6.3	16.7	1.1	120	Huang et al. (2013)
Industrial fats	*Yarrowia lipolytica*	8.7	44.0	3.8	120	Huang et al. (2013)
Cheese whey	*Mortierella isabellina*	23.1	17.3	4.0	120	Huang et al. (2013)
Glycerol	*Yarrowia lipolytica*	6.7	20.4	1.4	120	Huang et al. (2013)
Sewage sludge	*Lipomyces starkeyi*	9.4	68.0	6.4	120	Huang et al. (2013)
Tapioca starch	*Mucor SP*	28.0	17.8	5.0	120	Huang et al. (2013)
Sugar cane juice	ND	40.3	55.0	20.2	45	This work

sugarcane juice or sugarcane molasses, both extremely abundant and of low cost in Brazil. Sugarcane (*Saccharum* spp.) is one of the major cultures of the Brazilian agribusiness (Biosev 2013), and the leading producer of this crop in the world. According to Procana, the sugar-ethanol sector was responsible for about 2 % of the national GNP (gross national product) in 2011(FAOSTAT 2014). It is expected that 654 million tons will be harvested in 2015/2016 (Novacana 2008).

Its products and sub-products are widely used in the production of sugar, alcohol, heat, and electricity and, more recently, biodiesel and bioplastic (Globo Rural 2014). The bagasse from sugarcane juice extraction has been widely studied for the production of second generation ethanol, which can increase the production up to 30 % in the same planted area (Bajay et al. 2005). Each ton of sugarcane processed results in approximately 280 kg of bagasse containing 50 % humidity (Novacana 2008).

The production of bioproducts derived from sugarcane is only possible through the extraction of sugarcane juice, popularly known as "garapa", which contains approximately 15 % (p/p) of fermentable sugars (UNICA 2014). According to the UNICA (Union of Sugarcane Industry) the average productivity of sugar from sugarcane is approximately 70 ton of cane/ha (2013/2014 crop) (Rossetto 2014). Moreover, the average yield of sugar in Brazilian sugarcane in the referred years is 140 kg/ton (Rossetto 2014). With these numbers, it is possible to calculate the fermentable sugar (ATR) per hectare that is near 9800 kg. Over the past few years, the productivity of Brazilian sugarcane fields has increased constantly (UNICA 2014).

There are several sugarcane genetic improvement programmes underway in Brazil, such as those in Instituto Agronômico de Campinas (IAC), Centro de Tecnologia Canavieira (CTC), and Empresa Brasileira de Pesquisa Agropecuária (EMBRAPA). Each research group has a specific purpose, from the development of a drought resistant variety, with high sucrose content sugarcane, focusing on the increase of productivity. New techniques and harvesting equipments have been improved, with the common goal of increasing productivity of sucrose per hectare of cultivated sugarcane. Based on these information, it is clear that the current productivity (9800 kg per ha) may considerably increase in few years.

Numerous experiments were performed, including the choice and adequacy of strains, optimization of culture conditions, and process development and improvement. Formulation of the medium, substrate concentration, use of antibiotics, fermentation time, air flow, and/or oxygen, cell separation methods, methods of intracellular oil extraction, oil purification, biodiesel conversion, biodiesel purification, and even biodiesel performance tests on engines are examples of all experiments carried in this project. Low costs and minimization of environmental impacts were the main focus.

The following detailed methodologies are the result of many years of research. These conditions were established not only from the analysis and optimization of process conditions, considering the process costs, but also from the observation of the quality of oil and biodiesel obtained. The generated products contained around

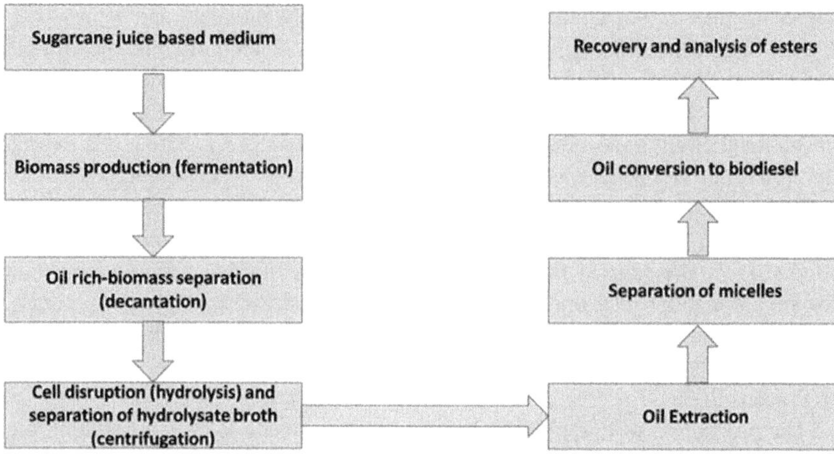

Fig. 1 Flowchart of biooil process

50 % of oleic acid in its composition which, according to (Sitepu et al. 2014), improves some properties of biodiesel.

The biooil process flow chart is presented in Fig. 1.

4.1 Biomass Production for SCO Synthesis

More than one hundred strains were screened for lipid production (SCO). These microorganisms can grow under different conditions and provide diverse products, having specific advantages and potential within the scope of the project. All tested strains tested were wild non-genetic-modified microorganisms (non-GMOs). The use of non-GMOs in this process represents a huge advance concerning the growing restricted health barriers. Both UFPR and Ouro Fino Company maintained the oil-producing strains stored in vegetative state through cryogenic techniques (10 % (v/v) glycerol solution at −80 °C or −196 °C).

The yeast strains were activated from cryogenic vials at 28 °C during 6 h. During this time, sugarcane-based medium was prepared for inoculation in order to reach a total soluble sugar concentration of 40 g/L (ideal). A nitrogen source was then added to the medium and the pH was adjusted to 5.0. Sterilization was then carried out to avoid contaminants and the need of antibiotics addition to the medium. The inoculated medium was kept at 32 °C and 180 rpm for approximately 12 h. A sample was withdrawn and analyzed for contaminants control. Serial inoculum propagations were then performed for different volumes of 15, 150, and 1500 ml and so on.

Table 2 Biomass, lipid production, and other fermentation parameters obtained in 15 L, 150 L, and 1500 L biorreactors

Time (h)	Cell concentration (g/L)	Nitrogen (g/L)	Sugar (g/L)	μ (h^{-1})	Lipid (g/L)	Productivity (g/L*h)
15 L reactor						
0	2.1	2.20	50.7	NA	NA	NA
24	27.4	0.01	4.9	0.038	NA	NA
150 L reactor						
0	4.1	2.20	42.1	NA	NA	NA
24	28.1	0.01	2.4	0.036	NA	NA
1500 L reactor						
0	3.6	2.16	30.8	NA	NA	NA
24	24.5	0.21	3.2	0.036	NA	NA
25	27.7	0.20	44.2	0.116	NA	NA
39	38.6	0.01	6.6	0.020	NA	NA
45	40.3	0.01	4.8	0.007	20.2	0.45

NA not available

In bioreactor scale, fermentation parameters are accurately controlled and monitored, such as oxygen demand and pH (which is not controlled in shaker cultures), for example. Fermentations was carried out in a 15 L reactor. The medium was prepared as described earlier (40 g/L of total soluble sugars from sugarcane per liter and a nitrogen source). When the reactor temperature reached 30 °C, pH was adjusted to 5.0 ± 0.2 and stirring was set to 200 rpm. After inoculation, eventual foaming may occur, which can result in culture leakage (an antifoaming agent can be used). The 15 L culture was then used, after 24 h of fermentation, as inoculum for the 15 L bioreactor. Fermentation conditions were identical to those used in the 15 L fermentation. The same operations were performed for the bioreactor of 150 L and 1500 L.

At the end of each process, at different scales, final concentrations of reducing sugars, nitrogen, cell, and oil concentration were determined. Fermentation was interrupted when the total sugar concentration was below 0.5 % (5 grams per liter) that happened from 24 to 48 h. In 1500 L, fermentation was carried out under fed-batch conditions. After 24 h, the reactor was fed with new concentration of the carbon source. Then, the process was continued till almost 48 h. Biomass concentration reached 40.3 g/L, which is considered very satisfactory.

Biomass, lipid production, and other fermentation parameters obtained from 15 L, 150 L, and 1500 L bioreactors are presented in Table 2.

After fermentation, decantation was then started and the biomass naturally settled down to the bottom of the fermentor for 4 h till the concentration reaches approximately 120 g/L. An efficiency of 99 % with cheap cell separation was attained.

4.2 Cell Disruption and Biooil Recovery

Biooil recovery from biomass is not easy, because the lipids of interest are accumulated intracellularly. An important step in oil recovery is cell disruption. The extraction without cell disruption is possible through the use of solvents (hexane, petroleum ether), which are toxic to humans and to the environment. Besides, they are derived from fossil sources, are difficult to separate and subsequent recovery, and results invariably in low purity lipids, requiring further purification steps (Sitepu et al. 2014). A diversity of methods for cell separation from the liquid medium was extensively studied: flocculation and sedimentation with and without additives, various types of filtration, the use of different types of filters (with and without the use of adjuvants) and centrifugation under various temperatures, times, and speeds.

Considering the economics of the process, the chosen technique for cell disruption was chemical hydrolysis with mineral acids. The duration of cell lysing step was determined by microscopic visualization of samples taken hourly from reactor, since it is very easy to distinguish intact cells from their debris. Cell disruption was complete when 99 % of the cells were ruptured. The aqueous phase containing lipids had low pH and approximately 10 g/L of soluble sugars and considerable amounts of HMF (hydroxymethylfurfural). The accumulated lipidproteic mass was then recovered with 30 % moisture content (w/w).

The medium with cells was sent to the separation unit, where lipidproteic mass was recovered for subsequent oil extraction. This mass consisted of a matrix of macromolecules formed by the vacuoles that store the oil (lipid fraction), and numerous proteins, including several binding proteins (protein fraction) that confers stability and homogeneous consistency.

4.3 Oil Extraction

Optimum extraction conditions were determined after a series of experiments involving at least five different variables including: the type of extraction, the degree of cell disruption, the moisture content of the lipidproteic mass, extraction time, and temperature. Best results were then reevaluated considering the toxicity of the solvent and the total cost of operation.

Oil extraction was carried out in three stages. The lipidproteic mass was transferred to the reactor at 35 °C under agitation. The approximate time of the extraction process was 15 min and the mixture contained in the reactor was allowed to stand for 5 min for phases' separation. The lower phase comprising the micelles was collected and sent to solvent evaporation unit.

The upper phase (extracted mass) still contained (approx. 20 %) of incorporated micelles, which could be discarded. This mass was then separated by centrifugation 6000 g during 40 s. The micelles obtained were sent to the extraction section. The centrifuged extracted mass still contained appreciable amount of oil that was reextracted with solvent. Sampling was carried out every 3 min to determine the oil concentration. The micelle was withdrawn from the reactor bottom and was reserved. The mass was again centrifuged at 6000 g during 40 s. The micelle obtained is reserved. The mass recovered from the centrifuge was extracted again at the same conditions. The final mass obtained was free from lipids, and thus consisted of a proteic mass that was dried.

4.4 Separation of Micelles

The separation of the micelles was carried out by distillation. The oil concentration in the micelle was determined. A 2 mL sample was withdrawn from the reboiler every 5 min. When three consecutive samples indicated the same concentration of oil in the micelle, solvent recovery was finished. This process lasts 30 min. The recovered solvent was reused in subsequent extractions. The oil was stored.

4.5 Analyses of the Oil

The analysis of the oil obtained from the oleaginous yeasts serves not only for characterization purposes, but also to predetermine the conditions of conversion steps to biodiesel production. This biooil is relatively new, and there is little information in the literature regarding the specific analysis for this product. Thus, adjustments to the analysis given by the American Oil Chemist's Society (AOCS) were applied, due to some characteristics of microbial oil that differ from vegetable oils and also due to the high content of antioxidative pigments in the obtained oil. These pigments give a very interesting visual identity to oil, as well as other very interesting properties such as high oxidative resistance and possible applications in other branches of industry, but they interfere in analysis made by titration. Some analysis were performed by gas chromatography, such as the acid number, which is the amount of free fatty acids (FFA) in the oil, the saponification number (SN). Most vegetable oils presents SN values from 180 to 200 mg KOH/g, as well as the microbial oil developed by the described methodology. Acid number of the obtained yeast oil during this project varied between 8 and 12 %.

The analysis of the peroxide value (PV) indicates if the oil suffered oxidation during the whole process (from hydrolysis to purification), caused by excessive exposure to air at high temperatures. Typically, PV is approx 5 meq/kg, close to the result of most vegetable oils.

Iodine index (II) provides the degree of insaturation of the oil, i.e., the amount of double bonds (which usually provide more noble properties to the oil). This parameter has strong variation according to the oil source. Palm oils present values

Table 3 Average profile of fatty acids from Biooil determined by gas chromatography

Fatty acid (% p/p)	Value	Maximum	Minimum
Miristic (C14:0)	1.0	0.0	1.1
Palmitic (C16:0)	21.5	19.9	22.6
Palmitoleic (C16:1)	0.7	0.0	1.1
Margaric (C17:0)	0.0	0.0	0.2
Heptadecenoic (C17:1)	0.0	0.0	0.2
Estearic (C18:0)	4.6	3.8	6.1
Oleic (C18:1)	62.1	60.4	63.7
Linoleic (C18:2) (Ш6)	7.6	4.4	9.6
Linolenic (C18:3) (Ш9)	0.7	0.0	1.2
Arachidic (C20:0)	0.1	0.0	0.3
Paulinic (C20:1)	0.4	0.0	0.7
Behenic (C22:0)	0.3	0.0	0.7
Eruciric (C22:1)	0.1	0.0	0.6
Lignoceric (C24:0)	0.7	0.0	1.6
Saturated (:0)	28.2	26.2	32.2
Insaturated (:1)	7.7	67.7	73.8

Analysis were performed in the CEMPEQC laboratory (Monitoring and Research Center of the Fuel Quality, Oils, Derivatives and Biofuels located in Araraquara, State of São Paulo, Brazil) On July 15, 2013

close to 50/100 g, whereas mamona oil is around 90/100 g. The biooil presented II of 80/100 g.

The fatty acid profile was also evaluated. The biooil profile is presented in Table 3. It was found that almost 70 % of esters were unsaturated, which provides optimum properties for biodiesel, such as low viscosity, cloud point, and freezing. By contrast, the oxidation time could present a low score, but as shown above, the oil has high antioxidants content which preserves much of its activity in the biodiesel.

4.6 Oil Conversion to Biodiesel

Many experiments were conducted to search an efficient and economically viable methodology. The chosen methodology consists of two stages. The first performs the conversion of all triglycerides into free fatty acids via acid esterification. When the reaction has reached the desired conversion, the transesterification was started (second stage).

The acid esterification was performed in an inox steel reactor with controlled agitation and temperature. Ethanol and sulfuric acid were added. When acid levels were higher than 98 %, the esterification reaction was interrupted. After the interruption of the esterification reaction, methanol and sodium hydroxide are mixed into the reactor. The mixture is kept stirring at the same speed and temperature for necessary time to convert at least 98 % of fatty acid esters. Usually after 15 min, separation reached the preestablished threshold. Methanol was then sent to a distillation unit for impurities removal. The purified methanol was then reused in further reactions.

4.7 Recovery and Analysis of Esters

Esters were purified by removing traces of methanol, water, catalysts used in esterification and transesterification reactions. Basically two operations were carried out: an extraction with water, and the use of selective adsorbents.

The first purification is called "washing". This operation was made at low agitation speed to avoid emulsions. Due to low agitation speed, its duration was 30 min. After this period, the agitation was stopped and water was drained through the reactor from the bottom and sent to the wastewater treatment system. The second purification step was the addition of selective adsorbent material. This second step was brief (approx. 10 min) and served to promote more intensive mixing between the phases. Then, sedimentation begins. The sobrenadant was centrifuged in a continuous centrifuge at 2500 g during one minute. From the top output, the purified esters were obtained, while the inside adsorbent material was retained. This adsorbent and the removed impurities were incinerated.

Esters were then analyzed according to the Brazilian National Petroleum Agency (*Agência Nacional do Petróleo - ANP*) (law no 45, August 25 2014). Figure 2 presents an analysis carried by the Greentech Laboratory, in Rio de Janeiro, State of Rio de Janeiro.

4.8 Engine Tests

The biodiesel that was produced through the described technology was tested with the support of a partnership with the Federal University of Rio de Janeiro (UFRJ)—Project, Research and Technological Studies Foundation—(COPPETEC).

Blind performance tests for emissions, consumption, and performance of the produced biodiesel and other biodiesel samples (referred as "reference") were carried out. The origin of the samples of biodiesel was not revealed before analysis. An engine, the Agrale single cylinder, which is considered as one of the most efficient, was used in emissions and performance tests that were conducted with different fuels. The engine characteristics are shown in Table 4.

An eddy (or absorption) current type dynamometer (DINAMATIC®), where the user can simulate the desired torque with the data acquisition system (Dinmont), was employed. The Dinmont software system has two analog output channels, one

SCHOOL OF CHEMISTRY

FEDERAL UNIVERSITY OF RIO DE JANEIRO

GreenTec

QUALITY REPORT OF B-100

Company: OUROFINO

Sample: HEDF 2012/01

PARAMETERS	RESULT	METHOD	UNIT	RESOLUTION N°7 - ANP	
				Min.	Max.
Aspect, at 25°C	Grade 2/Clear and free of impurities	NBR 16048	-	CFI*	CFI
Specific mass, at 20° C, min-max.	857.7	ASTM D4052	kg/m³	850	900
Kinematic viscosity, at 40°C, min-max.	5.24	ASTM D445	mm²/s	3	6
Water and sediments, max.	280	EN 12937	mg/kg		380
Total contamination, max.	14.9	EN 12662	mg/kg		24
Flash point, min.	172.0	ASTM D93	°C	100	
Ester content, min	97.4	EN 14103	% mass	96.5	
Carbon residue, max.	0.019	ASTM D4530	% mass		0.050
Sulphated ashes, max.	<0.01	ASTM D874	% mass		0.02
Total sulphur, max.	<0.01	NBR 15867	mg/kg		50
Sodium + potassium content max.	<0.01	NBR 15553	mg/kg		5
Calcium + magnesium content max.	<0.01	NBR 15553	mg/kg		5
Phosphorous content, max.	<0.01	NBR 15553	mg/kg		10
Corrosivity to copper, 3h at 50°C. max.	1	ASTM D130	-		1
Cold plugging point, max.	6	ASTM D6371	°C		19
Acid index, max.	0.28	ASTM D664	mg KOH/g		0.5
Free glycerin, max.	0.017	ASTM D6584	% mass		0.02
Total glycerin, max.	0.242	ASTM D6584	% mass		0.25
Monoglycerides, max.	0.429	ASTM D6584	% mass		0.80
Diglycerides, max.	0.129	ASTM D6584	% mass		0.20
Triglycerides, max.	0	ASTM D6584	% mass		0.20
Methanol/Ethanol, max.	0.02	EN 14110	% mass		0.2
Iodine index, max.	61.7	EN 14111	g/100g	Note	Note
Stability to oxidation at 110°C, min.	8.3	EN 14112	h	6	

* CFI = Clear and free of impures

Rio de Janeiro, 24th de October of 2012

Donato Aranda

Donato A. G. Aranda

(AC Certisign RFB G3 ICP Brasil - a8 96 9f 4e 8c 23 c4 ee 23 64 19 65 d0 2c 25 10 95 4a 2f 3b)

School of Chemitry/UFRJ

CRQ 03314401. 3ª. Zone E-mail: donato@eq.ufrj.br

Register N° 001 OUT 2012

Fig. 2 Analysis of biodiesel produced under the conditions described in this chapter

for spin and the other for torque control. It also has eight temperature channels compatible with thermocouple type K. Thus, the system is able to measure: speed, torque, power, and temperature (cooling water of the engine, exhaust gases). The

Table 4 Description of Agrale engine used in biodiesel tests

Type of engine	Diesel four-cycle
Brand/model	Agrale/M95W
Number of cylinders and type	1 vertical cylinder
Type of injection	Direct
Cylinder diameter	95 mm
Piston stroke	105 mm
Length of the rod	170 mm
Diameter of inlet valve	42 mm
Stroke of inlet valve	10.5 mm
Displacement	744 cm^3
Relationship rod/manivel	3.24
Compression ratio	21:01
Injection angle	−17° before PMS
Valve closing angle	−150° before PMS
Angle opening of the exhaust valve	160° before PMS

Fig. 3 Motor apparatus used in engine tests for biodiesel performance, consumption and emissions

motor apparatus, consisting of the motor, the dynamometer and the gas analyzer was used in engine tests of biodiesel performance, consumption and emissions analysis is presented in Fig. 3.

Two different mixtures of biodiesel:diesel were used for the tests: B5 (5 % biodiesel in mineral diesel) and B20 (20 % biodiesel in petroleum diesel). The motor performance test was carried out by running the motor with each fuel at 100, 75, 50, and 25 % of the maximum continuous power. Performance was analyzed by plotting Power x engine turn (rpm). Emission tests were based on the determination of NO_x, CO (Carbon Monoxide), CO_2 (Carbon Dioxide), HC (Hydrocarbons), and

O_2 content in the exhausting gas at 100, 75, 50, and 25 % of the maximum continuous power.

The specific consumption is the amount of fuel consumed to generate 1 kWh (about 860 kcal) energy. It is a parameter widely used to show how efficient an engine turns fuel into work. The use of this parameter is ideal for thermal efficiency because all involved variables are measured in standard unit: Time, Power, and Weight (Sitepu et al. 2014). Furthermore, the engine operating variables, such as temperature of the exhaust gases, air inlet, and fuel were also determined.

The main results of the complete biodiesel performance engine tests example are presented in the following in Fig. 4 for specific consumption, CO_2 emissions, CO emissions, O_2 emissions, NO_x emissions, and HC emissions in relation to engine spin (rpm).

As signaled previously, the specific consumption is the amount of fuel consumed to generate 1 kWh (about 860 kcal) energy. The more this value is low, more efficient is the fuel combustion, i.e., it generates more power for the same weight unit. In this case, an economy of 15 % can be achieved when the engine operates at 2500 rpm with the mixture B5 OF. For the blend B20 OF fuel, the economy at the same spin is 13 %. It is known that for bus and truck engines that are used in Brazil (with large internal volumes) this regime (2500 rpm) corresponds to an approximate speed of 95 km/h (speed constantly used by these vehicles). The use of microbial biodiesel represents a significant fuel economy if compared with other biodiesel consumed in Brazil (Fig. 4a).

CO_2 emissions (Fig. 4b) from microbial biodiesel are a half of the emissions from engines working with standard biofuels. The main reason for the development microbial biofuels is that high concentrations of CO_2 are liberated to the atmosphere and are associated with actual preoccupant weather imbalances. In all evaluated conditions, the microbial biodiesel presented lower CO_2 emissions, with reductions ranging from 11 up to 43 % in comparison to standard biodiesels. Engines operating at 2500 rpm with the mixture B20 OF were responsible for 220 % less CO_2 emissions, which was considered an exceptional result. Using B5 OF mixture, the emissions were 40 % lower than those produced by the standard biodiesel.

CO is a colorless, odorless gas produced by the incomplete combustion. In US, CO emissions were responsible for 3500 deaths per year that were caused by the smoke liberated from vehicles and malfunctioning of heating systems (Kaimen-Maciel et al. 2010). Due to its toxicity, the lower CO emissions by the produced microbial biodiesel, is an interesting alternative for the planet.

CO emissions are presented in Fig. 4c that technically indicates how complete the fuel was burned. There is again considerably higher performance of the fuel developed by the present technology, especially under high speed (2500 rpm) where the emission levels were 80 % lower compared to the standard fuel. When the engine operates at 2500 rpm, the emissions of microbial B20 OF are seven times lower than that observed when the engine worked with the standard B5. Smaller CO in the exhausting gas and the specific consumption curves reinforce the complete combustion of microbial OF biodiesel.

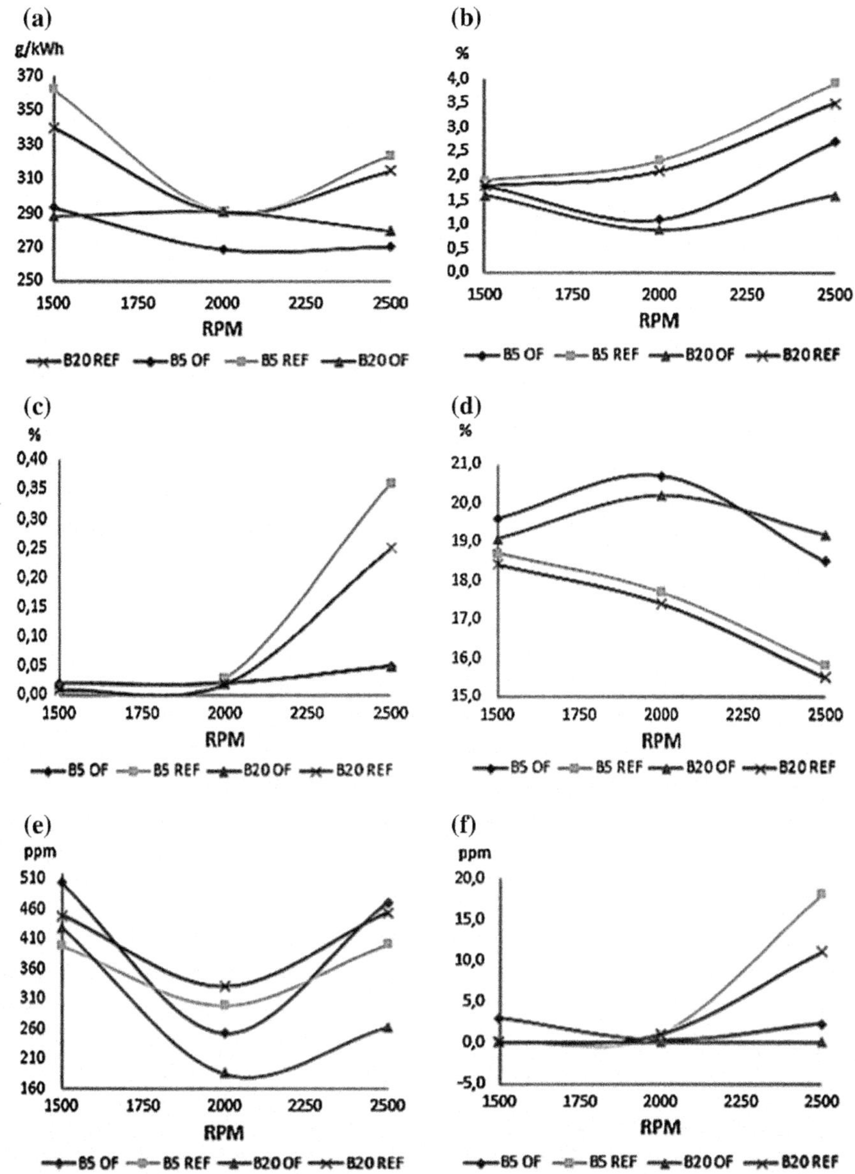

Fig. 4 Performance of biodiesel during engine tests for B5 OF—5 % microbial biodiesel produced by this technology; REF B5—5 % customary biodiesel "reference"; OF B20—20 % microbial biodiesel produced by this technology; REF B20—20 % customary biodiesel "reference". **a** Specific consumption (in g/kWh) versus RPM; **b** CO_2 emissions (in % at outlet gas) versus RPM; **c** CO emissions (in % at outlet gas) versus RPM; **d** O_2 emissions (in % at outlet gas) versus RPM; **e** NOx emissions (in ppm at out let gas) versus RPM and **f** HC emissions (is ppm at outlet gas) versus RPM

The emissions of O_2 in relation to the engine speed are presented in Fig. 4d. These results provide a clear view of the best burning of microbial OF biofuels. It appears that in all regimes of work, O_2 emissions by microbial OF biofuels were higher, i.e., it needs smaller amounts of oxygen for combustion. This reduction in consumption of O_2 can be very interesting in stationary engines, since it allows the use of blowers and/or smaller compressors, lower costs, and power consumption.

NO_x emissions of (nitrous oxides) are originated by three different mechanisms in the combustion chamber: (i) the reaction of atmospheric nitrogen with oxygen at high temperatures, (ii) hydrocarbon-free radical reaction with nitrogen molecules, and (iii) by the reaction of nitrogen with fuel. The nitrous oxide gases are harmful to health, causing eyes and respiratory system irritation. It also plays an important role in the environment, being partly responsible for acid rain and smog formation (Braun et al. 2003). So, a reduction in NO_x emissions can help to improve air quality and, consequently, improves peoples' health. NO_x emissions as a function of engine spin (rpm) are shown in Fig. 4e for the four types of analyzed biofuels. OF microbial biofuels were responsible for the lowest emissions. The comparison between microbial B20 OF and the standard B20 REF clearly shows that the developed microbial biofuel provoked 50 % less emissions. However, the microbial B5 OF showed the worst results at 1500 and 2500 rpm. Between 1800 and 2300 rpm, the NO_x emissions resulted from microbial B5 OF were below the standard B5 REF, however better results were obtained for other standard fuel rotations. It is difficult to clarify what may have caused such high discrepancy between the microbial B5 OF and the standard B5 REF. A more detailed study, involving the chemical composition of fuels, transport phenomena inside the combustion chamber could clarify this question and will be carried.

Most HC emissions are not directly harmful for peoples' health at concentrations found in the air. However, in chemical reactions in the troposphere it participates in the formation of NO_2 and ozone, which are dangerous for the environment and health (Braun et al. 2003). The HC's issued are extended to include a variety of other volatile organic compounds (VOC) such as alcohols and aldehydes. In Fig. 4f it is possible to see the hydrocarbons' concentration emitted from the engine tests for the four tested biofuels. In general, the microbial biodiesel emitted less HCs, with the exception of B5 OF that was running at 1500 rpm. In other concentrations and rotations, microbial OF biofuels provided better results than the standard biofuels. It is interesting to note that, for the microbial B20 OF, HC emissions were below the minimum that could be detected by the equipment (0.11 ppm).

5 Conclusions

The microbial sugarcane biodiesel idea can be a viable alternative for the diversification of the Brazilian energy matrix, and unlink it from variations in the soybean price. Taking in consideration the low cost of the culture medium (sugarcane juice) used in the production of microbial biodiesel, it is possible to say that it is an economically viable alternative in comparison with other vegetable oil sources that are normally used for biodiesel production. In this way, this process need, fewer inputs such as fertilizers, fungicides, and other details that are needed to soybean oil production, reducing costs, and preserving the environment. During sugarcane harvesting, less equipment and shorter distances are needed to be traveled to obtain the same amount of oil. Besides, the results obtained in engined tests show that microbial biodiesel, in terms of emission of pollutants, is much better than the traditional produced biodiesel. These results, when expanded to the Brazilian consumption of 54 billions of biodiesel in 2013, suggest an emission reduction of approximately 50 billions tons of CO_2 per year, or 40 % less of CO_2 emissions generated from microbial biodiesel.

References

Atabani A, Silitonga A, Badruddin I et al (2012) A comprehensive review on biodiesel as an alternative energy resource and its characteristics. Renew Sustain Energy Rev 16:2070–2093

Bajay SV, Berni MD, Lima CR (2005) Papel e Celulose. In: Rosillo-Cale, Rothman Bajay E (eds) Uso da Biomassa para Produção de Energia na Indústria Brasileira. Editora da Unicamp, Campinas

Biosev (2013) Biosev Relações com Investidores. http://ri.biosev.com/biosev/web/default_pt.asp?idioma=0&conta=28. Accessed 04 July 2015

Bonato ER, Bertagnolli PF, Lange CE, Rubin S (2000) Teor de óleo e de proteína em genótipos de soja desenvolvidas após 1990. Pesquisa Agropecuária Brasileira 35(12):2391–2398

Braun S, Appel LG, Schmal M (2003) A poluição gerada por máquinas de combustão interna movidas à diesel – A questão dos particulados. Estratégias atuais para a redução e controle das emissões e tendências futuras. Quim Nova 27(3):472–482

Cássia Albino J, Creste S, Figueira A (2008) Cana-de-Açúcar. Centro de Cana, Ribeirão Preto, Brazil

Dalmas Neto CJ (2012) Desenvolvimento de bioprocesso para produção de biocombustíveis à parir da pirólise rápida de microalgas. Dissertation, Universidade Positivo

FAOSTAT (2014) http://faostat.fao.org/site/339/default.aspx. Accessed 06 July 2015

Fischer H, Schmitt J, Schneider R et al (2015) Latest insights into past carbon cycle changes from CO_2 and δ13Catm. Nova Acta Leopoldina. http://epic.awi.de/37719/1/NAL_Bd121_Nr408_059-063_12_Fischer.pdf. Accessed 30 June 2015

Freitas DD, Thomé OSS, Bordeaux-Rego R (2013) Avaliação do impacto da concentração de biodiesel no diesel para rentabilidade do negócio mercado brasileiro. http://www.producao.uff.br/conteudo/rpep/volume132013/RelPesq_V13_2013_04.pdf.Accessed. 02 Aug 2015

Globo Rural (2014) http://g1.globo.com/economia/agronegocios/noticia/2014/12/usina-em-alagoas-comeca-produzir-o-etanol-de-segunda-geracao.html. Accessed 10 July 2015

Huang C, Zong M-H, Wu H, Liu Q-P (2009) Microbial oil production from rice straw hydrolysate by Trichosporon *fermentans*. Bioresour Technol 100(19):4535–4538

Huang C, Chen X-F, Xiong L et al (2013) Microbial oil production from corncob acid hydrolysate by oleaginous yeast Trichosporon coremiiforme. Biomass Bioenergy 49:273–278

Junior EC, Sacomano J, Neto MM (2014) Energia solar térmica: Inovação em aquecimento de água para processos industriais. Revista Brasileira de Engenharia de Biossistemas 8(3): 209–219

Kaimen-Maciel DR, Rocha FHB, Mancini DA (2010) Desmielinização tardia após intoxicação por monóxido de carbono. http://www.revistaneurociencias.com.br/edicoes/2010/RN1801/302%20relato%20de%20caso.pdf. Accessed 13 Aug 2015

Kalscheuer R, Stölting T, Steinbüchel A (2006) Microdiesel: *Escherichia coli* engineered for fuel production. Microbiology 152(9):2529–2536

Kiran EU, Trzcinski A, Webb C (2013) Microbial oil produced from biodiesel by-products could enhance overall production. Bioresour Technol 129:650–654

Li X, Lu B-S, Li H-L, Yin T (2012) Production of microbial oils fermented by two-step fermentation with corn stalk. Procedia Environ Sci 12:432–438

Meng X, Yang J, Xu X et al (2009) Biodiesel production from oleaginous microorganisms. Renew Energy 34(1):1–5

Miao X, Wu Q (2006) Biodiesel production from heterotrophic microalgal oil. Bioresour Technol 97:841–846

Ministério da Agricultura (2015) Cana-de-acucar. http://www.agricultura.gov.br/vegetal/culturas/cana-de-acucar. Accessed 20 July 2015

Ministério de Minas e Energia (2015) Boletim Mensal de Energia. http://www.mme.gov.br/web/guest/publicacoes-e-indicadores/boletins-de-energia. Accessed 12 Oct 2015

Novacana (2008) A produção de cana-de-açucar no Brasil (e no mundo). http://www.novacana.com/cana/producao-cana-de-acucar-brasil-e-mundo/. Accessed 25 July 2015

Pandey A, Lee D-J, Chisti Y, Soccol CR (eds) (2013) Biofuels from algae, 1st edn. Elsevier, Germany

Papanikolaou S, Komaitis M, Aggelis G (2004) Single cell oil (SCO) production by *Mortierella isabellina* grown on high-sugar content media. Bioresour Technol 95(3):287–291

Pometto A, Shetty K, Paliyath G, Levin RE (2014) Food biotechnology, 2nd edn. CRC Press, ISBN 9780824753290

Ratledge C (2004) Fatty acid biosynthesis in microorganisms being used for single cell oil production. Biochimie 86(11):807–815

Rodrigues SA, Batistela GC (2013) Uma revisão sobre a disponibilidade hídrica brasileira para geração de energia elétrica? Geoambiente (21). http://h200137217135.ufg.br/index.php/geoambiente/article/viewFile/27907/15761. Accessed 22 June 2015

Rossetto R (2014) BPUF Cana de açúcar. http://brasil.ipni.net/ipniweb/region/brasil.nsf/e0f085ed5f091b1b852579000057902e/716ec96288a9ea4383257cb0006f9ad8/$FILE/Palestra%20Raffaella%20Rossetto.pdf. Accessed 28 July 2015

Sitepu IR, Garay LA, Sestric R et al (2014) Oleaginous yeasts for biodiesel: current and future trends in biology and production. Biotechnol Adv 32(7):1336–1360

UNICA (2014) Moagem de cana-de-açúcar no centro-sul e atinge 41,33 milhões de toneladas na primeira quinzena de julho. unica.con [cited 20 May 2015]. http://www.unica.com.br/noticia/11196467920329114090/moagem-de-cana-de-acucar-no-centro-sul-desacelera-e-atinge-41-por-cento2C33-milhoes-de-toneladas-na-primeira-quinzena-de-julho. Accessed 27 July 2015

World Energy Outlook (2011) International Energy Agency, France. http:// www.worldenergyoutlook.org/media/weowebsite/energydevelopment/weo2011_energy_for_all.pdf. Accessed 23 June 2015

Xue FY, Miao JX, Zhang X, Luo H, Tan TW (2008) Studies on lipid production by *Rhodotorula glutinis* fermentation using monosodium glutamate wastewater as culture medium. Bioresour Technol 99(13):5923–5927

Biohydrogen

Saurabh Jyoti Sarma, Vinayak Laxman Pachapur, Satinder Kaur Brar, Mausam Verma and Carlos Ricardo Soccol

Abstract By virtue of its environmental friendly emissions, hydrogen has been considered as a potential green alternative of fossil fuels. Additionally, the gravimetric energy density of hydrogen is nearly three times higher than transportation fuels such as gasoline and diesel. Therefore, if hydrogen can be produced at commercial scale from renewable materials by using sustainable technologies, it will have enormous environmental benefits. In this context, research work has been focused on biohydrogen production from biomass, most preferably from the agro-industrial waste based feedstock. Nearly for the last three decades, investigators have been actively studying biohydrogen production from different renewable sources, and the trend of such investigations has been gradually changing from fundamental studies to technology development. Thus, in the present chapter, the development in biohydrogen research has been presented in a chronological order. The problems associated with different biohydrogen production methods and their potential solutions have been discussed. Finally, different recent government initiatives to promote a hydrogen-based economy have also been summarized.

Keywords Biohydrogen · Challenges · Government policies · Waste biomass · Historical overview

S.J. Sarma · V.L. Pachapur · S.K. Brar (✉)
Institut national de la recherche scientifique, Centre - Eau Terre Environnement, 490, Rue de la Couronne, Québec, QC G1K 9A9, Canada
e-mail: satinder.brar@ete.inrs.ca

M. Verma
CO2 Solutions Inc., 2300, rue Jean-Perrin, Québec, QC G2C 1T9, Canada

C.R. Soccol
Bioprocess Engineering and Biotechnology Department, Federal University of Paraná, Centro Politécnico, Usina Piloto B, Curitiba, Paraná CEP 81531-990, Brazil

C.R. Soccol et al. (eds.), *Green Fuels Technology*, Green Energy and Technology, DOI 10.1007/978-3-319-30205-8_16

1 Introduction

Fossil fuels fulfill most of the global transportation energy requirements. However, they are nonrenewable energy carriers with finite reserve. Additionally, overexploration and utilization of fossil fuel have adverse environmental consequences (Ntaikou et al. 2010). In this context, for the past few decades, biofuel production processes have been extensively investigated. As a result of these efforts, commercially successful biofuels such as biodiesel, bioethanol, and biomethane have been developed. Biohydrogen is another promising biofuel which, in comparison to these biofuels, has the potential to be a more efficient and environment-friendly alternative to fossil fuels. There is more than one reason to consider biohydrogen as a potential green alternative of petroleum hydrocarbon-based fuels. First, a range of agro-industrial wastes or nonfood biomass can be used as the feedstock for biohydrogen production. Second, during hydrogen combustion, only water is generated as the major emission. Although this process may release negligible amount of NOx, it does not release CO_2 or any other pollutants to the atmosphere. Likewise, as presented in Fig. 1, energy density of hydrogen (142 MJ/kg) is approximately three times that of gasoline (47 MJ/kg) or diesel (43 MJ/kg) (Şensöz et al. 2000; Sydney et al. 2014).

Based on the microorganisms used and the equipment involved, biohydrogen production processes can be broadly subdivided into three major categories as: (i) light-independent anaerobic fermentation, (ii) photofermentation, and (iii) electrochemical processes. All these techniques have their own advantages and disadvantages, and in the present chapter these points have been elaborately discussed. [NiFe]-hydrogenase, [FeFe]-hydrogenase, and nitrogenase are the enzymes directly associated with biohydrogen production. Investigations have been going on to understand the structural and functional properties of these enzymes. Similarly, different biochemical pathways involved in biohydrogen production have been

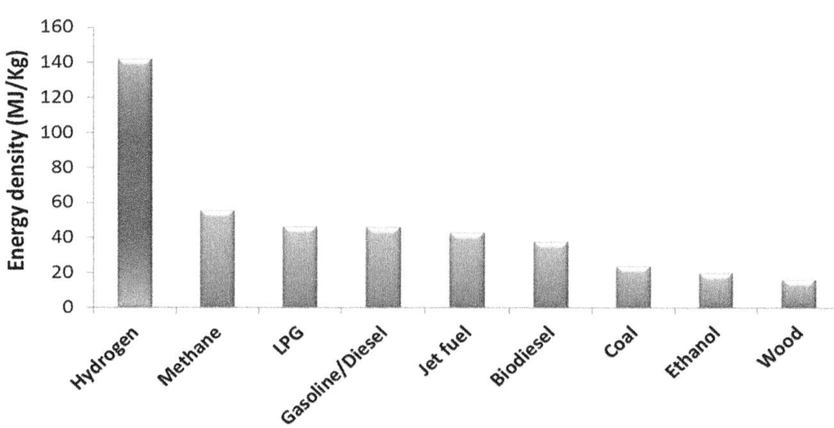

Fig. 1 Gravimetric energy density of different fuels

identified. Pyruvate: ferredoxin oxidoreductase (PFOR) pathway (Chen et al. 2006; Hallenbeck et al. 2012), Pyruvate: formate lyase (PFL) pathway (Hallenbeck et al. 2012; Yoshida et al. 2006), and Pentose phosphate (PP) pathways (Kim et al. 2011) are some of these pathways (Sarma et al. 2014b). The present chapter will cover a brief discussion on the enzymes and these pathways. Government policies are vital for successful biofuel production. Thus, the present chapter will discuss major government initiatives for encouraging industrial production of biohydrogen. Moreover, a short discussion on potential biohydrogen market will also be presented. In order to ensure maximum environmental benefit of the technology, biohydrogen production process can be integrated to organic waste treatment. Therefore, a critical evaluation of this approach will also be presented. Additionally, an attempt has been made to put forward an overview on chronological advancement in global biohydrogen production research.

2 Historical Overview of Global Biohydrogen Research

Hydrogen production potential of microorganism has been investigated for many decades. These investigations have facilitated better understanding of basic principles behind the phenomenon, and have created the possibility for large-scale production of biohydrogen as a renewable fuel. In Table 1, a chronological summary of the reports on biohydrogen production have been presented.

2.1 Initial Reports on Biohydrogen Research (Before 1990)

From Table 1, it is evident that investigations on biohydrogen processes have been started as early as 1942 (Gaffron et al. 1942). The authors have reported that in the presence of light, if anaerobic conditions are provided, microalgae, such as *Scenedesmus* or related species can utilize molecular hydrogen to reduce CO_2. However, in the dark, if nitrogen environment is provided, they can release hydrogen (Gaffron et al. 1942). The authors reported that by illumination the fermenting algae, hydrogen release could be enhanced (Gaffron et al. 1942). By using radioisotopic growth studies, Miller et al. (1973) reported that along with acetate, ethanol, and CO_2; *Ruminococcus albus* can produce hydrogen from glucose by Embden–Meyerhof–Parnas (EMP) pathway (Miller et al. 1973). In 1976, Harding et al. (1976) reported that certain *Bacteroides melaninogenicus* human isolates were capable of producing hydrogen in PY fructose medium. Joyner et al. (1977) studied hydrogen production by cell-free extracts of rumen bacteria. Likewise, hydrogen production during cellulose fermentation by anaerobic rumen fungus was investigated by Bauchop et al. (1981). According to the authors, this report was the first demonstration of hydrogen production by a fungus (Bauchop et al. 1981). Lactulose fermentation by colonic bacteria and production of hydrogen has been reported by

Table 1 A summary of different reports of biohydrogen production in chronological order

Period	Feedstock	Microorganism	Process and scale	H_2 production/yield	References
Before 1990	Glucose (0.2 %)	*Scenedesmus* species (Algae)	Fermentative and photochemical production	–	Gaffron et al. (1942)
	Water (photolysis)	*Clostridium pasteurianum* (hydrogenase only)	Photolysis of water in presence of chloroplast, ferredoxin and hydrogenase under continuous illumination	40 μmol/h/mg chlorophyll	Hall et al. (1978)
	Lactic acid-containing wastes	*Rhodospirillum rubrum (photosynthetic bacterium)*	Batch mode: 500 mL serum bottle	6 mL/h/g (dry weight) of cells	Zürrer et al. (1979)
			Continuous mode: 1 L	20 mL/h/g (dry weight) of cells	
	Alcohol factory's wastewater	*Clostridium butyricum (fermentative bacterium)*	5-L fermenter with immobilized whole cells	20 mL/min/kg wet gel	Karube et al. (1981)
	Synthetic medium	*Oscillatoria sp. Miami BG7 (Marine blue green algae)*	Photoproduction	260 μmol/mg chlorophyll/h	Kumazawa (1981)
1991–1999	Cane molasses stillage	*Citrobacter freundii*	Anaerobic fermentative hydrogen production at various scale ranging from 0.06 to 100 L	1.2 mol/mol	Vatsala (1992)
	Cellulose hydrolysate	*Clostridium sp.* (Strain No. 2)	Continuous hydrolysis of substrate in an aqueous two-phase system and continuous H_2 production in a separate fermenter (300 mL)	4.10 mmol of H_2/h	Taguchi et al. (1996)
	Baker's yeast wastewater	Mixed culture	Fluidized bed anaerobic digester at constant COD feed	800 ppm	Guwy et al. (1997)
	Starch	*Enterobacter aerogenes* HO-39a and *Clostridium butyricum* IFO13949 (coculture)	Batch fermentation using immobilization technique on glass porous beads. Working volume was 200 mL	2.6 mol/mol	Yokoi et al. (1998)
	Molasses	*Enterobacter aerogenes* strain E.82005	Batch fermentation with CO_2 removal. Working volume was 200 mL	1.58 mol-H_2/mol sugar	Tanisho et al. (1998)
	Whey dairy industry waste	*Rhodospirillum rubrum (photosynthetic bacterium)*	Batch/semi-continuous mode with modular outdoor reactor	40 L-H_2/mg^2 d	Modigell et al. (1998)

(continued)

Table 1 (continued)

Period	Feedstock	Microorganism	Process and scale	H_2 production/yield	References
	Organic fraction municipal solid waste (OFMSW)	Heat pretreated sludge and hydrogen-producing bacteria enriched from soybean meal silo	OFMSW digester of 100 mL working volume	180 mL-H_2/g TVS	Lay et al. (1999)
	ASN III medium	*Phormidium valderianum*/*Halobacterium halobium* and *Escherichia coli*	Photoproduction using cell immobilized packed-bed reactor	18.98 µmol-H_2/mg protein/h	Bagai et al. (1999)
	Sodium lactate and sodium glutamate	Mutant *Rhodobacter sphaeroides* (photosynthetic bacterium)	2 mL hydrogen electrode cell under monochromatic light	0.20 nmol-H_2/mg dcw	Vasilyeva et al. (1999)
2000–2010	Glucose, sucrose and cellobiase	*Enterobacter cloacae* IIT-BT 08	Batch fermentation using aerobic and anaerobic broth in 50-mL flask	2.2 mol-H_2/mol glucose, 6 mol-H_2/mol sucrose and 5.4 mol-H_2/mol cellobiose	Kumar et al. (2000)
	Slurry of naked or sensitized TiO semiconductor	*Rhodopseudomonas capsulate* (photosynthetic bacterium)	Pyrex glass photobioreactor	1.1–1.2 mL-H_2/h	Gurunathan (2000)
	Sweet potato starch residue	*C. butyricum* and *E. aerogenes*, and *Rhodobacter sp.* M-19	Repeated batch culture of two-step (dark and photofermentation). Working volume was 200 mL	7.0 mol-H_2/mol glucose	Yokoi et al. (2001)
	Glucose–mineral salts non-sterile medium	Mixed culture from soya bean meal	Continuous stirred-tank reactor using nitrogen sparging. Working volume was 2.3 L	1.43 mol-H_2/mol glucose	Mizuno et al. (2000)
	Glucose	Sludge mixed culture	Dark fermentation in 3 L fermentor with 6 h HRT	2.1 mol-H_2/mol glucose	Fang et al. (2002)
	Glucose and sucrose	Extreme thermophiles: *Caldicellulosiruptor saccharolyticus and Thermotoga elfii*	Dark fermentation in 100-mL crimp seal flasks at 70 and 65 °C, respectively	3.3 mol-H_2/mol glucose	Van Niel et al. (2002)
	Organic waste	*Rhodobacter capsulatus* and *Thermohydrogenium kirishi*	Active membrane system for separation of H_2/CO_2 mixture produced from aerobic and anaerobic reactors	900–1080 mL-H_2/h/mg protein	Teplyakov et al. (2002)
	Sucrose	Anaerobic microflora	60 mL working volume with C/N ratios ranged from 40 to 130	4.8 mol-H_2/mol sucrose	Lin et al. (2004)

(continued)

Table 1 (continued)

Period	Feedstock	Microorganism	Process and scale	H_2 production/yield	References
	Oxidation of CO and water to CO_2 and hydrogen	*Rhodospirillum rubrum* (phototrophic anaerobic bacterium)	Photoproduction of hydrogen from water–gas shift reaction in sealed serum bottle	0.98 mol-H_2/mol CO	Najafpour et al. (2004)
	Food waste and sewage sludge	Heat-treated seed sludge	Fermentation using 415 mL Wheaton media lab bottles	122.9 mL-H_2/g carbohydrate COD	Kim et al. (2004)
	Sucrose mineral medium	Heat-shocked mixed culture	Stirred-tank reactor by gas circulation with 24 h HRT	2.73 mol-H_2/mol sucrose	Zhang et al. (2005)
	Sucrose	Mixed microflora	Continuous reactor operations under varying sulfate concentration	3.6 mol-H_2/mol sucrose	Lin and Chen (2006)
	Olive pulp	Hydrogen-producing culture	500 mL CSTR-type digester at HRT of 30, 14.5 and 7.5 h	4–6 mol-H_2/g carbohydrates	Koutrouli et al. (2006)
	Glucose solution	*Escherichia coli* MC13-4	20 mL glass bottle connected to potable biofuel cell	1.27 mol-H_2/mol glucose	Ishikawa et al. (2006)
	Sucrose	Seed sludge from anaerobic digester	CSTR with working volume of 5 L, 12 h HRT followed with CO_2 sparging	1.68 mol-H_2/mol hexose	Kim et al. (2006)
	Artificial wastewater	Hydrogen-producing anaerobes	Batch experiments with 80 mL working volume using 5 nM gold particles	4.48 mol-H_2/mol sucrose	Zhang et al. (2007)
	Sucrose	Cattle dung and sludge as seed inoculum for dark and *Rhodobacter sphaeroides* SH2C for photofermentation	Dark fermentation of 150 mL followed with 36 mL anaerobic tubes under illumination	6.63 mol-H_2/mol sucrose	Tao et al. (2007)
	Glucose	*Escherichia coli*	80 mL compact stacked flat-bed reactor with immobilization of bacteria on agar supported on filter paper	1.2 mol-H_2/mol glucose	Ishikawa et al. (2008)
	Sucrose-laden wastewater	Sucrose-laden wastewater from reactor treating citrate-producing wastewater	Granule-based up-flow anaerobic sludge blanket of 4 L working volume with a HRT of 16.4 h	1.62 mol-H_2/mol hexose	Zhao et al. (2008)

(continued)

Table 1 (continued)

Period	Feedstock	Microorganism	Process and scale	H_2 production/yield	References
	Molasses-containing wastewater	Mixed microbial cultures	Expanded granular sludge bed (EGSB) process with 3.35 L working volume	3.47 mol-H_2/mol sucrose	Guo et al. (2008)
	Sodium acetate and yeast extract	Homogenized anaerobic sludge	Membraneless continuous flow microbial electrolysis cell (MEC)	3.9 mol-H_2/mol acetic acid	Tartakovsky et al. (2009)
	Sewage sludge from wastewater treatment process	Sterilized sewage sludge	Anaerobic self-fermentation with 150 mL working volume	16.26 mL-H_2/g VS	Xiao et al. (2009)
	Lignocellulosic materials	*Clostridium thermocellum*	Dark fermentation followed by electrohydrogenesis	9.95 mol-H_2/mol glucose	Lalaurette et al. (2009)
	Crude glycerol	*Rhodopseudomonas palustris*	Photofermentation using 125 mL serum bottles	6 mol-H_2/mol glycerol	Sabourin-Provost et al. (2009)
	Glucose	Anaerobically digested sludge	Continuously stirred reactor (5 L) and uncovered gravity settler (8 L)	2.8 mol-H_2/mol glucose	Hafez et al. (2010)
	Municipal solid wastes (MSW)	MSW endogenous bacteria	Bioreactor with a working volume of 4 L and supplied with current (0.06 A)	8.3 mmol-H_2/mg OC	Dictor et al. (2010)
	Poultry slaughterhouse sludge	Pretreated sludge	Aerobic thermophilic digestion using 100 mL serum bottle	136.9 mL-H_2/g treated sludge	Sittijunda et al. (2010)
2011–2014	Glucose	Anaerobically digested sludge	Sonicated biological hydrogen reactor (SBHR) with HRT of 12 h	2.1 mol-H_2/mol glucose	Elbeshbishy et al. (2011)
	Glucose	Recombinant *E. cloacae* CICC10017	Hydrogen-promoting gene was cloned and overexpressed into recombinant strain	2.55 mol-H_2/mol glucose	Song et al. (2011)
	Soft drink wastewater	Semisynthetic soft drink wastewater	Up-flow anaerobic packed-bed reactors with a working volume of 2.370 L	3.5 mol-H_2/mol sucrose	Peixoto et al. (2011)
	Rotten fruits of date palm	*Escherichia coli* EGY, *Clostridium acetobutylicum* ATCC 824, *Rhodobacter capsulatus* DSM 1710	Three fermentation stages: facultative anaerobe, strict anaerobic and photofermentation. Working volume 1.8 L	7.8 mol-H_2/mol sucrose	Abd-Alla et al. (2011)

(continued)

Table 1 (continued)

Period	Feedstock	Microorganism	Process and scale	H_2 production/yield	References
	Grass	*Clostridium pasteurianum*	Saccharification of grass by acid and alkaline treatment followed by batch fermentation in 120 mL serum bottles	72.21 mL-H_2/g dry grass	Cui et al. (2012)
	Beet molasses	*Rhodobacter capsulatus* JP91	Photofermentation using 125 mL serum bottles	10.5 mol-H_2/mol sucrose	Keskin et al. (2012)
	Synthetically prepared affluent	Inoculum obtained by the natural fermentation of the feed	Packed-bed bioreactor with altering the degree of backmixing using polynomial function	4.22 mol-H_2/mol sucrose	Fontes Lima et al. (2012)
	Molasses	–	Anaerobic CSTR with fuzzy controller	13.44 L-H_2/Day	Huang et al. (2012)
	Glucose	*Clostridium butyricum*	Nanoparticle encapsulation for enhancing hydrogen production	2.2 mol-H_2/mol glucose	Beckers et al. (2013)
	Glucose	Mixed culture from anaerobic digester	Electric field pretreatment of the seed inoculum for enhanced hydrogen production	1.43 mol-H_2/mol hexose	Jeong et al. (2013)
	Organic fraction of municipal solid waste (OFMSW)	Acclimated OFMSW	Solid substrate fermentation in semi-continuous and batch fermentation	1641 mmol-H_2/g VS	Escamilla-Alvarado et al. (2013)
	Crude glycerol	*Enterobacter aerogenes* NRRL B 407	Pretreatment of crude glycerol by salting out of soap	20 mmol-H_2/L medium	Sarma et al. (2014a)
	Beverage industry wastewater	Seed from compost of food waste	Temperature shift strategy to overcome inhibition in hydrogen production	1.68 mol-H_2/mol hexose	Sivagurunathan et al. (2014)
	Glucose and sucrose	*Enterobacter cloacae*	Application of $FeSO_4$ and synthesized iron oxide nanoparticles on fermentative hydrogen production	5.19 $\pm$ 0.12 mol-H_2/mol sucrose	Mohanraj et al. (2014)

Sahota et al. (1982). According to the authors, apart from hydrogen and carbon dioxide, acetic acid, lactic acid, and butyric acid were the other metabolites of the process (Sahota et al. 1982). Hydrogen production during fermentation or anaerobic respiration of *Rhodospirillum rubrum* and *Rhodopseudomonas capsulate* has been reported by Schultz et al. (1982). Karube et al. (1982) have reported that hydrogen productivity could be enhanced by immobilizing the *Clostridium butyricum* cells in a porous carrier so that it can be repeatedly used. The authors have mentioned that application of riboflavin can increase the hydrogen production (Karube et al. 1982). Application of anaerobic bacterial coculture for hydrogen from cellulose has been demonstrated by Odom et al. (1983). According to the authors, *Cellulomonas* strain ATCC 21399 can degrade cellulose, however, it does not produce hydrogen. *R. capsulate*, on the contrary, cannot utilize the cellulose directly, but it is capable of photoheterotrophically growing on organic acids or sugars. A coculture of *Cellulomonas* and hydrogenase-negative mutant *R. capsulate* has been reported to enhance the overall hydrogen yield (Odom et al. 1983). In a significant investigation, Mountfort et al. (1986) have demonstrated that unsaturated hydrocarbons can be reduced by hydrogen produced during anaerobic digestion of cellulose. According to the authors, upon addition of bromoethanesulfonic acid to the culture medium, methanogenesis was stopped and olefin, an unsaturated hydrocarbon, was subsequently reduced (Mountfort et al. 1986). Hydrogen metabolism has a role in the biodegradation of organic materials in anaerobic sediments (Goodwin et al. 1988). In an investigation, Goodwin et al. (1988) have studied the influence of pH on hydrogen metabolism in sediments. The authors have concluded that microbial hydrogen production and consumption in a sediment ecosystem are inhibited by gradual decrease in pH (Goodwin et al. 1988). Thus, from Table 1 and above discussion, it can be concluded that before 1990, microbial hydrogen production has been extensively investigated by different researchers and both light-dependent and independent-hydrogen production were explored. However, the research was mainly fundamental and the attempt to increase hydrogen production by process optimization or by genetic manipulation was rare in this period as the focus was mostly on feasibility of biohydrogen production.

2.2 Progress in Biohydrogen Research From 1991 to 1999

The advancement of biohydrogen production process during 1991–1999 has been summarized in Table 1. Wu et al. (1991) have reported the characterization of methanogenic granules of brewery wastewater treatment process. According to the authors, apart from methanogens, the granules contained hydrogen-producing organism (Wu et al. 1991). The granules were capable of methane production by using formate as the substrate. During this process, first hydrogen was produced and subsequently consumed to produce methane (Wu et al. 1991). Garcia-Lopez et al. (1996) have investigated in vitro methane and hydrogen production in ruminal fluid and the effect of 9,10-anthraquinone. The authors have reported that by increasing the levels of 9,10-anthraquinone, linear and quadratic decreases in methane production was achieved and correspondingly it was possible to increase the hydrogen production (Garcia-Lopez et al. 1996). Taguchi et al. (1996) have demonstrated an aqueous two-phase system for cellulose hydrolysis and continuous hydrogen production. The aqueous two-phase system used for cellulose hydrolysis was made up of 10 % polyethylene glycol and 5 % dextran, which can offer many advantages, such as recycling of the enzyme used in the process (Taguchi et al. 1996). According to the authors, by using a *Clostridium* strain, the system can produce hydrogen at a rate as high as 4.10 $mmol^{-1}$ h^{-1} (Taguchi et al. 1996). Hydrogen production using an anaerobic fluidized bed digester has been demonstrated by Guwy et al. (1997). The authors have reported that according to the operational parameters, hydrogen content of the biogas produced during the process changed. Thus, process parameters offer an opportunity to precisely control the anaerobic digestion process (Guwy et al. 1997). It can be found in the literature that by increasing the substrate loading rate from 40 to 63 kg COD m^{-3} day^{-1}, hydrogen concentration in the headspace can be increased from 290 to 640 ppm (Guwy et al. 1997). In a review, Nandi et al. (1998) have mentioned that hydrogen can be produced by certain photosynthetic bacteria, anaerobes, aerobes, facultative anaerobes, as well as methylotrophs. According to the authors, each of these groups has slightly different mechanism of hydrogen production and the subject has been extensively investigated from biochemistry, enzymology, and process technology point of view (Nandi et al. 1998). A low-cost and efficient bioreactor system has been developed for photosynthetic production of biohydrogen (Modigell et al. 1998). According to the authors, in the presence of sunlight, purple bacteria can produce hydrogen by photosynthesis (Modigell et al. 1998). During this process, the bacteria can use organic acids as the feedstock for hydrogen production. Compared to the traditional dark fermentation, hydrogen production by anoxygenic photoproduction technique used by these authors has relatively higher theoretical substrate energy conversion efficiency. Although hydrogen production by the bioreactor system designed by these authors will be costlier than traditional fossil fuel-based hydrogen production, but it will be less expensive than combined photovoltaic/electrolysis based process (Modigell et al. 1998). Yokoi et al. (1998) have demonstrated hydrogen production from starch by a coculture of *C. butyricum* and *Enterobacter aerogenes*. The process could remove O_2 from the culture medium to produce as high as 2 mol hydrogen per mol glucose without using any reducing agent. Continuous process involving these microorganisms immobilized in porous glass beads has been reported to produce nearly 2.6 mol hydrogen per mol glucose at a rate of 1.3 $L\text{-}H_2$ L^{-1} h^{-1} (Yokoi et al. 1998). Hydrogen production by a mutant strain of *Rhodobacter sphaeroides* has been reported by (Vasilyeva et al. 1999). The mutant strain used for this investigation was developed by subjecting the bacteria to UV irradiation, which resulted in an altered light-harvesting system. The authors have reported that compared to wild strain, the mutant could produce 1.5 folds more hydrogen at its absorption maxima, i.e., 800 and 850 nm (Vasilyeva et al. 1999). Bagai et al. (1999) have investigated photoevolution of hydrogen in a packed-bed reactor by using polyvinyl alcohol immobilized microbial strains such as *Halobacterium halobium, E. coli, and Phormidium valderianum*. The authors have reported that compared to a free cell system, the immobilized cell system was highly stable, and was able to produce hydrogen continuously for more than 60 days (Bagai et al. 1999).

Thus, from Table 1 and above discussion, it has been observed that during this decade, the studies were more focused on enhanced hydrogen production by process optimization.

2.3 Recent Developments in Biohydrogen Research (2000–2014)

In Table 1, the advancements in biohydrogen production research made during the period of 2000–2014 have been summarized. Mizuno et al. (2000) have demonstrated enhanced hydrogen production from glucose by driving away the accumulated hydrogen using nitrogen gas sparging. The authors have reported that by nitrogen sparging, specific hydrogen production rate could be increased from 1.446 mL hydrogen min^{-1} g^{-1} biomass to 3.131 mL hydrogen min^{-1} g^{-1} biomass (Mizuno et al. 2000). It was reported that the biogas mixture produced by this technique had 53.4 % of hydrogen and the yield was 0.85 mol/mol glucose (Mizuno et al. 2000). Photobiocatalytic hydrogen production using a system comprising of *Rhodopseudomonas capsulata* cells and sensitized TiO_2-MV^{2+} has been reported by Gurunathan (2000). According to the authors, compared to naked TiO_2, Rhodamine B and Ru $(bpy)_3^{2+}$ sensitized TiO_2 exhibited better hydrogen production (Gurunathan 2000). Fang et al. (2002) have investigated the effect of process pH on biohydrogen production. Based on 16S rDNA analysis, the authors have reported that corresponding to an increase in process pH, microbial diversity of the mixed culture used for hydrogen production was increased (Fang et al. 2002). At the optimal pH, i.e., 5.5, hydrogen content in the biogas mixture produced by the process has been reported to be 64 $\pm$ 2 %, and the yield was as high as 2.1 $\pm$ 0.1 mol hydrogen/mol glucose (Fang et al. 2002). Kim et al. (2004) have investigated the feasibility of co-digestion of food waste and sewage sludge for improved hydrogen production. The authors have reported that the food waste had superior specific hydrogen production potential to that of sewage sludge (Kim et al. 2004). Specific hydrogen production as high as 122.9 mL/g carbohydrate COD has been reported for optimum food waste: sewage sludge ratio of 87:13 (Kim et al. 2004). Based on the observation, the authors have mentioned that sewage sludge might have enriched the medium by supplying protein to enhance the hydrogen production (Kim et al. 2004). Thus, it could be considered as a good auxiliary substrate for fermentative hydrogen production. Najafpour et al. (2004) have demonstrated hydrogen production from synthesis gas by *R. rubrum*. Hydrogen production was achieved by water–gas shift reaction, where CO and water were oxidized by the microorganism to produce CO_2 and hydrogen (Najafpour et al. 2004). In order to grow the bacteria, liquid medium was prepared by using acetate as the carbon source and synthesis gas was used as the source of CO. The authors have reported that 1.5 g/L of acetate was optimum for the process, and hydrogen yield as well as CO conversion was close to theoretically possible maximum value (Najafpour et al. 2004). Effect of sulfate on fermentative hydrogen production by mixed culture has been studied by Lin and Chen (2006). Sulfate concentration ranging from 500 to 3000 mg SO_4^{2-} has been considered for this investigation. The investigators have reported that hydrogen production was inhibited by increased sulfate concentration mostly by shifting the metabolic pathway from butyrate to ethanol production (Lin and Chen 2006). Tao et al. (2007) have demonstrated a two-step dark and photofermentation process for enhanced hydrogen production from sucrose. Fermentation end products of the dark process have been further converted to hydrogen by photofermentation. It has been reported that by this sequential approach, cumulative hydrogen production can be increased from 3.67 mol-H_2/mol sucrose to 6.63 mol-H_2/mol sucrose (Tao et al. 2007). Zhao et al. (2008) have studied hydrogen production in a granule-based upflow anaerobic sludge blanket (UASB) reactor. The investigators have reported that hydrogen yield may change according to influent sucrose concentration and hydraulic retention time (HRT). An influent sucrose concentration of 14.5 g/L and an HRT 16.4 h was found to be optimum with a hydrogen yield of 1.62 mol-H_2/mol hexose (Zhao et al. 2008). Tartakovsky et al. (2009) have demonstrated continuous hydrogen production by using a membraneless microbial electrolysis cell. A gas phase cathode was used in the cell which was designed to avoid the need of any proton exchange membrane. Absence of proton exchange membrane and a short distance between the electrodes were helpful in reducing the internal resistance (Tartakovsky et al. 2009). Sittijunda et al. (2010) have investigated hydrogen production by self-fermentation of poultry slaughterhouse sludge. For this investigation, the authors have tested a dual digestion technique, where the sludge was first subjected to anaerobic thermophilic digestion followed by mesophilic digestion. Hydrogen yield as high as 136.9 mL-H_2/g total solid and a maximum production rate of 2.56 mL-H_2/L/h has been reported for this approach (Sittijunda et al. 2010). Song et al. (2011) have reported improved hydrogen production by overexpression of a gene encoding hydrogen-promoting protein. The hydrogen-promoting protein gene from *Enterobacter cloacae* IIT-BT 08 was cloned and overexpressed in *E. cloacae* CICC10017. Hydrogen yield as high as 2.55 ± 0.1 mol-H_2/mol glucose has been obtained for the recombinant strain, which was nearly 2-fold higher than the wild strain (Song et al. 2011). Fontes Lima et al. (2012) have studied the effect of backmixing on hydrogen production efficiency of an anaerobic packed-bed bioreactor. As a part of this study, different recycle ratios (*R*) were evaluated, and an *R* value of 0.6 was found to be optimum. According to the authors, at optimum recycle ratio hydrogen yield was as high as 4.22 mol-H_2 mol sucrose^{-1} (Fontes Lima et al. 2012). Escamilla-Alvarado et al. (2013) have studied biohydrogen production by solid substrate fermentation of municipal wastes. The authors have used semi-continuous and batch processes for the purpose and waste activated sludge has been evaluated as nitrogen-rich supplement (Escamilla-Alvarado et al. 2013). Cumulative hydrogen production as high as 1641 μmol-H_2/g volatile solid has been reported for this study (Escamilla-Alvarado et al. 2013).

Sarma et al. (2014a) have evaluated crude glycerol from biodiesel manufacturing plants as the feedstock for biohydrogen production. Addition of $MgSO_4$ to the crude glycerol-based medium has been shown to have beneficial effect on both hydrogen production and glycerol utilization by *E. aerogenes* (Sarma et al. 2014a). The effect of ferrous iron and iron oxide nanoparticles on fermentative hydrogen production has been evaluated by (Mohanraj et al. 2014). Likewise, a novel temperature shift strategy has been reported for enhanced hydrogen production by fermentation of beverage industry wastewater (Sivagurunathan et al. 2014). Thus, in the last decade, light-independent fermentative hydrogen production has been extensively investigated. In order to improve hydrogen yield, and to reduce the process cost, novel genetic approaches have been explored and waste-based feedstock have been tested.

3 Technical Challenges of Various Biohydrogen Production Processes

3.1 Light-Independent Anaerobic Fermentation

Technical challenges of hydrogen production by light-independent anaerobic fermentation have been listed in Table 2. In order to use it for hydrogen production, pretreatment of cellulose-based feedstock is necessary. In fact, cellulose hydrolysis is the rate-limiting reaction of cellulosic feedstock-based biohydrogen production (Lo et al. 2009). A less expensive, efficient, and yet sustainable method is desired for feedstock pretreatment for hydrogen production by this approach. Accumulation of fermentation end products such as volatile fatty acids may result in rapid depletion of process pH. It may result in metabolic pathway shifting and decrease in hydrogen production (Khanal et al. 2004). In order to maintain the hydrogen production ability of a process, therefore pH control is necessary. As already mentioned, pyruvate: ferredoxin oxidoreductase (PFOR) pathway, pyruvate: formate lyase (PFL) pathway, and pentose phosphate (PP) pathway are some of the metabolic pathways responsible for hydrogen production. Unfortunately, all these pathways ended up in different by-products. Therefore, suitable modification of these pathways to divert the substrate flow toward hydrogen production will be beneficial. Likewise, increase in hydrogen partial pressure due to accumulation of produced gas in the headspace of the reactor has negative effect on the process performance (Junghare et al. 2012). In a similar note, CO_2 accumulation in the fermentation medium has been shown to have inhibitory effect on hydrogen production (Tanisho et al. 1998). Due to accumulation of CO_2, succinate production may increase, which may be a reason of decrease in hydrogen production (Tanisho et al. 1998). Thus, continuous or periodic removal of accumulated gas may be necessary for enhanced hydrogen production by light-independent anaerobic fermentation.

Table 2 Technical challenges of various biohydrogen production processes

Process	Challenges	Potential solution	Remark(s)
Light-independent anaerobic fermentation	End product inhibition on enzyme activity	Use of aqueous two-phase system for simultaneous recovery of by-products	Process has been under development
	Complex substrates need saccharification	Substrate pretreatment	Different methods of substrate pretreatment for hydrogen has been studied
	Use of expensive reducing agents to maintain strict anaerobic environment	Use of facultative bacteria along with strict anaerobic bacteria: coculture approach	Facultative bacteria act as reducing agent with ability to remove oxygen
	Presence of hydrogen consumers, such as methanogens	Thermal and other pretreatment of the feedstock	The pretreated techniques have been reported to inhibit hydrogen consumers
	Significant biomass washout in continuous system due to high organic loading rate	Expanded granular sludge bed (EGSB) system with granular activated carbon (GAC) for development of biofilm to avoid costly immobilization technique	The EGSB system with high biomass retention at high organic loading rate proved to be a highly efficient biohydrogen production process
Photofermentation	Evolution of hydrogen by photosynthetic bacteria is inhibited by ammonium salts and molecular nitrogen	Biomass based feedstock low nitrogen content should be chosen	Cellulose of starch rich agro-industrial waste should be suitable for hydrogen production by photosynthetic bacteria
	Special requirements concerning pH value, temperature, light intensity, etc.	Modular outdoor bioreactor with very low need of external energy	New bioreactor concept has been tested to overcome these problems
	Two-stage dark–photofermentation have increased energy input and operational cost	Single stage fermentation using coculture of light-independent and light-dependent bacteria	Such strategy has been under development

(continued)

Table 2 (continued)

Process	Challenges	Potential solution	Remark(s)
Bioelectrochemical techniques	Loss of potential and reduced cell performance due to the membrane separating the anode and the cathode	Designing of membraneless microbial electrolysis cell with a liquid-phase anode and a gas phase cathode	Absence of the proton exchange membrane resulted in decreased internal resistance, reduced electrodes distance, minimized ohmic losses and increased volumetric hydrogen production
	Hydrogen production by microbial electrolysis cell may be inhibited by methanogenesis	UV irradiation of the system	Increase in hydrogen production

3.2 *Photofermentation*

It has been reported that purple photosynthetic bacteria such as, *R. capsulata*, can produce H_2 by anaerobic light-dependent electron transfer process involving nitrogenase enzyme (Hillmer et al. 1977). However, hydrogen production by nitrogenase may be inhibited by N_2 or ammonium salts (Hillmer et al. 1977). The authors have mentioned that maximum expression of the enzyme may be possible if the cells are photoheterotrophically grown in the presence of certain amino acids as nitrogen sources (Hillmer et al. 1977). Thus, proper selection of culture medium will be necessary for hydrogen production by this approach. A conventional pilot scale fermenter cannot be directly used for photosynthetic hydrogen production. An appropriate source of light is necessary for the approach, and the light should penetrate deep into the culture medium for efficient hydrogen production. Thus, compared to light-independent fermentative hydrogen production, scaling up of this method of hydrogen production may be more complicated.

3.3 *Bioelectrochemical Techniques*

Hydrogen production by microbial electrolysis cell may be inhibited by methanogenesis (Hou et al. 2014). By using an UV irradiation technique, hydrogen content of the gas mixture produced by the approach has been increased up to 91 %, whereas without UV irradiation methane content could be as high as 94 % (Hou et al. 2014). In order to separate the anode and the cathode of a microbial electrolysis cells, membrane is commonly used (Hu et al. 2008). However, it may result in loss of potential and reduced cell performance (Hu et al. 2008). Designing of single chambered microbial electrolysis cell could be a solution of this problem (Hu et al. 2008). Poor dimensional stability of sulfonated polyether ether ketone (SPEEK) proton conductor membrane conventionally used in microbial electrolysis cell has been reported (Chae et al. 2014). Recently, this type of membrane has been modified by introducing nanofiber-reinforced composite to overcome the problem (Chae et al. 2014). From above discussion, it can be concluded that microbial electrolysis cell technology for biohydrogen production has been going through a developmental phase and needs significant leads before coming to a scale-up level.

3.4 Common Challenges Associated with All Processes

In addition to the specific technical problems associated with the different methods of biohydrogen production, there are certain challenges common to all these techniques. High process cost is probably the most important challenge preventing large scale production of biohydrogen. Expensive chemicals needed for the process as well as high energy requirements are some of the factors directly associated with high process cost. Likewise, prior to their application for biohydrogen production, agro-industrial waste based feedstock generally needs hydrolysis or pretreatment. It adds up some additional steps as well as cost to biological hydrogen production. Presence of methanogens in the feedstock is other common problem of different biohydrogen production methods. Methanogens are known to use the hydrogen generated during the process and reduce the hydrogen yield. Overall, in order to increase the technoeconomic feasibility of large-scale production of biohydrogen, suitable solutions must be tailored to meet different production configurations.

4 Biohydrogen Production for Organic Waste Treatment

Almost all agro-industrial waste can be used as the feedstock for biohydrogen production, and this approach could be a sustainable method for management of these wastes. In a review, Kapdan et al. (2006) have mentioned that different food industry wastewaters such as olive mill wastewater and bakers yeast industry wastewaters can be simultaneously treated by using them as feedstock for biohydrogen production (Kapdan et al. 2006). According to the authors, carbohydrate-rich, low nitrogen solid waste materials such as cellulose and starch-containing wastes are ideal for hydrogen production (Kapdan et al. 2006). The authors have indicated that industrial waste with high COD or with potential environmental pollutants such as heavy metals and polycyclic aromatic hydrocarbons may need additional pretreatment prior to their application for hydrogen production (Kapdan et al. 2006). Similarly, Mamimin et al. (2011) have mentioned that proper disposal of agro-industrial waste is a burden for the industries; however, application of such waste for hydrogen production may eliminate the waste disposal problem of such industries as well as it could be helpful in reducing the raw material cost biohydrogen production (Mamimin et al. 2011). The authors have used palm oil mill effluent as the feedstock for hydrogen production, and reported that 46 % removal of COD was achieved by the approach (Mamimin et al. 2011). According to Kotay et al. (2006), if not disposed properly, sewage sludge could become an environmental hazard; however, it could be simultaneously treated by using it as the feedstock for hydrogen production (Kotay et al. 2006). The authors have demonstrated as high as 41.23 mL of H_2 could be produced by simultaneous reduction of 1 g COD of the sludge (Kotay et al. 2006). In a similar note, biohydrogen production and treatment of brown sugar wastewater has been demonstrated by Wang et al. (2010). Using an upflow anaerobic sludge bed reactor, a maximum hydrogen production rate of 5.98 L d^{-1} and 20 % removal of COD has been reported (Wang et al. 2010). Different similar reports on integrated hydrogen production and waste treatment have been summarized in Table 3. Based on the above discussion and the information presented in Table 3, it can be concluded that biohydrogen production from waste biomass can be a sustainable method for agro-industrial waste management; although further research should be considered to increase both hydrogen production and simultaneous COD removal.

5 Market Potential and Recent Government Policies on Biohydrogen Production

Considering the fact that hydrogen can be used as an alternative to fossil fuels, and that it is suitable for both fuel cells and internal combustion engines (Gillingham 2007), it could be postulated to have very high market potential. Nevertheless, supplying biohydrogen at a reasonable price will be the major challenge confronting potential commercial producers. In this context, government initiatives will be very crucial for economic feasibility of industrial biohydrogen production. The intention of different governments to promote hydrogen as a fuel is reflected by certain policies. Potential use of geothermal energy to produce hydrogen has been considered by US government in "Energy independence and security act of 2007" (http://www.hydrogen.energy.gov/pdfs/eisa_public_law_110-140_12-19-07.pdf; accessed on 17/09/2014). "International partnership for hydrogen and fuel cells in the economy" is an international institution with 18 partner countries/organizations, which was established in 2003 for realizing the transition to a hydrogen based economy (http://www.iphe.net/; accessed on 17/09/2014). Encouraging achievements made by each partner country in the field of hydrogen energy have been regularly communicated by the organization via meeting/workshops or by official website. Recently, The European Commission has lunched the second phase of its "Fuel Cells and Hydrogen (FCH) Joint Technology Initiative (JTI)." This phase of the initiative has a budget of €1.4 billion with 50:50 shares from EU and partner industries; where both partners will continue to invest until the fuel cell and

Table 3 A summary of different reports on biohydrogen production and simultaneous waste treatment

H_2 production process	Waste type	Microorganism/inoculum	H_2 production	COD removal	By-products	References
Light-independent anaerobic fermentation	Cellulose hydrolysate	*Clostridium sp.* (Strain No. 2)	4.46 mol-H_2/mol glucose	–	–	Taguchi et al. (1996)
	Organic fraction municipal solid waste	Heat pretreated sludge and hydrogen-producing bacteria	180 mL-H_2/g TVS	45 mL g/VSS/h	No methane	Lay et al. (1999)
	Food waste and sewage sludge	Heat-treated seed sludge	111.2 mL-H_2/g VSS/h	122.9 mL-H_2/g carbohydrate-COD reduced	Methane and alcohol	(Kim et al. 2004)
	Olive pulp	Hydrogen-producing culture	4–6 mmol-H_2/g carbohydrates	1.217 mg/mg of COD reduced	Acids and alcohol	Koutrouli et al. (2006)
	Molasses-containing wastewater	Mixed microbial cultures	3.47 mol-H_2/mol sucrose	–	89 % mass percentage of ethanol and acetate to the total volatile fatty acids %	Guo et al. (2008)
	Beverage industry wastewater	Seed from compost of food waste	1.68 mol-H_2/mol hexose	14.2 % removal	Acids and alcohol	Sivagurunathan et al. (2014)
Photofermentation	Lactic acid-containing wastes	Neutralized using NaOH	20 mL-H_2/h/g cells	92–99 % removal of lactate	–	Zürrer et al. (1979)
	Crude glycerol	*Rhodopseudomonas palustris*	6 mol-H_2/mol glycerol	–	–	Sabourin-Provost et al. (2009)
	Beet molasses	Beet molasses mixed with nutrients	10.5 mol-H_2/mol sucrose	–	–	Keskin et al. (2012)
Bioelectrochemical techniques	Sodium acetate and yeast extract	Homogenized anaerobic sludge	3.9 mol-H_2/mol acetic acid	90.6 % COD recovery	–	Tartakovsky et al. (2009)
	Municipal solid wastes (MSW)	MSW endogenous bacteria	8.3 mmol-H_2/mg OC	–	–	Dictor et al. 2010)

hydrogen technologies are introduced into the market. This initiative is expected to start within 2014 and its anticipated completion will be by 2024 (http://www.greencarcongress.com/2013/07/fch2jti-20130710.html; 18/09/2014). From the above discussion, it is evident that in order to make a transition to hydrogen-based economy, globally, remarkable initiatives have been taken by different governments. Thus, biohydrogen production processes are certain to have a political backup and good market in the future.

6 Concluding Statements

There are three reasons based on which biohydrogen can be considered as a green fuel for future transportation sector. First, hydrogen combustion does not result in harmful emissions, such as CO_2 or pollutant hydrocarbons. During this process, only water will be released to the atmosphere along with a negligible amount of NOx. Second, gravimetric energy density of hydrogen is nearly three times higher than that of commercial gasoline and diesel. Finally, biohydrogen can be produced from a range of agro-industrial waste materials resulting in simultaneous treatment of these wastes. Over the last 30 years, the focus of biohydrogen research has been gradually shifting from fundamental investigations to development of marketable technologies, including two-step and genetic interventions. Apart from the development of efficient technologies, government policies will be crucial for the overall success of hydrogen fuel-based economy.

Acknowledgments Authors are thankful to FRQNT, NSERC and INRS-ETE for financial assistance.

References

Abd-Alla MH, Morsy FM, El-Enany A-WE (2011) Hydrogen production from rotten dates by sequential three stages fermentation. Int J Hydrogen Energy 36(21):13518–13527

Bagai R, Madamwar D (1999) Long-term photo-evolution of hydrogen in a packed bed reactor containing a combination of *Phormidium valderianum*, *Halobacterium halobium*, and *Escherichia coli* immobilized in polyvinyl alcohol. Int J Hydrogen Energy 24(4):311–317

Bauchop T, Mountfort DO (1981) Cellulose fermentation by a rumen anaerobic fungus in both the absence and the presence of rumen methanogens. Appl Environ Microbiol 42(6):1103–1110

Beckers L, Hiligsmann S, Lambert SD, Heinrichs B, Thonart P (2013) Improving effect of metal and oxide nanoparticles encapsulated in porous silica on fermentative biohydrogen production by *Clostridium butyricum*. Bioresour Technol 133:109–117

Chae K-J, Kim K-Y, Choi M-J, Yang E, Kim IS, Ren X, Lee M (2014) Sulfonated polyether ether ketone (SPEEK)-based composite proton exchange membrane reinforced with nanofibers for microbial electrolysis cells. Chem Eng J 254:393–398

Chen X, Sun Y, Xiu Z, Li X, Zhang D (2006) Stoichiometric analysis of biological hydrogen production by fermentative bacteria. Int J Hydrogen Energy 31(4):539–549

Cui M, Shen J (2012) Effects of acid and alkaline pretreatments on the biohydrogen production from grass by anaerobic dark fermentation. Int J Hydrogen Energy 37(1):1120–1124
Dictor M-C, Joulian C, Touzé S, Ignatiadis I, Guyonnet D (2010) Electro-stimulated biological production of hydrogen from municipal solid waste. Int J Hydrogen Energy 35(19):10682–10692
Elbeshbishy E, Hafez H, Nakhla G (2011) Hydrogen production using sono-biohydrogenator. Int J Hydrogen Energy 36(2):1456–1465
Escamilla-Alvarado C, Ponce-Noyola T, Ríos-Leal E, Poggi-Varaldo HM (2013) A multivariable evaluation of biohydrogen production by solid substrate fermentation of organic municipal wastes in semi-continuous and batch operation. Int J Hydrogen Energy 38(28):12527–12538
Fang HH, Liu H (2002) Effect of pH on hydrogen production from glucose by a mixed culture. Bioresour Technol 82(1):87–93
Fontes Lima DM, Zaiat M (2012) The influence of the degree of back-mixing on hydrogen production in an anaerobic fixed-bed reactor. Int J Hydrogen Energy 37(12):9630–9635
Gaffron H, Rubin J (1942) Fermentative and photochemical production of hydrogen in algae. J Gen Physiol 26(2):219–240
Garcia-Lopez P, Kung L, Odom J (1996) In vitro inhibition of microbial methane production by 9, 10-anthraquinone. J Anim Sci 74(9):2276–2284
Gillingham K (2007) Hydrogen internal combustion engine vehicles: a prudent intermediate step or a step in the wrong direction? (Stanford)
Goodwin S, Conrad R, Zeikus J (1988) Influence of pH on microbial hydrogen metabolism in diverse sedimentary ecosystems. Appl Environ Microbiol 54(2):590–593
Guo W-Q, Ren N-Q, Wang X-J, Xiang W-S, Meng Z-H, Ding J, Qu Y-Y, Zhang L-S (2008) Biohydrogen production from ethanol-type fermentation of molasses in an expanded granular sludge bed (EGSB) reactor. Int J Hydrogen Energy 33(19):4981–4988
Gurunathan K (2000) Photobiocatalytic production of hydrogen using sensitized TiO_2–MV^{2+} system coupled *Rhodopseudomonas capsulata*. J Mol Catal A: Chem 156(1):59–67
Guwy A, Hawkes F, Hawkes D, Rozzi A (1997) Hydrogen production in a high rate fluidised bed anaerobic digester. Water Res 31(6):1291–1298
Hafez H, Nakhla G, El Naggar MH, Elbeshbishy E, Baghchehsaraee B (2010) Effect of organic loading on a novel hydrogen bioreactor. Int J Hydrogen Energy 35(1):81–92
Hall D, Rao K, Reeves S, Gogotov I (1978) Biocatalytic production of hydrogen by an in vitro system. Altern Energy Sour 8:3675–3696
Hallenbeck PC, Abo-Hashesh M, Ghosh D (2012) Strategies for improving biological hydrogen production. Bioresour Technol 110:1–9
Harding G, Sutter V, Finegold S, Bricknell K (1976) Characterization of *Bacteroides melaninogenicus*. J Clin Microbiol 4(4):354–359
Hillmer P, Gest H (1977) H_2 metabolism in the photosynthetic bacterium *Rhodopseudomonas capsulata*: H_2 production by growing cultures. J Bacteriol 129(2):724–731
Hou Y, Luo H, Liu G, Zhang R, Li J, Fu S (2014) Improved hydrogen production in the microbial electrolysis cell by inhibiting methanogenesis using ultraviolet irradiation. Environ Sci Technol 48(17):10482–10488
Hu H, Fan Y, Liu H (2008) Hydrogen production using single-chamber membrane-free microbial electrolysis cells. Water Res 42(15):4172–4178
Huang S-R, Chen H-T, Chung C-H, Wu C-C, Tsai T-Y, Chu C-Y, Lin C-Y (2012) Fermentative hydrogen production using a real-time fuzzy controller. Int J Hydrogen Energy 37(20):15575–15581
Ishikawa M, Yamamura S, Takamura Y, Sode K, Tamiya E, Tomiyama M (2006) Development of a compact high-density microbial hydrogen reactor for portable bio-fuel cell system. Int J Hydrogen Energy 31(11):1484–1489
Ishikawa M, Yamamura S, Ikeda R, Takamura Y, Sode K, Tamiya E, Tomiyama M (2008) Development of a compact stacked flatbed reactor with immobilized high-density bacteria for hydrogen production. Int J Hydrogen Energy 33(5):1593–1597

Jeong D-Y, Cho S-K, Shin H-S, Jung K-W (2013) Application of an electric field for pretreatment of a seeding source for dark fermentative hydrogen production. Bioresour Technol 139: 393–396
Joyner A, Winter W, Godbout D (1977) Studies on some characteristics of hydrogen production by cell-free extracts of rumen anaerobic bacteria. Can J Microbiol 23(3):346–353
Junghare M, Subudhi S, Lal B (2012) Improvement of hydrogen production under decreased partial pressure by newly isolated alkaline tolerant anaerobe, *Clostridium butyricum* TM-9A: optimization of process parameters. Int J Hydrogen Energy 37(4):3160–3168
Kapdan IK, Kargi F (2006) Bio-hydrogen production from waste materials. Enzym Microbial Technol 38(5):569–582
Karube I, Suzuki S, Matsunaga T, Kuriyama S (1981) Biochemical energy conversion by immobilized whole cells. Ann N Y Acad Sci 369(1):91–98
Karube I, Urano N, Matsunaga T, Suzuki S (1982) Hydrogen production from glucose by immobilized growing cells of *Clostridium butyricum*. Eur J Appl Microbiol Biotechnol 16(1):5–9
Keskin T, Hallenbeck PC (2012) Hydrogen production from sugar industry wastes using single-stage photofermentation. Bioresour Technol 112:131–136
Khanal SK, Chen W-H, Li L, Sung S (2004) Biological hydrogen production: effects of pH and intermediate products. Int J Hydrogen Energy 29(11):1123–1131
Kim S-H, Han S-K, Shin H-S (2004) Feasibility of biohydrogen production by anaerobic co-digestion of food waste and sewage sludge. Int J Hydrogen Energy 29(15):1607–1616
Kim D-H, Han S-K, Kim S-H, Shin H-S (2006) Effect of gas sparging on continuous fermentative hydrogen production. Int J Hydrogen Energy 31(15):2158–2169
Kim YM, Cho HS, Jung GY, Park JM (2011) Engineering the pentose phosphate pathway to improve hydrogen yield in recombinant *Escherichia coli*. Biotechnol Bioeng 108(12): 2941–2946
Kotay SM, Das D (2006) Feasibility of biohydrogen production from sewage sludge using defined microbial consortium. In: Proceedings of the 16th world hydrogen energy conference. Lyon, France, pp 209–210
Koutrouli EC, Gavala HN, Skiadas IV, Lyberatos G (2006) Mesophilic biohydrogen production from olive pulp. Process Saf Environ Prot 84(4):285–289
Kumar N, Das D (2000) Enhancement of hydrogen production by *Enterobacter cloacae* IIT-BT 08. Process Biochem 35(6):589–593
Kumazawa S (1981) Hydrogen metabolism by a marine blue-green alga, Oscillatoria Sp Miami Bg 7
Lalaurette E, Thammannagowda S, Mohagheghi A, Maness P-C, Logan BE (2009) Hydrogen production from cellulose in a two-stage process combining fermentation and electrohydrogenesis. Int J Hydrogen Energy 34(15):6201–6210
Lay J-J, Lee Y-J, Noike T (1999) Feasibility of biological hydrogen production from organic fraction of municipal solid waste. Water Res 33(11):2579–2586
Lin C-Y, Chen H-P (2006) Sulfate effect on fermentative hydrogen production using anaerobic mixed microflora. Int J Hydrogen Energy 31(7):953–960
Lin C, Lay C (2004) Carbon/nitrogen-ratio effect on fermentative hydrogen production by mixed microflora. Int J Hydrogen Energy 29(1):41–45
Lo YC, Huang C-Y, Fu T-N, Chen C-Y, Chang J-S (2009) Fermentative hydrogen production from hydrolyzed cellulosic feedstock prepared with a thermophilic anaerobic bacterial isolate. Int J Hydrogen Energy 34(15):6189–6200
Mamimin C, Prasertsan P (2011) Effect of temperature and initial pH on biohydrogen production from palm oil mill effluent: long-term evaluation and microbial community analysis. Electron J Biotechnol 14(5):9-9
Miller TL, Wolin M (1973) Formation of hydrogen and formate by *Ruminococcus albus*. J Bacteriol 116(2):836–846
Mizuno O, Dinsdale R, Hawkes FR, Hawkes DL, Noike T (2000) Enhancement of hydrogen production from glucose by nitrogen gas sparging. Bioresour Technol 73(1):59–65

Modigell M, Holle N (1998) Reactor development for a biosolar hydrogen production process. Renewable Energy 14(1):421–426

Mohanraj S, Kodhaiyolii S, Rengasamy M, Pugalenthi V (2014) Phytosynthesized iron oxide nanoparticles and ferrous iron on fermentative hydrogen production using *Enterobacter cloacae*: evaluation and comparison of the effects. Int J Hydrogen Energy 39:11920–11929

Mountfort DO, Kaspar HF (1986) Palladium-mediated hydrogenation of unsaturated hydrocarbons with hydrogen gas released during anaerobic cellulose degradation. Appl Environ Microbiol 52 (4):744–750

Najafpour G, Younesi H, Mohamed AR (2004) Effect of organic substrate on hydrogen production from synthesis gas using *Rhodospirillum rubrum*, in batch culture. Biochem Eng J 21(2): 123–130

Nandi R, Sengupta S (1998) Microbial production of hydrogen: an overview. Crit Rev Microbiol 24(1):61–84

Ntaikou I, Antonopoulou G, Lyberatos G (2010) Biohydrogen production from biomass and wastes via dark fermentation: a review. Waste Biomass Valorization 1(1):21–39

Odom JM, Wall JD (1983) Photoproduction of H_2 from cellulose by an anaerobic bacterial coculture. Appl Environ Microbiol 45(4):1300–1305

Peixoto G, Saavedra NK, Varesche MBA, Zaiat M (2011) Hydrogen production from soft-drink wastewater in an upflow anaerobic packed-bed reactor. Int J Hydrogen Energy 36(15): 8953–8966

Sabourin-Provost G, Hallenbeck PC (2009) High yield conversion of a crude glycerol fraction from biodiesel production to hydrogen by photofermentation. Bioresour Technol 100 (14):3513–3517

Sahota SS, Bramley PM, Menzies IS (1982) The fermentation of lactulose by colonic bacteria. J Gen Microbiol 128(2):319–325

Sarma SJ, Brar SK, Le Bihan Y, Buelna G, Soccol CR (2014a) Mitigation of the inhibitory effect of soap by magnesium salt treatment of crude glycerol–a novel approach for enhanced biohydrogen production from the biodiesel industry waste. Bioresour Technol 151:49–53

Sarma SJ, Dhillon GS, Brar SK, Bihan YL, Buelna G (2014b) Key enzymes in value-addition of glycerol to biohydrogen. In: Brar SK, Verma M (eds) Enzymes in value-addition of wastes. Nova Science Publishers, New York, pp 173–192

Schultz J, Weaver P (1982) Fermentation and anaerobic respiration by *Rhodospirillum rubrum* and *Rhodopseudomonas capsulata*. J Bacteriol 149(1):181–190

Şensöz S, Angın D, Yorgun S (2000) Influence of particle size on the pyrolysis of rapeseed (*Brassica napus* L.): fuel properties of bio-oil. Biomass Bioenergy 19(4):271–279

Sittijunda S, Reungsang A (2010) Biohydrogen production from dual digestion pretreatment of poultry slaughterhouse sludge by anaerobic self-fermentation. Int J Hydrogen Energy 35 (24):13427–13434

Sivagurunathan P, Sen B, Lin C-Y (2014) Overcoming propionic acid inhibition of hydrogen fermentation by temperature shift strategy. Int J Hydrogen Energy 39:19232–19241

Song W, Cheng J, Zhao J, Carrieri D, Zhang C, Zhou J, Cen K (2011) Improvement of hydrogen production by over-expression of a hydrogen-promoting protein gene in *Enterobacter cloacae*. Int J Hydrogen Energy 36(11):6609–6615

Sydney EB, Larroche C, Novak AC, Nouaille R, Sarma SJ, Brar SK, Letti LAJ, Soccol VT, Soccol CR (2014) Economic process to produce biohydrogen and volatile fatty acids by a mixed culture using vinasse from sugarcane ethanol industry as nutrient source. Bioresour Technol 159:380–386

Taguchi F, Yamada K, Hasegawa K, Taki-Saito T, Hara K (1996) Continuous hydrogen production by *Clostridium* sp. strain no. 2 from cellulose hydrolysate in an aqueous two-phase system. J Ferment Bioeng 82(1):80–83

Tanisho S, Kuromoto M, Kadokura N (1998) Effect of CO_2 removal on hydrogen production by fermentation. Int J Hydrogen Energy 23(7):559–563

Tao Y, Chen Y, Wu Y, He Y, Zhou Z (2007) High hydrogen yield from a two-step process of dark-and photo-fermentation of sucrose. Int J Hydrogen Energy 32(2):200–206

Tartakovsky B, Manuel M-F, Wang H, Guiot S (2009) High rate membrane-less microbial electrolysis cell for continuous hydrogen production. Int J Hydrogen Energy 34(2):672–677

Teplyakov V, Gassanova L, Sostina E, Slepova E, Modigell M, Netrusov A (2002) Lab-scale bioreactor integrated with active membrane system for hydrogen production: experience and prospects. Int J Hydrogen Energy 27(11):1149–1155

Van Niel EV, Budde M, De Haas G, Van Der Wal F, Claassen P, Stams A (2002) Distinctive properties of high hydrogen producing extreme thermophiles, *Caldicellulosiruptor saccharolyticus* and *Thermotoga elfii*. Int J Hydrogen Energy 27(11):1391–1398

Vasilyeva L, Miyake M, Khatipov E, Wakayama T, Sekine M, Hara M, Nakada E, Asada Y, Miyake J (1999) Enhanced hydrogen production by a mutant of *Rhodobacter sphaeroides* having an altered light-harvesting system. J Biosci Bioeng 87(5):619–624

Vatsala TM (1992) Hydrogen production from (cane-molasses) stillage by *Citrobacter freundii* and its use in improving methanogenesis. Int J Hydrogen Energy 17(12):923–927

Wang L, Li Y-F (2010) Biohydrogen production and wastewater treatment by anaerobic fermentation with UASB. In: International conference on mechanic automation and control engineering (MACE) 2010 IEEE, pp 1683–1686

Wu W-M, Hickey R, Zeikus J (1991) Characterization of metabolic performance of methanogenic granules treating brewery wastewater: role of sulfate-reducing bacteria. Appl Environ Microbiol 57(12):3438–3449

Xiao B, Liu J (2009) Biological hydrogen production from sterilized sewage sludge by anaerobic self-fermentation. J Hazard Mater 168(1):163–167

Yokoi H, Tokushige T, Hirose J, Hayashi S, Takasaki Y (1998) H_2 production from starch by a mixed culture of *Clostridium butyricum* and *Enterobacter aerogenes*. Biotechnol Lett 20 (2):143–147

Yokoi H, Saitsu A, Uchida H, Hirose J, Hayashi S, Takasaki Y (2001) Microbial hydrogen production from sweet potato starch residue. J Biosci Bioeng 91(1):58–63

Yoshida A, Nishimura T, Kawaguchi H, Inui M, Yukawa H (2006) Enhanced hydrogen production from glucose using ldh-and frd-inactivated *Escherichia coli* strains. Appl Microbiol Biotechnol 73(1):67–72

Zhang Y, Shen J (2007) Enhancement effect of gold nanoparticles on biohydrogen production from artificial wastewater. Int J Hydrogen Energy 32(1):17–23

Zhang Y, Liu G, Shen J (2005) Hydrogen production in batch culture of mixed bacteria with sucrose under different iron concentrations. Int J Hydrogen Energy 30(8):855–860

Zhao B-H, Yue Z-B, Zhao Q-B, Mu Y, Yu H-Q, Harada H, Li Y-Y (2008) Optimization of hydrogen production in a granule-based UASB reactor. Int J Hydrogen Energy 33(10): 2454–2461

Zürrer H, Bachofen R (1979) Hydrogen production by the photosynthetic bacterium *Rhodospirillum rubrum*. Appl Environ Microbiol 37(5):789–793

Biogas: An Evolutionary Perspective in the Indian Context

Shaikh Ziauddin Ahammad and T.R. Sreekrishnan

Abstract Anaerobic digestion is a well-known sustainable process for harnessing energy (in the form of biogas) from waste. India is among the pioneer countries to use AD processes for treating wastewater at larger scale. The first anaerobic digestion plant in India was built in 1859 in Bombay and biogas generated from the treatment facility was used for lighting streets. Use of AD to generate energy got promoted during and after World War II to meet higher energy demands. Most of the AD technologies in the world are primarily aimed at treating agricultural waste. The biogas generated from such big facilities can provide biogas for lighting and cooking to the local community. Biogas and subsequent energy generation gained popularity much later in Europe than in India. But, the biogas generation programme in Europe, especially in Germany, is implemented quite efficiently by utilizing agricultural residues in small scale (individual) as well as larger scale (community) reactors. There is a growing trend towards using larger, more sophisticated systems with better process control to produce electricity in India. Ministry of New and Renewable Energy (Govt. of India) has taken different initiatives (National Biogas and Manure Management Programme) to use AD systems to generate energy and to promote use of small-scale and large-scale anaerobic digesters in rural and semi-urban areas in India. Government has also given incentive to the families who have utilized their waste by setting up a family sized or community scale AD. National Programme on Biogas Development (NPBD) also promoted the use of cow dung slurry to produce energy from waste and as a result 4.17 million digesters have been set up to utilize cow dung generated by 289 million cattle with a gross capacity of 12 million family size biogas plants in India. Two different anaerobic digester designs extensively used in India under the NPBD are KVIC Model (floating drum type) and Deenbandhu model (fixed dome type). The major reason for the wide acceptance of this technology in India is the low-tech

S.Z. Ahammad · T.R. Sreekrishnan (✉)
Department of Biochemical Engineering and Biotechnology, Indian Institute of Technology Delhi, Hauz Khas, New Delhi 110016, India
e-mail: sree@iitd.ac.in

C.R. Soccol et al. (eds.), *Green Fuels Technology*,
Green Energy and Technology, DOI 10.1007/978-3-319-30205-8_17

design and the socio-economic benefits associated with the efficient utilization of waste. Lack of knowledge is the principal hurdle for not getting the technology utilized by a major portion of the rural population of India. The policies and implementation of appropriate support mechanisms can improve the AD utilization scenario in India. Increased public awareness programmes would help convincing people to adopt AD for recovering energy from different biodegradable waste residues. Use of clean fuel such as biogas instead of dung cake, agricultural residues and wood chips can substantially reduce the air pollution and help to maintain good environmental quality.

Keywords Anaerobic digestion · Agricultural residues · Biogas · Renewable energy

1 Introduction

Limited natural reserves of fossil fuels and increasing global warming are the major factors forcing us to look for alternate energy resources. Energy generation from wind power, hydropower, solar energy, geothermal power, tidal and wave power, biomass energy are gaining popularity and increasingly more researchers are looking at improving the efficiency of processes for their utilization. Energy generation from biomass, after its conversion to methane (major component of biogas) has a long history. During the tenth century BC in Assyria, biogas was used for heating water in bathrooms. Similar use was also found in Persia during the sixteenth century (Meynell 1976). The formal scientific discovery of the inflammable property of gases evolved from decaying organic matter was first determined by Jan Baptita Van Helmont in seventeenth century. In 1776, Count Alessandro Volta showed the correlation between the amount of degradable organic substances and the amount of marsh gas generated from it.

During the period 1804–1810, Dalton, Davy and Henry discovered the chemical composition of methane and they suggested that coal gas and marsh gas are very similar. They also showed methane production from decomposing cattle manure. At the end of nineteenth Century, it was discovered that the activity of a mixed microbial consortia, methanogens, is the key factor for methane production. In 1876, Herter reported the stoichiometric balance of methane (CH_4) and carbon dioxide (CO_2) formation from acetate obtained from sewage sludge (Meynell 1976). In 1906, Sohngen enriched two distinct acetate utilizing microbes, and he found that formate, CO_2 and hydrogen could act as precursors for its (methane) production. In 1930, Buswell and others identified anaerobic microorganisms and the conditions that promote their growth and subsequent methane production.

2 Anaerobic Digestion

Anaerobic digestion (AD) process is carried out by the syntrophic metabolic activities of different microbial consortia. Based on the metabolic activities, four distinct phases are observed in AD process. Hydrolysis, acidogenesis, acetogenesis and methanogenesis are the 4 major phases involved in AD process (Zenhder 1988). Figure 1 shows the anaerobic degradation pathway of organic substances.

Hydrolysis: It is the first stage of the organic matter decomposition process where bigger organic polymeric chains are broken down to smaller molecules like simple sugars, amino acids and fatty acids.

Acidogenesis: hydrolyzed organic matters are further degraded by the fermentative bacteria (acidogens) to produce different organic acids, long chain and short chain fatty acids, low alcohols and different degradation byproducts such as hydrogen (H_2), carbon dioxide (CO_2) ammonia (NH_3), and hydrogen sulphide (H_2S).

Acetogenesis: During this stage different organic acids are converted to acetic acid by different acetogens.

Metahnogenesis: The final stage of anaerobic digestion process where methane gas is produced from acetate, CO_2 and H_2 by the metabolic activities of acetoclastic

Fig. 1 Anaerobic degradation of organic compounds

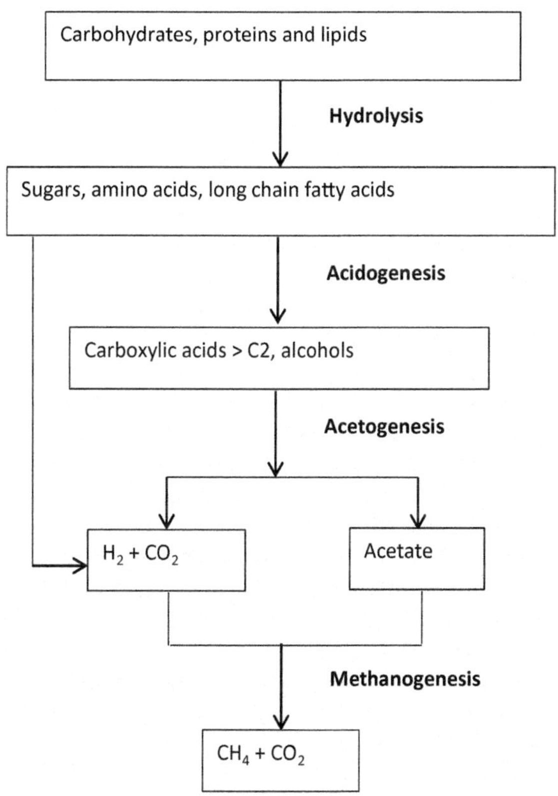

methanogens and hydrogenotrophic methanogens. Most often this step is considered as the rate limiting stage of AD process due to the slow growth rate of methanogens. The pathway for methane production is shown in Fig. 1.

2.1 Composition of Biogas

Biogas has approximately 60 % methane and 39 % carbon dioxide with small amounts of hydrogen sulphide, hydrogen, water vapour, and ammonia. It can be used as such to generate heat or electricity or enriched into bio-methane (>99 % methane). Calorific value of biogas is about 20–25 MJ/m^3 and energy potential is about 5.5–8 kWh/m^3.

2.2 Brief History of Anaerobic Digestion

In 1859, the first anaerobic digestion plant in India was built at a leper colony in Bombay. Few years later, in 1895, the first anaerobic digester in UK was built in Exeter in a sewage treatment facility to generate biogas from waste. The biogas generated from the treatment facility was used for lighting streets. Similar use of biogas was practiced in Bombay during 1897 to utilize the gas obtained during the treatment of human wastes at the Matunga Leper Asylum (Meynell 1976).

In 1904, Travis designed a two-stage process in which the suspended materials were separated and treated separately in hydrolyzing chamber. Most of the AD technologies in the world are primarily aimed to treat agricultural waste. The biogas generated from such big facilities can provide biogas for lighting and cooking to the local community. In China and India, there is a trend towards using larger, more sophisticated systems with better process control to produce electricity.

During 1914–1923, in Germany, digestion tanks were heated by burning methane produced by the anaerobic process.

Biogas and subsequent energy generation gained popularity much later in Europe than in developing countries, such as India, China and Nepal. But, the biogas generation programme in Europe, especially in Germany, is implemented quite efficiently by utilizing agricultural residues in small scale (individual) as well as larger scale (community) reactors.

In Europe, AD systems used for treating different wastes obtained from suitable farm, industries and municipalities have had a good record compared to North America. South American countries, such as Brazil are also following the footstep of India and China to adopt more AD-based treatment processes (i.e. vinasse, a byproduct of sugarcane-based ethanol production units).

Due to higher energy demand during and after World War II, options for different alternate energy sources are looked into. Biogas from AD systems was one of the most commonly used non-conventional energy resource. During World War II, biogas balloons were used to run public transport, buses and other vehicles. In a few

Fig. 2 Use of AD in various processes

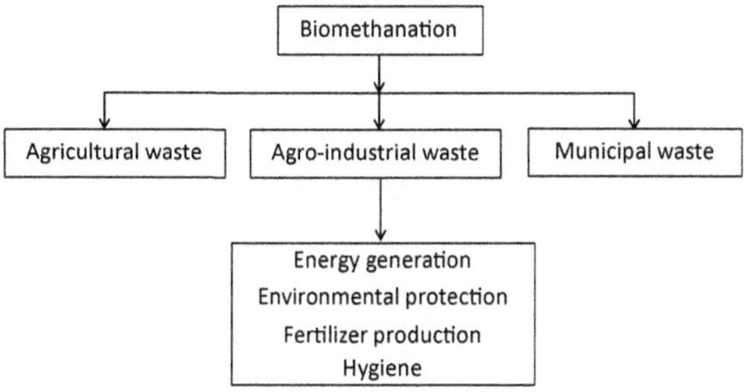

European countries those AD facilities have been in operation for more than 20–40 years. More than 600 farm-based AD systems are operational in Europe and among them the majority of the AD (around 250) systems are installed in Germany. Simplicity of design, local environmental regulations and strict implementation as well as different policies to promote renewable energy generation are the key factors for successful operation of these facilities.

Among other European countries, Denmark is using the large-scale AD facilities for quite some times and they have the greatest experience of using AD in different sectors. Denmark took the lead in using and promoting the benefits of AD technologies for treating co-digested manure, clean organic industrial wastes and source-separated municipal solid waste (MSW) in large-scale AD systems in Europe (Danish Ministry of Energy and Environment 1996). Around 18 large centralized plants are now in operation. Their energy initiative programme helped them to popularize the use of AD in different industries. 'Green Pricing' is one of the key policy tools used to encourage technology development and deployment. The policy allows the manufacturers to get some incentives while selling their products as they use biogas, a green fuel to generate electricity. Some industry uses biogas for heating process waters. The sale of hot water to the central heating systems is becoming a significant source of profits for project developers and different industries. Figure 2 shows the potential use of AD in different processes.

Worldwide, more than 1000 vendor-supplied systems are now in operation or are under construction and more than 35 classes of industries which use AD systems have been identified, including manufacturer and processors of chemicals, pharmaceuticals, fibre, food, meat and milk. Many use AD as a pre-treatment process to reduce the cost of sludge, control odours, and to reduce the final treatment costs at a municipal sewage treatment plant.

3 Process Configuration and Substrates

Reactor Configuration: Different reactors are used for treating various waste streams containing different organic matters. Depending upon the complexity/degradability of the organic matter, the reactor configuration is arrived at. If the hydrolysis step is the most difficult one, appropriate design of reactor and reactor operational strategies are considered to achieve desired hydrolysis and unhindered subsequent processes. Typically, low temperature anaerobic digestion suffers the problem of poor hydrolysis whereas mesophilic AD is predominantly controlled by the slow methanogenesis. As the methanogens are slow growing archaea, higher methanogen concentration is desired to improve the biogas production.

Based on the availability of resources, substrates and other operational conditions, two different reactor schemes such as suspended cell systems and attached cell systems are used. Attached cell systems are commonly used for treating high-strength wastewaters whereas slurry or suspended cell systems are used for stabilizing complex froth forming organics (PCP industries) and cow dung (Speece 1983).

Generally, composting is used to stabilize MSW, but (aerobic) composting is an *energy-consuming* process. It requires 50–75 kWh of electricity to stabilize one tonne of MSW. AD is a net *energy-producing* process and during anaerobic treatment of MSW, it can generate 75–150 kWh of electricity per tonne of MSW. The improved MSW treatment facilities have made significant progress towards its commercial use in recent years. There are reports of hog manure being used for biogas generation from Tulare, California. The biogas produced is being used to generate electricity. Apparently, the electricity generated is able to meet the electricity demands of the farm, including heating (Demirel and Yenigün 2002).

4 Indian Government Initiatives

Government of India has taken initiatives under the Ministry of New and Renewable Energy[1] to use AD systems to generate energy. Various programmes have been in operation to promote the use of small-scale and large-scale anaerobic digesters in rural and semi-urban areas in India. Government has also given incentive to the families who have utilized their waste by setting up a family sized anaerobic digester. Similar incentives have been given for community scale AD set-up. National Biogas and Manure Management Programme (NBMMP) (see Footnote 1) has been initiated to tackle the following programmes:

(a) Biogas-Based Distributed/Grid Power Generation Programme.
(b) Recovery of Energy from Industrial Wastes.
(c) Recovery of Energy from Urban Wastes.

The Ministry has also taken initiatives for efficient utilization of biogas.[2] In this regard, the Ministry has undertaken projects demonstrating integrated plants for

---

[1]National Biogas Manure Management Programme (NBMMP)—12th plan, No. 18-1/2014-BE (NBMMP), MNRE, Government of India, 2014.

[2]Evaluation report on National Project on Biogas Development Programme, Planning Commission, Government of India, 2002.

production, purification and bottling of biogas. It also has generation of smokeless villages using biogas systems in its agenda.

National programme on Biogas Development (NPBD) (see Footnote 2) also promotes the use of cow dung slurry to produce energy from waste. As a result of this programme, 4.17 million digesters have been set up to utilize cow dung generated by 289 million cattle with a gross capacity of 12 million family size biogas plants.

5 Reactor Configurations Used in Different AD Processes

Based on the hydrodynamic properties of microbial biomass and aqueous phase present in the reactor, the reactors are broadly classified into two groups; suspended cell system or slurry handing system and attached cell or immobilized cell reactors. Depending upon the requirement in the treatment process, appropriate configuration is used. Attached cell systems are considered as high rate reactors compared to suspended cell system due to higher biomass presence in the reactor. Upflow anaerobic sludge blanket (UASB), fluidized bed anaerobic reactors are the commonly used attached cell systems for treating wastewaters. Anaerobic continuous stirred tank reactor is one of the simplest designs of suspended cell system and is typically used with modified mixing conditions in treating waste slurry generated from domestic waste and cow dung. Anaerobic reactors are very delicate in nature due to the sensitive anaerobic microorganisms present in the reactor. Operation of complex reactors requires skilled operators and success of the reactor solely depends on strict monitoring, maintenance and troubleshooting of the reactors.

Different reactor designs have been developed, for carrying out the AD, with the objectives of improved biogas production. The success of the NPBD is very much dependent on the reactor design. Simple reactor configuration and minimum maintenance could make them attractive to use in rural or urban areas where skilled operators are not available.

Two different anaerobic digester designs extensively used in India under the NPBD are KVIC Model (Floating drum type) and Deenbandhu Model (Fixed dome type). Figures 3 and 4 show the schematic of the digesters (see Footnote 2).

Pragati Model is also developed by combining the concept of these two models where the lower part of the digester is semi spherical with a conical bottom and a floating drum acts as gas storage.

Deenbandhu model is the most widely used AD in India. It was developed by Action for Food Production (AFPRO), India in 1984. The design of the model considered two major factors; simplicity in design and low installation cost. Two concrete hemispheres are joined together to build the digester. Bottom half is used as the digestion chamber whereas the top is used for gas storage. Wastes come to the digester through inlet pipe from inlet tank and digested wastes go to the pit for further use as manure.

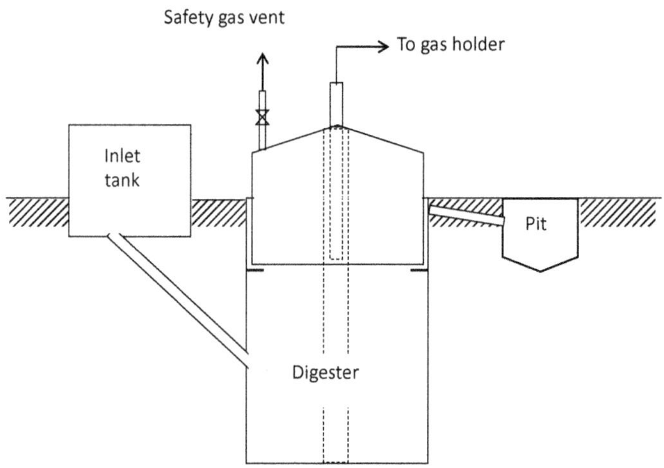

Fig. 3 Floating dome type anaerobic digester

Fig. 4 Fixed dome type anaerobic digester

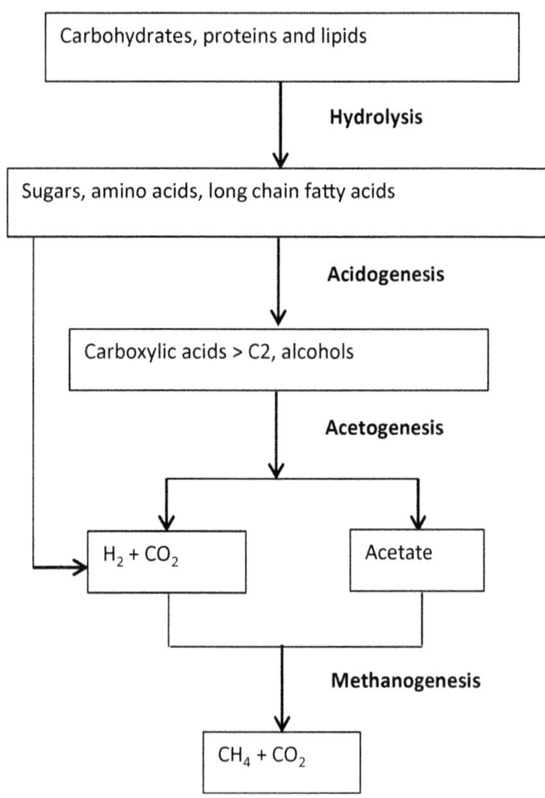

The size of the domestic digester depends upon the availability of feedstock or amount of waste generated by the family. About 25 kg of cattle dung is required for 1 m^3 of biogas production.

6 Economic Viability

Use of any technology and its acceptance always depends on its economic feasibility. Economic feasibility study of such AD technologies is generally carried out by considering the major cost components. The major expense heads for such AD systems are as follows (see Footnotes 1 and 2):

1. Installation Cost: The digester has three major components which are made of masonry structures. The inlet slurry mixing tank is the first compartment which is linked to the digester through connecting pipe and the digested slurry collection tank. Installation cost includes cost of construction materials, connecting pipes and gas delivery system, labour charges for digging the pit and making the digester.
2. Operating cost: Cost associated with mixing the feedstock and water in the inlet tank, removing the digested materials from outlet tank, drying the digested material are the major components of operating cost of such digester.
3. Maintenance cost: Consistent supply of the necessary quantity of feedstock and cattle dung is key factors for successful operation of such units. Requirement of feedstock, water and cattle dung depends on the size of digester. No major maintenance is required for such systems but the structure material and other associated installations have a fixed lifetime. The typical life of a Deenbandhu anaerobic digester is about 20 years.
   Annual operating cost includes the annual depreciation on construction work and other installations, maintenance charges and cost of dung.
4. Technology upgradation costs: Shredding of crop residues and pre-treatment with hot water to improve their digestibility is necessary when co-digested with cattle dung in the AD. Cost associated with mechanical shredding and pre-treatment and mixing the solids in the liquid can incur expenses for upgradation.
5. Transportation cost: Transportation of feedstock from the field and digested materials to the field/market involve the cost of transportation. It is more prominent in the case of large scale, community level digester.

6.1 *Profitability*

The low installation cost of anaerobic digester and minimal maintenance makes it very attractive and beneficial to replace the conventional unprocessed fuels like dung cakes, wood chips and crop residues commonly used in rural areas. It can also be used to replace LPG as 1 m^3 of biogas could give energy equivalent to 0.43 kg of LPG. A 1 m^3/d biogas producing digester is more than sufficient for any standard family to provide sufficient fuel for cooking.

The manure obtained from the digester is also another revenue generating material which can be used as a nitrogen-rich organic fertilizer. It provides a way to improve the soil fertility by supplementing essential nutrients to crops. It can replace expensive chemical fertilizers as well as reduce the pollution burden on the environment created by different synthetic chemicals.

Use of clean fuels such as biogas instead of dung cake, agriculture residues and wood chips can substantially reduce the air pollution and help to maintain good environmental quality.

The major reason for the wide acceptance of the technology in rural India is its low-tech design and the socio-economic benefits associated with the efficient utilization of waste.

6.2 Constraints

Lack of awareness is the principal hurdle for not getting the anaerobic digestion technology utilized to its full potential by a major portion of the rural population of India. More public campaigns and general awareness programmes need to be initiated by the Government and different NGOs to change the scenario. Implementation of different schemes and financial support for building the digester would promote the use of AD technology.

7 Present Interest in Anaerobic Digestion

There is an increased recognition, in both developing and industrialized countries, of the need for technical and economic efficiency in the allocation and exploitation of resources. A positive trend is growing to recover energy from waste and minimize the quantity of non-usable waste.

People are getting more interested about the effective use or disposal of household and community wastes in a manner which will reduce pollution burden as well as to generate revenue. We have moved ahead from the days when waste was considered as ‘garbage’ or ‘unusable’; it is presently considered as an important resource. Increasing popularity of AD in different sectors is the major driving force for developing advanced and energy-efficient AD systems which can be used for different purposes with some common goals such as:

(a) Treatment of organic wastes (organic or biodegradable part of solid waste) and wastewaters having a broad range of organic loads and substrate/waste concentrations.

(b) Energy generation and utilization.
(c) Improvement in sanitation and reduction/elimination of odours.
(d) Production of high-quality fertilizer.

R&D has shifted from basic studies to a different dimension where computational fluid dynamics and complex microbial models are being used for designing more sophisticated ADs to handle a wide variety of wastes (complex as well as simple) with different solid contents. The major concerns are to improve the rate of degradation and reduction of reactor size. Digester having capability to handle high organic loads as well as high rate digestion of diluted waste are other challenging areas to looked into. Other interesting research areas also include the improvement of fermentation and re-use of specific materials in integrated farming systems, biogas purification, simple but effective digester design/construction and development of digester for effective treatment of domestic wastewater.

Most often, increase in methanogenic population in the reactor is the key parameter to improve digestion rate and the capability of handing high organic load. Use of attached cell systems such as in fluidized bed reactors is one of the best solutions to tackle such issues. Separation of acidification and methanogenesis steps into two different reactors is another strategy to improve solids handling capacity and overall efficiency of AD.

Two-stage Anaerobic digestion: In conventional anaerobic digester, acidogenesis and methanogensis are carried out in a single reactor. A delicate balance is maintained between acidogens, acetogens (acid-formers) and methanogens in the reactor to achieve desired biogas production since the microorganisms involved in the two processes differ widely in terms of their metabolic needs, growth kinetics and sensitivity to different environmental conditions such as pH and temperature. Separation of acid-formers and methanogens into two reactors is desirable because this will make it possible to provide the optimum growth conditions for each group of microorganisms. This will enhance overall productivity, process stability and easiness for process monitoring and control. Different wastes are being treated using two-stage AD. The common configuration of the system is the fluidized bed or completely mixed acid phase digester followed by an UASB or high rate hybrid anaerobic reactor. Complex organic-containing wastes, such as waste food, food waste, oil mill effluent, dairy effluent, paper mill effluent, wood hydrolysate, cheese whey are preferably treated in two-stage AD system (Demirel and Yenigün 2002). Commonly, agricultural residues such as grass clippings, cannery waste, greenhouse waste and organic fractions of municipal refuse are treated in such reactors.

Many industries use UASB reactor to treat their effluents. UASB is an attached cell reactor system where the anaerobic microorganisms attach to one another to form a sludge blanket. Microorganisms present in the sludge blanket and wastewater interact while the wastewater passes through the reactor. Poor mixing (due to low liquid velocity), difficulty in developing the sludge bed and instability caused due to possible ‘wash-out’ of sludge bed are the major disadvantages of UASB.

To overcome the drawbacks of UASB, a hybrid system was developed by incorporating the advantages of UASB and fluidized bed reactors. The superior performance of hybrid anaerobic reactor (HAR) (Saravanan and Sreekrishnan 2005) due to better mixing and reactor stability make it more attractive to use over UASB. Substantial improvement in mixing reduces the time requirement to treat a given quantity of wastewater. It ultimately helps to reduce the reactor size without compromising on treatment capacity. Membrane-based anaerobic reactors have also been used in different industries to minimize discharge. The distillery industries in India follow a strict environmental regulation and as per the regulatory norms, all such industries should make their process comply with the 'zero discharge' policy. Commonly, these industries use AD followed by residual nitrogen removal and incineration of digested solids to make the process 'zero discharge. About 150 AD units are being used by different distillery industries to treat distillery effluents for generating biogas. Most of them use the biogas produced for heat (steam) and electrical power generation.

In India, almost all industrial processes use non-cow dung AD process to treat the generated wastewater. Different industrial wastes, such as poultry wastes, MSW, sewage, distillery effluents, dairy effluent, food processing waste, leather processing and rubber processing industries, slaughterhouse wastes, pulp and paper industry wastes, vegetable markets, kitchen and canteen wastes are using AD process to generate biogas and subsequently electricity from biogas.

At Namakkal district in Tamil Nadu, poultry wastes are used to generate 1.55 and 2.5 MW electricity and a similar process has been adopted at Panchkula, Haryana to produce 6 MW electrical power.

Use of AD has a major potential to generate energy from MSW. Current MSW generated by different states in India shows energy potential of 3670 MW of electricity. 6.6 MW project based on MSW are built at Hyderabad, Vijayawada (6 MW) and Lucknow (5 MW) (see Footnote 2).

8 Conclusions

Traditional methods of waste management are not very attractive any more as waste is regarded as a potential energy source and therefore considered as a valuable resource. Appropriate infrastructure and awareness can improve the waste to energy process in developed and developing countries. Substantial amount of energy can be gained from biogas generated during treatment of various kinds of waste materials in small scale as well as large-scale reactors. Anaerobic digestion is one of the attractive routes for renewable energy production. But, lack of proper understanding about its use and potential benefits limit its application in the rural areas of India. A more proactive approach by the government and putting in place necessary policy decisions can serve to propagate these technologies among the rural mass in India. There is also a need for public awareness programmes to convince the people to adopt such beneficial technologies for efficient resource utilization.

References

Danish Ministry of Energy and Environment (1996) Energy 21; The Danish Government's action plan for energy 1996. Denmark, Copenhagen
Demirel B, Yenigün O (2002) Two-phase anaerobic digestion processes: a review. J Chem Technol Biotechnol 77:743–755
Meynell P-J (1976) Methane: planning a digester. Schocken Books, New York, p 3
Saravanan V, Sreekrishnan TR (2005) Hydrodynamic study of biogranules obtained from an anaerobic hybrid reactor. Biotechnol Bioeng 91:715–721
Speece RE (1983) Anaerobic biotechnology for industrial wastewater treatment. Environ Sci Technol 17:416A–427A
Zenhder AJB (ed) (1988) Biology of anaerobic organisms. Wiley, New York

Biobutanol—"A Renewable Green Alternative of Liquid Fuel" from Algae

Sampa Maiti, Dulal Chandra Maiti, Mausam Verma and Satinder Kaur Brar

Abstract Increasing global energy demand, concern over global climate changes, and unstable and expensive petroleum resources, have led to the development of renewable energy sources that have driven research toward the utilization of biomass resources for the production of energy and fuels. In this context algae appear to be an emerging source of biomass for biobutanol that has the potential to give an alternative green solution to replace fossil fuel and reduce environmental issues. Biochemical production of butanol, a four carbon aliphatic alcohol, is promising due to its superior fuel properties as compared to ethanol. This chapter presents a comprehensive review on sustainable bioproduction and utilization of butanol as a biofuel and provides a glimpse on different potential biomass and microorganism for biochemical production of butanol. Main bottlenecks in biochemical production and recovery using conventional anaerobic acetone–butanol–ethanol (ABE) fermentation and corresponding recent counteractive steps to overcome these challenges were discussed systematically. A special emphasis has been given on the production of butanol a green alternative solution to fossil fuel using both micro- and macroalgae as potential biomass.

Keywords Biobutanol · Algae · ABE (acetone–ethanol–butanol) fermentation · Micro algae · Macro algae · Electrochemical production

S. Maiti · S.K. Brar (✉)
Institut national de la recherche scientifique, Centre—Eau Terre Environnement, 490, Rue de la Couronne, Québec, QC G1K 9A9, Canada
e-mail: satinder.brar@ete.inrs.ca

D.C. Maiti
Bajkul Milani Mahavidyalaya, Vidyasagar University, Medinipur, India

M. Verma
CO2 Solutions Inc., 2300, rue Jean-perrin, Québec, QC G2C 1T9, Canada

C.R. Soccol et al. (eds.), *Green Fuels Technology*,
Green Energy and Technology, DOI 10.1007/978-3-319-30205-8_18

1 Introduction

World energy is facing a large and diverse set of challenges as there is no proper balance between the demand and production of fossil fuel. Bursting population, short-term global politics and anxiety have raised the demand and price of liquid transport fuel. The detrimental impact on climate by CO_2 emissions and different components of greenhouse gas (GHGs) from the combustion of nonrenewable fuel has increased the urge to search renewable and green energy sources (Yang and Song 2008). Development of renewable chemicals and biofuel technologies is the current agenda in science and technology (Papoutsakis 2008). Biobutanol, an important chemical and solvent, had been recently considered as one such green alternative in liquid transportation fuel. As a fuel, biobutanol is emerging as third generation liquid biofuel. Compared to bioethanol, biobutanol has properties more comparable to gasoline due to higher energy content, octane improving power, low volatility, higher blending rate with gasoline without engine modification, convenient distribution using current pipeline infrastructure, and enhanced auto-emission performance (Cascone 2008).

Biobutanol has the potential to be produced from diversified biomass ranging from starchy food materials, lignocelluloses to renewable waste biomass by microbial acetone–ethanol–butanol (ABE) fermentation. However, the main bottlenecks in production are high feedstock cost which is about 60 % of total production cost, low productivity followed by recovery cost (Thang et al. 2010). First generation substrates are limited by high cost and scarcity of food, the second generation substrates compete for arable land and require additional cost to remove the lignin barrier. Hence, attention is drawn toward utilization of inexpensive, renewal, high carbon rich feedstock that has the potential to produce third generation biofuel. Fast growing, most abundant, high carbon rich algae have been considered as potential substrates for the production of third generation biofuel-butanol. This chapter describes briefly, biobutanol as a fuel and its biochemical production and challenges. Finally, particular emphasis is given to production of biobutanol from algae as potential substrate.

2 Butanol as Fuel

Butanol is four carbons containing, saturated aliphatic, alcohol having the molecular formula of C_4H_9OH (MW 74.12). In nature, four isomeric forms of butanol are abundant, *n*-butanol, 2-butanol, isobutanol, and tertiary butanol. Compared to other alcohols, such as propanol, ethanol, and methanol, all the isomers of butanol are superior as a biofuel considering all the parameters, such as octane number, heat of combustion, carbon dioxide production upon combustion, etc. A comparative study of these parameters, such as average octane (AKI rating/RON), cetane number, combustion energy (MJ/dm^3), evaporation heat (MJ/kg), carbon dioxide production (MJ/kg)) of different alcohol-based and other renewable biofuels are listed in Table 1.

Table 1 Chemical prosperities of biobutanol in comparison to other fuel

Fuel	Research octane number (RON)	Cetane number	Energy density (MJ/L)	Evaporation heat (MJ/kg)	Carbon dioxide production (MJ/kg)
Methanol	136	3	18.6	1.2	15
Ethanol	129	8	21.4	0.92	13
Biobutanol	96	25	29.2	0.43	15
Gasoline	91–99	–	32.5	0.36	14
Biohydrogen	–	–	2.8–12	–	N/A
Biodiesel	–	40–65	33.44	0.27	–

(Biofuel, The fuel of the future. http://biofuel.org.uk/bioalcohols.html)

Compared to other alcohols, less corrosive, less volatile, and less explosive butanol has higher flash point and lower vapor pressure. It can be used directly or blended with gasoline or diesel without any vehicle retrofit in higher proportion (Dürre 2007; Pfromm et al. 2010). Apart from these superior properties as fuel, butanol can be used as important chemical and solvent in different applications, such as in cosmetics, hydraulic fluids, detergent formulations, drugs, antibiotics, hormones and vitamins, and intermediate chemicals in different pharmaceuticals. Due to its superiority as fuel and application importance in versatile fields, there is a rising trend toward the production of butanol using both chemical and biochemical methods.

3 History of Biobutanol Production

Production of biobutanol using biochemical route is known as ABE fermentation. Biobutanol production using ABE fermentation was discovered by Pasteur in 1861 (Jones and Woods 1986). This fermentation was used commercially from the early part of twentieth century until WWII. In 1945, two thirds of industrially used biobutanol was produced by fermentation in U.S. However, the ABE fermentation process had lost competitiveness by 1960 s due to the increase of feedstock costs and advancement of the petrochemical industry except in Russia and in South Africa, where the substrate and labor costs were low (Lee et al. 2008). When cheaper butanol was produced from petrochemical sources using mainly oxo-process, the fermentative production lost attention and it became noncompetitive and was eventually discontinued. However, when oil price suddenly increased in 1973, intensive research interests returned to conventional production of butanol using the bioconversion of agricultural products, different industrial wastes, and other carbon rich sources into butanol. As the biochemical conversion of butanol is not economically attractive compared to petrobutanol, several inventions, modifications, and other efforts have been carried out at each step to make it more competitive vis-a-vis the conventional and old options. Brief summary of all these efforts during each step of the biochemical conversion are highlighted below.

4 Biochemical Production of Butanol

Biochemical production of biobutanol was carried out via ABE fermentation of sugar extract using solventogenic *Clostridia*. However, all the four isomers of butanol cannot be produced by biological process (Machado 2010). Only *n*-butanol and 2-butanol were produced by direct fermentation. Fermentation produced 2, 3-butanediol, which upon chemical conversion was transformed to secondary butanol. In literature, no such biological route was reported to produce tertiary butanol (Machado 2010). Biobutanol production via conventional anaerobic fermentation, involves different steps, such as biomass mobilization, bacterial strain selection and development, fermentation practices and recovery and purification from the fermentation broth.

4.1 Feedstock Selection

Selection of feedstock in biobutanol production is a major challenge as about 60 % of the total production cost is involved in this step. Production of biobutanol involves large variety of biomass, such as food crops, energy corps, agro-industrial waste, lignocelluloses, algae, glycerol, syngas, amino acids, CO_2, and other waste organic materials (Thang et al. 2010; Blanch et al. 2011; Huo et al. 2011; Köpke et al. 2010; Liew et al. 2006; Qureshi et al. 2008a, b; Taconi et al. 2009). First generation biofuel had negative impacts on food security and controversial energy balance. The second generation biofuels competed for arable land and required additional cost to remove lignin barrier. Attention was thus later shifted to third generation of biofuels using different photosynthetic microorganisms to create renewable fuels. Recently, algae (both micro and macro) have drawn attention of researchers as the third generation biomass for biofuel production (Ellis et al. 2012; Pittman et al. 2011; Demirbaş 2008; Harun et al. 2010; Jeong et al. 2010; Maeda et al. 1995; Nakas et al. 1983). Over the last three decades, there has been extensive research on algal biofuels production and the use of algae for CO_2 bioremediation (Borowitzka 2008). Both micro- and macroalgae have the potential to produce biobutanol—a third generation biofuel via sustainable bioproduction.

4.2 Microorganism Selection

Biobutanol generally can be produced by anaerobic fermentation by using solventogenic *Clostridia*; the rod-shaped, spore-forming Gram-positive bacteria which can use large variety of substrates from monosaccharides including many pentoses and hexoses to polysaccharides (Jones and Woods 1986). Among many solventogenic *Clostridia*, *Clostridia acetobutylicum*, *Clostridia beijerinckii*, *Clostridia saccharobutylicum*, and *Clostridia saccharoperbutylacetonicum* are primary solvent producers (Dürre 2007, 1998). However, the yield of biobutanol is very low as two moles of CO_2 is evolved per mole of glucose and by-products, such as acetone, ethanol, acetic acid, butyric acid, and hydrogen are also formed (Lütke-Eversloh and Bahl 2011) and solvent toxicity. To increase yield and solvent tolerance, several genetically engineered bacterial strains were developed which had been analyzed and reviewed recently (Papoutsakis 2008; Lütke-Eversloh and Bahl 2011). To enhance biobutanol titer by using different metabolic engineering and molecular biology of solventogenic *Clostridia* has been discussed vividly by several reviewers (Papoutsakis 2008; Lütke-Eversloh and Bahl 2011). However, most of the strains showed no remarkable increase in productivity except some hyper-butanol producing strains, such as *Clostridium beijerinckii* P260 and *C. beijerinckii* BA101 which produced ABE up to 33 g/L with a productivity threshold of about 12.5 g/L biobutanol (Formanek et al. 1997; Qureshi and Blaschek 2000). Hydrolysis of cellulosic biomass to sugar extract (saccharification) is required prior to fermentation by most of the strains. However, direct fermentation of cellulose to biobutanol was only made possible by using a single anaerobic *Clostridia* strain TU-103, *Clostridium cellulolyticum* (Higashide et al. 2011; Ray 2011). Some of the genetically modified non-*Clostridial* stains, such as *Escherichia coli*, *Saccharomyces cerevisiae* BY4742, *Ralstonia eutropha* H16 (LH74D), *Pseudomonas putida* S12, *Bacillus subtilis KS438 (BK1.0)* can convert wide range of substrates to biobutanol and have high solvent tolerance (Lee et al. 2008; Huo et al. 2011; Nielsen et al. 2009; Steen et al. 2008). Among the non-C*lostridial* strains, *P. putida* S12 can tolerate solvent at a concentration of about 6 % (v/v) (Nielsen et al. 2009). Moreover, higher production and tolerance of butanol by the microorganism would have a positive impact on the energy economy.

4.3 Fermentation Practices

Biphasic fermentation, substrate toxicity and product inhibition are major bottlenecks of ABE fermentation. Development of new fermentation practices, such as use of continuous immobilized cell reactor (Huang et al. 2010; Qureshi et al. 2004), cell recycle reactor (Jang et al. 2012; Tashiro et al. 2004), multistage continuous fermenter (Ramey 1998), free cell continuous reactor (Ezeji et al. 2007), vacuum fermentation (Mariano et al. 2011), and absorbent fermentation (He and Chen 2013) have increased productivity and concentration of biobutanol. Biofilm reactor (Qureshi et al. 2005; Qureshi and Maddox 1987) and electro-bioreactor (Atsumi et al. 2008; Li et al. 2012) are some of the recent additions in this regard. Optimization of the fermentation broth using different chemicals, such as glucose, hydrogen sparging in the headspace, butyric acid, butyrates and acetates, (Ezeji et al. 2007; Chen and Blaschek 1999; Tashiro et al. 2007; Pierrot et al. 1986) and coculture techniques had been reported (Tran et al. 2010, 2011) to obtain positive effects. These fermentation technologies used for first and second generation biofuel could enhance bioproduction of butanol to make the production comparable with petrobutanol.

4.4 Downstream Processing

The limitation of productivity and concentration is substantially related to the microorganisms. Although molecular technology has been used to improve solventogenic bacterial strains, it has yet been not easy to elevate the limitation to an acceptable high level, say, to be close to that of ethanol fermentation. However, biobutanol concentration in the fermenter still remains low (<0.5 %). Recovery of such low amount of biobutanol in combination with its high boiling point (118 °C; higher than water) requires a very high amount of energy for distillation, thus making the process of biobutanol production uneconomical. The energy requirement for distillation of 0.5 % (w/v) butanol is 79.5 MJ/kg of butanol (Matsumura et al. 1988), whereas the energy content of biobutanol itself is 34 MJ/kg of biobutanol (Ikegami et al. 2011). Thus, the energy consumed for biobutanol production is more than the energy released during biobutanol combustion. In the conventional distillation process, near about 40 % of the total production cost was involved. If the production of butanol can be enhanced from 12 to 19 g/L using some starch or corn as substrate, the recovery cost would be half (Papoutsakis 2008). However, to make biobutanol economically viable, emphasis has to be given mainly to the solvent recovery techniques because novel separation technology can fully exploit the productivity of a strain. Integration of fermentation with novel product recovery processes which separated ABE from the broth during fermentation retained biobutanol concentration below toxic level thus maintaining higher yield and productivity. During the last three decades, novel product recovery techniques, such as liquid–liquid extraction, ionic liquid extraction, adsorption, gas stripping, perstraction, pervaporation, and reverse osmosis have been developed (Ezeji et al. 2010; Green 2011; Vane 2005). Each method has its advantages and disadvantages and none can outperform all other methods. However, integration of fermentation with more than one recovery technique had been reported to be more efficient (Vane 2005) and it may have positive effect on energy economy due to its effectiveness toward higher solvent recovery.

Lots of efforts have already been reported to improve the production and amount of recovery of butanol from the fermentation broth. Nevertheless, there are still problems and challenges that need to be taken into account as detailed henceforth.

5 Challenges in Biochemical Production of Biobutanol

Biochemical production of butanol includes microbial ABE fermentation of sugar extract using solventogenic strains, such as *Clostridia* (Dürre 2007; Lee et al. 2008; Lütke-Eversloh and Bahl 2011). During the early 1980 s, a number of problems that have prevented biobutanol fermentation from being commercially viable were identified and these obstacles can be enumerated as: (1) high feedstock cost; (2) low product yield, usually of the order of 0.3 due to conversion of approximately 53 % of substrate into CO_2 and H_2; (3) high reducing sugar concentration to drive the process from acidogenic to solventogenic phase; (4) no concurrent sugar utilization (aldohexose followed by aldopentose); (5) low reactor productivity; (6) biphasic fermentation; (7) solvent toxicity; and (8) low product concentration in fermentation broth leads to high recovery cost. However, major bottleneck in biobutanol production is high feedstock cost which is about 60 % of the total production cost and which makes biobutanol less competitive than petrobutanol. Hence, use of fast growing, high carbon rich, and most abundant micro- and macroalgae as substrate to produce biobutanol can dilute the feedstock selection challenge to some extent.

6 Algae as Potential Source for Biobutanol and Other Biofuels

The US Department of Energy (DOE) devoted $25 million to algal fuels research in its aquatic species program at the National Renewable Energy Lab (NREL) in Golden, Colorado from 1978 to 1996 which set the platform for algal biofuel research today (Waltz 2009). Algae, a vast variety of photosynthetic species that are abundant in diverse environments (marine environments, fresh water and brackish water) (Mata et al. 2010; Nigam and Singh 2011), are responsible for $\sim$50 % of CO_2 sequestration and $\sim$50 % of O_2 genesis in the world (Lü et al. 2011). Algae can be classified based on the mode of nutrition, into three types, autotrophic, heterotrophic, and mixotrophic as shown in Fig. 1.

Autotrophic algae assimilate carbohydrates as a form of stored food via photosynthesis using sunlight and inorganic carbon from atmosphere. Heterotrophic algae use organic molecules to construct building blocks. However, mixtrophic algae use organic and inorganic carbon dioxide as source and convert them into carbohydrate and other building blocks in a short time period. All these metabolites have the potential to form sustainable renewable biofuel. The lipid part via biochemicals leads to production of sustainable biofuel biodiesel. Meanwhile, the carbohydrate part after lipid extraction can be used for production of different important potential biofuels, such as bioethanol, biobutanol, biohydrogen, biogas among others. Moreover, apart from these applications, they have utility in pharmaceuticals, cosmetics, food as shown in Fig. 2. Figure 2 depicts the versatility in

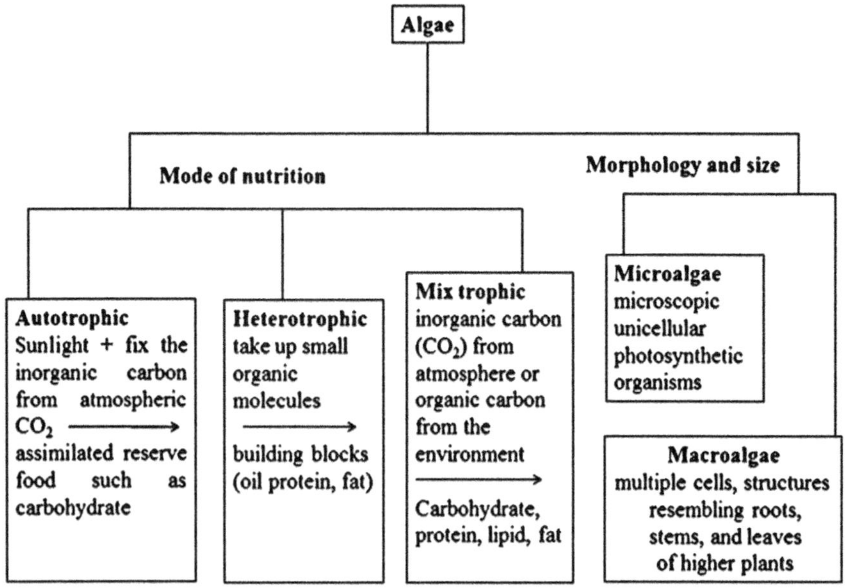

Fig. 1 Classification of algae according to **a** mode of nutrition; **b** morphology and size

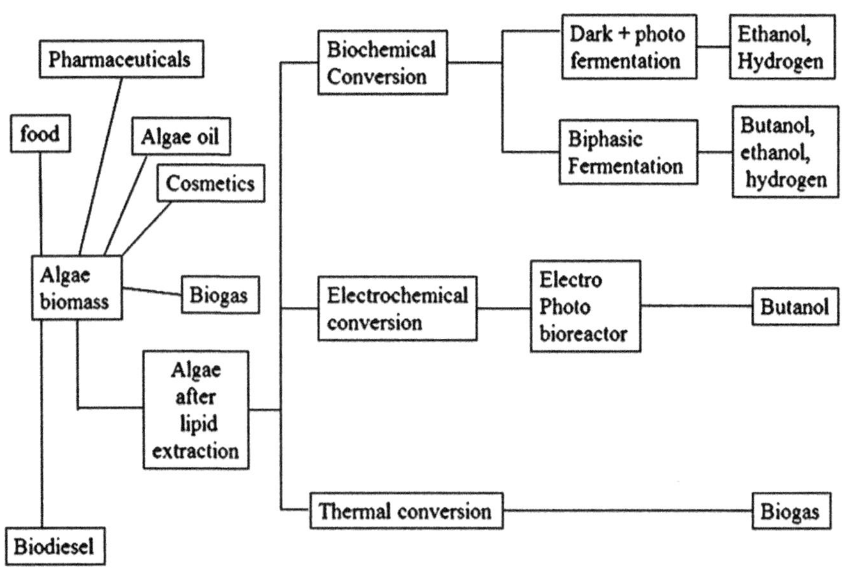

Fig. 2 Algae as a potential source of biofuels and other valuable products

utilization of algae as a potential source of biofuel and other commercial applications.

Hence, attention needs to be drawn to utilization of macro- and microalgae as biomass for biofuel production and various other applications. Based on their morphology and size, algae are categorized as micro- and macroalgae (shown in Fig. 1).

6.1 *Advantages of Algae as Feedstock for Biobutanol*

Algal biomass would serve as an advantageous substrate for biobutanol production due to its ubiquitous nature. Biofuels derived from microalgae are currently considered to be the most economical and technically viable route for producing biofuels to compete with petroleum-based fuels (Chisti 2010). Some of the advantages of using algal biomass over terrestrial biomass include no need for cultivation land, requires no fresh water rather algae refreshes wastewater by exploiting toxic materials as nutrients (Cantrell et al. 2008), ability to thrive in diverse ecosystems, higher growth rate and higher yield per area (www.nrel.gov/docs/legosti/old/4174.pdf), and do not compromise the production of food, fodder, and other products derived from terrestrial crops. Higher photon conversion efficiency, synthesis, and accumulation of large quantities of carbohydrate as inexpensive biomass for biobutanol production also make it more attractive (Subhadra and Edwards 2010). Several algal species can be metabolically engineered to produce more carbohydrates, which can be used as fermentation substrate for biobutanol production (Radakovits et al. 2010). Algae can be stored dry (<25 % moisture) and used for up to one year with higher conversion of carbohydrates to sugars using a 10 % w/v of dried algae (Jernigan et al. 2013). Moreover, higher tolerance microalgae can use CO_2 emitted from petroleum-based power station and other industrial source for their metabolism and reduce GHG emissions (Nigam and Singh 2011).

6.2 *Biobutanol from Microalgae*

Microalgae, the most abundant microorganism on the earth, can grow in diverse ecological marine environment, fresh water and brackish water all around the world. These are thought to be one of the earliest life forms on earth (Falkowski et al. 2004). They had the potential to grow in any adverse situations, such as extreme temperatures and pH conditions. Microalgae, such as *Chlorella, Dunaliella, Chlamydomonas, Scenedesmus, Spirulina* are known to contain a large amount (>50 % of the dry weight) of starch and glycogen, useful as raw materials for biobutanol production (Hirayama et al. 1996). Cellulose from microalgae can also be used for biobutanol production. As biobutanol is the end metabolite in

Table 2 Some higher carbohydrate containing microalgae already being used or to be used for biobutanol production

Microalgae	Percent starch or biomass after oil extraction (g/dry weight)	References
Laminaria hyperborea	55.0 (reserve food material)	Horn et al. (2000)
Chlamydomonas reinhardtii UTEX 90	53.0 (starch)	Kim (2009)
Green alga NKG 121701	>50.0 (starch)	Matsumoto et al. (2003)
Saccharina latissima	~50.0 (reserve food material)	Adams et al. (2009)
Spirulina fusiformis	37.3–56.1 (starch)	Rafiqul et al. (2003)
C. reinhardtii (UTEX2247)	45.0 (starch)	Hirano et al. (1997)
Spirogyra sp.	43.3 (biomass after oil extraction)	Hossain et al. (2008)
Chlorella sp.	21–27.0	Rodjaroen et al. (2007)
Scenedesmus sp.	13.0–21.0	Rodjaroen et al. (2007)
Ankistrodesmus sp.	_	Ellis et al. (2012)
Micromona sp.	_	Ellis et al. (2012)
Tetraselmis suecica	27.41 ± 2.08	Kassim et al. (2014)

metabolic cycle of solventogenic strains, required concentration of reducing sugar is high. Some microalgae which have higher carbohydrate content, they can be used for biobutanol production and are listed in Table 2.

In literature, composition of other microorganisms have been reported by several researchers, but higher carbohydrate containing microalgae were considered to be more potent as initial higher sugar concentration was required for biobutanol production. Many microalgae showed rapid growth under optimal conditions, such as some *Chlamydomanos* species required 6 h via asexual reproduction (Chen et al. 2010). Due to their high cell division rate, handling is often simpler in research application and it can be performed several times faster with microalgae than that of the terrestrial crop species (Packer 2009). Aqueous suspension of microalgae from Logan Lagoon was digested with acid and base to obtain 8.92 g/L soluble sugar solution. ABE fermentation of the sugar extract by *Clostridium saccharoperbutylacetonicum* produced 2.26 g/L biobutanol (yield: 0.201 g/g) (Ellis et al. 2012).

6.3 Biobutanol from Macroalgae

Akin to microalgae, macroalgae can also produce biofuel by converting their carbohydrate storage and cellulose from cell wall to fermentable sugars (Adams et al. 2009). Macroalgae which belong to brown and red algae group can have the potential to produce biofuel. Brown algae, such as *Laminaria, Saccorhiza, Alaria* grow up to meters and their main food reserves are laminarin and mannitol (Horn et al. 2000; Adams et al. 2009; Nobe et al. 2003). Red algae, such as *Gelidium amansii*, which is composed of cellulose, glucan, and galactan, also can serve as a potential feedstock for bioconversion to butanol (Kim 2009; Wi et al. 2009; Yoon et al. 2010). Macroalgae are also fast growing and produced large amounts of biomass as lower energy was required for the production of supporting tissue than terrestrial plants, and they had the capability to take up nutrients over their entire surface. *Ulva lactuca,* a species of green macroalgae, showed higher growth rate and high carbohydrate content and was used for biobutanol production. It produced 4 g/L butanol via fermentation using *C. beijerinckii* and *C. saccharoperbutylacetonicum*. At pilot scale, the production was higher and about 0.29 g butanol/g sugar was produced (Potts et al. 2012).

6.4 Culturing and Harvesting of Algae

There are two most commonly used techniques to cultivate microalgae, such as open pond system and shallow lagoons and closed photobioreactor system (Huesemann and Benemann 2009; Spolaore et al. 2006). For large-scale outdoor microalgae cultivation, one pond was generally widely used, as they are cheap and easy to build and handle (Brennan and Owende 2010). Depending on their size, shape, type of agitation and inclination, open ponds are mainly of three types, such as raceway pond, circular pond, and sloped pond (Shen et al. 2009). However, open pond system is less favorable due to limitation in controlling contamination from predators that leads to lower productivity, water loss through evaporation, and lower carbon dioxide use efficiency, high harvesting cost, and difficulties in temperature control (Shen et al. 2009; Chisti 2007; Lee 2001). *Chlorella vulgaris, Dunaliella salina*, *Spirulina*, *Haematococcus pluvialis* are produced in a variety of different open pond system by several industries. In Asia, much of the *Chlorella* production was performed in circular mixed open ponds. However, plastic tubes in ponds offer up to seven times the productivity of open ponds (http://www.biofuelstp.eu/algae.html). Photobioreactors (PBR) provide an easy system of controlling nutrients for growth and cultivation parameters and to prevent contamination leading to higher productivity, but require higher initial cost. However, higher biomass production may reduce the initial cost (Ugwu et al. 2008). Moreover, it also has another limitation instead of initial higher cost, such as pH gradient, dissolved oxygen and CO_2 along the tubes, wall growth, fouling, hydrodynamic stress, and higher scale-up expenses (Borowitzka 2008; Chen et al. 2010; Lee 2001; Ugwu et al. 2008). Therefore, microalgae production facility is an important factor to be considered for the optimum production of a specific microalgal species (Harun et al. 2010). *Haematococcus* is produced in several countries in closed photobioreactor. Recently, all the limitations of open pond system and PBR were overcome by hybrid method where the algae were first cultured in a photobioreactor to avoid contamination by unwanted strains and then transferred to open ponds where the algae were subjected to controlled nutrient conditions which enhanced the production of the desired product (Brennan and Owende 2010). Once desired algae were cultivated on a large scale, the next step was to harvest the algal biomass for biobutanol production.

6.5 Algal Biomass to Biobutanol

Algal biomass has been harvested and dewatered using several methods, such as flocculation (Sukenik et al. 1984), centrifugation (Brennan and Owende 2010), or filtration (Mohn 1980). Algae with high carbohydrate content were subjected to solventogenic fermentation to ABE (Ellis et al. 2012). Algae with high oil content were subjected to lipid extraction using any one of the methods, such as mechanical methods (https://www.uvm.edu/~epscor/pdfFiles/2010_algae_presentations/Ihab-vt100318uvmihf.pdf), solvent extraction (Ryckebosch et al. 2012), switchable solvents (Samorì et al. 2010), supercritical CO_2 (Demirbaş 2008), ultrasonic extraction (Cravotto et al. 2008), or flash depressurization (Couto et al. 2010). The spent aqueous algae extract remained enriched with sugars that can be used for anaerobic fermentation to produce biobutanol (Potts et al. 2012). Figure 3 shows the consecutive steps of harvesting of algae to biobutanol production.

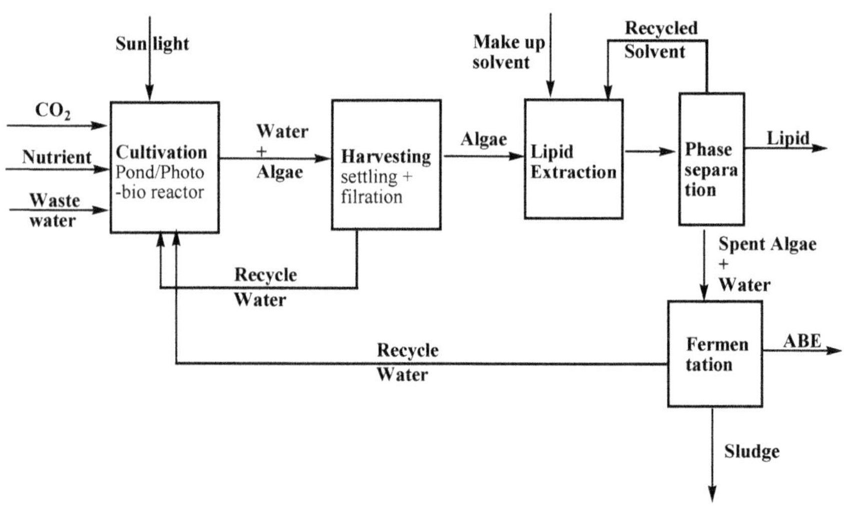

Fig. 3 Flow diagram starting from harvesting of algae to biobutanol production

6.5.1 Biobutanol from Algal Metabolites

Algae can assimilate considerable amounts of biomass in the form of starch/ cellulose, which can be converted to fermentable sugars and these sugars can be converted to biobutanol by a suitable butanol producer. Different steps and its corresponding substeps and required enzymes for conversion of starch/cellulose to biobutanol are depicted in Fig. 4.

List of enzymes involved in each steps in Fig. 4 comprise: (1) cellulose hydrolysis: cellulases, α-glucosidase; (2) starch hydrolysis: α-amylase, β-amylase, pullulanase, glucoamylase, α-glucosidase; (3) glucose uptake by the phosphorus transferase system (PTS) and conversion to pyruvate by the (Entner–Doudoroff pathway) EMP pathway; (4) pyruvate-ferrodoxin oxidoreductase; (5) phosphate acetyltransferase and acetate kinase, (6) acetaldehyde dehydrogenase and ethanol dehydrogenase; (7) acetoacetyl-CoA:acetate/butyrate:CoA transferase and acetoacetate decarboxylase; (8) phosphate butyltransferase and butyrate kinase; (9) butyraldehyde dehydrogenase and butanol dehydrogenase.

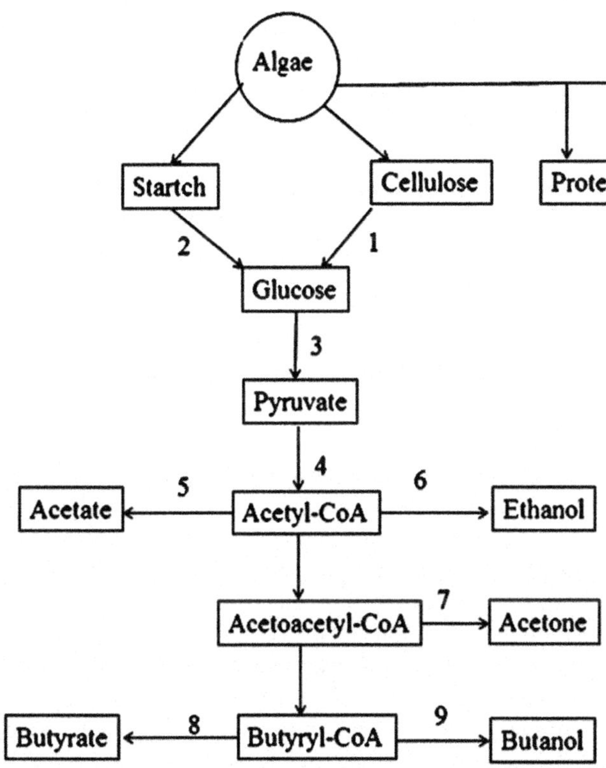

Fig. 4 Enzymes involved and metabolites production in each step during the conversion of algal biomass to biobutanol

Thus, when the starch or cellulose was extracted from the algae, it was converted to reducing sugar either by different pretreatment methods reported in the literature (Yamaguchi et al. 2009), or by sequential culturing by fungi (http://www.biofuelstp.eu/butanol.html). Later, the sugar was converted to biobutanol using conventional two-phase ABE fermentation mainly by solventogenic *Clostridia*. Five species of *Dunaliella* sp. were fermented with *Clostridium pasteurianum* to generate about 14–16 g/L of mixed solvents (*n*-butanol, propanediol and ethanol) (Nakas et al. 1983). Fermentation of the halophilic microalgae, *Dunaliella* with 4 % glycerol co-substrate was reported to produce 14–16 g/L of ABE. Pretreatment of algal biomass with 10 % acid and base followed by fermentation using *C. saccharoperbutylacetonicum* NI-4 for 96 h gave 23 g/L butanol at yield of 0.20 g/g and productivity 0.20 g/L/h (Klasson et al. 1991). Algal biobutanol production has been made easy by growing algae in race ways and dead zones water laden with overloading of nitrogen and phosphorous. The algae are scraped, dried, carbohydrates are extracted which are fermented to biobutanol via butyric acid, and sugars are converted into butyric, lactic, and acetic acids. This process was made faster by a special technique called electrode ionization. Thus, sugar extract obtained after pretreatment (acid followed by basic treatment) of algal biomass from the Logan Lagoon wastewater treatment facility was fermented using *C. saccharoperbutylacetonicum* N1–4 (ATCC 27021) to produce 2.74 g/L of total ABE. However, 7.27 g/L of ABE was produced when supplementing with glucose, and 9.74 g/L when supplementing with enzymes. Supplementation of enzymes produced the highest total ABE production yield of 0.311 g/g and volumetric productivity of 0. 102 g/L h. Non-pretreated algae with no supplementation produced 0.73 g/L total ABE (Ellis et al. 2012). The advances in production of biobutanol from microalgae using different technologies and substrates as carbon source are summarized in Table 3.

6.5.2 Electrochemical Production of Biobutanol Using Lithoautotrops

Photoautotrophs (plants, green algae, and other microorganisms) utilize solar energy to reduce CO_2 by electron from hydrogen of water producing biomass where energy is stored as C–C, C–H bonds liberating O_2. Lithoautotrophs utilize other forms of energy, such as electrical energy (or any form of energy converted to electrical energy) to reduce CO_2 using electrons from electricity or chemical sources (H_2O, H_2, H_2S, NH_3, Fe^{2+} or other reduced metal ions) to produce biomass that can be converted to transportation liquid fuel. Chemo-lithoautotrophic organisms, such as bacteria or archaea can fix CO_2 to grow and metabolize in the absence of light or reduced carbon source and thus can be used as a microbial factory for producing biobutanol. Solar electricity was used in the reactor to generate formic acid intermediate which enabled the microorganisms to grow and reduce CO_2 to alcohol (Lan and Liao 2012). Chemo-lithoautotrophic method requires no arable land and it has very low water usage, just to provide two electrons (as hydride) and molecular oxygen during CO_2 reduction. Moreover, it is a very good method of

Table 3 Production of biobutanol using either direct algal biomass or algae as a microorganism for biobutanol production

Microorganism	Carbon source	Production technology	Butanol production (g/L)	References
Synechococcus elongatus PCC 7942	CO_2	Photobioreactor	0.450 g/L (Isobutanol)	Atsumi et al. (2008)
Synechococcus elongatus PCC 7942	CO_2	Photobioreactor	0.030 g/L (*n*-butanol)	Lan and Liao (2012)
Clostridium saccharoperbutylacetonicum N1–4	(1) Carbohydrate from Lagoon city waste water microalgae	Batch	2.74 g/L (*n*-butanol)	Ellis et al. (2012)
	(2) Same biomass +1 % glucose supplement	Batch	2.727 g/L	
Clostridium beijerinckii, ATTC 35702	Carbohydrate from macroalgae (*Ulva lactuca*)	Batch	4 g/L	Potts et al. (2012)
Clostridium saccharoperbutyliticum	*(Tetraselmis suecica)*	Batch	0.07 g/L	Kassim et al. (2014)
Sargassum wightii	*Sargassum wightii*	Batch	15	Dubey et al. (2015)
Clostridium acetobutylicum ATCC824	*Chlorella vulgaris JSC-6*	Batch	0.89–0.93 g/h/L	Wang et al. (2014)

storing electrical energy as liquid fuel. A genetically modified microalga, such as *Synechococcus elongatus* PCC7942 was able to produce *n*-butanol and isobutanol using this method (Atsumi et al. 2008; Lan and Liao 2012).

6.5.3 Direct Production of Biobutanol by Engineered Microalgae

Nuclear transformation for control of metabolic pathways, chloroplastic transformation for high levels of protein expression, more clear-cut approaches to genetic alteration compared to higher plants (Zilber-Rosenberg and Rosenberg 2008) make algae more fascinating subjects for genetic modification to enhance or make it more useful for butanol production. However, only scarce information is available in this aspect. Nevertheless, after the success of metabolic engineering of *E. coli*, Liao group tried the same with photosynthetic organism. The *kivd* gene for ketoacid decarboxylase from *Lactococcus lactis* was transformed into *Synechococcus elongates* by an expression cassette under the control of the isopropyl-b-D-thiogalactoside (IPTG) inducible promoter *Ptrc* (Lü et al. 2011). Currently, this genetically modified *S. elongatus PCC7942* can produce directly butanol, isobutyraldehyde, and isobutanol from carbon dioxide. It can produce 1-butanol directly from light rather than harvesting glucose and converting glucose to 1-butanol using CO_2. About 48 mol of photons (of energy 173.5 kJ/mol, representing the average energy of 680 and 700 nm photons) in the presence of CO_2 as substrate had produced 1-butanol at 0.03 g/L (Lan and Liao 2012).

7 Future Outlook and Conclusion

Biobutanol production technology is a multidisciplinary process involving selective blending of environmental science, biochemistry, microbiology, genetic engineering, chemical engineering, process technology, and process and market economics. It involves different steps, such as biomass mobilization, bacterial strain selection and development, pretreatment of biomass, optimization of fermenting solution, fermentation practices, separation and purification, storage, transportation, and marketing. Any change in development of one or more steps had been shown to create new challenges to meet for cost effective production of biobutanol.

The potential of both micro- and macroalgae, for conversion of biobutanol has gained increased recognition. However, the production of biobutanol from algae is still at its natal stage. Moreover, considering it as an economically viable source for renewable biobutanol production, scientific and technological challenges are yet to be realized. To use algae as an alternative source of renewable energy, higher scale cultivation is required which has significant impact on the global energy economy. By overcoming few challenges, several uncertainties in all aspects of design and operation of these biomass production systems can bring out a positive effect. From the aspect of fermentation technology, in recent past, several new fermentation techniques, such as membrane cell cycle reactor, immobilized cell reactor, biofilm bioreactor, absorbent fermentation, and flash fermentation have been developed. There is scope of introducing novel adsorbents and porous support with increased selectivity to enhance production. In downstream processing, novel integration of more than one separation techniques commensurate with biofeed and fermentation will lower the cost of biobutanol production.

However there are no simple, certain, or sufficient solutions to the impending energy supply problem. Nevertheless, biobutanol production from algae is likely to be a major energy source and they could provide a solution of energy crisis.

Acknowledgments Financial support from the Natural Sciences and Engineering Research Council of Canada (NSERC, Discovery Grant 355254), MAPAQ (No. 809051) and Ministère des Relations Internationales du Québec (coopération Paraná-Québec 2010–2012; Quebec-Vietnam 2012–2015) are sincerely acknowledged.

References

Adams JM, Gallagher JA, Donnison IS (2009) Fermentation study on *Saccharina latissima* for bioethanol production considering variable pre-treatments. J Appl Phycol 21:569–574

Atsumi S, Cann AF, Connor MR, Shen CR, Smith KM, Brynildsen MP et al (2008) Metabolic engineering of *Escherichia coli* for 1-butanol production. Metab Eng 10:305–311

Blanch HW, Simmons BA, Klein-Marcuschamer D (2011) Biomass deconstruction to sugars. Biotechnol J 6:1086–1102

Borowitzka MA (2008) Marine and halophilic algae for the production of biofuels. J Biotechnol 136:S7

Brennan L, Owende P (2010) Biofuels from microalgae—a review of technologies for production, processing, and extractions of biofuels and co-products. Renew Sustain Energy Rev 14:557–577

Cantrell KB, Ducey T, Ro KS, Hunt PG (2008) Livestock waste-to-bioenergy generation opportunities. Bioresour Technol 99:7941–7953

Cascone R (2008) Biobutanol: a replacement for bioethanol? Chem Eng Prog 104(8)

Chen C-K, Blaschek H (1999) Acetate enhances solvent production and prevents degeneration in *Clostridium beijerinckii* BA101. Appl Microbiol Biotechnol 52:170–173

Chen P, Min M, Chen Y, Wang L, Li Y, Chen Q et al (2010) Review of biological and engineering aspects of algae to fuels approach. Int J Agric Bio Eng 2:1–30

Chisti Y (2007) Biodiesel from microalgae. Biotechnol Adv 25:294–306

Chisti Y (2010) Fuels from microalgae. Biofuels 1:233–235

Couto RM, Simões PC, Reis A, Da Silva TL, Martins VH, Sánchez-Vicente Y (2010) Supercritical fluid extraction of lipids from the heterotrophic microalga *Crypthecodinium cohnii*. Eng Life Sci 10:158–164

Cravotto G, Boffa L, Mantegna S, Perego P, Avogadro M, Cintas P (2008) Improved extraction of vegetable oils under high-intensity ultrasound and/or microwaves. Ultrason Sonochem 15:898–902

Demirbaş A (2008) Production of biodiesel from algae oils. Energy Sources Part A Recovery Utilization Environm Eff 31:163–168

Dubey KK, Dhingra AK, Rana S (2015) Optimisation of process parameters for enhanced biobutanol production from *Sargassum wightii* hydrolysate. Int J Energy Technol Policy 11:303–311

Dürre P (1998) New insights and novel developments in clostridial acetone/butanol/isopropanol fermentation. Appl Microbiol Biotechnol 49:639–648

Dürre P (2007) Biobutanol: an attractive biofuel. Biotechnol J 2:1525–1534

Ellis JT, Hengge NN, Sims RC, Miller CD (2012) Acetone, butanol, and ethanol production from wastewater algae. Bioresour Technol 111:491–495

Ezeji T, Qureshi N, Blaschek HP (2007) Production of acetone–butanol–ethanol (ABE) in a continuous flow bioreactor using degermed corn and *Clostridium beijerinckii*. Process Biochem 42:34–39

Ezeji T, Milne C, Price ND, Blaschek HP (2010) Achievements and perspectives to overcome the poor solvent resistance in acetone and butanol-producing microorganisms. Appl Microbiol Biotechnol 85:1697–1712

Falkowski PG, Katz ME, Knoll AH, Quigg A, Raven JA, Schofield O et al (2004) The evolution of modern eukaryotic phytoplankton. Science 305:354–360

Formanek J, Mackie R, Blaschek HP (1997) Enhanced butanol production by *Clostridium beijerinckii* BA101 grown in semidefined P2 medium containing 6 percent maltodextrin or glucose. Appl Environ Microbiol 63:2306–2310

Green EM (2011) Fermentative production of butanol—the industrial perspective. Curr Opin Biotechnol 22:337–343

Harun R, Singh M, Forde GM, Danquah MK (2010) Bioprocess engineering of microalgae to produce a variety of consumer products. Renew Sustain Energy Rev 14:1037–1047

He Q, Chen H (2013) Improved efficiency of butanol production by absorbed lignocellulose fermentation. J Biosci Bioeng 115:298–302
Higashide W, Li Y, Yang Y, Liao JC (2011) Metabolic engineering of *Clostridium cellulolyticum* for production of isobutanol from cellulose. Appl Environ Microbiol 77:2727–2733
Hirano A, Ueda R, Hirayama S, Ogushi Y (1997) CO_2 fixation and ethanol production with microalgal photosynthesis and intracellular anaerobic fermentation. Energy 22:137–142
Hirayama S, Nakayama H, Sugata K, Ueda R (1996) Process for the production of ethanol from microalgae. Google Patents
Horn S, Aasen I, Østgaard K (2000) Production of ethanol from mannitol by *Zymobacter palmae*. J Ind Microbiol Biotechnol 24:51–57
Hossain AS, Salleh A, Boyce AN, Chowdhury P, Naqiuddin M (2008) Biodiesel fuel production from algae as renewable energy. Am J Biochem Biotechnol 4:250–254
Huang H, Liu H, Gan Y-R (2010) Genetic modification of critical enzymes and involved genes in butanol biosynthesis from biomass. Biotechnol Adv 28:651–657
Huesemann M, Benemann J (2009) Biofuels from microalgae: review of products, processes and potential, with special focus on *Dunaliella* sp. The Alga Dunaliella: Biodiversity, Physiology, Genomics and Biotechnology' Science Publishers, New Hampshire
Huo Y-X, Cho KM, Rivera JGL, Monte E, Shen CR, Yan Y et al (2011) Conversion of proteins into biofuels by engineering nitrogen flux. Nat Biotechnol 29:346–351
Ikegami T, Negishi H, Sakaki K (2011) Selective separation of *n*-butanol from aqueous solutions by pervaporation using silicone rubber-coated silicalite membranes. J Chem Technol Biotechnol 86:845–851
Jang YS, Lee J, Malaviya A, Seung DY, Cho JH, Lee SY (2012) Butanol production from renewable biomass: rediscovery of metabolic pathways and metabolic engineering. Biotechnol J 7:186–198
Jeong TS, Choi W, Lee J, Oh KK (2010) Production of biobutanol from hydrolysate of red algae (*Gelidium amansii*) treated with acid saccharification. In: 32nd Symposium on biotechnology for fuels and chemicals, Hilton Clearwater Beach, Clearwater Beach, Florida, US
Jernigan A, May M, Potts T, Rodgers B, Hestekin J, May PI et al (2013) Effects of drying and storage on year-round production of butanol and biodiesel from algal carbohydrates and lipids using algae from water remediation. Environ Prog Sustain Energy 32:1013–1022
Jones DT, Woods DR (1986) Acetone-butanol fermentation revisited. Microbiol Rev 50:484
Kassim M, Potumarthi R, Tanksale A, Srivatsa S, Bhattacharya S (2014) Enzymatic saccharification of dilute alkaline pre-treated microalgal (*Tetraselmis suecica*) biomass for biobutanol production. Int J Biol Vet Agric Food Eng 6
Kim Y-S (2009) Pretreatment of *Gelidium amansii* for the production of bioethanol. In: The 31st Symposium on biotechnology for fuels and chemicals
Klasson K, Ackerson M, Clausen E, Gaddy J (1991) Bioreactors for synthesis gas fermentations. Resour Conserv Recycl 5:145–165
Köpke M, Held C, Hujer S, Liesegang H, Wiezer A, Wollherr A et al (2010) Clostridium ljungdahlii represents a microbial production platform based on syngas. Proc Natl Acad Sci 107:13087–13092
Lan EI, Liao JC (2012) ATP drives direct photosynthetic production of 1-butanol in cyanobacteria. Proc Natl Acad Sci 109:6018–6023
Lee Y-K (2001) Microalgal mass culture systems and methods: their limitation and potential. J Appl Phycol 13:307–315
Lee SY, Park JH, Jang SH, Nielsen LK, Kim J, Jung KS (2008) Fermentative butanol production by *Clostridia*. Biotechnol Bioeng 101:209–228
Li H, Opgenorth PH, Wernick DG, Rogers S, Wu T-Y, Higashide W et al (2012) Integrated electromicrobial conversion of CO_2 to higher alcohols. Science 335:1596
Liew S, Arbakariya A, Rosfarizan M, Raha A (2006) Production of solvent (acetone-butanol-ethanol) in continuous fermentation by *Clostridium saccharobutylicum* DSM 13864 using gelatinised sago starch as a carbon source. Malays J Microbiol 2:42–50

Lü J, Sheahan C, Fu P (2011) Metabolic engineering of algae for fourth generation biofuels production. Energy Environ Sci 4:2451–2466

Lütke-Eversloh T, Bahl H (2011) Metabolic engineering of *Clostridium acetobutylicum*: recent advances to improve butanol production. Curr Opin Biotechnol 22:634–647

Machado C (2010) Technical characteristics and current status of butanol production and use as biofuel. V Seminario Latinoamericano y del Caribe de Biocombustibles

Maeda K, Owada M, Kimura N, Omata K, Karube I (1995) CO_2 fixation from the flue gas on coal-fired thermal power plant by microalgae. Energy Convers Manage 36:717–720

Mariano AP, Qureshi N, Ezeji TC (2011) Bioproduction of butanol in bioreactors: new insights from simultaneous in situ butanol recovery to eliminate product toxicity. Biotechnol Bioeng 108:1757–1765

Mata TM, Martins AA, Caetano NS (2010) Microalgae for biodiesel production and other applications: a review. Renew Sustain Energy Rev 14:217–232

Matsumoto M, Yokouchi H, Suzuki N, Ohata H, Matsunaga T (2003) Saccharification of marine microalgae using marine bacteria for ethanol production. Biotechnology for fuels and chemicals. Springer, Berlin, pp 247–254

Matsumura M, Kataoka H, Sueki M, Araki K (1988) Energy saving effect of pervaporation using oleyl alcohol liquid membrane in butanol purification. Bioprocess Eng 3:93–100

Mohn F (1980) Experiences and strategies in the recovery of biomass from mass cultures of microalgae. Algae biomass: production and use/[sponsored by the National Council for Research and Development, Israel and the Gesellschaft fur Strahlen-und Umweltforschung (GSF), Munich, Germany]; Gedaliah S, Soeder CJ (eds)

Nakas J, Schaedle M, Parkinson C, Coonley C, Tanenbaum S (1983) System development for linked-fermentation production of solvents from algal biomass. Appl Environ Microbiol 46:1017–1023

Nielsen DR, Leonard E, Yoon S-H, Tseng H-C, Yuan C, Prather KLJ (2009) Engineering alternative butanol production platforms in heterologous bacteria. Metab Eng 11:262–273

Nigam PS, Singh A (2011) Production of liquid biofuels from renewable resources. Prog Energy Combust Sci 37:52–68

Nobe R, Sakakibara Y, Fukuda N, Yoshida N, Ogawa K, Suiko M (2003) Purification and characterization of laminaran hydrolases from *Trichoderma viride*. Biosci Biotechnol Biochem 67:1349–1357

Packer M (2009) Algal capture of carbon dioxide; biomass generation as a tool for greenhouse gas mitigation with reference to New Zealand energy strategy and policy. Energy Policy 37:3428–3437

Papoutsakis ET (2008) Engineering solventogenic clostridia. Curr Opin Biotechnol 19:420–429

Pfromm PH, Amanor-Boadu V, Nelson R, Vadlani P, Madl R (2010) Bio-butanol vs. bio-ethanol: a technical and economic assessment for corn and switchgrass fermented by yeast or *Clostridium acetobutylicum*. Biomass Bioenergy 34:515–524

Pierrot P, Fick M, Engasser J (1986) Continuous acetone-butanol fermentation with high productivity by cell ultrafiltration and recycling. Biotechnol Lett 8:253–256

Pittman JK, Dean AP, Osundeko O (2011) The potential of sustainable algal biofuel production using wastewater resources. Bioresour Technol 102:17–25

Potts T, Du J, Paul M, May P, Beitle R, Hestekin J (2012) The production of butanol from Jamaica bay macro algae. Environ Prog Sustain Energy 31:29–36

Qureshi N, Maddox I (1987) Continuous solvent production from whey permeate using cells of *Clostridium acetobutylicum* immobilized by adsorption onto bonechar. Enzym Microb Technol 9:668–671

Qureshi N, Blaschek H (2000) Butanol production using *Clostridium beijerinckii* BA101 hyper-butanol producing mutant strain and recovery by pervaporation. Twenty-first symposium on biotechnology for fuels and chemicals. Springer, Berlin, pp 225–235

Qureshi N, Lai L, Blaschek H (2004) Scale-up of a high productivity continuous biofilm reactor to produce butanol by adsorbed cells of *Clostridium beijerinckii*. Food Bioprod Process 82:164–173

Qureshi N, Annous BA, Ezeji TC, Karcher P, Maddox IS (2005) Biofilm reactors for industrial bioconversion processes: employing potential of enhanced reaction rates. Microb Cell Fact 4:24

Qureshi N, Ezeji TC, Ebener J, Dien BS, Cotta MA, Blaschek HP (2008a) Butanol production by *Clostridium beijerinckii*. Part I: use of acid and enzyme hydrolyzed corn fiber. Bioresour Technol 99:5915–5922

Qureshi N, Saha BC, Hector RE, Hughes SR, Cotta MA (2008b) Butanol production from wheat straw by simultaneous saccharification and fermentation using *Clostridium beijerinckii*: Part I —batch fermentation. Biomass Bioenergy 32:168–175

Radakovits R, Jinkerson RE, Darzins A, Posewitz MC (2010) Genetic engineering of algae for enhanced biofuel production. Eukaryot Cell 9:486–501

Rafiqul I, Hassan A, Sulebele G, Orosco C, Roustaian P, Jalal K (2003) Salt stress culture of blue green algae *Spirulina fusiformis*. Pak J Biol Sci 6:648–650

Ramey DE (1998) Continuous two stage, dual path anaerobic fermentation of butanol and other organic solvents using two different strains of bacteria. Google Patents

Ray KH (2012) Cars could run on recycled newspaper, Tulane scientists say. Tulane University news webpage Tulane University Retrieved March 2011, p 14

Rodjaroen S, Juntawong N, Mahakhant A, Miyamoto K (2007) High biomass production and starch accumulation in native green algal strains and cyanobacterial strains of Thailand. Kasetsart J (Nat Sci) 41:570–575

Ryckebosch E, Koenraad M, Imogen F (2012) Optimization of an analytical procedure for extraction of lipids from microalgae. J Am Oil Chem Soc 89:189–198.

Samorì C, Torri C, Samorì G, Fabbri D, Galletti P, Guerrini F et al (2010) Extraction of hydrocarbons from microalga *Botryococcus braunii* with switchable solvents. Bioresour Technol 101:3274–3279

Shen Y, Yuan W, Pei Z, Wu Q, Mao E (2009) Microalgae mass production methods. Trans ASABE 52:1275–1287

Spolaore P, Joannis-Cassan C, Duran E, Isambert A (2006) Commercial applications of microalgae. J Biosci Bioeng 101:87–96

Steen EJ, Chan R, Prasad N, Myers S, Petzold CJ, Redding A et al (2008) Metabolic engineering of *Saccharomyces cerevisiae* for the production of *n*-butanol. Microb Cell Fact 7:36

Subhadra B, Edwards M (2010) An integrated renewable energy park approach for algal biofuel production in United States. Energy Policy 38:4897–4902

Sukenik A, Shelef G (1984) Algal autoflocculation—verification and proposed mechanism. Biotechnol Bioeng 26:142–147.

Taconi KA, Venkataramanan KP, Johnson DT (2009) Growth and solvent production by *Clostridium pasteurianum* ATCC® 6013™ utilizing biodiesel-derived crude glycerol as the sole carbon source. Environ Prog Sustain Energy 28:100–110

Tashiro Y, Takeda K, Kobayashi G, Sonomoto K, Ishizaki A, Yoshino S (2004) High butanol production by *Clostridium saccharoperbutylacetonicum* N1-4 in fed-batch culture with pH-Stat continuous butyric acid and glucose feeding method. J Biosci Bioeng 98:263–268

Tashiro Y, Shinto H, Hayashi M, Baba S-I, Kobayashi G, Sonomoto K (2007) Novel high-efficient butanol production from butyrate by non-growing *Clostridium saccharoperbutylacetonicum* N1-4 (ATCC 13564) with methyl viologen. J Biosci Bioeng 104:238–240

Thang VH, Kanda K, Kobayashi G (2010) Production of acetone–butanol–ethanol (ABE) in direct fermentation of cassava by *Clostridium saccharoperbutylacetonicum* N1-4. Appl Biochem Biotechnol 161:157–170

Tran HTM, Cheirsilp B, Hodgson B, Umsakul K (2010) Potential use of *Bacillus subtilis* in a co-culture with *Clostridium butylicum* for acetone–butanol–ethanol production from cassava starch. Biochem Eng J 48:260–267

Tran HTM, Cheirsilp B, Umsakul K, Bourtoom T (2011) Response surface optimisation for acetone-butanol-ethanol production from cassava starch by co-culture of *Clostridium butylicum* and *Bacillus subtilis*. Maejo Int J Sci Technol 5(3)

Ugwu C, Aoyagi H, Uchiyama H (2008) Photobioreactors for mass cultivation of algae. Bioresour Technol 99:4021–4028

Vane LM (2005) A review of pervaporation for product recovery from biomass fermentation processes. J Chem Technol Biotechnol 80:603–629

Waltz E (2009) Biotech's green gold? Nat Biotechnol 27:15–18

Wang Y, Guo W-Q, Lo Y-C, Chang J-S, Ren N-Q (2014) Characterization and kinetics of bio-butanol production with *Clostridium acetobutylicum* ATCC824 using mixed sugar medium simulating microalgae-based carbohydrates. Biochem Eng J 91:220–230

Wi SG, Kim HJ, Mahadevan SA, Yang D-J, Bae H-J (2009) The potential value of the seaweed Ceylon moss (*Gelidium amansii*) as an alternative bioenergy resource. Bioresour Technol 100:6658–6660

Yamaguchi D, Masaaki K, Satoshi S, Kiyotaka N, Hideki K, Michikazu H (2009) Hydrolysis of cellulose by a solid acid catalyst under optimal reaction conditions. J Phys Chem-C 113:3181–3188

Yang Q, Song B (2008) Sustainability assessment of biofuels as alternative energy resources. IEEE International Conference on Sustainable energy technologies, 2008 ICSET 2008, IEEE; pp 1001–1006

Yoon JJ, Kim YJ, Kim SH, Ryu HJ, Choi JY, Kim GS et al (2010) Production of polysaccharides and corresponding sugars from red seaweed. Adv Mater Res 93:463–466

Zilber-Rosenberg I, Rosenberg E (2008) Role of microorganisms in the evolution of animals and plants: the hologenome theory of evolution. FEMS Microbiol Rev 32:723–735

Pyrolysis of Biomass for Biofuel Production

Maria Fernandez-Lopez, Antonio Avalos-Ramirez, Jose Luis Valverde and Luz Sanchez-Silva

Abstract Due to the depletion of fossil fuel reserves and the environmental issues derived from their use, biomasses have been proposed and used as one of the renewable energy sources for the replacement of fossil fuels. Biomass can be converted into energy by means of thermochemical conversion processes. The pyrolysis process is the thermal degradation of biomass under an inert atmosphere leading to three different products: solid char, liquid biofuel, and fuel gas. This thermochemical process involves complex and multiple reactions. In this chapter, the biomass pyrolysis is studied using one of the main analytical tools to evaluate the potential of biomass: the thermogravimetric analysis (TGA) coupled with mass spectrometry (MS). This tool lets to observe the four main stages of pyrolysis process: dehydration, devolatilization, char formation, and inorganic matter decomposition. Gases evolved were studied by means of mass spectrometry, and the solid fuel (char) derived from pyrolysis was characterized using different techniques, such as elemental analysis, TGA, bomb calorimetry, and inductively coupled plasma-atomic emission spectroscopy (ICP-AES). This chapter also discusses the kinetics of pyrolysis and the way to evaluate the kinetic parameters which are necessary for industrially scaling up.

Keywords Biomass · Manure · Thermochemical conversion · Pyrolysis · Kinetics

Abbreviations

TGA	Thermogravimetric Analysis
MS	Mass Spectrometry
ICP-AES	Inductively Coupled Plasma-Atomic Emission Spectroscopy

M. Fernandez-Lopez · J.L. Valverde · L. Sanchez-Silva (✉)
Department of Chemical Engineering, Univeristy of Castilla-La Mancha (UCLM), Avda. Camilo José Cela, 12, 13071 Ciudad Real, Spain
e-mail: marialuz.sanchez@uclm.es

A. Avalos-Ramirez
Centre National en Électrochimie et en Technologie Environnementales Inc., 2263, avenue du Collège, Shawinigan, (QC) G9N 6V8, Canada

C.R. Soccol et al. (eds.), *Green Fuels Technology*,
Green Energy and Technology, DOI 10.1007/978-3-319-30205-8_19

LSM	Livestock Manure
SWOT	Strengths, Weaknesses, Opportunities and Threats
DTG	Derivative Thermogravimetric
HHV	High Heating Value
T	Temperature
P	Pressure
α	Extent of conversion
A	Preexponential factor
E_a	Activation energy
R	Universal gas constant
FWO	Flynn-Wall-Ozawa
KAS	Kissinger-Akahira-Sunose

1 Introduction

Depletion of fossil fuel reserves and the environmental issues derived from their use are the main causes for the increased attention toward bioenergy. Currently, the consumption rates of main fossil fuels are about 91 million barrels per day of oil and 9 billion cubic meters per day of natural gas (BP 2013). Therefore, it is estimated that at the present consumption rates the reserves will satisfy 48 years of oil and 64 years of natural gas supply. In this context, biomass is considered to be one of the few viable replacement options (Shen et al. 2013), contributing to approximately 14 % to the world annual energy consumption (Shen et al. 2009).

The term biomass can be used to define all kinds of organic material including trees, crops, algae, and residues which are susceptible to be valorized into by-products or energy (McKendry 2002). The benefits of using biomass are related to its renewable nature which comes from the ability to use CO_2 for producing new biomass through photosynthesis cycle (Sharma et al. 2015). There are different types of biomass to produce biofuels. The use of crops to produce first-generation biofuels has generated controversy because it competes with human food production. This has caused several socioeconomic problems related with the increase of food prices and social revolts in pauper regions. Recently, the developments of second- and third-generation biofuels based on lignocellulosic biomass, marine biomass (especially microalgae), and organic wastes promise to alleviate these problems.

Thermochemical conversion of biomass is considered one of the most direct and promising routes for biomass valorisation, being Pyrolysis, combustion, and gasification the main thermal technologies to process biomass. Combustion is the conversion of biomass into several forms of useful energy in the presence of air or oxygen. Biomass gasification can be defined as the conversion of biomass into a gaseous fuel by heating it in a partial oxidation atmosphere. Pyrolysis is defined as the thermal decomposition of biomass under an inert atmosphere. It takes advantage of all the fractions of biomass producing three types of products: low calorific gas, pyrolytic oil, and solid char (López-González et al. 2014). The char is a high energy density solid fuel suitable for combustion and gasification processes. Among the different thermochemical conversion technologies, pyrolysis stands out for being a highly versatile and easily scale-up process.

From an industrial point of view, it is important to identify the volatile compounds released during the thermochemical processes before their industrial scale-up. Thermogravimetric analysis (TGA) has commonly been used to investigate the thermochemical conversion of solid raw material, including biomass. Although thermogravimetric analysis is a useful tool, it does not release qualitative information of biomass transformation during the analyses. Therefore, thermogravimetric analysis coupled with mass spectrometry (TGA–MS) is able to afford real-time and sensitive detection of evolved gases (López-González et al. 2013).

2 Biomass

The term *biomass* is referred to all organic materials including trees, crops, algae, and residues which are susceptible to be converted into bioproducts or energy (McKendry 2002). There are several definitions for the term biomasses. The United Nations Framework Convention on Climate Change defines biomasses as follows (Basu 2013): "*Non-fossilized and biodegradable organic material originated from plants, animals and micro-organisms. This shall also include products, by-products, residues and waste from agriculture, forestry and related industries as well as the non-fossilized and bio-degradable fractions of industrial and municipal wastes.*"

Biofuels can be classified as primary and secondary biofuels. Primary biofuels are used in an unprocessed form for heating, cooking, or electricity production, for example wood, chips, and pellets. The secondary biofuels are produced by processing biomass (e.g., ethanol and biodiesel) that can be used in vehicles and various industrial processes. The secondary biofuels are further divided into first-, second-, and third-generation biofuels depending on the basis of the raw material and the technology used to produce them (Nigam and Singh 2011).

The first-generation biofuels are generally produced from sugars, grains, or seeds and require relatively simple processes to obtain the finished fuel. The most well-known first-generation biofuel is ethanol produced by fermenting sugar obtained from crop plants. As mentioned before, the production of first-generation biofuels is questionable due to the conflicts generated with food competition.

Second-generation biofuels are generally produced by two different approaches: biological and thermochemical processes. The feedstock is mainly lignocellulosic biomasses which can be agro-industrial residues, nonedible crops, or forestry products. Third-generation biofuels are specifically derived from microorganism, being microalgae the most developed feedstock (Nigam and Singh 2011).

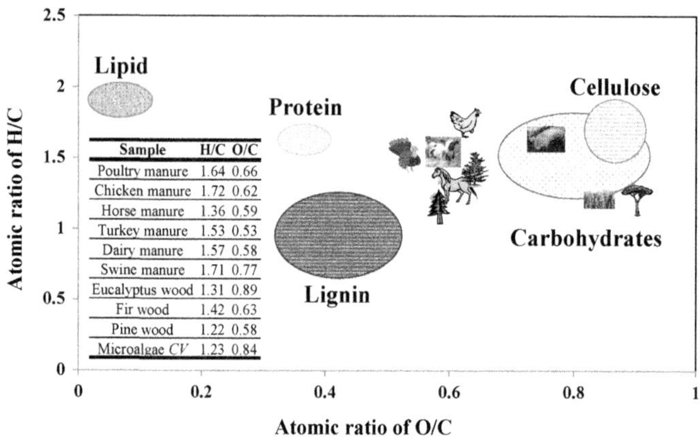

Fig. 1 Van Krevelen diagram with different kinds of terrestrial, marine, and manure biomass (Fernandez-Lopez et al. 2015)

Figure 1 shows the classification of different biomasses based on a Van Krevelen diagram where H/C versus O/C atomic ratios on a dry ash free basis (daf). This kind of diagram is useful to compare and select different types of fuels according to their H/C and O/C atomic ratios. The lower these two ratios, the higher the content of carbon in the biomass structure, being more suitable for biological or thermochemical processes. Figure 1 shows the main groups of biomolecular compounds including lipids, proteins, carbohydrates, lignin, and cellulose (Hammes et al. 2008). Figure 1 compares some lignocellulosic biomasses (pine, fir, and eucalyptus), the microalgae *Chlorella Vulgaris*, and animal manures (chicken, turkey, horse, seabird, dairy, and swine) (Fernandez-Lopez et al. 2015).

2.1 Lignocellulosic Biomasses

Among the different types of biomasses, lignocellulosic biomass represents a renewable and largely available source of raw feedstock that can be transformed into liquid and gas fuels, thermochemical products, and other energy-related end products (Naik et al. 2010).

It is recognized the high potential of woody crops (hardwoods and pines) and non-woody, high-yielding annual and perennial crops (Miscanthus, switchgrass, and sweet sorghum) due to multiple advantages. Some of the benefits of energy crops include less capital-intensive conversion technologies, attractive opportunity for local and regional self-sufficiency, reduction of greenhouse gas emissions, and viable alternative to fossil fuel use (López-González et al. 2015).

Lignocellulose biomass is mainly composed of cellulose, hemicellulose, and lignin, along with smaller amounts of pectin, protein, extractives (nonstructural sugars, nitrogenous material, chlorophyll, and waxes), and ash. The composition of these constituents can vary according to the different plant species. For example, hardwood has greater amounts of cellulose whereas wheat straw and leaves have more hemicellulose. In addition, ratios between various constituents within a single plant vary with age, stage of growth, and other conditions (Kumar et al. 2009).

Cellulose is the main structural constituent in plant cell walls and is found in an organized fibrous structure. This linear polymer consists of B-glucose subunits linked to each other by β-1,4-glycosidic bonds. Cellulose in biomass is present in both crystalline and amorphous forms. Crystalline cellulose comprises the major proportion of cellulose, whereas a small percentage of unorganized cellulose chains forms amorphous cellulose (Kumar et al. 2009; Hendriks and Zeeman 2009).

Hemicellulose is a complex carbohydrate structure that consists of different monomers like pentoses, hexoses, and sugar acids. The dominant component of hemicellulose from hardwood and agricultural plants is xylan. Hemicellulose has lower molecular weight than cellulose, and branches with short lateral chains that consist of different sugars. Hemicellulose is the connection between the lignin and the cellulose fibers, constituting the whole cellulose–hemicellulose–lignin network (Hendriks and Zeeman 2009). The structure of hemicellulose is either a homopolymer or a heteropolymer with short branches linked by β-1,4-glycosidic bonds and occasionally β-1,3-glycosidic bonds.

Lignin is a complex, large molecular structure containing cross-linked polymers of phenolic monomers. It is also, after cellulose and hemicellulose, one of the most abundant polymers in nature and is present in the cellular wall. It is an amorphous heteropolymer consisting of three different phenylpropane units (*p*-coumaryl, coniferyl, and sinapyl alcohol) that are held together by different kinds of linkages. The main purpose of lignin is to give the plant structural support, impermeability, and resistance against microbial attack and oxidative stress. The amorphous heteropolymer is also non-water soluble and optically inactive. All this makes the lignin structure very resistant (Kumar et al. 2009; Hendriks and Zeeman 2009).

2.2 Manure

The accumulation of livestock manure (LSM) has been associated with some hygienic and environmental problems due to its high pollution potential and high production. In this sense, contaminations of subterranean and surface water, odors, greenhouse gases, and ammonia emissions are some of the problems of LSM accumulation. Traditional uses of LSM, as a fertilizer and landfill, have to be changed due to land saturation with phosphorous and more strict regulations (Cao et al. 2015). Currently, there is an increasing interest in the valorization of manure as a solid fuel. Therefore, the utilization of manure for waste-to-bioenergy generation could be a sustainable choice since it is considered a zero-cost feedstock (Fernandez-Lopez et al. 2015). To assess the viability of manure as a feedstock for thermochemical conversion processes, the strengths, weaknesses, opportunities, and threats (SWOT analysis) could be used to evaluate it based on a wide knowledge of

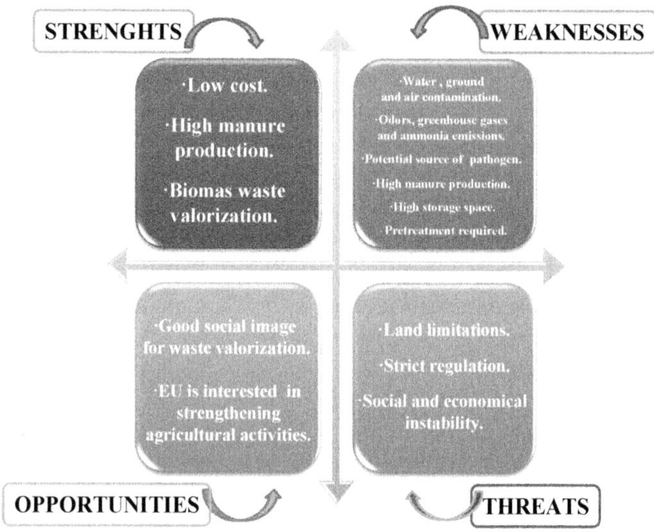

Fig. 2 SWOT matrix concerning the use of manure as fertilizer

the present situation and future trends of the market. The SWOT matrix concerning the use of manure as fertilizer is shown in Fig. 2.

In this case, the SWOT analysis was applied to decide whether the use of manure as fertilizer is the most appropriate one to this type of biomass. According to the SWOT matrix, there are more weaknesses than strengths in the use of manure as a fertilizer. As can be seen, the item "high manure production" was placed in both the strengths and the weaknesses. The high production is a strength since the viewpoint of a continuous raw material but also is a weakness because there is more production than the fertilizer market can assume. Moreover, land limitations are one of the threats.

Therefore, by considering manure as a raw material for thermochemical conversion processes, the majority of the weaknesses and the threats (pollution, contamination, and the strict regulation) could be solved.

3 Thermochemical Conversion of Biomass: Pyrolysis Process

Thermochemical conversion processes can be used for the transformation of dry biomass samples. Depending on the amount of oxygen present in the reaction atmosphere, three different types of processes can be defined: pyrolysis, combustion, and gasification. Combustion is the oxidation of biomass with air or under an oxidizing atmosphere with excess of oxygen. In the case of gasification process, the atmosphere is poor in oxygen. Pyrolysis can be defined as the degradation of the biomass by heating it in a non-oxidant atmosphere, leading to three different products: solid char, bio-oil, and fuel gas. The ratio of these products could vary by changing the operating conditions during the pyrolysis process. Low temperature and long vapor residence time favor the production of charcoal. High temperature and long residence time increase the production of gas, and moderate temperature and short vapor residence time are optimum for producing liquids (Bridgwater 2012).

The pyrolysis process can be generally represented by the following reaction:

$$C_nH_mO_p(\text{biomass}) \rightarrow \sum_{\text{liquid}} C_xH_yO_z + \sum_{\text{gas}} C_aH_bO_c + H_2O + C(\text{char})$$

Pyrolysis process could be divided into four general stages:

- **Drying**. In this stage, the moisture present in the biomass fuel is removed at temperatures below 100 °C.
- **Devolatilization**. It is the main pyrolytic stage in which the volatile compounds are converted into vapors and oxidized. This stage usually takes place from 100 to 300 °C.
- **Char formation**. Reactions such as thermal cracking and dehydrogenation take place in this stage, leading to the char formation. The temperature range for this step is from 300 to 600 °C.
- **Inorganic matter decomposition**. At temperatures above 600 °C, the inorganic matter present in the biomass sample could be decomposed, whereas the secondary cracking of the char formed could also take place.

4 Characterization and Simulation of Pyrolysis Process Using Thermal Analysis

4.1 Thermogravimetric Analysis Coupled to Mass Spectrometric Analysis (TGA–MS)

Thermogravimetric analysis (TGA) is a thermoanalytical technique which has been commonly used for the study of the solid-state thermal degradation (White et al. 2011). Thermoanalytical methods are crucial in order to obtain the fundamental experimental data for the characterization of the solid phase (Mura 2015). TGA measures the weight loss of a sample caused by the release of volatile compounds or devolatilization, during the thermal degradation. The maximum reaction rate can be obtained by taking the first derivative of such thermogravimetric curves (leading to the derivative thermogravimetric (DTG) curves).

Pyrolysis has been widely studied by means of TGA (López-González et al. 2014; Worasuwannarak et al. 2007; Wu et al. 2012; Guerrero et al. 2014). In this chapter, the pyrolysis of a manure dairy sample is presented as an example of the utilization of TGA technique for the thermal degradation study.

Figure 3 shows the thermogravimetric (TGA) and derivative thermogravimetric (DTG) analysis at a heating rate of 10 °C/min for the pyrolysis process of the dairy manure sample. Mass spectrometry profile for the pyrolysis process of the dairy sample is shown in Fig. 4.

The pyrolysis process of dairy sample could be divided into five stages as shown in Fig. 3. The first one appears at temperatures below 125 °C, which represents the drying process of the sample. Therefore, water is mainly detected in this stage as it can be seen in Fig. 4. The second step occurs at temperatures between 200 and

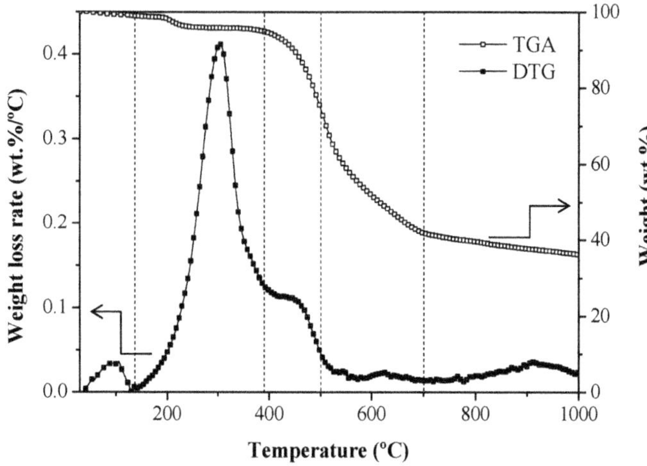

Fig. 3 Thermogravimetric (TGA) and derivative thermogravimetric (DTG) curves for the pyrolysis process of the manure sample (dairy sample)

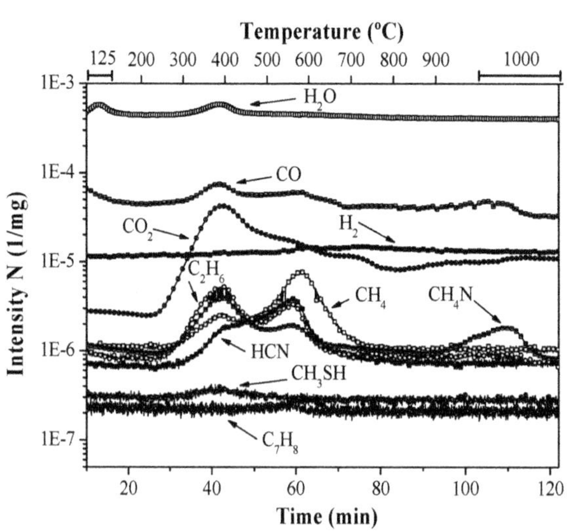

Fig. 4 Mass spectra for the pyrolysis process of the manure sample (dairy sample)

390 °C. In this stage, the maximum decomposition rate takes place at 300 °C, being associated with hemicellulose and glucoside linkage depolymerisation (Wnetrzak et al. 2013). The maximum weight loss is approximately 50 wt%, being considered the main pyrolytic stage. In this stage, most gaseous products are detected, obtaining the highest yields for CO_2 and CO. Sulfur products are only detected in this stage as CH_3SH. In addition, nitrogen compounds such as amines and cyanides are identified. The third stage is characterized by a shoulder appearing between 390 and 500 °C in the DTG curves. This shoulder is ascribed to the decomposition of lipids and other *N*-containing compounds (Wnetrzak et al. 2013). The main emission peaks corresponding to nitrogen compounds and CH_4 are found in this stage.

Furthermore, aromatic compounds such as toluene are also detected. H_2 starts evolving in this stage, indicating that char transformation reactions such as thermal cracking and dehydrogenation are taking place (Sanchez-Silva et al. 2012). The last two stages are identified by two peaks between 550–700 and 800–1000 °C. These stages are associated with the decomposition of the formed char and the decomposition of inorganic matter (López-González et al. 2013). Additionally, CO, CO_2, and CH_4N are detected at 1000 °C.

4.2 *Characterization of Solid Fuel Char Obtained from Pyrolysis*

Char obtained from the pyrolysis process of dairy manure sample was characterized by elemental analysis, TGA, bomb calorimetry, and inductively coupled plasma-atomic emission spectroscopy (ICP-AES).

The ultimate analysis is used to measure the carbon (C), hydrogen (H), nitrogen (N), oxygen (O), and sulfur (S) content of a sample. Moisture, ash, fixed carbon, and volatile matter content of the samples are obtained by proximate analysis. These are the four most important chemical characteristics in any type of fuel. The heat of combustion (high heating value, HHV) was determined using a Parr 1356 bomb calorimeter according to UNE 164001:2005 EX at constant volume and a reference temperature of 25 °C. ICP-AES elemental analysis was used to obtain the weight percentage of metallic elements in the char coming from the manure sample. Table 1 shows the characterization of dairy manure char described in this section.

5 Kinetics of Biomass Pyrolysis

5.1 *Kinetic Models*

Generally speaking, the thermal decomposition of biomass can be expressed as follows:

Table 1 Ultimate and proximate analysis, HHV, and mineral content of manure dairy char

	Ultimate analysis (wt%)[a]					Proximate analysis (wt%)				Bomb calorimeter	
Sample	C	H	N	S	O^{diff}	Moisture	Ash	Volatile matter	Fixed carbon^{diff}	HHV (MJ/kg)	
Dairy manure char	40.19	0.57	0.22	0.16	2.01	2.85	56.85	10.33	33.55	16.3	
Mineral content (ppm)											
Sample	Al	Ca	Cr	Cu	Fe	K	Mg	Na	Ni	P	Si
Dairy manure char	5864	28,322	–	–	8696	8793	9744	11,377	–	25,818	9961

[a] *daf* dry ash free; O^{diff} obtained by difference of C, H, N, S, and ash; *Fixed carbon* ^{diff} calculated from difference of moisture, ash, and volatile matter

$$A(\text{solid}) \rightarrow B(\text{solid}) + C(\text{liquid}) + D(\text{gas})$$

The kinetic study of the thermal decomposition reactions is based on the measure and the parameterization of the reaction rate, being this reaction rate dependent on three main variables: temperature (*T*), pressure (*P*), and the extent of conversion (α) as follows:

$$\frac{\mathrm{d}\alpha}{\mathrm{d}t} = k(T)f(\alpha)h(P) \tag{1}$$

The term *h*(*P*) represents a pressure dependence. In general, the gaseous products are removed from the reaction medium by the carrier gas flow. Therefore, the pressure could be kept constant during all the thermochemical processes. Consequently, the majority of the kinetic methods consider the reaction rate as a function of only two variables, *T* and α:

$$\frac{\mathrm{d}\alpha}{\mathrm{d}t} = k(T)f(\alpha) \tag{2}$$

Equation 2 represents the reaction rate of a single-step process. Although the process mechanism involves more than one reaction, one of them can determine the overall kinetics (Vyazovkin et al. 2011). Therefore, Eq. 2 can be used to describe the overall reaction rate of the thermochemical processes.

The temperature dependence *k*(*T*) is typically parameterized by the Arrhenius equation:

$$k(T) = A\mathrm{e}^{\frac{-E_a}{RT}} \tag{3}$$

where *A* is the preexponential factor, E_a is the activation energy, and *R* is the universal gas constant. These kinetic parameters, which are experimentally determined, are called "effective" or "apparent" parameters, because they represent the overall process and not the individual parameters of each step.

The function which describes the conversion degree can be expressed using different reaction models, *f*(α), which are described elsewhere (López-González et al. 2013). The extent of conversion α is defined as follows:

$$\alpha = \frac{m_0 - m_t}{m_0 - m_f} \tag{4}$$

where m_0, m_t, and m_f represent the masses at time $t = 0$, $t = t$, and $t = tf$, respectively.

Combining Eqs. 2 and 3 yields

$$\frac{\mathrm{d}\alpha}{\mathrm{d}t} = A\mathrm{e}^{\frac{-E_a}{\mathrm{RT}}} f(\alpha) \tag{5}$$

Integrating Eq. 5, the integral function $g(\alpha)$ can be defined as follows:

$$g(\alpha) \equiv \int_0^\alpha \frac{\mathrm{d}\alpha}{f(\alpha)} = A\int_0^t \mathrm{e}^{\frac{-E_a}{\mathrm{RT}}} \mathrm{d}t \tag{6}$$

Defining a constant heating rate as $\beta = \mathrm{d}T/\mathrm{d}t$, Eq. 6 is rearranged as follows:

$$g(\alpha) \equiv \int_0^\alpha \frac{\mathrm{d}\alpha}{f(\alpha)} = \frac{A}{\beta}\int_{T_0}^T e^{\frac{-E_a}{\mathrm{RT}}} \mathrm{d}T \tag{7}$$

Equation 7 is the general equation necessary to obtain the kinetic parameters of the biomass thermal decomposition. It can be observed that this equation has not an analytical solution. Therefore, a number of approximate solutions were given in the past (Vyazovkin et al. 2011). There are different methods to solve and obtain the kinetic parameters from Eq. 7. On the one hand, the fitting models allow to obtain the kinetic parameters by selecting the adequate reaction model $f(\alpha)$ for each biomass decomposition considered during thermochemical processes. On the other hand, isoconversional methods are used to obtain these kinetic parameters without the requirement of a reaction model specification. The fitting process requires experimental data obtained at different heating rates. Isoconversional methods consider that the reaction rate at constant extent of conversion is only a function of temperature.

This chapter illustrates the use of different isoconversional methods to obtain and compare the activation energy of the pyrolysis process using a manure dairy sample.

5.2 Differential Isoconversional Methods

The most common differential isoconversional method is that of Friedman (Vyazovkin et al. 2011; Friedman 1964) and is based on the following equation obtained from Eq. 6:

$$\ln \beta\left(\frac{\mathrm{d}\alpha}{\mathrm{d}t}\right)_\alpha = \ln[f(\alpha)A] - \frac{E_a}{\mathrm{RT}_\alpha} \tag{8}$$

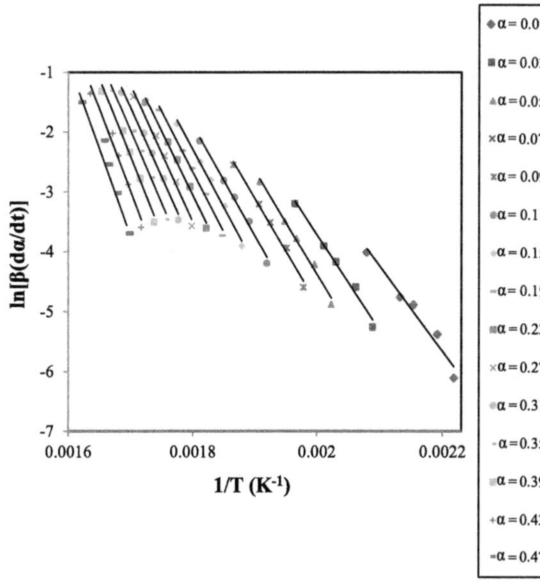

Fig. 5 Friedman experimental data fitting at different α for the pyrolysis process of dairy manure sample

For each given α, activation energy E_a can be easily obtained by the slope of the representation of $\ln \beta\left(\frac{d\alpha}{d_t}\right)_\alpha$ against $\frac{1}{T_\alpha}$. The dependence of the activation energy on the extent of conversion gives important information about the mechanism of the reaction. If the activation energy does not vary with the extent of conversion, the mechanism is the same for the whole process. If so, it means that the mechanism changes.

Figure 5 shows the Friedman experimental data fitting at different extents of conversion (α) for the pyrolysis process of the dairy manure sample. Activation energy and correlation coefficient for each extent of conversion selected are presented in Table 2. It can be observed that, in all the cases, the correlation coefficients correspond to linear fittings and they were higher than 0.9.

5.3 *Integral Isoconversional Methods: Flynn–Wall–Ozawa (FWO) and Kissinger–Akahira–Sunose (KAS)*

As aforementioned, Eq. 7 has not an analytical solution. For this reason, a number of integral isoconversional methods have been proposed, which differ each other in the different temperature integral approximations used. Many of these approximations give rise to linear equations of the general form (Vyazovkin et al. 2011; Starink 2003):

Table 2 Activation energy and correlation coefficient at each conversion for dairy manure sample pyrolysis

Dairy manure (temperature range: 170–360 °C)					
Friedman		FWO		KAS	
r^2	Ea (kJ/mol)	r^2	Ea (kJ/mol)	r^2	Ea (kJ/mol)
0.964	118	0.977	125	0.973	111
0.986	133	0.984	141	0.982	126
0.990	143	0.988	152	0.986	136
0.991	146	0.989	155	0.987	139
0.993	152	0.992	161	0.991	144
0.990	153	0.993	164	0.992	147
0.992	161	0.993	171	0.992	153
0.992	168	0.992	175	0.991	157
0.991	176	0.993	182	0.992	164
0.987	189	0.991	191	0.990	172
0.989	191	0.989	195	0.988	176
0.987	200	0.988	202	0.987	182
0.984	215	0.989	214	0.988	194
0.975	228	0.983	221	0.981	200
0.962	242	0.974	238	0.971	216

$$\ln\left(\frac{\beta}{T_{\alpha}^{B}}\right) = \text{Const} - C * \left(\frac{E_{\alpha}}{\text{RT}_{\alpha}}\right) \tag{9}$$

where B and C are the parameters determined by the type of the temperature integral approximation. For example, Eq. 9 will take the following form if the approximation of Ozawa and/or Flynn and Wall (FWO) is used:

$$\ln(\beta_i) = \text{Const} - 1.052 * \left(\frac{E_{\alpha}}{\text{RT}_{\alpha}}\right) \tag{10}$$

Another widespread approximation is that by Kissinger–Akahira–Sunose (KAS) equation:

$$\ln\left(\frac{\beta_i}{T_{\alpha,i}^{2}}\right) = \text{Const} - \left(\frac{E_{\alpha}}{\text{RT}_{\alpha}}\right) \tag{11}$$

In both cases, the activation energy is easily obtained from the slope of the representation of $\ln\beta_i$ and $\ln\left(\frac{\beta_i}{T_{\alpha,i}^{2}}\right)$ for FWO and KAS equations, respectively, versus $1/T_{\alpha}$.

In Fig. 6, experimental data fittings with these two methods (FWO and KAS) at different extents of conversion (α) for the pyrolysis process of the dairy manure sample are shown. Activation energy and correlation coefficient for each extent of conversion selected are listed in Table 2.

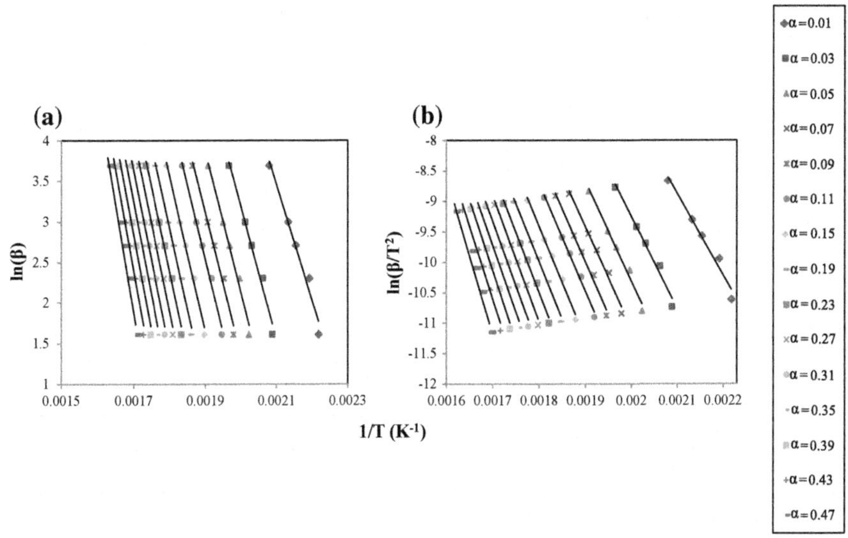

Fig. 6 **a** FWO and **b** KAS experimental data fitting at different α for the pyrolysis process of dairy manure sample

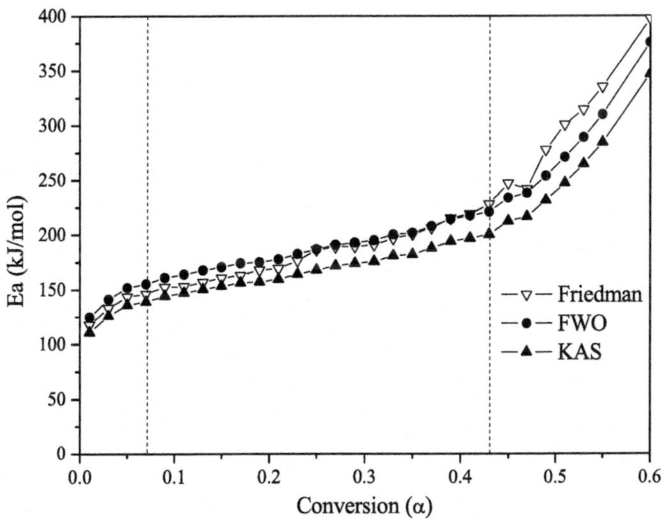

Fig. 7 Comparison of the activation energies calculated by Friedman, FWO, and KAS methods for the pyrolysis process of dairy manure sample

The temperature range at which the activation energy was derived from these methods ranges from 170 to 330 °C, which corresponds with a range of conversion from 0.05 to 0.47. This temperature range, for which the activation energy is constant, corresponds to the devolatilization stage of the dairy sample (Fig. 3). Figure 7 shows a comparison among the activation energies obtained by the three methods considered.

This way, the average values of activation energy obtained from Friedman, FWO, and KAS methods are 170, 175, and 157 kJ/mol, respectively (Table 2), which are comparatively very similar in magnitude.

To sum up, TGA is an excellent tool to study the thermal behavior of a biomass sample during pyrolysis process, providing the experimental data necessary to carry out the kinetic analysis. Concerning this kinetic analysis, isoconversional methods can be used to easily obtain the activation energy of the pyrolysis process, therefore avoiding more complex mathematical analysis associated to the decomposition process.

References

Basu P (2013) Biomass gasification, pyrolysis and torrefaction: practical design and theory: Elsevier science

BP (2013) BP statistical review of world energy, June 2014. Available online at: bp.com/statisticalreview

Bridgwater AV (2012) Review of fast pyrolysis of biomass and product upgrading. Biomass Bioenergy 38:68–94

Cao JP, Huang X, Zhao XY, Wei XY, Takarada T (2015) Nitrogen transformation during gasification of livestock compost over transition metal and Ca-based catalysts. Fuel 140: 477–483

Fernandez-Lopez M, Puig-Gamero M, Lopez-Gonzalez D, Avalos-Ramirez A, Valverde J, Sanchez-Silva L (2015) Life cycle assessment of swine and dairy manure: pyrolysis and combustion processes. Bioresour Technol 182:184–192

Friedman HL (1964) Kinetics of thermal degradation of char-forming plastics from thermogravimetry. Application to a phenolic plastic. J Polymer Sci Part C: Polymer Symp 6(1): 183–195

Guerrero MRB, Marques da Silva Paula M, Zaragoza MM, Gutiérrez JS, Velderrain VG, Ortiz AL et al (2014) Thermogravimetric study on the pyrolysis kinetics of apple pomace as waste biomass. Int J Hydrogen Energy 39(29):16619–16627

Hammes K, Smernik RJ, Skjemstad JO, Schmidt MWI (2008) Characterisation and evaluation of reference materials for black carbon analysis using elemental composition, colour, BET surface area and 13C NMR spectroscopy. Appl Geochem 23(8):2113–2122

Hendriks ATWM, Zeeman G (2009) Pretreatments to enhance the digestibility of lignocellulosic biomass. Bioresour Technol 100(1):10–18

Kumar P, Barrett DM, Delwiche MJ, Stroeve P (2009) Methods for pretreatment of lignocellulosic biomass for efficient hydrolysis and biofuel production. Ind Eng Chem Res 48(8):3713–3729

López-González D, Fernandez-Lopez M, Valverde JL, Sanchez-Silva L (2013) Thermo-gravimetric-mass spectrometric analysis on combustion of lignocellulosic biomass. Bioresour Technol 143:562–574

López-González D, Fernandez-Lopez M, Valverde JL, Sanchez-Silva L (2014) Pyrolysis of three different types of microalgae: kinetic and evolved gas analysis. Energy. 73:33–43

López-González D, Avalos-Ramirez A, Giroir-Fendler A, Godbout S, Fernandez-Lopez M, Sanchez-Silva L et al (2015) Combustion kinetic study of woody and herbaceous crops by thermal analysis coupled to mass spectrometry. Energy

McKendry P (2002) Energy production from biomass (part 1): overview of biomass. Bioresour Technol 83(1):37–46

Mura P (2015) Analytical techniques for characterization of cyclodextrin complexes in the solid state: a review. J Pharm Biomed Anal 113:226–238

Naik S, Goud VV, Rout PK, Jacobson K, Dalai AK (2010) Characterization of Canadian biomass for alternative renewable biofuel. Renew Energ. 35(8):1624–1631

Nigam PS, Singh A (2011) Production of liquid biofuels from renewable resources. Prog Energ Combust. 37(1):52–68

Sanchez-Silva L, López-González D, Villaseñor J, Sánchez P, Valverde JL (2012) Thermogravimetric–mass spectrometric analysis of lignocellulosic and marine biomass pyrolysis. Bioresour Technol 109:163–172

Sharma A, Pareek V, Zhang D (2015) Biomass pyrolysis—a review of modelling, process parameters and catalytic studies. Renew Sust Energ Rev. 50:1081–1096

Shen DK, Gu S, Luo KH, Bridgwater AV, Fang MX (2009) Kinetic study on thermal decomposition of woods in oxidative environment. Fuel 88(6):1024–1030

Shen D, Hu J, Xiao R, Zhang H, Li S, Gu S (2013) Online evolved gas analysis by thermogravimetric-mass spectroscopy for thermal decomposition of biomass and its components under different atmospheres: Part I. Lignin Bioresour Technol 130:449–456

Starink MJ (2003) The determination of activation energy from linear heating rate experiments: a comparison of the accuracy of isoconversion methods. Thermochim Acta 404(1–2):163–176

Vyazovkin S, Burnham AK, Criado JM, Pérez-Maqueda LA, Popescu C, Sbirrazzuoli N (2011) ICTAC kinetics committee recommendations for performing kinetic computations on thermal analysis data. Thermochim Acta 520(1–2):1–19

White JE, Catallo WJ, Legendre BL (2011) Biomass pyrolysis kinetics: a comparative critical review with relevant agricultural residue case studies. J Anal Appl Pyrol 91(1):1–33

Wnetrzak R, Kwapinski W, Peters K, Sommer SG, Jensen LS, Leahy JJ (2013) The influence of the pig manure separation system on the energy production potentials. Bioresour Technol 136:502–508

Worasuwannarak N, Sonobe T, Tanthapanichakoon W (2007) Pyrolysis behaviors of rice straw, rice husk, and corncob by TG-MS technique. J Anal Appl Pyrol 78(2):265–271

Wu H, Hanna MA, Jones DD (2012) Thermogravimetric characterization of dairy manure as pyrolysis and combustion feedstocks. Waste Manage Res J Int Solid Wastes Public Clean Assoc ISWA 30(10):1066–1071

Life-Cycle Assessment of Biofuels

Luiz Alberto Junior Letti, Júlio César de Carvalho, Sérgio José da Costa, Thatyana Santiago Martins, Nádia da Silva Ramos, Marciane Cristina Dotto, Adenise Lorenci Woiciechowski and Carlos Ricardo Soccol

Abstract Life-cycle assessment (LCA) is a powerful tool to evaluate economic efficiency and environmental impacts of a product or processes. One of its most important applications in the last decades—considering the debates around energetic matrix changes and the impacts of human activity over the environment—is on biofuels life cycle, where the main target is to compare the use of a certain biofuel (a renewable source of energy) with a fossil fuel. In the first part of this chapter, we briefly discuss LCA theory, scope and objectives, besides some of the commonly used parameters to evaluate and compare different processes efficiencies. In the sequence, we focus on some specificities of biofuels life cycle, and finally we present some case studies concerning the most important biofuels—biodiesel, bioethanol, and biohydrogen, among others.

Keywords Life cycle · Bioethanol · Biodiesel · Biohydrogen · Biomethane · Microalgae · GHG

Abbreviations

EgLCA	Energetic life-cycle assessment
EvLCA	Environmental life-cycle assessment
FC-HEV	Fuel cell—hydrogen electric vehicle
FER	Fuel energy ratio
GHG	Greenhouse gases
ICEV	Internal combustion engine vehicle
LCA	Life-cycle assessment
LUC	Land-use change
LULUCF	Land use, land-use change and forestry

L.A.J. Letti · J.C. de Carvalho · A.L. Woiciechowski · C.R. Soccol (✉)
Federal University of Paraná, Curitiba, Brazil
e-mail: soccol@ufpr.br

S.J. da Costa · T.S. Martins · N. da Silva Ramos · M.C. Dotto
Federal University of Tocantins, Gurupi, Brazil

C.R. Soccol et al. (eds.), *Green Fuels Technology*,
Green Energy and Technology, DOI 10.1007/978-3-319-30205-8_20

NEG	Net energy gain
TTW	Tank to wheel
WTT	Wheel to tank
WTW	Wheel to wheel

1 Introduction

Over the years, society has undergone numerous changes of habits, leading to a significant increase in consumption of varied and innovative products. The global population growth in the last three decades has resulted in a need for increased raw materials and energy generation to fulfill the current consumption demands. But the increase of the consumption may cause more environmental impacts at any stage of their development or manufacturing.

Currently, much is being discussed about environmental issues and the responsibilities of companies and organizations. The main target is to seek for less impactful and more transparent energy technologies, and also the acquisition of detailed knowledge about the stages of production of new products or services. So it becomes important to use tools that evaluate the contribution of a particular technology and its sustainability.

In this context it becomes important to study a tool called "Life-Cycle Analysis"—LCA, as it allows the analysis of how a particular product or service may interact with the environment, and at what stage, whether during the production, use, disposal or final disposal lays and its degree of interaction and/or interference over the environment.

2 Life-Cycle Assessment (LCA)

According to NREL, Life-cycle assessment (LCA) is "a standardized technique that tracks all material, energy, and pollutant flows of a system—from raw material extraction, manufacturing, transport, and construction to operation and end-of-life disposal." As an example, the first analyses carried out by the Coca-Cola Company, in 1969, in order to comply with the emissions and waste generated by packaging their drinks used the LCA tool (Hunt and Franklin 1996).

Caldeira-Pires et al. (2005), signals that the LCA is a process that aims to evaluate the impacts to the environment and health, associated to a product, process, or service.

Robles Junior and Bonelli (2006), state that every product may cause in any form some kind of environmental impact, not only at the moment of their discharge, but also during their entire life cycle, beginning at the raw material extraction until the final disposal, impacting to the environmental pollution, residues and emissions.

According to Lima et al. (2007), the interest of industries and specialists, environmental organizations and the public in general to LCA is increasing each day, due to their disposition to know mainly the quality of the industrial products. In many developing countries, the LCA of products is not yet a very well known and used tool, only large companies and institutions such as Mercedez-Benz, Volvo, and ITAL (Instituto Técnico de Alimentação), Food Technical Institute of Brazil use this tool.

However, according to Lemos and Barros (2006), in Brazil and at most of the Latin American countries this actions are beyond the desirable when compared to what is being done at the more developed economic regions: European Union, North American Free Tread Agreement (NAFTA), and Asian-Pacific Economic Cooperation (APEC).

Particularly for Brazil and for the countries with a large surface, covering different climates, and with a diversity of people traditions the development of a methodology suitable to all regions, faces serious problems due to LCA is hardly dependent of regional circumstances; in these cases all studies are based in standards, ISO 14040 in the case of Brazil.

The methodology to proceed the LCA is described in NBR ISO 14040 standard (ABNT 2001) that establish that the study must include all phases such as definition, objective and scope, inventory analysis, environmental impacts evaluation and interpretation.

In this way the standards ISO 14044 (2006) defines the LCA as a comprehensive method to evaluate the environmental impact of a product or process along its life span. In other words, LCA is an environmental tool that evaluates the life of a product or process. The evaluation of Life Cycle (LCA—Life-cycle Assessment) can be seen as an environmental management tool that gathers data from processes and uses environmental models to evaluate the impact of a technology or process in a global way. This is a more complete analysis when compared to analyzing one factor of the system or one portion at a time. It is through this kind of analysis that is possible to evaluate how sustainable a biofuel production can be such as soybean biodiesel compared to palm biodiesel or bioethanol from sugar cane or from corn, for example.

3 Steps of a LCA

The LCA studies begun in 1960 and aroused great technical interest during the oil crisis that happened in the 1970s. After the crisis, there was a development slowdown in the energy sector. In the end of 1980s, concerning about the utilization of inconsistent tools to justify sustainable declarations led to the LCA standardization through the ISO 14000 series, and in the last 20 years it was possible to identify a huge growth at the area together with the sustainable concerning, global warming, and the development of renewable fuels (SAIC 2006).

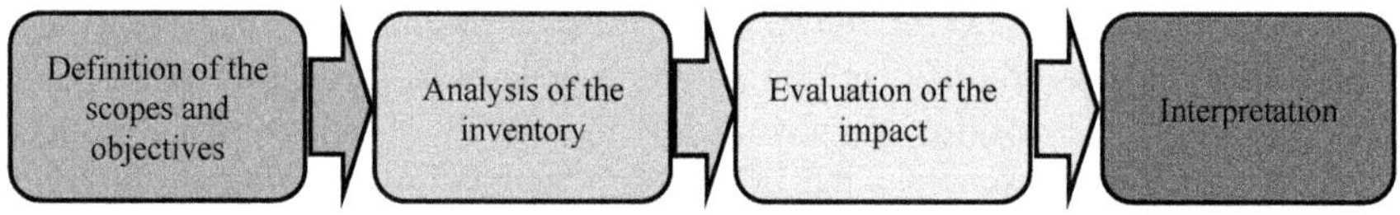

Fig. 1 Steps of a product of a generic life cycle evaluation

The life-cycle evaluation involves four basic steps that are conducted according to the specific objectives—if LCA will be performed to evaluate environmental impacts, to compare products or production routes, or to compare industrial processes impacts, etc.

Although the steps are more or less sequential (as shown in Fig. 1), revisit each step is necessary to adapt the model to new realities and novelties that are improved along the evaluation process.

4 Definition of the Scope and Objectives

To develop any model it is essential to define the model's objectives, and what it must illustrate (which is the scope). As LCA covers a great quantity of different data (consumed energy and matter quantity, quantities of generated residues and the different kinds of impacts caused by these residues, etc.), the way to collect these data, describe them and list them must be established previously. At this point, it is necessary to define the functional unit to be used, which kind of analysis in terms of origin and destination of the products, and which are the borders of the system.

Depending on the objectives of the study, the limits of the system can be shortened or expanded—these are the borders of the system. The most common approaches adopted in an LCA are:

- Cradle to Grave: considers the whole life cycle from the extraction of the raw materials to the disposal of the end products.
- Cradle to gate: begins with the extraction of raw materials and ends in the entrance of the factory.
- Gate to gate: considers just the industrial phase of transformation of a product. It is useful to study unitary processes. The adequate joint of several gate-to-gate processes, plus the preindustrial processes and the disposal of the products, can be used to develop the whole cradle to grave assessment.
- Gate to grave: considers the product from the end gate of the factory to its final disposal.

For biofuels, the analysis is usually performed in a "*cradle-to-grave*" base because the analysis complexity does not improve the quality of the impact evaluation and sustainability that is in general the final objective of the LCA. In these cases, the analysis is much more of the system than the product, but calculated based on a functional unit, such as: tons of raw material at the factory entrance, tons of product, production area, etc.

Specifically for transport fuels and vehicles, the LCA used are:

- Wheel to Tank (WTT): considers the stages of feedstock, production (processing) of fuel and its delivery (or transmission in the case of energy).
- Tank to Wheel (TTW): considers the use of the fuel in the vehicle, or the use of the energy.
- Wheel to Wheel (WTW): considers the whole cycle (wheel to tank plus tank to wheel) and is usually the most important way to assess the net energy gain (NEG) or loss of a process whose product is a fuel or energy.

5 Analysis of the Inventory

At this step, it is possible to quantify the entrances and exits of each subsystem that compose the global process. In the case of biofuel LCA, the inventory analysis includes the evaluation of quantities of agricultural inputs, planted area, energy uses, harvesting, transportation and production labor, and data of raw matter quantities, energy consumption, and residues generated at the various industrial processing steps. At this point, in order to improve the model confidence, it is important to do an accurate data collection and this is the point that justifies the necessity of the realization of a biofuel LCA in accordance to the local reality: the input costs, productivity, and energy consumption at the biodiesel production. In Brazil, for example, the scenario is different than that in Indonesia or Malaysia and radically different than what happens at the biodiesel production with another agricultural base, such as colza in Europe. Similarly, the cultures can have different performances in different regions used for plantation and it is essential to obtain real data from the field to evaluate the real potential of the established cultures.

On the other hand, if the inventory analysis is extended "strain above" ad infinitum, the analysis becomes impracticable. Due to this, in general, some simplifications are carefully defined at the scope, in order to establish the use of previous LCA studies, for example, to the fertilizer production. There is a series of European and American tables of data that not always reflect the reality of a specific case because they express average values or they were generated in a different reality. Fortunately, LCA studies of inputs are being developed to reflect the reality of many specific cases, adapted to specific realities, for example, the Brazilian reality, and the use of modular models allow the adaptation of a model in development for more refined data.

The inventory analysis is translated into an inventory table, which summarizes the relevant elementary flows (separated, for example, in fertilizers, diesel, and fuel consumption, carbon emission, etc.) to one or more products or cases. This table is not an impact evaluation "*per se*", being basically a summary of inputs and exits of material and energy that presents behind a series of submodels and references. An example is given in Fig. 2.

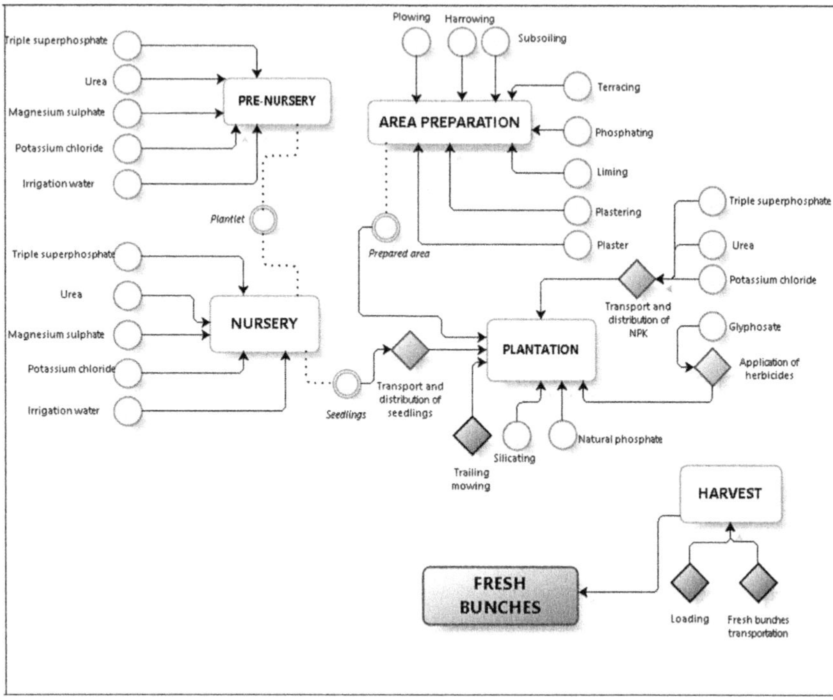

Fig. 2 Flowchart containing some information (mass input and output, WTG) to be used in inventory analysis for the first steps of second generation biofuels production from lignocellulosic material—other data, such as energy input and output would be also required for compilation of the complete table

6 Sustainability and Impact Assessment

The inventory analysis data and the analysis of the models involved are translated in terms of potential effects, according to the methodologies currently accepted from IPCC (to the greenhouse gases (GHG) emissions), or CalTox (to ecotoxicity), among others. Besides the model selection and classification and characterization of the environmental impacts, it is necessary to translate the results in a form that could allow the comparison with other impact evaluations. To perform the comparisons, it is important to use a common data basis, for example a certain scenario or referential product standardizing the results based at the reference scenery, applied to different products.

This step escapes a little from what is measurable (at the inventory analysis) and depends on the models and references chosen. For example, today it is considered that the greenhouse effect caused by methane (CH_4) is 21 times bigger than the carbonic gas (CO_2). This kind of consideration must be clearly defined at this step and again the modularity of the model is an advantage for model revision and refining.

7 Interpretation of the Results

How to compare biofuel or different scenarios? The data interpretation is always a little subjective, but it is always possible to compare different scenarios from graphics that illustrate standardized data and aggregated impacts—for example, to greenhouse effect gases, acidification, ecotoxicity, etc. Besides, two important results can be obtained working on the inventory table and models of the process and of the SIA: comparing impacts for different scenarios, a *sensibility analysis* allows the identification of critical points at the process or associate the environmental impacts to extended deadlines and even the confidence degree (or error) associated to the model predictions. The interpretation of the results includes also the possibility of reporting specific and desirable factors to the planning and the decision-making, such as the impact on the fossil fuel consumptions, natural resources consumptions (such as minerals for fertilizers) or electrical energy, to beyond the environmental results explored at SIA.

8 Biomass as Alternative Raw Material for Sustainable Biofuel Production

The use of biomass as alternative source of energy has been widely used around the world, as one of the many alternatives for substitution of fossil fuels. The most recent debates are on which extent biofuels are better than fossil fuels, both in economical and environmental aspects. The sustainability of biofuels production can be performed by the so called LCA. As said before, LCA is a tool to estimate environmental and economic net flux impacts of a part of a process, a whole process, or a chain of processes. It can aim on environmental impacts (EvLCA), or in energetic impacts (EgLCA). Both analyses are important, but for biofuels the last one is more commonly employed to evaluate and compare the efficiency of different fuels, considering both environmental and economic factors (Davis et al. 2009).

Energy from biomass can be obtained directly via burning, but the focus nowadays is the liquid fuels that can be obtained from biomass. Ethanol and biodiesel are the main examples of liquid fuels obtained from fermentation of simple sugars (first generation biofuels), while the second generation biofuels are produced via saccharification of lignocellulosic material followed by fermentation of the resulting sugars. Some authors also define another class of biofuels (third generation), produced via transesterification of microalgae biomass and/or via reusing of wastes from production of second generation biofuels (Menten et al. 2013).

Corn is the principal feedstock used in North America to produce bioethanol, while sugarcane is the main raw material in Brazil, and beet sugar is used in Europe. The challenge for scientists researching second and third generation biofuels is: greater energy output than corn and sugarcane-based ethanol; smaller land-use change (LUC); positive nutrient cycling in soil.

The main candidates for substituting corn are switchgrass and *Miscanthus*. Literature data regarding environmental and economic effects of biofuels production are controversial. Many factors are responsible for discrepancies in final LCA results, such as differences in the considered pathways and methodologies employed. Specifically for biofuels LCA, there is divergence on the description of biomass, data to describe conversion from biomass to fuel, definition of system boundaries, and the impact of coproducts (Davis et al. 2009).

Regarding the boundaries in an LCA for biofuels, they can vary from a study to other. Depending on the scale considered, the results may vary significantly. Energy and material flows must be taken in account, and the methodology adopted by the researcher to estimate the parameters can also change the results. Also, the inventory data regarding other processes required to produce or build other parts of the system will impact in the final results.

9 Important Parameters for Biofuels LCA

Some important parameters given as result for biofuels LCA are: fuel energy ratio (FER), GHG balance, and LUC.

Also called NEG, the FER is defined as the ratio between the energy obtained by burning a fuel obtained in a process and the required energy in the said process. FER of ethanol and other biofuels are usually compared to that of fossil fuels.

Greenhouse gases (GHG) are responsible for greenhouse effect. These gases are capable to absorb and emit radiation in the infrared range. GHG balance in a process is the net difference between generated and consumed GHG during the said process.

Land use, LUC, and forestry (LULUCF) is defined by the United Nations Climate Change Secretariat as "A greenhouse gas inventory sector that covers emissions and removals of GHG resulting from direct human-induced land use, LUC, and forestry activities."

10 Biofuels Versus Fossil Fuels

In 1998, US Department of Agriculture and US Department of Energy performed a complete study to compare the Life Cycles of biofuels with fossil fuels. The results shown the use of B100 (100 % biodiesel) in substitution of fossil fuels for urban buses reduces in 95 % the consumption of petroleum in its life cycle. Concerning energetic efficiency, the production process of biodiesel produces 3.2 units of energy for every unit of fossil fuel consumed during its life cycle, while this number drops to 0.83 for conventional diesel. The most controversial result found out in this study is the net reduction of around 78 % of CO_2 emission in the biofuel life cycle when compared to the fossil fuel life cycle. Some authors have criticized the model adopted by the agencies, concerning Indirect LUCs.

The impacts caused by production of biofuels (or fuels in general) can be classified as direct or indirect. Direct impacts are related to emissions in each step of the production process, such as the machinery used for harvesting, growing of the feedstock, transport of the feedstock, final use of the fuel. Indirect impacts are related to consequences caused by land substitution, for example: changes in livestock costs and numbers and changes between pasture, forest, and cropland. Several debates have taken place concerning this topic. For example, while some authors (Searchinger et al. 2008) argue that the production of biofuels from soybean is not sustainable, almost duplicating GHG emissions over a period of 30 years, others (Kim et al. 2009) argue that is difficult to establish indirect LUC real impacts.

Humpeno et al. (2013) selected EU as scenario to evaluate the impacts in terms of GHG emission for first generation biofuels. They demonstrated that there is a 50 % reduction in emissions when compared to fossil fuels if LUC is not considered—in this scenario, the target of 35 % reduction established by EU laws would be achieved. Although, in the most realistic scenario considering LUC impacts, the reduction range was found to be between 2 and 13 %, which means the targets would not be fully achieved, and with a risk of even enhancing GHG emissions.

11 Study Cases

11.1 Biohydrogen

Biohydrogen has a very large energy content and generates water when it is burnt—the ideal biofuel. However, the bioenergetics of its synthesis is not that favorable, and the theoretical amount of biohydrogen that can be produced by anaerobic digestion is 4 mol/mol of glucose:

$$C_6H_{12}O_6 + 2H_2O \rightarrow 4H_2 + 2C_2H_4O_2 + 2CO_2$$

The acetic acid formed can be recovered or converted into methane in a second stage:

$$2C_2H_4O_2 \rightarrow 2CH_4 + 2CO_2$$

If only hydrogen is generated as biofuel, then less than 40 % of the energy of the feedstock (if it is carbohydrates) can be recovered. If methane is also produced, then the overall theoretical energy recovery raises to 96 %. This explains why the technology is being actively researched nowadays.

More hydrogen can be produced if an extra input of energy is used, and that is what Manish and Banerjee proposed in an article from 2008. They evaluated the GHG emission and the energy efficiency of producing biohydrogen from sugarcane, evaluating the efficiency of a dark fermentation, a photofermentation, and a two-stage process (Table 1).

Table 1 GHG emissions and energy efficiency in the production of biohydrogen (also compared with steam reforming). The functional unit is 1 MJ of hydrogen thermal energy

Process	Without by-products				With by-products		
	GHG g CO_2	Nonrenewable energy use (MJ)	Energy efficiency (%)	Net energy ratio	GHG g CO_2	Nonrenewable energy use (MJ)	Energy efficiency (%)
Steam methane reforming	89.5	1.31	64	0.64	89.5	1.31	64
Dark fermentation	38.5	0.43	9.6	1.9	−608.4	−7.41	89.1
Photofermentation	24.5	0.28	25.6	3	−153.1	−1.73	82.3
Two-stage process	23.8	0.27	27.2	3.1	−136.4	−1.53	81.6
Biocatalyzed electrolysis	37.1	0.45	25.7	1.8	−122.4	−1.26	76.8

Adapted from Manish and Banerjee (2008)

Table 2 Fuel life cycle energy and CO_2 WTT, TTW, and WTW results for selected biofuels

Technology—fuel	WTT, MJ/MJ fuel	WTT, gCO_2/MJ fuel	TTW, MJ/km	TTW, gCO_2/km	WTW, MJ/km	WTW, gCO_2/km
ICEV-diesel	0.16	14.00	1.67	124.40	0.27	147.78
ICEV-B10-sunflower	0.22	16.50	1.63	110.7	0.36	137.60
ICEV-B10-microalgae	0.22	16.10	1.63	110.7	0.36	136.94
FC-HEV-hydrogen centralized reforming	0.72	99.10	1.08	0	0.78	107.03
FC-HEV-hydrogen electrolysis (on-site)	3.62	184.21	1.08	0	3.91	198.95
FC-HEV-hydrogen potato peels	0.34	11.72	1.08	0	0.36	12.66

Adapted from Ferreira et al. (2010)
WTT Wheel to Tank, *TTW* Tank to Wheel, *WTW* Wheel to Wheel, *ICEV* Intern Combustion Engine Vehicle, *FC-HEV* Fuel Cell–Hydrogen Electric Vehicle

The authors found that, compared to steam methane reforming, the energy efficiencies of biohydrogen production are lower, but the GHG emission is also lower, especially for dark fermentation. They also found that the overall process efficiency is enhanced by the use of the coproducts, showing that this may be the key for sustainable production: the integration in biorefineries.

A study by Ferreira et al. (2010) compares several biofuels in a WTT LCA (as shown in Table 2), where it is shown that biohydrogen (from potato waste) compares favorably with sunflower biodiesel: the consumption of energy is half that required for hydrogen production by reforming methane (natural gas), with 11.72 g of CO_2/MJ of fuel (see Table 2—LCA made with SimaPRO7.1). The authors also determined that most of the energy consumption occurs during the pretreatment and fermentation, and GHG emissions are higher in farming, pretreatment, and treatment—while distribution and compression generate less than 5 % of the emissions.

11.2 Biomethane

Biomethane is produced from biogas, which has around 40–50 % of CO_2 and 50–60 % of methane. Biogas has a long history of use. Produced naturally in anaerobic digestion plants, its use has been growing steadily in cities and in farms (from manure digestion), and the LCA of this biofuel has been done by several researchers. The process occurs in two steps with multiple microorganisms hydrolyzing and converting the organic matter into hydrogen, several organic acids, and ammonia, and other microorganisms reusing the hydrogen and splitting the organic acids into methane and CO_2. The overall reaction is:

Table 3 GHG emissions and yield for selected feedstocks, for small-scale biogas production

	GHG, g/MJ biogas	DM content, %	Yield (MJ/t residue)	Raw energy input (MJ/MJ biogas)
Cattle manure	17.1	8	469	–
Straw	16.0	86	5376	–
Corn silage	9.3	35	3763	–
Municipal solid wastes	15.1	40	2649	0.389
Grease separator sludge	4.6	5	1098	–

Adapted from data of Poeschl et al. (2012)

$$C_6H_{12}O_6 \rightarrow 3CH_4 + 3CO_2$$

From that reaction, it seems that the usual proportion of methane to CO_2 in the gas is 50:50; however, substrates with a relatively lower content of oxygen, such as proteins and lipids, give more methane.

Biomethane production is widespread, because its production is a way of using the energy in organic wastes that must be disposed or treated anyway—both industrial wastes and municipal solid wastes and wastewater. However, the benefits of producing methane depend on the integration of the process. If the only objective is the production of energy, it is easy to develop a system that economically produces gas, with a secondary treatment. However, the process has also its impacts, and the complete treatment of the residue usually requires large area or energy-consuming secondary effluent treatment, which makes the overall process highly dependent on the system integration and overall systems size, as shown in Table 3, with data adapted from Poeschl et al. (2012).

11.3 Bioethanol

Wang et al. (2014) performed LCA and GHG emission balances for two scenarios in Brazil: (1) Conventional production of ethanol from sugarcane and (2) Coupled first and second generation ethanol production from sugarcane and sugarcane bagasse. The main conclusion concerning economy is: conventional ethanol production is more favorable then combined conventional plus lignocellulosic ethanol. However, if technological improvements in lignocellulosic ethanol are performed, the production cost would be compared to conventional production, reaching 0.31U $/L in a prospective scenario for 2020. About GHG savings, the saves range from 86 to 150 % and from 80 to 95 % for scenarios 1 and 2 respectively, also in a prospective scenario for 2020.

Luo et al. (2009) evaluated two scenarios involving bioethanol production: (1) Base case—coproduction of sugar and ethanol from sugarcane, and electricity from bagasse; (2) Future scenario—coproducing sugar and ethanol from sugarcane and bagasse, and electricity from wastes of the process (filter cake from the step of juice clarification and biomass generated during the fermentation steps). The results indicated GHG emissions for both scenarios are lower when compared to gasoline (respectively 87 and 24 %). In terms of economicity, the future scenario is the best one (0.26 U$/kg), followed by the base case (0.30 U$/kg), and both being better than gasoline (0.50 U$/kg). However, for other environmental indicators, such as ecotoxicity, acidification and eutrophization, both base case and future scenarios are worse for the environment when compared to the fossil fuel.

Wiloso et al. (2012) corroborate the results shown above. The authors conclude, based on an extent literature review, that 2nd bioethanol has two very important indicators in its favor: net energy output and reduction in GHG emissions. About other indicators, such as acidification and eutrophization, the authors call attention to the fact that the results are not conclusive, and depend on the assumptions made by the researchers for each case, and on the methodologies adopted.

11.4 Biodiesel

Biodiesel, the second most important biofuel, a mixture of fatty acid esters. It is usually derived from vegetable oils (soybean, rapeseed, Jatropha, palm) which are transesterified with methanol:

$$\begin{aligned} &R_1COO(R_2COO)CHCH_2R_3COO + 3CH_3OH \rightarrow R_1COOCH_3 \\ &\quad + R_2COOCH_3 + R_3COOCH_3 + HOCH_2CH(OH)CH_2OH \end{aligned}$$

The source of fat can also be animal or microbial, and the alcohol can be ethanol or even a higher alcohol. The radicals in oils and fats are usually different, and the source of the oil is paramount for defining the quality of the derived biofuel and the sustainability of the production. More specifically, biodiesel produced using residual fats or oils (fat from abattoirs or reclaimed cooking oil) may be partially hydrolyzed and/or oxidized, and will have a more difficult and costly processing. At the other side, oils produced from vegetable oils will compete with the use as food (e.g., soybean or rapeseed) or base chemicals, (e.g., palm). The technology for production is quite straightforward, with several companies producing catalyzers for transesterification.

After bioethanol, biodiesel is the most important biofuel, with a world production around 19.7 Mt/year. (WorldWatch Institute 2015) The popularity is in part due to the fact that biodiesel can burn efficiently in diesel engines, without further adaptation, and that the production process involves only the modification of oils in relatively compact industrial plants, as compared to fermentation based plants. Finally, various countries have governmental policies and incentives for substitution of part of the diesel for biodiesel, as a strategy for immediate reduction of emissions and incentive for the development of the industry. Therefore, it is no wonder that there are several studies regarding biodiesel LCA. A study from 1998 from USA-DOE (Sheenan et al. 1998) showed that soybean biodiesel produced in USA (one of the top producers of soybeans in the world) and used as a fuel in buses needs and input of 0.2318 MJ/MJ of biodiesel energy produced, which is comparable to the 0.2009 MJ/MJ of petroleum diesel needed. In this wheel-to-wheel LCA, soybean oil agriculture, crushing and conversion correspond to 95 % of the energy demand, while transportation (soybeans, oil and diesel) correspond to 5 %. However, in terms of emissions, biodiesel generates 18.2 g fossil CO_2/MJ, much less than the 84.4 g fossil CO_2/MJ diesel. The overall production of CO_2 is similar for both biofuels, and the production of SO_x and CO are smaller for biodiesel (by 8 and 34 %), but that of NO_x is higher (by 13.4 %), as compared to diesel.

In a more recent evaluation of biodiesel, Achten et al. (2010) analyzed *Jatropa* biodiesel for use in cars in India. They estimated that the energy input is around 0.222 MJ/MJ of biodiesel energy, and a production of 123gCO_2 equivalent/MJ, 85 % of which due to N_2O emissions in the plantation. The authors postulate that the integration of the process with production of biogas can enhance its energy efficiency.

11.5 Microalgae

Microalgae are photosynthetic microorganisms widely regarded as a possible source of lipids for biodiesel production. The lipid content of this rather heterogeneous group of organisms ranges usually from 10 to 40 %, with some authors reporting productions reaching up to 70 % of lipids. However, only recently a few production plants have produced enough microalgal biodiesel for a good evaluation of the life cycle of the biofuel. On the bright side, these microorganisms have very large productivities compared to land plants: an annualized production of 50–70t/ha, which even with a low content of lipids, far exceeds the production of oleaginous crops such as palm. However, microalgae are produced in very low concentrations (around 1–5 g/L), and the recovery of the biomass and the extraction of the oil are energy-demanding steps. Clarens et al. (2010) evaluated the production of microalgae for biofuels and found that the energy demand for production of 1 MJ of microalgal energy is 0.946 MJ, a result four times worse than canola and 10 times worse than switchgrass. However, the CO_2 emissions are around 6 g/MJ, not considering the CO_2 production in burning the biomass. The eutrophication was lesser than canola, corn or switchgrass. The authors went on evaluating the influence of factors such as irradiance and the use of nutrients from wastewaters, finding that this is indeed the way to go: effluents from activated sludge systems gave an energy input of 0.008 MJ/MJ algal energy, with emissions of 3.7 gCO_2/MJ. With source-separated urine, which provides high concentrations of N and P, the overall process becomes even more attractive. Lardon et al. (2009) also concluded that the balance for energy production using microalgae is not favorable, unless high-lipid microalgae are produced continuously in low N conditions, and that downstream must be efficient. They also state, and we agree, that LCA is a good tool for indication of the bottlenecks and challenges of this relatively new technology.

11.6 Solid Biofuels

Wood is presumably the oldest fuel, and yet it remains an important source of energy, especially for industrial processes that are conducted in regions where wood is produced for pulping or production of lumber. Industrial processes also burn other solid biomasses, such as wheat straw or sugarcane bagasse. These are fuels impractical for use as vehicle fuel, although they are important in industry.

The production of 1 MJ of heat energy using forest wood causes the emission of 4.2 g of CO_2 and overall 6 g of GHG, while wheat straw contributes with 4.3 and 8.0 g, respectively. Both fuels have a low energy demand around 0.06 MJ/MJ of heat energy (data from Pehnt 2006). The difference in emissions is due to the use of fertilizers in the intensive production of wheat, and the higher sulfur content of the residual straw—giving higher NO_X and SO_X emissions.

12 Conclusions

It is clear that LCA is a powerful tool to show the benefits and difficulties involved in biofuel production. For example, a 10 % mix of biodiesel and diesel has lower GHG emissions that pure diesel; however, the use of microalgae for biodiesel production can be only marginally better than the use of sunflower oil. At the other side, a technology that may seem hard to defend such as methane reforming may prove adequate because it compares favorably with fossil fuels while having a zero emission of CO_2 in cities. This prompts us to evaluate how CO_2 can, for example, be captured on-site, making the process more attractive. In all cases, it seems that the integration of residue processing in biorefineries is the way to reduce emissions and increase overall efficiency.

References

ABNT—Associação Brasileira de Normas Técnicas (2001) *NBR ISO* 14040: Gestão ambiental—Avaliação do Ciclo de Vida—Princípios e Estrutura. Rio de Janeiro, 10 p

Achten WMJ, Almeida J, Fobelets V, Bolle E, Mathijs E, Singh VP, Tewari DN, Verchot LV, Muys B (2010) Life cycle assessment of jatropha biodiesel as transportation fuel in rural India. Appl Energ 87:3652–3660

Caldeira-Pires A, Souza-Paula MC, Villas Bôas RC (2005) A Avaliação do Ciclo de Vida: a ISO 14040 na América Latina. Abipti, Brasília, p 337

Clarens AF, Ressurecion EP, White MA, Colosi LM (2010) Environmental life cycle comparison of algae to other bioenergy feedstocks. Environ Sci Technol 44:1813–1819

Davis SC, Anderson-Teixeira KJ, DeLucia EH (2009) Life cycle analysis and the ecology of biofuels. Trends Plant Sci 14(3):140–146

Ferreira AF, Baptista P, Silva C, Stolten D, Grube T (2010) Analysis of energy Consumption and CO_2 emissions of the life cycle of biohydrogen applied to the portuguese road transportation sector. In: Proceedings of the world hydrogen energy conference. Essen, Germany, pp 16–21

Humpeno F, Schaldach R, Cikovani Y, Schebek L (2013) Effects of land-use change on the carbon balance of 1st generation biofuels: an analysis for the european union combining spatial modeling and LCA. Biomass Bioenerg 56:166–178

Hunt R, Franklin E (1996) LCA—how it came about. Personal reflections on the origin and the development of LCA in the USA. Int J Life Cycle Ass 1(1)

ISO 14044 (2006) Environmental management and life cycle assessment and requirements and guidelines

Kim H, Kim S, Dale BE (2009) Biofuels, land use change, and greenhose gas emissions: some unexplored variables. Environ Sci Technol 43(3):961–967

Lardon L, Hélias A, Sialve B, Steyer JP, Bernard O (2009) Life-cycle assessment of biodiesel production from microalgae. Environ Sci Technol 43(17):6475–6481

Lemos HM, Barros RLP (2006) Ciclo de Vida dos Produtos: Certificação e Rotulagem Ambiental. Rio de Janeiro: Comitê Brasileiro das Nações Unidas para o Meio ambiente, SEBRAE

Lima AMF, Caldeira-Pires A, Kiperstok A (2007) Evolução dos Trabalhos de Avaliação do Ciclo de Vida nas Instituições Acadêmicas Brasileiras. In Conferência Internacional Ciclo de Vida, 2. São Paulo

Luo L, Van der Voet E, Huppes G (2009) Life cycle assessment and life cycle costing of bioethanol form sugarcane in Brazil. Renew Sust Energ Rev 13:1613–1619

Manish S, Banerjee R (2008) Comparison of biohydrogen production processes. Int J Hydrogen Energ 33(1):279–286

Menten F, Chèze B, Petouillard L, Bouvart F (2013) A review of LCA green house gas emissions results for advanced biofuels: the use of meta-regression analysis. Renew Sust Energ Rev 26:108–134

Pehnt M (2006) Dynamic life cycle assessment (LCA) of renewable energy technologies. Renew Energ 31:55–71

Poeschl M, Ward S, Owende P (2012) Environmental impacts of gas deployment–Part II: life cycle assessment of multiple production and utilization pathways. J Cleaner Prod 24:184–201

Robles Junior A, Bonelli VV (2006) Gestão da Qualidade e do Meio ambiente: Enfoque Econômico, Financeiro e Patrimonial. Atlas, São Paulo

Searchinger T, Heimlich R, Houghton RA, Dong FX, Elobeid A, Fabiosa J, Tokgoz S, Hayes D, Yu TH (2008) Use of US croplands for biofuels increase greenhouse gases through emissions from land-use change. Science 319(5867):1238–1240

Sheehan J, Camobreco V, Duffield J, Graboski M, Shapouri H (1998) An overview of biodiesel and petroleum diesel life cycles. USA-DOE

Wang L, Quiceno R, Price C, Malpas R, Woods J (2014) Economic and GHG emissions analyses for sugarcane ethanol in Brazil: looking forward. Renew Sust Energ Rev 40:571–582

Wiloso I, Heijungs R, de Snoo GR (2012) LCA of second generation bioethanol: a review and some issues to be resolved for good LCA practice. Renew Sust Energ 16:5295–5308

Worldwatch Institute (2015) Biofuels poised to displace oil. Washington, Nov 2015

Patents on Biofuels

Eduardo Scopel Ferreira da Costa, Flavia Thomaz Soccol and Carlos Ricardo Soccol

Abstract Global energy matrix is mainly based on fossil fuels. As the amount of non-renewable fuels diminishes, a need for alternative sources of fuels and ways to make them feasible arise, in an effort to turn the global energy matrix in the next decades more diverse. Besides, the unbridled use of these fuels may harm the environment and cause several problems, such as air pollution, and unexpected changes in entire ecosystems. Therefore, renewable fuels come in handy to solve those problems. Biofuels, such as biodiesel, ethanol, are fuels derived somehow from biomass. Even though biofuels production have been increasing consistently, it is only feasible due to incentives from governmental policies. So, in order for them to compete with other forms of energy, such as oil-derived fuels, there is a continuous need for innovation. One way to measure innovation in any area, is to assess information from patent applications. The present review consists in assessing biofuels (biodiesel and ethanol) patent information from its main drivers in the past decade.

Keywords Patents · Biofuels · Bioethanol · Biodiesel · Butanol · Biogas · Innovation

1 Introduction

This chapter compares patent applications in biofuel technologies focused priority in the biodiesel and ethanol, in the last 10 years (from 2005 to 2015). Patent applications statistics are used as indicators for technological development, as the main objective of a patent is to protect a new technology. The chapter covers every aspect of the production process of biofuels, focusing on biotechnological aspects and its most recent trends.

E.S. Ferreira da Costa · F.T. Soccol · C.R. Soccol (✉)
Department of Bioprocesses Engineering and Biotechnology (DEBB),
Federal University of Paraná (UFPR), Curitiba, Brazil
e-mail: soccol@ufpr.br

C.R. Soccol et al. (eds.), *Green Fuels Technology*,
Green Energy and Technology, DOI 10.1007/978-3-319-30205-8_21

Alongside this review, it will be possible to identify, by the analysis of patent applications made by inquiry in research institutes and companies, on their national patent offices

- Evolution of the technological development in the most relevant countries in the area;
- Technological routes that have been receiving more attention;
- Most important drivers.

2 Biofuels

Fossil fuels, mainly oil-based, form the basis of the global energy matrix—especially for transportation uses and heat generation. However, with the amount of available oil decreasing, volatility of prices, political instability at producing regions and environmental problems caused by its unsustainable usage, demand for alternative fuels has grown, in order to diversify this matrix. In this context, several kinds of alternative fuels have emerged, among which, biofuels stand out. Biofuels are renewable fuels derived from biomass.

It is highly likely that until 2050, bioenergy will respond for almost one third of the world's energy demand, due to the influence of present renewable energy and climate change policies. Just to meet the goal established by the World Energy Council for carbon emission reduction, the consumption of bioenergy would have to triplicate. Biofuels, such as biodiesel and bioethanol are going to be very relevant in this scenario (World Energy Council 2013; Guo et al. 2015).

Even with the recent increase in global production capacity of oil, natural gas and coal—the main sources of non-renewable energy—their growth rate did not keep pace with global consumption. Biofuels global production grew more than 6.0 % in 2013, reaching over 65 million tons oil-equivalent, representing 2.7 % of global energy consumption, from which United States (43.5 %) and Brazil (24.2 %) accounted for 67.7 % (British Petroleum 2015).

In 2006, biofuels represented about 1.8 % of the fuel used for transportation globally, mainly due to its use in European countries and developing countries, such as Brazil. Ethanol had its production tripled from 2000 to 2007 reaching 52 billion liters, while biodiesel production expanded from 1 to about 11 billion liters, on the same period. This growth was motivated by the increase of the ethanol amending in gasoline fuel and of biodiesel in diesel. USA (ethanol from corn), Brazil (ethanol from sugar cane) and the European Union (biodiesel from rapeseed) lead the production of transport biofuels. China is also increasing its production of biofuels (Demirbas 2009; UNEP 2009).

In 2010, investments in renewable energy reached US$211 billion, 32 % more than 2009. Developing country investments in biofuel projects exceeded those of developed countries. Almost 3 % of global road transport fuels in 2010 were liquid biofuels. Regarding ethanol, Brazil and USA together represented 88 % of the global production, being the USA the world's leading ethanol exporter (REN21 2011; UNEP 2011).

So, it is safe to assume that those investments sponsor a lot of technological development, not only in universities and government institutions, but also in energy companies. With increased technological development new inventions will certainly arise, and may increment biofuels production, quality, and efficiency. As the best available way for creators to protect their inventions is the patent system, studying these patents are a good way to assess technological development in any area, including bioenergy and biofuels.

3 Patent Statistics

Patents are a relevant source of information about any kind of technology. Its content comprises important information about its novelty, inventive step processes, location of emerging technologies, inventive networks, emerging technologies, state of art problems, most contributing inventors, among others. When combined with complementary data, they provide basis for a broad analysis of various dimensions of innovations, as the role of intellectual property of economic performance, entrepreneurship and on search for links in science and technological development. As many others, these indicators also have their pros and cons, and reflect several stages of the innovation process (OECD 2009; WTO 1994).

According to WTO (1994), for a patent to be granted 4 main requirements must be taken into consideration: novelty, inventiveness, industrial application and sufficient descriptiveness. Because of this, most of patent documents will have the newest information regarding a specific technology, in order to fulfill the novelty and inventiveness requirements. The industrial application requirement leads us to the conclusion that most patents have relevant information to be used in industries, regardless if they are granted, because, either way, that information will remain in public domain, aiding someone to solve a technical problem the invention was supposed to solve. Also, a patent document must have sufficient and complete information, in order for a person skilled in the art to be able to carry out the invention properly. It is in this concept that we rely to identify the importance of patent documents as source of technological information.

From 2002 and 2008, almost 3,000 biofuel-related patents were published, with considerable growth in the last two years of the referred period. In 2007, the number of biofuel patents (1045) was more than the combined total of solar power (555) and wind power (282). Assuming that biofuel, solar power, and wind power are the leading renewable energy technologies, then, in 2007, biofuel patents, more specifically biodiesel, clearly dominate renewable energy in terms of new technologies development. It is expected that legislation directed to climate change will continue influencing biofuel patents. In 2022, at least 16 billion gallons of USA transportation fuel will have to be cellulosic biofuel. Also, 21 billion gallons of

USA transportation fuel will have to be derived from sources different than traditional ethanol fuel (Kamis and Joshi 2008).

4 Ethanol

The potential of ethanol as alternative fuel has raised worldwide attention. Among the advantages of using ethanol blended with gasoline or alone, are lower quantities of emissions of carbon monoxide, nitrogen oxides and reactive hydrocarbons, after combustion. Even the blended one reduces those emissions, since ethanol acts as an oxidizing agent (Takahashi et al. 2000; Badger 2002; Mielenz 2001).

The production process of ethanol from sugar cane begins with the pretreatment of the raw materials. This process consists of the removal of organic material and shredding the cane into smaller pieces. After this process, the feedstock is fed and milled to extract juice and separate it from the cane bagasse, which is a fiber residue of the process. A great part of the bagasse is composed of cellulosic material, which can be used to produce second generation ethanol. It is also used as feedstock for boilers, to produce low-cost energy, which supply electricity and steam for the process. The milling process aims, mainly, to extract sucrose from the cane. The sugarcane juice is, then, filtered, chemically treated and filtered and pasteurized. The next step is evaporation of the juice, in order to increase its concentration of sugar. Following this step, we reach the crystallization of the syrup, by boiling or cooling, what leads to a mixture of clear crystals with molasses, with high sugar concentrations. Molasses are removed by centrifugation, after which is pretreated with pasteurization and addition of lime. This process will sterilize the molasse, leaving it ready to be fermented. The sugars from molasses or sugarcane juice are then fermented into ethanol by yeasts, during the fermentation step, which can vary from 4 to 12 h. The resulting wine is centrifuged, to recover the yeast. Then, the alcohol in the fermented wine is distillated, using different boiling points, to separate it from other materials. The obtained product is hydrated ethanol (Smeets et al. 2006; Kumar et al. 2009).

To produce second generation ethanol, using cellulosic material as raw material, it is necessary to hydrolyze cellulosic material, especially cellulose, to glucose. This is usually performed with acid, enzymatic or microbial hydrolysis. However, there are several difficulties to achieve success in those processes. One of those comes from the nature and function of cellulose. This molecule is known to act as a wall to protect terrestrial plant cell wall from intruders and degradation. In order to achieve that, cellulose has strong molecular hydrogen bonds, that difficult enzymatic hydrolysis. Also, the coating of crystalline cellulose by hemicellulose makes it difficult for enzymes to perform hydrolysis. In addition to that, there is a need for lignin, another molecule that is bound to cellulose in its crystalized form, to be removed so that hydrolysis can be performed. Because of all that, there is an utter need for pretreatment of cellulosic material that is not always easily performed. So, there is room for innovation in the production of second generation ethanol (Viikari et al. 2012).

In the Web of Knowledge Derwent Innovation Index, almost 9,000 patent applications were found when searched for ethanol related keywords and International Patent Classification (IPC), since 2005. With this total, some important statistics were established, such as history of applications, most significant countries and applicants, and the fields where most of these applications have been made. This review will also describe some of the most cited patents on each area.

4.1 Ethanol Patents History

In Fig. 1 it is possible to see the trend in ethanol-related patent applications.

The first patent applications boom dates back to the late 1970s, when the first public incentives to ethanol producing companies were created—such as tax incentives and obligation of blending ethanol into gasoline fuel—mainly in Brazil (*Pro-Álcool* Program) and USA (Energy Tax Act of 1978), as a response to the 1973 oil crisis. Those policies were modified several times throughout the 1980s and the 1990s, but in USA they were removed after the oil barrel prices went down in 1991, causing the bankruptcy of several ethanol industries there (Sissine 2008; Bastos 2007). This may be one of the causes for the low number of patent applications during the 1990s decade.

The second patent boom occurred after 2003, when the first flexible-fuel car, which runs with either gasoline or ethanol, was launched in Brazil. As that technology became available, ethanol use as the main fuel for vehicles in Brazil grew

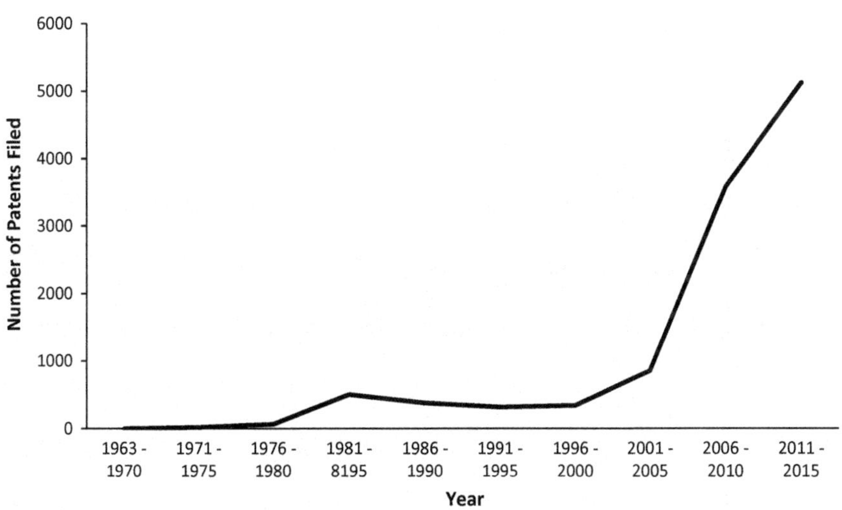

Fig. 1 Evolution of patent applications related to ethanol in the world, from 1963 to 2015

rapidly. Brazil's flexible-fuel car percentage from the total light vehicles went from 2.9 in 2003 to 83.4 in 2012 (Anfavea 2012). At the same time, crude oil prices rose rapidly, forcing USA to become more aware of their dependability of foreign countries on that matter (RFA 2014). It is evident the growth in the number of applications after 2000, with the increase in ethanol use for transportation (threefold from 2000 to 2007) (UNEP 2009). After 2011, it is possible to see a reduction in the growth rate of application numbers, which can relate to a decrease in ethanol production since 2011, mainly in US (RFA 2014). Also, this data can be explained by the development of new techniques to obtain crude oil, which increased oil production in the world and, consequently, brought down its prices.

4.2 *Most Relevant Drivers*

In Fig. 2 is shown the countries that filed more patent applications related to ethanol, since 2005, via Patent Cooperation Treaty (PCT), a sort of global patent. Those numbers relate only to patent applications filed through PCT and not in the residence countries.

As expected, USA, the biggest producer and exporter of ethanol, is the country that filed most patents in the area in any analyzed period, followed by Europe, China and Canada. One reason to explain why Brazil, the second biggest ethanol producer, is not on this list is that patents filed in the residence countries of the applicant were not considered. Brazilian companies and universities are not used to protect their inventions outside their boarders, which may hinder Brazil's effort in technological development of ethanol. Another factor that can be considered is that Brazilian ethanol is derived from sugarcane, which is the raw material with most concentration of sucrose, being considerably easier to transform to ethanol, when compared to corn (USA) or rapeseed (Europe).

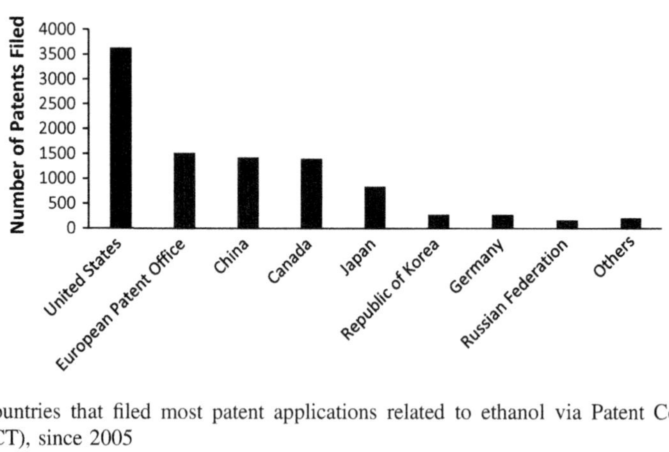

Fig. 2 Countries that filed most patent applications related to ethanol via Patent Cooperation Treaty (PCT), since 2005

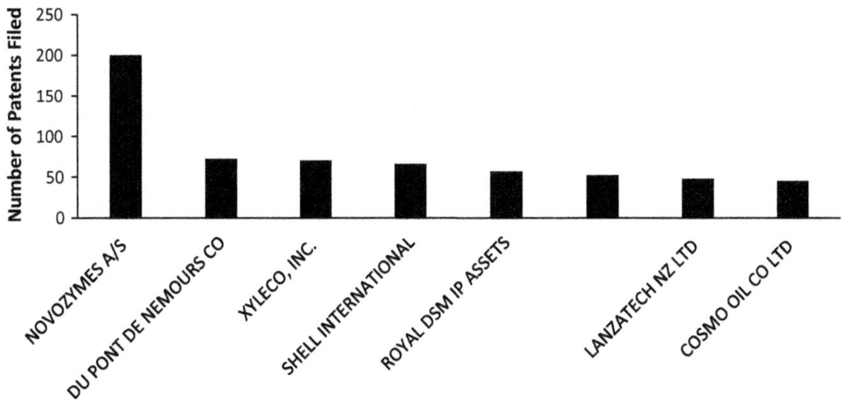

Fig. 3 Companies that filed most patent applications related to ethanol via Patent Cooperation Treaty (PCT), since 2005

The eight companies that filed most patent applications via PCT since 2005 are shown in Fig. 3.

Novozymes, a Denmark-based company, is the most relevant driver in the world, regarding ethanol technological development. It is a biotechnology company that has a strong global leadership in enzyme production. This is an interesting data, as one of the most—if not the most—researched topics regarding ethanol is the pretreatment of cellulosic material with enzymes, such as cellulases, which is one of the main products produced by Novozymes.

4.3 *Most Cited Patents*

In order to assess the technologies that have been receiving most attention in the past 10 years, the most cited patents regarding ethanol production were reviewed.

Schreck and Balmer (2009a) describe an integrated method to produce ethanol and hydrogen, the first by indirect fermentation and the second by pyrolysis. Their main goal is to perform homoacidogenic fermentation, using acidogenic microorganisms, with an aqueous solution solvent, comprising carbohydrates, to obtain organic acid. After that, the obtained organic acid is esterified with an esterifying alcohol, with longer carbon chain than the acid, in order to provide an ester that will be, thereafter, hydrogenated, obtaining a solution containing ethanol. This solution is then distilled into a low boiling fraction, containing ethanol, and a high boiling fraction, containing ester and esterifying alcohol, which is recycled. They also protect pyrolyzis of the biomass feed, in order to obtain gases, containing hydrogen. This patent application is in accordance with the concept of biorefinery. In a second similar invention, the same authors (Schreck and Balmer 2009b) claim a process for the separation steps from the previous invention, in each part of the process. The main methods used for separation is liquid–liquid extraction, for separating the organic phases of each solution.

The process patent filed by Czartoski et al. (2011) consists in the fractionation of microalgae biomass into three parts—organic, aqueous and insoluble. In the organic fraction, biocrude oil, containing lipids, hydrocarbons and semi volatile organic compounds, is obtained; in the aqueous fraction sugars, organic acids, protein products, glycerol, ethanol, phosphate, and other compounds are present; from the insoluble fraction can be found microalgae debris, insoluble sugars and proteins, among other compounds. The biomass fractioning process consists in cell lysis by addition of an acidic solution, to reduce the solution pH to less than 5, and heat. Cell lysate is then subjected to a treatment with a polar (e.g., alcohol) and a non-polar solvent (e.g., hexane), generating an organic and a water-soluble phase. The organic phase contains lipid components, while the water-soluble contains cell debris, carbohydrates and proteins. The organic phase is distilled with a solvent, separating the biocrude oil from other lipids, which is then transesterified, with alcohol and a catalyst. After that the oil phase is separated, resulting in glycerol, hydrocarbons, biodiesel and high value lipids. The water-soluble phase is subjected to an enzymatic hydrolysis, followed by a yeast fermentation to produce ethanol and protein products (from yeast debris). Every residue from these steps can be reused in the beginning of the process, finishing the concept of a biorefinery.

Verser and Eggeman (2008) patent application describes a process that comprises the reaction of biomass pyrolysis, such as syngas (hydrogen, carbon monoxide, carbon dioxide or methanol), with a fermentation product, such as carboxylic acid. The fermentation process consists in the conversion of a carbohydrate containing substrate into carboxylic acid by at least one microorganism, such as *Clostridium formicoaceticum, Clostridium butyricum, Moorella thermoacetica, Thermoanaerobacter kivui, Lactobacillus delbrukii, Propionibacterium acidipropionici, Propionispera arboris, Anaerobiospirillum succinicproducens, Bacteriodes amylophilus* or *Bacteriodes ruminicola*.

A process for the fermentation of pyrolysis products, such as syngas, by a claimed recombinant microorganism, *Clostridium ragsdalei*, to produce ethanol is claimed by Huhnke et al. (2008).

The patent application filed by Geir et al. (2009) claims a process for removal of water from a water/ethanol solution, comprising several steps, such as evaporation, distillation, compression, heat exchange and vapor permeation or molecular sieve. The novelty of this patent resides in the vapor permeation or molecular sieve step.

Harris et al. (2005) provide a method for the saccharification of a cellulosic material with recombinant enzymes with cellulolytic activity, supposed to enhance the cellulolytic activity. The invention comprises (a) a process for hydrolysis of cellulosic material with the recombinant enzymes and; (b) a fermentative process with a microorganism capable of producing said recombinant enzymes.

In the invention protected by Vehmaanperae et al. (2007), a process for treatment of cellulosic material is described. This process uses at least one recombinant enzyme from the list comprising cellobiohydrolase, endoglucanase and beta-glugosidases, all obtained from microorganisms, such as *Thermoascus aurantiacus*, *Acremonium thermophilium* and *Chaetomium thermophilium*. The recombinant enzymes may be produced by *Trichoderma* or *Aspergillus*. These enzymes may also have xylanase activity.

Dunson et al. (2006) invented a technology where cellulosic material is treated with a saccharification enzymes consortium, in an aqueous solution containing ammonia, so that the solution pH is alkaline. They suggest that presence of ammonia at low concentration increases the effectiveness of the pretreatment, because it competes with hydrolysis via ammonolysis of acetyl esters in biomass to form acetamide, which is less toxic to the used microorganisms, *Zymomonas mobilis*, thus making the removal of acetic acid from the solution unnecessary.

Bartek et al. (2010) disclose an invention based on a process to transform wood biomass to ethanol, including the saccharification of the cellulosic material and the removal of lignin and other non-fermentable sugar from the substrate. Also, it is claimed a process to add hydrogen from the same refinery to the fermentation of biomass, in order to avoid the formation of undesirable products, such as oxygenated acidic compounds and aromatics.

Nguyen (2008) protected an invention that comprises the conversion of wood biomass into ethanol, by saccharification of biomass carbohydrates. The removal of lignin from wood is also claimed. Hydrolysis of the wood chips is performed under low pH values (from 2 to 3) and high temperatures and pressure values (from 424.15 to 573.15 K and 1034 to 6894 kPa, respectively) for up to 30 min. The heat is added through vapor injection. To avoid toxic compounds, and improve fermentation, neutralization step is performed, with high pH values (from 8 to 10) to precipitate such compounds. To remove lignin and other lignocellulosic solids, the prehydrolysate solution is washed with a solution of sodium chlorite/acetic acid anhydrous or chlorine dioxide or sodium hypochlorite, and/or acidic hydrogen peroxide in a delignification reactor. After that, the filtrate is contacted with solutions of sodium chlorite/glacial acetic acid (anhydrous). After a washing step and filtration, the lignin is then separated from the cellulosic solution. Cellulose can, then, be transformed into glucose by enzymatic hydrolysis, or directly into ethanol, with cellulose, yeast, or bacteria.

Justin et al. (2009) protect an invention claiming a nucleotide sequence encoding a protein with lignocellulosic activity, such as glycosyl hydrolase, cellulase, endocellulase, glucanase, cellobiohydrolase, beta-glucosidadse or mannanase, useful to produce ethanol from cellulosic material.

Leschine and Warnick (2008) describe a process for ethanol production from a lignocellulosic material, needless of any steps to remove lignin or other toxic compounds, being the fermentation performed by anaerobic bacteria, *Clostridium phytofermentans* that can be done in co-culture, with *Zymomonas mobilis*. The lignocellulosic material must be hydrolyzed by either acid or enzymatic way, after having its size diminished, before fermentation.

The invention proposed by Penttilae et al. (2005) describes a process for production of ethanol from cellulosic material, such as softwood. Before fermentation for ethanol production, the substrate must be pretreated with steam, generating two fractions, a solid (mainly cellulose) and a soluble one (mainly hemicellulose). The solid fraction is transformed into a fiber suspension, after filtration, and pre-hydrolyzed with cellulase and beta-glucosidase at optimal conditions, returning 20 % of the original material as glucose. The final solution is then fermented with *Saccharomyces cerevisiae* to produce ethanol. The soluble fraction is added, gradually, to the fermentation at its latter stage, improving the yield and production rate.

Galvez and Richards (2010) developed a process in which biomass, such as corn stover, is milled to a powder by a milling device, such as a hammer mill, and pretreated with a cocktail of enzymes. This cocktail must be able to convert cellulose into glucose and hemicellulose into xylose and arabinose. Before subjecting the substrate to the enzyme catalysis, the milled grains must be grinded to a uniform particle size, to provide a better exposure to the catalytic process. After the conversion of cellulose into fermentable sugars, the fermentation step begins to produce biofuels. About 95 % of the pretreated particles should have a size of 150–300 microns and the enzyme pool should consist of cellulases, xylanases and ligninases.

Kim et al. (2008) describe a method for production of ethanol from polysaccharides extracted from marine algae. The method comprises extraction of polysaccharides, such as agar, starch, carrageenan, alginic acid and fibrin, by acid hydrolysis. In the first step, agar, carrageenan and alginic acid are separated from starch and fibrin. Any of those polyssacharides may be hydrolyzed with an enzyme, such as beta-agarase, beta-galactosidase, beta-glucosidase, endo-1,4-beta-glucanase, alpha-amylase, glucoamylase and cellulase, under optimal conditions, to produce fermentable sugars, such as galactose, 3,6-anydrogalactose, glucose, fucose, rhamnose, xylose and mannose. To obtain fermentable sugars from agar, which is mainly formed of galactan, a two-step process is used, comprising both acid and enzymatic hydrolysys. The enzyme used in the enzymatic catalysys may be beta-agarose or beta-galactosidase. If the polyssacharide is fibrin, the recommended process is to use an acid and/or an enzymatic hydrolysis (any beta-glucosidase and endo-1,4-beta-glucanase may be used). When starch is the polyssacharide, the same process can be performed, but the enzymes used are preferably amylases. Also, the patent protects a method for direct fermentation of marine algae material, containing any of the aforementioned polyssacharides. In this case, two or more enzymes must be used simultaneously, depending on the algae used. If the used algae is green algae, the enzymes used must be able to convert starch and fibrin into fermentable sugars. If original marine algae are used, eliminate impurities before the hydrolysis is recommended, by a process that comprises a washing, a drying and a pulverizing steps.

Taira et al. (2005) claims a process for production of a recombinant hybrid enzyme, similar to alpha-amylase, useful for liquefying starch used to produce ethanol by fermentation with yeast.

In Abbas et al. (2006), a process for extracting, liquefying, saccharifying and fermenting starch from corn kernel into ethanol is described. The method provided in the invention consists in processing corn kernels, removing its pericarp and germ, by alkali or acid debranning, or ammonia addition to the kernel, before conditioning it with steam or hot water in high temperatures (from 323.15 to 372.15 K), and milling it.

The patent application filed by Foody et al. (2006) describes an enzymatic hydrolysis of cellulose, consisting of adding aqueous slurry, with at least 20 % of cellulose, to the bottom of a hydrolysis reactor, and with limited slurry flow velocity controlling the suspended solids upward at a rate slower than that of the liquid. After that, cellulase is added to the aqueous slurry, obtaining a product comprising glucose and cellobiose. It is claimed an enzyme composition, including cellulase enzymes and flocculants; a kit for the aforementioned process; the hydrolysis process; and a method for the preparation of the enzyme composition. The cellulase may be produced by microorganisms, of such genus as *Aspergillus*, *Humicola*, *Trichoderma*, *Bacillus* and/or *Thermohifida*. The pretreatment of this material is performed at high temperatures (433.15–533.15 K) for up to 30 min, at an alkaline pH, prior to the enzymatic hydrolysis.

A method for production of ethanol and biogas using plant-derived carbohydrate and aminoacids (leucine, isoleucine, valine, or a mixture) as biocatalysts for the yeast fermentation is described in the invention protected by Golubkov (2005). These aminoacids are supposed to increase the yield of low chain alcohols production for obtaining biogas. Also, it doubles the productivity of the fermentation of the low chain alcohols.

Those patent applications are the most cited ones, and they refer to diverse areas of technology, although it is clear that second generation ethanol has been receiving most attention, mainly technologies related to cellulosic material hydrolysis, as expected.

5 Biodiesel

The chemical reaction of a vegetable oil with an alcohol (methanol or ethanol) will result in an ester (methyl or ethyl, depending on the alcohol used), that is called biodiesel. Since ethanol already is widely produced in Brazil and USA, ethyl ester would be the most interesting option, but commercially it is easier to find methyl ester. The use of biodiesel is limited to blending with diesel oil, up to now, what already helps reducing air pollutant emissions. Using biodiesel helps to restrain the increase of the greenhouse effect, and may reach strong results when used in a large global scale. Biodiesel is not aggressive to the environment and it is a renewable source of energy. It does not carry the same drawbacks that fossil fuels do, such as high market prices and negative impact on the environment (UNEP 2009; Rovere et al. 2011).

Biodiesel has a lot of advantages, such as the capacity to reduce the carbon emissions, add more job opportunities, decrease the use of fossil fuels, and increases vegetable oil supplies, which will reduce its costs. It is one of the most environmentally friendly, from the fuels available in the market, carrying the potential to reduce up to 80 % the CO_2 life cycle (Sambodo 2008; Mekhilefa et al. 2011).

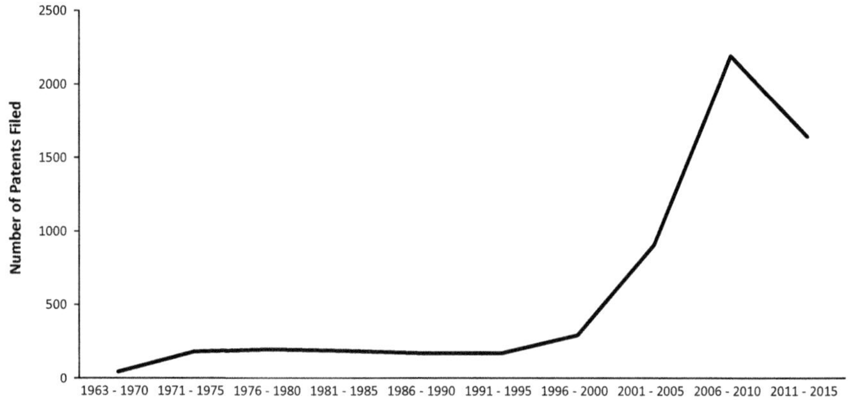

Fig. 4 Evolution of patent applications related to biodiesel in the world, from 1963 to 2015

Biodiesel current global production is of about 11 billion liters and is expected to grow up to 24 billion liters, until 2017, since a lot of countries will increase the minimum amount of blending on diesel fuels (IEA 2007; EBTP 2011).

5.1 *Biodiesel Patent History*

The Fig. 4 shows the patent application of inventions related to biodiesel history in the world.

Different from ethanol, the boom of technological development of biodiesel started on the 1990s, mainly due to the increase in biodiesel production and consumption in Europe, initially to be mixed with diesel fuel. At the end of this decade and at the beginning of the 2000s, biodiesel mixing in diesel fuel started to become mandatory in several countries, due to government regulations. From 2001 to 2006, biodiesel world production grew six times, for instance (Worldwatch Institute 2007). As its production rose, the same happened with related patent applications. The decrease that can be noted in the last 5 years may be due to the unaccounted 2015 patent applications and those applications that are still in the secrecy phase (18 months after filing date).

5.2 *Most Relevant Drivers*

In Fig. 5 is shown the countries that filed more patent applications related to biodiesel, since 2005, via Patent Cooperation Treaty (PCT). Those numbers relate only to patent applications filed through PCT and not in the residence countries.

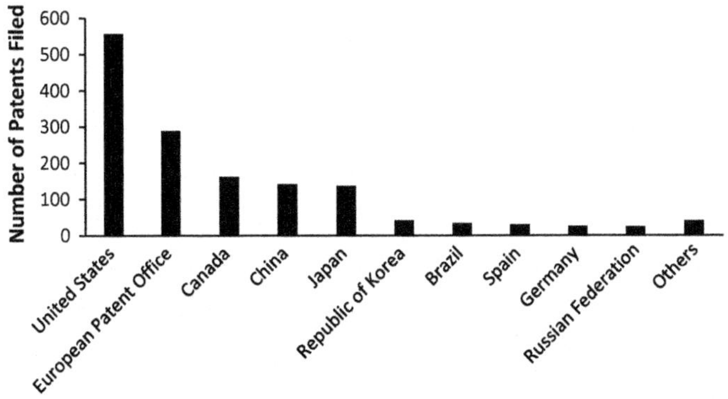

Fig. 5 Countries that filed most patent applications related to Biodiesel via Patent Cooperation Treaty (PCT), since 2005

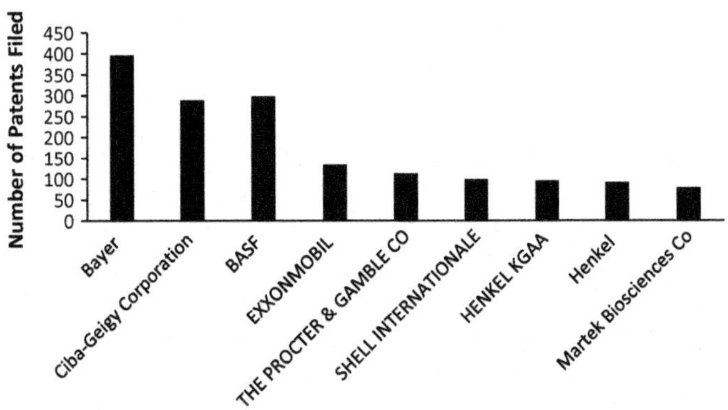

Fig. 6 Companies that filed most patent applications related to biodiesel via Patent Cooperation Treaty (PCT), since 2005

Europe, mainly via France and Germany, is the continent that produce most biodiesel, but the country that tops the ranking is United States (64 thousand barrels of oil-equivalent), followed by Germany (54.7), Argentina (47.9) and Brazil (46.7) (EIA 2012). As expected, USA also leads the ranking of patent applications in biodiesel, followed by the European Patent Office (EPO). Brazil also appears on the ranking at the 9th place.

In Fig. 6, the nine companies that filed most patent applications in the world, since 2005, are shown.

From the nine most active companies, three are based in Germany (Bayer, Basf, and Henkel) and 3 in USA (ExxonMobil, Procter & Gamble and Martek). From those companies, only two are primarily energy companies. Some companies, such as Martek, work exclusively with biosciences, and others such as Bayer, Basf, and

Procter & Gamble are groups that act in a variety of fields, including energy and fuels.

5.3 Most Cited Patents

The most cited patents regarding bodiesel production were reviewed, to make possible an assessment of the technologies are trending in the area.

Trimbur et al. (2008) present an invention comprising a method for producing biodiesel from oil accumulated in a microorganism, such as microalgae or yeast. The referred microorganism is capable of expressing a lipid pathway enzyme at high rates. These microorganisms may be recombinant, having exogenous genes, or selected from the nature. The aforementioned exogenous genes encode proteins selected from a fatty acyl–acyl carrier protein thioesterase, a fatty acyl-CoA reductase, a fatty aldehyde reductase, a fatty acyl coenzyme A, a fatty aldehyde decarbonylase, and an acyl carrier protein, lipase, sucrose transporter, sucrose invertase, fructokinase or polyssaccharide-degrading enzyme. Oil extracted from those microorganisms is then transesterified to biodiesel, by any chemical or enzymatic catalysis.

In the invention proposed by Chua et al. (2010) a process for the production of biodiesel from cells of the genus *Prototheca*, mainly from the species *Prototheca moriformis*, *Prototheca krugani*, *Prototheca stagnora* or *Prototheca zopfii* is described. The used microorganism must have a gene that encodes at least one protein of the group that comprises sucrose invertase, fatty acyl-ACP thioesterase, fatt acyl-CoA/aldehyde reductase, fatty acyl-CoA reductase, fatty aldehyde reductase, fatty aldehyde decarbonylase, acyl carrier protein and an antibiotic. With these proteins, the cell is able to metabolize fatty acid or oil into esters (biodiesel).

Caspari et al. (2008) protected an invention consisting in the cultivation of microalgae, such as *Dunaliella salina*, to accumulate oil. The microalgae cultivated are harvested by centrifugation, and the lipids are separated by contacting the cellular solution with hexane. The hexane/lipid phase, after filtration to exclude biomass fragments, contains pigments that are separated by contacting with graphene. After centrifugation, clear liquid oil is released and the hexane is removed by evaporation. With this process, the final lipid extract is free of pigments and solvent, with very few separation processes.

Herskowirz et al. (2006) invented a technology for the single-step production of liquid biodiesel, by hydrodeoxygenation and hydroisomeration of any kind of oil, derived from plants or animals. The process consists in deoxygenating, removing oxygen in the form of water or CO_2, oil to form C_{14} to C_{18} paraffins, which are hydroisomerized, in a fixed-bed reactor with gas and liquid running downflow, using one or more catalysts, such as metal or acidic component (i.e., platinum or palladium, silica or fluoride alumina and zeolites) in the same stage to form isoparaffins, so that they can be used as diesel-like fuels. The invention claims that the process is able to produce a liquid fuel that comprises 2–10 % lighter naphta products that have a boiling temperature of up to 423.15 K. The hydrogen separated from the effluent obtained in the one-step reaction may be used as supplement hydrogen for the reaction, with a recirculation system.

Oyler (2008) propose a process for the accumulation of oil in microalgae in a two-step fermentation process. The first step comprises cultivation of cyanobacteria in an autotrophic phase, using light and CO_2, in order to maximize microorganism growth. The second step consists in a heterotrophic growth of the same cyanobacteria with a stress induction mechanism. The heterotrophic growth is provided with a sugar feed and the stress induction may be with light or nutrient deprivation, and/or the injection of a reactive oxygen source of chemical additives. The second step aims for the production of oil by the algae, which is hydrogenated or transesterified afterwards to produce biodiesel.

Huntley and Redalje (2007) describe a process for oil production using photosynthetic microorganisms capable of fermenting carbon dioxyde from fossil fuel combustion. Produced oil can be transformed into biodiesel. This process is divided in two stages, one in an opened system and the other one in a closed one. The first stage is a continuous closed system that aims the production of biomass, using carbon dioxyde as substrate, and adding light and other nutrients to improve biomass growth. The microorganism used in this step may be from the species *Tetrasilmis suecica*, *Isochysis galbana*, *Phaeodactylum tricornutum*, *Nannochloropsis* sp., *Dunaliella primolecta* or *Nitzschia closterium*. The closed systems must be designed to occupy less than 20 % of the total area of the cultures. The innoculum rate of the second step, in the opened system, should be at least 5 % of its total load, and the fermentation period must be inferior to 5 days. These parameters, when observed, make the production process easier, the construction less complex and less prone to contaminations in the opened system. In the closed system the microorganism must duplicate at least once every 16 h, while in the closed system the duplication rate should be up to 8 per day. The growth rate of the microorganism is controlled by limiting the carbon dioxyde availability.

The patent application filed by Hammond and Wang (2005) claim a process of biodiesel production from acidic lipids, with small amounts of excess methanol at its end, thus eliminating recycling steps. Purification of the lipid phase is said to be eased, as well. This is performed by the addition of methyl alcohol to the acidic lipid solution at 270 % of the theoretical amount needed to full conversion in acid oil or methyl esters. After the addition of methanol as a catalyst, sulfuric acid, hydrogen chloride or p-toluenesulfonic acid is added to start the esterification reaction that will enable the production of the fatty acid methyl esters and glycerol. For best performance of this process, the determination of free fatty acid concentration in the acidic lipid solution is necessary.

O'Connor et al. (2007) filed a patent for a process that aims to separate the components, lignin, hemicellulose and crystalline cellulose, of non-edible substrates, such as bagasse, straw, corn stover, corn husk, sugar beet pulp, chopped straw, cotton linter, corn stalk, corn cob, wood chip, saw dust, tree bark and grass, and develop its susceptibility to be converted into liquid fuels. The individual components are then converted into fuels by appropriated methods. The process consists in biomass activation, in order to sensitize the material to hydrothermal conversion, followed by a solvent addition (optional), partial biomass conversion to form solubilized material, and a final step of recycling the unconverted biomass. The activation of biomass is performed by the addition of an inorganic material in its particulate form, such as cationic, anionic or natural clays, hydrotalcite-like materials, layered materials, ores, minerals, metal oxides, hydroxides of metals and mixtures thereof, and an organic material, both with catalytic activity. After this, the activated biomass is homogenized and extruded in a kneader, milling device or fluid or spouted bed, at elevated temperature, with the presence of steam. The conversion step may be performed either by hydrothermal, pyrolysis, gasification thermal cracking, catalytic cracking, acid hydrolysis or enzymatic hydrolysis. The separation step of the unconverted biomass is performed by a filtration device, with the addition of a solvent, steam and under elevated pressure. The final product is biodiesel.

Gilbeau (2006), Krafft et al. (2005) and O'Connor et al. (2008) describe process for use of glycerol obtained in biodiesel production. Gilbeau's invention aims to produce epoxi. Krafft et al. describe a process for production of dichloropropanol by reacting glycerol with a chlorinating agent. And O'Connor et al. use glycerol to produce olefins, alkanes, coke or syngas (hydrogen and carbon monoxide) by fluid catalytic cracking technology.

Luxem and Troy (2004) describe a process for single-step production of biodiesel, by heating to a temperature between 423.15 and 443.15 K a mixture of a vegetable oil source containing free fatty acids or glycerides, with up to 5 equivalents of methanol, and an acid catalyst. This mixture is subjected to high pressure conditions (670 to 2068 kPa). At the end of the process, biodiesel is obtained, either by transesterification of glycerides and conversion of free fatty acid, with no need for neutralization or recovery of residual free fatty acids.

Jones et al. (2010) developed a process for cultivating microalgae (*Cyanobacteria*, *Amphiprora*, *Chaetoceros*, *Isochrysis*, *Scenedesmus*, *Chlorella*, *Spirulina*, *Coelastrum*, *Micractinium*, *Euglena*, or *Dunaliella*) in an environment that comprises a plurality of fish species, that are used to control several parameters of the algae cultivation, such as nitrogen, phosphorous and carbon dioxide concentration, oxygen level, zooplankton level, mollusk and crustacean population, and temperature uniformity. To induce oil accumulation by the microalgae, they are stressed by nutrient limiting performed by controlling the fish population.

Portnoff et al. (2005) propose a process for biodiesel production with enhanced productivity and better conversion rates by applying to the transesterification reaction at least one of the following: modulated microwave or radio-frequency energy, with a potency that can be as high as 100 Watts and with a frequency between 400 MHz and 5 GHz (either amplitude, frequency or pulse width may be modulated); use of a heterogeneous catalyst mixed in a velocity high enough to provoke high shear rates; use of a homogenous catalyst (acid or base) in emulsions; perform the reaction in high pressure values (up to 689 kPa higher than the autogenous pressure).

Khalil and Leite (2005) invented an integrated process to produce biodiesel, among other products, from castor beam seeds. The oleaginous seeds are reacted with ethyl alcohol and an alkaline catalyst to obtain an alkyl ester, used as biodiesel. The reaction product is then filtrated, obtaining solid and liquid phases. After drying and sieving the solid phase, two fractions are obtained, one containing starchy carbohydrates that can be used to produce ethanol, via fermentation, and another one comprising cattle feed or hull that can be used to produce fertilizers. The liquid phase is distilled to obtain ethanol, and the remaining fraction is decanted into a heterogeneous solution, comprising esters and glycerin. After separation, both glycerin and esters are neutralized, and the esters formulated to obtain biodiesel.

Most of the biodiesel patents applications mentioned in this chapter refer to production of this fuel using microalgae to accumulate oil. This technology is especially interesting due to its capability of producing biodiesel with almost no residues or carbon emissions. Microalgae are autotrophic organisms that can metabolize carbon dioxide, using light as catalyst, into oxygen and water. With some stress factors, some microalgae are able to accumulate oils that, after extraction, can be transformed into biodiesel. In this scenario, no edible substrate must be used and very few pollutants are produced. Therefore, the demanding interest in this technology is justified.

6 Other Biofuels

There are other biofuels receiving attention recently, from which butanol and biogas are highlighted. Patent statistics from these fuels are briefly described in the following sections.

6.1 Butanol

ABE (Acetone, Butanol and Ethanol) fermentation, using Clostridia as microorganism and a starchy or lignocellulosic feedstock, can be used to produce butanol as fuel, even though synthetic production, derived from a petrochemical route, is still the most used process. China is the only country currently producing butanol from fermentation. ABE fermentation is not economically feasible, but the increasing interest in non-fossil fuels brought attention to bioprocesses for producing butanol. Besides its use as fuel, butanol is used as a chemical precursor for paints, polymers and plastics. Global production in 2012 was estimated to be almost 4.0 million tons with an expected growth of almost 5.0 % until 2018 (Jiang et al. 2014; European Biofuels Technology Platform 2015).

In Fig. 7 it is possible to see the trend in butanol related patent applications. As more than 90 % of the total butanol related patents were filed after 2005, only the annual trend from this year on is presented. In Fig. 8 is shown the countries that filed more patent applications related to butanol. Those numbers relate only to patent applications filed through PCT and not in the residence countries. United States leads by a considerable margin. The eight companies that filed most patent applications via PCT are shown in Fig. 9, being Butamax, a joint venture between DuPont and BP, the main producers of butanol, the biggest applicant, by far.

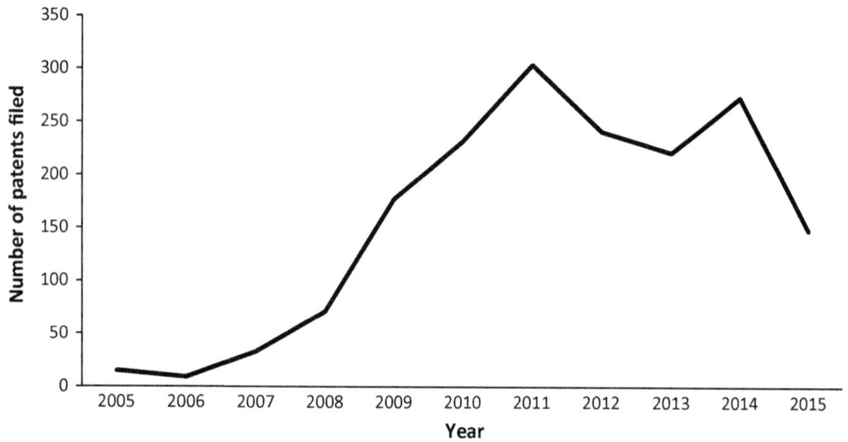

Fig. 7 Evolution of patent applications related to butanol in the world, from 2005 to 2015

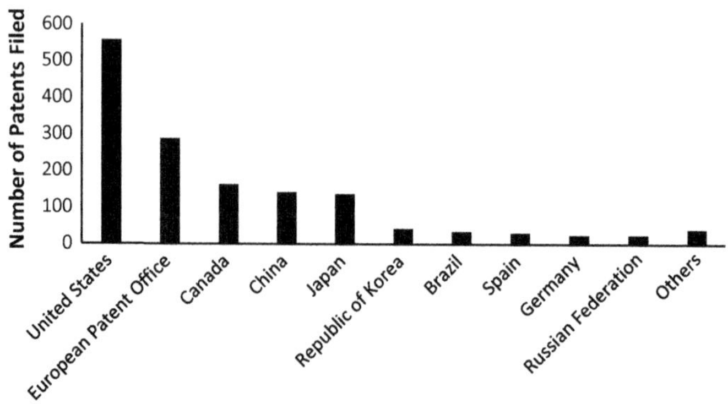

Fig. 8 Countries that filed most patent applications related to butanol via Patent Cooperation Treaty (PCT), since 2005

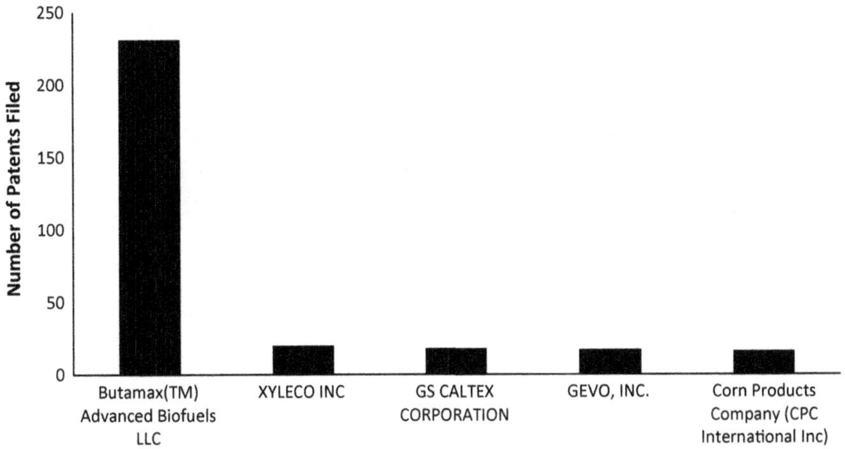

Fig. 9 Companies that filed most patent applications related to butanol via Patent Cooperation Treaty (PCT), since 2005

6.2 Biogas

Biogas is usually referred to as a gaseous fuel produced from anaerobic fermentation of organic products, and is mainly composed of methane and carbon dioxide. It is very common to produce it from sewage feedstock. Production of biogas in 2012 was over 56 billion m^3, and its biggest producers are Germany, UK, The Netherlands, Korea, and Brazil (WBA 2015).

Biogas related patent application growth from the past 10 years is shown in Fig. 10. As happens with butanol, most inventions regarding biogas production were made in the past 2 decades. United States, China, EU and Japan are the most active countries regarding the filing of patents regarding biogas, with no significant difference amongst them, as presented in Fig. 11. Figure 12 show that Mistui Oil &

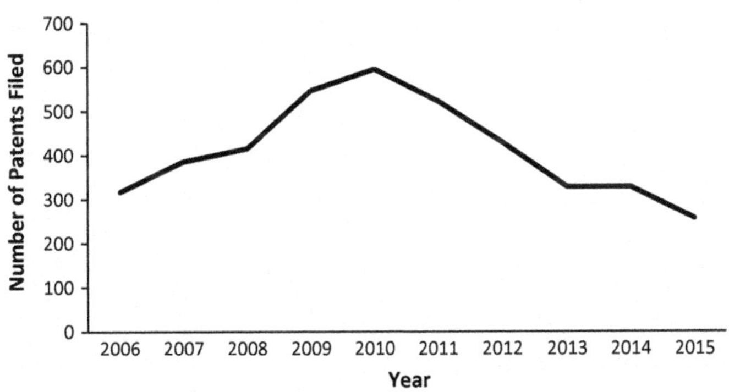

Fig. 10 Evolution of patent applications related to biogas in the world, from 2005 to 2015

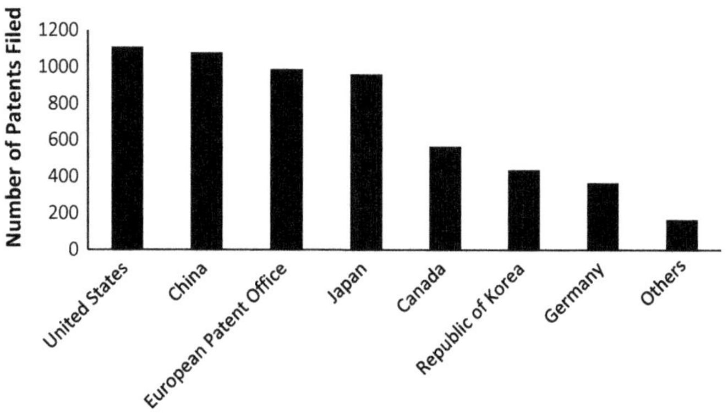

Fig. 11 Countries that filed most patent applications related to biogas via Patent Cooperation Treaty (PCT), since 2005

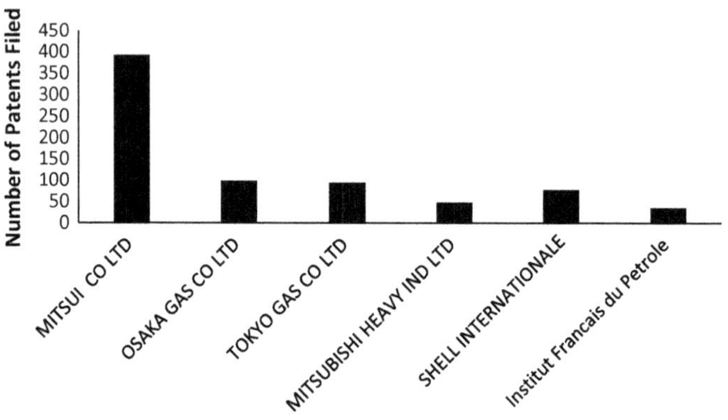

Fig. 12 Companies that filed most patent applications related to biogas via Patent Cooperation Treaty (PCT), since 2005

Gas, one of the companies spun-off from Mitsui Co. Ltd., is the biggest patent applicant regarding biogas. This company is also one of the biggest producers of natural gas in the world.

7 Conclusion

Patent applications in biofuels have been growing constantly in the past decade. Due to a soaring interest for renewable and cleaner forms of energy, investments will continue to arise for better, more efficient and cheaper technologies for production of biofuels, since, up to now, these products are only feasible thanks to governmental policies. Also, there are concerns that production of biodiesel and ethanol use edible raw material as substrate. To avoid that most researched technologies regarding biofuels are the ones using industrial sub-products, such as cellulosic materials, or microorganisms that grow without need of edible sources. There are still challenges to make those technologies more available, as can be seen from the patent documents cited in this chapter. So, the expectation is that patent applications will continue to grow in the near future to provide a more diversified global energy matrix, with less dependency on fossil fuels.

References

Abbas C, Binder TP, Beery KE, Cecava MJ, Doane PH, Holzgraefe DP, Solheim LP (2006) Process for the production of animal feed and ethanol and novel feed. PCT Patent Application WO2006113683, 26 Oct 2006

ANFAVEA (2012) Anúario da Industria Automobilistica Brasileira 2012. ANFAVEA - Associação Nacional dos Fabricantes de Veículos Automotores (Brasil)

Badger PC (2002) Ethanol from cellulose: a general review. In: Janick J, Whipkey A (eds) Trends in new crops and new uses. SHS Press, Alexandria, pp 17–21

Bartek R, Yanik S, O'Connor P (2010) Processing biomass with a hydrogen source. PCT Patent Application WO2010135734. 25 Nov 2010

Bastos MB (2007) Brazil's ethanol program—an insider's view. In: Economides MJ (ed) Energy Tribune

British Petroleum (2015) BP statistical review of world energy. London

Caspari M, Vick B, Radaelli G (2008) Methods and compostions for production and purification of biofuel from plants and microalgae. PCT Patent Application WO2008060571. 22 May 2008

Chua P, Espina K, Franklin S, Rudenko G, Somanchi A (2010) Manufacturing of tailored oils in recombinant heterotrophic microorganisms. PCT Patent Application WO2010063031. 03 Jun 2010

Czartoski TJ, Perkins R, Villanueva JL, Richards G (2011) Algae biomass fractionation. PCT Patent Application WO2012078852. 14 Jun 2012

Demirbas A (2009) Political, economic and environmental impacts of biofuels: a review. App Energy 86:S108–S117

Dunson JB, Tucker M, Elander R, Hennessey SM, Elander RT (2006) Treatment of biomass to obtain ethanol. PCT Patent Application WO2006110900. 19 Oct 2006

EBTP (2011) Growth in production and use of biofuels worldwide. In: Global biofuels—an overview. http://www.biofuelstp.eu/global_overview.html#prod. Accessed 27 Sept 2011

EIA (2012) International energy statistics. In: US Department of Energy, Washington. http://www.eia.gov/cfapps/ipdbproject/iedindex3.cfm?tid=79&pid=81&aid=1&cid=regions&syid=2008&eyid=2012&unit=TBPD. Accessed 08 Oct 2015

European Biofuels Technology Platform (2015) Butanol. http://www.biofuelstp.eu/butanol.html. Accessed 13 Nov 2015

Foody B, Rahme Z, Tolan JS (2006) Upflow settling reactor for enzymatic hydrolysis of cellulose. PCT Patent Application WO2006063467. 22 Jun 2006

Galvez A, Richards G (2010) Materials and methods for converting biomass to biofuel. PCT Patent Application WO2010025171. 04 Mar 2010

Geir H, Kjetil E, Ivar GC (2009) Method for dewatering a mixture of mostly ethanol and water. PCT Patent Application WO2009048335. 16 Apr 2009

Gilbeau P (2006) Method for making an epoxide. PCT Patent Application WO2006100311. 28 Sept 2006

Golubkov I (2005) Method for producing hydrocarbons and oxygen-containing compounds, from biomass. PCT Patent Application WO2005040392. 06 May 2005
Guo M, Song W, Buhain J (2015) Bioenergy and biofuels: history, status, and perspective. Renew Sust Energ Rev 42:712–725
Hammond EG, Wang T (2005) Method of converting free fatty acids to fatty acid methyl esters with small excess of methanol. US Patent 6,965,044. 15 Nov 2005
Harris P, Rey M, Ding H (2005) Methods for enhancing the degradation or conversion of cellulosic material. US Patent Application US2007077630. 05 Apr 2007
Herskowirz M, Landau M, Reizner I, Kaliya M (2006) Production of diesel fuel from vegetable and animal oils. US Patent Application US2006207166. 21 Sept 2006
Huhnke RL, Lewis RS, Tanner RS (2008) Isolation and characterization of novel Clostridial species. PCT Patent Application WO2008028055. 06 Mar 2008
Huntley ME, Redalje DG (2007) Continuous-batch hybrid process for production of oil and other useful products from photosynthetic microbes. PCT Patent Application WO2007013899. 01 Feb 2007
IEA (2007) Biodiesel production. Energy Technology Essentials, International Energy Agency
Jiang Y, Liu J, Jiang W, Yang Y, Yang S (2014) Current status and prospects of industrial bio-production of n-butanol in China. Biotechnol Adv. doi:10.1016/j.biotechadv.2014.10.007
Jones DV, Morgenthaler GE, Stephen D, Wu BC (2010) Systems and methods for producing biofuels from algae. PCT Patent Application WO2010036334. 01 Apr 2010
Justin ST, Nahla A, Gordana D, Ellen B, Peter L, Dycaico M, Toby R, Poland J, Hefner Y, Marie MS (2009) Enzymes for the treatment of lignocellulosics, nucleic acids encoding them and methods for making and using them. PCT Patent Application WO2008095033. 07 Ago 2008
Kamis R, Joshi M (2008) Biofuel patents are booming. Baker & Daniels. Washington DC
Khalil CN, Leite LCF (2005) Process for producing biodiesel. US Patent Application US2005011112,. 20 Jan 2005
Kim GS, Shin M, Kim YJ, Oh KK, Kim JS, Ryu HJ, Kim KH (2008) Method of producing biofuel using sea algae. PCT Patent Application WO2008105618. 04 Sept 2008
Krafft P, Gilbeau P, Gosselin B, Claessens S (2005) Process for producing dichloropropanol from glycerol, the glycerol coming eventually from the conversion of animal fats in the manufacture of biodiesel. PCT Patent Application WO2005054167. 16 Jun 2005
Kumar S, Singh SP, Mishra IM, Adhikari DK (2009) Recent advances in production of bioethanol from lignocellulosic biomass. Chem Eng Technol 32:517–526
Leschine S, Warnick TA (2008) Systems and methods for producing biofuels and related materials. PCT Patent Application WO2007089677. 09 Aug 2007
Luxem FJ, Troy WM (2004) Method of making alkyl esters using pressure. US Patent 6,768,015. 27 July 2004
Mekhilefa S, Sigaa S, Saidurb R (2011) A review on palm oil biodiesel as a source of renewable fuel. Renew Sust Energ Rev 15:1937–1949
Mielenz JR (2001) Ethanol production from biomass: technology and commercialization status. Curr Opin Microbiol 4:324–329
Nguyen XN (2008) Integrated process for separation of lignocellulosic components to fermentable sugars for production of ethanol and chemicals. US Patent Application US2008057555. 06 Mar 2008
O'Connor P, Huber GW, Corma CAA, Savanaud LLA (2008) Fluid catalytic cracking of oxygenated compounds. PCT Patent Application WO2008020047. 21 Feb 2008
O'Connor P, Stamires D, Daamen S (2007) Process for the conversion of biomass to liquid fuels and specialty chemicals. PCT Patent Application WO2007128800. 15 Nov 2007
OECD (2009) Patent statistics manual. 2nd edn. Organisation for Economic Co-operation and Development (OECD)
Oyler JR (2008) Two-stage process for producing oil from microalgae. US Patent Application US2008160593. 03 July 2008

Penttilae M, Siika-Aho M, Uusitalo J, Viikari L (2005) Process for producing ethanol. PCT Patent Application WO2005118828. 15 Dec 2005
Portnoff MA, Purta DA, Nasta MA, Zhang J, Pourarian F (2005) Methods for producing biodiesel. US Patent Application US2005274065. 15 Dec 2005
REN21 (2011) Renewables 2011 global status report. Renewable energy policy network for the 21st century. Paris
RFA (2014) 2014 Production. In: World fuel ethanol production. Renewable Fuels Association (RFA). http://www.ethanolrfa.org/pages/statistics. Accessed 02 Oct 2015
Rovere EL, Pereira AS, Simões AF (2011) Biofuels and sustainable energy development in Brazil. World Dev 39:1026–1036
Sambodo MT (2008) Sustainable biofuel development in Indonesia, Hoping the Dream Come True. Paper presented at the Sustainable Biofuel Development Research Workshop pp. 3–6 Jakarta, October 2008
Schreck DJ, Balmer NL (2009a) Integrated facility for producing alcohol using homoacidogenic fermentation. PCT Patent Application WO2009009322. 15 Jan 2009
Schreck DJ, Balmer NL (2009b) Improved indirect process for producing ethanol. PCT Patent Application WO2009009320. 15 Jan 2009
Sissine F (2008) Renewable energy: background and issues for the 110th congress. In CBS Reposrt for Congress. 25 Mar 2008
Smeets E, Junginger M, Faaij A, Walter A, Dolzan A (2006) Sustainability of Brazilian bio-ethanol. Universiteit Utrecht & Unicamp, Utrecht
Taira R, Tkagi S, Hjort C, Vikso-Nielsen A, Allain E, Udagawa H, Fukuyama S, Matsui T, (2005) Enzymes for starch processing. PCT Patent Application WO2005003311. 13 Jan 2005
Takahashi CM, Lima KGC, Takahashi DF, Altehthum F (2000) Fermentation of sugar cane bagasse hemicellulosic hydrolysate and sugar mixtures to ethanol by recombinant Escherichia coli KO11. World J Microb Biotechnol 16:829–834
Trimbur DE, Im C, Dillon HF, Day AG, Franklin S, Coragliotti A (2008) Production of oil in microorganisms. PCT Patent Application WO2008151149. 11 Dec 2008
UNEP (2009) Towards sustainable production and use of resources. United Nations Energy Programme (UNEP), Paris
UNEP (2011) Global trends in renewable energy investment. United Nations Energy Programme, Paris
Vehmaanperae J, Alapuranen M, Puranen T, Siika-Aho M, Kallio J, Hooman S, Voutilainen S, Halonen T, Viikari L (2007) Treatment of cellulosic material and enzymes useful therein. PCT Patent Application WO2007071818. 28 June 2007
Verser DW, Eggeman TJ (2008) Energy efficient methods to produce products. PCT Patent Application WO2008098254. 14 Aug 2008
Viikari L, Vehmaanperä J, Koivula A (2012) Lignocellulosic ethanol: from science to industry. Biomass Bioenerg 46:13–24
WBA (2015) WBA global bioenergy statistics 2015. http://www.worldbioenergy.org/sites/default/files/WBA%20Global%20Bioenergy%20Statistics%202015%20%28low%29.pdf. Accessed 07 Jan 2016
World Energy Council (2013) World energy resources: bioenergy. London
Worldwatch Institute (2007) Biofuels for transport: global potential and implications for sustainable energy and agriculture in the 21st centruy. Earthscan, London
WTO (1994) Agreement on trade-related aspects of intellectual property rights (TRIPS). World Trade Organization (WTO), Marrakesh

Economic and Environmental Aspects of Biofuels

Emmanuel Bertrand, Marilys Pradel and Claude-Gilles Dussap

Abstract Within the framework of sustainable development, a quadruple transition (demographic, food, energetic and environmental) is necessary. The integrated biomass valorization, including the production of biofuels via the biorefinery concept lies at the cross section of these global challenges. Biofuels production competes for water and land use with respect to feed and food production. Securing these resources in an uncertain geopolitical environment is a strategic issue for many countries. It also has favorable (climate change mitigation) and adverse implications (fertilizers consumption, local pollutions). Therefore, the topic leads to huge political and economic debates at national and international scales. The results obtained from the implementation of supportive strategies for the development of this "new" bioeconomy indicate, at least in the near future, mixed results in comparison to the objectives that have been set. This may be explained by windfall effects created on the international markets and underline two imperative needs. The development and improvement of environmental methods, such as life cycle assessment expanded to the scope of international trade and not restricted to the regional or national economies will be necessary in order to install ambitious and unambiguous agreements for policymaker's in the future international conferences such as the ones on climate change.

Keywords Biofuels · Bioeconomy · Environment · Life cycle analysis

E. Bertrand · C.-G. Dussap (✉)
Université Clermont Auvergne, Université Blaise Pascal, Institut Pascal, UMR CNRS 6602, BP 10448, 63000 Clermont-Ferrand, France
e-mail: C-Gilles.Dussap@univ-bpclermont.fr

E. Bertrand
LabEx IMobS3, Université Blaise Pascal, BP 80026, 63171 Aubière Cedex, France

M. Pradel
IRSTEA, UR TSCF, Domaine des Pallaquins, 03150 Montoldre, France

C.R. Soccol et al. (eds.), *Green Fuels Technology*,
Green Energy and Technology, DOI 10.1007/978-3-319-30205-8_22

1 Introduction

The concept of a bioeconomy is an economy where the sources for materials, chemicals, and energy come from renewable biological sources (Organization for Economic Co-operation and Development 2009; European Commission 2012). The development of the bioeconomy concept would require continuous public investments in research and the establishment of regulatory framework coupled with financial incentives to lead the private sector investments toward the development and commercialization of new bio-based products (Zilberman et al. 2013). At the end of the year 2014, at least 164 countries have set renewable energy targets; 145 have already engaged renewable energy policies. More than 60 of them have also engaged strategies toward the development of a bioeconomy. Of course, it includes major economic players such as the United States, Canada, Brazil, China, India, most of the countries of the European Union, but also smaller ones such as Kenya, Zimbabwe, or Sri Lanka (REN21 2015). Therefore, the development of a bioeconomy approach is spreading worldwide, no matter of the size and strength of the national economy and can no longer be considered as marginal.

Biofuels are the most visible output for this existing bioeconomy. The biorefinery concept aims at replacing at least part of the petroleum-based products. A focus on the energetic sector at the end of 2014 (Fig. 1) indicates that about 80 % of the global energy is still coming from fossil origin. The transportation sector is particularly pointed out because of its quasi-exclusive dependence on liquid fuels. In year 2011, more that 93 % of the 2500 millions of tons of oil equivalent used for transportation came from nonrenewable feedstock (International Energy Agency 2012). Moreover, the sector is charged with about 15 % of the global emissions of greenhouse gases worldwide and this proportion increases close to a third in some countries.

Furthermore, in an effort to add value to the whole fraction of the biomass, 12 building blocks for bio-based chemical activities have been proposed: four carbons 1,4 di-acids (succinic, fumaric and malic), 2,5-furan dicarboxylic acid (FDCA), 3-hydroxypropionic acid (3-HPA), aspartic acid, glucaric acid, glutamic acid, itaconic acid, levulinic acid, 3-hydroxybutyrolactone, glycerol, sorbitol, and xylitol (Werpy and Petersen 2004). Most of these chemicals are now coming from the production processes and for some of them with significant amount. For instance, the production of glutamic acid via the sodium glutamate (MSG) pathway already reached 3 millions of tons produced in 2014, with an increase of approximately 70 % over the last decade. In this context, the development of biorefineries appears as a credible alternative for a more sustainable future (Menten et al. 2013).

Five mainstreams energy technologies are from renewable origins: wind power, (370 GWy = 12,000 PJ), hydropower (1055 GWy = 33,000 PJ), geothermal (18.8 GWy = 590 PJ), solar (177 GWy = 5600 PJ for photovoltaic plus 4.4 GWy = 139 PJ for thermal power), and biomass energy (93 GWy = 2930 PJ) for a total installed capacity of 1712 GWy = 54,000 PJ of renewable origin during the year 2014. Considering 7 billion people, it represents an average power of 234 W of renewable origin or 7.7 GJ per inhabitant (REN21 2015). If these sources are not

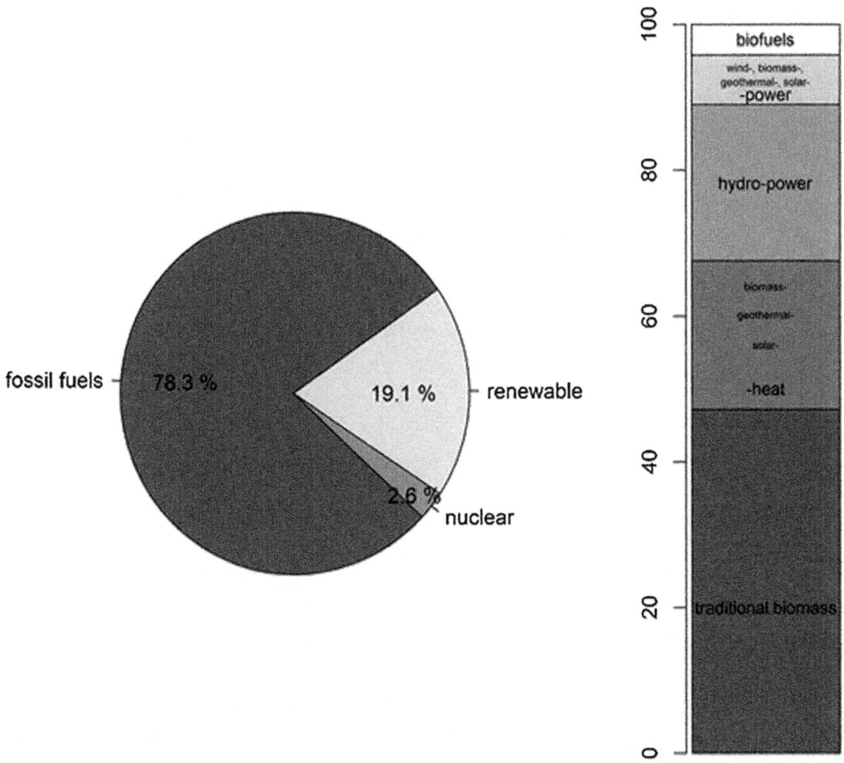

Fig. 1 Global energy consumption outlook in year 2014

in competition, they are at least adaptable (with some technical adjustments that are already available or still in the development phase) for real applications. In order to understand the potential contribution of biofuels for the energetic mix in general, the economy including the transportation sector in particular, one has to compare biofuels-based products with their fossils and renewable alternatives.

The chapter mainly examines the economic and environmental impacts of the first-generation biofuel that is the only one largely available on the market today. Based on this experience, the process of continuous improvement for the future generations of biofuels is investigated.

2 Overview of the Global Market for First-Generation Biofuels

One difficulty to assess and compare several production processes is to evaluate the performances of several processes on a common basis, i.e., with a suitable conversion table. The key values used throughout this chapter are given in Table 1.

Table 1 Key values for comparison of the energy units with volume production units, e.g. 1 billion liter ethanol equivalent to 23.4 × 10^{15} J = 23.4 PJ

Compound	Calorific value (MJ/kg)	Calorific value (MJ/L)	Specific mass (kg/L)
Petrol	47.3	35.7	0.755
Diesel	44.8	37.3	0.833
Crude oil	39.5	34.3–36.4	0.87–0.91
Ethanol	29.7	23.4	0.789
Methanol	19.9	15.8	0.792
Butanol	33.1	26.8	0.810
Wood	15		0.24–0.52
Charcoal	15–27		0.208
Lignite (dry)	16–20		0.801
Coal (Anthracite)	30–33		1.506
Methane	50		
Dihydrogen	120.5		

First, it must be outlined that biofuels are strongly linked with the history of modern mobility. In 1860, Nicholas Otto, began his experiments with ethanol-powered engines while Rudolf Diesel in 1893 first designed his motor to run on vegetable oils. Brazil started to use bioethanol from sugarcane as a fuel since the mid 1920s. However, the discovery of new oil wells, the improvement of drilling, and safe storage techniques make gasoline-based products the most competitive fuel available on the market. The renewed interest for biofuel production began in the 1970s in response to the two oil shocks and its related energy security issues. Governments started national programs to develop biofuel production. Mandates were particularly strong in Brazil and in the United States to promote bioethanol (Timilsina and Shrestha 2014). It was also driven by agricultural policies in the industrialized countries, as they experienced overproduction and their associated low prices issues for most of the crops. In this context, biofuels were considered as an ecological way to diversify the valuations opportunities (Gnansounou and Dauriat 2011).

2.1 Ethanol

The commercial production of ethanol as a biofuel is depicted Fig. 2 (left). It has increased in average at a rate of 14 % per year in the period 2004–2010. However, the production level was stabilized for the period 2010–2013. According to the World Bank, a slight decrease was caused by high corn prices in the United States originated from the mid-year drought. In 1989, 22 % of the 28.9 million of tons of oil equivalent used for the Brazilian transportation sector were from bioethanol, i.e., 10.7 billion liters of ethanol (251 PJ) (Timilsina and Shresta 2014). Although the

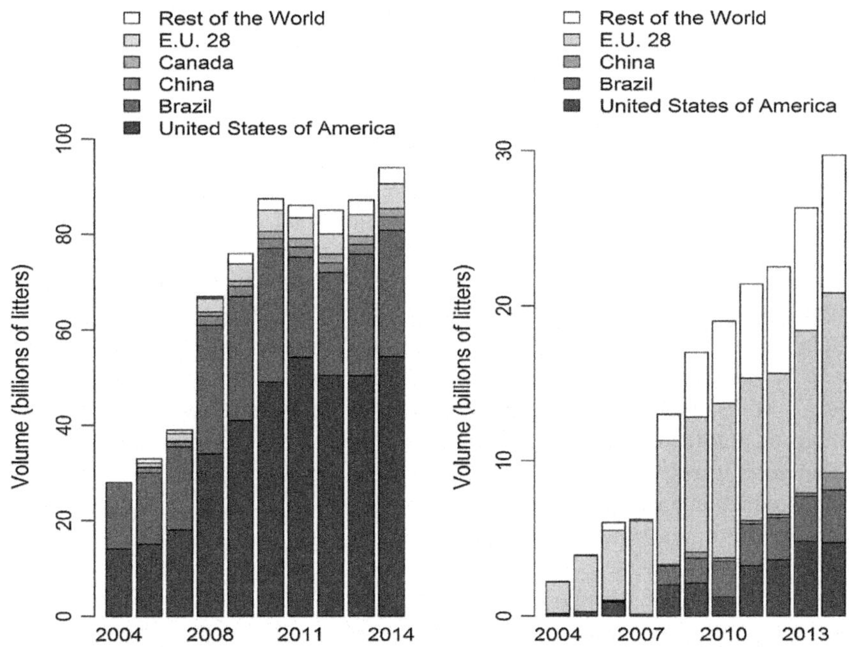

Fig. 2 Ethanol (*left*) and biodiesel (*right*) production worldwide. Adapted from REN21 (2015)

consumption increased in Brazil to reach 15.46 billion liters ethanol (360 PJ) in 2014, the share decreased to represent only 16 % of the 72.3 million of tons of oil equivalent used in year 2012 in the transportation sector. Thus, the development of the first-generation ethanol industry was not sufficient alone to support a rapidly growing economy in Brazil and the priority was recently given to the exportations with an approximate export capacity of 10 billion of liters (or 234 PJ) available for the export. Indeed, the discovery and exploitation of new offshore oil fields in Brazil allowed the country to fully meet its energetic needs with 1.93 million barrels of oil produced each day since 2006.

Brazil was leader in ethanol production from sugarcane until 2006 but was then significantly outdated by the United States of America and its very powerful corn sector. It should be outlined that the ethical question of the acceptability of genetically engineered crops being cultivated in opened fields, as well as their positive and adverse effects on the environment and the biodiversity remain an open question in many countries of the world, including the European Union. Anyway, the two leaders are still representing more than 85 % of the 94 billion liters bioethanol produced worldwide in 2014. Recent ethanol players include France, China, and Canada, while others such as India are on their way toward a significant commercial production.

Most recently, the production in 2014 appears to start rising again. It is certainly linked to the good harvests and consequently to the abundance of agricultural commodities available at an affordable price. It is noteworthy to indicate that a second and recent driver is probably the introduction of advanced biofuels as the first production plants for the second-generation producing ethanol from lignocellulosic materials are now coming on the market. Such plants have begun to produce at industrial scale in late 2013 and 2014 in the USA. They have the potential to transform significantly the biofuel sector. For instance, three facilities are already beginning operations in the Midwest of the United States (Abengoa's 95 million liters per year (2.2 PJ) in Hugoton, Kansas, Dupont's 115 million liters per year (2.7 PJ) in Nevada, Iowa and the POET-DSM plant in Emmetsburg, Iowa). Two more facilities are located in Brazil: Granbio's 308 million liters per year (8.9 PJ) in Sao Miguel dos Campos, Alagoas, and Raizen Energia S/A's 158 million liters per year (3.6 PJ) plant. The current production of second-generation ethanol is thus more than 100 times lower than the actual production from the first-generation. Therefore, it will be of great interest for the coming years to monitor and evaluate the production increase, its repartition between the main technologies and the emergence of new producers worldwide.

2.2 *Biodiesel*

The production of biodiesel as a biofuel is presented Fig. 2 (right). In comparison with ethanol, the production of biodiesel is still four times lower in volume. However, the sector is growing at a higher annual rate, 35 % in average from 2004 to 2012, and this increase is still 18 % strong after 2010, when the ethanol production started stagnation. One of the reasons for this rapid increase is that the biodiesel production scheme is not only producing energy but also proteins as a coproduct. These proteins are of great significance for the animal feed sector. Recent prospective analyses at the horizon 2030 indicate that it would be possible to sustain the global oil demand, but that the protein supply for animal feed will be under significant stress.

The sector used to be the exclusive preserve of the European Union in general and France and Germany in particular until 2004. The production was essentially based on rapeseed and sunflower as the main feedstock. At this time, Argentina (soybean), Indonesia, Malaysia (palm oil) but also the United States (soybean) reached significant production levels. The market is therefore segmented between more than six different countries. In 2012, the United States and Argentina with 3.6 and 3.1 billion liters (135 and 115 PJ), respectively, took first and second place. One of the reasons for such a drastic increase in the production is the introduction of governmental blending mandates toward the refiners. For instance, in the United States, the Environmental Protection Agency (EPA) set a target of 4.8 billion liters (180 PJ) to be blended in diesel fuel in 2013 under the Federal Renewable Fuel standard (RFS). These targets roughly correspond to the production level reached by the United States for the years 2013 and 2014. Similarly, in order to meet its commitments to reduce greenhouse gases emissions, the European Union has set a target of a 10 % market share in 2020 (Directive 2009/28/CE 2009). The current incorporation rate in volume being approximately of 7.5 % is slightly below the roadmap strategy. However, to achieve these objectives, the European Union has become net importer of biodiesel since 2001. For instance, 31 PJ were already imported in 2009 from the United States and Argentina. Similarly, concerning the first-generation bioethanol 16 PJ (0.75 billion liters) were imported mainly from Brazil (Lamers et al. 2011).

2.3 Solid Biomass Used as Fuel

The use of solid biomass as a fuel is always difficult to assess because of its domestic use. However, according to Lamers et al. analysis (2012), the international trade for biomass has grown from approximately 56–300 PJ between 2000 and 2010. The trade for wood pellets reached the most pronounced increase from 8.5 to 120 PJ compared to wood waste (77 PJ), fuelwood (76 PJ), wood chips (17 PJ), residues (9 PJ), and round wood (2.4 PJ). Approximately two-third of the market was conducted inside the European Union in 2010. The economical profitability is in the case of solid fuels the major explanation. The main exporting countries, such as Canada are characterized, by a high stockpile available, low feedstock costs, and have already a strong wood industry running. As a commodity, solid fuels are used to be regionally outsourced but prices differences and a new framework as well have favored a growing international trade.

2.4 Prices

The average prices for ethanol and biodiesel are globally following the same trends since 2006. Ethanol has increased steadily from 0.41 US$ L^{-1} (17.52 US$/GJ) in 2006 to 0.85 US$ L^{-1}(36.4 US$/GJ) in 2012 before starting to decrease again up to 0.62 US$ L^{-1} (26.5 US$/GJ) in 2013. Biodiesel prices leveled up from 0.90 US $ L^{-1} (23.6 US$/GJ) to 1.55 US$ L^{-1} (41.4 US$/GJ) before coming down to 1.51 US$ L^{-1} (40.1 US$/GJ) in 2013 (Timilsina and Shresta 2014). It must be outlined that the prices are almost equivalent when reported to their calorific content. This overall picture must be modulated by the regional specificities. For instance, in the European Union, which has now a reduced capacity to produce ethanol, prices are growing steadily. However, in Brazil or in the United States, the prices exhibit a huge volatility principally caused by the price of feedstocks (sugarcane and corn). For instance, due to climatic issues, Brazil experienced poor harvests between 2009 and 2011 and for the first time, corn-based ethanol produced on in the USA become cheaper than the Brazilian ethanol. Regarding biodiesel, the fluctuations are lower in the European Union than in other countries, probably because of a more diversified feedstock (sunflower, rapeseed) for its production than in the United States where soybean oil is used in majority.

2.5 Critical Analysis of the First-Generation Biofuels

The first environmental assessment studies of the first-generation biofuels show a positive energy balance (energy produced is about twice as much than the energy consumed) and a potential for reducing the greenhouse gases that is considerable (Ometto et al. 2009). The authors also concluded that the fuel ethanol life cycle is responsible for the consumption of a high quantity and diversity of nonrenewable resources. These resources are linked to the mechanization level of the rural activities, the intensive use of pesticides, fertilizers, and diesel. The ethanol life cycle process is also responsible to negative and adverse impacts at the local and regional scale such as ozone formation, acidification, ecotoxicity, and human toxicity. The main causes for these impacts are the use of fertilizers and the burning after harvest that is in the case of the first-generation ethanol traditionally realized to cogenerate electricity during the process. Similarly, Querini (2012) has found distributed environmental consequences for the first-generation biofuels. These impacts are broken down by their origins in the production cycle of biodiesel from sunflower and first- generation ethanol from wheat. They are presented on Fig. 3 and globally compared to other fuels for the transportation sector in Table 2. In particular, whatever the feedstock being considered, the agricultural practices are responsive for a significant part of these production impacts. Depending on the good agricultural practices, even the global warming potential is in favor of fossil fuels. Ethanol from sugarcane is confirmed as the most environmental-friendly biofuel for the first-generation.

Figure 4 represents the global warming potential for different fuels in a Well-to-Wheel life cycle assessment analysis (Querini 2012). The environmental impacts of fossil fuels occur for a small part during the extraction and refining processes and essentially during the usage phase. Biofuels are characterized by their double distributed environmental impacts: (i) throughout the production chain of the added value, (ii) in most of the environmental compartments considered by the major life cycle assessment methodologies. The question of the good agricultural practices, the sustainable management of land, water, and fertilizers inputs as well as the processing technologies and the final use of the produced resource becomes the central issue. In this context, the economic and ecological performances from one biofuel to another are fundamentally challenged. However, some initiatives are gradually introduced in order to provide sustainability certifications to the most environmentally virtuous biofuels, such as the guideline to evaluate the sustainability performances of biofuels published by the Natural Resources Defense Council (NRDC 2014). At first, the electric mobility from renewable origin presents the most favorable balance, while first-generation biofuels have contrasted results.

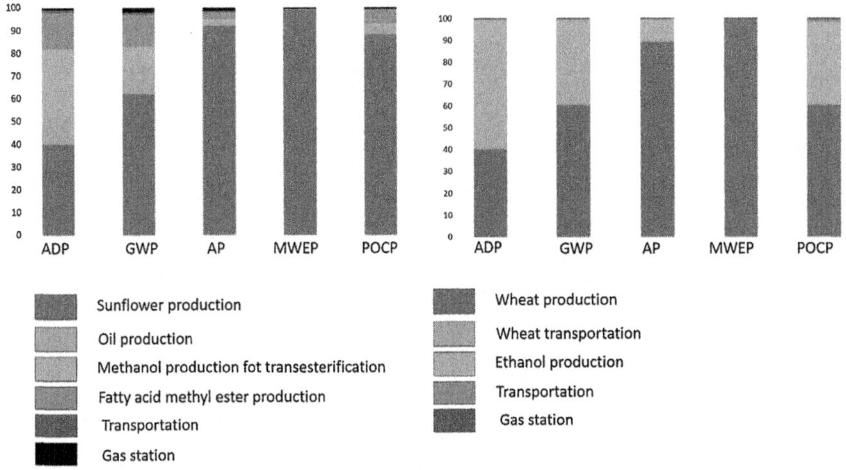

Fig. 3 Relative share of the environmental impacts according to the production stage of 1 MJ biodiesel originated from sunflower (*left*), or 1 MJ of bioethanol originated from the first-generation ethanol from wheat (*right*) distributed at the gas station. *ADP E* abiotic energetic depletion (fossil fuels use); *GWP* global warming potential according to RECIPE 2008 methodology; *AP* acidification potential according to RECIPE 2008 methodology; *MWEP* marine water eutrophication potential according to RECIPE 2008 methodology; *POCP* photochemical oxidation creation potential according to CML 2001 methodology. Reproduced from Querini (2012) with permission

However, as it is the case for the greenhouse gases, almost all of the impacts are for the electric mobility from renewable origin, reported on the abiotic depletion potential as their construction require precious metals and rare earths. This might be the cause of other sustainability issues such as international land grabbing. Furthermore, the data necessary for these evaluations are not easily accessible and the storage technologies necessary for these intermittent energies remain a challenging issue for the engineers and the attributional life cycle assessment methodologies. Recent studies taking into account the various energy carrier clearly demonstrate that they are susceptible to modify these emissions by a factor three (Yazdanie et al. 2014) and even cause some environmental rebound effects on categories such as the freshwater eutrophication potential (Font Vivanco et al. 2014).

Electricity distribution networks are able to absorb approximately 25 % of energy from renewable origin. Beyond that level, they are neither able to compensate the fluctuations associated with the intermittency of the production nor to answer to the daily consumption peaks. In this context, the adaptation of the existing distribution networks with the smart grid and with the development of high capacity storage systems would require a significant level of investments. Studies about these storage technologies show that onshore and offshore wind can be the support of effective and quantitative energy storage over for 300 h, when it is limited to only 24 h for the photovoltaic systems (Carbajales-Dale et al. 2014). On

Table 2 Impacts for 1 MJ of first-generation biofuel distributed at the gas station in the European Union

Impact for 1 MJ distributed at the gas station originated from	Diesel	Biodiesel from soybean	Biodiesel from Sunflower	Biodiesel from palm Oil	Gasoline	Ethanol from wheat	Ethanol from sugar beet	Ethanol from sugarcane
Energetic depletion (CML 2001) [MJ]	1.19 [1.15–1.23]	0.37 [0.29–0.46]	0.27 [0.22–0.34]	0.35 [0.27–0.44]	1.16 [1.13–1.20]	0.40 [0.33–0.49]	0.25 [0.08–0.27]	0.10 [0.07–0.13]
Global warming potential (RECIPE 2008) [g CO_2 eq]	20 [17–24]	43 [33–58]	23 [19–33]	28 [21–38]	15 [12–18]	49 [34–66]	28 [14–40]	11 [8–14]
Acidification potential (RECIPE 2008) [mg SO_2 eq]	48 [42–72]	566 [396–821]	242 [169–351]	352 [246–511]	45 [38–65]	496 [263–793]	279 [158–439]	204 [157–289]
Eutrophication potential (RECIPE 2008) [mg N eq]	23 [16–31]	447 [220–110]	1700 [835–4139]	264 [130–652]	22 [15–30]	532 [412–693]	121 [69–184]	121 [60–247]
Photochemical oxidation potential (CML 2001) [mg $C_2H_{4\ eq}$]	6 [5–8]	28 [25–33]	24 [21–28]	49 [44–58]	5 [4–7]	15 [11–20]	19 [15–24]	12 [8–17]
PM10 (RECIPE 2008) [mg PM_{10} eq]	16 [13–22]	114 [84–155]	55 [41–76]	98 [72–133]	16 [13–22]	103 [63–171]	68 [38–126]	58 [55–234]

Data reproduced from Querini (2012) with permission

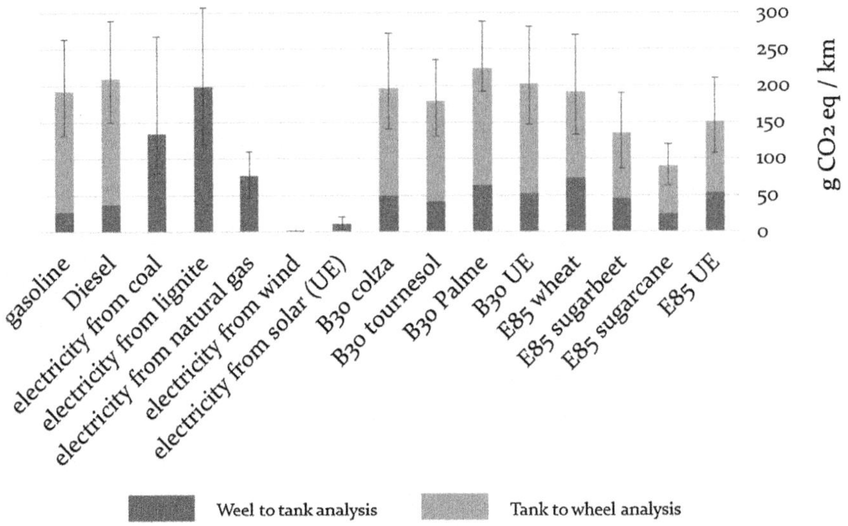

Fig. 4 Comparison of the global warming potential for vehicles based on their respective source of energy. Reproduced from Querini (2012) with permission

the contrary, the distribution schemes for the energy originated from the biomass, with the notable exception of bio hydrogen, are readily available in most of the countries, since they can use the same network than the one for fossil fuels. Moreover, for the countries that already own a gas distribution network; this network might be adapted with a reasonable investment to enable biogas injection at multiple inputs points. Such a network already permits to be used for energy storage purposes. These past investments in distribution networks for some countries are considerable assets, which must be considered in the selection and development of renewable technology in general and for the transportation sector in particular.

This benchmarking exercise has the advantage to identify the impacts and the sources for progress. It might be tricky if it turns out to discuss what the best technology is, and where the investments should be made. It seems unfair to compare technologies that have very different degree of maturity in a statically way. Therefore, it is concluded that life cycle analysis tools should rather be considered dynamically with an engineering purpose of identifying weaknesses and helping toward the implementation of better production processes. Recent works are taking full use of these potentialities for the improvement of biofuel production pathways in a combined life cycle assessment/chemical engineering approach (Gillani et al. 2013).

The improvement and the standardization of life cycle analysis methods (Jolliet et al. 2005) make it possible to evaluate complex systems with extended boundaries. In fact, in the life cycle assessment methodology, the limits of the system studied are defined in relation to the problem to solve. Therefore, these boundaries will not be the same to improve technically a process or to assess the possible

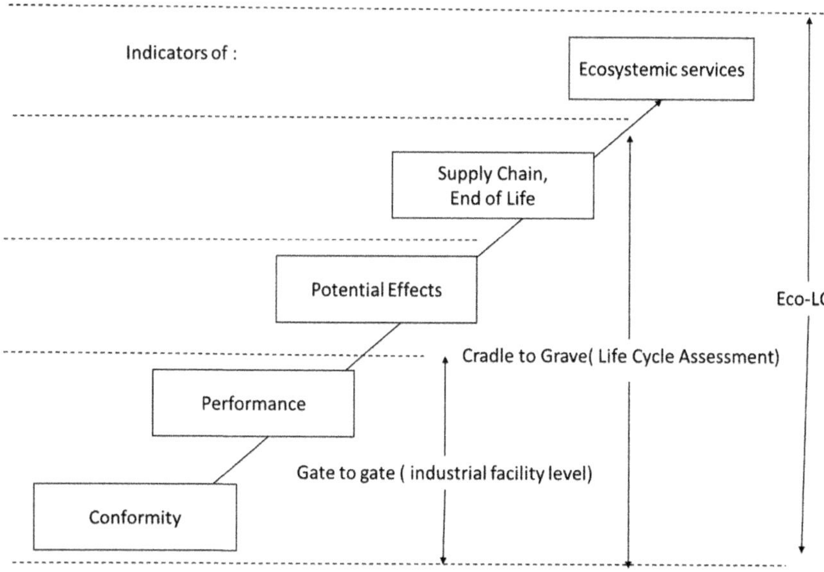

Fig. 5 Five level indicator frameworks promoting the idea of an evolutionary process adapted from Veleva and Ellenbecker (2001)

impacts on the regional biodiversity. Figure 5 represents the diversity of situations that could be covered with the life cycle assessment methodology.

Unfortunately, wider boundaries are to date still associated with greater uncertainties. The inclusion of these indirect effects such as the competition for feed and food (Koh and Gahzoul 2008), the land use changes (Lapola et al. 2010), the use of irrigation, and long distance trade flows make the environmental report of first-generation biofuels significantly less favorable (Panichelli 2012; Panichelli and Gnansounou 2015). The social question gave also mixed results for the development of local agricultural cooperatives. It seems that Brazil has failed at least at the present time to genuinely integrate the farmers in the production chain of the benefit (Stattman and Mol 2014). Concerning the production of palm oil in Indonesia, some authors even show that the impact on the environment and biodiversity of the palm plantation are directly related to the type of company responsible for the operation (Lee et al. 2014). Prospective studies show that the dietary changes in most of the emerging countries, with the notable exception of India would be responsible for two-third of the need of new agricultural land against one-third for the demographic growth (Khoury et al. 2014; Valin et al. 2014). These dietary changes are characterized by an increase in the total calories ingested in general and the one originated from red meat in particular. Currently approximately 72 % of the cultivated land is already dedicated to animal feeding purposes. Without any change in food consumption patterns and improvement in the production of proteins from animal origin, it will be unsustainable to produce food and biofuels simultaneously.

All biofuels generations considered, the United Nations Environment Program estimates that demand for agricultural land for the production of biofuels for transport will rise from 30 million hectares in 2009 to over 110 in 2050 to meet the commitments already taken with respect to the reduction of greenhouse gas emissions in the transportation sector (Unep 2014). Such a large-scale deployment would never be sustainable with the sole use of the first-generation biofuels (Mohr and Raman 2013).

2.6 New Investments in Renewable Energies

The new investments in capacity in renewable energy according to the Renewable Energy Network (REN21 2015) are depicted Fig. 6. Since the economic crisis in 2008, these investments are being reequilibrated between developed, which investments started to stagnate, and developing countries in which the investments remain steadily growing (Fig. 6 left). The sectorial breakdown per sector, as well as the net growth for 2014 in comparison to 2013 [Fig. 6 (right)] clearly show that these redistribution is beneficial for the wind- and solar- electricity sectors but not for the biomass and biofuel sectors. Drivers for these redistributions of funding are to search among the criticism that arise against the first-generation biofuel, and in particular, the fuel versus food debate that reached its paroxysm in 2008. In this context, the biofuel support polices remain under review in the United States and Europe.

According to Timilsina and Shresta (2014), the investments for biofuel refineries were approximately 16 billion US$ in 2008 and suffered a threefold drop to 6.8

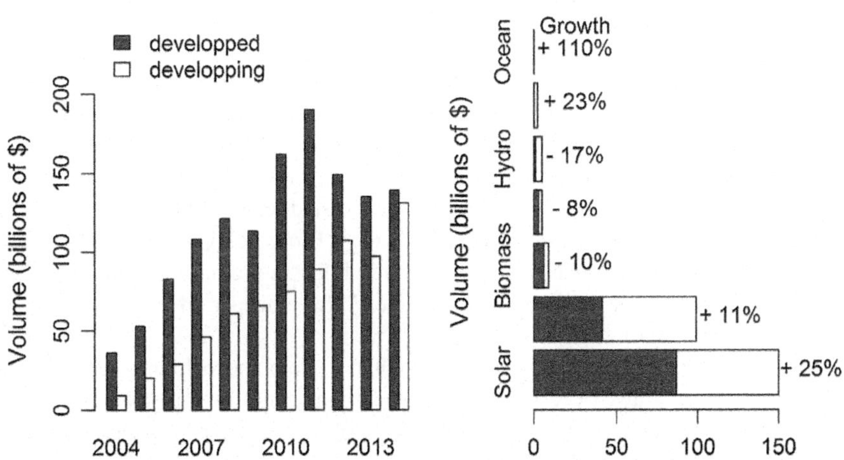

Fig. 6 Global investments in renewable energy from 2004 to 2014 (*left*) and sectorial ventilation in 2014. Growth represents the variation in investments from 2013 to 2014. Adapted from REN21 (2015)

billion in 2011, and even 5 billion in 2012. In this period, many of the 650-ethanol plants worldwide were operating below their theoretical capacity while others experienced temporary closed down because of the very high volatility of the prices, the fluctuating demand and the many reserves that has been made on first-generation ethanol. Brazil had the capacity to produce approximately 37 billion liters (860 PJ) in more than 440 plants in 2012. However, because of the specificity of the feedstock used, sugarcane that has a short storage life, Brazil has an excess of 30 % in sugarcane milling capacity and the production facilities are generally oversized. In reality, the production reached 26 billion liters (610 PJ). As the United States are primarily using corn, which as a longer shelf-life, the USA with approximately 210 ethanol plants had the capacity to produce 56 billion liters (1310 PJ) in 2012. Therefore, US plants have on average about three times the annual capacity of Brazilian plants. Feedstock differentiation and implementation of integrated first- and second- generation's bioethanol refineries might be an opportunity for the existing Brazilian facilities (Dias et al. 2013).

The European Commission has already proposed to limit the proportion of biofuels from the first-generation biofuel used for the transportation sector to 5 % and to remove the subsidies for food crop-based biofuels by 2020 (Timilsina and Shresta 2014). If these commitments are confirmed, the second-generation biofuels are expected to produce the other 5 % in volume within the next five years in the European Union. Thus, once the last technological and economic uncertainties are controlled, investments at the industrial scale should be on the raise again.

3 Economic and Environmental Considerations for the Development of Second and Third-Generation Biofuels

The second-generation processes extend the resource to the whole plant and to organic wastes of diverse origins. The first industrial commercial plants are already coming on the market, but these technologies are still under development to make the various technological pathways (biochemical, or thermochemical) more sustainable and competitive. We encourage the interested reader to refer in this book to Chapters "Second-generation bioethanol" and "Biodiesel and Bioethanol from Microalgae" for further technical and scientific information on these processes and to Chapter "Life Cycle Assessment of Biofuels", for a deeper understanding of this reference methodology.

The production of second-generation ethanol from the cellulosic agricultural residues (wheat straw, rice, corn stover, and sugar cane) or forestry (wood from eucalyptus, poplar, pine) has the advantage to be not directly in competition with food production and to ensure additional opportunities and sustainable income to the agroforestry sector in difficulty. It would be too long to describe all the production processes in details. However, two main pathways for the synthesis of second-generation fuels are competing: the biochemical and thermochemical pathways (Nanda et al. 2014). The bioconversion of lignocellulose requires several sequential steps: (i) pretreatment of the biomass and removing the lignin residues permits the release of the cellulose and hemicellulose. (ii) It allows the saccharification by the depolymerization of these polysaccharides that become available as hexoses and pentoses. (iii). The simultaneous fermentation of glucose and xylose mixture produces ethanol, (iv) that is further distilled and rectificated. These steps are for instance explained extensively in the book Biofuels: alternative feedstocks and conversion processes (Pandey et al. 2011). The third-generation processes use microalgae either heterotrophically or autotrophically. They are still under intensive development at a laboratory and demonstration pilot plant scale (Pruvost et al. 2015).

3.1 Impact of Pretreatment Technologies

The pretreatment of biomass can be conducted according to physico-chemical, chemical, hydrothermal, and biological methods. A large diversity of these pretreatments technologies are still being tested at a laboratory scale, knowing they are the most important bottleneck toward competitive second-generation fuels. These treatments are described extensively in the book Chapters "Pretreatment Processes for Cellulosic Ethanol Production: Processes Integration and Modeling for the Utilization of Lignocellulosics Such as Sugarcane Straw" and "Fungal Enzymatic Degradation of Cellulose".

3.1.1 Physico-chemical, Chemical, and Hydrothermal Pretreatments

Among the physico-chemical pretreatment methods, alkaline wood fibers explosion with ammonia (AFEX) is probably the most commonly used. It is carried out at high temperature and pressure (60–200 °C and 1.4–4.8 MPa). A sudden pressure reduction induces an explosive evaporation of ammonia causing the removal of lignin and hemicellulose without the production of any inhibitor compounds for the fermentation (Chiaramonti et al. 2012). Ozonolysis has the same effect at room temperature. However, the amount of ozone being required is considerable (Prasad et al. 2007); this makes the technology nonprofitable for the production of commodities. Others physico-chemical methods are taking advantage from gamma rays, pulsed electric fields, electron beams, ultrasound (Nanda et al. 2014), or microwaves (Zhu et al. 2006) but they are not industrially developed.

The chemical pretreatment is carried out in three ways. The sulfuric or concentrated hydrochloric acid hydrolysis is not recommended since the concentrated acids are a source of corrosion for the reactors and a risk to the operators. These acids need to be recovered to make the method economically competitive and to limit the burdens on the environment. In addition, they are the cause of furfurals formation, a well-known fermentation inhibitor. This pretreatment was replaced by the combination of dilute acid hydrolysis with subsequent enzymatic treatment. In this case, the operating temperature is 130–150 °C. A recent energy study shows that the majority of the energy necessary is supplied for the heating of biomass. Although the sugar yields obtained are comparable, the energy needed increases considerably (Mafe et al. 2015). Alkaline hydrolysis with ammonium hydroxide or sodium hydroxide allows less sugar degradation than with the acid hydrolysis. However, it causes saponification and intermolecular reactions between ester bonds that leads to cross-links between hemicellulose and other components. In addition, the recycling of the bases is often a difficult and costly process at the industrial scale (Wyman et al. 2005). The organosolv process uses the direct action of water and a dissolved organic solvent such as ethanol or methanol to solubilize the lignin and the hydrolyzed hemicellulose. The temperature varies from 20 to 200 °C depending on the organic solvent used. However, the furfurals are produced in sufficient concentrations (3–15 mM) to inhibit fermentation (Zhao et al. 2009). More recently, the use of ionic liquids, such as acetate 1-ethyl-3-methylimidazolium found applications to extract the lignin and solubilize the cellulose by reducing its crystallinity. Thus, it gives an easier access to the side chains of the polysaccharide for the cellulases (Lee et al. 2009).

Steam explosion is the first form of hydrothermal pretreatment. The wood is treated with high-pressure saturated vapor steam (160–260 °C) causing the decomposition of the hemicellulose and the degradation of lignin. However, inhibitors are also generated and an additional washing step with water is generally necessary (Garrote et al. 1999). At a pressure and temperature above its critical point (22.1 MPa, 374 °C), the water behaves as a weak solvent and acts as a catalyst, via its hydrogen and hydroxide ions, in the hydrolysis of many nonpolar substances including lignin, hemicellulose, and cellulose. The delignification can be further improved by using mixed co-solvents (water–ethanol at 190 °C and 16 MPa). However, considering that fermentation inhibitors are once again formed, it requires post-detoxification treatments. Furthermore, as the hydrolysis starting point for the cellulose (230 °C) and for the hemicellulose (100 °C) are quite low, the neo-formed monomers of xylose and glucose will continue to degrade into organic acids such as formic, acetic, glycolic, or pyruvic acids. These organic acids may be useful as a substrate for methane and hydrogen production via specific anaerobic fermentation pathways, but are unwelcome substances for ethanol production (Kumar 2013).

3.1.2 Biochemical Pretreatments

The biochemical pretreatment also called saccharification is the enzymatic hydrolysis that is carried out to convert the cellulose into glucose and the hemicellulose into xylose. It is generally performed after a first pretreatment of the biomass so that the cellulolytic and hemicellulolytic enzymes have facilitated access to polymers. In comparison to other pretreatments, the operational cost of the enzyme is low. In addition, the operating conditions are relatively mild, pH 4.8 and 45–50 °C. This step can be performed ahead of the fermentation or simultaneously. In this second case, the accumulation of sugars in the reactor is lower and induced higher saccharification and fermentation yields (Nanda et al. 2014). The saccharification yields can be further improved by using a cocktail of several well-selected enzymes working synergistically (Van Dyk and Pletschke 2012). More recently, research in synthetic biology is being conducted with the aim at introducing labile linkage into the lignin backbone of the crops in order to facilitate its chemical depolymerization during the process (Wilkerson et al. 2014). The successful engineering of plant cell walls offers consequently opportunities toward enhanced biofuel production but adverse effects on plant growth and development have also to be considered with appropriate strategies, such as the selection of adequate promoters and right spatiotemporal expression profiles (Loqué et al. 2015).

3.2 Thermochemical or Biochemical Treatments?

3.2.1 Thermochemical Treatments

Four major thermochemical processes are readily available on the market. Pyrolysis, used to convert organic material into oil at temperatures below 700 °C and gasification, used to obtain a syngas at temperatures ranging from 800 to 1400 °C are traditionally the technologies of choice. The gasification might be carried out either in a fixed bed (but it gives a poor gas and a significant tar rate) or in a fluidized bed with a significant improvement in the process performance. The fluidized bed reactors make it possible to operate at high pressure and are producing the richer syngas available but at a demonstration stage only (Ruiz et al. 2013). However, only the syngas way allows the production of ethanol by a catalytic synthesis (Nanda et al. 2014). Besides the syngas-ethanol pathway appears as a crossroads for the synthesis of particularly interesting green chemistry components by opening the way toward other molecules such as butadiene (Makshina et al. 2014). Finally, a hybrid thermochemical–biochemical pathway makes it possible to convert syngas into ethanol through anaerobic mixed cultures for instance with *Alkalibaculum bachii* in coculture with *Clostridium propionicum* (Liu et al. 2014). Although this pathway has to be intensified, it is considered as a credible alternative for the catalytic synthesis pathway and a possible way for the recovery of organic wastes from household origin (Latif et al. 2014). The alcohols of higher molecular weight that have been produced such as *n*-butanol and *n*-hexanol have better combustion properties in comparison with ethanol because of their higher energy density (ethanol 23.4 MJ L^{-1}; *n*-butanol 27 MJ L^{-1}; *n*-hexanol 32 MJ L^{-1}) and a lower solubility in water. Therefore, they are good candidates to be incorporated in fuels for aviation (Sarathy et al. 2014). *n*-butanol can also be obtained from the sole biochemical pathway via the acetone–butanol–ethanol or ABE fermentation but will need more improvements from the molecular biology to become economically competitive in the near future (Wang et al. 2014).

Finally, hydrothermal liquefaction further enables the transformation of organic matter into oil at about 100 bar and 300 °C (Duan and Savage 2011). Similarly, the gasification in supercritical water permits the transformation of organic matter into syngas (mixture of CO, H_2, CO_2 and H_2S). They are used to treat high moisture biomass (>50 % water) without going through a pre-drying step (Yakaboylu et al. 2015). For this reason, both methods seem particularly well suited for the development of third-generation biofuels from microalgae (Santana et al. 2012; Delrue et al. 2013). Due to the high pressure required, they are, however, neither energetically nor economically effective in the treatment of organic materials with lower water content such as those used for the second-generation biofuels.

3.2.2 Biochemical Treatments

The bioconversion of the produced monosaccharides to ethanol is realized by fermentation. The solid state fermentation could be advantageous as compared to the submerged fermentation because of a greater ethanol yield, reduced investment and operating costs due to the smaller size of the fermentation units and lower costs of purification (Couto and Sanrormán 2006). Many fungi, bacteria and yeasts have the ability to produce ethanol. *Saccharomyces cerevisiae* does not possess the ability to use xylose natively as a substrate. Metabolic engineering techniques have, however, allowed the expression of the missing enzymes xylose reductase, xylodeshydrogenase, and xylulokinase. It leads to an improvement of about 15 % of the ethanol yield (Katahira et al. 2008). The bacterium *Zymomonas mobilis* has the potential to produce ethanol with theoretical yields higher that 97 % but suffers from very high specificity for the substrates: glucose, fructose, and sucrose only (Fukuda et al. 2009). However, metabolic engineering approaches make it a natural candidate as a microorganism platform for the future biorefineries (He et al. 2014). The resulting mixture is then distilled and rectified in accordance with the conventional chemical engineering technologies on distillation column. The removal of the water residues could be further achieved on molecular sieved absorbers. Current research shows that it is possible to reduce significantly the energetic cost associated with these steps at the condition of a slightly higher initial investment (Palacios-Bereche et al. 2015). Energy optimization routines conducted throughout the downstream processing of ethanol indicate that the energy efficiency of the existing facilities can be improved by 23 % for the ethanol production and up to 49 % if coproducts are being considered (Lassmann et al. 2014).

3.2.3 Comparative Assessment

The biochemical pathway has suffered for a long time from its comparison with the thermochemical pathway. Techno-economic studies emphasize the major role of the biomass pretreatment costs and the potential generation of inhibitors of the alcoholic fermentation. Furthermore, the cost of the saccharification enzymes was also a fundamental bottleneck to be economically competitive without any incentive policy implemented by the governments (Klein-Marcuschamer and Blanch 2015). However, recent scientific advances in the cellulose (Aden and Foust 2009) and hemicellulose hydrolysis (Sella Kapu and Trajano 2014) have helped to establish effective and competitive conversion processes. These advances are limiting the production of furfural with the concomitant development of new, more tolerant strains (Fujitomi et al. 2012) or able to simultaneously use xylose in addition to glucose as a substrate (Ito et al. 2010).

The latest techno-economic studies demonstrate a minimum ethanol selling price of approximately 0.58–0.70 US$ L. The breakdown analysis indicated that the cost is divided into 35 % for the feedstock materials (corn stover), 16 % for the production of enzymes and 49 % for the other manufacturing costs. This is nearly one US$ per liter (in constant $) down as compared to 2000 (NREL 2011, Gabriel and El-Halwagi 2013). The cost of ethanol is consequently at approximately 115 US$ for the same energy content than one barrel of Brent. However, prices for fossil fuels are very volatile and are a concern for the economical profitability of the biofuel-based sector. Due to geopolitical considerations, the price for a barrel of Brent was for example at a historically low level of approximately 60$ on March 2015 and even 31$ in January 2016.

The technical comparison between the biochemical and thermochemical pathway, gave comparable performances in terms of ethanol yields. It is presently approximately 280 L per ton of poplar wood and it is planned to reach 360 per ton within the next 10 years. The environmental balance is in favor of the biochemical pathway for the reduction of greenhouse gas emissions: 0.3 against 0.8 kg eq. CO_2 eq per liter of oil equivalent and the depletion of fossil resources: −8 MJ against − 2 MJ per liter of oil equivalent. However, the water consumption and the use of sulfuric acid are in its disfavor (Mu et al. 2010). 14 L of water were used to produce 1 L of oil equivalent with the dilute acid sulphuric pathway against only 2.5 L for the thermochemical pathway. In the long term, the economic comparison of the biochemical versus the thermochemical pathway might be favorable on the thermochemical pathway again thanks to possible scale-up gains on the energetic performances (Mu et al. 2010).

The greenhouse gases reduction potential for the second-generation ethanol of perineal crops was estimated at 90–103 % for corn stover, 77–97 % for switchgrass and up to 90–130 % for *Miscanthus giganteus*. This performance is almost twice improved over the first-generation: 19–48 % for corn and 40–62 % for sugarcane (Wang et al. 2012). Regarding poplar, its greenhouse gas reduction potential varies from 10 to 90 % depending on the farming practices and the transformation processes used (Guo et al. 2014). These observed values and the causes responsible for their variations are confirmed for other wood species (Morales et al. 2015).

The potential for reducing the greenhouse gases is only slightly affected by the choice of biomass pretreatment technology (Pourbafrani et al. 2014). However, this technological choice is crucial for the recovery of other valuable bio-products in a biorefinery strategy and they can significantly improve the balance sheet (FitzPatrick et al. 2010). For example, xylitol can be extracted from bark and recovered as a sucrose substitute. Lignin may be recovered not only energetically in cogeneration but also in fine chemistry toward the production of vanillin or bioplastics for example (Eerhart et al. 2014).

A critical point in the development of the sector concerns the capacity of the production unit and the associated supply chain to deliver the raw materials. Studies show that a production unit in the range of 80–100 million liters per year, collecting its raw material in an approximate radius of 50 km presents the best balance between return on investment, minimization of the environmental impacts, creation of local employment, and support for the local agroforestry (Yue et al. 2014). It is for instance equivalent to approximately 8 % of all the lignocellulosic materials produced in a 50 km range from the facility in France. These second-generation development works are currently resulting in the establishment of the first pre-commercial demonstration plants in Europe (e.g., Futurol project, France, 3.5 million liters per year) and the first commercial production units in the US. Prospective analysis indicates that the European Union can potentially feed 70–80 biorefineries using straw and wood. The biochemical pathway has the potential to create more employment per unit of raw materials than the thermochemical pathway (Thornley et al. 2014).

3.3 Future Challenges to Improve the Environmental Balances of Biofuels

Figure 7 (top) presents the greenhouse gases reduction potential for different feedstock of first-generation. These effects are further amplified with better yields per hectare. In the case of second-generation *Miscanthus x giganteus*, *Arundo donax,* and *Pennisetum purpureum* seem the most promising feedstock for achieving high-energy yields (Laurent et al. 2015). Unfortunately, there is neither ideal feedstock nor universal process for the biofuel production process and the performances are dependent on the good adequacy between the feedstock properties and the whole process. In the case of second-generation fuels, the pretreatment step is of particular importance. For instance, depending on the composition of the plant cell wall and in particular the amount of ramifications on the polysaccharides fractions, either the dilute acidic pretreatment for wheat straw or the ammonia fiber expansion for corn stover (Wyman et al. 2005) will perform best (Fig. 7, bottom). Furthermore, in the situation of intensive farming, a high input of nitrogen fertilizer might lead to adverse effects. In fact, ammonia is produced with the Haber reaction at 200–400 bars and 450 °C in the reaction of the nitrogen from air and natural gas. This is both a high energetic costly process and part of the explanation on the relationship between fossil fuels and food crops prices (Esmaeili and Shokoohi 2011).

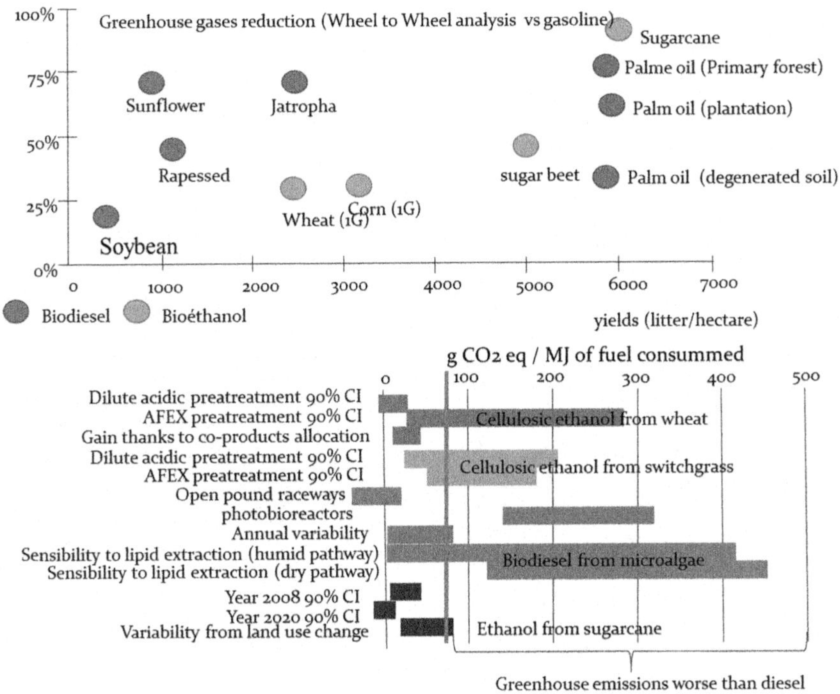

Fig. 7 Comparative assessment of the greenhouse gases reduction potential according to the type of feedstock used (*top*) and to the type of pretreatment and amelioration potential of the existing technologies. AFEX stands for ammonia fiber expansion. Data compiled from Maurya et al. (2015), Mafe et al. (2015) and Pandey et al. (2014)

Therefore, the nitrogen cycle needs to be closed in a nonartificial way at the agricultural parcel. Although, the bioelectricity produced in cogeneration from sugarcane bagasse in Brazil certainly helps to make the whole process economically viable even without any subsidy, it also leads to less nitrogen fixed on the soils and sustainability issues (Mitchell et al. 2000). The practice of Green Cane Trash Blanket (Fig. 8) proves its potential to increase the chemical, biological, and physical fertility of the soils (Thorburn et al. 2005). Alternative agricultural practices such as cocultivation of annual cereal-legume intercrops prove its potential to increase feedstocks yield without additional fertilizers inputs (Pelzer et al. 2014). The right balance has to be found between the plant proportion used for electric and heat generation and those returning to the ground, between earnings in the short-term and long-term management of the resource.

Similarly, for vinasses, the final by-product of the biomass distillation from sugar crops and one of the main wastes of the fermentation into ethanol, raises much attention because of environmental problems associated with the inadequate and indiscriminate disposal in soils and water (Christofoletti, et al. 2013). Many studies proved its ecotoxicity. For instance, the effect on seed germination depends

Fig. 8 Fresh trash blanket in a cane block soon after harvest in Brazil

on the concentration and the crop (Pant and Adholeya 2007). It may induce a decrease in the germination rate of peas or sunflowers (Algur and Kadioglu 1992) but vinasses at 10 % has a positive effect on the germination of onions seeds (Ramana et al. 2002). Yesalida (1999) demonstrated genotoxic effects of vinasses on fecundity and longevity of *Drosophila melanogaster*. Furthermore, vinasses also contribute to greenhouse gases emissions during its storage, transportation, and application to soils. For instance, de Oliveira and coworkers (2013) estimate that the application of 200 m^3 ha^{-1} of vinasse triples the CO_2 emissions and increases the release of N_2O by the soil. The fluxes of N_2O and CH_4, converted into CO_2 equivalent, indicate that each m^3 ha^{-1} of vinasse applied to the soil emits 0.491 and 0.314 kg of CO_2 eq in the burnt and unburnt sugarcane areas. However, rather than being a waste, vinasses might also be used as a valuable resource toward the production of biogas via anaerobic fermentation pathways (Syaichurrozi et al. 2013). This production of biogas might be further used to produce the necessary N-fertilizer in an attempt to decorrelate the agricultural practices from the massive use of fossil fuels. Similarly, the selection of specific anaerobic consortia makes it also possible to produce biohydrogen and volatile fatty acids (Sydney et al. 2014).

Hence, the principles of industrial ecology (Erkman 1997, 1998) characterized by a strong industrial and political will to minimize the negative externalities caused by the production processes, applied to the biorefinery concept demonstrate their potential. Taking into account, the whole process rather than segmented unit operations is even more critical in the case of third-generation biofuels.

Microalgae are a promising feedstock for a sustainable production of biofuels. The lipid productivity may be high and the production can be done using seawater, without any competition for food or land use. The technology must be transferred from high-value small niche markets toward industrial scale of commodities production (Barbosa and Wijffels 2013). However, it seems in the short to medium timescale (until 2050) not economically feasible to produce biofuels from microalgae at a significant extent. Depending on the techno-economic studies and the underlying technological options; photo bioreactors, open-pound raceways, auto- or heterotrophy; the prices for crude fatty acid methyl esters have been established between 3 and 30 US\$ L^{-1}, which is even for the most optimistic scenario still highly unfavorable as compared to fossil fuels (Grima et al. 2013). However, in a bio-based economy, the valorization of by-products for food and fine chemical in an integrated biorefinery strategy may lead to feasibility (Downes and Hu 2013; Barbosa and Wijffels 2013). In this situation, there may be a competition between the different uses of the microalgae since the mode of cultivation strongly influences the yields for the various interesting compounds such as produced hydrogen, lipid content or valuable pigments, and antioxidants. Once again, it appears of prime importance to consider all the production steps of the benefit as a whole rather than in an individualized way.

More recently, projects are on the way to take full use of the bioremediation potential of microalgae (Renuka et al. 2015) in order to improve the environmental impact of first and second-generation biorefineries. In fact, in accordance to Gay Lussac's reaction, the maximum efficiency establishes that 100 kg of glucose will produce 48.4 kg of ethanol and 22 kg of carbon dioxide. The carbon dioxide associated with nitrogen and phosphorus supplied from wastewater from the distillation or from residential area might become a substrate for the microalgae. These kinds of projects integrating first, second and third-generation biorefineries will make it possible to gain the experience needed to deploy the third-generation facilities in an economically and competitive way.

The fundamental challenge concerning the life cycle assessment methodology is the allocation of the impacts to the old garbage uses as new resources. In the current situation, the state of the art states that all the environmental impacts are systematically allocated to the products of economic interest. The modalities of this allocation, according to the economic value, mass, energy content, or any other possibilities are to be defined by the analyst according to the purpose of his study. Therefore, when a waste is produced in the process, it has by definition no environmental impact. Only the associated costs for its ultimate treatment can be attributed to this waste. However, when a specific waste becomes a resource, due to some technical improvement's or paradigm shift, it seems appropriate to reassess the way the environmental impacts are allocated. This is a critical step to make sure that this life cycle methodology can be used dynamically for the continuous improvement of existing production processes and that the observed environmental gains are not simply due to some accounting refinements.

4 Impact of Policies

Many strategies are designed in order to mitigate the impact of human activities on the environment. It is the case for the large-scale deployment of cleaner energy technologies. An environmental assessment of these impacts is necessary. Life cycle assessment (LCA) is up to date the method of choice for these evaluations. However, one of the difficulties with LCA methodology is that the definition of the system boundaries is chosen depending of the aim of the study. Indeed, it has to be applied to different types of decisions from the evaluation of a very specific product (considering a given process scheme) to the potential indirect effects caused by the introduction of this new product on the market. Consequently, the deeper its penetration on the market, in terms of volume sales and territory covered, the likelier these indirect effects will be. In other terms, Timilsina (2014) define the problem in this way: "Does the increasing oil price have an impact on demand for biofuels? And reciprocally, does an expansion of biofuels put pressure on oil price to fall down?" Despite a large number of econometric studies analyzing the relationship between the oil prices and the biofuels markets, no consensus can be made (Cha and Bae 2011). Computable General Equilibrium models are able to assess inter-sectoral linkages. They found out a global penetration rate for biofuel of 5.4 % in 2020. A 25 % increase in oil prices from the baseline enhances this penetration rate to 7 %. In addition, a 100 % increase in oil prices leads to a 13 % penetration rate. Middle and lower income countries will experience more rapid biofuel penetration in response to these high oil prices than the one with higher income (Timilsina 2014). Agricultural commodities are the primary raw materials for biofuel production. An increased price for the feedstock will lead to the rise of production costs and to a reduced demand. In this case, the correlation between oil price and agricultural commodities is a major concern. To the best of our knowledge, there is hardly any study evaluating the feasibility to decouple the agricultural crops production from fossil fuels with the help of biofuels and other renewable energies.

Fossil fuels are partly cheaper because of the pollution and the other negative externalities that are not taken into account in their prices. To create the conditions for a competition between fossils and biofuels one has to tax the first and/or subsidy the second (Timilsina et al. 2014). Therefore, in the absence of any regulatory mechanism giving the benefit to the ecological, social, political, and maybe technological intelligence aiming at preserving the natural capital, the large-scale development of "green fuels" will be just another additional energy to support the global growth and economic activity and will never be a real alternative associated with the energetic transition.

It is striking that in their prospective analysis report Deep Pathways to Decarbonization (SDSN and IDDRI 2014), whatever the scenario being considered (sobriety, efficiency, diversity, or decarbonization), the economists attribute this mechanism to the sole carbon tax (obviously for practical reasons); that is to say that the carbon impact is the only one of the many environmental impacts associated with human activities. There is to date no prospective study showing what would be the possible pollution report from the greenhouse gas emissions toward the other environmental compartments of such a system. Furthermore, it should be noticed that the potential impacts on environment increase inequalities both globally, between countries, and locally, between people. A final questioning seems necessary: why is the value systematically given to the remediation of the environmental damages rather than to the rational management of the common good? In this situation, the development of regionalized solutions taking into account the environmental parameters, such as climate, local feedstocks and biodiversity is an axis of research that shall not be overlooked. These aspects are all more critical that the consequential life cycle analyses; they were recently reported to mislead the economic and political decision-makers (Plevin et al. 2014). This is not only to assess what the best final state is, but also to evaluate what is the best way to get there (McKone et al. 2011) and many reflections are being conducted in this way (Milder et al. 2008; Davis et al. 2009; Wiloso et al. 2012; Gobert and Brullot 2013).

5 Conclusion

Biofuels production processes, whatever the generation, are at the center of industrial ecology principles. In this context, the development of integrated bio refineries combining the best of the first, second, and third-generation processes in one same geographical unit is a promising line to work on.

However, even an ideally environmental-friendly energy resource, though large investments and governmental subsidies may have negative indirect effects. This leads to a drop in the prices of fossil resources and makes them even more attractive and used in other parts of the planet. Without any regulation, these financials mechanisms have the side effect of making the efforts obsolete and even counterproductive, at least on the global environmental indicators, such as the emissions of greenhouse gases. Therefore, in the absence financial regulatory, the large-scale development of "green fuels" will be just another additional energy to support the global economic activity and never a real alternative associated with the energetic transition.

To conclude, the environmental impact of the production of biofuels cannot be tackled without a transparent and standardized collaboration of all stakeholders. They have a common interest to join forces in a holistic research approach that has to be necessarily conducted in a transdisciplinary approach, and not dictated by the sole economic aspects. The development of the biorefinery lies now at a crossroad. It must be refocused on the territorial aspects, especially with the agricultural and forestry sector. It is mandatory to evaluate the actual environmental impacts in terms of waste generation, raw material consumption and regeneration, water availability, and waste recycling processes. This must be done in relation with a geographical context, including all socioeconomic backgrounds, resources availability, and short-range markets possibilities. But the virtuous modes of energy generation for the future remain largely to be reinvented.

Acknowledgments This work has been sponsored by the French government research program "Investissements d'avenir" through the IMobS3 Laboratory of Excellence (ANR-10-LABX-16-01), by the European Union through the program Regional competitiveness and employment 2007–2013 (ERDF—Auvergne region), and by the Auvergne region.

References

Aden A, Foust T (2009) Techno economic analysis of the dilute sulphuric acid and enzymatic hydrolysis process for the conversion of corn stover to ethanol. Cellulose 16(4):535–545

Alghur OF, Kadioglu A (1992) The effects of vinasse on the growth, biomass and primary productivity in pea (*Pisum sativum*) and sunflower (*Helianthus anus*). Agric Ecosyst Environ 39(3,4):139–144

Barbosa MJ, Wijffels RH (2013) Biofuels from microalgae. In: Richmond A, Hu Q (eds) Handbook of microalgal culture, applied phycology and biotechnology, 2nd edn. Wiley, Hoboken, pp 566–577

Carbajales-Dale M, Barnhart CJ, Benson SM (2014) Can we afford storage? A dynamic net energy analysis of renewable electricity generation supported by energy storage. Energy Environ Sci 7 (5):1538–1544

Cha KS, Bae JS (2011) Dynamic impacts of high oil prices on the bioethanol and feedstock markets. Energy Policy 39(2):753–760

Chiaramonti D, Prussi M, Ferrero S et al (2012) Review of pretreatment processes for lignocellulosic ethanol production, and development of an innovative method. Biomass Bioenergy 46:25–35

Christofoletti CA, Escher JP, Correia JE et al (2013) Sugarcane vinasse: environmental implications of its use. Waste Manage 33(12):2752–2761

Couto SR, Sanromán MÁ (2006) Application of solid-state fermentation to food industry—A review. J Food Eng 76(3):291–302

Davis SC, Anderson-Teixeira KJ, DeLucia EH (2009) Life-cycle analysis and the ecology of biofuels. Trends Plant Sci 14(3):140–146

de Oliveira BG, Carvalho JLN, Cerri CEP et al (2013) Soil greenhouse gas fluxes from vinasse application in Brazilian sugarcane areas. Geoderma 200–201:77–84

Delrue F, Li-Beisson Y, Setier PA et al (2013) Comparison of various microalgae liquid biofuel production pathways based on energetic, economic and environmental criteria. Bioresour Technol 136:205–212

Dias MOS, Junqueira TL, Rossell CEV et al (2013) Evaluation of process configurations for second generation integrated with first generation bioethanol production from sugarcane. Fuel Process Technol 109:84–89

Directive 2009/28 CE of the European Parliament (2009) On the promotion of the use of energy from renewable. Available online at http://eur-lex.europa.eu/legal-content/fr/ALL/?uri=CELEX:32009L0028. Last accessed on Sept 2015

Downes CM, Hu Q (2013) First principles of techno-economic analysis of algal mass culture. In: Richmond A, Hu Q (eds) Handbook of microalgal culture, applied phycology and biotechnology, 2nd edn. Wiley, Hoboken, pp 310–326

Duan P, Savage PE (2011) Hydrothermal liquefaction of a microalga with heterogeneous catalysts. Ind Eng Chem Res 50(1):52–61

Eerhart AJJE, Huijgen WJJ, Grisel RJH et al (2014) Fuels and plastics from lignocellulosic biomass via the furan pathway; a technical analysis. RSC Adv 4(7):3536–3549

Erkman S (1997) Industrial ecology: an historical view. J Cleaner Prod 5(1,2):1–10

Erkman S (1998) Vers une Ecologie Industrielle. Editions Charles Leopold Meyer, Paris

Esmaeili A, Shokoohi Z (2011) Assessing the effect of oil price on world food prices: application of principal component analysis. Energy Policy 39(2):1022–1025

European Commission (2012) Innovating for sustainable growth: a bioeconomy for Europe; final, Brussels. Available online at http://ec.europa.eu/research/bioeconomy/pdf/bioeconomy communicationstrategy_b5_brochure_web.pdf. Last accessed on Sept 2015

FitzPatrick M, Champagne P, Cunningham MF et al (2010) A biorefinery processing perspective: treatment of lignocellulosic materials for the production of value-added products. Bioresour Technol 101(23):8915–8922

Font Vivanco D, Freire-González J, Kemp R et al (2014) The remarkable environmental rebound effect of electric cars: a microeconomic approach. Environ Sci Technol 48:1063–12072

Fujitomi K, Sanda T, Hasunuma T et al (2012) Deletion of the PHO13 gene in *Saccharomyces cerevisiae* improves ethanol production from lignocellulosic hydrolysate in the presence of acetic and formic acids, and furfural. Bioresour Technol 111:161–166

Fukuda H, Kondo A, Tamalampudi S (2009) Bioenergy: sustainable fuels from biomass by yeast and fungal whole-cell biocatalysts. Biochem Eng J 44(1):2–12

Gabriel KJ, El-Halwagi MM (2013) Modeling and optimization of a bioethanol production facility. Clean Technol Environ Policy 15(6):931–944

Garrote G, Dominguez H, Parajó JC (1999) Hydrothermal processing of lignocellulosic materials. Holz Als Roh- Und Werkstoff 57(3):191–202

Gillani ST, Belaud JP, Sablayrolles C et al (2013) A cape based life cycle assessment for evaluating the environmental performance of non-food agro-processes. Chem Eng Trans 3:211–216

Gnansounou E, Dauriat A (2011) Life-cycle assessment of biofuels. In: Pandey A, Larroche C, Ricke SC et al (eds) Biofuels, alternative feedstocks and conversion processes. Academic Press, Oxford, pp 25–50

Gobert J, Brullot S (2013) Bioraffineries, biocarburants, produits bio-sourcés: l'écologie comme prétexte? In: Diemer A, Figuiere C, Pradel M (eds) Ecologie politique vs Ecologie Industrielle: Quelles stratégies pour le développement durable?. Editions Oeconomia, Clermont-Ferrand, pp 232–252

Grima EM, Acién Fernández FG, Robles Medina A (2013) Downstream processing of cell mass and products. In: Richmond A, Hu Q (eds) Handbook of microalgal culture, applied phycology and biotechnology, 2nd edn. Wiley, Hoboken, pp 267–309

Guo M, Littlewood J, Joyce J et al (2014) The environmental profile of bioethanol produced from current and potential future poplar feedstocks in the EU. Green Chem 16:4680–4695

He MX, Wu B, Qin H et al (2014) *Zymomonas mobilis*: a novel platform for future biorefineries. Biotechnol Biofuels 7(1):101

International Energy Agency (2012) World energy outlook 2012. Available online at www.iea.org/publications/freepublications/publication/WEO2012_free.pdf. Last accessed on Sept 2015

Ito T, Hotta A, Uchida A et al (2010) Intensification of bioethanol fermentation by recombinant yeast with xylose isomerase pathway. Chem Eng Trans 20:103–108

Jolliet O, Dubreuil A, Gloria T et al (2005) Progresses in life cycle impact assessment within the UNEP/SETAC life cycle initiative. Int J Life Cycle Assess 10(6):447–448

Katahira S, Ito M, Takema H et al (2008) Improvement of ethanol productivity during xylose and glucose co-fermentation by xylose-assimilating S. cerevisiae via expression of glucose transporter Sut1. Enzyme Microb Technol 43(2):115–119

Khoury CK, Bjorkman AD, Dempewolf H et al (2014) Increasing homogeneity in global food supplies and the implications for food security. Proc Natl Acad Sci 111(11):4001–4006

Klein-Marcuschamer D, Blanch HW (2015) Renewable fuels from biomass: technical hurdles and economic assessment of biological routes. AIChE J 61(9):2689–2701

Koh LP, Ghazoul J (2008) Biofuels, biodiversity and people: understanding the conflicts and finding opportunities. Biol Conserv 141(10):2450–2460

Kumar S (2013) Sub- and Supercritical Water Technology for Biofuels. In: Lee JW (ed) Advanced biofuels and bioproducts. Springer, New York, pp 147–183

Lamers P, Hamelinck C, Junginger M et al (2011) International bioenergy trade—a review of past developments in the liquid biofuel market. Renew Sustain Energy Rev 15(6):2655–2676

Lamers P, Junginger M, Hamelinck C et al (2012) Developments in international solid biofuel trade—an analysis of volumes, policies, and market factors. Renew Sustain Energy Rev 16 (5):3176–3199

Lapola DM, Schaldach R, Alcamo JA et al (2010) Indirect land-use changes can overcome carbon savings from biofuels in Brazil. Proc Natl Acad Sci 107(8):3388–3393

Lassmann T, Kravanja P, Friedl A (2014) Simulation of the downstream processing in the ethanol production from lignocellulosic biomass with ASPEN Plus® and IPSEpro. Energy Sustain Soc 4(1):4–27

Latif H, Zeidan AA, Nielsen AT et al (2014) Trash to treasure: production of biofuels and commodity chemicals via syngas fermenting microorganisms. Curr Opin Biotechnol 27:79–87

Laurent A, Pelzer E, Loyce C et al (2015) Ranking yields of energy crops: a meta-analysis using direct and indirect comparisons. Renew Sustain Energy Rev 46:41–50

Lee SH, Doherty TV, Linhardt RJ et al (2009) Ionic liquid-mediated selective extraction of lignin from wood leading to enhanced enzymatic cellulose hydrolysis. Biotechnol Bioeng 102 (5):1368–1376

Lee JSH, Abood S, Ghazoul J et al (2014) Environmental impacts of large-scale oil palm enterprises exceed that of smallholdings in Indonesia forst loss from Sumatra's oil palm industry. Conserv Lett 7(1):25–33

Liu K, Atiyeh HK, Stevenson BS et al (2014) Continuous syngas fermentation for the production of ethanol, n-propanol and n-butanol. Bioresour Technol 151:69–77

Locqué D, Scheller HV, Pauly M (2015) Engineering of plant cell walls for enhanced biofuel production. Curr Opin Plant Biol 25:151–161

Mafe OAT, Davies SM, Hancock J et al (2015) Development of an estimation model for the evaluation of the energy requirement of dilute acid pretreatments of biomass. Biomass Bioenergy 72:28–38

Makshina EV, Dusselier M, Janssens W et al (2014) Review of old chemistry and new catalytic advances in the on-purpose synthesis of butadiene. Chem Soc Rev 43(22):7917–7953

Maurya DP, Singla A, Negi S (2015) An overview of key pretreatment processes for biological conversion of lignocellulosic biomass to bioethanol. 3 Biotech 5(5):597–609

McKone TE, Nazaroff WW, Berck P et al (2011) Grand challenges for life-cycle assessment of biofuels. Environ Sci Technol 45(5):1751–1756

Menten F, Chèze B, Patouillard L et al (2013) A review of LCA greenhouse gas emissions results for advanced biofuels: the use of meta-regression analysis. Renew Sustain Energy Rev 26: 108–134

Milder JC, McNeely JA, Shames SA et al (2008) Biofuels and ecoagriculture: can bioenergy production enhance landscape-scale ecosystem conservation and rural livelihoods? Int J Agric Sustain 6(2):105–121

Mitchell RDJ, Thorrn PJ, Larsen P (2000) Quantifying the loss of nutrients from the immediate area when sugarcane residues are burnt. In: Proceedings of the 2000 conference of the Australian Society of Sugar Cane Technologies, pp 206–211

Mohr A, Raman S (2013) Lessons from first generation biofuels and implications for the sustainability appraisal of second-generation biofuel. Energy Policy 63:114–122

Morales M, Quintero J, Conejeros R et al (2015) Life cycle assessment of lignocellulosic bioethanol: environmental impacts and energy balance. Renew Sustain Energy Rev 42: 1349–1361

Mu D, Seager T, Rao S et al (2010) Comparative life cycle assessment of lignocellulosic ethanol production: biochemical versus thermochemical conversion. Environ Manage 46(4):565–578

Nanda S, Mohammad J, Reddy SN et al (2014) Pathways of lignocellulosic biomass conversion to renewable fuels. Biomass Convers Biorefinery 4(2):157–191

NRDC (2014) Report R: 14-04-A. Biofuel sustainability performance guidelines. Available online at www.nrdc.org/energy/files/biofuels-sustainability-certification-report.pdf. Last accessed on Sept 2015

NREL (2011) Technical report NREL/TP-5100-47764 Process design and economics for biochemical conversion of lignocellulosic biomass to ethanol. Dilute acid pretreatment and enzymatic hydrolysis of corn stover., Oak Ridge, U.S. Department of Energy, Office of Scientific and Technical Information

Ometto AR, Hauschild MZ, Lopes Roma WN (2009) Lifecycle assessment of fuel ethanol from sugarcane in Brazil. Int J Life Cycle Assess 14(3):236–247

Organization for Economic Co-operation and Development (2009) The bioeconomy to 2030: designing a policy agenda, Paris

Palacios-Bereche R, Ensinas AV, Modesto M et al (2015) Double-effect distillation and thermal integration applied to the ethanol production process. Energy 82:512–523

Pandey A, Larroche C, Ricke SC et al (eds) (2011) Biofuels: alternative feedstocks and conversion processes. Academic Press, Burlington

Pandey A, Negi S, Binod P et al (eds) (2014) Pretreatment of biomass, process technologies. Elsevier, Amsterdam

Panichelli LA (2012) Impact of biofuels production on land-use change and greenhouse gas emissions/methodological framework and system dynamics modeling applied to soybean based biodiesel production in Argentina, Ecole Polytechnique de Lausanne

Panichelli LA, Gnansounou E (2015) Impact of agricultural base biofuel production on greenhouse gas emissions from land use change: key modelling choices. Renew Sustain Energy Rev 42:344–360

Pant D, Adholeya A (2007) Biological approaches for treatment of distillery wastewater: a review. Bioresour Technol 98:2321–2334

Pelzer E, Hombert N, Jeuffroy MH et al (2014) Meta-analysis of the effect of nitrogen fertilization on annual cereal-legume intercrop production. Agron J 106(5):1775–1786

Plevin RJ, Delucchi MA, Creutzig F (2014) Using attributional life cycle assessment to estimate climate-change mitigation benefits misleads policy makers: attributional LCA can mislead policy makers. J Ind Ecol 18(1):73–83

Pourbafrani M, McKechnie J, Shen T et al (2014) Impacts of pre-treatment technologies and co-products on greenhouse gas emissions and energy use of lignocellulosic ethanol production. J Clean Prod 78:104–111

Prasad S, Singh A, Joshi HC (2007) Ethanol as an alternative fuel from agricultural, industrial and urban residues. Resour Conserv Recycl 50(1):1–39

Pruvost J, Cornet JF, Le Borgne F et al (2015) Theoretical investigation of microalgae culture in the light changing conditions of solar photobioreactor production and comparison with cyanobacteria. Algal Res 10:87–99

Querini F (2012) Analyse de cycle de vie des énergies alternatives pour l'automobile et proposition méthodologiques pour une meilleure évaluation des impacts locaux. Thèse de Doctorat ISAE-ENSMA, Poitiers

Ramana A, Biswas AK, Kundu S et al (2002) Effect of distillery effluent on seed germination in some vegetable crops. Bioresour Technol 82:273–275

REN21 (2015) Renewables 2015 global status report. Available online at http://www.ren21.net/status-of-renewables/global-status-report/. Last accessed on Sept 2015

Renuka N, Sood A, Prasanna R et al (2015) Phycoremediation of wastewaters: a synergistic approach using microalgae for bioremediation and biomass generation. Int J Environ Sci Technol 12(4):1443–1460

Ruiz JA, Juárez MC, Morales MP et al (2013) Biomass gasification for electricity generation: review of current technology barriers. Renew Sustain Energy Rev 18:174–183

Santana AA, Jesus S, Larrayoz MA et al (2012) Supercritical carbon dioxide extraction of algal lipids for the biodiesel production. Procedia Eng 42:1755–1761

Sarathy SM, Oßwald P, Hansen N et al (2014) Alcohol combustion chemistry. Prog Energy Combust Sci 44:40–102

SDSN, IDDRI (2014) Pathways to deep decarbonization report. Available online at www.unsdsn.org/wp-content/uploads/2014/09/DDPP_Digit_updated.pdf. Last accessed Sept 2015

Sella Kapu N, Trajano HL (2014) Review of hemicellulose hydrolysis in softwoods and bamboo. Biofuels Bioprod Biorefin 8(6):857–870

Stattman SL, Mol APJ (2014) Social sustainability of Brazilian biodiesel: the role of agricultural cooperatives. Geoforum 54:282–294

Syaichurrozi I, Budiyono, Sumardiono S (2013) Predicting kinetic model of biogas production and biodegradability organic materials: biogas production from vinasse at variation of COD/N ratio. Bioresour Technol 149:390–397

Sydney EB, Larroche C, Novak AC et al (2014) Economic process to produce biohydrogen and volatile fatty acids by a mixed culture using vinasse from sugarcane ethanol industry as nutrient source. Bioresour Technol 159:380–386

Thorburn PJ, Meier EA, Probert ME (2005) Modelling nitrogen dynamics in sugarcane systems: recent advances and applications. Field Crops Res 91:337–351

Thornley P, Chong K, Bridgwater T (2014) European biorefineries: implications for land, trade and employment. Environ Sci Policy 37:255–265

Timilsina GR (2014) Oil price and biofuels. In: Timilsina GR, Zilberman D (eds) The impacts of biofuels on the economy, environment, and poverty. Springer, New York, pp 103–110

Timilsina GR, Shrestha A (2014) An overview of global markets and policies. In: Timilsina GR, Zilberman D (eds) The impacts of biofuels on the economy, environment, and poverty. Springer, New York, pp 1–14

Timilsina GR, Csordás S, Mevel S (2014) Biofuel policies: subsidy vs. carbon tax. In: Timilsina GR, Zilberman D (eds) The impacts of biofuels on the economy, environment, and poverty. Springer, New York, pp 123–129

UNEP (2014) Assessing global land use: balancing consumption with sustainable supply. A Report of the Working Group on Land and Soils of the International Resource Panel

Valin H, Sands RD, van der Mesnbrugghe D et al (2014) The future of food demand: understanding differences in global economic models. Agric Econ 45(1):51–67

Van Dyk JS, Pletschke BI (2012) A review of lignocellulose bioconversion using enzymatic hydrolysis and synergistic cooperation between enzymes—factors affecting enzymes, conversion and synergy. Biotechnol Adv 30(6):1458–1480

Veleva V, Ellenbecker M (2001) Indicators of sustainable production: framework and methodology. J Clean Prod 9(6):519–554

Wang M, Han J, Dunn JB et al (2012) Well-to-wheels energy use and greenhouse gas emissions of ethanol from corn, sugarcane and cellulosic biomass for US use. Environ Res Lett 7(4):045905

Wang Y, Janssen H, Blaschek HP (2014) Fermentative biobutanol production: an old topic with remarkable recent advances. In: Bisaria VS, Kondo A (eds) Bioprocessing of renewable resources to commodity bioproducts. Wiley, Hoboken

Werpy T, Petersen G (2004) Top value added chemical from biomass, vol 1. Results of screening for potential candidates from sugars and synthetic gas. No. DOE/GO-102004-1992, 15008859, Office of the Biomass Program, Oak Ridge

Wilkerson CG, Mansfield SD, Lu F et al (2014) Monolignol ferulate transferase introduces chemically labile linkages into the lignin backbone. Science 344(6179):90–93

Wiloso EI, Heijungs R, de Snoo GR (2012) LCA of second generation bioethanol: a review and some issues to be resolved for good LCA practice. Renew Sustain Energy Rev 16(7): 5295–5308

Wyman CE, Dale BE, Elander RT et al (2005) Coordinated development of leading biomass pretreatment technologies. Bioresour Technol 96(18):1959–1966

Yakaboylu O, Harinck J, Smit K et al (2015) Supercritical water gasification of biomass: a literrature and technology overview. Energies 8(2):859–894

Yazdanie M, Noembrini F, Dossetto L et al (2014) A comparative analysis of well-to-Wheel primary energy demand and greenhouse gas emissions for the operation of alternative and conventional vehicles in Switzerland, considering various energy carrier production pathways. J Power Sources 249:333–348

Yeşilada E (1999) Genotoxic activity of vinasse and its effect on fecundity and longevity of *Drosophila melanogaster*. Bull Environ Contam Toxicol 63(5):560–566

Yue D, You F, Snyder SW (2014) Biomass-to-bioenergy and biofuel supply chain optimization: overview, key issues and challenges. Comput Chem Eng 66:36–56

Zhao X, Cheng K, Liu D (2009) Organosolv pretreatment of lignocellulosic biomass for enzymatic hydrolysis. Appl Microbiol Biotechnol 82(5):815–827

Zhu S, Wu Y, Yu Z (2006) Comparison of three microwave/chemical pretreatment processes for enzymatic hydrolysis of rice straw. Biosyst Eng 93(3):279–283

Zilberman D, Kim E, Kirschner S et al (2013) Technology and the future bioeconomy. Agric Econ 44:95–102

Printed by Printforce, the Netherlands